Animal Diversity

Animal Diversity

SIXTH EDITION

Cleveland P. Hickman, Jr.
Washington and Lee University

Larry S. Roberts
Florida International University

Susan L. Keen
University of California, Davis

Allan Larson
Washington University

David J. Eisenhour
Morehead State University

Original Artwork by
William C. Ober, M.D., and Claire W. Garrison, R.N.

Connect
Learn
Succeed™

ANIMAL DIVERSITY, SIXTH EDITION

Published by McGraw-Hill, a business unit of The McGraw-Hill Companies, Inc., 1221 Avenue of the Americas, New York, NY 10020. Copyright © 2012 by The McGraw-Hill Companies, Inc. All rights reserved. Previous editions © 2009, 2007, and 2003. Printed in the United States of America. No part of this publication may be reproduced or distributed in any form or by any means, or stored in a database or retrieval system, without the prior written consent of The McGraw-Hill Companies, Inc., including, but not limited to, in any network or other electronic storage or transmission, or broadcast for distance learning.

Some ancillaries, including electronic and print components, may not be available to customers outside the United States.

This book is printed on acid-free paper.

1 2 3 4 5 6 7 8 9 0 QDB/QDB 1 0 9 8 7 6 5 4 3 2 1

ISBN 978-0-07-131543-2
MHID 0-07-131543-8

www.mhhe.com

Brief Contents

Contents

Chapter 20

Mammals 408

About the Authors

Cleveland P. Hickman, Jr.

Cleveland P. Hickman, Jr., Professor Emeritus of Biology at Washington and Lee University in Lexington, Virginia, has taught zoology and animal physiology for more than 30 years. He received his Ph.D. in comparative physiology from the University of British Columbia, Vancouver, B.C., in 1958 and taught animal physiology at the University of Alberta before moving to Washington and Lee University in 1967. He has published numerous articles and research papers on fish physiology, in addition to co-authoring these highly successful texts: *Integrated Principles of Zoology, Biology of Animals, Animal Diversity, Laboratory Studies in Animal Diversity, and Laboratory Studies in Integrated Principles of Zoology.*

Over the years, Dr. Hickman has led many field trips to the Galápagos Islands. His current research is on intertidal zonation and marine invertebrate systematics in the Galápagos. He has published three field guides in the Galápagos Marine Life Series for the identification of echinoderms, marine molluscs, and marine crustaceans. (To read more about these field guides, visit *http://www.galapagosmarine.com.*)

His interests include scuba diving, woodworking, and participating in chamber music ensembles.

Dr. Hickman can be contacted at
hickmanc@wlu.edu

Larry S. Roberts

Larry S. Roberts, Professor Emeritus of Biology at Texas Tech University and an adjunct professor at Florida International University, has extensive experience teaching invertebrate zoology, marine biology, parasitology, and developmental biology. He received his Sc.D. in parasitology at the Johns Hopkins University and is the lead author of Schmidt and Roberts' *Foundations of Parasitology.* Dr. Roberts is also co-author of *Integrated Principles of Zoology, Biology of Animals, and Animal Diversity*, and is author of *The Underwater World of Sport Diving.*

Dr. Roberts has published many research articles and reviews. He has served as president of the American Society of Parasitologists, Southwestern Association of Parasitologists, and Southeastern Society of Parasitologists, and is a member of numerous other professional societies. Dr. Roberts also serves on the Editorial Board of the journal *Parasitology Research.* His hobbies include scuba diving, underwater photography, and tropical horticulture.

Dr. Roberts can be contacted at
Lroberts1@compuserve.com

Susan Keen

Susan Keen is a senior lecturer in the Section of Evolution and Ecology at the University of California at Davis. She received her Ph.D. in zoology from the University of California at Davis, following an M.Sc. from the University of Michigan at Ann Arbor. A native of Canada, she obtained her undergraduate education at the University of British Columbia in Vancouver.

Dr. Keen is an invertebrate zoologist fascinated with jellyfish life histories. She has a particular interest in the life cycles of organisms that have both asexual and sexual phases, as do most jellyfishes. Her other research has included work on sessile marine invertebrate communities, spider populations, and Andean potato evolution.

Dr. Keen has been teaching evolution and animal diversity within the Introductory Biology series for 15 years. She enjoys all facets of the teaching process, from lectures and discussions to the design of effective laboratory exercises. She offers seminars for the Davis Honors Challenge program and for undergraduate and graduate students interested in teaching methods for biology. She was a National Academies Education Fellow in the Life Sciences for 2005-2006. She received the Academic Federation Award for Excellence in Teaching in 2006 and an Excellence in Education Award from the Associated Students group at Davis in 2004. Her interests include weight training, horseback riding, gardening, travel, and mystery novels.

Dr. Keen can be contacted at
slkeen@ucdavis.edu

Allan Larson

Allan Larson is a professor at Washington University, St. Louis, MO. He received his Ph.D. in genetics at the University of California, Berkeley. His fields of specialization include evolutionary biology, molecular population genetics and systematics, and amphibian systematics. He teaches courses in zoology, genetics, macroevolution, molecular evolution, and the history of evolutionary theory, and has organized and taught a special course in evolutionary biology for high-school teachers.

Dr. Larson's research uses DNA sequences to examine evolutionary relationships among vertebrate species, especially salamanders and lizards. The students in Dr. Larson's laboratory have participated in zoological field studies around the world, including in Africa, Asia, Australia, Madagascar, North America, South America, and the Caribbean Islands. Dr. Larson has authored numerous scientific publications, and has edited for the journals *The American Naturalist, Evolution, Journal of Experimental Zoology, Molecular Phylogenetics and Evolution,* and *Systematic Biology*. Dr. Larson serves as an academic advisor to undergraduate students and supervises the undergraduate biology curriculum at Washington University.

Dr. Larson can be contacted at
larson@wustl.edu

David J. Eisenhour

David J. Eisenhour is a professor of biology at Morehead State University in Morehead, Kentucky. David, known as "Dr. E" to his students, delights in teaching, especially in laboratory activities. In 2001 David received the Master Teacher Award from the student Greek Councils at MSU. He received his Ph.D. in zoology from Southern Illinois University, Carbondale. He teaches courses in human anatomy and physiology, general zoology, comparative anatomy, ichthyology, and vertebrate zoology. David has an active research program that focuses on systematics, conservation biology, and natural history of North American freshwater fishes. He has a particular interest in the diversity of Kentucky's fishes and is writing a book about that subject. He and his graduate students have authored several publications. David serves as an academic advisor to prepharmacy students.

His interests include fishing, landscaping, home remodeling, and entertaining his three young children, who, along with his wife Lynn, are enthusiastic participants in fieldwork.

Dr. Eisenhour can be contacted at
d.eisenhour@morehead-st.edu

Preface

Animal Diversity is tailored for the restrictive requirements of a one-semester or one-quarter course in zoology, and is appropriate for both nonscience and science majors of varying backgrounds. This sixth edition of *Animal Diversity* presents a survey of the animal kingdom with emphasis on diversity, evolutionary relationships, functional adaptations, and environmental interactions.

Organization and Coverage

The sixteen survey chapters of animal diversity are prefaced by four chapters presenting the principles of evolution, ecology, taxonomy, and animal architecture. Throughout this revision, we updated references and worked to streamline the writing.

Chapter 1 begins with a brief explanation of the scientific method—what science is (and what it is not)—and then introduces evolutionary principles. Following a historical account of Charles Darwin's life and discoveries, we present the five major components of Darwin's evolutionary theory, the important challenges and revisions to his theory, and an assessment of its current scientific status. This approach reflects our understanding that Darwinism is not a single, simple statement easily confirmed or refuted. It also prepares the student to dismiss the arguments of creationists who misconstrue scientific challenges to Darwinism as contradictions to the validity of organic evolution. The chapter ends with discussion of micro- and macroevolution.

Chapter 2 explains the principles of ecology, with emphasis on populations, community ecology, and variations in the life-history strategies of natural populations. The treatment includes discussions of niche, population growth and its regulation, limits to growth, competition, energy flow, nutrient cycles, and extinction.

Chapter 3, on animal architecture, is a short but important chapter that describes the organization and development of the body plans that distinguish major groups of animals. This chapter includes a picture essay of tissue types and a section explaining important developmental processes responsible for the evolutionary diversification of the bilateral animals.

Chapter 4 treats taxonomy and phylogeny of animals. We present a brief history of how animal diversity has been organized for systematic study, emphasizing current use of Darwin's theory of common descent as the major principle underlying animal taxonomy. Our summary of continuing controversies over concepts of species and higher taxa includes discussion of how alternative taxonomic philosophies guide our study of evolution. We give special attention to phylogenetic systematics (cladistics) and the interpretation of cladograms. Chapter 4 also emphasizes that current issues in ecology and conservation biology depend upon our taxonomic system.

The sixteen survey chapters provide comprehensive, current, and thoroughly researched coverage of the animal phyla. We emphasize the unifying phylogenetic, architectural, and functional themes of each group, and illustrate them with detailed coverage of representative forms. Each chapter includes succinct statements of the diagnostic characteristics and major subgroups of the focal taxa.

The Linnean classifications in each chapter are positioned following other coverage of a particular group, in most cases immediately preceding the summary at the end of the chapter. Discussions of phylogenetic relationships take a cladistic viewpoint, with cladograms showing the structure of each group's history and the origin of the principal shared derived characters. Phylogenetic trees add temporal evolutionary hypotheses to the cladistic analyses.

Changes in the Sixth Edition

We continue in a revised format the major new structural feature of the previous edition: a cladogram depicting phylogenetic relationships among animal taxa appears in the inside front cover and serves to order our coverage of animal diversity in Chapters 5–20. We introduce a new graphic format for this and all other cladograms in the book, depicting branching events in a rectangular rather than triangular form; educational studies reveal that this format is a more intuitive one for students learning phylogenetics for the first time. The reformatted cladogram from the inside front cover appears in small form at the start of each taxonomic chapter, with the chapter's taxonomic coverage highlighted on it.

Our unifying cladogram also includes some substantive revisions from the previous edition. Revised hypotheses of the phylogenetic structure of lophotrochozoan protostomes appear in our cladogram and guide a corresponding restructuring of taxonomic coverage in Chapters 8 and 9. Three new taxa, Platyzoa (Platyhelminthes, Gastrotricha, and Gnathifera), Polyzoa (Cycliophora, Entoprocta, and Ectoprocta), and Kryptochozoa (Brachiopoda, Phoronida, and Nemertea), serve to formalize new hypotheses of lophotrochozoan phylogeny. In addition, new phylum Xenoturbellida now appears on the animal cladogram and in Chapter 14.

Many additional systematic updates appear in the detailed coverage of Chapters 5–20. Chapter 5 presents a revised taxonomy of the single-celled eukaryotes as 16 clades. Amoebae, formerly grouped as a polyphyletic taxon, are now distributed among several new taxa. Chapter 6 introduces a new taxonomic class of sponges, Homoscleromorpha. Chapter 10 adds discussion of the contrasting Conchifera versus Seralia hypotheses of molluscan phylogenetic diversification. Chapter 11 now places the former phylum Echiura within Annelida, and it includes new discussion of phylogenetic diversification within polychaetes. Chapter 13 is revised to include more discussion on phylogenetic relationships among myriapods, chelicerates, and pancrustaceans. Revisions to early chordate evolution appear in Chapter 15, and new hypotheses of hagfish phylogeny in Chapter 16. Chapter 20 incorporates major recent revisions to human evolution as revealed by fossil discoveries, including *Ardipithecus*. A substantial revision to mammal ordinal taxonomy also appears in Chapter 20. Many of these updates include expanded coverage of critical group characters, such as fiber optic properties of sponge spicules (Chapter 6), feeding structures of ctenophores (Chapter 7), and the visual system of cephalopod molluscs (Chapter 10). We provide additional photographs of some taxa not formerly well illustrated, such as caudofoveates and solenogasters (Chapter 10).

Because Darwin's theory of common descent is, more than any other single theory, the primary guiding principle of the book, we use it to illustrate the hypothetico-deductive method of science in a new essay that appears early in Chapter 1. We emphasize that an explanatory theory must allow us continually to test and potentially to falsify its predictions. Although Darwin formulated the theory of common descent using morphological, paleontological, and geographic data, it now guides investigation of molecular and cellular characteristics that were unknown and inconceivable to Darwin. We show how molecular and chromosomal data have confirmed predictions of Darwin's theory that humans and apes form a branch on the tree of life. We also show how new data often falsify predictions derived from earlier results. For example, we add new results from DNA-sequence comparisons among flightless birds (Chapter 1) to show that these results are not completely concordant with earlier hypotheses based on morphological characters, and that further testing and

reanalysis of available data must be done to resolve these contradictions. Expansion of this example also allows students to test their understanding of how characters are mapped onto the branches of a phylogenetic tree and how hypotheses of character homoplasy are generated during this process. This exercise prepares students for our more detailed coverage in this edition of the evolutionary origins of vertebrate jaws (Chapter 15), turtle shells (Chapter 18), and avian feathers (Chapter 19). New features of Chapter 1 prepare students to understand our discussions in Chapters 5–20 of phylogenetic and taxonomic revisions made possible by new discoveries that reject prior phylogenetic hypotheses.

We add or expand examples to illustrate critical theoretical concepts that some reviewers found initially too abstract. For example, we illustrate the complementarity of comparative and experimental sciences using the role of limb muscles in animal locomotion (Chapter 1). Experimental methods reveal the utility of limb muscles in locomotion for various animal populations; comparative methods using a phylogenetic hypothesis of species relationships allow us to synthesize such results across many species to test hypotheses of evolutionary changes in locomotion and associated changes in animal diversity. We emphasize more strongly in this edition the fact that evolution emerges at the populational level of complexity, although its consequences are often measured in molecular and organismal characters. We highlight the geographic theories of species-level diversification as the critical aspect of Darwin's theory of multiplication of species that distinguishes it from his theory of common descent (Chapter 1). We present in Chapter 1 a completely revised and updated depiction of the stratigraphic record of horse evolution, and a new photograph that better illustrates the macroevolutionary theory of impact crises. A new synthesis of evolutionary and developmental principles underlies revised coverage of the origins of vertebrate jaws (Chapter 15), amniote characteristics (Chapter 18), turtle shells (Chapter 18), and feathers (Chapter 19), as well as revised interpretations of the evolutionary diversifications of chordates (Chapter 15) and of marine fishes (Chapter 16).

Our coverage of the population-genetic principles of evolution now includes an explanatory graph of the relative frequencies of dominant versus recessive phenotypes in a randomly mating population as a function of its allelic frequencies (Chapter 1). This figure allows us to revise our introduction of Hardy-Weinberg equilibrium in the main text, removing statements that presupposed a student's confoundment of allelic dominance with allelic frequency. We add a new figure and text to demonstrate how scoring of protein polymorphisms permits measuring allelic frequencies in a population. We cover in greater detail the genetical theory of natural selection, first by defining the critical parameters of fitness and the average effect of an allelic substitution; we then illustrate these measurements for a hemoglobin polymorphism in some human populations of Africa (Chapter 1). This example reveals interactions between natural selection and genetic drift in the evolutionary adaptation of human populations to a malarial environment.

Ecological principles feature new explanations and examples, beginning in Chapter 2. To illustrate the concepts of niche differences, ecological competition, and their consequences for geographic distributions of species, we add an example from the ecology of Appalachian woodland salamanders (Chapter 2), including a new figure. Using a hypothetical example based on a cichlid fish species, we illustrate the possible consequences of a trophic polymorphism on metapopulation dynamics in a species (Chapter 2). We completely revise our illustration of age structures of populations using newly compiled date that show how the human population of Mexico changed from being actively growing in 1975 to having a stable demographic structure in 2006 (Chapter 2). We expand our coverage of commensalisms and mutualism to show how these phenomena intergrade with each other (Chapter 2). An exciting new discovery reveals that episodic occurrence of sunspots can influence the timing of predator-prey cycles in lynx and snowshoe hare populations in Canada (Chapters 2 and 20). We expand our coverage of Müllerian mimicry to note that the participating species are not necessarily equally distasteful, and that Müllerian mimicry thus intergrades with Batesian mimicry (Chapter 2). We update our graphical illustration of trophic levels (Chapter 2).

Extensions of these ecological principles and related conservation issues occur throughout the book. These include a photograph of a red tide (Chapter 5), discussion of medicinal compounds from sponges and their symbionts (Chapter 6), a photograph and discussion of coral bleaching (Chapter 7), anticancer properties of bryozoans (Chapter 9), revised coverage of life-history ecology of cycliophorans (Chapter 9) and of nematomorphs (Chapter 12), descriptions of worms that consume carcasses of whales (Chapter 11), greater coverage and a photograph of the anomalous camel spider of Middle Eastern deserts (Chapter 13), conservation biology of tuataras (Chapter 18), DDT and lead poisoning in avian populations (Chapter 19), and polyandry and extra-pair copulations in birds (Chapter 19). Also receiving new coverage are some economic threats to animal conservation, including commercial harvesting of shark fins (Chapter 16), hagfish skins (Chapter 16), and rhinoceros horns (Chapter 20).

Chapter 3 on animal architecture is extensively reordered, with one figure updated and a new figure added. This chapter now begins a discussion of the evolutionary origins and hallmarks of multicellularity, which continues through Chapters 5 and 6.

Chapter 4 includes some important clarifications and extensions of systematic principles. Because many cladistic systematists are exploring rank-free taxonomies as an alternative to Linnean principles, we discuss the critical contrasts between ranked and rank-free taxonomies and explain why the latter are gaining popularity. We extend our coverage of species concepts by adding the cohesion concept of species as a refinement of the evolutionary species concept. Added discussion clarifies the distinction between these concepts and the phylogenetic species concept, and it provides greater understanding of the criteria that taxonomists consider most important in recognizing species. We introduce the concept of convexity to permit more precise definitions and explanations of the contrasting concepts of monophyletic, paraphyletic, and polyphyletic groups. We greatly expand our boxed essay on deriving phylogenies from aligned DNA sequences to show how statistical approaches (maximum likelihood and Bayesian methods) can use evolutionary models of base substitution to make phylogenetic inference more powerful. We revise this chapter's coverage of the major subdivisions of the animal kingdom to delete an archaic system and to illustrate a rank-free, indented presentation of hierarchically nested taxa.

Each taxonomic chapter (5–20) retains a summary of groups characteristics in list form, but we enhance comparability of these lists among chapters by adopting the following standard ordering of kinds of features listed: (1) unique features; (2) habitats; (3) symbiosis with other organisms; (4) symmetry and metamerism; (5) triploblastic characters; (6) coelomic characters; (7) epithelial characters, cuticular characters, skin/scales; (8) digestive system; (9) body musculature; (10) nervous system; (11) sensory systems; (12) asexual reproduction/regeneration; (13) sexual characters and hermaphroditism; (14) excretory system; (15) respiratory system; (16) circulatory system. Because the number of characteristics listed as well as relevance of particular categories varies among chapters, only the ordering and not the exact numbering of characteristics is consistent across chapters.

As with prior revisions, we strive for greater clarity of wording based on specific comments received from reviewers of past editions of the textbook. Of particular utility in this revision were suggestions on rewording the structure/function relationships of characters to avoid the appearance of teleology. For example, in describing the fossil *Ichthyostega* in Chapter 17, our earlier wording noted that it had "a stronger backbone to support the body in air" and "new muscles to elevate the head" as if these structures evolved for a preordained purpose of terrestrial locomotion; we revise such statements to retain relevant descriptions without suggesting a purposeful goal.

We have added at the end of each chapter a special thought question labeled "For Further Thought." These questions are more open ended than the other questions and would be good stimuli for group discussions.

Teaching and Learning Aids

Vocabulary Development

Key words are boldfaced, which serves as a cross-reference to the glossary for definition, pronunciation, and derivation of each term. Derivations of generic names of animals are given where they first appear in the text. In addition, derivations of many technical and zoological terms are provided, allowing students to recognize the more common roots that recur in many technical terms.

Chapter Prologues

A distinctive feature of this text is an opening essay at the beginning of each chapter. Each essay presents a theme or topic relating to the subject of the chapter to stimulate interest. Some present biological, particularly evolutionary, principles; others illuminate distinguishing characteristics of the animal group treated in the chapter.

Chapter Notes

Chapter notes, which appear throughout the book, augment the text material and offer interesting sidelights without interrupting the narrative.

For Review

Each chapter ends with a concise summary, review questions, and a list of annotated selected references. The review questions enable students to test themselves for retention and understanding of the more important chapter material.

Art Program

The appearance and usefulness of this text are much enhanced by numerous full-color paintings by William C. Ober and Claire W. Garrison. Bill's artistic skills, knowledge of biology, and experience gained from an earlier career as a practicing physician have enriched the authors' zoology texts through many editions. Claire practiced pediatric and obstetric nursing before turning to scientific illustration as a full-time career. Texts illustrated by Bill and Claire have received national recognition and won awards from the Association of Medical Illustrators, American Institute of Graphic Arts, Chicago Book Clinic, Printing Industries of America, and Bookbuilders West. Bill and Claire also are recipients of the Art Directors Award.

Web Pages

At the end of each survey chapter is a selection of related Internet links. The URLs for the pages are in the text's website at *www.mhhe.com/hickmanad6e*.

Supplements

Instructor's Manual

Each chapter of the Instructor's Manual provides a detailed chapter outline, lecture enrichment suggestions, a commentary, and critical thinking questions. This material should be particularly helpful for first-time users of the text, although experienced teachers also may find much of value. The Instructor's Manual is available on this text's Online Learning Center at *www.mhhe.com/hickmanad6e*.

Animal Diversity Website www.mhhe.com/hickmanad6e

For the Student

- This convenient website takes studying to a whole new level. Students will find animations, key term flashcards, interactive web links, and more. What a great way to get a better grade!
- The *Animal Diversity* website is also home to the **Zoology Essential Study Partner**. This unique learning tool allows students to test their understanding of important zoological concepts through the use of animations, learning activities, quizzing, and interactive diagrams.

For the Instructor

- Instructors will appreciate a password-protected site that offers an Instructor's Manual, test questions, PowerPoint lecture outlines, guide to teaching animal molecular phylogenetics, and more!
- Instructors can also access the **Presentation Center** through the website. This online digital library contains all the illustrations, photos, and tables from *Animal Diversity*, sixth edition, which can be used to create customized lectures, visually enhanced tests and quizzes, compelling course websites, or attractive printed supplements. In addition, you have access not only to the assets specific to *Animal Diversity*, but also to thousands of assets from many McGraw-Hill titles. The Presentation Center's dynamic search engine allows you to explore by discipline, course, textbook chapter, asset type, or keyword.

Animal Diversity Laboratory Manual

The laboratory manual by Cleveland P. Hickman, Jr., and Lee B. Kats, *Laboratory Studies in Animal Diversity*, is designed specifically for a survey course in zoology.

McGraw-Hill: *Digitized Biology Video Clips*

McGraw-Hill is pleased to offer digitized biology video clips on DVD. Licensed from some of the highest-quality science video producers in the world, these brief segments range from about five seconds to just under three minutes in length and cover all areas of general biology, from cells to ecosystems. Engaging and informative, McGraw-Hill's digitized biology videos will help capture students' interest while illustrating key biological concepts and processes. Topics include mitosis, ameba locomotion, rain forest diversity, Darwin's finches, tarantula defense, nematodes, bird/water buffalo mutualism, poison-dart frogs, echinoderms, and much more! ISBN-13: 978-0-07-312155-0 (ISBN-10: 0-07-312155-X)

Electronic Textbook

CourseSmart is a new way for faculty to find and review eTextbooks. It is also a great option for students who are interested in accessing their course materials digitally and saving money. CourseSmart offers thousands of the most commonly adopted textbooks across hundreds of courses from a wide variety of higher education publishers. It is the only place for faculty to review and compare the full text of a textbook online, providing immediate access without the environmental impact of requesting a print exam copy. At CourseSmart, students can save up to 50% off the cost of a print book, reduce their impact on the environment, and gain access to powerful Web tools for learning, including full text search, notes and highlighting, and e-mail tools for sharing notes between classmates. *www.CourseSmart.com*

Acknowledgments

We wish to thank the following individuals for reviewing the sixth edition of this text:

Frank Anderson, *Southern Illinois University*
James Coggins, *University of Wisconsin—Milwaukee*
James Doyle, *College of the Holy Cross*
Harold Heatwole, *North Carolina State University*
Bradford Hollingsworth, *San Diego State University*
Carey Krajewski, *Southern Illinois University*
Alvin Kristopher Lappin, *California State Polytechnic University (Pomona)*
Jon Scales, *Midwestern State University*

The authors express their gratitude to the able and conscientious staff of McGraw-Hill who brought this book to its present form. We extend special thanks to sponsoring editor Rebecca Olsen, developmental editor Wendy Langerud, project manager Jane Mohr, designer Brenda Rolwes, and executive marketing manager Patrick Reidy. All played essential roles in shaping the sixth edition.

Science of Zoology and Evolution of Animal Diversity

Evolutionary diversification of Hawaiian honeycreepers.

Kauai akialoa

Laysan finch

Kona finch

Akepa

Maui parrotbill

Nukupuu

Alauwahio

Anianiau

Amakihi

Palila

Ou

Akiapolaau

Crested honeycreeper

-ai-hawane

Apapane

Mamos

iwi

A Legacy of Change

Life's history is a legacy of perpetual change. Despite the apparent permanence of the natural world, change characterizes all things on earth and in the universe. Countless kinds of animals and plants have flourished and disappeared, leaving behind an imperfect fossil record of their existence. Many, but not all, have left living descendants that resemble them to varying degrees.

We observe and measure life's changes in many ways. On a short evolutionary timescale, we see changes in the frequencies of different genetic traits within populations. For example, evolutionary changes in the relative frequencies of light- and dark-colored moths occurred within a single human lifetime in polluted areas of industrial England. On the other hand, formation of new species and dramatic changes in organismal appearance, as shown by evolutionary diversification of Hawaiian birds, require longer timescales covering 100,000 to 1 million years. Major evolutionary trends and episodic mass extinctions occur on even larger timescales, covering tens of millions of years. The fossil record of horses through the past 50 million years shows a series of different species replacing older ones. The fossil record of marine invertebrates shows episodic mass extinctions separated by intervals of approximately 26 million years.

Organic evolution is the irreversible, historical change that we observe in living populations and in the earth's fossil record. Because every feature of life is a product of evolutionary processes, biologists consider organic evolution the keystone of all biological knowledge.

Zoology (Gr. *zōon,* animal, + *logos,* discourse on, study of) is the scientific study of animals. It is part of biology (Gr. *bios,* life, + *logos*), the study of all life. Explaining the panorama of animal diversity—how animals function, live, reproduce, and interact—is exciting and challenging.

To explain the diversity of animal life, we must study its long history, which began more than 600 million years ago. From the earliest animals to the millions of animal species living today, this history demonstrates extensive and ongoing change, which we call **evolution.** We depict the history of animal life as a branching genealogical tree, called a **phylogeny** or **phylogenetic tree.** We place the earliest species ancestral to all animals at the trunk; all living animal species fall at the growing tips of the branches. Each successive branching event represents the historical splitting of an ancestral species to form new ones. Newly formed species inherit many characteristics from their immediate ancestor, but they also evolve new features that appear for the first time in the history of animal life. Each branch therefore has its own unique combination of characteristics and contributes a new dimension to the spectrum of animal diversity.

The scientific study of animal diversity has two major goals. The first is to reconstruct a phylogeny of animal life and to find where in evolutionary history we can locate the origins of major characteristics—multicellularity, a **coelom, spiral cleavage,** vertebrae, **homeothermy**—and all other dimensions of animal diversity as we know it. The second major goal is to understand historical processes that generate and maintain diverse species and adaptations throughout evolutionary history. Darwin's theory of evolution allows us to apply scientific principles to attain both goals.

■ Principles of Science

A basic understanding of zoology requires understanding what science is, what it excludes, and how one gains knowledge using the scientific method. In this section we examine the methodology that zoology shares with science as a whole. These features distinguish the sciences from other disciplines, such as art and religion.

Despite an enormous impact of science on our lives, many people have only a minimal understanding of science. Public misunderstanding of scientific principles as applied to animal diversity revealed itself to us on March 19, 1981, when the governor of Arkansas signed into law the Balanced Treatment for Creation-Science and Evolution-Science Act (Act 590 of 1981). This act falsely presented creation-science as a valid scientific endeavor. Further legal scrutiny revealed that creation-science was not science, but rather a religious position advocated by a minority of America's religious community.

Enactment of this law incited a historic lawsuit tried in December 1981 in the court of Judge William R. Overton, U.S. District Court, Eastern District of Arkansas. The American Civil Liberties Union filed the suit on behalf of 23 plaintiffs, including religious leaders and groups representing several denominations, individual parents, and educational associations. Plaintiffs contended that this law violated the First Amendment to the U.S. Constitution, which prohibits establishment of religion by government. This amendment prohibits passing a law that would favor one religious position over another one. On January 5, 1982, Judge Overton permanently prohibited Arkansas from enforcing Act 590.

Considerable testimony during the trial clarified the nature of science. On the basis of testimony by scientists, Judge Overton stated explicitly these essential characteristics of science:

1. It is guided by natural law.
2. It must be explanatory by reference to natural law.
3. Its conjectures are testable against the empirical world.
4. Its conclusions are tentative and not necessarily the final word.
5. It is falsifiable.

Pursuit of scientific knowledge is guided by physical and chemical laws that govern the state of existence. Scientific knowledge must explain observations by reference to natural law without intervention of any supernatural being or force. We must record observations that directly or indirectly test hypotheses about nature. We must discard or modify any conclusion if further observations contradict it. As Judge Overton stated, "While anybody is free to approach a scientific inquiry in any fashion they choose, they cannot properly describe the methodology used as scientific, if they start with a conclusion and refuse to change it regardless of the evidence developed during the course of the investigation." Science lies outside religion, and scientific knowledge does not favor one religious position over another.

Unfortunately, the religious position formerly called creation-science later reappeared in American politics with the name "intelligent design theory." We once again defended science education against this scientifically meaningless doctrine. On December 20, 2005, Judge John E. Jones III of the U.S. District Court for the Middle District of Pennsylvania ruled unconstitutional the teaching of intelligent design, which had been mandated by the Dover school board. The local voters already had rejected the eight board members who supported the intelligent-design requirement, replacing them with candidates who actively opposed teaching intelligent design as science.

Scientific Method

The essential criteria of science form the **hypothetico-deductive method.** One begins this process by generating **hypotheses,** or potential explanations of a phenomenon of nature. These hypotheses are usually based on prior observations of nature (figure 1.1) or on theories derived from such observations. Scientific hypotheses often constitute general statements that might explain a large number of diverse observations. The hypothesis of natural selection, for example, explains our observations that many different species have accumulated

A

C

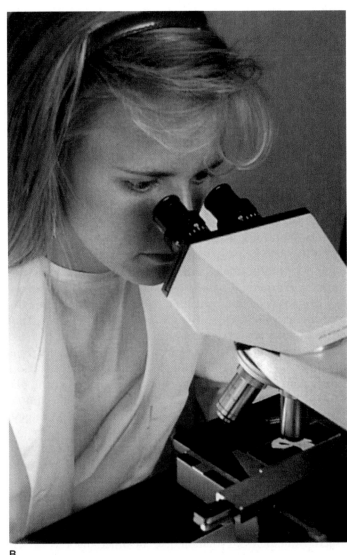

B

figure 1.1

Examples of observation in zoological research. **A,** Observing coral in the Caribbean Sea. **B,** Observing nematocyst discharge, **C,** from hydrozoan tentacles (see p. 137).

favorable characteristics that adapt them to their environments. Based on a hypothesis, a scientist must say, "If my hypothesis correctly explains past observations, then future observations must match specific expectations."

The scientific method comprises a series of steps:

1. Observation
2. Question
3. Hypothesis
4. Empirical test
5. Conclusions
6. Publication

Observations form a critical first step in evaluating the life histories of natural populations. For example, the observation of crab larvae of varying sizes might cause an observer to question whether the rate of larval growth is higher in undisturbed populations than in ones exposed to a chemical pollutant. A null hypothesis is then generated to permit an empirical test. A null hypothesis is worded so that the data can reject it if it is false. The null hypothesis pertaining to larval growth rates of crabs in undisturbed versus polluted conditions is that the growth rates are the same in both kinds of populations. The investigator then performs an empirical test by gathering data on larval growth rates in a set of undisturbed crab populations and in a set of populations exposed to the chemical pollutant. Ideally, the undisturbed populations and the chemically treated populations are equivalent for all conditions except for the chemical in question. If measurements show consistent differences in growth rate between the two sets of populations, the null hypothesis is rejected. The investigator then concludes that the chemical pollutant does alter larval growth rates. A statistical test is usually needed to ensure that the differences between the two groups are greater than would be expected from chance fluctuations alone. If the null hypothesis cannot be rejected, the investigator concludes that the data do not reveal any effect of the chemical treatment. Results of the study are then published to communicate findings to other researchers, who may repeat the procedure, perhaps using additional populations of the same or a different species. Conclusions of the initial study then serve as the observations for further questions and hypotheses to reiterate the scientific process.

Note that a null hypothesis cannot be proved correct using the scientific method. If the available data are compatible with it, the hypothesis serves as a guide for collecting additional data that potentially might reject it.

Our most successful hypotheses are the ones that make specific predictions that can be confirmed by large numbers of empirical tests. If a hypothesis is very powerful in explaining a large variety of related phenomena, it attains the status of a **theory**. Evolution by natural selection is a good example; it provides a potential explanation for the occurrence of many different traits observed among animal species. Each of these instances constitutes a specific hypothesis generated from the theory of evolution by natural selection. Natural selection explains many disparate features of the animal kingdom, but we have learned that it is not the exclusive means of evolution. The most useful theories are those that can explain the largest array of different natural phenomena.

We emphasize that the word "theory," when used by scientists, is not arbitrary speculation as is often implied by nonscientific usage. Creationists make this mistake when they call evolution "only a theory" to imply that it is little better than a random guess. In fact, evolutionary theory is supported by such massive evidence that its scientific validity is well established. Nonetheless, evolution, like all other theories in science, cannot be proved correct by mathematical logic, but is testable, tentative, and falsifiable.

Experimental and Evolutionary Sciences

One can group the many questions raised about animal life into two major categories. The first category seeks to explain **proximate causes** (also called immediate causes) that guide biological systems at all levels of complexity. It includes explaining how animals perform their metabolic, physiological, and behavioral functions at molecular, cellular, organismal, and even population levels. For example, how is genetic information expressed to guide the synthesis of proteins? What signal causes cells to divide to produce new cells? How does population density affect the physiology and behavior of organisms?

We call the biological sciences that investigate proximate causes **experimental sciences** because they use the **experimental method** to test mechanistic explanations of a biological system. This method has three steps: (1) predicting from a tentative explanation how a system being studied would respond to a disturbance, (2) making the disturbance,

and (3) comparing observed results to predicted ones. An investigator repeats the experiment multiple times to eliminate chance occurrences that might produce errors. **Controls** (repetitions of an experimental procedure that lack the disturbance) eliminate any unperceived conditions that might bias an experiment's outcome.

Processes by which animals maintain their body temperature under different environmental conditions, digest food, migrate to new habitats, or store energy are additional examples of phenomena studied by experimentation. Experimental sciences in biology include molecular biology, cell biology, endocrinology, immunology, physiology, developmental biology, and community ecology.

In contrast to proximate causes, **evolutionary sciences** address **ultimate causes** that have molded biological systems and their properties through evolutionary time. For example, what evolutionary factors cause some birds to make complex seasonal migrations between temperate and tropical regions? Why do different species of animals have different numbers of chromosomes in their cells? Why do some animal species maintain complex social systems, whereas individuals of other species remain largely solitary?

A scientist's use of the phrase "ultimate cause," unlike Aristotle's usage, does not imply a preconceived goal for natural phenomena. An argument that nature has a predetermined goal, such as evolution of the human mind, is termed teleological. **Teleology** is the mistaken notion that the evolution of living organisms is guided by purpose toward an optimal design. A major success of Darwinian evolutionary theory is its rejection of teleology in explaining biological diversification.

Evolutionary sciences require the **comparative method** rather than just experimentation. Characteristics of molecular biology, cell biology, organismal structure, development, and ecology are compared among related species to identify patterns of variation. Scientists then use patterns of similarity and dissimilarity to test hypotheses of relatedness and thereby to reconstruct the phylogenetic tree that relates the species being compared. Systematics is the ordering of organisms according to their inferred evolutionary relationships for comparative study. Recent advances in DNA sequencing technology permit precise tests of evolutionary relationships among all animal species. Comparative studies also serve to test hypotheses of evolutionary processes that have molded diverse animal species. Clearly, evolutionary sciences use results of experimental sciences as a starting point. Experimental results reveal how limb muscles typically perform in locomotion in various animal species, and comparative studies based on evolutionary relationships among those species reveal evolutionary changes in locomotion that underlie current species diversity. Evolutionary sciences include comparative biochemistry, molecular evolution, comparative cell biology, comparative anatomy, comparative physiology, and phylogenetic systematics.

The Power of a Theory

Darwin's theory of common **descent** (p. 16) illustrates the scientific importance of general theories that give unified explanations to diverse kinds of data. Darwin proposed his theory of descent with modification of all living forms because it explained the patterns of similarity and dissimilarity among organisms in anatomical structures and cellular organization.

Anatomical similarities between humans and apes led Darwin to propose that humans and apes share more recent common ancestry with each other than they do with any other species. Darwin was unaware that his theory, a century later, would provide the primary explanation for similarities and dissimilarities among species in the structures of their chromosomes, sequences of amino acids in homologous proteins, and sequences of bases in homologous genomic DNA.

The accompanying figure shows photographs of a complete haploid set of chromosomes from each of four ape species: human (*Homo sapiens*), bonobo (the pygmy chimpanzee, *Pan paniscus*), gorilla (*Gorilla gorilla*), and orangutan (*Pongo pygmaeus*). Each chromosome in the human genome has a corresponding chromosome with similar structure and gene content in the genomes of other ape species. The most obvious difference between human and ape chromosomes is that the large second chromosome in the human nuclear genome was formed evolutionarily by a fusion of two smaller chromosomes characteristic of the ape genomes. Detailed study of the human and other ape chromosomes shows remarkable correspondence between them in genic content and organization. Ape chromosomes are more similar to each other than they are to chromosomes of any other animals.

Comparison of DNA and protein sequences among apes likewise confirms their close genetic relationships, with humans and the two chimpanzee species being closer to each other than any of these species are to other apes. DNA sequences from the nuclear and mitochondrial genomes independently support the close relationships among ape species and especially the grouping of humans and chimpanzees as close relatives. Homologous DNA sequences of humans and chimpanzees are approximately 99% similar in base sequence.

Studies of variation in chromosomal structure, mitochondrial DNA sequences, and nuclear DNA sequences produced multiple independent data sets, each one potentially capable of rejecting Darwin's theory of common descent. Darwin's theory would be rejected, for example, if the chromosomal structures and DNA sequences of apes were no more similar to each other than to those of other animals. The data in this case support rather than reject predictions of Darwin's theory. The ability of Darwin's theory of common descent to make precise predictions of genetic similarities among these and other species, and to have those predictions confirmed by numerous empirical studies, illustrates its great strength. As new kinds of biological data have become available, the scope and strength of Darwin's theory of common descent have increased enormously. Indeed, nothing in biology makes sense in the absence of this powerful explanatory theory.

Comparative Karyotype of Great Apes

The human haploid genome contains 22 autosomes (I–XXII) and a sex chromosome (X or Y). The human chromosome is shown first in each group of four, followed by the corresponding chromosomes of bonobo, gorilla, and orangutan, in that order. Note that the chromatin of human chromosome II corresponds to that of two smaller chromosomes (marked p and q) in other apes.

Origins of Darwinian Evolutionary Theory

Charles Robert Darwin and Alfred Russel Wallace (figure 1.2) were the first to establish evolution as a powerful scientific theory. Today, evolution can be denied only by abandoning reason. As the English biologist Sir Julian Huxley wrote, "Charles Darwin effected the greatest of all revolutions in human thought, greater than Einstein's or Freud's or even Newton's, by simultaneously establishing the fact and discovering the mechanism of organic evolution." Darwinian theory allows us to explain both the genetics of populations and long-term trends in the fossil record. Darwin and Wallace did not originate the basic idea of organic evolution, which has an ancient history. We review first the history of evolutionary thinking as it led to Darwin's theory and then discuss critical evidence supporting Darwin's theory.

Pre-Darwinian Evolutionary Ideas

Early Greek philosophers, notably Xenophanes, Empedocles, and Aristotle, developed the idea that life has a long history of evolutionary change. They recognized fossils as evidence for former life, which they thought had been destroyed by natural catastrophe. Despite their intellectual inquiry, ancient Greeks failed to establish an evolutionary concept that could guide a meaningful study of life's history. Opportunities for evolutionary thinking were even more restricted as the metaphorical biblical account of earth's creation became accepted as requiring no mechanistic explanation. The year 4004 B.C. was fixed by Archbishop James Ussher (mid-seventeenth century) as the time of life's creation.

Evolutionary views were considered heretical, but they refused to die. The French naturalist Georges Louis Buffon (1707–1788) stressed environmental influences on modifications of animal form and extended the earth's age to 70,000 years.

Lamarckism: The First Scientific Hypothesis for Evolution

The first complete hypothesis for evolution was authored by the French biologist Jean Baptiste de Lamarck (1744–1829) (figure 1.3) in 1809, the year of Darwin's birth. He made the first convincing argument that fossils were remains of extinct animals. Lamarck's evolutionary mechanism, **inheritance of acquired characteristics**, was engagingly simple: Organisms, by striving to meet the demands of their environments, acquire adaptations and pass them by heredity to their offspring. According to Lamarck, giraffes evolved a long neck because their ancestors lengthened their necks by stretching to obtain food and then passed the lengthened neck to their offspring. Lamarck proposed that over many generations, these changes accumulated to produce the long necks of modern giraffes.

We call Lamarck's concept of evolution *transformational,* because it claims that, as individual organisms transform their characteristics through the use and disuse of parts, heredity makes corresponding adjustments to produce evolution. We now reject transformational theories because genetic studies show that traits acquired during an organism's lifetime, such as strengthened muscles, are not transmitted to offspring.

Darwin's evolutionary theory differs from Lamarck's in being a *variational* theory. Evolution occurs at the level of the **population,** and it includes changes across generations in the organismal characteristics that prevail in the population. Darwin argued that organisms possessing hereditary characteristics that conferred an advantage for survival or reproduction would leave more offspring than would other organisms, thereby causing populations to accumulate across generations the characteristics most favorable to the survival and reproductive success of their bearers.

A   **B**

figure 1.2

Founders of the theory of evolution by natural selection. **A,** Charles Robert Darwin (1809–1882), as he appeared in 1881, the year before his death. **B,** Alfred Russel Wallace (1823–1913) in 1895. Darwin and Wallace independently developed the same theory. A letter and essay from Wallace written to Darwin in 1858 spurred Darwin into writing *On the Origin of Species*, published in 1859.

figure 1.3

Jean Baptiste de Lamarck (1744–1829), French naturalist who offered the first scientific explanation of evolution. Lamarck's hypothesis that evolution proceeds by inheritance of acquired characteristics was rejected by genetic research.

Charles Lyell and Uniformitarianism

The geologist Sir Charles Lyell (1797–1875) (figure 1.4) established in his *Principles of Geology* (1830–1833) the principle of **uniformitarianism**. Uniformitarianism encompasses two important assumptions that guide scientific study of the history of nature. These assumptions are (1) that the laws of physics and chemistry have not changed throughout earth's history, and (2) that past geological events occurred by natural processes similar to those that we observe in action today. Lyell showed that natural forces, acting over long periods of time, could explain the formation of fossil-bearing rocks. Lyell's geological studies convinced him that earth's age must be many millions of years. These principles were important because they discredited miraculous and supernatural explanations of the history of nature and replaced them with scientific explanations. Lyell also stressed the gradual nature of geological changes that occur through time, and he argued that such changes have no inherent directionality. Both of these claims left important marks on Darwin's evolutionary theory.

Darwin's Great Voyage of Discovery

"After having been twice driven back by heavy southwestern gales, Her Majesty's ship *Beagle,* a ten-gun brig, under the command of Captain Robert FitzRoy, R.N., sailed from Devonport on the 27th of December, 1831." Thus began Charles Darwin's account of the historic five-year voyage of the *Beagle* around the world (figure 1.5). Darwin, not quite 23 years old, had asked to accompany Captain FitzRoy on the *Beagle,* a small vessel only 90 feet in length, which was about to make an extensive surveying voyage to South America and the Pacific (figure 1.6). It was the beginning of the most important scientific voyage of the nineteenth century.

During this voyage (1831–1836), Darwin endured seasickness and erratic companionship from Captain FitzRoy, but his endurance and early training as a naturalist equipped him for his work. The *Beagle* made many stops along the coasts of South America and adjacent islands. Darwin made extensive collections and observations of the faunas and floras of these regions. He unearthed numerous fossils of animals long extinct and noted a resemblance between fossils of South American pampas and known fossils of North America. In the Andes, he encountered seashells embedded in rocks at 13,000 feet. He experienced a severe earthquake and watched mountain torrents that relentlessly wore away the earth. These observations and his reading of Lyell's *Principles of Geology* during the voyage strengthened Darwin's conviction that natural forces could explain geological features of the earth.

figure 1.4

Sir Charles Lyell (1797–1875), English geologist and friend of Darwin. His book *Principles of Geology* greatly influenced Darwin during Darwin's formative period. This photograph was made about 1856.

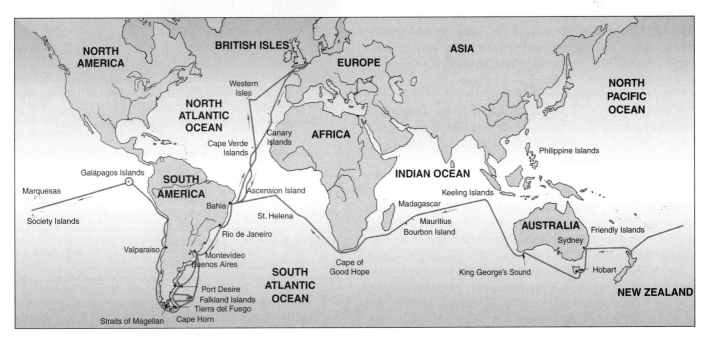

figure 1.5

Five-year voyage of H.M.S. *Beagle.*

A

B

figure 1.6

Charles Darwin and H.M.S. *Beagle*. **A**, Darwin in 1840, four years after the *Beagle* returned to England, and a year after his marriage to his cousin, Emma Wedgwood. **B**, H.M.S. *Beagle* sails in *Beagle* Channel, Tierra del Fuego, on the southern tip of South America in 1833. This watercolor was painted by Conrad Martens, one of two official artists during the voyage of the *Beagle*.

In mid-September of 1835, the *Beagle* arrived at the Galápagos Islands, a volcanic archipelago straddling the equator 600 miles west of Ecuador (figure 1.7). The fame of these islands stems from their oceanic isolation and rugged volcanic terrain. Circled by capricious currents, surrounded by shores of twisted lava, bearing skeletal brushwood baked by equatorial sunshine, almost devoid of vegetation, inhabited by strange reptiles and by convicts stranded by the Ecuadorian government, these islands had few admirers among mariners. By the middle of the seventeenth century, Spaniards called these islands "Las Islas Galápagos"—the tortoise islands. The giant tortoises, used for food first by buccaneers and later by American and British whalers, sealers, and ships of war, were the islands' principal attraction. At the time of Darwin's visit, these tortoises already were heavily exploited.

During the *Beagle*'s five-week visit to the Galápagos, Darwin documented the unique character of the Galápagos plants and animals, including the giant tortoises, marine iguanas, mockingbirds, and ground finches. Darwin later described these studies as the "origin of all my views."

Darwin was struck by the fact that, although the Galápagos Islands and Cape Verde Islands (visited earlier in this voyage) were similar in climate and topography, Galápagos plants and animals resembled most closely those of the South American mainland, and they were entirely different from the African-derived forms of the Cape Verde Islands. Each Galápagos island often contained a unique species that nonetheless resembled forms on other Galápagos islands. In short, Galápagos life must have originated in continental South America and then undergone modification in various environmental conditions of different islands. He concluded that living forms were neither divinely created nor immutable; they were, in fact, products of a long history of evolutionary change.

figure 1.7

The Galápagos Islands viewed from the rim of a volcano.

"Whenever I have found that I have blundered, or that my work has been imperfect, and when I have been contemptuously criticized, and even when I have been overpraised, so that I have felt mortified, it has been my greatest comfort to say hundreds of times to myself that 'I have worked as hard and as well as I could, and no man can do more than this.'"

—*Charles Darwin, in his autobiography, 1876.*

On October 2, 1836, the *Beagle* returned to England, where Darwin conducted most of his scientific work (figure 1.8). Most of Darwin's extensive collections had preceded him there, as had his notebooks and diaries kept during the cruise. Darwin's journal, published three years after the *Beagle*'s return to England,

figure 1.8

Darwin's study at Down House in Kent, England, is preserved today much as it was when Darwin wrote *On the Origin of Species*.

was an instant success and required two additional printings within its first year. Darwin's entertaining book, *The Voyage of the Beagle,* became one of the most popular travel books of all time.

The main product of Darwin's voyage, his theory of evolution, would continue to develop for more than 20 years after the *Beagle*'s return. In 1838, he "happened to read for amusement" an essay on populations by T. R. Malthus (1766–1834), who stated that animal and plant populations, including human populations, tend to increase beyond the capacity of their environment to support them. Darwin already had been gathering information on artificial selection of animals under domestication. He was especially fascinated by artificial breeds of pigeons. Many pigeon breeds differed so much in appearance and behavior that they would be considered different species if found in nature. All clearly had been derived from a single wild species, the rock pigeon (*Columbia livia*). After reading Malthus's article, Darwin realized that a process of selection in nature, driven by a "struggle for existence" because of overpopulation, could be a powerful force for evolution of wild species.

Darwin allowed the idea to develop in his own mind until 1856, when he began to assemble his voluminous data into a work on origins of species. He expected to write four volumes, a very big book, "as perfect as I can make it." However, his plans took an unexpected turn.

In 1858, he received a manuscript from Alfred Russel Wallace (1823–1913), an English naturalist in Malaya with whom he corresponded. Darwin was stunned to find that in a few pages, Wallace summarized the main points of the natural selection theory on which Darwin had worked for two decades. Rather than withhold his own work in favor of Wallace as he was inclined to do, Darwin was persuaded by two close friends, Lyell and a botanist, Hooker, to publish his views in a brief statement that would appear together with Wallace's paper in the *Journal of the Linnean Society.*

Portions of both papers were read before an unimpressed audience on July 1, 1858.

For the next year, Darwin worked urgently to prepare an "abstract" of the planned four-volume work. This book was published in November 1859, with the title *On the Origin of Species by Means of Natural Selection, or the Preservation of Favoured Races in the Struggle for Life.* The 1250 copies of the first printing were sold the first day! The book instantly generated a storm that has never abated. Darwin's views were to have extraordinary consequences on scientific and religious beliefs, and they remain among the greatest intellectual achievements of all time.

Once Darwin's caution had been swept away by publication of *On the Origin of Species,* he entered an incredibly productive period of evolutionary thinking for the next 23 years, producing book after book. He died on April 19, 1882, and was buried in Westminster Abbey. The little *Beagle* had already disappeared, having been retired in 1870 and sold for scrap.

■ Darwin's Theory of Evolution

Darwin's theory of evolution is now over 150 years old. Biologists frequently are asked, "What is Darwinism?" and "Do biologists still accept Darwin's theory of evolution?" These questions do not have simple answers because Darwinism encompasses several different, although mutually connected, theories. Professor Ernst Mayr of Harvard University argued that Darwinism should be viewed as five major theories. These five theories have somewhat different origins and fates and cannot be treated as only a single hypothesis. The theories are (1) **perpetual change,** (2) **common descent,** (3) **multiplication of species,** (4) **gradualism,** and (5) **natural selection.** The first three theories are generally accepted as having universal application throughout the living world. Gradualism and natural selection remain somewhat controversial among evolutionists; they are clearly important evolutionary processes, but they might not explain as much of animal evolution as Darwin thought. Creationists often misrepresent legitimate controversies regarding gradualism and natural selection as challenges to the first three theories, whose validity is strongly supported by all relevant facts.

1. **Perpetual change.** This is the basic theory of evolution on which the others are based. It states that the living world is neither constant nor perpetually cycling, but is always changing. Characteristics of organisms undergo modification across generations throughout time. This theory originated in antiquity but did not gain widespread acceptance until Darwin advocated it in the context of his other four theories. Perpetual change is documented by the fossil record, which clearly refutes any claims for a recent origin of all living forms. Because it has withstood repeated testing and is supported by an overwhelming number of observations, we now regard perpetual change as fact.

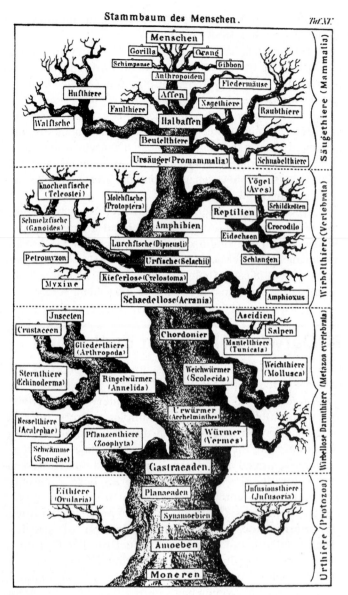

Stammbaum des Menschen.   *Taf.XI.*

figure 1.9

An early tree of life drawn in 1874 by the German biologist Ernst Haeckel, who was strongly influenced by Darwin's theory of common descent. Many phylogenetic hypotheses shown in this tree, including a unilateral progression of evolution toward humans (= Menschen, *top*), subsequently have been refuted.

2. **Common descent.** The second Darwinian theory, common descent, states that all forms of life propagated from a common ancestor through a branching of lineages (figure 1.9). An opposing argument, that different forms of life arose independently and descended to the present in linear, unbranched genealogies, is refuted by comparative studies of organismal form, cellular structure, and macromolecular structures (including those of the genetic material, DNA). All of these studies confirm the theory that life's history has the structure of a branching evolutionary tree, called a phylogeny. Species that share relatively recent common ancestry (within the past several million years) have more similar features at all levels than do species whose most recent common ancestor occurred tens or hundreds of millions of years ago. Darwin's theory of common descent guides much ongoing research to reconstruct life's phylogeny using patterns of similarity and dissimilarity observed among species. The resulting phylogeny provides the basis for our taxonomic classification of animals (see Chapter 4).

3. **Multiplication of species.** Darwin's third theory states that evolution produces new species by splitting and transforming older ones. This theory adds a spatial dimension to evolutionary processes. When populations of a species become isolated from each other by geographic barriers, the isolated populations undergo separate evolutionary change and can diverge from each other. For example, when sea level was higher in the past than it is now, low areas of Cuba were inundated, dividing its land area into multiple isolates. Lizard populations that were formerly parts of a single species evolved species-level differences in isolation before another lowering of sea level reconsolidated Cuba as we know it today.

   Species are now generally viewed as reproductively distinct populations of organisms that usually but not always differ from each other in organismal form. Once species are fully formed, they propagate as separate evolutionary lineages, and interbreeding does not occur freely among members of different species, or the resulting hybrid offspring do not persist. Evolutionists generally agree that splitting and transformation of lineages produce new species, although much controversy remains concerning the details of this process and the precise meaning of the term "species"(see Chapter 4). Biologists are actively studying evolutionary processes that generate new species.

4. **Gradualism.** Darwin's theory of gradualism states that large differences in anatomical traits that characterize different species originate by accumulation of many small incremental changes over very long periods of time. This theory opposes the notion that large anatomical differences arise by sudden genetic changes. This theory is important because genetic changes having very large effects on organismal form are usually harmful to an organism. It is possible, however, that some genetic variants that have large effects on an organism are nonetheless sufficiently beneficial to be favored by natural selection. Therefore, although we know that gradual evolution occurs, it does not necessarily explain the origins of all structural differences observed among species. Scientists are studying this question actively.

5. **Natural selection.** Natural selection explains why organisms are constructed to meet the demands of their environments, a phenomenon called **adaptation.** This theory describes a natural process by which populations accumulate favorable characteristics throughout long periods of evolutionary time. Adaptation formerly was considered strong evidence against evolution. Darwin's theory

Darwin's Explanatory Model of Evolution by Natural Selection

Observation 1
Organisms have great potential fertility, which permits exponential growth of populations. (*Source:* Thomas Malthus)

Observation 2
Natural populations normally do not increase exponentially but remain fairly constant in size. (*Source:* Charles Darwin and many others)

Observation 3
Natural resources are limited. (*Source:* Thomas Malthus)

Inference 1
A struggle for existence occurs among organisms in a population. (*Source:* Thomas Malthus)

Observation 4
Variation occurs among organisms within populations. (*Source:* animal breeding and systematics)

Observation 5
Variation is heritable. (*Source:* animal breeding)

Inference 2
Varying organisms show differential survival and reproduction, favoring advantageous traits (= natural selection). (*Source:* Charles Darwin)

Inference 3
Natural selection, acting over many generations, gradually produces new adaptations and new species. (*Source:* Charles Darwin)

Source: E. Mayr, One Long Argument, 1991, Harvard University Press, Cambridge, MA.

of natural selection was therefore important for convincing people that a natural process, amenable to scientific study, could produce new adaptations and new species. Demonstration that natural processes could produce adaptation was important to the eventual acceptance of all five Darwinian theories. Darwin developed his theory of natural selection as a series of five observations, and he made three inferences from them (see box on this page):

Observation l: Organisms have great potential fertility. All populations produce large numbers of gametes and potentially large numbers of offspring each generation. Population size would increase exponentially at an enormous rate if all of the individuals produced each generation survived and reproduced. Darwin calculated that, even for slow-breeding organisms such as elephants, a single pair breeding from age 30 to 90 and having only six offspring could produce 19 million descendants in 750 years.

Observation 2: Natural populations normally remain constant in size, except for minor fluctuations. Natural populations fluctuate in size across generations and sometimes go extinct, but no natural populations show the continued exponential growth that their reproductive capacity theoretically could sustain.

Observation 3: Natural resources are limited. Exponential growth of a natural population would require unlimited natural resources to provide food and habitat for an expanding population, but natural resources are finite.

> **Inference 1: A continuing *struggle for existence* occurs among members of a population.** Survivors represent only a portion, often a very small portion, of all individuals produced each generation. Darwin wrote in *On the Origin of Species* that "it is the doctrine of Malthus applied with manifold force to the whole animal and vegetable kingdoms." Struggle for food, shelter, and space becomes increasingly severe as overpopulation develops.

Observation 4: All populations show organismal *variation*. No two individuals are exactly alike. They differ in size, color, physiology, behavior, and many other ways.

Thomas Henry Huxley (1825–1895), one of England's greatest zoologists, on first reading the convincing evidence of natural selection in Darwin's *On the Origin of Species*, is said to have exclaimed, "How extremely stupid not to have thought of that!" He became Darwin's foremost advocate and engaged in often bitter debates with Darwin's critics. Darwin, who disliked publicly defending his own work, was glad to leave such encounters to his "bulldog," as Huxley called himself.

Observation 5: Variation is heritable. Darwin noted that offspring tend to resemble their parents, although he did not understand how. Many years later, the hereditary mechanism discovered by Gregor Mendel would be applied to Darwin's theory.

Inference 2: Varying organisms show *differential survival and reproduction* favoring advantageous traits (= natural selection). Survival in a struggle for existence is not random with respect to the hereditary variation in a population. Some traits give their possessors an advantage in using their environmental resources for effective survival and reproduction. Survivors transmit their favored traits to offspring, thereby causing those traits to accumulate in the population.

Inference 3: Over many generations, natural selection gradually produces new adaptations and new species. Differential reproduction of varying organisms gradually transforms species and causes their long-term "improvement." Darwin knew that people often use hereditary variation to produce useful new breeds of livestock and plants. *Natural* selection acting over millions of years should be even more effective in producing new types than *artificial* selection imposed during a human lifetime. Natural selection acting independently on geographically separated populations would cause them to diverge from each other, thereby generating reproductive barriers that lead to speciation.

Natural selection can be considered a two-step process with a random component and a nonrandom component. Production of variation among organisms is the random part. Mutational processes have no inherent tendency to generate traits that are favorable to an organism. The nonrandom part is differential persistence of different traits, determined by the effectiveness of traits in permitting their possessors to use environmental resources to survive and to reproduce.

Differential survival and reproduction among varying organisms is called **sorting** and should not be equated with natural selection. We now know that even random processes (genetic drift, see p. 29) can produce sorting. If a garden planted with equal numbers of red- and white-flowered plants suffers severe damage from a hurricane, it is unlikely that equal numbers of red- and white-flowered plants will survive to produce seeds. If red-flowered plants constitute 70% of the survivors, sorting has occurred in favor of the red-flowered plants. In this case, flower color provided no advantage in withstanding the hurricane damage. Most likely, a larger number of red-flowered plants happened to be growing in better-protected locations, permitting their differential survival. This sorting therefore cannot be attributed to natural selection because the character being sorted had no causal influence on the outcome. If in the same garden, white-flowered plants produced more seeds and offspring because they were more visible to a nocturnal moth pollinator, we would observe sorting favoring the white flowers and could attribute this sorting to selection; white flower color in this case provided a reproductive advantage over red color, leading the white-flowered plants to increase in frequency in the next generation. Darwin's theory of natural selection states that sorting occurs *because certain traits give their possessors advantages in survival and reproduction* relative to others that lack those traits. Therefore, selection is one specific cause of sorting.

The popular phrase "survival of the fittest" predates the publication of Darwin's theory of natural selection; it was introduced by the British social philosopher Herbert Spencer, and later applied to Drawin's theory. Unfortunately, this phrase often implies unbridled aggression and violence in a bloody, competitive world. In fact, natural selection operates through many other characteristics of living organisms. For example, many Russian evolutionists argued that animals practicing mutual aid enjoyed the greatest survival advantages in harsh climates.

Evidence for Darwin's Five Theories of Evolution

Perpetual Change

Perpetual change in the form and diversity of animal life throughout its 600- to 700-million-year history reveals itself most directly in the fossil record. A **fossil** is a remnant of past life uncovered from the earth's crust (figure 1.10). Some fossils

A

B

C

D

figure 1.10

Four examples of fossil material. **A**, Stalked crinoids (sea lilies, class Crinoidea, phylum Echinodermata; see p. 304) from 85-million-year-old Cretaceous rocks. The fossil record shows that these echinoderms reached their greatest diversity millions of years earlier and began a slow decline to the present. **B**, The fossil of an insect that got stuck in the resin of a tree 40 million years ago, after which the resin hardened into amber. **C**, Fish fossil from rocks of the Green River Formation, Wyoming. Such fish swam here during the Eocene epoch, approximately 55 million years ago. **D**, Electron micrograph of tissue from a fly fossilized as shown in **B**; the nucleus of a cell is marked in red.

constitute complete remains (insects in amber and mammoths), actual hard parts (teeth and bones), or petrified skeletal parts infiltrated with silica or other minerals (ostracoderms and molluscs). Other fossils include molds, casts, impressions, and fossil excrement (coprolites). In addition to documenting organismal evolution, fossils reveal profound changes in the earth's physical environments, including major changes in the locations of lands and seas. Because many organisms left no fossils, a complete record of past history is always beyond our reach; nonetheless, discovery of new fossils and reinterpretation of familiar ones expand our knowledge of how the forms and diversity of animals changed through geological time. Evolutionary study of the fossil record is called paleontology.

On rare occasions, fossil remains include soft tissues preserved so well that electron microscopy reveals recognizable cellular organelles! Insects and even small vertebrates, such as lizards, can be entombed in amber, the fossilized resin of trees. One study of a fly entombed in 40-million-year-old amber revealed structures corresponding to muscle fibers, nuclei, ribosomes, lipid droplets, endoplasmic reticulum, and mitochondria (figure 1.10D). This extreme case of mummification probably occurred because chemicals in the plant sap diffused into the embalmed insect's tissues. A fictional extraction and cloning of DNA from embalmed insects that had bitten and then sucked the blood of dinosaurs was the technical basis for Michael Crichton's best-seller *Jurassic Park*.

Interpreting the Fossil Record

The fossil record is biased because preservation is selective. Vertebrate skeletal parts and invertebrates with shells and other hard structures left the best record (figure 1.10). Soft-bodied animals, including jellyfishes and most worms, are fossilized only under very unusual circumstances, such as those that formed the Burgess Shale of British Columbia (figure 1.11). Exceptionally favorable conditions for fossilization produced a Precambrian fossil bed in South Australia, tar pits at Rancho La Brea (Hancock Park, Los Angeles), great dinosaur beds (Alberta, Canada, and Jensen, Utah; figure 1.12), the Olduvai Gorge of Tanzania, and the early Cambrian Chengjiang beds of China.

Pikaia, an early chordate

Amiskwia, from an extinct phylum

Odontogriphus, from an extinct phylum

Eldonia, a possible echinoderm

A   Fossil trilobite from Burgess Shale

Anomalocaris canadensis, from an extinct phylum

Halichondrites, a sponge

Aysheaia, an onychophoran

Opabinia, from an extinct phylum

Xianguangia, an anenome-like animal

Yohoia, a unique arthropod

Sidneyia, a unique arthropod

Canadia, a polychaete

Marrella splendens, a unique arthropod

Wiwaxia, from an extinct phylum

Ottoia, a priapulid

Dinomischus, from an extinct phylum

Hallucigenia, from an extinct phylum

B

figure 1.11

A, Fossil trilobite visible at the Burgess Shale Quarry, British Columbia. **B**, Animals of the Cambrian period, approximately 540 million years ago, as reconstructed from fossils preserved in the Burgess Shale of British Columbia, Canada. The major body plans of living animals appeared rather abruptly at this time.

Fossils are deposited in stratified layers, with new deposits forming above older ones. If left undisturbed, which is rare, a sequence is preserved so that the ages of fossils are directly proportional to their depth in stratified layers. Stratigraphy is the study of fossil-bearing rocks. Characteristic fossils often serve to identify particular layers. Certain widespread marine invertebrate fossils, including various foraminiferans (see p. 111) and echinoderms (see p. 292), are such good indicators of specific geological periods that we call them "index," or "guide," fossils. Unfortunately, layers are usually tilted or folded or show faults (cracks). Old deposits exposed by erosion might be covered by new deposits in a different plane. When exposed to tremendous pressures or heat, stratified sedimentary rock metamorphoses into crystalline quartzite, slate, or marble, thereby destroying fossils.

Stratigraphy for two major groups of African antelopes and its evolutionary interpretation are shown in figure 1.13. These antelope species are identified by the characteristic sizes and shapes of their horns, which form much of the fossil record of this group. Solid vertical lines in figure 1.13 denote the temporal distributions of species as determined by the presence of their characteristic horns in rock strata of various ages. Red lines denote the fossil records of living species, and gray lines denote the fossil records of extinct species. The dotted gray lines show the inferred relationships among living and fossil species based upon their sharing of homologous structural features.

figure 1.12

A fossil skeleton from Dinosaur Provincial Park, Alberta, Canada.

Geological Time

Long before the earth's age was known, geologists divided its history into a table of succeeding events based on ordered layers of sedimentary rock. The "law of stratigraphy" produced a relative dating, with the oldest layers at the bottom and the most recent at the top of a sequence. Time was divided into eons, eras, periods, and epochs as shown on the endpapers inside the back cover of this book. Time during the last eon (Phanerozoic) is expressed in eras (for example, Cenozoic), periods (for example, Cambrian), epochs (for example, Paleocene), and sometimes smaller divisions of an epoch.

In the late 1940s, radiometric dating methods were developed for determining the absolute ages (in years) of rock formations. Several independent methods are now used, all based on radioactive decay of naturally occurring elements into other elements. These "radioactive clocks" are independent of pressure and temperature changes and therefore not affected by often violent earth-building activities.

One method, potassium-argon dating, uses the decay of potassium-40 (^{40}K) to argon-40 (^{40}Ar) (12%) and calcium-40 (^{40}Ca) (88%). The half-life of potassium-40 is 1.3 billion years,

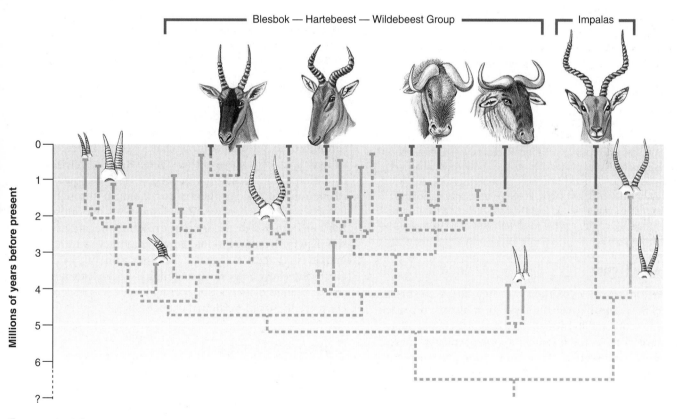

figure 1.13

Stratigraphic record and inferred evolutionary relationships among alcelaphine (blesboks, hartebeests, wildebeests) and aepycerotine (impalas) antelopes in Africa. Species in this group are identified by characteristic sizes and shapes of horns found in rock strata of various ages. Solid vertical lines show the temporal distribution of species in rock strata whose ages are shown on the scale at the left side of the figure. Red lines show the temporal distributions of living species, and gray lines show the temporal distributions of extinct species in rock strata. Dotted gray lines show the inferred relationships among species based on their sharing of homologous structural features. The relative constancy of horn structure within species through geological time is consistent with the theory of punctuated equilibrium (see p. 25). This fossil record shows that rates of speciation and extinction are higher for alcelaphine antelopes than for impalas.

meaning that half of the original atoms will decay in 1.3 billion years, and half of the remaining atoms will be gone at the end of the next 1.3 billion years. This decay continues until all radioactive potassium-40 atoms are gone. To measure the age of a rock, one calculates the ratio of remaining potassium-40 atoms to the amount of potassium-40 originally there (the remaining potassium-40 atoms plus the argon-40 and calcium-40 into which they have decayed). Several such isotopes exist for dating purposes, some of which help to determine the age of the earth itself. One of the most useful radioactive clocks depends on decay of uranium into lead. With this method, rocks over 2 billion years old can be dated with a probable error of less than 1%.

The fossil record of macroscopic organisms begins near the start of the Cambrian period of the Paleozoic era, approximately 600 million years before present (BP). Geological time before the Cambrian period is called the Precambrian era or the Proterozoic eon. Although the Precambrian era occupies 85% of all geological time, it receives much less attention than do later eras, partly because oil, which provides a commercial incentive for much geological work, seldom exists in Precambrian formations. The Precambrian era contains well-preserved fossils of bacteria and algae, and casts of jellyfishes, sponge spicules, soft corals, segmented flatworms, and worm trails. Most, but not all, are microscopic fossils.

The more well-known carbon-14 (^{14}C) dating method is of little help in estimating ages of geological formations because its short half-life restricts the use of ^{14}C to quite recent events (less than about 40,000 years). It is especially useful, however, for archaeological studies. This method is based on the production of radioactive ^{14}C (half-life of approximately 5570 years) in the upper atmosphere by bombardment of nitrogen-14 (^{14}N) with cosmic radiation. Radioactive ^{14}C enters the tissues of living animals and plants, and an equilibrium is established between atmospheric ^{14}C and ^{14}N in living organisms. At death, ^{14}C exchange with the atmosphere stops. In 5570 years, only half of the original ^{14}C remains in a preserved fossil. One estimates a fossil's age by comparing the ^{14}C content of the fossil with that of living organisms.

Evolutionary Trends

The fossil record reveals evolutionary change across the broadest scale of time. Throughout the geological history recorded by the fossil record, millions of species have arisen and almost as many have gone extinct. Animal species typically survive approximately 1 million to 10 million years, although their durations are highly variable. When we study patterns of species or taxon replacement through time, we observe **trends.** Trends are directional changes in characteristic features or patterns of diversity in a group of organisms. Fossil trends clearly demonstrate Darwin's principle of perpetual change.

A well-studied fossil trend is the evolution of horses from the Eocene epoch to the present. Looking back to the Eocene epoch, we see many different genera and species of horses that replaced each other through time (figure 1.14). George Gaylord Simpson (see p. 80) showed that this trend is compatible with Darwinian evolutionary theory. Three characteristics that show the clearest trends in horse evolution are body size, tooth structure, and foot structure. Compared to modern horses, those of extinct genera were small, their teeth had a relatively small grinding surface, and their feet had a relatively large number of toes (four). Throughout the subsequent Oligocene, Miocene, Pliocene, and Pleistocene epochs, new genera arose and old ones went extinct. In each case, there was a net increase in body size, expansion of the grinding surface of teeth, and reduction in number of toes. As the number of toes declined, the central digit became increasingly more prominent in the foot, and eventually only this central digit remained.

The fossil record shows a net change not only in the characteristics of horses but also in the numbers of different horse genera (and numbers of species) that exist through time. Many horse genera of past epochs have been lost to extinction, leaving only a single survivor. Evolutionary trends in diversity are observed in fossils of many different groups of animals (figure 1.15).

Our use of "evolutionary trend" does not imply that more recent forms are superior to older ones or that the changes represent progress in adaptation or organismal complexity. Although Darwin predicted that such trends would show progressive adaptation, many contemporary paleontologists consider progressive adaptation rare among evolutionary trends. Observed trends in the evolution of horses do not imply that contemporary horses are superior in any general sense to their Eocene predecessors.

Trends in fossil diversity through time are produced by different rates of species formation versus extinction through time. Why do some lineages generate large numbers of new species whereas others generate relatively few? Why do different animal groups undergo higher or lower rates of extinction (of species, genera, or families) throughout evolutionary time? To answer these questions, we must turn to Darwin's other four theories of evolution. Regardless of how we answer these questions, however, observed trends in animal diversity clearly illustrate Darwin's principle of perpetual change. Because Darwin's remaining four theories rely on perpetual change, evidence supporting these theories strengthens Darwin's theory of perpetual change.

Common Descent

Darwin proposed that all plants and animals have descended from some one form into which life was first "breathed." Life's history forms a branching tree, called a phylogeny, that gives all of life a unified evolutionary history. Pre-Darwinian evolutionists, including Lamarck, advocated multiple independent origins of life, each of which gave rise to lineages that changed through time without the extensive branching required by Darwin's theory. Like all good scientific theories, common descent makes several important predictions that can be tested and potentially used to reject it. According to this theory, one should be able to trace the genealogies of all modern

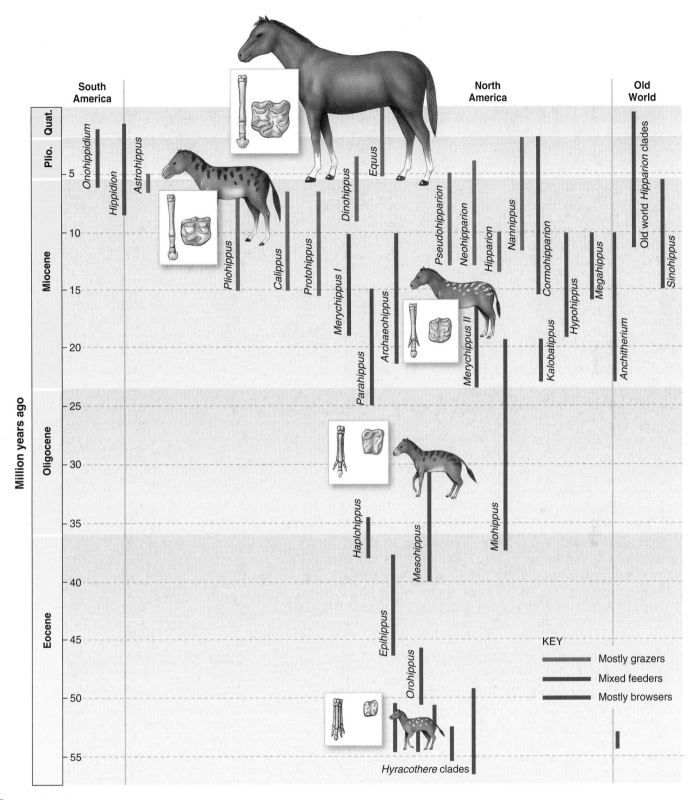

figure 1.14

Stratigraphy of genera of horses from the Eocene epoch to the present. Evolutionary trends toward increased size, elaboration of molars, and loss of toes are shown along with bars denoting temporal durations and continental locations of genera.

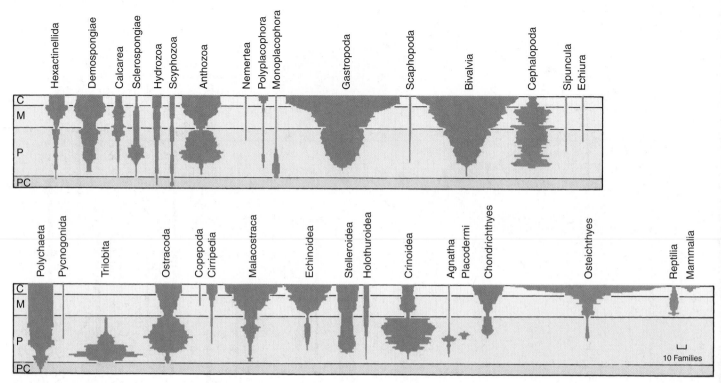

figure 1.15

Diversity profile s of taxonomic families from different animal groups in the fossil record. A scale marks the Precambrian (PC), Paleozoic (P), Mesozoic (M), and Cenozoic (C) eras. The number of families is indicated by the width of each profile.

species backward until they converge on ancestral lineages shared with other species, both living and extinct. We should be able to continue this process, moving farther backward through evolutionary time to reach a primordial ancestor of all life on earth. All forms of life, including many extinct forms that represent dead branches, connect to this tree somewhere. Although reconstructing a history of life in this manner may seem almost impossible, it has in fact been extraordinarily successful. How has this difficult task been accomplished?

Homology and Reconstruction of Phylogeny

Darwin recognized a major source of evidence for common descent in the concept of **homology**. Darwin's contemporary, Richard Owen (1804–1892), used this term to denote "the same organ in different organisms under every variety of form and function." A classic example of homology is the limb skeleton of vertebrates. Bones of vertebrate limbs maintain characteristic structures and patterns of connection despite diverse modifications for different functions (figure 1.16). According to Darwin's theory of common descent, structures that we call homologies represent characteristics inherited with some modification from a corresponding feature in a common ancestor.

Darwin devoted an entire book, *The Descent of Man and Selection in Relation to Sex,* largely to the idea that humans share common descent with apes and other animals. This

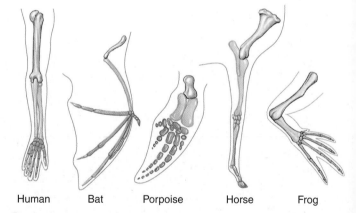

Human   Bat   Porpoise   Horse   Frog

figure 1.16

Forelimbs of five vertebrates show skeletal homologies: brown, humerus; orange, radius and ulna; purple, "hand" (carpals, metacarpals, and phalanges). Homologies of bones and patterns of connection are evident despite evolutionary modification for various uses.

idea was repugnant to many Victorians, who responded with outrage (figure 1.17). Darwin built his case mostly on anatomical comparisons that revealed homology between humans and apes. To Darwin, the close resemblances between apes and humans could be explained only by common descent.

Throughout the history of all forms of life, evolutionary processes generate new characteristics that are transmitted across generations. Every time a new feature becomes established

figure 1.17

This 1873 advertisement for Merchant's Gargling Oil ridicules Darwin's theory of the common descent of humans and apes, which was widely doubted by the general public during Darwin's lifetime.

in a lineage destined to be ancestral to others, a new homology originates. The pattern formed by these homologies provides evidence for common descent and allows us to reconstruct a branching evolutionary history of life. We can illustrate such evidence using a phylogenetic tree of ground-dwelling flightless birds (figure 1.18). A new skeletal homology (see figure 1.18) arises on each of the lineages shown (descriptions of these homologies are not included because they are highly technical). The different groups of species located at the tips of the branches contain different combinations of these homologies, thereby revealing common ancestry. For example, ostriches show homologies 1 through 5 and 8, whereas kiwis show homologies 1, 2, 13, and 15.

The tree structure inferred from analysis of skeletal structures of flightless birds can be tested by data gathered independently from DNA-sequence information (see Chapter 4). The phylogeny of flightless birds inferred from DNA-sequence data does not agree completely with the one inferred from skeletal structures (figure 1.18); if we choose the hypothesis favored by DNA-sequence data, then we must hypothesize that some of the skeletal structures either arose

multiple times or were lost on some lineages as shown on figure 1.18B. The conflict between the phylogenetic hypotheses derived from skeletal structures and from DNA sequences requires that systematists examine their phylogenetic characters and analyses for sources of error in inferring the detailed phylogenetic relationships among these species. All of the phylogenetic data support the hypothesis that these flightless birds are more closely related to each other than they are to any other living species.

Branches of an evolutionary tree combine species into a **nested hierarchy** of groups within groups (see Chapter 4). Smaller groups (species grouped near terminal branches) are contained within larger ones (species grouped by basal branches, including the trunk of the tree). If we erase the tree structure but retain the patterns of homology observed in the terminal groups of species, we can reconstruct the branching structure of the entire tree. Evolutionists test the theory of common descent by observing patterns of homology present in all groups of organisms. The pattern formed by all homologies taken together should specify a single branching tree that represents the evolutionary genealogy of all living organisms.

The nested hierarchical structure of homology is so pervasive in animals that it forms the basis for our systematic groupings (genera grouped into families, families grouped into orders, orders into classes, classes into phyla, and phyla into the animal kingdom). Plants and fungi show similar hierarchical patterns of homology and corresponding systematic groupings. Hierarchical classification even preceded Darwin's theory because this pattern was so evident, but it was not explained scientifically before Darwin. Once common descent was understood, biologists began investigating structural, molecular, and/or chromosomal homologies of animal groups to infer evolutionary relationships. Taken together, the nested hierarchical patterns uncovered by these studies permit us to reconstruct the evolutionary trees of many groups and to continue testing our phylogenetic hypotheses with new data. Methods presented in Chapter 4 show how we use Darwin's theory of common descent to reconstruct the evolutionary history of life and to construct a taxonomic system that summarizes evolutionary relationships among species.

Characters of different organisms that perform similar functions are not necessarily homologous. The wings of bats and birds, although homologous as vertebrate forelimbs, are not homologous as wings. The most recent common ancestor of bats and birds had forelimbs, but the forelimbs were not in the form of wings. The wings of bats and birds evolved independently and have only superficial similarity in their flight structures.

Bat wings are formed by skin stretched over elongated digits, whereas bird wings are formed by feathers attached along the forelimb. Such functionally similar but nonhomologous structures are often termed *analogues*.

Note that the earlier evolutionary hypothesis that life arose many times, forming unbranched lineages, predicts linear

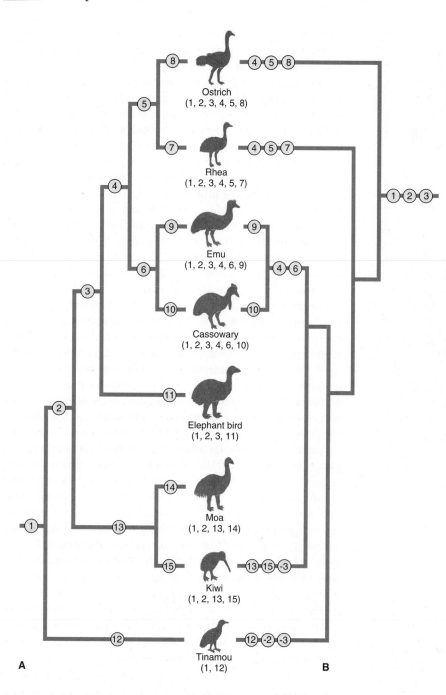

A, The phylogenetic pattern specified by 15 homologous structures in the skeletons of a group of flightless birds. Homologous features are numbered 1 through 15 and are marked both on the branches of the tree on which they arose and on the birds that have them. If you were to erase the tree structure, you would be able to reconstruct it without error from the distributions of homologous features shown for the birds at the terminal branches. **B**, A recent analysis of molecular data suggests a different pattern of relationships among the living flightless birds (all except moas and elephant birds). If the molecular analysis is correct, then one must reinterpret evolution of skeletal characters 2, 3, 4, and 5. In the interpretation shown, character 2 is lost by tinamous (−2), and character 3 is lost by both kiwis and tinamous (−3). Character 4 arises independently in ostriches, rheas, and a common ancestor of cassowaries and emus. Character 5 originates separately in ostriches and rheas. Multiple origins and losses complicate phylogenetic analysis, as explained in Chapter 4.

figure 1.18

sequences of evolutionary change with no nested hierarchy of homologies among species. Because we do observe nested hierarchies of homologies, that hypothesis is rejected. Note also that a supernatural creationist argument can make no testable predictions about any pattern of homology and therefore fails the criteria of a scientific theory of animal diversity.

Ontogeny, Phylogeny, and Recapitulation

Zoologists find in animal development important clues to an animal's evolutionary history. **Ontogeny** is the history of an organism's development through its entire life. Early developmental and embryological features contribute greatly to our knowledge of homology and common descent. Comparative studies of ontogeny show how evolutionary alteration of developmental timing generates new **phenotypes** (expressed characteristics, or the appearance of an organism), thereby causing evolutionary divergence among lineages.

Comparisons of gene expression among animals show that in forms as dissimilar as insects and humans, homologous genes may guide developmental differentiation of anterior versus posterior body segments. For example, mutations of such genes, termed **homeotic genes,** in fruit flies can cause awkward developmental changes such as legs in the place of antennae or an extra pair of wings. Such genes provide an evolutionary "tool kit" that can be used to construct new body parts by relocating patterns of gene expression to different parts of a developing embryo. Perhaps the most famous homeotic genes are those containing a sequence of 180 base pairs, called the **homeobox,** which encodes a protein sequence that binds to other genes, thereby altering their expression. For example, evolution of the paired limbs of terrestrial vertebrates occurred by activating a caudal "tool kit" of homeobox genes at the paired pectoral and pelvic regions of the body where limbs develop. Genetic and developmental tools evolved for constructing a tail were thus available for the new role of constructing paired limbs.

The German zoologist Ernst Haeckel, a contemporary of Darwin, proposed the influential hypothesis that each successive stage an in organism's development represented an adult form present in its evolutionary history. For example, a human embryo with gill depressions in its neck was considered to signify a fishlike ancestor. On this basis, Haeckel gave his generalization: *ontogeny (individual development) recapitulates (repeats) phylogeny (evolutionary descent).* This notion later was simply called **recapitulation,** or the **biogenetic law.** Haeckel based his biogenetic law on a flawed premise that evolutionary change occurs by successively adding stages onto the end of an unaltered ancestral ontogeny, condensing the ancestral ontogeny into earlier developmental stages. This notion was based on Lamarck's concept of inheritance of acquired characteristics (see p. 6).

A nineteenth-century embryologist, K. E. von Baer, gave a more satisfactory explanation of the relationship between ontogeny and phylogeny. He argued that early developmental features were simply more widely shared among different animal groups than were later ones. For example, figure 1.19 shows early embryological similarities between organisms whose adult forms are very different. Adults of animals with relatively short and simple ontogenies often resemble the pre-adult stages of other animals whose ontogeny is more elaborate, but the embryos of descendants do not necessarily resemble the adults of their ancestors. However, even early development undergoes evolutionary divergence among groups, and it is not as stable as von Baer thought.

We now know many parallels between ontogeny and phylogeny, but features of an ancestral ontogeny can be shifted to either earlier or later stages in descendant ontogenies.

Evolutionary change in the timing of development is called **heterochrony,** a term initially used by Haeckel to denote exceptions to recapitulation. Because the lengthening or shortening of ontogeny can change different parts of an organism independently, we often see a mosaic of different kinds of developmental evolutionary change in a single lineage. Therefore, cases in which an entire ontogeny recapitulates phylogeny are rare.

Despite many changes in scientific thinking about the relationships between ontogeny and phylogeny, one important fact remains clear. Darwin's theory of common descent is strengthened enormously by the many homologies found among developmental stages of organisms belonging to different species.

Multiplication of Species

Multiplication of species through time is a logical corollary to Darwin's theory of common descent. A branch point on a phylogenetic tree means that an ancestral species has split into two different ones. Darwin's theory postulates that variation present within a species, especially variation between geographically separated populations, provides material from which new species are produced. Because evolution is a branching process, the total number of species produced by evolution increases through time, although most of these species eventually go extinct without leaving descendant species. A major challenge for evolutionists is to discover the processes by which an ancestral species "branches" to form two or more descendant species.

Before we explore multiplication of species, we must decide what we mean by "species." No consensus exists regarding the definition of species (see Chapter 4). Most biologists would agree, however, on three important criteria for recognizing a species:

1. Individuals of the same species descend from a common ancestral population and form an unbranched lineage of ancestral-descendant populations.
2. Individuals of the same species exhibit reproductive compatibility (ability to interbreed) among individuals and reproductive incompatibility between species (for sexually reproducing organisms).
3. Individuals within the same species maintain genotypic and phenotypic cohesion (lack of abrupt differences among populations in allelic frequencies [see p. 28] and in organismal appearance).

The criterion of reproductive compatibility has received the greatest attention in studies of species formation, also called **speciation**.

Biological factors that prevent different species from interbreeding are called **reproductive barriers**. A primary problem of speciation is to discover how two initially compatible populations evolve reproductive barriers that cause them to become distinct, separately evolving lineages. How

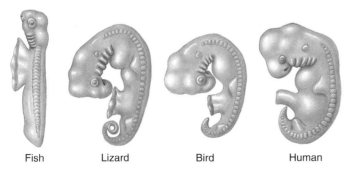

| Fish | Lizard | Bird | Human |

figure 1.19

Comparison of gill arches (purple) of different embryos. All are shown separated from the yolk sac. Note the remarkable similarity of these four embryos at this early stage in development.

do populations diverge from each other in their reproductive properties while maintaining complete reproductive compatibility within each population?

Geographical barriers between populations are not equivalent to reproductive barriers. Geographical barriers refer to spatial separation of two populations. They prevent gene exchange and are usually a precondition for speciation. Reproductive barriers are various morphological, physiological, ecological, and behavioral factors that prevent interbreeding between different species. Geographical barriers do not guarantee that reproductive barriers will evolve. Reproductive barriers are most likely to evolve under conditions that include small population size, a favorable combination of selective factors, and long periods of geographical isolation. One or both of a pair of geographically isolated populations may become extinct prior to evolution of reproductive barriers between them. Over the vast span of geological time, however, conditions sufficient for speciation have occurred many millions of times.

Reproductive barriers between populations usually evolve gradually. Evolution of reproductive barriers requires that diverging populations be kept physically separate ("isolated") for long periods of time. If the diverging populations were reunited before reproductive barriers were completely formed, interbreeding would occur between the populations and they would merge. Speciation by gradual divergence in animals usually requires perhaps 10,000 to 100,000 years or more. Geographical isolation followed by gradual divergence is the most effective way for reproductive barriers to evolve, and many evolutionists consider geographical separation a prerequisite for branching speciation. Speciation that results from evolution of reproductive barriers between geographically separated populations is called **allopatric speciation**, or geographical speciation. For example, geological formation of the Isthmus of Panama fragmented an ancestral population of *Eucidaris* sea urchins, leading to formation of the pair of species shown in figure 1.20.

Evidence for allopatric ("in another land") speciation occurs in many forms, but perhaps most convincing is an occurrence of geographically separated but adjoining, closely related populations that illustrate the gradual origin of reproductive barriers. Populations of a salamander, *Ensatina eschscholtzii*, in California are a particularly clear example (figure 1.21). These populations show evolutionary divergence in color pattern and collectively form a geographical ring around California's central valley. Genetic exchange between differentiated, geographically adjoining populations is evident through the formation of hybrids and occasionally regions of extensive genetic exchange (called zones of introgression). Two populations at the southern tip of the geographical range (called *E. e. eschscholtzii* and *E. e. klauberi*) make contact but do not interbreed. A gradual accumulation of reproductive differences among contiguous populations around the ring is visible, with the two southernmost populations separated by strong reproductive barriers.

figure 1.20

Geographic formation of species in the sea urchin (*Eucidaris*). Formation of the Isthmus of Panama separated an ancestrel population, leading to evolution of separate Caribbean (*E. tribuloides*) and Pacific (*E. thouarsi*) species.

Additional evidence for allopatric speciation comes from observations of animal diversification on islands. Oceanic islands formed by volcanoes are initially devoid of life. Plants and animals from a continent or from other islands occasionally colonize new islands in separate invasions. Invaders often encounter situations ideal for evolutionary diversification, because environmental resources that were exploited heavily by other species on the mainland are free for colonization on a sparsely populated island. Because colonization of oceanic islands is rare, populations established on islands are effectively isolated geographically from their parental populations and can undergo divergent evolution, leading to reproductive barriers and speciation. Archipelagoes, such as the Galápagos Islands, greatly increase opportunities for speciation in this manner.

Evolutionists have often wondered whether the isolation of populational gene pools needed for reproductive barriers to evolve might sometimes occur without geographical isolation. Populations that are reproductively active at different seasons or on different substrates could, in theory, achieve gene-pool isolation without geographic isolation. The term *sympatric speciation* is used to denote species formation not involving geographical isolation. Intermediate between allopatric and sympatric speciation is a condition in which the diverging populations are geographically separate but make contact along a borderline; speciation arising in this manner is called *parapatric speciation*. Many zoologists doubt that sympatric and parapatric speciation have been important in animal evolution.

figure 1.21

Speciation in progress: geographic variation of color patterns in the salamander genus *Ensatina*. Populations of *Ensatina eschscholtzii* form a geographic ring around the Central Valley of California. Adjacent, differentiated populations throughout the ring can exchange genes except at the bottom of the ring where the sub-species *E. e. eschscholtzii* and *E. e. klauberi* overlap without interbreeding. These two subspecies would be recognized as distinct species if intermediate populations linking them across the ring were extinct. This example demonstrates that reproductive barriers between populations can evolve gradually.

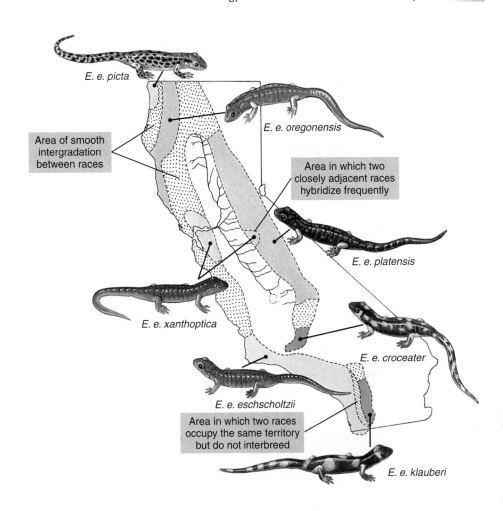

Area of smooth intergradation between races

Area in which two closely adjacent races hybridize frequently

Area in which two races occupy the same territory but do not interbreed

E. e. picta
E. e. oregonensis
E. e. platensis
E. e. xanthoptica
E. e. croceater
E. e. eschscholtzii
E. e. klauberi

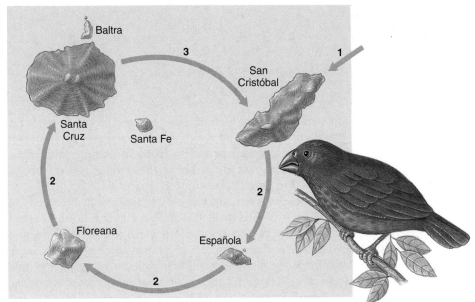

Baltra
San Cristóbal
Santa Cruz
Santa Fe
Floreana
Española

figure 1.22

Tentative model for evolution of the 13 species of Darwin's finches on the Galápagos Islands. This model postulates three steps: (1) Immigrant finches from South America reach the Galápagos and colonize an island; (2) after a population becomes established, finches disperse to other islands where they adapt to new conditions and change genetically; and (3) after a period of isolation, secondary contact is established between different populations. Different populations are then recognized as separate species if they cannot interbreed successfully.

Production of many ecologically diverse species from a common ancestral species is called **adaptive radiation**, especially when these species arise within a short interval of geological time (a few million years). Galápagos finches clearly illustrate adaptive radiation on an oceanic archipelago (figures 1.22 and 1.23).

Galápagos finches (the name "Darwin's finches" was popularized in the 1940s by the British ornithologist David Lack) are close relatives, but each species differs from others in the size and shape of its beak and in its feeding habits. Darwin's finches descended from a single ancestral population that arrived

figure 1.23

A, Adaptive radiation in 10 species of Darwin's finches from Santa Cruz, one of the Galápagos Islands. Differences in bills and feeding habits are shown. All apparently descended from a single common ancestral finch from South America. **B,** Woodpecker finch, one of 13 species of Galápagos Islands finches, using a slender twig as a tool for feeding. This finch worked for about 15 minutes before spearing and removing a wood roach from a break in the tree.

from South America and subsequently colonized different islands of the Galápagos archipelago. These finches underwent adaptive radiation, occupying habitats that in South America were denied to them by other species better able to exploit those habitats. Galápagos finches thus acquired characteristics of mainland binds as diverse and unfinchlike as warblers and woodpeckers (figure 1.23B). The founding of new island populations by a small number of migrants may have accelerated evolutionary divergence among these island finches (see p. 29). A fourteenth species of finch, found on isolated Cocos Island far north of the Galápagos archipelago, represents yet another speciation event in this impressive adaptive radiation.

Gradualism

Darwin's theory of gradualism opposed arguments for a sudden origin of species. Small differences, resembling those that we observe among organisms within populations today, are the raw material from which different major forms of life evolved. This theory shares with Lyell's uniformitarianism a notion that we must not explain past changes by invoking unusual catastrophic events that are not observable today. If new species originated in single, catastrophic events, we should be able to see these events happening today, and we do not. What we observe instead are small, continuous changes in phenotypes occurring in natural populations. Such continuous changes can produce major differences among species only by accumulating over many thousands to millions of years. A simple statement of Darwin's theory of gradualism is that accumulation of quantitative changes leads to qualitative change.

Phenotypic gradualism was controversial when Darwin first proposed it, and it is still controversial. Not all phenotypic changes are small, incremental ones. For example, some mutations that appear during artificial breeding, traditionally called "sports," change a phenotype substantially in a single mutational step. Sports that produce dwarfing occur in many species, including humans, dogs, and sheep, and have been used by animal breeders to achieve desired results; for example, a sport that deforms limbs was used to produce ancon sheep, which cannot jump hedges and are therefore easily contained (figure 1.24). Many colleagues of Darwin who accepted his other theories considered phenotypic gradualism too extreme. If sports can be used in animal breeding, why must we exclude them from our evolutionary theory? In favor of gradualism, some have replied that sports always have negative side effects that would prevent affected organisms from surviving in natural populations. Indeed, it is questionable whether the ancon sheep in figure 1.24, despite its attractiveness to farmers, would propagate successfully in the presence of its long-legged relatives without human intervention. Naturalists nonetheless report mutations of large effect that appear adaptive in natural populations; a mutation responsible for a large difference in beak size in the African finch species *Pyrenestes ostrinus* allows large-beaked birds to eat hard seeds, whereas small-beaked forms eat softer seeds.

When we view Darwinian gradualism on a geological timescale, we may expect to find in the fossil record a long series of intermediate forms bridging phenotypes of ancestral and descendant populations (figure 1.25). This predicted pattern is called **phyletic gradualism**. Darwin recognized that phyletic gradualism is not often revealed by the fossil record. Studies conducted since Darwin's time likewise have failed to produce a continuous series of fossils as predicted by phyletic gradualism. Is the theory of gradualism therefore refuted by the fossil record? Darwin and others claim that it is not, because the fossil record is too imperfect to preserve each minor change in animal form that occurs across generations in a species lineage. Although evolution is a slow process by our standards, it is rapid relative to the rate at which good fossil deposits accumulate. Others have argued, however, that abrupt origins and extinctions of species in the fossil record force us to conclude that phyletic gradualism is rare.

Niles Eldredge and Stephen Jay Gould proposed a model called **punctuated equilibrium** in 1972 to explain discontinuous evolutionary changes observed throughout geological time. Punctuated equilibrium states that phenotypic evolution is concentrated in relatively brief events of branching speciation, followed by much longer intervals of morphological evolutionary stasis (figure 1.26). Speciation is an episodic event, occurring over a period of approximately 10,000 to 100,000 years. Because species may survive for 5 million to 10 million years, the speciation event is a "geological instant," representing 1% or less of a species' existence. Ten thousand years is plenty of time, however, for Darwinian evolution to accomplish dramatic changes. Therefore, a small fraction of the total evolutionary history of a group accounts for most of the morphological evolutionary change that occurs.

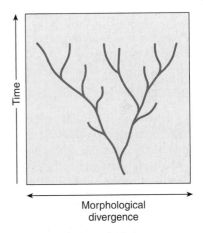

figure 1.25

A gradualist model of evolutionary change in morphology, viewed as proceeding more or less steadily through geological time (millions of years). Bifurcations followed by gradual divergence led to speciation.

figure 1.24

The ancon breed of sheep arose from a "sport" mutation that caused dwarfing of their legs. Many of Darwin's contemporaries criticized him for claiming that such large mutations are not important for evolution by natural selection.

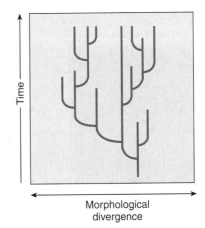

figure 1.26

A punctuated equilibrium model shows evolutionary change concentrated in relatively rapid bursts of branching speciation (*lateral lines*) followed by prolonged periods of little change throughout geological time (millions of years).

Evolutionists who lamented the imperfect state of the fossil record were treated in 1981 to the opening of an uncensored page of fossil history in Africa. Peter Williamson, a British paleontologist working in fossil beds 400 m deep near Lake Turkana, documented a remarkably clear record of speciation in freshwater snails. Geological strata of the Lake Turkana basin reveal a history of instability. Earthquakes, volcanic eruptions, and climatic changes caused waters to rise and to fall episodically, sometimes by hundreds of feet. Thirteen lineages of snails show long periods of stability interrupted by relatively brief periods of rapid change in shell shape when snail populations were fragmented by receding waters. These populations diverged to produce new species that then remained unchanged through thick deposits before becoming extinct and being replaced by descendant species. The transitions occurred within 5000 to 50,000 years. In the few meters of sediment where speciation occurred, transitional forms were visible. Williamson's study conforms well to a punctuated equilibrium model, which remains an important challenge to gradualism on a geological timescale.

Natural Selection

Many examples show how natural selection alters populations in nature. Sometimes selection can proceed very rapidly as happened, for example, when flies and mosquitoes quickly developed high resistance to insecticides. Doses that at first killed almost all pests later failed to control them. As more insects were exposed to insecticides, those most sensitive were killed, leaving more space and less competition for resistant strains to multiply. Thus, as a result of selection, mutations bestowing high resistance, but previously rare in these populations, increased in frequency. Likewise, widespread medical use of some antibiotics, such as penicillin, has caused bacterial strains to evolve resistance to these antibiotics. Some staphylococci secrete an enzyme, called penicillinase, that inactivates penicillin and confers bacterial resistance to it.

Perhaps the most famous instance of rapid evolution by natural selection is that of industrial melanism (dark pigmentation) in the peppered moths of England (figure 1.27). Before 1850, peppered moths were white with black speckling on their wings and body. In 1849, a mutant black form of the species appeared. It became increasingly common, reaching frequencies of 98% in Manchester and other heavily industrialized areas by 1900. Peppered moths, like most moths, are active at night and rest in exposed places during daytime, depending upon matching their backgrounds using cryptic coloration to conceal themselves from potential predators. Experimental studies have shown that, consistent with the hypothesis of natural selection, birds often eat moths that do not match their surroundings, but birds in the same area frequently fail to find moths that match their surroundings. The mottled pattern of the normal white form blends well with lichen-covered tree trunks. With increasing industrialization, soot from thousands of chimneys darkened the bark of trees for miles around centers such as Manchester. Against this dark background, white moths were conspicuous to predatory birds, whereas mutant black forms were camouflaged. When pollution diminished, the frequency of lightly pigmented individuals increased in moth populations (figure 1.27C), just as the hypothesis of natural selection predicted. However, using the hypothesis of natural selection to explain industrial melanism requires further testing, particularly more detailed studies of the behavioral ecology of peppered moths and their predators.

A recurring criticism of natural selection is that it cannot generate new structures or species but can only modify existing ones. Most structures in their early evolutionary stages could not have performed roles that the fully formed structures perform, and therefore it is unclear how natural selection could have favored them. What use is half a wing or the rudiment of a feather for a flying bird? To answer this criticism, Darwin proposed that many structures evolved initially for purposes different from the uses that they have today. Rudimentary feathers could have been useful in thermoregulation, for example. Their role in flying would have evolved later after they incidentally acquired aerodynamic properties that

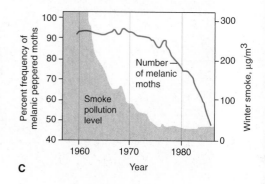

A

B

C

figure 1.27

Light and melanic forms of peppered moths, *Biston betularia*, on **A**, an unpolluted lichen-covered tree and **B**, a soot-covered tree near industrial Birmingham, England. These color variants have a simple genetic basis. **C**, Recent decline in frequency of the melanic form due to diminished air pollution in industrial areas of England. Frequency of the melanic form exceeded 90% in 1960, when smoke and sulfur dioxide emissions were still high. Later, as emissions fell and light-colored lichens began to grow again on tree trunks, the melanic form became more conspicuous to predators. By 1986, only 50% of the moths were melanic, the rest having been replaced by the light form.

subjected them to selection for improved flying. **Exaptation** denotes the utility of a structure for a biological role that was not part of the structure's evolutionary origin. Exaptation contrasts with adaptation, which implies that a structure arose by natural selection for the utility in question. Bird feathers are therefore adaptations for thermoregulation but exaptations for flight. Because the anatomical differences observed among organisms from different, closely related species resemble variation observed within species, it is unreasonable to propose that selection cannot lead beyond a species boundary.

Revisions of Darwinian Evolutionary Theory

Neo-Darwinism

The most serious weakness in Darwin's argument was his failure to identify correctly a mechanism for inheritance. Darwin saw heredity as a blending phenomenon in which the hereditary factors of parents melded in their offspring. Darwin also invoked the Lamarckian hypothesis that an organism could alter its heredity through use and disuse of body parts and through direct environmental influence. Darwin did not realize that hereditary factors could be discrete and nonblending, and that a new genetic variant therefore could persist unaltered from one generation to the next. The German biologist August Weismann (1834-1914) rejected Lamarckian inheritance by showing experimentally that modifications of an organism during its lifetime do not change its heredity, and he revised Darwinian evolutionary theory accordingly. We now use the term **neo-Darwinism** to denote Darwinian evolutionary theory as revised by Weismann. The genetic basis of neo-Darwinism eventually became what we now call the **chromosomal theory of inheritance**, a synthesis of Mendelian genetics and cytological studies of the segregation of chromosomes into gametes.

Gregor Mendel published his theories of inheritance in 1868, 14 years before Darwin died. Darwin presumably never read this work, although it was found in Darwin's extensive library after his death. Had Mendel written Darwin about his results, it is possible that Darwin would have modified his theory accordingly. However, it is also possible that Darwin could not have seen the importance of hereditary mechanisms to the continuous variations and gradual changes that represent the hub of Darwinian evolution. It took other scientists many years to establish the relationship between Mendelian genetics and Darwin's theory of natural selection.

Emergence of Modern Darwinism: A Synthetic Theory

In the 1930s, a new generation of geneticists reevaluated Darwinian evolutionary theory from a mathematical perspective. These were population geneticists, scientists who studied variation in natural populations of animals and plants and who had a sound knowledge of statistics. Gradually, a new comprehensive theory emerged that brought together population genetics, paleontology, biogeography, embryology, systematics, and animal behavior in a Darwinian framework.

Population geneticists study evolution as change in the genetic compositions of populations. With the establishment of population genetics, evolutionary biology became divided into two subfields. **Microevolution** pertains to evolutionary changes in frequencies of variant forms of genes within populations. **Macroevolution** refers to evolution on a grand scale, encompassing origins of new organismal structures and designs, evolutionary trends, adaptive radiation, phylogenetic relationships of species, and mass extinction. Macroevolutionary research is based in systematics and the comparative method (see p. 4). Following this evolutionary synthesis, both macroevolution and microevolution have operated firmly within a tradition of neo-Darwinism, and both have expanded Darwinian theory in important ways.

Microevolution: Genetic Variation and Change Within Species

Microevolution is the study of genetic change occurring within natural populations. A population is a reproductively cohesive group of organisms of the same species; populations of sexually reproducing forms show interbreeding among their members. Variant forms of a single gene are called **alleles**. For example, in human populations, three different allelic forms occur for the gene encoding ABO blood types: I^A, I^B, and i. A person has two copies of this gene, one copy inherited from each parent. A person's **genotype** denotes the pair of allelic forms present for a particular gene. For the ABO blood type, the possible genotypes are I^AI^A, I^AI^B, I^Ai, I^BI^B, I^Bi, ii. A genotype is *homozygous* if it contains two copies of one allele (for example, I^AI^A, I^BI^B, ii) and *heterozygous* if it contains copies of two alleles (for example, I^AI^B, I^Ai, I^Bi). An organismal trait associated with a particular genotype is called the *phenotypic effect* of that genotype. A capital letter in the allelic designation indicates that an allele is genetically dominant, being expressed in both homozygous and heterozygous genotypes. A lower-case letter indicates that an allele is genetically recessive, revealing its characteristics only in its homozygous genotype. Therefore, two genotypes specify blood types A (I^AI^A, I^Ai) and B (I^BI^B, I^Bi), but blood types AB (I^AI^B) and O (ii) have single genotypes. Many genetic traits show this standard pattern of inheritance, called Mendelian inheritance.

Occurrence of different alleles of a gene in a population is called **polymorphism**. All alleles of all genes possessed by members of a population collectively form its **gene pool**. Polymorphism is potentially enormous in large populations because at observed mutation rates, many different alleles are expected for all genes.

Population geneticists study polymorphism by identifying allelic forms of a gene present in a population and then measuring their relative frequencies in that population. The relative frequency of a particular allele of a gene in a population is called its **allelic frequency**. Because each individual genotype contains two copies of this gene, the total number of copies present in a population is twice the number of individuals. What fraction of this total is represented by each different allelic form? In France, we find the following allelic frequencies: $I^A = 0.46$, $I^B = 0.14$, and $i = 0.40$. In Russia, the corresponding allelic frequencies differ ($I^A = 0.38$, $I^B = 0.28$, and $i = 0.34$), demonstrating microevolutionary divergence between these populations (figure 1.28). Genetically, alleles I^A and I^B are dominant to i, but i is nearly as frequent as I^A and exceeds the frequency of I^B in both populations. Dominance describes the *phenotypic effect* of an allele in heterozygous individuals, not its relative abundance in a population of individuals. In many human populations, genetically recessive traits, including the O blood type, blond hair, and blue eyes, are very common. In the following discussion, we demonstrate that Mendelian inheritance and dominance do not alter allelic frequencies directly or produce evolutionary change in a population.

Genetic Equilibrium

A mathematical theorem called **Hardy-Weinberg equilibrium** (see box, p. 30) permits us to estimate the relationship between the frequency of an allele in a population and the frequency of a phenotype influenced by that allele. We apply the theorem to one locus at a time, tracing the alleles through the formation of gametes followed by fertilization of gametes to produce individuals of the next generation. The frequency of an allele in the gametes produced by a population equals its frequency in the gene pool as defined in the preceding paragraph. Random mating consists mathematically of drawing pairs of gametes at random from the gene pool, uniting each random pair, and then quantifying the frequencies of the resulting organismal genotypes. Mating is usually random with respect to most molecular genetic traits, such as blood types, because they are unlikely to influence mate choice or recognition. We can estimate from Hardy-Weinberg equilibrium the ratios of genotypes and phenotypes that should occur in the following generation.

Consider a human population that contains a dominant allele for normal pigmentation (A) and a recessive allele for albinism (a), which is rare in the gene pool. Figure 1.29 shows the expected frequency of the dominant phenotype in a population as a function of the frequency of the dominant allele in the gene pool. When an allele is rare, copies of that allele occur almost entirely in heterozygous genotypes (Aa in our example), which express the dominant allele phenotypically; only a tiny fraction of the copies of a rare allele occurs in homozygous form (aa). The frequency of the recessive phenotype in the population is thus much less than the frequency of the recessive allele in the gene pool. Using conventional symbols, p denotes the frequency of allele A and q ($= 1 - p$) denotes the frequency of allele a. Under random mating, the genotypic frequencies are p^2 (AA), $2pq$ (Aa) and q^2 (aa). The frequency of albinism in humans is approximately 1/20,000. Assuming that mating is random with respect to genotype at this locus, we calculate using Hardy-Weinberg equilibrium:

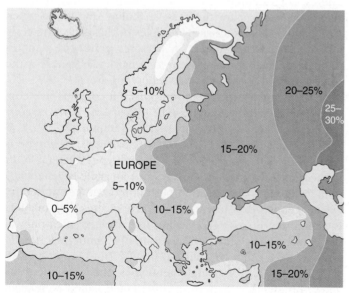

figure 1.28

Frequencies of the blood-type B allele (I^B) among humans in Europe. This allele is more common in the east and rarer in the west. The allele may have arisen in eastern Europe and gradually diffused westward through genetically continuous populations. This allele has no known selective advantage, and its changing frequency probably represents random genetic drift.

figure 1.29

The ratio of dominant to recessive phenotypes in a randomly mating population (in Hardy-Weinberg equilibrium) plotted as a function of the frequency of the dominant allele. Frequency of the recessive allele equals 1−p. The 3:1 ratio of dominant to recessive phenotypes in the second generation of Mendel's crosses occurred because all individuals of the preceding generation were heterozygous thus, $p = q = 0.5$, a condition not common in natural populations. As the dominant allele approaches a frequency of one, virtually no individuals express the recessive phenotype, thus permitting recessive lethal alleles to persist in a population at very low frequencies (0.001).

$$q^2 = \frac{1}{20,000}$$

$$q = \sqrt{\frac{1}{20,000}} = \frac{1}{141}$$

$$p = 1 - q$$

$$= \frac{141}{141} - \frac{1}{141} = \frac{140}{141}$$

The frequency of carriers is:

$$A/a = 2\,pq = 2 \times \frac{140}{141} \times \frac{1}{141} = \frac{1}{70}$$

One person in every 70 is a carrier!

Tay-Sachs disease in humans is associated with homozygosity for a recessive lethal allele; individuals homozygous for the lethal allele die early in childhood. Natural selection keeps recessive lethal alleles rare in the population, because individuals homozygous for such alleles never reproduce. Natural selection does not eliminate recessive lethal alleles from the population, however, because virtually all copies of those alleles occur in heterozygous genotypes, which are phenotypically normal. Mating is thus random with respect to whether individuals carry the lethal allele. For a recessive lethal allele present in 2 of every 100 persons (but homozygous in only 1 in 1000 fertilizations), 50 generations of selection are required to reduce the frequency of the allele to 1 of every 100 persons.

Eugenics deals with improvement of hereditary qualities in humans by social control of mating and reproduction. While the genetic argument against eugenics in the preceding paragraph is compelling, a eugenics program lingered in the United States (and elsewhere) until it was finally dispatched following Adolph Hitler's attempt to "purify" the races in Europe by genocide.

Processes of Evolution: How Genetic Equilibrium Is Upset

Genetic equilibrium is disturbed in natural populations by (1) random genetic drift, (2) nonrandom mating, (3) migration, and (4) natural selection, as well as by interactions among these factors. Recurring mutation is the ultimate source of variability in all populations, but upsetting genetic equilibrium usually requires interaction with one or more other factors. We consider these other factors individually.

Genetic Drift

Some species, such as the cheetah (figure 1.30), contain very little genetic variation, probably because their ancestral lineages were sometimes restricted to very small populations. A small population clearly cannot contain large amounts of genetic variation. Each individual organism has at most two different allelic forms of each gene, and a single breeding pair contains at most four different allelic forms of a gene. Suppose that we have such a breeding pair. We know from Mendelian genetics that chance decides which allelic forms of a gene get passed to offspring. It is therefore possible by chance alone that one or two parental alleles in this example are not passed to any offspring. It is highly unlikely that the different alleles present in a small ancestral population are all passed to descendants without any change in allelic frequency. This chance fluctuation in allelic frequency from one generation to the next, including loss of alleles from a population, is called **genetic drift.**

Genetic drift occurs to some degree in all populations of finite size. Perfect constancy of allelic frequencies, as predicted by Hardy-Weinberg equilibrium, occurs only in infinitely large populations, and such populations occur only in mathematical models. All populations of animals are finite and therefore experience some effect of genetic drift, which becomes greater, on average, as population size declines. Genetic drift erodes the genetic variability of a population. If population size remains small for many generations in a row, genetic variation can be greatly depleted. This loss is harmful to the evolutionary success of a species because it restricts potential genetic responses to environmental change. Indeed, biologists are concerned that cheetah populations may have insufficient variation for continued survival.

The term *population bottleneck* denotes a large reduction in the size of a population, causing loss of genetic variation, by genetic drift, followed by an increase in population size. The loss of variation is proportional to the number of generations that population size remains small before the population expands. A bottleneck associated with the founding of a new geographic population is called a founder effect and sometimes initiates the formation of a new species.

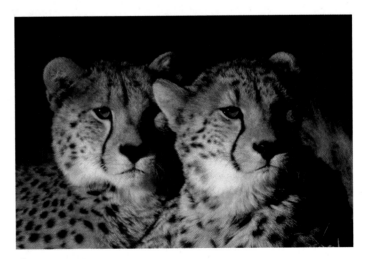

figure 1.30

Cheetahs, a species whose genetic variability has been depleted to very low levels because of small population size in the past.

Hardy-Weinberg Equilibrium: Why Mendelian Heredity Does Not Change Allelic Frequencies

The Hardy-Weinberg law is a logical consequence of Mendel's first law of inheritance, also called the law of segregation, and expresses a tendency toward equilibrium inherent in Mendelian heredity. Mendel's first law of inheritance states that each organism contains a pair of genetic factors for each variable trait. Different forms of these factors are called alleles. The paired factors in a given organism may be copies of the same allele (homozygous) or different alleles (heterozygous). In either case, the formation of gametes involves "segregation" of the paired factors so that each gamete receives only one factor for a given trait. Each gamete from a homozygous organism contains a copy of the same allele. A gamete from a heterozygous organism contains one of its two alleles, and that organism produces its two kinds of gametes in equal frequency.

Let us select for our example a population having a single locus bearing just two alleles, T and t. The phenotypic expression of this gene might be, for example, ability to taste a chemical compound called phenylthiocarbamide. Individuals in a population can be of three genotypes for this locus, T/T, T/t (both tasters), and t/t (non-tasters). In a sample of 100 individuals, suppose that 20 have the genotype T/T, 40 are T/t, and 40 are t/t. The following table shows the allelic frequencies (remember that every individual's genotype has two copies of a gene):

Genotype	Number of Individuals	Copies of the T Allele	Copies of the t Allele
T/T	20	40	
T/t	40	40	40
t/t	40		80
TOTAL	100	80	120

Of the 200 copies, the proportion of the T allele is $80/200 = 0.4$ (40%), and the proportion of the t allele is $120/200 = 0.6$ (60%). It is customary to use p and q to represent the two allelic frequencies. The genetically dominant allele is represented by p, and the genetically recessive by q. Thus:

$$p = \text{frequency of } T = 0.4$$
$$q = \text{frequency of } t = 0.6$$
$$\text{Therefore, } p + q = 1$$

Having calculated allelic frequencies in this sample, let us determine whether these frequencies change spontaneously in a new generation of the population. Assuming that mating is random (gametes are sampled independently in pairs), each individual is expected to contribute an equal number of gametes to a common pool from which the next generation is formed. Frequencies of gametes in the pool are proportional to allelic frequencies in the sample: 40% of the gametes are T, and 60% are t (ratio of 0.4:0.6). Both ova and sperm will, of course, show similar frequencies. The next generation is formed as shown here:

	Ova	
Sperm	$T = 0.4$	$t = 0.6$
$T = 0.4$	$T/T = 0.16$	$T/t = 0.24$
$t = 0.6$	$T/t = 0.24$	$t/t = 0.36$

Collecting genotypes, we have:

$$\text{frequency of } T/T = 0.16$$
$$\text{frequency of } T/t = 0.48$$
$$\text{frequency of } t/t = 0.36$$

Next, we determine the values of p and q from randomly mated populations. From the table, we see that the frequency of T is the sum of genotypes T/T, which is 0.16, and one-half of the genotype T/t, which is 0.24:

$$T(p) = 0.16 + .5(0.48) = 0.4$$

Similarly, the frequency of t is the sum of genotypes t/t, which is 0.36, and one-half the genotype T/t, which is 0.24:

$$t(p) = 0.36 + .5(0.48) = 0.6$$

The new generation bears exactly the same allelic frequencies as its parent population! Note that no increase has occurred in the frequency of the genetically dominant allele T. Thus *in a freely interbreeding, sexually reproducing population, the frequency of each allele would remain constant generation after generation in the absence of natural selection, migration, recurring mutation, and genetic drift* (see p. 29). A mathematically minded reader would recognize that the genotype frequencies T/T, T/t, and t/t are actually a binomial expansion of $(p = q)^2$:

$$(p + q)^2 = p^2 + 2pq + q^2 = 1$$

Note that the equilibrium calculations give *expected* frequencies, which are unlikely to be realized exactly in a population of limited size. For this reason, limits to population size cause evolutionary change.

Most genes have more than just a single pair of alleles, especially when we measure genetic variation at the DNA sequence level. The binomial expansion shown above can be used for any number of alleles. Suppose that we have three alleles (T_1, T_2, T_3) whose frequencies are denoted p, q, and r, respectively. We now have six possible genotypes with the following Hardy-Weinberg equilibrium frequencies:

$$(p + q + r)^2 = \underset{p^2}{T_1 T_1} + \underset{2pq}{T_1 T_2} +$$
$$\underset{q^2}{T_2 T_2} + \underset{2pr}{T_1 T_3} + \underset{2qr}{T_2 T_3} + \underset{r^2}{T_3 T_3}$$

As the number of alleles at a gene increases, the proportion of the population having heterozygous genotypes also increases.

Nonrandom Mating

If mating is nonrandom, genotypic frequencies deviate from Hardy-Weinberg expectations. For example, if two different alleles of a gene are equally frequent ($p = q = .5$), we expect half of the genotypes to be heterozygous ($2pq = 2 [.5] [.5] = .5$) and one-quarter to be homozygous for each allele ($p^2 = q^2 [.5]^2 = .25$). In **positive assortative mating,** individuals mate preferentially with others of the same genotype, as when albinos mate with other albinos, for example. Matings among homozygous parents generate offspring that are homozygous like themselves. Matings among heterozygous parents produce on average 50% heterozygous offspring and 50% homozygous offspring (25% of each alternative type) each generation. Positive assortative mating increases the frequency of homozygous genotypes and decreases the frequency of heterozygous genotypes in a population but does not change allelic frequencies.

Preferential mating among close relatives also increases homozygosity and is called **inbreeding.** Whereas positive assortative mating usually affects one or a few traits, inbreeding simultaneously affects all variable traits. Strong inbreeding greatly increases the chances that rare recessive alleles become homozygous and thereby are expressed.

Inbreeding is a serious problem in zoos holding small populations of rare animals. Matings of close relatives tend to bring together genes from a common ancestor and increase the probability that two copies of a deleterious gene come together in the same organism. The result is "inbreeding depression." The management solution is to enlarge genetic diversity by bringing together captive animals from different zoos or by introducing new stock from wild populations if possible. Paradoxically, where zoo populations are extremely small and no wild stock can be obtained, deliberate inbreeding is recommended. This procedure selects for genes that tolerate inbreeding; deleterious genes disappear if they kill animals homozygous for them.

Migration

Migration prevents different populations of a species from becoming genetically dissimilar. If a species is divided into many small populations, genetic drift and selection acting separately in different populations can produce evolutionary divergence among them. A small amount of migration among populations each generation keeps different populations from diverging strongly. For example, the French and Russian populations show some genetic divergence (see figure 1.28), but continuing migration prevents them from becoming completely distinct.

Natural Selection

Natural selection can change both allelic frequencies and genotypic frequencies in a population. Although effects of selection are often reported for particular polymorphic genes, it is important to remember that natural selection acts on whole animals, not on isolated traits. The organism that possesses a superior combination of traits is favored. An animal may have some traits that confer no advantage or even a disadvantage, but the animal is successful overall if its combination of traits is favorable. When we claim that a particular genotype has a higher relative **fitness** than do alternative genotypes, we state that the genotype on average confers an advantage in survival and reproduction in a population. If alternative genotypes have unequal probabilities of survival and reproduction, Hardy-Weinberg equilibrium is upset.

Using the genetic theory of natural selection, one can measure relative fitness values associated with different genotypes in a population. Geneticists often use W to denote the expected average fitness of a genotype in a population, with the genotype of highest fitness given a value of one and fitnesses of other genotypes indicated as fractions.

We illustrate measurement of fitness using genetic variation associated with the disease sickle cell anemia in human populations. Considering only the alleles for normal hemoglobin (A) and sickle cell hemoglobin (S) for the beta-hemoglobin gene in human populations, the possible genotypes are AA, AS, and SS. Measurements of viability of individuals of these three genotypes in nonmalarial environments give a fitness value of 1 to genotypes AA and AS and a fitness of 0.2 to genotype SS. People having the SS genotype, who are susceptible to severe anemia, are expected to contribute only 20% as many offspring to the next generation on average as are individuals having the AA or AS genotypes. In malarial environments, genotype AS has the highest fitness (= 1); genotype AA has a slightly decreased fitness (= 0.9) because these individuals have a greater incidence of malaria than do AS individuals, and SS has a low fitness (= 0.2) because of anemia. From these measured fitness values and knowledge of the frequencies of alleles in a population and its system of mating, one can calculate the **average effect** that an allele has on the phenotype of relative fitness in that population. In the example of sickle cell anemia, the average effect of allele S on fitness in a malarial environment is a balance between the strongly negative effect it has when homozygous and the positive effect that it has when heterozygous with allele A.

Some traits and combinations of traits are advantageous for certain aspects of an organism's survival or reproduction and disadvantageous for others. Darwin used the term **sexual selection** to denote selection of traits that are advantageous for obtaining mates but perhaps harmful for survival. For example, bright colors and elaborate feathers might enhance a male bird's competitive ability to obtain mates while simultaneously increasing his vulnerability to predators (figure 1.31). Environmental changes can alter the selective values of different traits.

Selection is often studied using quantitative traits, those that show continuous variation with no obvious pattern of Mendelian segregation in their inheritance. The values of a trait in offspring often are intermediate between the values of their parents. Such traits are influenced by variation at many genes, each of which follows Mendelian inheritance and contributes a small, incremental amount to a phenotype. Traits that show quantitative variation include tail length in mice, length of a leg segment in grasshoppers, number of gill rakers in sunfishes, number of peas in pods, and height of adult humans. When values are graphed with respect to frequency distribution, they often approximate a normal, bell-shaped curve (figure 1.32A). Most individuals fall near the average, fewer fall well above or below the average, and extreme values form "tails" of the frequency curve with increasing rarity.

Selection can act on quantitative traits to produce three different kinds of evolutionary response. **Stabilizing selection**

figure 1.31

A pair of wood ducks. Brightly colored feathers of male birds probably confer no survival advantage and might even be harmful by alerting predators. Such colors nonetheless confer advantage in attracting mates, which overcomes, on average, any negative consequences of these colors for survival. Darwin used the term "sexual selection" to denote traits that give an individual an advantage in attracting mates, even if these traits are neutral or harmful for survival.

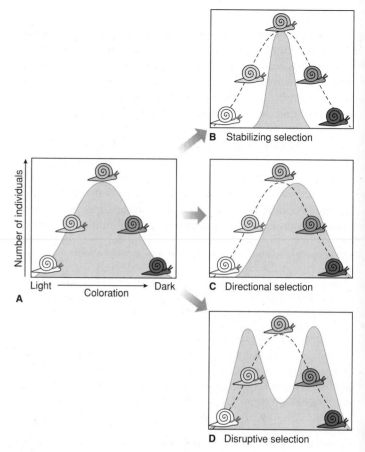

figure 1.32

Responses to selection on a continuous (polygenic) character, coloration in a snail. **A,** Frequency distribution of coloration before selection. **B,** Stabilizing selection culls extreme variants from the population, in this case eliminating unusually light or dark individuals and thereby stabilizing the mean. **C,** Directional selection shifts the population mean, in this case by favoring darkly colored variants. **D,** Disruptive selection favors both extremes but not the mean; the mean is unchanged, but the population no longer has a bell-shaped distribution of phenotypes.

favors average values of a trait and disfavors extreme ones (figure 1.32B). **Directional selection** favors an extreme value of a phenotype and causes a population average to shift toward it (figure 1.32C). When we think about natural selection producing evolutionary change, we usually have directional selection in mind, although that is not the only possibility. A third alternative is **disruptive selection,** in which two different extreme phenotypes are favored simultaneously, but their average is disfavored (figure 1.32D). If inbreeding or positive assortative mating accompanies disruptive selection, this can cause a population to become bimodal, meaning that two very different phenotypes predominate.

Interactions of Selection, Drift, and Migration

Subdivision of a species into small populations that exchange migrants is an optimal way to promote rapid adaptive evolution of a species. Interactions of genetic drift and selection in different populations permit many different genetic combinations of many polymorphic genes to be tested against natural selection. Migration among populations permits particularly favorable new genetic combinations to spread throughout the species as a whole. Interactions of selection, genetic drift, and migration in this example produce evolutionary change qualitatively different from what would result if any of these three factors acted alone. The geneticist Sewall Wright called this interaction *shifting balance* because it permits a population to switch from an initial adaptive combination of traits to a new and possibly better one. Genetic drift, nonrandom mating, migration, and natural selection interact in natural populations to create an enormous opportunity for evolutionary change; the perpetual stability predicted by Hardy-Weinberg equilibrium almost never lasts across any substantial amount of evolutionary time.

The importance of interactions between natural selection and genetic drift in adaptive evolution is illustrated by the beta-hemoglobin variation discussed in the preceding section. We noted that in malarial environments, natural selection retains both the *A* and *S* alleles in the population because individuals with the *AS* genotype uniquely combine the benefits of malarial resistance and normal respiration. A rare third allele called hemoglobin *C* also occurs in some malarial regions of western Africa. Like hemoglobin *S*, allele *C* was derived from allele *A* by a single mutation changing the sixth codon; in allele *C*, lysine (AAG) substitutes for glutamic acid (GAG). In malarial areas, natural selection acts against allele *C* in heterozygous genotypes because *AC* individuals are susceptible to malaria (as are *AA* individuals), and *CS* individuals have severe anemia. Natural selection favors the *C* allele in homozygous form; *CC* individuals have malarial resistance, normal respiration, and much higher fitness than do *AS* individuals in malarial areas. If the population were fixed for the *C* allele, all individuals would benefit from both malarial resistance and normal respiration.

Why has the *C* allele not become the most frequent one in malarial regions of western Africa? We know from Hardy-Weinberg equilibrium that in a randomly mating population, a rare allele occurs almost exclusively in heterozygous genotypes with the more common alleles. Because selection favors *AS* individuals over *AC* and *CS* individuals in malarial areas, selection acts to eliminate allele *C* from those populations.

In a few local populations of western Africa, genetic drift caused the *C* allele to reach a relatively high frequency prior to the onset of malaria. In these populations alone, the *CC* genotype occurs with sufficient frequency that the positive action of natural selection on these individuals exceeds the negative effects of selection on the *AC* and *CS* individuals in the same population. The *C* allele has increased in frequency by natural selection in these local populations. One expects that natural selection eventually would cause the *C* allele to become fixed in these local populations, and that gene flow from them to others would allow the *C* allele to become fixed across malarial Africa. One hopes, of course, that effective treatment of malaria will overcome the need for an evolutionary solution to this problem, as the latter would require many generations and much intervening illness. Nonetheless, this example shows how interactions between genetic drift, natural selection, and gene flow can shift a population's response to malarial selection from maintaining a polymorphism of the *A* and *S* alleles to fixing the *C* allele.

Protein Polymorphism

Beginning in the 1960s, studies of protein variation provided the first unambiguous evidence that animal populations typically contain large amounts of genetic variation. Studies of protein polymorphism are now largely superseded by studies of variation in DNA sequences sampled from both the nuclear and mitochondrial genomes. DNA studies reveal even larger amounts of variation than did studies of proteins. We focus here on protein variation, for both its historical importance and its more straightforward interpretation using the principles of Hardy-Weinberg equilibrium (p. 28).

Different allelic forms of genes encode proteins that often differ slightly in their amino acid sequence. This phenomenon is called **protein polymorphism.** If these differences affect the protein's net electric charge, the different allelic forms can be separated using protein electrophoresis (figure 1.33). We can identify the genotypes of particular individuals for protein-coding genes and measure allelic frequencies in a population.

Over the last 45 years, geneticists using this approach have discovered far more variation than was previously expected. Despite the high levels of polymorphism discovered using protein electrophoresis, these studies underestimate both protein polymorphism and the total genetic variation present in a population. For example, protein polymorphism that does not involve charge differences is not detected. Furthermore, because the genetic code is degenerate (more than one codon for most amino acids), protein polymorphism does not reveal all of the genetic variation present in protein-coding genes. Genetic changes that do not alter protein structure sometimes alter patterns of protein synthesis during development and can be very important to an organism. When all kinds of variation are considered, it is evident that most species have an enormous potential for further evolutionary change.

figure 1.33

Study of genetic variation in proteins using gel electrophoresis. **A,** An electrophoretic apparatus separates allelic variants of proteins that differ in charge because of differences in their sequence of amino acids. **B,** Genetic variation in the protein leucine aminopeptidase for nine brown snails, *Helix aspersa.* Two different sets of allelic variants are revealed. The top set contains two alleles [denoted fast (F) and slow (S) according to their relative movement in the electric field]. Individuals homozygous for the fast allele show only a single fast band on the gel (FF), those homozygous for the slow allele show only a single slow band (SS), and heterozygous individuals have both bands (FS). The lower set contains three different alleles denoted fast (F), medium (M), and slow (S). Note that no individuals shown are homozygous for the medium (M) allele.

Macroevolution: Major Evolutionary Events and Processes

Macroevolution describes large-scale events and processes in organic evolution. Speciation links macroevolution and microevolution. The major trends in the fossil record described earlier (see figures 1.13 through 1.15) fall clearly within the realm of macroevolution. Patterns and processes of macroevolutionary change emerge from those of microevolution, but they acquire some degree of autonomy in doing so. The emergence of new adaptations and species, and the varying rates of speciation and extinction observed in the fossil record, transcend the fluctuations of allelic frequencies within populations.

Speciation and Extinction Through Geological Time

Evolutionary change at the macroevolutionary level provides a new perspective on Darwin's theory of natural selection. Although a species can persist for many millions of years, it ultimately has two possible evolutionary fates: it may give rise to new species or become extinct without leaving descendants. Rates of speciation and extinction vary among lineages, and lineages that have the highest speciation rates and the lowest extinction rates produce the greatest diversity of living forms. The characteristics of a species might make it more or less likely than others to undergo speciation or extinction. Because many characteristics are passed from ancestral to descendant species analogous to heredity at the organismal level, lineages whose properties enhance the probability of speciation and confer resistance to extinction should dominate the living world. This species-level process that produces differential rates of speciation and extinction among lineages is analogous

in many ways to natural selection. It represents an expansion of Darwin's theory of natural selection.

Species selection denotes the differential survival and multiplication of species through geological time based on variation among lineages in species-level properties. These species-level properties include mating rituals, social structuring, migration patterns, and geographic distribution. Descendant species usually resemble their ancestors for these properties. For example, a "harem" system of mating in which a single male and several females compose a breeding unit characterizes some mammalian lineages but not others. We expect speciation rates to be enhanced by social systems that promote the founding of new populations by small numbers of individuals. Certain social systems might increase the probability that a species survives environmental challenges through cooperative action. Such properties would be favored over geological time by species selection.

Mass Extinctions

When we study evolutionary change on an even larger timescale, we observe episodic events in which large numbers of taxa become extinct nearly simultaneously. These events are called **mass extinctions** (figure 1.34). The most cataclysmic of these extinction episodes happened about 250 million years ago, when at least half of the families of shallow-water marine invertebrates and fully 90% of marine invertebrate species disappeared within a few million years. This was the **Permian extinction.** The **Cretaceous extinction,** which occurred about 65 million years ago, marked the end of dinosaurs as well as numerous marine invertebrates and many small reptilian species.

Causes of mass extinctions and their occurrence at intervals of approximately 26 million years throughout the past 250 million years are difficult to explain. Some researchers have proposed biological explanations for these mass extinctions, and others consider them artifacts of our statistical and taxonomic

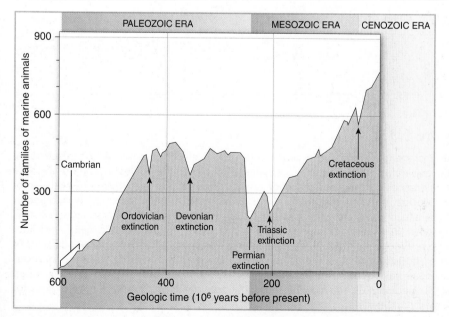

figure 1.34

Changes in numbers of taxonomic families of marine animals through time, from the Cambrian period to the present. Sharp drops represent five major extinctions of skeletonized marine animals. Note that despite these extinctions, the overall number of marine families has increased over the past 540 million years. (Results follow the analysis of David Raup and J. John Sepkoski, Jr. as reported in 1982. Subsequent revisions to geological dating move the start of the Cambrian forward from 600 million years ago as shown to 542 million years ago, requiring compression of the left 60 million years of the figure.)

analyses. Walter Alvarez proposed that the earth was occasionally bombarded by asteroids, causing the mass extinction at the end of the Cretaceous and perhaps others (figure 1.35). The drastic effects of such a bombardment were observed in July 1994 when fragments of Comet Shoemaker-Levy 9 bombarded Jupiter. The first fragment to hit Jupiter was estimated to have the force of 10 million hydrogen bombs. Twenty additional fragments hit Jupiter within the following week; one of them was 25 times more powerful than the first fragment. This bombardment was the most violent event in the recorded history of our solar system. Bombardments by asteroids or comets could change the earth's climate drastically, sending debris into the atmosphere and blocking sunlight. Temperature changes would challenge ecological tolerances of many species. This hypothesis is being tested in several ways, including searches for impact craters left by asteroids and for altered mineral content of rock strata where mass extinctions occurred. Unusual concentrations of the rare-earth element iridium in strata at the end of the Cretaceous indicate that this element entered the earth's atmosphere through asteroid bombardment. So far, the end-Cretaceous mass extinction is the only one that coincides with strong geological evidence for asteroid bombardment.

Sometimes lineages favored by species selection are unusually at risk during a mass extinction. Climatic changes produced by hypothesized asteroid bombardments could produce selective challenges very different from those encountered at other times in the earth's history. Selective discrimination of particular biological traits by mass extinction events is termed **catastrophic species selection.** For example, mammals survived the end-Cretaceous mass extinction that destroyed dinosaurs and other prominent vertebrate and invertebrate groups. The ability of small mammals to hibernate in underground burrows might have permitted them to withstand climatic stresses that killed more exposed animals.

figure 1.35

Meteor crater of the Arizona desert is among the youngest and best-preserved impact craters. Geologists estimate that it formed approximately 50,000 years ago when a meteor measuring 30 meters wide and weighing 100,000 tons landed at a speed of 20 kilometers per second. Asteroid impacts associated with the end-Cretaceous mass extinctions would have greatly exceeded the power of this one, but geological changes accumulating over the past 65 million years have lessened the preservation of those craters.

Following this event, mammals were able to use environmental resources that previously had been denied them, leading to their adaptive radiation.

Natural selection, species selection, and catastrophic species selection interact to produce macroevolutionary trends that we see in the fossil record. Intensive study of these interacting causal processes has made modern paleontology an active and exciting field.

■ Summary

Zoology is the scientific study of animals. Science is the acquisition of knowledge guided by hypotheses that are explanatory with reference to natural law, testable, tentative, and falsifiable. These criteria form the basis for the scientific method, which is called hypothetico-deductive. On the basis of prior observations, a scientist formulates an explanatory hypothesis. Predictions about future observations based on the hypothesis guide researchers to collect data that can support or falsify the hypothesis. A testable hypothesis for which there is a large amount of supporting data, particularly a hypothesis that explains a very large number of observations, is called a theory.

Experimental sciences seek to identify proximate or immediate causes in biological systems and use the experimental method. Hypotheses are tested by experiments, which must include controls. Evolutionary sciences seek ultimate causes in biological systems using the comparative method.

Organic evolution explains the diversity of living organisms as the historical outcome of gradual change from previously existing forms. Evolutionary theory is strongly identified with Charles Robert Darwin, who presented the first credible explanation for evolutionary change.

Darwin derived much of the material used to construct his theory from his experiences on a five-year voyage around the world aboard H.M.S. *Beagle.*

Darwin's evolutionary theory has five major components. Its most basic proposition is *perpetual change,* the theory that the world is neither constant nor perpetually cycling but is steadily undergoing irreversible change. The fossil record amply demonstrates perpetual change in continuing fluctuations of animal form and diversity following the Cambrian appearance of animals 600 million years ago.

Darwin's theory of *common descent* states that all organisms descend from a common ancestor through branching genealogical lineages. This theory explains morphological homologies among organisms as characteristics inherited with modification from a corresponding feature in their common evolutionary ancestor. Patterns of homology formed by common descent with modification permit us to classify organisms according to their evolutionary relationships.

A corollary of common descent is *multiplication of species* through evolutionary time. Allopatric speciation describes evolution of reproductive barriers between geographically separated populations to generate new species. Adaptive radiation denotes proliferation of many adaptively disparate species from a single ancestral lineage. Oceanic archipelagoes, such as the Galápagos Islands, are particularly conducive to adaptive radiation of terrestrial organisms.

Darwin's theory of *gradualism* states that large phenotypic differences between species are produced by accumulation through evolutionary time of many individually small changes. Gradualism is still controversial. Mutations that have large effects on phenotype have been useful in animal breeding, leading some to dispute Darwin's claim that such mutations are not important in evolution. From a macroevolutionary perspective, punctuated equilibrium states that most evolutionary change occurs in relatively brief events of branching speciation, separated by long intervals in which little phenotypic change accumulates.

Darwin's fifth major statement is that *natural selection* is the major guiding force of evolution. This principle is founded on observations that all species overproduce their kind, causing a struggle for limited resources. Because no two organisms are exactly alike, and because variable traits are at least partially heritable, those organisms whose hereditary endowment enhances their use of resources for survival and reproduction contribute disproportionately to the next generation. Over many generations, the sorting of variation by selection produces new species and new adaptations.

Population geneticists have discovered principles by which the genetic properties of populations change through time. A particularly important discovery, called Hardy-Weinberg equilibrium, shows that Mendelian heredity does not change the genetic composition of populations. Important sources of evolutionary change include genetic drift, non-random mating, migration, natural selection, and their interactions. Mutations are the ultimate source of all new variation on which these sources of evolutionary change act.

Macroevolution denotes evolutionary change on a geological timescale. Macroevolutionary studies measure rates of speciation, extinction, and changes in diversity through time. These studies have expanded Darwinian evolutionary theory to include higher-level processes that regulate rates of speciation and extinction among lineages, including species selection and catastrophic species selection.

■ Review Questions

1. What are the essential characteristics of science? Describe how evolutionary studies fit these characteristics. Why are arguments based on supernatural or miraculous explanations, such as "creation-science" or "intelligent design," not valid scientific hypotheses?
2. What is the relationship between a hypothesis and a theory?
3. Explain how biologists distinguish between experimental and evolutionary sciences.
4. Briefly summarize Lamarck's hypothesis for evolution. What evidence rejects this hypothesis?
5. What is "uniformitarianism?" How did it influence Darwin's evolutionary theory?
6. Explain how Darwin's journey on the *Beagle* and his study of pigeon breeding affected his thinking.
7. What key idea, contained in Malthus's essay on populations, helped Darwin formulate his theory of natural selection?
8. Explain how each of the following contributes to Darwin's evolutionary theory: fossils, geographic distributions of closely related animals, homology, animal classification.
9. How do modern evolutionists view the relationship between ontogeny and phylogeny?

10. What major evolutionary lesson is illustrated by Darwin's finches on the Galápagos Islands?
11. How does observation of "sports" in animal breeding challenge Darwin's theory of gradualism? Why did Darwin consider such mutations to have little evolutionary importance?
12. What does the theory of punctuated equilibrium state about the occurrence of speciation throughout geological time?
13. Describe the observations and inferences that compose Darwin's theory of natural selection.
14. Identify the random and nonrandom components of Darwin's theory of natural selection.
15. Describe some recurring criticisms of Darwin's theory of natural selection. How are these criticisms refuted?
16. A common but mistaken notion is that because some alleles are genetically dominant and others are recessive, dominants eventually replace all recessives in a population. How does the Hardy-Weinberg equilibrium refute this notion?
17. Assume that you are measuring frequencies of alternative conditions of a variable trait from two animal populations. The trait is controlled by a single allelic pair, A and a, and you can distinguish all three phenotypes, AA, Aa, and aa (intermediate inheritance). Your sample includes:

Population	AA	Aa	aa	TOTAL
I	300	500	200	1000
II	400	400	200	1000

Calculate the distribution of phenotypes in each population as expected under Hardy-Weinberg equilibrium. Is population I in equilibrium? Is population II in equilibrium?
18. After studying a population for a trait determined by a single pair of alleles, you find that this population is not in equilibrium. What possible reasons might explain its departure from equilibrium?
19. Explain why genetic drift is more powerful in small populations.
20. Is it easier for selection to remove a deleterious recessive allele from a randomly mating population or from a highly inbred population? Why?
21. Contrast microevolution and macroevolution.

For Further Thought Why will our knowledge of the origin of life always be more tentative than our knowledge of the subsequent evolution of life's diversity?

■ Selected References

See also general references on page 432.

Avise, J. C. 2004. Molecular markers, natural history, and evolution, ed. 2. Sunderland, Massachusetts, Sinauer Associates. *An exciting account of using molecular genetics to understand evolution.*

Browne, E. J. 2002. Charles Darwin: a biography. New York, Knopf. *A comprehensive two-volume biography of Darwin completed in 2002.*

Conner, J. K., and D. L. Hartl. 2004. A primer of ecological genetics. Sunderland, Massachusetts, Sinauer Associates. *An introductory text on population genetics.*

Coyne, J. A., and H. A. Orr. 2004. Speciation. Sunderland, Massachusetts, Sinauer Associates. *A detailed coverage of species formation.*

Darwin, C. 1859. On the origin of species by means of natural selection, or the preservation of favoured races in the struggle for life. London, John Murray. *There were five subsequent editions by the author.*

Freeman, S., and J. C. Herron. 2007. Evolutionary analysis, ed. 4. Upper Saddle River, New Jersey, Pearson/Prentice-Hall. *An introductory textbook on evolutionary biology designed for undergraduate biology majors.*

Futuyma, D. J. 2005. Evolution. Sunderland, Massachusetts, Sinauer Associates. *A challenging introductory textbook on evolution.*

Gould, S. J. 2002. The structure of evolutionary theory. Cambridge, Massachusetts, Belknap Press of Harvard University Press.

A provocative account of what fossils tell us about evolutionary theory.

Kitcher, P. 1982. Abusing science: the case against creationism. Cambridge, Massachusetts, MIT Press. *A treatise on how knowledge is gained in science and why creationism does not qualify as science. The argument refuted as creation-science in this book is equivalent to recent arguments termed "intelligent design theory."*

Mayr, E. 2001. What evolution is. New York, Basic Books. *A general survey of evolution by a leading evolutionary biologist.*

Templeton, A. R. 2006. Population genetics and microevolutionary theory. Hoboken, New Jersey, Wiley-Liss. *A very thorough and exciting coverage of population-genetic theory, methods, and discoveries.*

■ Custom Website

The *Animal Diversity* Online Learning Center is a great place to check your understanding of chapter material. Visit www.mhhe.com/hickmanad6e for access to key terms, quizzes, and more! Further enhance your knowledge with Web links to chapter-related material.

Explore live links for these topics:

Introductory Sites
Writing Papers and Study Tips
Glossaries and Dictionaries
Careers in Science
Utility and Organizational Sites
Scientific Method

Evolution
Darwinian Evolutionary Theory
Evidences for Evolution
Theories of Evolutionary Patterns

2

Animal Ecology

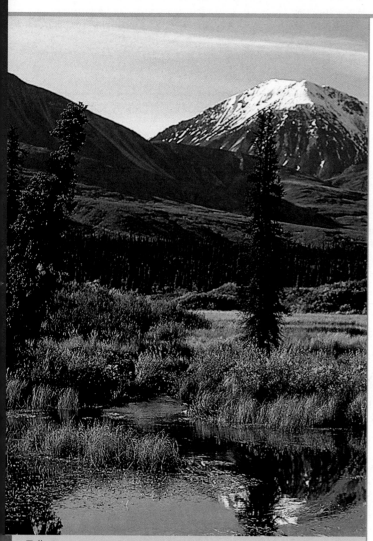

Talkeetna Mountain Range, Alaska.

Ecological Niche and Animal Species Diversity

Life's diversity is organized into a hierarchy of interacting units: an individual organism, a population, a community, and most inclusively an ecosystem. Central to ecological study of animal diversity is habitat, the spatial location where an animal lives. What an animal does in its habitat is its niche: how it gets its food, how it achieves reproductive perpetuity—in short, how it survives and stays adapted in a Darwinian sense.

The niche of a species evolves, and no other species in the community can evolve to exploit exactly the same resources at the same place and time. This illustrates the "competitive exclusion principle": no two species can occupy the same niche if they are to live together in a stable ecological community. Different species are therefore able to form an ecological community in which each has a different role in their shared environment.

In the mid-nineteenth century, German zoologist Ernst Haeckel introduced the term **ecology**, defined as the relation of the animal to its organic as well as inorganic environment. Environment here includes everything external to the animal but most importantly its immediate surroundings. Although we no longer restrict ecology to animals alone, Haeckel's definition remains basically sound. Animal ecology is now a highly synthetic science that incorporates the behavior, physiology, genetics, and evolution of animals to study interactions between populations of animals and their environments. The major goal of ecological studies is to explain how these diverse interactions influence the geographical distribution and abundance of animal populations. Ecological knowledge is a crucial part of understanding animal diversity and for ensuring the continued survival of many populations.

E cology encompasses a hierarchy of biological systems in interaction with their environments. At the base of the ecological hierarchy is an **organism** such as an individual animal. To understand why animals live where they do, ecologists must examine the varied physiological and behavioral mechanisms that animals use to survive, to grow, and to reproduce. For example, a near-perfect physiological balance between production and loss of heat is required for success in certain endothermic species (birds and mammals, whose body temperature is regulated internally) under extreme temperatures that occur in the Arctic or a desert. Other species succeed in these situations by escaping the most extreme conditions through migration or hibernation. Insects, fishes, and other ectotherms (animals whose body temperature responds to environmental heat) compensate for fluctuating temperatures through behavioral responses (figure 2.1) or by altering biochemical and cellular processes involving enzymes, lipid organization, and the neuroendocrine system. Thus an animal's physiological capacities permit it to occupy changing and often adverse environments. Behavioral responses are important also for obtaining food, finding shelter, escaping enemies and unfavorable environments, finding a mate, courting, and tending to offspring. Physiological mechanisms and behaviors that improve adaptability to environments help organisms to survive. Ecologists who focus their studies at the organismal level are called physiological ecologists or behavioral ecologists.

Animals in nature coexist with others of the same species as reproductive units called **populations**. Populations have properties that cannot be discovered by studying individual animals alone; these properties include genetic variability among individuals (polymorphism), a gene pool (p. 27), growth in numbers over time, and factors that limit the density of individuals in a given area. Ecological studies at the population level help us to predict the future success of endangered species and to discover controls for pest species.

Just as individuals are not alone in nature, populations of different species co-occur in more complex associations called **ecological communities**. One measure of an ecological community's complexity is the number of species present, called **species richness**. The populations of species in a community interact with each other in many ways, the most prevalent ones being **predation**, **parasitism**, and **competition**. **Predators** obtain energy and nutrients by killing and eating prey. **Parasites** derive similar benefits from their host organism, but they live on or in the host and usually do not kill the host. A **parasitoid** lives on or in a host but eventually kills its host organism. Competition occurs when food and/or space are limited and members of different species interfere with each other's use of their shared resources. Communities are complex because all of these interactions occur simultaneously, and their individual effects on the community often cannot be isolated.

Ecological communities are biological components of even larger, more complex entities called **ecosystems**. An

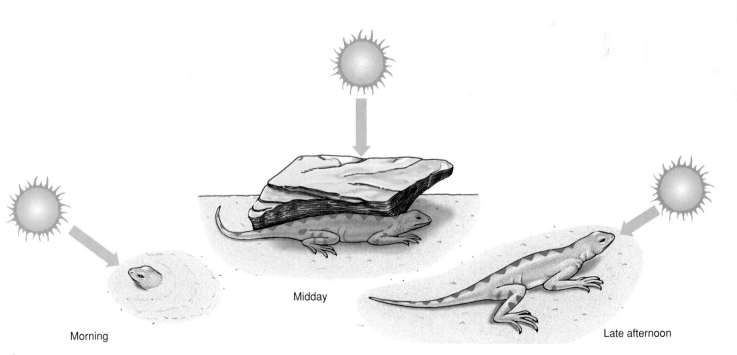

figure 2.1

How a lizard (an ectotherm) regulates its body temperature behaviorally. In the morning, the lizard absorbs the sun's heat through its head while keeping the rest of its body protected from cool morning air. Later it emerges to bask. At noon, with its body temperature high, it seeks shade from the hot sun. When the air temperature drops in the late afternoon, it emerges and lies parallel to the sun's rays.

ecosystem consists of all populations in an ecological community and their physical environment. The study of ecosystems reveals two key processes in nature: the flow of energy and the cycling of materials through biological channels. The largest ecosystem is the **biosphere**, the thin veneer of land, water, and atmosphere that envelops the earth and supports all life.

■ Environment and the Niche

An animal's *environment* comprises all conditions that directly affect its chances for survival and reproduction. These factors include space; forms of energy, such as sunlight, heat, wind, and water currents; and materials, such as soil, air, water, and numerous chemicals. Environment also includes other organisms, which can be an animal's food or its predators, competitors, hosts, or parasites. Environment thus includes both abiotic (nonliving) and biotic (living) factors. **Resources**, such as space and food, are the factors that an animal uses directly.

A resource is either expendable or nonexpendable, depending on how an animal uses it. Food is expendable, because once eaten it is no longer available. Food therefore must be continuously replenished. Space, whether total living area or a subset such as the number of suitable nesting sites, is not exhausted by being used, and thus is nonexpendable.

The physical space where an animal lives is its **habitat**. Size of habitat varies. A rotten log is a normal habitat for carpenter ants. Such logs occur in larger habitats called forests where deer also live. However, deer forage in open meadows, so their habitat is larger than the forest. On a larger scale, some migratory birds occupy forests of the north temperate region during summer and move to the tropics during winter. Thus, habitat is defined by an animal's normal activity rather than by arbitrary physical boundaries.

Animals of any species demonstrate environmental limits of temperature, moisture, and food within which they can grow, reproduce, and survive. A suitable environment therefore must meet all requirements for life. A freshwater clam inhabiting a tropical lake could tolerate the temperature of a tropical ocean, but the ocean's salinity would be fatal. A brittle star living in the Arctic Ocean could tolerate the salinity of a tropical ocean but not its temperature. Thus temperature and salinity are two separate dimensions of an animal's environmental limits. If we add another variable, such as pH, we increase our description to three dimensions (figure 2.2). If we consider all environmental conditions that permit members of a species to survive and to multiply, we define a role for that species in nature distinct from those of other species. This unique, multidimensional fingerprint of a species is called its **niche** (see p. 38). Dimensions of niche vary among members of a species, making the niche subject to evolution by natural selection. The niche of a species undergoes evolutionary changes over successive generations.

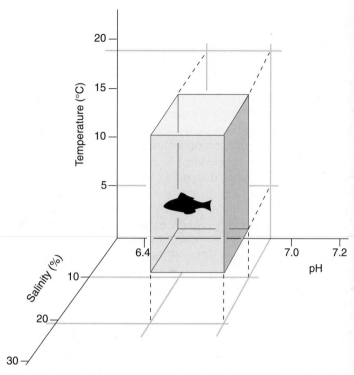

figure 2.2

Three-dimensional niche volume of a hypothetical animal showing three tolerance ranges. This graphic representation is one way to show the multidimensional nature of environmental relations. This representation is incomplete, however, because additional environmental factors also influence growth, reproduction, and survival.

One can classify particular animals as either generalists or specialists with respect to their tolerance of environmental conditions. For example, most fish species are adapted to either freshwater or seawater, but not both. However, those that live in salt marshes, such as the killifish, *Fundulus heteroclitus,* easily tolerate changes in salinity that occur over tidal cycles in these estuarine habitats as fresh water from land mixes with seawater. Similarly, although most snakes are capable of eating a wide variety of animal prey, others have narrow dietary requirements; for example, the African snake *Dasypeltis scaber* is specialized for eating bird eggs "(figure 18.16, p. 379)."

However broad its tolerance limits, an animal experiences only a single set of conditions at a time. An animal probably does not experience in its lifetime all of the environmental conditions that it potentially can tolerate. Thus, we must distinguish an animal's **fundamental niche**, which describes its potential role, and its **realized niche**, the subset of potentially suitable environments that an animal actually experiences. Likewise, we must distinguish fundamental niche from realized niche at the population and species levels. For example, competition within a community may limit the realized niche of a species to a

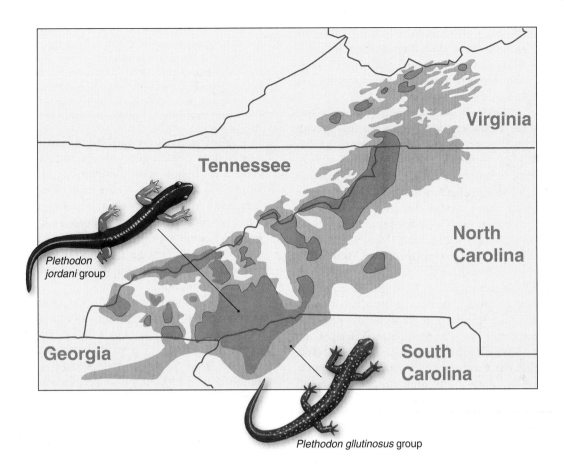

figure 2.3

Geographic consequences of competitive exclusion. The fundamental niches of woodland salamanders of the *Plethodon jordani* group restrict them to the coldest environments located at mountaintops in the southern Appalachian Mountains (red shading). Salamanders of the *Plethodon glutinosus* group (which includes the species *P. teyahalee* mentioned in the text) have a broader temperature tolerance, usually occupying warmer woodlands at lower elevations (including areas shown in gray on the map), but they can occupy the colder habitats favored by the *P. jordani* group provided that members of the *P. jordani* group are absent. Where they occur, populations of the *P. jordani* group aggressively exclude members of the *P. glutinosus* group from their geographic territory. Climatic cycles occurring on tens of thousands of years alter the geography of this competitive exclusion; the gray areas on the map would be cold enough to support *P. jordani*–group populations at times of maximum glaciation, and *P. glutinosus*–group populations thus would be driven from these areas during colder parts of the climatic cycle. Competition from the *P. jordani* group thus limits the realized niche of many *P. glutinosus*–group populations to a subset of their fundamental niche in the dimensions of temperature and elevation.

From Weisrock & Larson, 2006. Biological Journal of the Linnean Society 89:25–51.

much smaller range of conditions than those predicted by its fundamental niche. For example, the lungless slimy salamander *Plethodon teyahalee* occupies forest-floor habitats in the southern Appalachian Mountains. It occupies the coolest, mountaintop forests only where species of the *Plethodon jordani* group are absent (figure 2.3). Climatic requirements restrict the *P. jordani*-group species to cool mountaintop isolates, from which they exclude *P. teyahalee* through aggressive encounters. Competition with the *P. jordani*-group species thereby constrains the realized niche of some *P. teyahalee* populations to a smaller and higher average range of temperatures than its fundamental niche would permit.

■ Populations

Animals exist in nature as members of populations. A population is a reproductively interactive group of animals of a single species (see p. 6). A species might comprise a single, cohesive population or many geographically disjunct populations. A geographically and genetically cohesive population that is separable from other such populations is called a **deme**. Because members of a deme regularly interbreed, they share a common gene pool (see p. 27). A deme of cichlid fishes in Cuatro Cienegas, Mexico, was mistakenly diagnosed as two separate species because some individuals have strong "molariform" jaws capable of crushing snails, whereas others have weaker "papilliform" jaws capable of processing only soft food items. Molecular genetic studies show that despite their morphological differences, these fishes interbreed and share a common gene pool, thereby forming a single deme of the species *Cichlasoma minckleyi.*

Movement of individuals among demes within a species can impart some evolutionary cohesion to the species as a whole. Local environments can change unpredictably, sometimes depleting or eliminating a local deme. Immigration is therefore a crucial source of replacement among demes within a region. A species can avoid extinction if risk of extinction is spread among many demes, because simultaneous destruction of the environments of all demes is unlikely unless a catastrophe is geographically widespread. Interaction among demes in this manner is called **metapopulation dynamics**, with "metapopulation" denoting a population subdivided into multiple genetically interacting demes. In some species, gene flow and recolonization among demes are nearly symmetrical. If some demes are stable and others more susceptible to extinction, the more stable ones, termed **source demes**, differentially supply emigrants to the less stable ones, called **sink demes**. Suppose that the freshwater fish *Cichlasoma minckley* is subdivided geographically into multiple demes that differ from each other in the relative proportions of molariform and papilliform individuals. A deme that contains mostly molariform individuals should be more stable in times of food scarcity than demes having mostly "papilliform" individuals, because the molariform jaws permit a fish to exploit snails as food when the preferred softer food items are depleted. Such a deme could function as a source deme following a decline in the fish's favored prey species.

Each population or deme has a characteristic **age structure**, **sex ratio**, and **growth rate**. **Demography** is the study of these properties and the factors that influence them. Demographic characteristics (individual size, age, number of offspring) vary according to the lifestyles of the species under study. For example, some animals (and many plants) are **modular**. Modular animals, such as sponges, corals, and ectoprocts (see p. 178), form colonies of genetically identical organisms. Reproduction is by asexual **cloning**, as described for hydrozoans in Chapter 7 (see p. 139). Colonies propagate also by fragmentation, as seen on coral reefs during severe storms. Pieces of coral can be scattered by wave action on a reef, forming propagules for new reefs. For these modular animals, age structure and sex ratio are difficult to measure. Changes in colony size are used to measure growth rate as an alternative to counting individuals and measuring their ages and numbers of offspring.

Most animals are **unitary**. However, even some unitary species reproduce by **parthenogenesis**. Parthenogenetic species occur in many animal taxa, including insects, fishes, salamanders, and lizards. Usually such groups contain only females, which lay unfertilized eggs that hatch into daughters genotypically identical to their mothers. The praying mantis *Bruneria borealis,* common in the southeastern United States, is a unitary parthenogenetic animal.

Parthenogenesis ("virgin origin") is the development of an embryo from an unfertilized egg or from one in which the male and female nuclei fail to unite following fertilization. Parthenogenesis takes many forms and is surprisingly widespread ecologically and taxonomically in the animal kingdom.

Most animals are biparental, and reproduction follows a period of organismal growth and maturation. Each new generation begins with a **cohort** of individuals born at the same time. Of course, individuals of any cohort do not all survive to reproduce. For a population to retain constant size from generation to generation, each adult female must replace herself on average with one daughter that survives to reproduce. If females produce on average more than one viable daughter, the population grows; if fewer than one, the population declines.

Animal species have different characteristic patterns of **survivorship** from birth until death of the last member of a cohort. The three principal types of survivorship are illustrated in figure 2.4. Curve I, in which all individuals die at the same time, probably occurs rarely in nature. Curve II, in which rate of mortality as a proportion of survivors is constant over all ages, characterizes some animals that care for their young, as do many birds. Human populations generally fall somewhere between curves I and II, depending on nutrition and medical care.

Survivorship of most invertebrates, and of vertebrates such as fish that produce great numbers of offspring, resembles curve III. For example, a mature female marine snail, *Ilyanassa obsoleta,* produces thousands of eggs each reproductive period.

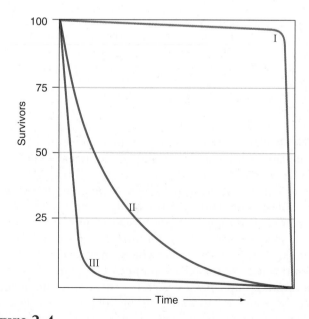

figure 2.4

Three types of theoretical survivorship curves. See text for explanation.

Zygotes become free-swimming planktonic veliger larvae, which are scattered far from the mother's location by oceanic currents. They form part of the plankton and experience high mortality from numerous animals that consume plankton. Furthermore, larvae require a specific, sandy substrate on which to settle and then to metamorphose into an adult snail. The probability of a larva surviving long enough to find a suitable habitat is very low, and most of the cohort dies during the veliger stage. One therefore sees a rapid drop in survivorship in the first part of the curve. The few larvae that do survive to become snails have better odds of surviving further, as shown by the more gentle slope of the curve for older snails. Thus, high reproductive output balances high juvenile mortality.

Most animals do not survive to reach reproductive age, and those that do might typically reproduce only once before they die. In many insect species of the temperate zone, adults reproduce once before the onset of winter and die, leaving only their eggs to overwinter and to repopulate their habitat the following spring. Similarly, Pacific salmon after several years return from the ocean to fresh water to spawn only once, after which all the adults of a cohort die. This condition in which an organism reproduces only once during its life history is termed **semelparity**. However, other animals survive long enough to produce multiple cohorts of offspring that may mature and reproduce while their parents are still alive and reproductively active. **Iteroparity** denotes the occurrence of more than one reproductive cycle in an organism's life history.

Populations of animals containing multiple cohorts, such as robins, box turtles, and humans, exhibit **age structure**. Analysis of age structure reveals whether a population is actively growing, stable, or declining. Figure 2.5 shows age-structure profiles of a human population at two different points in time.

Population Growth and Intrinsic Regulation

Population growth is the difference between rates of birth and death. As Darwin recognized from an essay by Thomas Malthus (see p. 9), all populations have an inherent ability to grow exponentially. This ability is called the intrinsic rate of increase or **intrinsic growth rate,** denoted by the symbol **r.** The steeply rising curve in figure 2.6 shows this kind of growth. If species continually grew in this fashion, earth's resources soon would be exhausted. A bacterium dividing three times per hour could produce a colony a foot deep over the entire earth after 36 hours, and this mass would be over our heads only one hour later. Animals have much lower potential growth rates than bacteria, but they could achieve the same kind of result over a longer period of time, given unlimited resources. Many insects lay thousands of eggs each year. A single codfish can spawn 6 million eggs in a season, and a field mouse can produce 17 litters of five to seven offspring each year. Obviously, unrestricted growth is not prevalent in nature.

Even in the most benign environment, a growing population eventually exhausts food or space. Exponential increases such as locust outbreaks or planktonic blooms in lakes must end when food or space is expended. Among all the resources that could limit a population, the one in shortest supply relative to the needs of the population is depleted before the others.

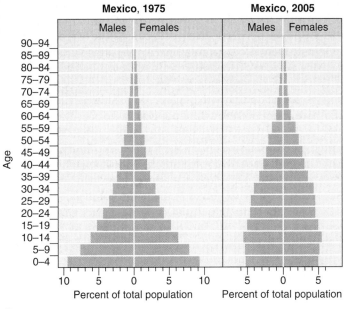

figure 2.5

Age structure profiles of the human populations of Mexico in 1975 and 2005 contrast the rapidly growing, youthful population of 1975 with the stable population of 2005, where the fertility rate approximates replacement.

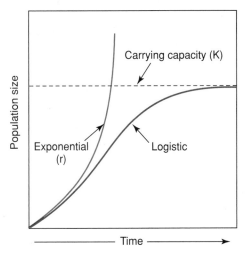

figure 2.6

Population growth, showing exponential growth of a species in an unlimited environment, and logistic growth in a limited environment. K = population size at carrying capacity; r = intrinsic rate of population growth.

Exponential and Logistic Growth

We describe the sigmoid growth curve (see figure 2.6) by a simple model called the logistic equation. The slope at any point on the growth curve is the growth rate, how rapidly the population size changes with time. If **N** represents the number of organisms and **t** the time, we can express growth in the language of calculus as an instantaneous rate:

$$d\text{N}/d\text{t} = \text{the rate of change in number of organisms per time at a particular instant in time.}$$

When populations experience unlimited resources (unlimited food and space, and no competition from other organisms), growth is limited only by the inherent capacity of the population to reproduce itself. Under these ideal conditions, growth is expressed by the symbol **r,** defined as the intrinsic rate of population growth per capita. The index **r** is actually the difference between birth rate and death rate per individual in the population at any instant. The growth rate of the population as a whole is then

$$d\text{N}/d\text{t} = \text{rN}$$

This expression describes the rapid, exponential growth illustrated by the early upward-curving portion of the sigmoid growth curve (see figure 2.6).

Growth rate for a natural population slows as the upper limit is approached, and eventually stops. At this point, N has reached its maximum density because the space being studied has become "saturated" with animals. This limit is called the carrying capacity of the environment and is expressed by the symbol **K.** The sigmoid population growth curve can now be described by the logistic equation, written as

$$d\text{N}/d\text{t} = \text{rN}\left(\left[\text{K} - \text{N}\right]/\text{K}\right)$$

This equation states the rate of increase per unit of time ($d\text{N}/d\text{t}$ = rate of growth per capita [**r**] × population size [**N**] × unutilized freedom for growth [**K** − **N**]/**K**). One can see from the equation that when the population approaches carrying capacity, **K** − **N** approaches 0, $d\text{N}/d\text{t}$ also approaches 0, and the curve flattens.

Populations occasionally overshoot the carrying capacity of the environment so that **N** exceeds **K.** The population then exhausts some resource (usually food or shelter). The rate of growth, $d\text{N}/d\text{t}$, then becomes negative, and the population must decline.

This one is termed the **limiting resource**. The largest population that the limiting resource can support in a habitat is called the **carrying capacity** of that environment, symbolized by **K.** Ideally, a population slows its growth rate in response to diminishing resources until it just reaches K, as represented by the sigmoid (S-shaped) curve (blue line) in figure 2.6. The box on this page compares the mathematical expressions of exponential and sigmoid (or logistic) growth curves. Sigmoid growth occurs when population density exerts negative feedback on growth rate. This phenomenon is called **density dependence**, and it is the mechanism for intrinsic regulation of populations by negative feedback from population size. If the limiting resource is expendable, as is food, carrying capacity is reached when the rate of resource replenishment equals the rate of depletion by the population; the population is then at K for that limiting resource. According to the logistic model, when population density reaches K, rates of birth and death are equal, and growth of the population ceases. If food is replaced at a rate that supports the current population but no more, a population of grasshoppers in a green meadow, for example, may be at carrying capacity even though we see plenty of unconsumed food.

Although experimental populations of protozoa often fit a logistic growth curve closely, most populations in nature tend to fluctuate above and below carrying capacity. For example, after sheep were introduced to Tasmania around 1800, their numbers changed logistically, with small oscillations around an average population size of about 1.7 million; we thereby infer the carrying capacity of the environment to be 1.7 million sheep (figure 2.7A). Ring-necked pheasants introduced on an island in Ontario, Canada, exhibited wider oscillations (figure 2.7B).

Why do intrinsically regulated populations oscillate this way? First, the carrying capacity of an environment can change over time, causing a change in population density as dictated by a limiting resource. Second, animals always experience a lag between the time that a resource becomes limiting and the time that the population responds by reducing its rate of growth. Third, **extrinsic factors** occasionally limit a population's growth below carrying capacity. We consider extrinsic factors on page 45.

On a global scale, humans have the longest record of exponential population growth (figure 2.7C). Although famine and war have restrained growth of populations locally, global human growth declined only when bubonic plague ("black death") decimated much of Europe during the fourteenth century. What then is the carrying capacity for the human population? The answer is unclear, because technological advances have increased our ability to extract useful resources from our environments.

Agriculture increased the carrying capacity of the environment, and the human population grew steadily from 5 million around 8000 B.C., when agriculture was introduced, to 16 million around 4000 B.C. Despite the toll of terrible famines, disease, and war, the population reached 500 million by 1650. With the Industrial Revolution in Europe and England in the eighteenth century, followed by a medical revolution, discovery of new lands for colonization, and better agricultural practices, the human carrying capacity increased dramatically. The population doubled to 1 billion around 1850. It doubled again to 2 billion by 1927, reached 4 billion in 1974, passed 6 billion in October 1999, and is expected to reach 8.9 billion

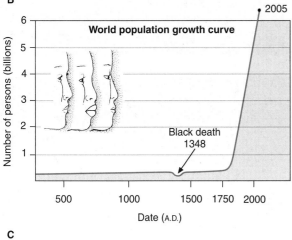

figure 2.7

Growth curves for **A,** sheep, **B,** ring-necked pheasant, and **C,** world human populations throughout history. Note that the sheep population on an island is stable because people control the population, but the ring-necked pheasant population oscillates greatly, probably due to large changes in carrying capacity. Where would you place the carrying capacity for the human population?

by 2040. Thus, growth has been exponential and remains high (figure 2.7C). What would constitute the maximum sustainable agricultural output is uncertain, but food production cannot sustain population growth indefinitely.

Between 1970 and 2004, the annual growth rate of the human population decreased from 1.9% to 1.23%. At 1.23%, it will take nearly 57 years for the world population to double, rather than 36.5 years at the higher annual growth rate. The decrease is credited to better education for women and family planning. Nevertheless, half the global population is under 25 years old. Thus, despite the drop in growth rate, the greatest surge in population lies ahead, with a projected 3 billion people to be added within the next five decades, the most rapid increase in human numbers ever.

Extrinsic Limits to Growth

We have seen that the intrinsic carrying capacity of a population for an environment prevents unlimited exponential growth of the population. Population growth also can be limited by extrinsic biotic factors, such as predation, parasitism (including disease-causing pathogens), and interspecific competition, or by abiotic influences, such as floods, fires, and storms. Although abiotic factors certainly reduce populations in nature, they cannot truly regulate population growth because their effect is wholly independent of population size; abiotic limiting factors are thus **density-independent**. A single hailstorm might kill most of the young wading birds, or a forest fire might eliminate entire populations of many animals, regardless of how many individuals exist.

In contrast, biotic factors can and do act in a **density-dependent** manner. Predators and parasites respond to changes in density of their prey and host populations, respectively, to maintain those populations at fairly constant sizes. These sizes are below carrying capacity, because populations regulated by predation or parasitism are not limited by their resources. Competition between species for a common limiting resource lowers the effective carrying capacity for each species below that of either one alone. Salamanders of the species *Plethodon teyahalee* occur together with single species of the *Plethodon jordani* group at some scattered intermediate elevations in the southern Appalachian Mountains; where the species overlap, both species have lower densities than are typical for them outside the small area of co-occurrence. Experimental removal of either species from a site of co-occurrence increases population density of the remaining species.

▪ Community Ecology

Interactions Among Populations in Communities

Populations of animals that form an ecological community interact in various ways that can be detrimental (−), beneficial (+), or neutral (0) to each species, depending on the nature of the interaction. For instance, a predator's effect on prey is (−), because survival of the prey animal is reduced. However, the same interaction benefits the predator (+) because the food obtained from prey increases the predator's ability to survive and to reproduce. Thus, the predator-prey interaction is + −. Ecologists use this shorthand notation to characterize interspecific interactions because it shows the direction in which the interaction affects each species.

We see other kinds of + − interactions. One of these is **parasitism**, in which the parasite benefits by using the host as a home and a source of nutrition, and the host is harmed. **Herbivory**, in which an animal eats a plant, is another + − relationship. **Commensalism** is an interaction that benefits one species and neither harms nor benefits the other (0 +). For example, most bacteria that normally inhabit our intestinal tracts do not affect us (0), but the bacteria benefit (+) by having food and a place to live. Another example of commensalism is the association of pilot fishes and remoras with sharks (figure 2.8). These fishes get the "crumbs" remaining when the host shark makes its kill.

Organisms engaged in mutualism (+ +) both benefit from their ecological interaction (figure 2.9). A distinction between commensalism and mutualism is sometimes difficult to make in practice. For example, if the harmless bacteria residing in the human gut prevent entry of harmful bacteria, the apparent commensalism becomes a mutualism. Likewise, if remoras remove parasites from their shark hosts (see figure 2.8), then this commensalism likewise become a mutualism.

Some mutualistic relationships are not only beneficial but necessary for the survival of one or both species. Such relationships are termed *obligatory mutualisms*; by contrast, *facultative mutualisms* are not required for a species' survival. The relationship between a termite and protozoa inhabiting its

gut illustrates an obligatory mutualism. Bacterial symbionts of the protozoa can digest wood eaten by the termite because the protozoa produce an enzyme, lacking in the termite, that digests cellulose; the termite lives on waste products of protozoan-bacterial metabolism. In return, the protozoans and their bacteria gain a habitat and food supply. Such absolute interdependence among species is a liability if one of the participants is lost. For example, *Calvaria* trees native to the island of Mauritius have not reproduced successfully for over 300 years, because their seeds germinate only after being eaten and passed through the gut of a dodo bird, now extinct.

Competition between species reduces fitnesses of both (− −). Many biologists, including Darwin, considered competition the most common and important interaction in nature. Ecologists have constructed most of their theories of community

figure 2.8

Four remoras, *Remora* sp., attached to a shark. In this classic example of commensalism, remoras feed on fragments of food left by their shark host, as well as on pelagic invertebrates and small fishes. Although they actually are good swimmers, remoras prefer to be pulled through the water by marine creatures or boats. The shark host may benefit by having embedded copepod skin parasites removed by remoras.

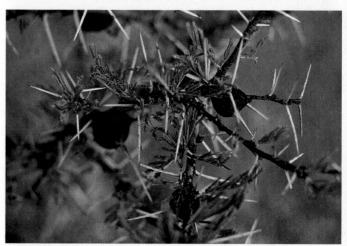

figure 2.9

Among the many examples of mutualism that abound in nature is that of the whistling thorn acacia of the African savanna and the ants that make their homes in the acacia's swollen galls. The acacia provides both protection for the ants' larvae (*lower photograph of opened gall*) and honeylike secretions used by ants as food. In turn, ants protect the tree from herbivores by swarming out as soon as the tree is touched. Giraffes, however, which love the tender acacia leaves, seem immune to the ants' fiery stings.

structure based on the premise that competition is the chief organizing factor in species assemblages. Sometimes the effect on one species in a competitive relationship is negligible. This condition is called **amensalism**, or **asymmetric competition** (0 −). For example, two species of barnacles, *Chthamalus stellatus* and *Balanus balanoides,* compete for space on rocky intertidal habitats. A famous experiment by Joseph Connell[1] demonstrated that *B. balanoides* excluded *C. stellatus* from a portion of the habitat, whereas *C. stellatus* had no effect on *B. balanoides.*

We have treated interactions as occurring between pairs of species. However, in natural communities containing populations of many species, a predator often has more than one prey species, and several animal species often compete for the same resource. Thus, ecological communities are complex and dynamic, a challenge to ecologists who study this level of natural organization.

Competition and Character Displacement

Competition occurs when two or more species share a **limiting resource**. Simply sharing food or space with another species does not produce competition unless the resource is in short supply relative to the needs of the species that share it. Thus, we cannot assume that competition occurs in nature simply by showing that two species share a resource. However, we can find evidence of competition by investigating the different ways that species exploit a resource.

Competing species reduce conflict by reducing overlap of their niches. **Niche overlap** is the portion of resources shared by the niches of two or more species. For example, if two species of birds eat seeds of exactly the same size, competition eventually excludes the species less able to exploit the resource. This example illustrates the principle of **competitive exclusion**: strongly competing species cannot coexist indefinitely. This principle was discovered in 1932 by the Soviet microbiologist G. F. Gause, who performed experiments on the mechanism of competition between two species of yeast. To coexist in the same habitat, species must specialize by partitioning a shared resource and using different portions of it. We call ecological specialization of this kind **character displacement**.

Character displacement usually appears as differences in organismal morphology or behavior related to exploitation of a resource. For example, in his classic study of the Galápagos finches (see p. 24), the English ornithologist David Lack noticed that beak sizes of these birds depended on whether they occurred together on the same island (figure 2.10). On the islands Daphne and Los Hermanos, where *Geospiza fuliginosa* and *G. fortis* occur separately and therefore do not

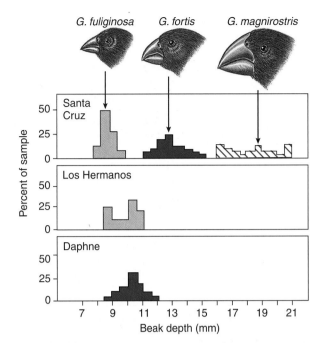

figure 2.10

Displacement of beak sizes in Darwin's finches from the Galápagos Islands. Beak depths are given for the ground finches *Geospiza fuliginosa* and *G. fortis* where they occur together (sympatric) on Santa Cruz Island and where they occur alone on the islands Daphne and Los Hermanos. *G. magnirostris* is another large ground finch that lives on Santa Cruz.

compete with each other, beak sizes are nearly identical; on the island Santa Cruz, where both *G. fuliginosa* and *G. fortis* coexist, their beak sizes do not overlap. These results suggest resource partitioning, because beak size determines the size of seeds selected for food. Recent work by the American ornithologist Peter Grant confirms what Lack suspected: *G. fuliginosa* with its smaller beak selects smaller seeds than does *G. fortis* with its larger beak. Where the two species coexist, competition between them led to evolutionary displacement of beak sizes to diminish competition. Absence of competition today has been called "the ghost of competition past."

Character displacement promotes coexistence by reducing niche overlap. When several species share the same general resources by such partioning, they form a **guild**. Just as a guild in medieval times constituted a brotherhood of men sharing a common trade, species in an ecological guild share a common livelihood. The term guild was introduced to ecology by Richard Root in his 1967 paper on niche patterns of the blue-gray gnatcatcher.[2] A classic example of a bird guild is Robert MacArthur's study of a feeding guild comprising five species of warblers in spruce woods of the northeastern United States.[3] At first glance, one might ask how five birds, very similar in

[1]Connell, J. H. 1961. The influence of interspecific competition and other factors on the distribution of the barnacle *Chthamalus stellatus*. Ecology **42:** 710–723.

[2]Root, R. B. 1967. The niche exploitation pattern of the blue-gray gnatcatcher. Ecological Monographs **37:** 317–350.

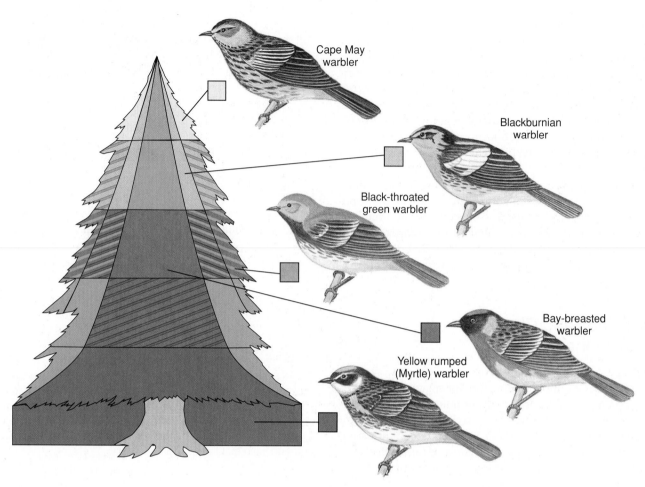

figure 2.11

Distribution of foraging effort among five species of wood warblers in a northeastern spruce forest. The warblers form a feeding guild.

size and appearance, could coexist by feeding on insects in the same tree. However, on close inspection, MacArthur found subtle differences in sites of foraging among these birds (figure 2.11). One species searched only on the outer branches of spruce crowns; another species used the top 60% of the tree's outer and inner branches away from the trunk; another species concentrated on inner branches closer to the trunk; another species used the midsection from the periphery to the trunk; and still another species foraged in the bottom 20% of the tree. These observations suggest that structural differences in habitat separate each warbler's niche within this guild.

Guilds are not limited to birds. For example, a study done in England on insects associated with Scotch broom plants revealed nine different guilds of insects, including three species of stem miners, two gall-forming species, two that fed on seeds, and five that fed on leaves. Another insect guild comprises three species of praying mantids that avoid both competition and predation by differing in size of prey, timing of hatching, and height of vegetation in which they forage.

[3]MacArthur, R. H. 1958. Population ecology of some warblers of northeastern coniferous forests. Ecology **39**: 599–619.

Predators and Parasites

The ecological warfare waged by predators against their prey causes coevolution: predators get better at catching prey, and prey get better at escaping predators. This is an evolutionary race that a predator cannot afford to win. If a predator became so efficient that it exterminated its prey, the predator species would become extinct. Because most predators feed on more than a single species, specialization on a single prey to the point of extermination is uncommon.

When a predator does rely primarily on a single prey species, both populations tend to fluctuate cyclically. First, prey density increases; then predator density increases until prey become scarce. At that point, predators must adjust their population size downward by leaving the area, lowering reproduction, or dying. When the density of a predator population falls enough to allow reproduction by prey to outpace mortality from predation, the cycle begins again. Thus, populations of both predators and prey show cycles of abundance, but increases and decreases in predator abundance are slightly delayed relative to those of prey because of the time lag in a predator's response to changing prey density.

Laboratory experiments with protozoa reveal this process (figure 2.12).

Perhaps the longest documented natural example of a predator-prey cycle is between Canadian populations of snowshoe hares and lynxes (see figure 20.20, p. 422). Abundance of lynx (predator) follows that of showshoe hares (prey) on a 10-year cycle. Interestingly, the timing of these predator-prey cycles appears to track 10-year cycles in abundance of sunspots, which increase the solar energy that reaches earth and possibly also growth of vegetable matter on which hares feed. A global climatic variable thus influences timing of the density-dependent predator-prey cycles.

The war between predators and prey reaches high art in the evolution of defenses by potential prey. Potential prey can escape detection because their bodies match their background or resemble some inedible feature of the environment (such as a stick). Such defenses are called *cryptic*. In contrast to cryptic defenses, animals that are toxic or distasteful to predators actually advertise their strategy with bright colors and conspicuous behavior, a type of defense called *aposematic*. These species are protected because predators learn to avoid them after distasteful encounters. A predator can use mimicry aggressively by matching another organism that attracts its prey. For example, the crab spider shown in figure 13.9A (p. 252) mimics flower petals among which it sits to draw as prey insects that visit the flowers in search of nectar or pollen.

When distasteful prey adopt warning coloration, advantages arise for palatable prey. Palatable prey can deceive potential predators by mimicking distasteful prey, a phenomenon called *Batesian mimicry*. For example, monarch butterflies are brightly colored, and they are noxious as prey because caterpillars store poison (cardiac glycoside) obtained from eating milkweeds. Birds learn from experience not to try to eat monarch butterflies. Monarchs serve as **models** for other species, called **mimics**, that do not possess toxins of their own but look deceptively like the toxic model species. A Batesian mimic essentially parasitizes the model population; a predator that encounters the edible mimic first might later harm models, expecting them to be edible (figure 2.13 A and B).

In another form of mimicry, termed *Müllerian mimicry*, two or more toxic species resemble each other (figure 2.13C). What might an animal that has its own poison gain by evolving resemblance to another poisonous animal? The answer is that a predator needs only to experience the toxicity of one species to avoid all similar prey. A predator can learn one warning signal more easily than many! The different species participating in a case of Müllerian mimicry often are not equally distasteful. If a mildly distasteful species shares warning coloration with a highly toxic species, the mildly distasteful species gains more advantage from the mimicry than does the highly toxic species. Batesian mimicry is thus an extreme case in a continuum of differential toxicity among species that share warning coloration.

Sometimes one population's influence on others is so pervasive that its absence drastically changes the entire community. We call such a population a **keystone species**.[4] On rocky intertidal shores of western North America, the sea star *Pisaster ochraceous* is a keystone species. Sea stars are a major predator of the mussel *Mytilus californianus*. When sea stars were removed experimentally from a patch of Washington State coastline, mussel populations expanded, occupying all of the space previously used by 25 other invertebrate and algal species (figure 2.14). Keystone predators act by reducing prey populations below a level where resources, such as space, are limiting. The original notion that all keystone species are predators has been broadened to include any species whose removal causes the extinction of others.

By reducing competition, keystone species allow more species to coexist on a resource. Consequently, they contribute to maintaining diversity in a community. Keystone species illustrate a more general phenomenon, disturbance. Periodic natural disturbances, such as fires and hurricanes, also can prevent monopolization of resources and competitive exclusion by a few broadly adapted competitors. Disturbances can permit more species to coexist in such highly diverse communities as coral reefs and rain forests.

Parasites are often considered freeloaders because they appear to benefit from their hosts at no expense. **Ecto-parasites**, such as ticks and lice, infest many different kinds of animals. The host provides nutrition from its body and aids dispersal of the parasite. However, we must consider that the evolutionary pathway to parasitism from free-living forms often has costs as well as benefits. **Endoparasites**, such as tapeworms (see Chapter 8), have lost their ability to choose habitats. Also, because they must move among hosts to complete their life

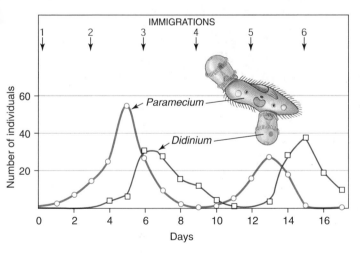

figure 2.12

Classic predator-prey experiment by the Soviet biologist G. F. Gause in 1934 shows the cyclic interaction between predator (*Didinium*) and prey (*Paramecium*) in laboratory culture. When the *Didinium* find and eat all the *Paramecium*, the *Didinium* themselves starve. Gause could keep the two species coexisting only by occasionally introducing one *Didinium* and one *Paramecium* to the culture (*arrows*); these introductions simulated migration from an outside source.

[4]Paine, R. T. 1969. A note on trophic complexity and community stability. American Naturalist **103:** 91–93.

figure 2.13

Artful guises abound. **A,** A palatable viceroy butterfly (*top*) mimics a distasteful monarch butterfly (*below*). **B,** A harmless clearwing moth (*top photograph*) mimics a yellowjacket wasp, which is armed with a stinger (*lower photograph*). Both **A** and **B** illustrate Batesian mimicry. **C,** Two unpalatable tropical butterflies of different families resemble each other, an example of Müllerian mimicry.

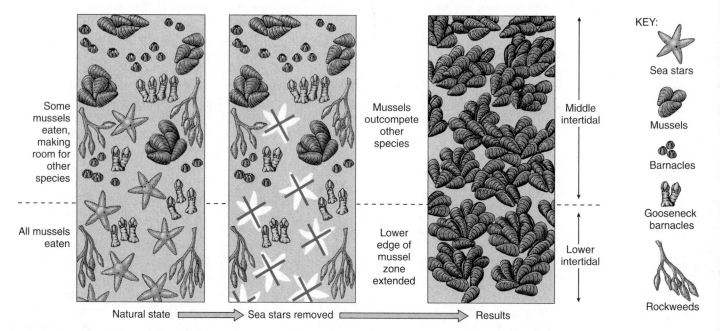

figure 2.14

Experimental removal of a keystone species, the predatory sea star *Pisaster ochraceus,* from an intertidal community completely changes the structure of the community. With their principal predator missing, mussels form dense beds by outcompeting and replacing other intertidal species.

cycle, the chance that a single individual will live to reproduce is very low. The more intermediate hosts involved in a parasite's life cycle, the lower the likelihood of success, and the greater reproductive output must be to balance mortality.

Biologists often are puzzled by the complexity of parasite-host relationships. For example, a trematode parasite (p. 161) of the marine gastropod *Ilyanassa obsoleta* actually changes its host's behavior to complete its life cycle. These snails live in sandy-bottomed intertidal habitats in eastern North America. If the snails are exposed to air when the tide recedes, they normally burrow into sand to avoid desiccation. If, however, a snail is infected with the trematode *Gynaecotyla adunca*, it moves shoreward on high tides preceding low night tides to be left on the beach on the receding tide. Then, as in the legend of the Trojan Horse, the snail sheds larval forms called cercariae into the sand, where they can infect the next intermediate host, a beach-living crustacean. The crustacean can be eaten by a gull or other shorebird, the definitive hosts for this trematode. The life cycle is completed when the bird defecates into water, releasing eggs from which hatch larvae that will infect more snails.

Coevolution between parasite and host is expected to generate an increasingly benign, less virulent relationship if host organisms are uncommon and/or difficult for a parasite to infest. Selection favors a benign relationship, because a parasite's fitness is diminished if its host dies. Virulence is correlated, at least in part, with availability of new hosts. When alternative hosts are common and transmission rates are high, continued colonization of new hosts makes any particular host's life less valuable to the parasite, so that high virulence may have no disadvantage.

Ecosystems

Transfer of energy and materials among organisms within ecosystems is the ultimate level of organization in nature. Energy and materials are required to construct and to maintain life, and their incorporation into biological systems is called **productivity**. Ecologists subdivide productivity into component **trophic levels** based on how organisms obtain energy and materials. Trophic levels are linked together into **food chains**, which denote movement of energy from plant compounds to organisms that eat plants, then to other organisms that eat the plant feeders, and possibly further through a linear series of organisms that feed and are then eaten by others.

Steve Stinson and Roanoke Times and World-News

Life Without the Sun

For many years, ecologists thought that all animals depended directly or indirectly on primary production from solar energy. However, in 1977 and 1979, dense communities of animals were discovered living on the sea floor adjacent to vents of hot water issuing from rifts (such as the Galápagos Rift and East Pacific Rise) where tectonic plates on the sea floor are slowly spreading apart. Such communities were subsequently found on numerous ocean ridges (Juan de Fuca, East Pacific, Southeast Pacific, mid-Atlantic, and Indian Ocean), back-arc basins (mostly West Pacific), hydrocarbon or hypersaline cold seeps (Gulf of Mexico, West Coast of Africa, Asia), and subduction zones (Mediterranean, west coasts of North and South America). These communities (see photo) include several species of molluscs, some crabs, polychaete worms, enteropneusts (acorn worms), and giant pogonophoran-siboglinid tubeworms. The temperature of seawater above and immediately around vents is 7° to 23°C where

A population of giant pogonophoran-siboglinid tubeworms grows in dense profusion near a Galápagos Rift thermal vent, photographed at 2800 m (about 9000 feet) from the deep submersible *Alvin*. Also visible in the photograph are mussels and crabs.

it is heated by basaltic intrusions, whereas the surrounding normal seawater is 2°C.

The producers in these vent communities are chemoautotrophic bacteria that derive energy from oxidation of large amounts of hydrogen sulfide or methane in vent water and fix carbon dioxide into organic carbon. Some animals in vent communities—for example, bivalve molluscs—are filter feeders that ingest bacteria. Others, such as the giant pogonophoran-siboglinid tubeworms, which lack mouths and digestive tracts, harbor colonies of symbiotic bacteria in their tissues and use the organic carbon that these bacteria synthesize.

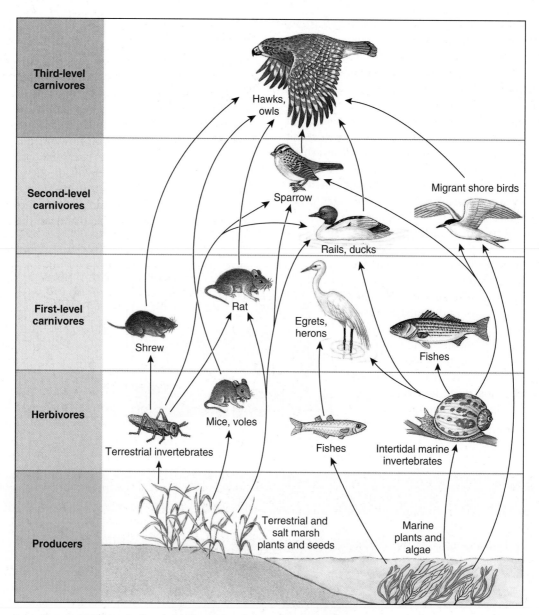

figure 2.15

Primary producers and consumers (herbivores, carnivores) in a midwinter food web in *Salicornia* salt marsh of San Francisco Bay area.

Food chains connect to form **food webs** (figure 2.15), which are branching pathways for transfer of energy and materials among organisms within an ecosystem.

　　Primary producers are organisms that begin productivity by fixing and storing energy from outside the ecosystem. Primary producers usually are plants or algae that capture solar energy through **photosynthesis** (but see an exception in the box on p. 51). Powered by solar energy, plants assimilate and organize minerals, water, and carbon dioxide into living tissue. All other organisms survive by consuming this tissue, or by consuming organisms that consumed this tissue. **Consumers** include **herbivores**, which eat plants directly, and **carnivores**, which eat other animals. The most important consumers are **decomposers**, mainly bacteria and fungi that break dead organic matter into its

mineral components, returning it to a soluble form available to plants at the base of the nutrient cycle (see p. 54). Although important chemicals such as nitrogen and carbon are reused endlessly through biological cycling, all energy ultimately dissipates from the ecosystem as heat and is not recycled. Thus, no ecosystems are truly closed; all require input of new energy from the sun or from hydrothermal vents (see box on previous page).

Energy Flow and Productivity

Every organism in nature has an **energy budget**. Just as we each must partition our income for housing, food, utilities, and taxes, each organism must obtain enough energy to meet its metabolic costs, to grow, and to reproduce.

Ecologists divide the budget into three main components: **gross productivity**, **net productivity**, and **respiration**. Gross productivity is like gross income; it is the total energy assimilated, analogous to your pay before deductions. When an animal eats, it digests food and absorbs nutrients. Most energy assimilated from these nutrients serves the animal's metabolic demands, which include cellular metabolism and regulation of body heat in endotherms. Energy used for metabolic maintenance constitutes respiration, which is deducted from gross productivity to yield net productivity, an animal's take-home pay. Net productivity is energy stored by an animal in its tissues as **biomass**. Animals use some of this energy for growth and for reproduction, which constitutes population growth.

Laws of Thermodynamics

Scientists' observations of how heat and energy interact within a system revealed the laws of thermodynamics. The first law of thermodynamics states that energy cannot be created or destroyed. Although energy can change from one form to another, the total amount of energy in an isolated system remains unchanged.

In the context of an animal's energy budget, the first law of thermodynamics can be expressed by the following equation, in which gross and net productivity are represented by P_g and P_n, respectively, and respiration is R:

$$P_n = P_g - R$$

This equation conveys the important messages that (1) the energy budget of every animal is finite and may be limiting and (2) energy is available for growth of individuals and populations only after satisfying maintenance.

The second law of thermodynamics, which states that the total disorder or randomness of a system always increases, is important when we study energy transfers between trophic levels in food webs. Energy for maintenance, R, usually constitutes more than 90% of the assimilated energy (P_g) for animal consumers. More than 90% of the energy in an animal's food is lost as heat, and less than 10% is stored as biomass. Each succeeding trophic level therefore contains only 10% of the energy in the next lower trophic level. Most ecosystems are thereby limited to five or fewer trophic levels.

Our ability to feed a growing human population is influenced profoundly by the second law of thermodynamics. Humans, who are at the end of the food chain, may eat grain, fruits, and vegetables of plants that fix the sun's energy in chemical bonds; this very short chain represents an efficient use of potential energy. When humans eat beef from cattle, which in turn eat grass that fixes the sun's energy, the addition of a trophic level decreases available energy by a factor of 10. Ten times as much plant biomass is needed to feed humans who eat beef as to feed humans who eat grain, although cattle grazing is possible on land that is too marginal to sustain agriculture. Consider a person who eats a bass that eats a sunfish that eats zooplankton that eats phytoplankton that fixes the sun's energy. The tenfold loss of energy at each trophic level in this five-step chain requires that the pond must produce 5 tons of phytoplankton for a person to gain a pound by eating bass. If the human population depended on bass for survival, we would quickly exhaust this resource.

The productivity of oceans is very low and limited largely to estuaries, marshes, reefs, and upwellings where nutrients are available to phytoplankton producers. Marine fisheries supply approximately 18% of the world's protein, but much of this protein is used to feed livestock and poultry. If we remember the rule of 10-to-1 loss in energy with each transfer of material between trophic levels, we can see that using fish as food for livestock rather than for humans is poor use of a valuable resource in a protein-deficient world. Fishes that we prefer to eat include flounder, tuna, and halibut, which are three or four levels up the food chain. Every 125 g of tuna requires one metric ton of phytoplankton to produce. The less desirable fishes at lower trophic levels are used to feed farm animals or as fertilizer and are therefore consumed by humans only indirectly.

Ecological Pyramids

When we examine the food chain in terms of biomass at each level, we can construct **ecological pyramids** either of numbers or of biomass. A pyramid of numbers (figure 2.16A), also called an **Eltonian pyramid**, depicts numbers of organisms that are transferred between each trophic level. This pyramid provides a vivid impression of the great difference in numbers of organisms involved in each step of the chain, and it supports the observation that large predatory animals are rarer than the small animals on which they feed. However, a pyramid of numbers does not indicate actual mass of organisms at each level.

More instructive are pyramids of biomass (figure 2.16B), which depict the total bulk, or "standing crop," of organisms at each trophic level. Such pyramids usually slope upward because mass and energy are lost at each transfer. However, in aquatic ecosystems whose producers are algae, which have short life spans and rapid turnover, the pyramid is inverted. Algae tolerate heavy exploitation by zooplankton consumers. Therefore, the base of the pyramid (biomass of phytoplankton) is smaller than the biomass of zooplankton it supports. This inverted pyramid is analogous to a person who weighs far more than the food in a refrigerator, but who can be sustained from the refrigerator because food is constantly replenished.

Oxford ecologist Charles Elton invented the concepts of food chains and ecological pyramids in 1923. Working for a summer on a treeless arctic island, Elton watched arctic foxes as they roamed, noting what they ate and, in turn, what their prey had eaten, until he traced the complex cycling of nitrogen in food throughout the animal community. Elton realized that life in a food chain comes in discrete sizes, because each form has evolved to be much bigger than what it eats. He thus explained the common observation that large animals are rare while their smaller prey are common; the ecological pyramids illustrating this phenomenon now bear Elton's name.

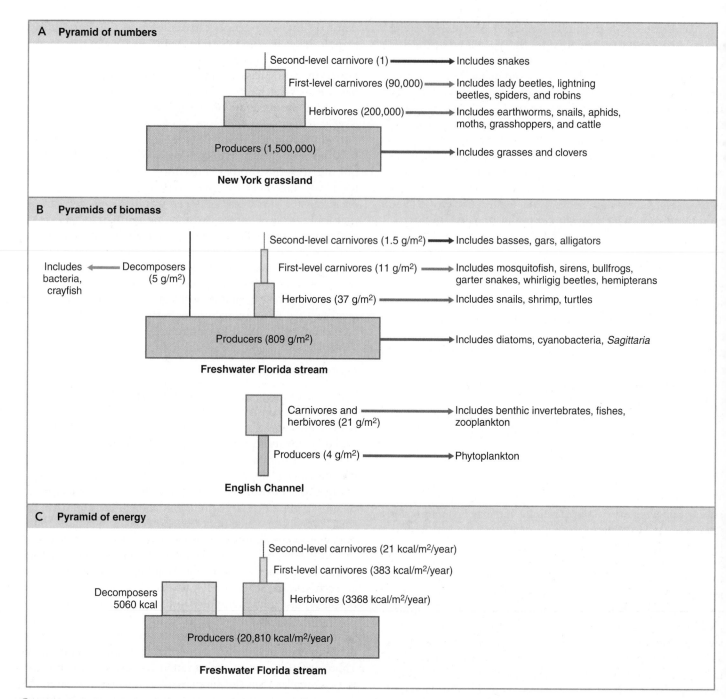

figure 2.16

Ecological pyramids with representative organisms from each trophic level. **A,** Pyramid of numbers of organisms counted for 100 square feet of grassland in New York. **B,** Pyramids of biomass for a freshwater Florida stream (*top*) and plankton of the English Channel (*inverted pyramid at bottom*). **C,** Pyramid of energy for the same Florida stream shown in **B.** (*From the work of E. P. Odum and H. T. Odum.*)

A third type of pyramid is a pyramid of energy, which shows rate of energy flow between levels (figure 2.16C). An energy pyramid is never inverted because the amount of energy transferred from each level is less than the amount that entered it. A pyramid of energy gives the best overall picture of community structure because it depicts the loss of energy from a community as energy flows from primary producers through the food web. In the English Channel, energy of phytoplankton exceeds that of zooplankton, even though the biomass of phytoplankton is less than the biomass of zooplankton (because grazing by zooplankton consumers is heavy).

Nutrient Cycles

All elements essential for life come from environmental air, soil, rocks, and water. When plants and animals die and their

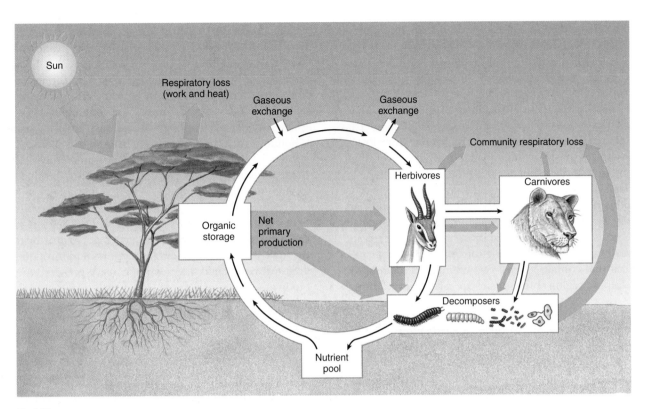

figure 2.17

Nutrient cycles and energy flow in a terrestrial ecosystem. Note that nutrients are recycled (black arrows), whereas energy flow (*orange*) is one way.

bodies decay, or when organic substances are burned or oxidized, elements and inorganic compounds essential for life (nutrients) are released and returned to the environment. Decomposers fulfill an essential role in this process by feeding on the remains of plants and animals and on fecal material. The result is that nutrients flow in a perpetual cycle between biotic and abiotic components of the ecosystem. Nutrient cycles are often called **biogeochemical cycles** because they involve exchanges between living organisms (*bio-*) and the rocks, air, and water of the earth's crust (*geo-*). Continuous input of energy from the sun keeps nutrients flowing and the ecosystem functioning (figure 2.17).

We tend to think of biogeochemical cycles in terms of naturally occurring elements, such as water, carbon, phosphorus, and nitrogen. For example, the nutrient pool of the soil shown in figure 2.17 contains a store of phosphorus, some of which is in compounds that can be extracted by living plants. Phosphorus obtained by plants can be stored in the plants' tissues, returned to the soil, or consumed by herbivorous animals. The herbivores likewise store phosphorus in their bodies, return it to the soil via waste products, or provide it to carnivores that consume them. These carnivores likewise retain some phosphorus, return some to the soil via waste products, and may themselves be the prey of other carnivores. The amount of phosphorus available in the nutrient pool can be a limiting factor in the nutrient cycle of a local community.

Nutrient pools can be altered greatly by fertilizers and other compounds that people synthesize and add to the soil. Our synthetic compounds often challenge nature's nutrient cycling because decomposers have not evolved ways to degrade them. Among the most harmful synthetic compounds are pesticides. Pesticides in natural food webs can be insidious for three reasons. First, many pesticides become concentrated as they travel up succeeding trophic levels. The highest concentrations occur in top carnivores, such as hawks and owls, diminishing their ability to reproduce. Second, many species that are killed by pesticides are not pests, but merely innocent bystanders, called nontarget species. Nontarget effects happen when pesticides move out of the agricultural field to which they were applied, through rainwater runoff, leaching through soil, or dispersal by wind. The third problem is persistence; some chemicals used as pesticides have a long life span in the environment, so that nontarget effects persist long after the pesticides have been applied. Genetic engineering of crop plants aims to improve their resistance to pests to lessen the need for chemical pesticides.

■ Biodiversity and Extinction

Biodiversity exists because rates of speciation on average slightly exceed rates of extinction in earth's evolutionary history. Approximately 99% of all species that have lived are extinct. If we divide the fossil history of animal life into million-year intervals throughout the past 540 million years, each interval shows at least 5% extinction of the species that entered it. Extinction events killing at least 30% of existing species have occurred episodically approximately every 10 million years, and truly catastrophic events that killed at least 65% of existing species have occurred, on average, every 100 million years. Although the earth's ecosystems have recovered from loss of biodiversity in the past, the time for recovery has been enormously long from the perspective of a human life span.

At rates unprecedented during human history, we are losing biodiversity, genetic diversity within species, and habitat diversity. Fragmentation of populations, as seen especially on islands, leads to loss of genetic diversity and makes a species unusually susceptible to extinction. About half of all areas in the world that contain at least two endemic (confined to a small geographic region) bird species are islands, even though islands constitute less than 10% of the earth's terrestrial habitats. Island species are often particularly prone to destruction by introduction of invasive exotic species. For example, diverse land snails of the genus *Partula* on the Tahitian island of Moorea were displaced by introduced exotic snails. Mainland habitats, such as forests, are fragmented into virtual islands when development clears vast areas of habitat and when introduced species invade these areas. Because tropical regions have high species endemism, human fragmentation of these environments is particularly likely to produce species extinction.

Human activity clearly has induced numerous species extinctions, and we must avoid making the present time rival the great extinction crises of earth's geological history. Only through significant conservation efforts and through laws that limit environmental degradation can we preserve the planet's biodiversity.

Summary

Ecology is the study of relationships between organisms and their environments to explain the distribution, abundance, and diversity of life on earth. The physical space where an animal lives, and which contains its environment, is its habitat. Within the habitat are physical and biological resources that an animal uses to survive and to reproduce, which constitute its niche.

Animal populations consist of demes of interbreeding members sharing a common gene pool. Cohorts of animals have characteristic patterns of survivorship that represent adaptive trade-offs between parental care and number of offspring. Animal populations comprising overlapping cohorts have an age structure that indicates whether they are growing, declining, or at equilibrium.

Every species in nature has an intrinsic rate of increase that gives it the reproductive potential for exponential growth. The human population is growing exponentially at about 1.23% each year, and it is expected to increase from 6.4 billion to 8.9 billion by the year 2040. Population growth can be regulated intrinsically, by the carrying capacity of the environment; or extrinsically, by competition between species for a limiting resource, or by predators or parasites. Density-independent abiotic factors can limit, but not truly regulate, population growth.

Communities consist of populations that interact with one another in any of several ways, including competition, predation, parasitism, commensalism, and mutualism. These relationships cause coevolution among populations within communities. Guilds of species avoid competitive exclusion by character displacement, the partitioning of limited resources by morphological specialization. Keystone species of predators are those that control community structure and reduce competition among prey, which increases species diversity. Parasites and their hosts evolve a benign relationship that ensures their coexistence.

Ecosystems consist of communities and their abiotic environments. Animals occupy the trophic levels of herbivorous and carnivorous consumers within ecosystems. All organisms have an energy budget consisting of gross and net productivity, and respiration. For animals, respiration usually is at least 90% of this budget. Thus, transfer of energy from one trophic level to another is limited to about 10%, which in turn limits the number of trophic levels in an ecosystem. Ecological pyramids of energy depict how productivity decreases in successively higher trophic levels of food webs.

Ecosystem productivity is described by measuring energy flow and nutrient cycles within ecosystems. All energy is lost as heat, but nutrients and other materials, including pesticides, are recycled. No ecosystem, including the global biosphere, is closed because they all depend upon imports of energy from the sun or hydrothermal vents.

Review Questions

1. The term *ecology* is derived from a Greek word meaning "house" or "place to live." However, as used by scientists, the terms ecology and environment do not mean the same thing. How do these terms differ?
2. How would you distinguish the concepts of ecosystem, community, and population?
3. What is the distinction between habitat and environment?
4. Define a niche. How does the *realized niche* of a population differ from its *fundamental niche?* How does the concept of niche differ from the concept of guild?
5. Populations of independently living (unitary) animals have a characteristic age structure, sex ratio, and growth rate. However, these properties are difficult to measure for modular animals. Why?
6. Explain which of the three survivorship curves in figure 2.4 best fits the following: (a) a population in which mortality as a proportion of survivors is constant; (b) a population in which few individuals die early and most

individuals live to old age; (c) a population in which the very young experience heavy mortality but the survivors live to old age. Offer an example from the real world of each survivorship pattern.

7. Contrast exponential and logistic growth of a population. Under what conditions might you expect a population to exhibit exponential growth? Why cannot exponential growth be perpetuated indefinitely?

8. Growth of a population can be limited by either density-dependent or density-independent mechanisms. Define and contrast these two mechanisms. Offer examples of how growth of the human population might be curbed by either agent.

9. Herbivory is beneficial for an animal (+) but harmful to the plant it eats (−). What are some + − interactions among animal populations? What is the difference between commensalism and mutualism?

10. Explain how character displacement can ease competition between coexisting species.

11. Define predation. How does the predator-prey relationship differ from the parasite-host relationship? Why is the evolutionary race between predator and prey one that the predator cannot afford to win?

12. Mimicry of monarch butterflies by viceroys is an example of a palatable species resembling a toxic one. What is the advantage to the viceroy of this form of mimicry? What is the advantage to a toxic species of mimicking another toxic species?

13. A keystone species is one whose removal from a community causes extinction of other species. How does this extinction happen?

14. What is a food chain? How does a food chain differ from a food web?

15. What is a trophic level, and how does it relate to a food chain?

16. Define *productivity* as the word is used in ecology. What is a primary producer? What is the distinction between gross productivity, net productivity, and respiration? How is net productivity related to biomass?

17. What conditions produce an inverted pyramid of biomass in which the consumers have a greater biomass than the producers? Describe an example of an inverted pyramid of numbers of organisms, in which there are, for example, more herbivores than plants on which they feed.

18. The pyramid of energy has been offered as an example of the second law of thermodynamics. Why?

19. Animal communities surrounding deep-sea thermal vents apparently exist in total independence of solar energy. How is this existence possible?

For Further Thought Natural processes of species formation and extinction emerge on a timescale measured in millions of years, whereas human-mediated extinctions are measured on a scale of years to decades. How does this discrepancy complicate our attempts to conserve biodiversity?

■ Selected References

See general references on page 432.

Chase, J. M., and M. A. Liebold. 2003. Ecological niches: linking classical and contemporary approaches. Chicago, University of Chicago Press. *An insightful coverage of niche concepts in community ecology.*

Levin, S. A. (ed.). 2009. The Princeton Guide to Ecology. Princeton, New Jersey, Princeton University Press. *A comprehensive review of ecological topics with contributions from many authorities.*

Molles, M. C., Jr. 2010. Ecology: concepts and applications, ed. 5. New York, McGraw-Hill. *A concise and well-illustrated survey of ecology.*

Sax, D. F., J. J. Stachowicz, and S. D. Gaines. 2005. Species invasions: insights into ecology, evolution, and biogeography. Sunderland, Massachusetts, Sinauer Associates. *The establishment of alien species in an area can reveal ecosystem properties and the structuring of ecological communities.*

Sibly, R. M., J. Hone, and T. H. Clutton-Brock. 2003. Wildlife population growth rates. Cambridge, U.K., Cambridge University Press. *Contributions by many ecologists illustrate current methods for measuring the growth of natural populations.*

Sinclair, A. R. E., J. M. Gosline, G. Holdsworth, C. J. Krebs, S. Boutin, J. N. M. Smith, R. Boonstra, and M. Dale. 1993. Can the solar cycle and climate synchronize the snowshoe hare cycle in Canada? Evidence from tree rings and ice cores. American Naturalist 141: 173–198. *An example of how a global climatic variable can influence the timing of a density-dependent, predator-prey cycle.*

Zimmer, C. 2000. Parasite Rex. New York, The Free Press (a division of Simon & Schuster). *A fascinating book for nonspecialists by a talented science writer. Many examples of parasites that affect or control their host's behavior and drive host evolution.*

■ Custom Website

The *Animal Diversity* Online Learning Center is a great place to check your understanding of chapter material. Visit www.mhhe.com/hickmanad6e for access to key terms, quizzes, and more! Further enhance your knowledge with Web links to chapter-related material.

Explore live links for these topics:

General Ecology Sites
Animal Population Ecology
Life Histories
Population Growth
Human Population Growth
Community Ecology

Competition
Parasitism, Predation, Herbivory
Coevolution
Food Webs
Primary Productivity

3

Animal Architecture

Cnidarian polyps have radial symmetry and the cell-tissue level of organization (*Dendronephthya* sp.).

New Designs for Living

Zoologists today recognize 34 animal phyla, each phylum characterized by a distinctive body plan and array of biological properties that set it apart from all other phyla. Nearly all are the survivors of perhaps 100 phyla that existed 600 million years ago during the Cambrian period. How long it took for these taxa to evolve is the subject of active debate, but virtually all of the major body plans that we see today, together with many other novel plans that we know only from the fossil record, were established by the Cambrian. Entering a world sparse in species and mostly free of competition, these new life forms began widespread experimentation, producing new themes in animal architecture. Nothing since has equaled the Cambrian explosion. Later bursts of speciation that followed major extinction events produced only variations on established themes.

Established themes, in the form of distinctive body plans, are passed down a lineage from an ancestral population to its descendants. Most molluscs carry a hard shell. Bird forelimbs become wings. These ancestral traits limit the morphological scope of descendants no matter what their lifestyle. Although penguin bodies are modified for an aquatic life, the bird wings and feathers adapt to aquatic environments differently than do fish fins and scales. Despite structural and functional evolution, new forms are constrained by the architecture of their ancestors.

T he English satirist Samuel Butler proclaimed that the human body was merely "a pair of pincers set over a bellows and a stewpan and the whole thing fixed upon stilts." While human attitudes toward the human body are distinctly ambivalent, most people less cynical than Butler would agree that the body is a triumph of intricate, living architecture. Less obvious, perhaps, is that humans and most other animals share an intrinsic material design and fundamental functional plan despite vast differences in structural complexity. This essential uniformity of biological organization derives from the common ancestry of animals and from their basic cellular construction. In this chapter, we consider the limited number of body plans that underlie the diversity of animal form and examine some of the common architectural themes that animals share.

The Hierarchical Organization of Animal Complexity

The eukaryotic cell is the basic building block of animals. There are five major levels of organization within the eukaryotes. Our description of levels begins with a single cell and encompasses four types of cell-cell interactions in multicellular bodies (table 3.1).

At the first level of organization, the *protoplasmic level*, cells vary in terms of their subcellular structures, including organelles. Organisms at the next level of organization, *the cellular level*, may be colonial or multicellular. Cells at this level communicate with other cells in the aggregation

Table 3.1   Levels of Organization in Organismal Complexity

1. *Protoplasmic level of organization.* Protoplasmic organization characterizes unicellular organisms. All life functions are confined within the boundaries of a single cell. Within a cell, living substance is differentiated into organelles that perform specialized functions.

2. *Cellular level of organization.* Colonial organization denotes an aggregation of undifferentiated cells, whereas multicellular organization denotes an aggregation of cells that are functionally differentiated. A division of labor is evident, so that some cells are concerned with, for example, reproduction, while others function in nutrition. Such cells have little tendency to become organized into tissues. Some protozoan colonial forms that have distinct somatic and reproductive cells might be placed at the cellular level of organization. Many authorities also place sponges at this level.

3. *Cell-tissue level of organization.* At this level similar cells aggregate into definite patterns or layers, thus becoming a tissue, a group of similar cells organized to perform a common function. Some authorities assign sponges to this level, although jellyfishes and their relatives (Cnidaria) more clearly demonstrate the tissue plan. Both groups are still largely of the cellular level of organization because most of the cells are scattered and not organized into tissues. An excellent example of a tissue in cnidarians is the nerve net, a definite tissue structure formed from nerve cells and their processes and functioning in coordination. Three distinct types of junctions between cells are present (adherens, septate, and gap junctions) in animals.

4. *Tissue-organ level of organization.* Organs usually contain more than one kind of tissue and have a more specialized function than tissues. This is the organizational level of the flatworms (Platyhelminthes), which have well-defined organs such as eyespots, a digestive tract, and reproductive organs. Only the reproductive organs are well organized into a system.

5. *Organ-system level of organization.* When organs work together to perform some function, they form the most complex level of organization—the organ system. Systems are associated with basic body functions—circulation, respiration, digestion, and others. The simplest animals that show this type of organization are nemertean worms, which have a complete digestive system distinct from the circulatory system. Most animal phyla demonstrate this type of organization, illustrated by chordates such as frogs, by arthropods such as crabs, and by annelids such as earthworms.

and may secrete an extracellular matrix, called an **ECM**, to which cells adhere. Collagen is a familiar component of ECM found in animals. The distinction between coloniality and multicellularity lies in the degree of specialization. Colonial organisms consist of aggregations of unspecialized cells, all able to reproduce. In multicellular organisms, cells are specialized for particular functions and reproduction is limited to some cells in the aggregation. The evolution of multicellularity occurred independently in perhaps as many as 25 lineages of unicellular eukaryotes (see Chapter 5 for some examples). One such event within a lineage called Opisthokonta led to the animals (metazoans).

Animals evolved greater structural complexity by combining cells into larger units. The simplest metazoans show the cellular level of organization, in which cells demonstrate division of labor but are not strongly associated to perform a specific collective function (table 3.1). In the more complex *cell-tissue* level, similar cells are grouped together and perform their common functions as a highly coordinated unit. In animals of the *tissue-organ level* of organization, tissues are assembled into still larger functional units called *organs*. Usually one type of tissue carries the burden of an organ's chief function, as muscle tissue does in the heart; other tissues—epithelial, connective, and nervous—perform supportive roles. The chief functional cells of an organ are called its **parenchyma** (pa-ren´ka-ma; Gr. *para*, beside, + *enchyma*, infusion). The supportive tissues are its **stroma** (Gr. bedding). For instance, in the vertebrate pancreas the secreting cells are the parenchyma; the capsule and connective tissue framework represent the stroma.

Most metazoa have an additional level of complexity in which different organs operate together; this is known as the *organ-system level* of organization. Eleven different organ systems are observed in metazoans: skeletal, muscular, integumentary, digestive, respiratory, circulatory, excretory, nervous, endocrine, immune, and reproductive. The great evolutionary diversity of these organ systems is covered in Chapters 8–20.

■ Animal Body Plans

As described in the prologue to this chapter, ancestral body plans constrain the form of descendant organisms. Animal body plans differ in level of organization, in number of embryonic germ layers, in form and number of body cavities, and in body symmetry. Symmetry can generally be determined from the external appearance of an animal, but other features require more detailed study.

Animal Symmetry

Symmetry refers to balanced proportions, or correspondence in size and shape of parts on opposite sides of a median plane. *Asymmetrical* bodies are not balanced; there is no plane through which they can be divided into identical halves (figure 3.1A). *Spherical symmetry* means that any plane passing through the center divides the body into equivalent, or mirrored, halves (figure 3.1B). This type of symmetry occurs chiefly among some protozoan groups and is rare in metazoans. Spherical forms are best suited for floating and rolling.

Radial symmetry applies to forms that can be divided into similar halves by more than two planes passing through the longitudinal axis (figure 3.1C). These are tubular-, vase-, or bowl-shaped animals, in which one end of the longitudinal axis is usually the mouth (the oral surface). Examples are the hydras, and jelly fishes in phylum Cnidaria, sea urchins (phylum Echinodermata), and some sponges; in sessile forms, the **aboral** surface is used in attachment. A variant form is biradial symmetry, in which, because of some part that is single or paired rather than radial, only one or two planes passing through the longitudinal axis produce mirrored halves. Sea walnuts or comb jellies (phylum Ctenophora), which are roughly globular in form but have a pair of tentacles, are an example. Radial and biradial animals are usually sessile, freely floating, or weakly swimming. The two phyla that are primarily radial, Cnidaria and

A Asymmetry

B Spherical symmetry

C Radial symmetry

D Bilateral symmetry

figure 3.1

Asymmetrical bodies **A,** lack planes of symmetry. Planes of symmetry are illustrated by spherically **B,** radially **C,** and bilaterally **D,** symmetrical animals.

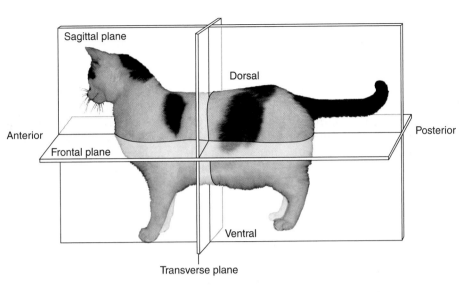

figure 3.2

Terms used to describe positions on the body of a bilaterally symmetrical animal.

Ctenophora, are called the **Radiata**. Echinoderms (sea stars and their kin) are primarily bilateral animals (their larvae are bilateral) that have become secondarily radial as adults.

Bilateral symmetry applies to animals that can be divided along a sagittal plane into two mirrored portions, right and left halves (figure 3.1D). Some convenient terms used for locating regions of animal bodies are **anterior**, to designate the head end; **posterior**, the opposite or tail end; **dorsal**, the back side; and **ventral**, the front or belly side (figure 3.2). **Medial** refers to the midline of the body; **lateral** refers to the sides. **Distal** parts are farther from the middle of the body; **proximal** parts are nearer. Figure 3.2 also illustrates three planes of the body. A **frontal plane** (also sometimes called a coronal plane) divides a bilateral body into dorsal and ventral halves by running through the anteroposterior axis and the right-left axis at right angles to the **sagittal plane**, the plane dividing an animal into right and left halves. A **transverse plane** (also called a cross section) would cut through a dorsoventral and a right-left axis at right angles to both the sagittal and frontal planes, separating the body into anterior and posterior portions.

The appearance of bilateral symmetry in animal evolution was a major innovation because bilateral animals are much better fitted for directional (forward) movement than are radially symmetrical animals. Bilateral animals are collectively called **Bilateria**. Bilateral symmetry is strongly associated with **cephalization**, the differentiation of a head end. Concentration of nervous tissue and sense organs in a head bestows obvious advantages to an animal moving through its environment head first. This is the most efficient positioning of instruments for sensing the environment and responding to it. An animal's mouth usually occurs on the head where sensory structures guide forward locomotion toward food.

Cephalization is always accompanied by differentiation along an anteroposterior axis (**polarity**). Each end of the body forms a pole. The release of substances from each pole may produce gradients of chemical signals along an anteroposterior axis. Such gradients are important during development as regions of the body develop differently in response to the chemical signals they experience.

Development of Animal Body Plans

The body plan of an animal forms as part of an inherited developmental sequence. This sequence begins after fertilization of an egg to form a **zygote**. The zygote is a single large cell until the process of cleavage divides it into a large number of smaller cells called blastomeres. Cleavage is an orderly sequence of cell divisions.

Cleavage occurs in several different ways: sponges and cnidarians (sea anemones and their kin) lack a distinct cleavage pattern, but bilateral animals typically exhibit either radial or spiral cleavage (figure 3.3). Although there is much debate about which type of cleavage occurred first, biologists assume that these two types of cleavage each evolved only once. Thus, animals with spiral cleavage form a monophyletic group, as do those with radial cleavage (see p. 65).

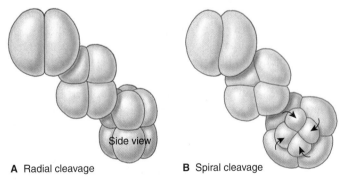

A Radial cleavage   **B** Spiral cleavage

figure 3.3

Radial and spiral cleavage patterns shown at two-, four-, and eight-cell stages; all diagrams are top views unless otherwise labeled. **A,** Radial cleavage, typical of echinoderms, chordates, and hemichordates. **B,** Spiral cleavage, typical of molluscs, annelids, and other protostomes. Arrows indicate clockwise movements of small cells following division of large cells.

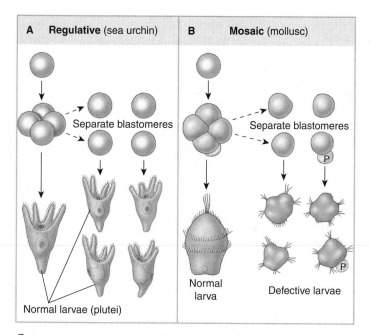

figure 3.4

Regulative and mosaic development. **A,** Regulative development. If the early blastomeres of a sea urchin embryo are separated, each develops into a complete larva. **B,** Mosaic development. When the blastomeres of a mollusc embryo are separated, each gives rise to a partial, defective larva. P indicates polar bodies which are shed.

In **radial cleavage**, cleavage produces tiers, or layers, of cells on top of each other in an early embryo (figure 3.3A). Radial cleavage occurs with **regulative development** where each blastomere of the early embryo, if separated from the others, can adjust or "regulate" its development into a complete and well-proportioned (though possibly smaller) embryo (figure 3.4A).

Spiral cleavage, found in several phyla, differs from radial in several ways. By the eight-cell stage cleavage produces two quartets of cells that come to lie not on top of each other but in furrows between the cells (see figure 3.3B). In addition, spirally cleaving eggs tend to pack their cells tightly together much like a cluster of soap bubbles, rather than just lightly contacting each other as do those in many radially cleaving embryos. Spirally cleaving embryos also differ from radial embryos in that most have a form of **mosaic development**, in which the organ-forming determinants in the egg cytoplasm become strictly localized in the egg, even before the first cleavage division. The result is that if the early blastomeres are separated, each will continue to develop for a time as though it were still part of the whole. Each forms a defective, partial embryo (figure 3.4B).

A special feature of most spirally cleaving embryos is that, at about the 29-cell stage, a blastomere called the 4d cell is formed that will give rise to all mesoderm of the embryo. Mesoderm is one of three germ layers usually present in an embryo; the other two germ layers are ectoderm and endoderm.

Ultimately, all features of an adult animal are derived from one of these three germ layers, but adult structures do not form until well after cleavage.

Cleavage proceeds until the zygote is divided into many small cells, typically surrounding a fluid-filled cavity (figure 3.5). At this point, the embryo is called a **blastula**, and the fluid-filled cavity is a **blastocoel**. In animals other than most sponges, the blastula becomes a two-layered stage called a **gastrula** (figure 3.5A). The gastrula has two germ cell layers, ectoderm and endoderm. The outer germ cell layer, ectoderm, surrounds the blastocoel. The endoderm surrounds and defines an inner body cavity called the **gastrocoel**. The gastrocoel will become the gut cavity in an adult animal. Some body cavities, such as the gut, persist into adulthood, but other cavities are lost, and still others appear only later in development. Adult sea anemones and their kin are two-layered animals, so they do not develop beyond the gastrula stage, but most animals develop three embryonic germ layers and additional body cavities.

Body Cavities

The most obvious internal space or body cavity is a gut developing from an embryonic gastrocoel (figure 3.5A). The gut always has at least one opening, the blastopore. In adult sea anemones and their kin, this opening ingests food and releases digestive wastes. A gut with only one opening is called a two-way, blind, or incomplete gut. Most animals develop a second opening to the gut, creating a tube (figure 3.5B). Tubular guts permit sequential movement of food from mouth to anus and are called one-way or complete guts.

In many animals, the gut is surrounded by a fluid-filled cavity. This cavity can form in several ways. For animals with only two germ layers, such as cnidarians (including sea anemones, corals, and jelly fishes), the embryonic blastocoel persists into adulthood as a cavity surrounding the blind gut (figure 3.5A). The blastocoel may also persist in animals with a third germ layer, mesoderm, but the addition of a layer of mesoderm inside the blastocoel causes structural changes.

In some animals, mesoderm lines the outer edge of the blastocoel, lying next to the ectoderm (figure 3.5C); when this occurs, the blastocoel is renamed the **pseudocoelom** in adult animals. "Pseudocoelom" means "false coelom," and the descriptive nature of this name will be clear once a coelom is understood.

In other animals, mesoderm completely fills the blastocoel (figure 3.5D). These animals are **acoelomate**, meaning that the body is without a coelom. The only body cavity is the space inside the gut tube, the gut being surrounded by a mass of tissue derived from mesoderm.

In the development of most bilaterally symmetrical animals, the blastocoel fills with mesoderm, and then a new cavity forms *inside* the mesoderm (figure 3.5E). This new cavity, completely surrounded by mesoderm, is a **coelom** (Gr. *koilos*, hollow or cavity). A coelom differs from a pseudocoelom, which has mesoderm on only the outer

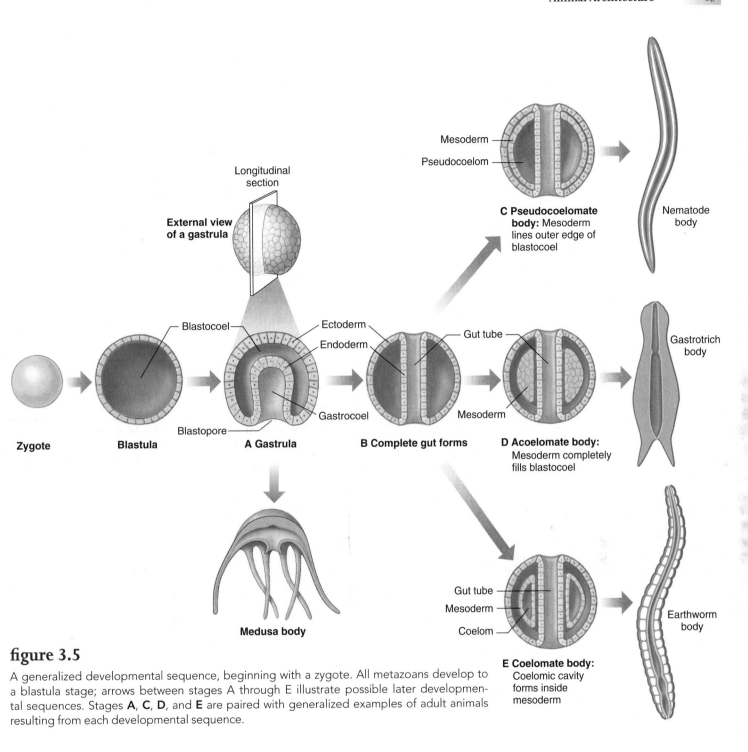

figure 3.5

A generalized developmental sequence, beginning with a zygote. All metazoans develop to a blastula stage; arrows between stages A through E illustrate possible later developmental sequences. Stages **A**, **C**, **D**, and **E** are paired with generalized examples of adult animals resulting from each developmental sequence.

edge of the cavity. A coelom may form by one of two methods: **schizocoely** or **enterocoely** (figure 3.6), or by modifying these methods. In schizocoelous formation (Gr. *schizein,* to split), the coelom arises from splitting of mesodermal bands that originate when cells in the blastopore region migrate into the blastocoel. In enterocoelous formation (Gr. *enteron,* gut), the coelom comes from pouches of the archenteron, or primitive gut, that push outward into the blastocoel (figure 3.6).

Once development is complete, the results of schizocoelous and enterocoelous formations are indistinguishable. Both

produce a true coelom lined with a mesodermal peritoneum (Gr. *peritonaios,* stretched around) and having mesenteries in which the visceral organs are suspended.

The evolution of a coelom, a fluid-filled body cavity between the outer body wall and the gut, was a major developmental event affecting the body plans of many bilateral animals. A coelom provides a tube-within-a-tube arrangement that allows much greater body flexibility than is possible in animals lacking an internal body cavity. A coelom also provides space for visceral organs and permits greater size and complexity by exposing more cells to surface exchange. A

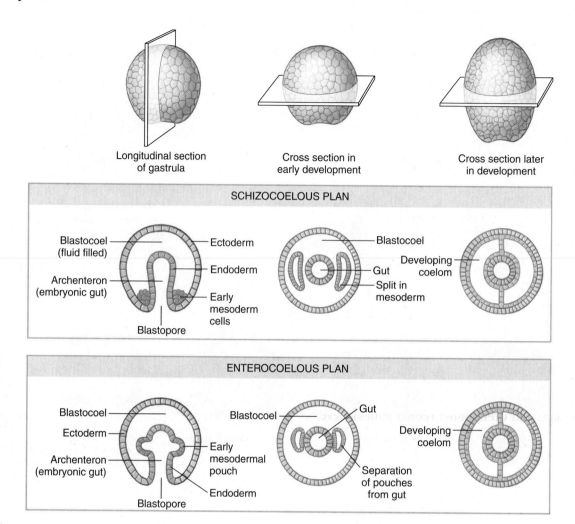

Longitudinal section
of gastrula

Cross section in
early development

Cross section later
in development

SCHIZOCOELOUS PLAN

Blastocoel
(fluid filled)

Ectoderm

Endoderm

Archenteron
(embryonic gut)

Early
mesoderm
cells

Blastopore

Blastocoel

Developing
coelom

Gut

Split in
mesoderm

ENTEROCOELOUS PLAN

Blastocoel

Ectoderm

Archenteron
(embryonic gut)

Early
mesodermal
pouch

Endoderm

Blastopore

Blastocoel

Gut

Developing
coelom

Separation
of pouches
from gut

figure 3.6

Types of mesoderm and coelom formation. In schizocoelous formation, the mesoderm originates from the wall of the embryonic gut near the blastopore and proliferates into a band of tissue that splits to form the coelom. This method of coelom formation may not be homologous in all phyla where it occurs. In enterocoelous formation, most mesoderm originates as a series of pouches from the embryonic gut; these pinch off and enlarge to form the coelom. In both formations, the coeloms expand to obliterate the blastocoel.

fluid-filled coelom additionally serves as a hydrostatic skeleton in some forms, especially many worms, aiding in such activities as movement and burrowing.

Metameric or Segmented Body Plans

Metamerism, also called segmentation, is serial repetition of similar body segments along the longitudinal axis of the body (figure 3.7). Each segment is called a metamere, or somite. In forms such as earthworms and other annelids, in which metamerism is most clearly represented, the segmental arrangement includes both external and internal structures of several systems. There is repetition of muscles, blood vessels, nerves, and the setae of locomotion. Some other organs, such as those of sex, may be repeated in only a few segments.

Evolutionary changes have obscured much of the segmentation in many animals, including humans. True metamerism characterizes only three phyla: Annelida, Arthropoda, and

Chordata (figure 3.7), although many diverse animals display superficial segmentation of the ectoderm and the body wall.

How Many Body Plans Are There?

Animals comprise 34 phyla as shown on the cladogram inside the front cover. Members of all phyla develop from cleavage through the blastula stage, but the number of embryonic germ layers formed after the blastula varies among phyla. The number of germ layers affects the tissues present in adults. Sponges, in phylum Porifera, have no tissue layers in the adult body, and were originally thought to develop only to a single-layered blastula. However, detailed study has demonstrated gastrula formation and the presence of two tissue layers in the larval stages of a few

Annelida

Arthropoda

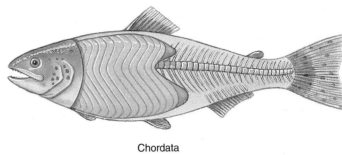

Chordata

figure 3.7

Segmented phyla. These three phyla illustrate an important principle in nature—metamerism, or repetition of structural units. Segmentation brings more varied specialization because segments, especially in arthropods, have become modified for different functions.

species. Reorganization of the body after larval settlement causes tissue layers not to persist in the adult body plan. Members of a few other phyla, such as Cnidaria (sea anemones and their kin) and Ctenophora (comb jellies), develop two embryonic germ layers and have two adult body layers. These animals are typically radially symmetrical.

A vast majority of phyla contain bilaterally symmetrical animals developing from the three embryonic germ layers mentioned previously: ectoderm, endoderm, and mesoderm. The presence of three germ layers is called "triploblasty," so these animals are collectively the triploblastic Bilateria. Within the Bilateria, biologists recognize two groups of animals: **Protostomia** (called protostomes) and **Deuterostomia** (called deuterostomes) (find these groups on the cladogram inside front cover). These two evolutionary groups, or clades, differ in the pattern of cleavage, in the formation of mesoderm, and in the way that a coelom is made, if a coelom is present.

Deuterostome Body Plans

Deuterostomes share an inherited common developmental sequence that begins with radial regulative cleavage and progresses from a blastula to a gastrula. The blastopore becomes the anus (figure 3.8). The name Deuterostomia means "second mouth" and refers to formation of the mouth from a second opening in the embryo. Other distinguishing hallmarks of the deuterostomes are summarized in figure 3.8 and include formation of a coelom by enterocoely in most group members. A notable exception is the vertebrate chordates. Invertebrate chordates form the coelom by enterocoely as expected, but vertebrates develop via a modified, and independently derived, version of schizocoely.

There are four deuterostome phyla. Members of phylum Echinodermata, which includes the sea stars and sea urchins,

undergo typical deuterostome development to form a free-swimming larval stage (figure 3.9). This larval stage metamorphoses into a small version of an adult echinoderm. Other deuterostomes include acorn worms and related animals in phylum Hemichordata, strange wormlike animals with a central band of cilia that were recently classified as deuterostomes and placed in phylum Xenoturbellida, as well as tunicates, lancelets (see p. 291), and vertebrates in phylum Chordata. Metamerism occurs in the chordates.

Protostome Body Plans

Protostomes share an inherited developmental sequence that begins with spiral mosaic cleavage and proceeds through a blastula and gastrula stage (see figure 3.5). Protostome means "mouth first" and refers to the formation of the mouth from the embryonic blastopore. The anus forms secondarily in protostomes (see figure 3.8).

There is no body plan common to protostomes. Some protostomes, such as flatworms (phylum Platyhelminthes), have acoelomate bodies where a spongy mass of cells called parenchyma, derived from mesoderm, completely fills the region between the epidermis and the digestive tract (figure 3.10, *middle*). Readers should note that in this case "parenchyma" does not have the same meaning as when used for cells within an organ on page 60. Other protostomes, such as roundworms in phylum Nematoda, are pseudocoelomate (figure 3.10, *top*). A pseudocoelom surrounds the gut, giving the body a tube-within-a-tube arrangement similar to that in animals with a true coelom. However, a pseudocoelom is persistent blastocoel and is not lined with a peritoneum, the thin cellular membrane derived from mesoderm that lines the body cavity in animals with a true coelom (figure 3.10, *bottom*). **Coelomate** animals such as earthworms (phylum Annelida) and insects (phylum Arthropoda) are also protostomes;

figure 3.8

Developmental tendencies of protostomes and deuterostomes. These tendencies are much modified in some groups—for example, the vertebrates. Cleavage in mammals, reptiles, birds, and many fishes is neither radial nor spiral in form. Within deuterostomes, vertebrates have evolved a mode of coelom formation that is schizocoelous, but not homologous to protostome schizocoely. Coelom formation may have evolved several times within the protostomes.

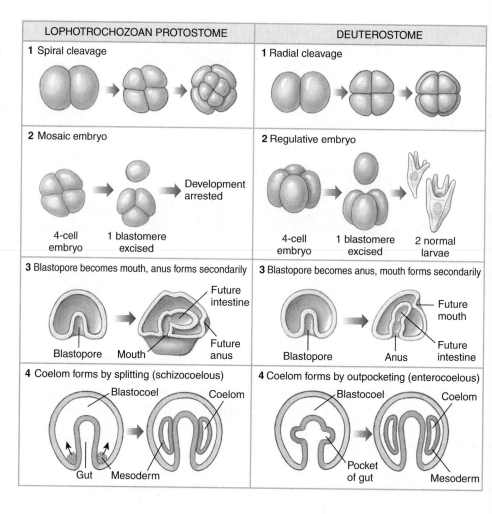

LOPHOTROCHOZOAN PROTOSTOME	DEUTEROSTOME
1 Spiral cleavage	**1** Radial cleavage
2 Mosaic embryo — 4-cell embryo, 1 blastomere excised → Development arrested	**2** Regulative embryo — 4-cell embryo, 1 blastomere excised → 2 normal larvae
3 Blastopore becomes mouth, anus forms secondarily — Blastopore, Mouth, Future intestine, Future anus	**3** Blastopore becomes anus, mouth forms secondarily — Blastopore, Anus, Future mouth, Future intestine
4 Coelom forms by splitting (schizocoelous) — Blastocoel, Coelom, Gut, Mesoderm	**4** Coelom forms by outpocketing (enterocoelous) — Blastocoel, Coelom, Pocket of gut, Mesoderm

when a coelom is present, it forms by schizocoely in protostomes.

Protostome Subgroups

Protostomia comprises two subgroups that were initially identified using molecular characters. One subgroup, the ecdysozoan protostomes or **Ecdysozoa**, includes all animals that molt their cuticles. The name comes from the root *ekdysis*, meaning "to strip off." This group comprises arthropods, nematodes, and members of six other phyla (see cladogram inside front cover). Metamerism is present in the arthropods.

The other protostome subgroup, **Lophotrochozoa**, includes a very diverse collection of animals. Part of the name comes from *lophos*, meaning "crest or tuft," in reference to a horseshoe-shaped whorl of tentacles called a **lophophore** (see p. 180) that occurs in some members of the group, such as lampshells (phylum Brachiopoda). Another part of the name comes from *trochos*, meaning "wheel," in reference to the band of cilia on a larval form called a **trochophore** (see p. 192). A trochophore larval stage occurs in the life cycles of many lophotrochozoans, including snails (phylum Mollusca) and worms (phylum Annelida). There are 17 lophotrochozoan phyla (see cladogram inside front cover). Metamerism occurs in phylum Annelida.

Components of Metazoan Bodies

Metazoan bodies consist of cellular components derived from the three embryonic germ layers, and some extracellular components. Extracellular on noncellular components are fluids that surround cells and structural substance made by cells, but lying outside the cells.

Cellular Components: Tissues

A tissue is a group of cells specialized for performing a common function. The study of tissues is called **histology** (Gr. *histos*, tissue, + *logos*, discourse). During embryonic development, the three germ layers become differentiated into different tissues. In animals at or above the tissue level of organization, all cells exist as part of a tissue. Some tissues are composed of several kinds of cells, and some tissues, such as connective tissue, include extracellular materials made by tissue cells, as well as noncellular fluids.

There are four kinds of tissues: epithelial, connective (including vascular), muscular, and nervous tissues (figure 3.11). This surprisingly short list of basic tissue types is able to meet the diverse requirements of animal life.

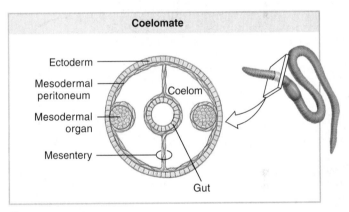

figure 3.10

Pseudocoelomate, acoelomate, and coelomate body plans are shown as cross sections of representative animals. Note the relative positions of mesodermally derived parenchyma, peritoneum, and body organs.

Epithelial Tissue

An **epithelium** (pl., epithelia) is a sheet of cells that covers an external or internal surface. Outside the body, epithelium forms a protective covering. Inside, an epithelium lines all the organs of the body cavity, as well as ducts and passageways through which various materials and secretions move. On many surfaces, epithelial cells are often modified into glands that produce lubricating mucus or specialized products such as hormones or enzymes.

Epithelia are classified by cell form and number of cell layers. Simple epithelia (figure 3.12) occur in all animals that have tissues, whereas stratified epithelia (figure 3.13) are mostly restricted to vertebrates. All types of epithelia are supported by an underlying basement membrane, an extracellular matrix or

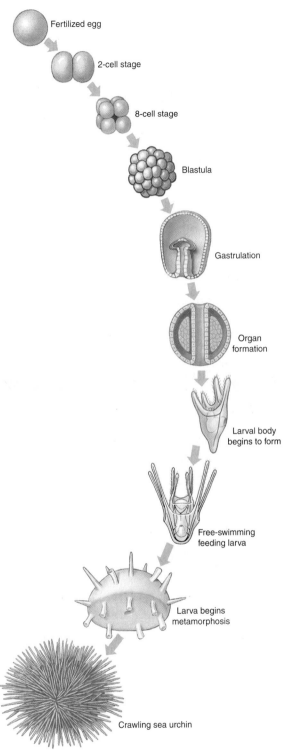

figure 3.9

Development in a sea urchin. Fertilization of an urchin egg is followed by cleavage, producing a mass of cells. This mass rearranges to form a single cell layer surrounding a cavity (blastula stage). In the next stage, a gut and more tissue layers form (gastrulation). Once a gut is present, the urchin embryo develops a larval body. The free-swimming larva feeds and grows in ocean surface waters. The larva metamorphoses into a tiny bottom-dwelling sea urchin; the urchin feeds and grows, reaching sexual maturity in this body form.

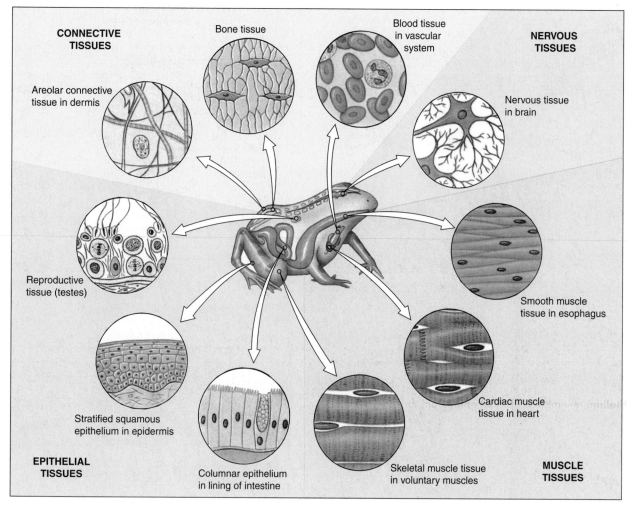

figure 3.11

Metazoan tissue types illustrated within the body of a frog.

ECM, which is a thin mat containing **collagen** (Gr. *kolla*, glue, + *genos*, descent), a protein material of great tensile strength. Blood vessels never enter epithelial tissues, so they depend on diffusion of oxygen and nutrients from underlying tissues.

Connective Tissue

Connective tissues are a diverse group of tissues that serve various binding and supportive functions. They are so widespread in the body that removal of other tissues would still leave the complete form of the body clearly apparent. Connective tissue is composed of relatively few cells, a great many *extracellular fibers,* and a fluid, gel-like or rigid **ground substance** (also called **matrix**), in which the fibers are embedded. Fibers and ground substance together form an *ECM.* We recognize several different types of connective tissue. Two kinds of connective tissue proper occur in vertebrates. *Loose connective tissue* is composed of fibers and both fixed and wandering cells suspended in a syrupy ground substance (figure 3.14A). *Dense connective tissue*, which includes tendons and ligaments, is composed largely of densely packed fibers (figure 3.14B).

Much of the fibrous tissue of connective tissue is composed of collagen, which is the most abundant protein in the animal kingdom and occurs in animal bodies wherever both flexibility and resistance to stretching are required. The connective tissue of invertebrates, as in vertebrates, consists of cells, fibers, and extracellular matrix, but usually it is not as elaborately developed.

Other types of connective tissue include *blood*, **lymph**, and **interstitial fluid**, composed of distinctive cells in a watery ground substance, the plasma. As plasma leaks from capillaries into the spaces between cells, it becomes interstitial fluid. Excess interstitial fluid is collected in lymphatic ducts, where it joins lymph. Lymph consists of lymphocytes and fluids. Some lymph is returned to the circulatory system when lymph ducts meet veins going to the heart. Blood, lymph, and interstitial fluid are collectively considered vascular tissue. Vascular tissue lacks fibers under normal conditions. **Cartilage** is a semirigid form of connective tissue with closely packed fibers embedded in a gel-like ground substance (figure 3.14C). *Bone* is a calcified connective tissue containing calcium salts organized around collagen fibers (figure 3.14D).

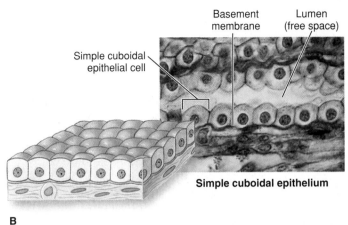

figure 3.12

Types of simple epithelium. **A, Simple squamous epithelium** is composed of flattened cells that form a continuous delicate lining of blood capillaries, lungs, and other surfaces. It permits the passive diffusion of gases and tissue fluids into and out of cavities. **B, Simple cuboidal epithelium** is composed of short, boxlike cells. Cuboidal epithelium usually lines small ducts and tubules, such as those of the kidney and salivary glands, and may have active secretory or absorptive functions. **C, Simple columnar epithelium** resembles cuboidal epithelium, but the cells are taller and usually have elongated nuclei. This type of epithelium lines highly absorptive surfaces, such as the intestinal tract of most animals. The cells often bear minute, fingerlike projections called microvilli that greatly increase the absorptive surface. In some organs, such as the female reproductive tract, the cells may be ciliated.

Muscular Tissue

Muscle is the most common tissue of most animals. It originates (with few exceptions) from mesoderm, and its unit is the cell, or *muscle fiber*, which is specialized for contraction. When viewed with a light microscope, *striated muscle* appears transversely striped (striated), with alternating dark and light bands (figure 3.15A, B). In vertebrates, we recognize two types of striated muscle: *skeletal* and *cardiac muscle*. A third kind of muscle is *smooth* (or visceral) *muscle*, which lacks the characteristic alternating bands of the striated type (figure 3.15C). Unspecialized cytoplasm of muscles is called *sarcoplasm*, and contractile elements within the fiber are **myofibrils**.

Nervous Tissue

Nervous tissue is specialized for receiving stimuli and conducting impulses from one region to another. Two basic types of cells in nervous tissue are **neurons** (Gr. nerve), the basic functional unit of the nervous system, and **neuroglia** (nurō-gleə Gr. nerve + *glia*, glue), a variety of nonnervous cells that insulate neuron membranes and serve various supportive functions. Figure 3.16 shows the functional anatomy of a typical neuron.

Extracellular Components of the Metazoan Body

In addition to the hierarchically arranged cellular structures discussed, metazoan animals contain two important noncellular components: body fluids and extracellular structural elements. In all metazoans, body fluids are subdivided into two fluid "compartments": those that occupy **intracellular space**, within the body's cells, and those that occupy *extracellular space*, outside the cells. In animals with closed vascular systems (such as segmented worms and vertebrates), extracellular fluids are subdivided further into *blood plasma*, **lymph**, and **interstitial fluid**. Interstitial fluid, sometimes called tissue fluid, occupies the space surrounding cells. However, many invertebrates have open vascular systems, with no true separation of blood plasma from interstitial fluid.

If we were to remove all specialized cells and body fluids from the interior of the body, we would be left with the third element of the animal body: extracellular structural elements. This is the supportive material of the organism, forming the noncellular parts of loose connective tissue (especially well developed in vertebrates but present in all metazoa), cartilage (molluscs and chordates), bone (vertebrates), and cuticle

Free surface

Stratified squamous
epithelial cell

Nuclei

Basement
membrane

Stratified squamous epithelium

Basement
membrane

Free surface

Connective   Nucleus   Transitional
tissue   epithelial cell

Transitional epithelium—unstretched

Transitional epithelium—stretched

Stratified squamous epithelium consists of two to many layers of cells adapted to withstand mild mechanical abrasion. The basal layer of cells undergoes continuous mitotic divisions, producing cells that are pushed toward the surface, where they are shed and replaced by new cells beneath them. This type of epithelium lines the oral cavity, esophagus, and anal canal of many vertebrates, and the vagina of mammals.

Transitional epithelium is a type of stratified epithelium specialized to accommodate great stretching. This type of epithelium lines the urinary tract and bladder of vertebrates. In the relaxed state, it appears to be four or five cell layers thick, but when stretched out, it has only two or three layers of extremely flattened cells.

figure 3.13

Types of stratified epithelium.

(arthropods, nematodes, annelids, and others). These elements provide mechanical stability and protection. In some instances, they also act as a depot for exchange of materials and serve as a medium for extracellular reactions. We describe the diversity of extracellular skeletal elements characteristic of the different groups of animals in Chapters 10–20.

The term "intercellular," meaning "between cells," should not be confused with the term "intracellular," meaning "within cells."

■ Complexity and Body Size

The most complex levels of metazoan organization permit, and to some extent even promote, evolution of large body size (figure 3.17). Large size confers several important physical and ecological consequences for an organism. As animals become larger, the body surface increases much more slowly than does body volume because surface area increases as the square of body length (length^2), whereas volume (and usually mass)

increases as the cube of body length (length^3). In other words, a large animal will have less surface area relative to its volume than will a small animal of the same shape. The surface area of a large animal may be inadequate for respiration and nutrition by cells located deep within the body. There are two possible solutions to this problem. One solution is to fold or invaginate the body surface to increase surface area or, as exploited by flatworms, to flatten the body into a ribbon or disc so that no internal space is far from the surface. This solution allows the body to become large without internal complexity. However, most large animals evolved a second solution; they developed internal transport systems to shuttle nutrients, gases, and waste products between the cells and the external environment.

Larger size buffers an animal against environmental fluctuations, provides greater protection against predation and enhances offensive tactics, and permits more efficient use of metabolic energy. A large mammal uses more oxygen than a small mammal, but the cost of maintaining its body temperature is less per gram of weight for a large mammal than for a small one. Large animals also can move at less energy cost than small animals can. A large mammal uses more oxygen

Nucleus   Collagen fiber   Elastic fiber

A

Loose connective tissue, also called areolar connective tissue, is the "packing material" of the body that anchors blood vessels, nerves, and organs. It contains fibroblasts that synthesize the fibers and ground substance of connective tissue and wandering macrophages that phagocytize pathogens or damaged cells. The different fiber types include strong collagen fibers (thick and violet in micrograph) and elastic fibers (thin and branching in micrograph) formed of the protein elastin. Adipose (fat) tissue is considered a type of loose connective tissue.

Nucleus   Fibers

B

Dense connective tissue forms tendons, ligaments, and fasciae (fásha), the latter arranged as sheets or bands of tissue surrounding skeletal muscle. In a tendon (shown here), the collagenous fibers are extremely long and tightly packed together.

Chondrocyte   Lacuna   Matrix

C

Cartilage is a vertebrate connective tissue composed of a firm gel matrix containing cells (chondrocytes) living in small pockets called lacunae, and collagen fibers or elastic fibers (depending on the type of cartilage). In hyaline cartilage shown here, both collagen fibers and ground substance are stained uniformly purple, and cannot be distinguished from one another. Because cartilage lacks a blood supply, all nutrients and waste materials must diffuse through the ground substance from surrounding tissues.

Central canal   Osteocytes   Mineralized
(Haversian canal)   in lacunae   matrix

D

Bone, strongest of the vertebrate connective tissues, contains mineralized collagen fibers. Small pockets (lacunae) within the matrix contain bone cells, called osteocytes. The osteocytes communicate with blood vessels that penetrate bone by means of a tiny network of channels called canaliculi. Unlike cartilage, bone undergoes extensive remodeling during an animal's life, and can repair itself following even extensive damage.

figure 3.14

Types of connective tissue.

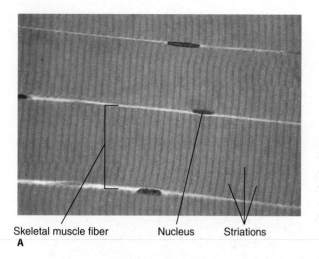

Skeletal muscle fiber   Nucleus   Striations

A

Skeletal muscle is a type of striated muscle found in both invertebrates and vertebrates. It is composed of extremely long, cylindrical fibers, which are multinucleate cells that may reach from one end of the muscle to the other. Viewed through the light microscope, the cells appear to have a series of stripes, called striations, running across them. Skeletal muscle is called voluntary muscle (in vertebrates) because it contracts when stimulated by nerves under conscious cerebral control.

B

Striations   Nucleus of cardiac   Intercalated discs
                    muscle cell   (special junctions
                                                between cells)

Cardiac muscle is another type of striated muscle found only in the vertebrate heart. The cells are much shorter than those of skeletal muscle and have only one nucleus per cell (uninucleate). Cardiac muscle tissue is a branching network of fibers with individual cells interconnected by junctional complexes called intercalated discs. Cardiac muscle is called involuntary muscle because it does not require nerve activity to stimulate contraction. Instead, heart rate is controlled by specialized pacemaker cells located in the heart itself. However, autonomic nerves from the brain may alter pacemaker activity.

C

Nuclei of smooth muscle cells

Smooth muscle is nonstriated muscle found in both invertebrates and vertebrates. Smooth muscle cells are long, tapering strands, each containing a single nucleus. Smooth muscle is the most common type of muscle in invertebrates, where it serves as body wall musculature and lines ducts and sphincters. In vertebrates, smooth muscle cells are organized into sheets of muscle circling the walls of the gut tube, blood vessels, respiratory passages, and urinary and genital ducts. Smooth muscle is typically slow acting and can maintain prolonged contractions with very little energy expenditure. Its contractions are involuntary and unconscious. The principal functions of smooth muscles are to push the material in a tube, such as the intestine, along its way by active contractions or to regulate the diameter of a tube, such as a blood vessel, by sustained contraction.

figure 3.15

Types of muscle tissue.

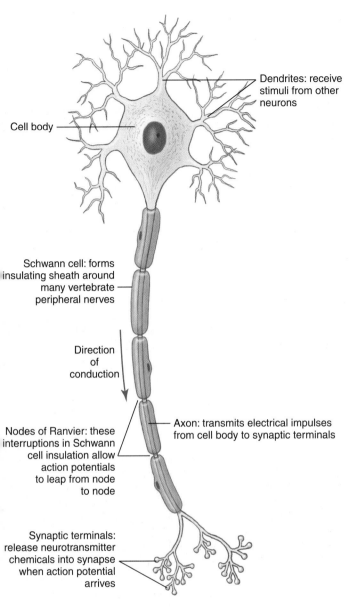

figure 3.16

Functional anatomy of a neuron. From the nucleated body, or soma, extend one or more **dendrites** (Gr. *dendron*, tree), which receive electrical impulses from receptors or other nerve cells, and a single **axon** that carries impulses away from the cell body to other nerve cells or to an effector organ. The axon is often called a **nerve fiber**. Schwann cells are examples of neuroglia. Neurons are separated from other neurons or from effector organs by specialized junctions called synapses.

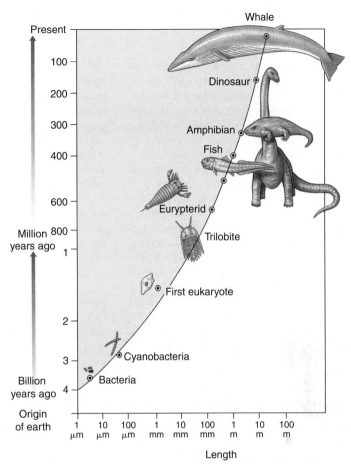

figure 3.17

Graph showing the largest organisms (absent trees and fungi) present at different periods of life on earth. Note that both scales are logarithmic.

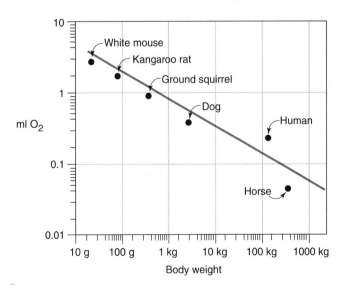

figure 3.18

Net cost of running for mammals of various sizes. Each point represents the cost (measured in rate of oxygen consumption) of moving 1 g of body over 1 km. The cost decreases with increasing body size.

in running than does a small mammal, but the energy cost of moving 1 g of its body over a given distance is much less for a large mammal than for a small one (figure 3.18). For all of these reasons, the ecological opportunities of larger animals are very different from those of small ones. The earth is home to organisms across a wide range of sizes and types. The next chapter describes the astonishing diversity of unicellular forms.

Summary

From the relatively simple organisms that mark the beginnings of life on earth, animal evolution has produced many more intricately organized forms. However, adaptive modifications for different lifestyles within a phyletic line of ancestry are constrained by the ancestral pattern of body architecture.

Whereas a unicellular organism performs all life functions within the confines of a single cell, a complex multicellular animal is an organization of subordinate units that are united at successive levels. Cells of metazoans combine into various tissues performing common functions. The basic tissue types are nervous, connective, epithelial, and muscular. Tissues are organized into larger functional units called organs, and organs are associated to form systems.

Every organism has an inherited body plan that may be described in terms of broadly inclusive characteristics, such as symmetry, presence or absence of body cavities, partitioning of body fluids, presence or absence of segmentation, degree of cephalization, and type of nervous system. Symmetrical bodies may be spherical, radial, or bilateral. Bilateral symmetry tends to be correlated with cephalization and movement in one direction. Bilaterally symmetrical animals typically have three tissue layers and most have a complete gut with a mouth and anus. Many radially symmetrical animals have incomplete guts. The gut is the primary body cavity of an animal, but it may be surrounded by another cavity such as a coelom or pseudocoelom.

Based on several developmental characteristics, bilateral metazoans are divided into two major groups. The Deuterostomia have radial cleavage and regulative development, and the mouth forms secondarily to the anus rather than from the blastopore. The Protostomia are characterized by spiral cleavage, mosaic development, and the mouth forming at or near the embryonic blastopore. Protostomia contains two subgroups: the lophotrochozoan protostomes and the ecdysozoan protostomes.

In addition to cells, a metazoan body contains fluids, divided into intracellular and extracellular fluid compartments, and extracellular structural elements, which are fibrous or formless elements that serve various structural functions in the extracellular space.

One correlate of increased body complexity in animals is increased body size, which offers certain advantages, such as more effective predation, reduced energy cost of locomotion, and improved homeostasis. However, the most species-rich and abundant organisms are not large.

Review Questions

1. Name the five levels of organization in organismal complexity, and explain how each successive level differs from the one preceding it.
2. Distinguish between an asymmetrical and a symmetrical body.
3. Distinguish among spherical, radial, and bilateral symmetry.
4. Use the following terms to identify regions on your body and on the body of a frog (see p. 61): anterior, posterior, dorsal, ventral, lateral, distal, proximal.
5. How would frontal, sagittal, and transverse planes divide your body?
6. What is the difference between radial cleavage and spiral cleavage?
7. Aside from the gut cavity, what types of body cavities exist and where do these cavities lie, relative to the gut cavity?
8. Match each description with at least one illustrative organism:

2 EMBRYONIC TISSUES _____	A. SEA STAR
3 EMBRYONIC TISSUES _____	B. SEA ANEMONE
PROTOSTOME _____	C. FLATWORM
DEUTEROSTOME _____	D. EARTHWORM
METAMERIC BODY _____	E. LOBSTER
_____	F. FISH
_____	G. FROG
_____	H. INSECT

9. What is metamerism? Name three phyla showing metamerism.
10. What are the distinguishing developmental hallmarks of the two major groups of bilateral animals, the Protostomia and the Deuterostomia?
11. What are the four major types of tissues in the body of an animal?
12. How would you distinguish simple and stratified epithelium? What characteristic of stratified epithelium might explain why it, rather than simple epithelium, lines the oral cavity, esophagus, and vagina?
13. What three elements are present in all connective tissue? Give some examples of the different types of connective tissue.
14. What are the three kinds of muscle found in animals? Explain how each is specialized for particular functions.
15. Describe the principal structural and functional features of a neuron.
16. Explain why plasma, lymph, and interstitial fluid are considered extracellular.

For Further Thought Calculate the surface area and volume for two spherical organisms, one with a radius of 1 mm and the other with a radius of 10 mm. Determine the surface area-to-volume ratio for each organism. What problems might occur for each organism because of its size? How would these problems be solved?

■ Selected References

See also general references on page 432.

Baguña, J., and M. Ruitort. 2004. The dawn of bilaterian animals: the case of acoelomorph flatworms. Bioessays 26:1046–1057. *Goes through the hypotheses for the transition between radial and bilateral body plans and the evidence for them.*

Bonner, J. T. 1988. The evolution of complexity by means of natural selection. Princeton, New Jersey, Princeton University Press. *Levels of complexity in organisms and how size affects complexity.*

Kessel, R. G., and R. H. Kardon. 1979. Tissues and organs: a text-atlas of scanning electron microscopy. San Francisco, W. H. Freeman & Company. *Collection of excellent scanning electron micrographs with text.*

Kirk, D. L. 2005. A twelve-step program for evolving multicellularity and a division of labor. BioEssays 27:299–310. *A comparison of the degree of cellular independence found across different species of the volvocine flagellates; members of one genus evolved a multicellular body.*

McGowan, C. 1999. A practical guide to vertebrate mechanics. New York, Cambridge University Press. *Using many examples from his earlier book,* Diatoms to dinosaurs, 1994, *the author describes principles of biomechanics that underlie functional anatomy. Includes practical experiments and laboratory exercises.*

Radinsky, L. B. 1987. The evolution of vertebrate design. Chicago, University of Chicago Press. *A lucid functional analysis of vertebrate body plans and their evolutionary transformations over time.*

Rokas, A. 2008. The origins of multicellularity and the early history of the genetic toolkit for animal development. Annual Review of Genetics 42:235–251. *This fascinating review discusses the genetic and morphological basis of multicellular bodies.*

Willmer, P. 1990. Invertebrate relationships: patterns in animal evolution. Cambridge, Cambridge University Press. *Chapter 2 is an excellent discussion of animal symmetry, developmental patterns, origin of body cavities, and segmentation.*

■ Custom Website

The *Animal Diversity* Online Learning Center is a great place to check your understanding of chapter material. Visit www.mhhe.com/hickmanad6e for access to key terms, quizzes, and more! Further enhance your knowledge with Web links to chapter-related material.

Explore live links for these topics:

Architectural Pattern and Diversity of Animals
Animal Systems
Cells and Tissues: Histology
Basic Tissue Types
Vertebrate Laboratory Exercises

Principles of Development/Embryology in Vertebrates

Taxonomy and Phylogeny of Animals

Molluscan shells from the collection of Jean Baptiste de Lamarck (1744–1829).

Order in Diversity

Evolution has produced a great diversity of species in the animal kingdom. Zoologists have named more than 1.5 million species of animals, and thousands more are described each year. Some zoologists estimate that the species named so far constitute less than 20% of all living animals and less than 1% of all those that have existed.

Despite its magnitude, the diversity of animals is not without limits. Many conceivable forms do not exist in nature, as our myths of minotaurs and winged horses show. Animal diversity is not random but has a definite order. Characteristic features of humans and cattle never occur together in a single organism as they do in mythical minotaurs; nor do the characteristic wings of a bird and body of a horse occur together naturally as they do in the mythical horse Pegasus. Humans, cattle, birds, and horses are distinct groups of animals, and yet they do share some important features, including vertebrae and a skull, that separate them from even more dissimilar forms, such as insects and flatworms.

All human cultures classify familiar animals according to patterns in animal diversity. These classifications have many criteria. Some societies classify animals according to their usefulness or destructiveness to human endeavors; others may group animals according to their roles in mythology. Biologists organize animal diversity in a nested hierarchy of groups within groups according to their evolutionary relationships. Ordered patterns in the sharing of homologous features among species reveal these evolutionary relationships to us. This ordering is called a "natural system" because it reflects relationships that exist among animals in nature, outside the context of human activity. A systematic zoologist has three major goals: to discover all species of animals, to reconstruct their evolutionary relationships, and to communicate those relationships by a precise taxonomic system.

Darwin's theory of common descent (see Chapter 1) is the underlying principle that guides our search for order in the diversity of animal life. Our science of **taxonomy** ("arrangement law") produces a formal system for naming and grouping species to communicate this order. Animals that have very recent common ancestry share many features in common and are grouped most closely in our taxonomic system; dissimilar animals that share only very ancient common ancestry are placed in different taxonomic groups except at the "highest" or most inclusive levels.

Taxonomy is part of a broader science of **systematics,** which studies variation among animal populations to reveal their evolutionary relationships. Taxonomy predates evolutionary biology, however, and thus many taxonomic practices are relics of pre-evolutionary worldviews. Adjusting our taxonomic system to accommodate evolution has produced many problems. Taxonomy has reached an unusually active and controversial point in its development, with several alternative taxonomic systems competing for use. To explain this controversy, we must review the history of animal taxonomy.

■ Linnaeus and Taxonomy

The Greek philosopher and biologist Aristotle was the first to classify organisms based on their structural similarities. Following the Renaissance in Europe, the English naturalist John Ray (1627–1705) introduced a more comprehensive system of classification and a new concept of species. Rapid growth of systematics in the eighteenth century culminated in the work of Carolus Linnaeus (1707–1778; figure 4.1), who produced our current scheme of classification.

Linnaeus was a Swedish botanist at the University of Uppsala. He had a great talent for collecting and classifying organisms, especially flowering plants. Linnaeus produced an extensive system of classification for both plants and animals. This scheme, published in his great work, *Systema Naturae,* used **morphology** (the comparative study of organismal form) for arranging specimens in collections. He divided the animal kingdom into species and gave each one a distinctive name. He grouped species into genera, genera into orders, and orders into "classes". (We use quotation marks or a capital letter to distinguish "class" as a formal taxonomic rank from its broader meaning as a group of organisms that share an essential property.) Because Linnaeus's knowledge of animals was limited, his lower categories, such as genera, were very broad and included animals that are only distantly related. Much of his classification has been drastically altered, but its basic principles are still in use.

Linnaeus's scheme of arranging organisms into an ascending series of groups of increasing inclusiveness is a **hierarchical system** of classification. Groupings of organisms called **taxa** (sing., **taxon**) are given one of several standard taxonomic ranks to indicate the general inclusiveness of each group. The hierarchy of taxonomic ranks has been expanded considerably since Linnaeus's time (table 4.1). It now includes seven mandatory ranks for the animal kingdom, in descending series: kingdom, phylum, "class," order, family, genus, and species. Any organism being classified must be placed into at least seven taxa, one at each of these mandatory ranks. Taxonomists have the option of subdividing these seven ranks even further to recognize more than seven taxa (superfamily, subfamily, superorder, suborder, and others) for any particular group of organisms. More than 30 taxonomic ranks are used. For very large and complex groups, such as fishes and insects, these additional ranks are needed to express different degrees of evolutionary divergence.

A taxonomist's choice of a group of species for recognition as a formally ranked taxon always has some arbitrariness. For example, should the taxonomic family Hominidae be restricted to genus *Homo* (humans) and all fossil genera that are closer to *Homo* than to genus *Pan* (bonobos and chimpanzees), or should it comprise the more inclusive grouping of genera *Homo, Pan, Gorilla,* and *Pongo* (orangutans) plus fossils closer to these genera than to gibbons? Within the last few decades, anthropologists have shifted Hominidae from the former usage primarily to the latter one (p. 423). Does the confusing arbitrariness of ranking taxa outweigh the usefulness of ranks for reminding us which taxa are more inclusive than others? Could we devise a rank-free taxonomy that encodes the positions of species on an evolutionary tree of common descent? As we write, taxonomists are actively trying to answer these questions. Meanwhile, both ranked and rank-free taxonomies of animals are in use.

Systematization Versus Classification

Introduction of evolutionary theory into animal taxonomy changed the taxonomist's task from classification to **systematization.** Classification denotes the construction of

figure 4.1

Carolus Linnaeus (1707–1778). This portrait was made of Linnaeus at age 68, three years before his death.

Table 4.1 Examples of Taxonomic Categories to Which Representative Animals Belong

	Human	Gorilla	Southern Leopard Frog	Bush Katydid
Kingdom	Animalia	Animalia	Animalia	Animalia
Phylum	Chordata	Chordata	Chordata	Arthropoda
Subphylum	Vertebrata	Vertebrata	Vertebrata	Uniramia
Class	Mammalia	Mammalia	Amphibia	Insecta
Subclass	Eutheria	Eutheria	—	Pterygota
Order	Primates	Primates	Anura	Orthoptera
Suborder	Anthropoidea	Anthropoidea	Neobatrachia	Ensifera
Family	Hominidae	Hominidae	Ranidae	Tettigoniidae
Subfamily	Homininae	Homininae	Raninae	Phaneropterinae
Genus	*Homo*	*Gorilla*	*Lithobates*	*Scudderia*
Species	*Homo sapiens*	*Gorilla gorilla*	*Lithobates sphenocephalus*	*Scudderia furcata*
Subspecies	—	—	—	*Scudderia furcata furcata*

Hierarchical taxonomy of four species (human, gorilla, southern leopard frog, and bush katydid). Higher taxa generally are more inclusive than lower-level taxa, although taxa at two different levels may be equivalent in content. Closely related species are united at a lower point in the hierarchy than are distantly related species. For example, humans and gorillas are united at the subfamily (Homininae) and above; they are united with southern leopard frogs at the subphylum level (Vertebrata) and with bush katydids at the kingdom (Animalia) level. Mandatory Linnean ranks are shown in bold type.

classes, groupings of organisms possessing a common feature, called an essence, that is used to define the class. Organisms that possess the essential feature are members of the class by definition, and those that lack it are excluded. Because evolving species are always subject to change, the static nature of classes makes them a poor basis for a taxonomy of living systems. In systematization, on the other hand, the taxonomist groups species that form units of common evolutionary descent. Species placed into a taxonomic group include the most recent common ancestor of the group and its descendants and thus form a branch of the phylogenetic tree of life. The species of such a group thus constitute a system of common descent, not a class defined by possession of an essential characteristic. It remains common, although technically erroneous, for systematists to call their taxonomic systems "classifications."

Because organismal characteristics are inherited from ancestral to descendant species, character variation is used to diagnose systems of common descent, but there is no requirement that an essential character be maintained throughout the system for its recognition as a taxon. The role of morphological or other features in systematization is therefore fundamentally different from the role of such characters in classification. In classification, a taxonomist asks whether a species being classified contains the defining feature(s) of a particular taxonomic class; in systematization, a taxonomist asks whether the characteristics of a species confirm or reject the hypothesis that it descends from the most recent common ancestor of a particular taxon. For example, tetrapod vertebrates descend from a common ancestor that had four limbs, a condition retained in most but not all of its descendants. Although they lack limbs, caecilians (see p. 356) and snakes (see p. 376) are tetrapods because they are parts of this system of common descent; other morphological and molecular characters group them respectively with living amphibians and lizards.

Although the hierarchical structure of Linnean classification is retained in current taxonomy, the taxa are groupings of species related by evolutionary descent with modification, as diagnosed by sharing of homologous characters. Moving up the taxonomic hierarchy from a species toward more inclusive groups, each taxon represents the descendants of an earlier ancestor, a larger branch of the tree of life.

Binomial Species Nomenclature

Linnaeus's system for naming species is called **binomial nomenclature.** Each species has a Latinized name composed of two words (hence binomial) written in italics (underlined if handwritten or typed). The first word is the name of the **genus,** written with a capital initial letter; the second word is the **species epithet,** which identifies the species within the genus and is written with a lowercase initial letter (table 4.1). The great communicative value of Latin species names is that they are used consistently by scientists in all countries and languages; they are much more precise than "common names," which vary culturally and geographically.

The name of a genus is always a noun, which is either feminine or masculine. The species epithet is usually an adjective that must agree in gender with the genus. For instance, the scientific name of a common robin is *Turdus migratorius* (L. *turdus,* thrush; *migratorius,* of the migratory habit). A species epithet never stands alone; the complete binomial must be used to name a species. Names of genera must refer only to single groups of organisms; a single name cannot be given to two different genera of animals. However, the same species epithet may be used in different genera to denote different species. For example, the scientific name of a white-breasted nuthatch is *Sitta carolinensis*. The species epithet *"carolinensis"* is used in other genera, including *Poecile carolinensis* (Carolina chickadee) and *Anolis carolinensis* (green anole, a lizard) to mean "of Carolina." All ranks above species are designated using a capitalized noun, such as Iguanidae (the lizard family that contains *Anolis*).

The person who first names and describes a species is called the authority. The authority deposits in a museum a formally designated type specimen, which formally carries the name of the species. The authority's name and date of publication of the species description often appear after the species name. Thus, *Didelphis marsupialis* Linnaeus, 1758, tells us that Linnaeus was the first person to publish the species name of opossums. The authority citation is not part of the scientific name but rather an abbreviated bibliographical reference. If the genus name of a species is changed following its initial description, the authority's name appears in parentheses.

■ Species

While discussing Darwin's book *On the Origin of Species* in 1859, Thomas Henry Huxley (see p. 12) asked, "In the first place, what is a species? The question is a simple one, but the right answer to it is hard to find, even if we appeal to those who should know most about it." So far in this chapter, we have used the term "species" as if it had a simple and unambiguous meaning, but actually, Huxley's commentary is as valid today as it was over 140 years ago. Our concepts of species have become more sophisticated, but the different concepts and disagreements surrounding their use are as evident now as in Darwin's time.

Criteria for Recognition of Species

Despite widespread disagreement about what constitutes a species, biologists agree that certain criteria are important for recognizing species. First, **common descent** is central to nearly all modern concepts of species. Members of a species must trace their ancestry to a common ancestral population, although not necessarily to a single pair of parents. Species are thus historical entities. Variation among organisms serves to test the hypothesis that a particular grouping of organisms forms a population evolving separately from other such populations.

A second criterion is that species must be the *smallest distinct groupings* of organisms sharing patterns of ancestry and descent; otherwise, it would be difficult to separate species from higher taxa whose members also share common

descent. Morphological characters traditionally have been used to identify such groupings, but chromosomal and molecular characters are now extensively used for this purpose.

A third important criterion is the *reproductive community,* which pertains to sexually reproducing organisms; members of a species must form a reproductive community that excludes members of other species. This criterion is very important to many modern concepts of species. For organisms whose reproduction is strictly **asexual,** reproductive community entails occupation of a particular ecological niche in a particular place so that a reproducing population responds as a unit to evolutionary processes such as natural selection and genetic drift (see p. 29).

Any species has a distribution through space, called its *geographic range,* and a distribution through time, called its *evolutionary duration.* Species differ greatly from each other in both of these dimensions. Throughout the evolutionary duration of a species, its geographic range might change many times. A geographic range could be either continuous or disjunct, having breaks within it where the species is absent. Suppose that we find two similar but not identical populations living 300 miles apart with no closely related populations between them. Are we observing a single species with a disjunct distribution, or two different but closely related species? Suppose that these populations have been separated historically for 50,000 years. Is this enough time for them to have evolved separate reproductive communities, or do they form parts of a single reproductive community? Species concepts differ in how they answer these questions.

Concepts of Species

Before Darwin, a species was considered a distinct and immutable entity. The concept that species are defined by fixed, essential features (usually morphological) is called the **typological species concept.** Taxonomists discarded this concept following establishment of Darwinian evolutionary theory.

The most influential concept of species inspired by Darwinian evolutionary theory is the **biological species concept** formulated by Theodosius Dobzhansky and Ernst Mayr. In 1983, Mayr stated the biological species concept as follows: *"A species is a reproductive community of populations (reproductively isolated from others) that occupies a specific niche in nature."* Note that a species is identified here according to reproductive properties of populations, groups of related organisms inhabiting a particular geographic area, not according to organismal morphology. This definition establishes that a species is an *interbreeding* population of individuals having *common descent.* By adding the criterion of **niche,** an ecological concept denoting an organism's role in its ecological community (p. 40), we recognize that members of a reproductive community constitute an ecological entity in nature. Because a reproductive community should maintain genetic cohesiveness, organismal variation should be relatively smooth and continuous within species and discontinuous between them. Although a biological species is based on reproductive properties of populations rather than on organismal morphology, morphology nonetheless can help us to diagnose biological species.

The biological species concept has been strongly criticized for several reasons. First, it refers to contemporary populations and provides little guidance for tracing the temporal duration of a species lineage through its past history. How far back must we trace a species lineage before we cross its historical boundary with its ancestral species? Second, according to the biological species concept, species do not exist in groups of organisms that reproduce only asexually. It is common taxonomic practice, however, to describe species in all groups of organisms. A third problem is that systematists using the biological species concept often disagree on the amount of reproductive divergence necessary for considering two populations separate species. Reproductive barriers between populations range from strong to weak, and some barriers, such as differences in the preferred habitat or time of mating, could be eliminated by natural selection.

The **evolutionary species concept** was proposed by the mammalian paleontologist George Gaylord Simpson (figure 4.2) in the 1940s to add an evolutionary time dimension to the biological species concept. This concept persists in a modified form today. A current definition of the evolutionary species is *a single lineage of ancestor-descendant populations that maintains its identity from other such lineages and that has its own evolutionary tendencies and historical fate.* Note that the criterion of common descent is retained here in the need for a species to have a distinct historical identity. Unlike the biological species concept, the evolutionary species concept applies both to sexually and asexually reproducing forms. As long as continuity of diagnostic features is maintained by an evolving lineage, it is recognized as a single species. Abrupt changes in diagnostic features mark a boundary between different species in evolutionary time.

An ability of geographic populations to evolve collectively as a single, genetically cohesive unit through evolutionary time is critical to the evolutionary species concept. Population geneticist Alan Templeton updated this concept in 1989 to make explicit the expectation that populations of a species evolve as a genetically cohesive unit by natural selection and genetic drift. Templeton defined his **cohesion species concept** as follows: *the most inclusive population of individuals having the potential for phenotypic cohesion through intrinsic cohesion mechanisms.* The cohesion mechanisms include gene flow across the geographic expanse of the species, shared loss of alleles through genetic drift, and shared genetic changes caused by natural selection. Another way of stating the cohesion criterion is that any individual in a species is a possible common ancestor of the entire species at some future time. For example, a new allele that arises by mutation in a single person could spread throughout the human population over many generations, ultimately to become shared by all members of the species at some future time. New mutations arising in other species, even in our closest relatives of the genus *Pan*, could not enter the human gene pool.

The last concept that we present is the **phylogenetic species concept.** Ornithologist Joel Cracraft defines the phylogenetic species concept as *an irreducible (basal) grouping of organisms diagnosably distinct from other such groupings and within which there is a parental pattern of ancestry and descent.* This concept also emphasizes common descent, and both asexual and sexual groups are covered. Any population that has become spatially separated from others and has undergone character evolution that distinguishes it would be recognized as a species. The criterion of irreducibility requires that no more than one such population be placed in a single species. The phylogenetic species concept emphasizes recognizing as species the smallest groupings of organisms that have undergone independent evolutionary change. The evolutionary and cohesion species concepts would group into a single species geographically disjunct populations that demonstrate some genetic divergence but share major "evolutionary tendencies," whereas the phylogenetic species concept would treat them as separate species. In general, the phylogenetic species concept would describe a larger number of species than would any other concept. The phylogenetic species concept is intended to encourage us to reconstruct patterns of evolutionary common descent on the finest scale possible. The evolutionary and cohesion concepts place greater emphasis on whether historically separated populations have the biological potential to merge into a single lineage in the future.

Current disagreements concerning concepts of species should not be considered trivial or discouraging. Whenever a field of science enters a phase of dynamic growth, old concepts are reevaluated and either refined or replaced with newer, more progressive ones. Active debate among systematists

figure 4.2

Mammalian paleontologist George Gaylord Simpson (1902–1984) formulated the evolutionary species concept and principles of evolutionary taxonomy.

shows that this field has acquired unprecedented activity and importance in biology. Just as Thomas Henry Huxley's time was one of enormous advances in biology, so is the present time. Both times are marked by fundamental reconsiderations of the meaning of species.

Some species are divided into subspecies, in which case a trinomial nomenclature is employed (see katydid example, table 4.1, and salamander example, figure 1.21); such species are called **polytypic.** Generic, specific, and subspecific names are printed in italics (underlined if handwritten or typed). A polytypic species contains one subspecies whose subspecific name is a repetition of the species epithet and one or more additional subspecies whose names differ. Thus, to distinguish geographic variants of *Ensatina eschscholtzii*, one subspecies is named *Ensatina eschscholtzii eschscholtzii*, and different subspecies names are used for each of the six other subspecies (see figure 1.21). Both the genus name and the species epithet may be abbreviated as shown in figure 1.21. Formal recognition of subspecies has lost popularity among taxonomists because boundaries between subspecies rarely are distinct. Recognition of subspecies usually highlights one or a few superficial characters that do not always diagnose an evolutionarily distinct unit. Subspecies, therefore, should be viewed as tentative statements indicating that the species status of the populations needs further investigation.

DNA Barcoding of Species

DNA barcoding is a technique for diagnosing organisms to species using sequence information from a standard gene present in all animals. The mitochondrial gene encoding cytochrome *c* oxidase subunit 1 (*COI*), which contains about 650 nucleotide base pairs, is a standard "barcode" region for animals. DNA sequences of *COI* usually vary among individuals of the same species but not extensively, so that variation within a species is much smaller than differences among species. DNA barcoding is applied to specimens in nature by taking a small DNA sample from blood or another expendable tissue. The method is useful also for specimens in natural-history museums, zoos, aquaria, and frozen-tissue collections. DNA sequences from such sources are checked against a public reference library of species identifiers to assign unknown specimens to known species. DNA barcoding does not solve the controversies regarding use of different species concepts, but it often permits the origin of a specimen to be identified to a particular local population, which is valuable information regardless of the species status that a taxonomist assigns to that population.

Taxonomic Characters and Reconstruction of Phylogeny

A major goal of systematics is to infer an evolutionary tree or **phylogeny** that relates all extant and extinct species. One constructs a phylogenetic tree by studying organismal features, formally called **characters,** that vary among species.

A character is any feature that a taxonomist uses to study variation within or among species. Potentially useful taxonomic characters include morphological, chromosomal, and molecular features. Taxonomists find characters by observing patterns of similarity among organisms. If two organisms possess similar features, they may have inherited these features from an equivalent one in a common ancestor.

Character similarity that results from common ancestry is called **homology** (see p. 18). The character of feathers in birds illustrates homology; all birds have feathers that were inherited with various modifications from those of their most recent common ancestor. However, similarity does not always reflect common ancestry. Independent evolutionary origins of similar features on different lineages produce patterns of similarity among organisms that misrepresent common descent; this occurrence complicates the work of taxonomists. Character similarity that does not accurately represent patterns of common descent is called nonhomologous similarity, or **homoplasy.** For example, independent evolution of image-forming eyes by cephalopod molluscs and by vertebrates illustrates homoplasy; the most recent common ancestor of molluscs and vertebrates did not have such eyes. For an example of molecular homoplasy, see the interpretation of character 41 in the boxed essay, Phylogenies from DNA Sequences (see p. 86).

Using Character Variation to Reconstruct Phylogeny

To reconstruct the phylogeny of a taxon using characters that vary among its members, the first step is to determine which variant form of each character occurred in the most recent common ancestor of the entire taxon of interest. This form is called the **ancestral character state** for the taxon as a whole. We presume that all contrasting forms of the character arose later within the group, and these forms are called evolutionarily **derived character states.** The ancestral/descendant relationships among the alternative states of a character in the taxon of interest constitute the **polarity** of the character.

For example, suppose that our study group is the tetrapod vertebrates, which includes all amphibians, reptiles (including birds), and mammals. Two contrasting character states of egg structure are presence versus absence of an extraembryonic membrane called the **amnion** in the egg. An amnion occurs in the eggs of all reptiles and mammals, but it is absent from all eggs of amphibians. Which of these alternative states most likely characterized the most recent common ancestor of all tetrapod vertebrates?

The method used to examine polarity of a variable character is called **outgroup comparison.** We begin by selecting an additional group of organisms, called an **outgroup,** that is phylogenetically close but not within the group being studied. The various groups of bony fishes constitute appropriate outgroups for polarizing variation in egg structure of tetrapods. Next, we infer that any character state found within *both* the study group and its outgroup most likely characterized the most recent common ancestor of the study group. All bony fishes lack an amnion in their eggs; therefore, we infer that the absence of an amnion is ancestral for the tetrapods and presence of an amnion is derived. Polarity of this character indicates that an amnion first evolved in an ancestral lineage of all modern reptiles and mammals. Polarity of characters is evaluated most effectively when several different outgroups are used. All character states found in a study group that are absent from appropriate outgroups are considered derived within the study group (see figure 4.3 for additional derived characters of vertebrates).

Organisms or species that share a derived character state are hypothesized to form a **clade** (Gr. *klados*, branch), formally defined as a group that includes the most recent common ancestor of all group members and all descendants of that ancestor. A derived character shared by members of a clade is formally called a **synapomorphy** (Gr. *synapsis*, joining together, + *morphē*, form) of that clade. Taxonomists use synapomorphies as evidence of homology to infer that a particular group of organisms forms a clade. Within tetrapods, presence of an amnion is a synapomorphy that identifies all reptiles and mammals as a clade called the **amniotes**. Within amniotes, absence of teeth and presence of feathers are synapomorphies that identify birds as a clade. A clade corresponds to a unit of evolutionary common descent. The pattern formed by the sharing of derived states of all characters within

figure 4.3

A cladogram as a nested hierarchy of taxa. *Amphioxus* (see p. 318) is the outgroup, and the study group comprises four vertebrates (bass, lizard, horse, and monkey). Four characters that vary among vertebrates are used to generate a simple cladogram: presence versus absence of four limbs, amniotic eggs, hair, and mammary glands. For all four characters, absence is considered the ancestral state in vertebrates because this condition is shared by the outgroup, *Amphioxus*, and at least one ingroup member; for each character, presence is derived within vertebrates. Because they share presence of four limbs and amniotic eggs as synapomorphies, lizards, horses, and monkeys form a clade relative to bass. This clade is subdivided further by two synapomorphies (presence of hair and mammary glands) that unite horses and monkeys relative to lizards. We know from comparisons involving even more distantly related animals that vertebrae and jaws constitute synapomorphies of vertebrates and that *Amphioxus*, which lacks these features, falls outside the vertebrate clade.

our study group reveals a **nested hierarchy** of clades within clades. The goal is to identify all clades nested within the study group, which would reveal patterns of common descent among all species in the group. Each clade may be formally recognized as a taxon and perhaps given a rank in the Linnean hierarchy.

A nested hierarchy of clades is presented as a branching diagram called a **cladogram** (figure 4.3; see also figure 1.18). The branches on a cladogram are a formal device for indicating a nested hierarchy of clades within clades. The same nested hierarchy is sometimes presented as an indented list of taxon names; for example, the cladogram for major taxa of living gnathostome vertebrates shown in figure 16.2 (see p. 330) could be represented as:

Chondrichthyes
    Holocephali
    Elasmobranchii

Osteichthyes
    Actinopterygii
    Sarcopterygii
        Dipnoi (lobe-finned fishes)
    Tetrapoda

Taxonomists often make a technical distinction between a cladogram and a **phylogenetic tree.** A cladogram is not strictly equivalent to a phylogenetic tree, whose branches represent real lineages that occurred in the evolutionary past. To obtain a phylogenetic tree, one must add to a cladogram information concerning ancestors, durations of evolutionary lineages, or amounts of evolutionary change that occurred on lineages. However, because the branching order of a cladogram matches that of the corresponding phylogenetic tree, a cladogram often serves as a first approximation of the structure of the phylogenetic tree.

Sources of Phylogenetic Information

We find characters used to construct cladograms in comparative morphology (including embryology), comparative cytology, and comparative biochemistry. **Comparative morphology** examines varying shapes and sizes of organismal structures, including their developmental origins. As we discuss in later chapters, variable structures of skull bones, limb bones, and integument (scales, hair, feathers) are particularly important for reconstructing the phylogeny of vertebrates. Comparative morphology uses specimens obtained from both living organisms and fossilized remains. **Comparative biochemistry** uses the sequences of amino acids in proteins and the sequences of nucleotides in nucleic acids to identify variable characters for constructing a cladogram or phylogenetic tree (figure 4.4). Recent work shows that some fossils retain enough DNA for comparative biochemical studies. **Comparative cytology** (also called **karyology**) uses variation in the numbers, shapes, and sizes of chromosomes and their parts to obtain variable characters for constructing cladograms. Comparative cytology is used almost exclusively on living rather than fossilized organisms because chromosomal structure is not well preserved in fossils (see figure 1.10D).

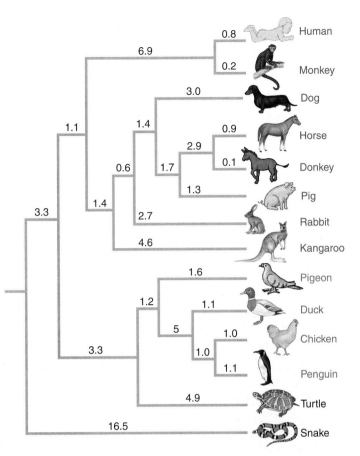

figure 4.4

An early phylogenetic tree of representative amniotes based on inferred base substitutions in the gene that encodes the respiratory protein, cytochrome c. The numbers on the branches are estimated minimum numbers of mutational changes needed to explain amino acid substitutions along different evolutionary lineages. Publication of this tree by Fitch and Margoliash in 1967 was influential in convincing systematists that molecular sequences contain phylogenetic information. Subsequent work confirms some hypotheses, including the hypotheses that mammals (blue) and birds (red) form nonoverlapping clades, while rejecting others; the kangaroo, for example, should be outside a branch containing all other mammals sampled.

To add the evolutionary timescale needed for producing a phylogenetic tree, we must consult the fossil record. We look for the earliest appearance of derived morphological characters in fossils to estimate ages of clades distinguished by those characters. The ages of fossils showing derived characters of a particular clade are determined by radioactive dating (see p. 15) to estimate the age of the clade.

■ Theories of Taxonomy

A theory of taxonomy establishes principles that we use to recognize and to rank taxonomic groups. There are two currently popular theories of taxonomy: (1) evolutionary taxonomy and (2) phylogenetic systematics (cladistics). Both are based on evolutionary and phylogenetic principles but differ on how those evolutionary principles are used. These differences have important implications for how we use a taxonomy to study evolutionary processes. Evolutionary taxonomy predates phylogenetic systematics and retains many aspects of Linnean taxonomy; for this reason, it is sometimes called "traditional evolutionary taxonomy." Evolutionary taxonomy was well established by the 1940s; phylogenetic systematics arose in the 1960s as a replacement for evolutionary taxonomy, which some systematists considered too subjective and misleading.

The relationship between a taxonomic group and a phylogenetic tree or cladogram is important for both theories. This relationship takes one of three forms: **monophyly, paraphyly,** or **polyphyly.** A taxon is monophyletic if it includes the most recent common ancestor of all members of a group and all descendants of that ancestor (figure 4.5A). The terms monophyletic group and clade are synonymous. A taxon is paraphyletic if it includes the most recent common ancestor of all members of a group and some but not all descendants of that ancestor (figure 4.5B). A taxon is polyphyletic if it does not include the most recent common ancestor of all members of a group; this situation requires the group to have had at least two separate evolutionary origins, usually requiring independent evolutionary acquisition of a diagnostic feature (figure 4.5C). For example, if birds and mammals were grouped in a taxon called Homeothermia, it would be a polyphyletic taxon because birds and mammals descend from two quite separate amniotic lineages that have evolved homeothermy independently. The most recent common ancestor of birds and mammals is not homeothermic and does not occur in the polyphyletic Homeothermia just described.

Monophyletic and paraphyletic groups share the property of **convexity,** which distinguishes them from polyphyletic groups. A group is convex if you can trace a path between any two members of the group on a cladogram or phylogenetic tree without leaving the group. For example, on figure 4.5 you could trace a connection between any pair of points in the blue areas of parts A or B without leaving the blue area. For the polyphyletic grouping in part C of figure 4.5, one cannot trace the path between species C and E without leaving the group designated by blue shading. In figure 4.5C, if a systematist added the full path connecting species C and E to the group shown but continued to omit the paths leading to species A, B, and H, then the new grouping thus formed would be convex and paraphyletic rather than polyphyletic. Demonstration that a group is not convex is the formal criterion for considering the group polyphyletic. Both evolutionary and cladistic taxonomies accept monophyletic groups and reject polyphyletic groups. They differ regarding acceptance of paraphyletic groups.

Evolutionary Taxonomy

Evolutionary taxonomy incorporates two different evolutionary principles for recognizing and ranking higher taxa: (1) common descent and (2) amount of adaptive evolutionary change, as

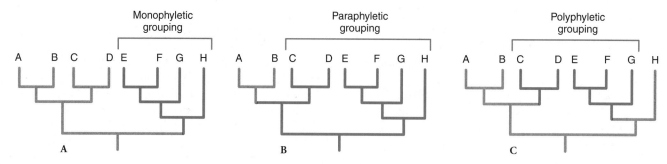

figure 4.5

Relationships between phylogeny and taxonomic groups illustrated for a hypothetical phylogeny of eight species (A through H). **A,** A monophyletic group contains the most recent common ancestor of all members of the group and all of its descendants. **B,** A paraphyletic group contains the most recent common ancestor of all members of the group and some but not all of its descendants. **C,** A polyphyletic group does not contain the most recent common ancestor of all members of the group, thereby requiring the group to have at least two separate phylogenetic origins. Monophyletic and paraphyletic groups are *convex*, meaning that one can trace a path from any member of the group to any other member without leaving the group; any group that fails the convexity criterion is considered polyphyletic.

A

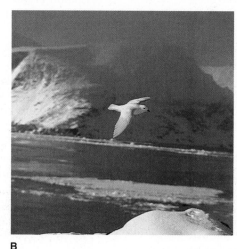

B

figure 4.6

A, Penguin. **B,** Diving petrel. Penguins (avian family Spheniscidae) were considered by George G. Simpson a distinct adaptive zone within birds because of their adaptations for submarine flight. Simpson believed that the adaptive zone ancestral to penguins resembled that of diving petrels, which display adaptations for combined aerial and aquatic flight. Adaptive zones of penguins and diving petrels are distinct enough to be recognized taxonomically as different families within a common order (Ciconiiformes).

shown on a phylogenetic tree. Evolutionary taxa must have a single evolutionary origin, and must show unique adaptive features.

George Gaylord Simpson (see figure 4.2) was highly influential in developing and formalizing principles of evolutionary taxonomy. According to Simpson, a particular branch on an evolutionary tree is given the status of a higher taxon if it represents a distinct **adaptive zone.** Simpson describes an adaptive zone as "a characteristic reaction and mutual relationship between environment and organism, a way of life and not a place where life is led." By entering a new adaptive zone through a fundamental change in organismal structure and behavior, an evolving population can use environmental resources in a completely new way.

A taxon forming a distinct adaptive zone is termed a **grade.** Simpson gives the example of penguins as a distinct adaptive zone within birds. The lineage immediately ancestral to all penguins underwent fundamental changes in form of the body and wings to switch from aerial to aquatic locomotion (figure 4.6). Aquatic birds that can fly both in air and underwater are somewhat intermediate in habitat, morphology, and

behavior between aerial and aquatic adaptive zones. Nonetheless, obvious modifications of a penguin's wings and body for swimming represent a new grade of organization. Penguins are therefore recognized as a distinct taxon within birds, family Spheniscidae. The Linnean rank of a taxon depends upon the breadth of its adaptive zone: the broader the adaptive zone when fully realized by a group of organisms, the higher the taxon is ranked.

Evolutionary taxa may be either monophyletic or paraphyletic. Recognition of paraphyletic taxa requires, however, that taxonomies distort patterns of common descent. An evolutionary taxonomy of anthropoid primates provides a good example (figure 4.7). This taxonomy places humans (genus *Homo*) and their immediate fossil ancestors in family Hominidae and places chimpanzees (genus *Pan*), gorillas (genus *Gorilla*), and orangutans (genus *Pongo*) in family Pongidae. However, the pongid genera *Pan* and *Gorilla* share more recent common ancestry with Hominidae than they do with the remaining pongid genus, *Pongo*. Family Pongidae is therefore paraphyletic because it does not include humans, who also descend from

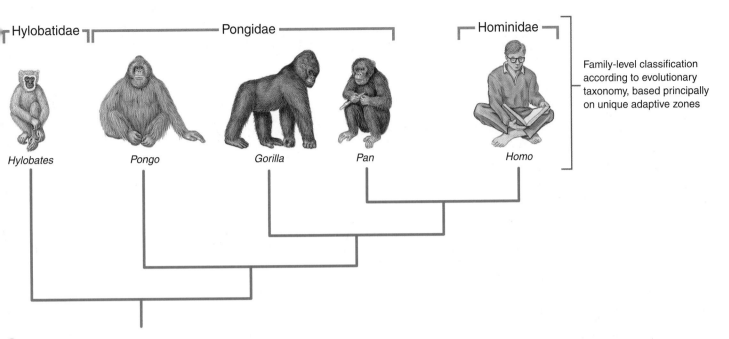

figure 4.7

Phylogeny and family-level classification of anthropoid primates. Evolutionary taxonomy groups the genera *Gorilla, Pan,* and *Pongo* into the paraphyletic family Pongidae because they share the same adaptive zone or grade of organization. Humans (genus *Homo*) are phylogenetically closer to *Gorilla* and *Pan* than any of these genera are to *Pongo,* but humans are placed in a separate family (Hominidae) because they represent a different grade of organization. Cladistic taxonomy eliminates the paraphyletic family Pongidae and groups *Pongo, Gorilla, Pan,* and *Homo* into a single monophyletic family, Hominidae. Gibbons (genus *Hylobates*) form the monophyletic family Hylobatidae, which is compatible with both evolutionary and cladistic classifications.

its most recent common ancestor (figure 4.7). Evolutionary taxonomists nonetheless recognize pongid genera as a single, family-level grade of **arboreal** (tree-dwelling), herbivorous primates having limited mental capacity; they show a family-level adaptive zone. Humans are terrestrial, omnivorous primates who possess greatly expanded mental and cultural attributes, thereby forming a distinct adaptive zone at the taxonomic level of a family.

Evolutionary taxonomy has been challenged from two opposite directions. One challenge states that because phylogenetic trees can be very difficult to obtain, it is impractical to base our taxonomic system on common descent and adaptive evolution. We are told that our taxonomy should represent a more easily measured feature, overall similarity of organisms evaluated without regard to phylogeny. This principle is called **phenetic taxonomy.** Phenetic taxonomy contributed some useful analytical methods but did not have a strong impact on animal taxonomy, and scientific interest in this approach is in decline.

Despite the difficulties of reconstructing phylogeny, zoologists still consider this endeavor a central goal of their systematic work, and they are unwilling to compromise this goal for methodological simplicity.

Phylogenetic Systematics/Cladistics

A second and stronger challenge to evolutionary taxonomy is called **phylogenetic systematics,** or **cladistics.** As the first name implies, this approach emphasizes the criterion of

figure 4.8

Willi Hennig (1913–1976), German entomologist who formulated the principles of phylogenetic systematics/cladistics.

common descent, and as the second name implies, it is based on the cladogram of a group being classified. This approach to taxonomy was first proposed in 1950 by the German entomologist Willi Hennig (figure 4.8) and therefore is sometimes called

Phylogenies from DNA Sequences

Most phylogenetic reconstruction currently underway comes from collecting genomic DNA sequences for the species of interest and analyzing those data using statistical approaches, especially maximum-likelihood and Bayesian methods. Such analysis requires sophisticated computer algorithms and cannot be done by hand. Nonetheless, statistical inference of phylogenies is designed to separate patterns of homology from homoplasy in the data, and to use the former to trace common descent of the species being studied. To explain statistical phylogenetics, we first must examine the basic phylogenetic principles of character analysis that underlie statistical phylogenetics.

A simple example illustrates cladistic analysis of DNA sequence data to examine phylogenetic relationships among species. Note that this procedure of character analysis follows the general principles described on pages 81–82. The study group in this example contains three species of chameleons, two from the island of Madagascar (*Brookesia theili* and *B. brygooi*) and one from Equatorial Guinea (*Chamaeleo feae*). The outgroup is a lizard of the genus *Uromastyx*, which is a distant relative of chameleons. Do the molecular data in this example confirm or reject the prior taxonomic hypothesis that the two Madagascan chameleons are more closely related to each other than either one is to the Equatorial Guinean species?

The molecular information in this example comes from a piece of the mitochondrial DNA sequence (57 bases) for each species. Each sequence encodes amino acids 221–239 of a protein called "NADH dehydrogenase subunit 2" in the species from which it was obtained. These DNA base sequences are aligned and numbered as

Each column in the aligned sequences constitutes a character that takes one of four states: A, C, G, or T (a fifth possible state, absence of the base, is not observed in this example). Only characters that vary among the three chameleon species potentially contain information on which pair of species is most closely related. Twenty-three of the 57 aligned bases show variation *among chameleons,* as shown here in bold letters:

To be useful for constructing a cladogram, a character must demonstrate sharing of derived states (= synapomorphy). Which of these 23 characters demonstrate synapomorphies for chameleons? For each of the 23 variable characters, we must ask whether one of the states observed in chameleons is shared with the outgroup, *Uromastyx*. If so, this state is judged ancestral for chameleons and the alternative state(s) derived. Derived states are identified for 21 of the 23 characters just identified; derived states are shown here in blue:

Note that polarity is ambiguous for two variable characters (at positions 23 and 54) whose alternative states in chameleons are not observed in the outgroup.

Of the characters showing derived states, 10 show synapomorphies among chameleons. These characters are marked here with numbers 1, 2, or 3 below the appropriate column.

The eight characters marked "1" show synapomorphies grouping the two Madagascan species (*Brookesia theili* and *B. brygooi*) to the exclusion of the Equatorial Guinean species, *Chamaeleo feae*. We can represent these relationships as a cladogram:

We can explain evolution of all characters favoring this cladogram by placing a single mutational change on the branch ancestral to the two *Brookesia* species. This is the simplest explanation for evolutionary changes of these characters.

Characters marked "2" and "3" disagree with our cladogram and favor alternative relationships:

To explain evolutionary changes in characters favoring cladograms 2 or 3 using cladogram 1, we need at least two changes per character. Likewise, if we try to explain evolution of characters favoring cladogram 1 on cladogram 2 or 3, we need at least two changes for each of these characters. These two diagrams show the minimum numbers of changes required for character 5 (which favors cladogram 1) and character 41 (which favors cladogram 3) on cladogram 1; the ancestral state of each character appears at the root of the tree and the nucleotide states observed in each species at the tips of the branches:

character 5 (1 change)

character 41 (2 changes)

A principle called **parsimony** can be used to resolve conflicts among taxonomic characters, as seen here. Parsimony represents the simplest limiting case of the maximum-likelihood and Bayesian methods for statistical inference of phylogeny, so we present it first and then show how the statistical methods build on this framework. We choose as our best working hypothesis the cladogram that requires the smallest total amount of character change. In our example, parsimony favors cladogram 1. For all 10 phylogenetically informative characters, cladogram 1 requires a total of 12 changes of character state (one for each of the 8 characters favoring it and two for each of the other 2 characters); cladograms 2 and 3 each require at least 19 character-state changes, 7 steps longer than cladogram 1. By choosing cladogram 1, we claim that characters favoring cladograms 2 and 3 show homoplasy in their evolution.

The molecular sequences shown in this example therefore confirm predictions of the prior hypothesis, based on the appearance and geography of these chameleons, that the *Brookesia* species shared a common ancestor with each other more recently than either one did with *Chamaeleo feae*.

As a further exercise, you should convince yourself that the 12 characters that vary among chameleons but do not demonstrate unambiguous sharing of derived states are equally compatible with each of the three possible cladograms shown. For each character, find the minimum total number of changes that must occur to explain its evolution on each cladogram. You will see, if you do this exercise correctly, that the three cladograms do not differ in minimum numbers of changes required for each of these characters. For this reason, the characters are phylogenetically uninformative by the parsimony criterion.

The parsimony method just illustrated makes some assumptions: base substitutions are equally likely to occur at any of the 57 sites, all 3 possible substitutions are equally likely at any site (for example, at site 5 one assumes that C would be equally likely to change to A, G, or T given that a substitution occurs), and the expected amount of molecular evolution on any branch is proportional to its temporal duration. Suppose that site 16 is unusually subject to mutational change, that a change from C to T is ten times more likely than one from C to A or G, and that the internal branch on the tree (a branch is called "internal" if it groups two or more of the species being studied to the exclusion of other species) is very short compared to the tip branches (those that terminate in a living species whose DNA sequence was determined). The likelihood of getting the observed data for that site by parallel changes C to T in the tip branches of the two *Brookesia* species might equal or exceed the likelihood of a single change C to T in the internal branch of hypothesis 1. The data at site 16 then would not necessarily favor hypothesis 1 over the alternatives. If we can obtain detailed knowledge of the evolutionary properties of the sites in this DNA sequence (for example, by studying its evolution in a large number of lizards), we might favor a phylogenetic method that incorporates a detailed model of DNA-sequence evolution.

Maximum-likelihood and Bayesian approaches to phylogenetic inference are

---

*Ancestral nucleotide state inferred by outgroup comparison.

(continued)

efficient means for using detailed models of molecular evolution to test phylogenetic hypotheses from aligned DNA sequences. First, we estimate an evolutionary model for the aligned sequences: how much do sites differ from each other in their tendencies to vary, which kinds of substitutions (C to A, C to G, C to T, etc.) are most likely to occur. Second, each site is evaluated with respect to each of the alternative possible trees to determine which tree has the highest likelihood of producing the observed data (such as the CTTC pattern shown for site 16). In this second step, branches of the tree can vary in length, relaxing the assumption that expected amounts of substitution are proportional to the temporal duration of the branch. For our data, the maximum-likelihood method would evaluate the probability of observing the results in each of the 57 columns considering all possible trees, and measure the likelihood of the entire data set for each contrasting tree. The tree with the highest likelihood of producing the observed data is the favored phylogenetic hypothesis. We can reject contrasting trees whose likelihoods are judged much lower than would be expected by chance alone. The Bayesian approaches operate in a similar manner, but they permit an investigator to evaluate the contributions of a new data set relative to prior results. For example, we could evaluate the probability that the data favor hypothesis 1 after incorporating phylogenetic results for these same species obtained from another data set. Calculating likelihoods for contrasting trees is part of the Bayesian operation, and in most phylogenetic applications this is the critical factor in favoring one tree over alternatives. For this reason, results of maximum-likelihood and Bayesian analyses are usually the same, with Bayesian analyses currently more popular because of computational efficiency.

Note that the maximum-likelihood and Bayesian approaches use more of the data set than our parsimony analysis did. To estimate branch lengths (numbers of substitutions occurring on them), sites whose derived states arose on a single terminal branch contribute useful information. Given hypothesis 1, the terminal branch connecting *C. feae* from its common ancestor with *Brookesia* is longer than the terminal branches connecting the *Brookesia* species to their most recent common ancestor with each other. Using sites for which we identified derived states in the preceding exercise but which were not parsimony informative, we see that the terminal branch leading to *C. feae* requires seven substitutions (sites 10, 15, 30, 32, 38, 56, 57), whereas the terminal branch leading to *B. brygooi* requires only three changes (sites 32, 50, 55), and the terminal branch leading to *B. theili* requires only two changes (sites 12, 14). Using the likelihood method, these sites collectively would favor hypothesis 1 even though none of the sites is parsimony informative. We thus expect parallel substitutions to occur more frequently in the *C. feae* lineage and one of the two *Brookesia* lineages (as probably occurred at sites 35 and 41) than in both *Brookesia* lineages (such changes would produce parsimony-informative sites favoring hypothesis 1, indistinguishable in our data from sites undergoing one change in the lineage immediately ancestral to the two *Brookesia* species).

The previous paragraph should make clear why inferring phylogenies using maximum-likelihood and Bayesian approaches would be very hard to do by hand, even for the data in our example. Finding the optimal tree topologies, branch lengths, relative probabilities of substitution at different sites, and relative probabilities of different kinds of substitution requires testing many alternative conditions and comparing their likelihoods. Computer algorithms can explore this parameter space in an efficient manner, but it would be prohibitively tedious to do by hand. Nonetheless, sharing of derived states at a site as predicted by hypotheses of homology remains the primary reason why the tree that requires only a single evolutionary change to explain the observed variation has a higher likelihood than do trees that require two or more parallel substitutions. Statistical inference of phylogeny using maximum-likelihood and Bayesian methods thus remains grounded in the basic cladistic principles. When parameters of the maximum-likelihood and Bayesian analyses are set to the simple conditions used for parsimony analysis, their results correspond to those that we obtained using parsimony.

Data from Townsend, T., and A. Larson. 2002. Molecular phylogenetics and mitochondrial genomic evolution in the Chamaeleonidae (Reptilia, Squamata). Molecular Phylogenetics and Evolution 23:22–36.

"Hennigian systematics." All taxa recognized by Hennig's cladistic system must be monophyletic. We saw previously how evolutionary taxonomists' recognition of the primate families Hominidae and Pongidae distorts genealogical relationships to emphasize the adaptive uniqueness of the Hominidae. Because the most recent common ancestor of the paraphyletic family Pongidae is also an ancestor of family Hominidae, recognition of Pongidae is incompatible with cladistic taxonomy. To avoid paraphyly, cladistic taxonomists have discontinued use of the traditional family Pongidae, placing chimpanzees, gorillas, and orangutans with humans in the family Hominidae (as shown for *Gorilla gorilla* in table 4.1).

Disagreement regarding the validity of paraphyletic groups may seem trivial at first, but its important consequences become clear when we discuss evolution. For example, claims that amphibians evolved from bony fish, that birds evolved from reptiles, or that humans evolved from apes might be made by an evolutionary taxonomist but are meaningless to a cladist. We imply by these statements that a descendant group (amphibians, birds, or humans) evolved from part of an ancestral group (bony fish, reptiles, or apes, respectively) from which the descendant is excluded. This usage automatically makes the ancestral group paraphyletic, and indeed bony fish, reptiles, and apes as traditionally recognized are paraphyletic groups. How are such paraphyletic groups recognized? Do they share distinguishing features not shared by a descendant group?

Paraphyletic groups are usually defined in a negative manner. They are distinguished only by the absence of features found in a particular descendant group, because any traits shared from their common ancestry are present also in the excluded descendants (unless secondarily lost). For example, apes are those "higher" primates that are not humans. Likewise, fish are those vertebrates that lack the distinguishing characteristics of tetrapods (amphibians and amniotes). What does it mean then to say that humans evolved from apes? To an evolutionary taxonomist, apes and humans are different adaptive zones or grades of organization; to say that humans evolved from apes states that bipedal organisms of large brain capacity evolved from arboreal, quadrupedal organisms of smaller brain capacity. To a cladist, however, the statement that humans evolved from apes says essentially that humans evolved from an arbitrary grouping of species that lack the distinctive characteristics of humans; such a statement contains no useful information. An extinct ancestral group is always paraphyletic because it excludes a descendant that shares its most recent common ancestor. Although many such groups have been recognized by evolutionary taxonomists, none are recognized by cladists.

Cladists denote the common descent of different taxa by identifying **sister taxa.** Sister taxa share more recent common ancestry with each other than either one does with any other taxon. For example, the sister group of humans appears to be chimpanzees (including the pygmy chimpanzee or bonobo), with gorillas forming a sister group to humans and chimpanzees combined. Orangutans are the sister group of the clade that comprises humans, chimpanzees, and gorillas; gibbons form the sister group of the clade that comprises orangutans, chimpanzees, gorillas, and humans (see figure 4.7). These apes and humans share a common ancestor that is not shared with any other living species.

Evolutionary paleontologists traditionally use the family-level adjective **hominid** to denote humans and fossil species closer to humans than to other apes; however, a cladist's use of "hominid" refers to any member of the expanded family Hominidae, which includes humans, chimpanzees, gorillas, orangutans, and fossil forms that are phylogenetically closer to these species than to other living primates.

Current State of Animal Taxonomy

The formal taxonomy of animals that we use today was established using principles of evolutionary systematics and has been revised recently in part using principles of cladistics. Introduction of cladistic principles initially replaces paraphyletic groups with monophyletic subgroups while leaving the remaining taxonomy mostly unchanged. A thorough revision of taxonomy along cladistic principles, however, will require profound changes. A new taxonomic system called PhyloCode is being developed as an alternative to Linnean taxonomy; this system replaces Linnean ranks with codes that denote the nested hierarchy of monophyletic groups conveyed by a cladogram. In our coverage of animal taxonomy, we emphasize taxa that are monophyletic and therefore consistent with criteria of both evolutionary and cladistic taxonomy. We continue, however, to use Linnean ranks. Some new groups, especially ones comprising single-celled organisms, encompass multiple phyla and are therefore designated as named clades without a Linnean rank. For familiar taxa that are clearly paraphyletic grades, we note this fact and suggest alternative taxonomic schemes that contain only monophyletic taxa.

When discussing patterns of descent, we avoid statements such as "mammals evolved from reptiles" that imply paraphyly and instead specify appropriate sister-group relationships to convey the order of branching in a phylogeny. We avoid calling groups of organisms primitive, advanced, specialized, or generalized because all groups of animals contain combinations of primitive, advanced, specialized, and generalized features; these terms are best restricted to describing specific characteristics and not an entire group.

Revision of taxonomy according to cladistic principles can cause confusion. In addition to new taxonomic names, we see old ones used in unfamiliar ways. For example, cladistic use of "bony fishes" includes amphibians and amniotes (including reptilian groups, birds, and mammals) in addition to the finned, aquatic animals that evolutionary taxonomists normally group under the term "fish." Cladistic use of "reptiles" includes birds in addition to snakes, lizards, turtles, and crocodilians; however, it excludes some fossil forms, such as synapsids, that evolutionary taxonomists traditionally place in Reptilia (see Chapter 18). When using these seemingly familiar terms, taxonomists must be very careful to specify whether they are referencing traditional evolutionary taxa or newer cladistic taxa.

■ Major Divisions of Life

Since Aristotle's time, people have tried to assign every living organism to one of two kingdoms: plant or animal. Unicellular forms were arbitrarily assigned to one of these kingdoms, although recognition of the kingdoms was based primarily on properties of multicellular organisms. This system has outlived its usefulness. It does not convey common descent among organisms accurately. Under the traditional, two-kingdom system, neither animals nor plants constitute monophyletic groups.

Several alternative systems have been proposed to classify unicellular forms. In 1866, Haeckel proposed a new kingdom, Protista, to include all single-celled organisms. At first, bacteria and cyanobacteria (blue-green algae), forms that lack nuclei bounded by a membrane, were included with nucleated unicellular organisms. Later, systematists recognized important differences between the anucleate bacteria and cyanobacteria (prokaryotes) and all other organisms that have membrane-bounded nuclei (eukaryotes). In 1969, R. H. Whittaker proposed a five-kingdom system that incorporated a basic prokaryote-eukaryote distinction. Kingdom Monera contained prokaryotes. Kingdom Protista contained unicellular eukaryotic organisms (protozoa and unicellular eukaryotic algae). Multicellular organisms were split into three kingdoms based on mode of nutrition and other fundamental differences in organization. Kingdom Plantae included multicellular photosynthesizing organisms (higher plants and multicellular algae). Kingdom Fungi contained molds, yeasts, and fungi, which obtain their food by absorption.

Invertebrates (except the protozoa) and vertebrates form the kingdom Animalia. Most of these forms ingest their food and digest it internally, although some parasitic forms are absorptive.

All of these systems were proposed without regard to the phylogenetic relationships needed to construct evolutionary or cladistic taxonomies. The oldest phylogenetic events in the history of life have been obscure, because different forms of life share very few characters that can be compared among these higher taxa to reconstruct their phylogeny. More recently, however, a cladistic classification of all life forms has been proposed based on phylogenetic information obtained from molecular data (especially the nucleotide base sequence of ribosomal RNA). According to this tree (figure 4.9), Carl Woese and coworkers recognized three monophyletic **domains** above the kingdom level: Eucarya (all eukaryotes), Bacteria (the true bacteria), and Archaea (prokaryotes differing from bacteria in structure of membranes and in ribosomal RNA sequences). They did not divide Eucarya into kingdoms, but if we retain Whittaker's kingdoms Plantae, Animalia, and Fungi, Protista is paraphyletic because this group does not contain all the descendants of its most recent common ancestor (all eukaryotes shown on figure 4.9, except for the animals, plants, and fungi, are protistan groups). To maintain a cladistic classification, Protista must be discontinued by recognizing as separate kingdoms all of the labeled branches of Eucarya shown in figure 4.9.

Until recently, heterotrophic **protists** were traditionally studied in zoology courses as the animal phylum Protozoa. Given current knowledge and principles of phylogenetic systematics, this practice commits a taxonomic error because "protozoa" do not form a monophyletic group. Protists, now often arrayed in numerous separate phyla, are nonetheless of interest to students of zoology because of their shared eukaryotic ancestry with multicellular animals.

Major Subdivisions of the Animal Kingdom

The phylum is the largest category in Linnean classification of the animal kingdom. Animal (= metazoan) phyla are often ordered to produce additional, informal taxa intermediate between phylum and kingdom. Taxon Eumetazoa includes all animal phyla except Porifera and Placozoa, with the placement of phylum Mesozoa in Eumetazoa remaining controversial. Taxon Eumetazoa is divided into Radiata (phyla Cnidaria and Ctenophora) and Bilateria (all remaining eumetazoan phyla).

Bilateral animals are customarily divided into **Protostomia** and **Deuterostomia** based on their embryological development (figure 4.10). Note, however, that the individual characters listed in figure 4.10 are not completely diagnostic in separating protostomes from deuterostomes. Some phyla

figure 4.9

Phylogenetic overview of the three domains of life—Archaea, Bacteria, and Eucarya—based on analysis of genes encoding ribosomal RNA. Because of their endosymbiotic origins, organellar genomes of domain Eucarya (mitochondria, chloroplasts) are phylogenetically within the Bacteria rather than the clade that includes all eucaryotic nuclear genomes. Organisms of domain Eucarya therefore include cellular components of disparate evolutionary origins.

PROTOSTOMES			DEUTEROSTOMES		
	Spiral cleavage	Cleavage mostly spiral	Cleavage mostly radial	Radial cleavage	
	Cell from which mesoderm will derive	Endomesoderm usually from a particular blastomere designated 4d	Endomesoderm from enterocoelous pouching (except chordates)	Endomesoderm from pouches from primitive gut	
	Primitive gut / Mesoderm / Coelom / Blastopore	In coelomate protostomes, the coelom forms as a split in mesodermal bands (schizocoelous)	All coelomate; coelom from fusion of enterocoelous pouches (except chordates, which are schizocoelous)	Coelom / Mesoderm / Primitive gut / Blastopore	
	Anus / Annelid (earthworm) / Mouth	Mouth from, at, or near blastopore; anus a new formation	Anus from, at, or near blastopore; mouth a new formation	Mouth / Anus	
		Embryology mostly determinate (mosaic)	Embryology usually indeterminate (regulative)		
		Includes phyla Platyhelminthes, Nemertea, Annelida, Mollusca, Arthropoda, minor phyla	Includes phyla Echinodermata, Hemichordata, and Chordata, and formerly Chaetognatha, Phoronida, Ectoprocta, Brachiopoda		

figure 4.10

Basis for distinctions between divisions of bilateral animals.

are difficult to place into one of these two categories because they possess characteristics of each group.

Molecular phylogenetic studies have challenged traditional classification of Bilateria, but results are not yet conclusive enough to present a precise hypothesis of phylogenetic relationships among bilaterian phyla. Molecular phylogenetic results place four phyla traditionally classified as deuterostomes (Brachiopoda, Chaetognatha, Ectoprocta, and Phoronida) in Protostomia. Furthermore, the traditional major groupings of protostome phyla (acoelomates, pseudocoelomates, and coelomates) appear not to be monophyletic. Instead, most protostomes are divided into two major monophyletic groups called Lophotrochozoa and Ecdysozoa. Reclassification of Bilateria, including some changes in the phyla recognized, is

Bilateria
    Protostomia: phyla Acoelomorpha, Chaetognatha
        Lophotrochozoa: phyla Platyhelminthes,
            Mesozoa, Nemertea, Rotifera, Gastrotricha,
            Acanthocephala, Mollusca, Annelida, Echiura,
            Sipuncula, Phoronida, Ectoprocta, Entoprocta,
            Gnathostomulida, Micrognathozoa, Brachiopoda
        Ecdysozoa: phyla Kinorhyncha, Nematoda,
            Nematomorpha, Priapulida, Arthropoda,
            Tardigrada, Onychophora, Loricifera
    Deuterostomia: phyla Chordata, Hemichordata,
        Echinodermata

Although further study is needed to test these new groupings, we use them to organize our survey of animal diversity.

■ Summary

Animal systematics has three major goals: (1) to identify all species of animals, (2) to evaluate evolutionary relationships among animal species, and (3) to group animal species hierarchically in taxonomic groups (taxa) that convey evolutionary relationships. Taxa are ranked to denote increasing inclusiveness as follows: species, genus, family, order, "class," phylum, and kingdom. All of these ranks can be subdivided to signify taxa that are intermediate between them. The names of species are binomial, with the first name designating the genus to which the species belongs (first letter capitalized) followed by a species epithet (lowercase), both written in italics. Taxa at all other ranks are given single, nonitalicized names.

The biological species concept has guided recognition of most animal species. A biological species is defined as a reproductive community of populations (reproductively isolated from others) that occupies a specific niche in nature. A biological species is not immutable through time but changes during its evolution. Because the biological species concept may be difficult to apply in spatial and temporal dimensions, and because it excludes asexually reproducing forms, alternative concepts have been proposed. These alternatives include the evolutionary species concept, the cohesion species concept, and the phylogenetic species concept. No single concept of species is universally accepted by all zoologists, but zoologists agree that a species should constitute a population lineage whose evolutionary descent is separate from those of other such lineages. Because species lineages are expected to differ from each other in the DNA sequence of the rapidly evolving mitochondrial gene *COI*, this gene sequence serves as a diagnostic barcode to assign specimens to species.

Two major schools of taxonomy are currently active. Evolutionary taxonomy groups species into higher taxa according to joint criteria of common descent and adaptive evolution; such taxa have a single evolutionary origin and occupy a distinctive adaptive zone. A second approach, called phylogenetic systematics or cladistics, emphasizes common descent exclusively in grouping species into higher taxa. Only monophyletic taxa (those having a single evolutionary origin and containing all descendants of the group's most recent common ancestor) are used in cladistics. In addition to monophyletic taxa, evolutionary taxonomy recognizes some paraphyletic taxa (those having a single evolutionary origin but excluding some descendants of the group's most recent common ancestor). Both schools of taxonomy exclude polyphyletic taxa (those having more than one evolutionary origin).

Both evolutionary taxonomy and cladistics require that common descent among species be assessed before higher taxa are recognized. Comparative morphology (including development), cytology, and biochemistry serve to reconstruct the nested hierarchical relationships among taxa that reflect the branching of evolutionary lineages through time. The fossil record provides estimates of the ages of evolutionary lineages. Comparative studies and the fossil record jointly permit us to reconstruct a phylogenetic tree representing the evolutionary history of the animal kingdom.

Molecular systematists infer phylogenetic relationships among species from variation in aligned DNA sequences of homologous genes sampled from each species being studied. We illustrate molecular phylogenetic analysis using the criterion of maximum parsimony and show how models of molecular evolution are applied to phylogenetic inference using maximum-likelihood and Bayesian approaches.

Taxonomists recognize three monophyletic domains above the kingdom level: Eucarya (all eukaryotes), Bacteria (the true bacteria), and Archaea (prokaryotes differing from bacteria in membrane structure and ribosomal RNA sequences). Traditionally, all living forms were placed into two kingdoms (animal and plant), but more recently a five-kingdom system (animals, plants, fungi, protistans, and monerans) has been followed. Neither system conforms to the principles of evolutionary or cladistic taxonomy because both place single-celled organisms into either paraphyletic or polyphyletic groups. Based on our current knowledge of the phylogenetic tree of life, "protozoa" do not form a monophyletic group and they do not belong within the animal kingdom, which comprises multicellular forms (metazoa).

■ Review Questions

1. List in order, from most inclusive to least inclusive, the principal categories (ranks of taxa) in Carolus Linnaeus's system of classification.
2. Explain why the system for naming species that originated with Linnaeus is binomial.
3. How does the biological species concept differ from earlier typological concepts of species? Why do evolutionary biologists prefer it to typological species concepts?
4. What problems have been identified with the biological species concept? How do other concepts of species attempt to overcome these problems? What problems do these other concepts entail?
5. What are the properties of phylogenetically useful (or informative) characters? How are such characters used to construct a cladogram?
6. What is the difference between a cladogram and a phylogenetic tree? Given a cladogram for a group of species, what additional information is needed to obtain a phylogenetic tree?
7. How do monophyletic, paraphyletic, and polyphyletic taxa differ? How do these differences affect the validity of such taxa for both evolutionary and cladistic taxonomies?
8. How do cladists and evolutionary taxonomists differ in their classifications of humans and apes? Contrast their respective interpretations of the statement that humans evolved from apes, which evolved from monkeys.
9. What are the five kingdoms distinguished by Whittaker? How does their recognition conflict with the principles of cladistic taxonomy?
10. For the human, gorilla, and leopard-frog taxa in table 4.1, construct a rank-free ordering using indentations to denote nesting of groups within groups, as shown on p. 91.

For Further Thought If a taxonomist constructs a rooted phylogentic tree for a group of living species, the structure of the tree alone can be used to distinguish hypotheses of monophyly versus nonmonophyly of a particular subgroup. If monophyly is rejected for a particular subgroup, tree topology alone cannot distinguish paraphyly from polyphyly. What additional information is needed to distinguish paraphyly from polyphyly?

■ Selected References

See also general references on page 432.

Aguinaldo, A. M. A., J. M. Turbeville, L. S. Linford, M. C. Rivera, J. R. Garey, R. A. Raff, and J. A. Lake. 1997. Evidence for a clade of nematodes, arthropods and other moulting animals. Nature **387:** 489–493. *This molecular phylogenetic study challenges traditional classification of the Bilateria.*

Avise, J. C. 2006. Evolutionary pathways in nature: a phylogenetic approach. Cambridge, U.K., Cambridge University Press. *A survey of phylogenetic knowledge.*

de Queiroz, K. 1988. Systematics and the Darwinian Revolution. Philosophy of Science 55: 238–259. *Criticism of evolutionary taxonomy by a phylogenetic systematist, including explanation of the contrast between classification and systematization.*

Ereshefsky, M. 2001. The poverty of the Linnaean hierarchy. Cambridge, UK, Cambridge University Press. *A philosophical study of biological taxonomy, emphasizing problems with Linnean taxonomy.*

Ereshefsky, M. (ed.). 1992. The units of evolution. Cambridge, Massachusetts, MIT Press. *A thorough coverage of concepts of species, including reprints of important papers on the subject.*

Felsenstein, J. 2002. Inferring phylogenies. Sunderland, Massachusetts, Sinauer Associates. *A thorough coverage of phylogenetic methods.*

Freeman, S., and J. C. Herron. 2007. Evolutionary analysis, ed. 4. Upper Saddle River, New Jersey, Pearson/Prentice Hall. *See chapter 4, "Estimating Evolutionary Trees," for further coverage of phylogenetic methods.*

Hull, D. L. 1988. Science as a process. Chicago, University of Chicago Press. *A study of the working methods and interactions of systematists, containing a thorough review of the principles of evolutionary, phenetic, and cladistic taxonomy.*

Maddison, W. P., and D. R. Maddison. 2008. MacClade 4.08. Sunderland, Massachusetts, Sinauer Associates, Inc. *A program for MacIntosh computers that conducts phylogenetic analyses of systematic characters. The instruction manual stands alone as an excellent introduction to phylogenetic procedures. The computer program is user-friendly and excellent for instruction in addition to serving as a tool for analyzing real data.*

Mayr, E., and P. D. Ashlock. 1991. Principles of systematic zoology. New York, McGraw-Hill. *A detailed survey of systematic principles as applied to animals.*

Swofford, D. 2002. Phylogenetic analysis using parsimony (and other methods) PAUP* version 4. Sunderland, Massachusetts, Sinauer Associates. *A powerful computer package and instructive manual for constructing phylogenetic trees from data.*

Valentine, J. W. 2004. On the origin of phyla. Chicago, Illinois, University of Chicago Press. *Phylogenetic analysis of animal phyla in a paleontological perspective.*

Wagner, G. P. (ed.). 2001. The character concept in evolutionary biology. San Diego, Academic Press. *A thorough coverage of evolutionary character concepts.*

■ Custom Website

The *Animal Diversity* Online Learning Center is a great place to check your understanding of chapter material. Visit www.mhhe.com/hickmanad6e for access to key terms, quizzes, and more! Further enhance your knowledge with Web links to chapter-related material.

Explore live links for these topics:

Classification and Phylogeny of Animals
Methods of Classification
Trees of Life
Speciation

Protozoan Groups (Unicellular Eukaryotes)

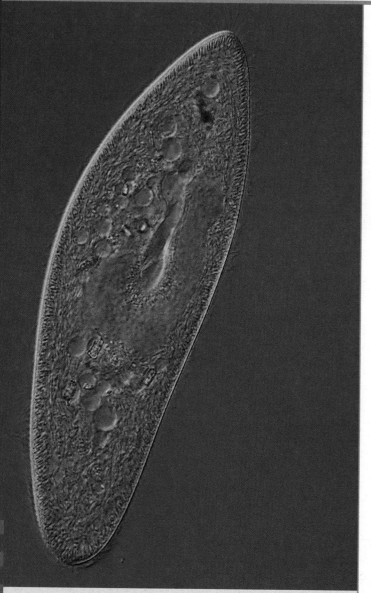

A paramecium.

Emergence of Eukaryotes and a New Life Pattern

The first reasonable evidence for life on earth dates from approximately 3.5 billion years ago. The first cells were prokaryotic bacteria-like organisms. The early prokaryotes diversified over an enormous time span: their prokaryotic descendants now belong to two groups, the Bacteria and the Archaea. An ability to photosynthesize evolved within bacteria; the cyanobacterial invention of oxygen-releasing photosynthesis changed the earth. The ancient prokaryotes also gave rise to the common ancestor of eukaryotes when two lineages merged by symbiogenesis. In this process, cells of one prokaryotic lineage engulfed, but did not digest, cells of another prokaryotic lineage. The engulfed cell was eventually reduced to an organelle inside the host cell. The eukaryotic products of symbiogenesis include mitochondria and plastids.

A mitochondrion originated from an aerobic prokaryote capable of deriving energy from carbon compounds using environmental oxygen. An anaerobic bacterium that engulfed such an aerobic form gained a capacity to grow in the oxygen-enriched environment that developed as cyanobacteria became more abundant. The engulfed aerobic bacterium persisted inside the host cell as a mitochondrion with its own genetic material. Over evolutionary time, most, but not all, genes from the mitochondrion came to reside in the host cell nucleus. Almost all present-day eukaryotes have mitochondria and are aerobic.

The eukaryotic plastid originated when a cell engulfed a photosynthetic cyanobacterium. When a prokaryote is engulfed and modified to become a eukaryotic organelle, the organelle is said to develop by primary endosymbiosis. Chloroplasts in clade Plantae (red algae, green algae, and land plants, among others) arose in this way. However, a eukaryotic cell may also obtain plastids from another eukaryote. This is secondary endosymbiosis. Two similar cells with plastids may have formed very differently, so it can be hard to untangle the evolutionary relationships among the diverse array of unicellular eukaryotes alive today.

The assemblage of eukaryotic unicellular organisms is collectively called protista or protozoa. The inclusion of "zoa" in the name originally referred to two animal-like features: the absence of a cell wall and the presence of at least one motile stage in the life cycle. However, these are not homologous features and we now recognize protozoans as belonging to many different clades whose ages exceed those of animals, fungi, and plants. As this chapter makes clear, many protozoans impact human life; they feed on corneas or brain tissue, cause diseases such as malaria or dysentery, and impact livestock. Our ability to treat protozoan disease agents depends on our ability to classify them correctly because susceptibility to particular drugs is often lineage-dependent. This is an exciting time to study these tiny organisms.

A protozoan is a eukaryotic organism in which all life activities proceed within the limits of a single plasma membrane. Protozoa do demonstrate a basic body plan or grade—the single eukaryotic cell—and they amply demonstrate the enormous adaptive potential of that grade. Over 64,000 species have been named, and more than half of these are known only from fossils. Some researchers estimate that there may be 250,000 species of protozoa.

Protozoa occur wherever life exists. They are highly adaptable and easily distributed from place to place. They require moisture, whether they live in marine or freshwater habitats, soil, decaying organic matter, or plants and animals. They may be sessile or free-swimming, and they form a large part of the floating plankton. Some species may have spanned geological eras of more than 100 million years.

Protozoa play an enormous role in the economy of nature. Their fantastic numbers are evidenced by the gigantic oceanic and soil deposits formed by their skeletons. About 10,000 species of protozoa are symbiotic in or on animals or plants, or sometimes even on other protozoa. The symbiotic relationship may be **mutualistic** (both partners benefit), **commensalistic** (one partner benefits without affecting the other), or **parasitic** (one partner benefits at the expense of the other) (see Chapter 2). Some of the most important diseases of humans and domestic animals are caused by parasitic protozoa.

Protozoans are often described by their body type: ciliates have numerous cilia covering the cell membrane; flagellates have one or more flagella used to propel the cell; amebas have irregular shapes caused by flowing protoplasm inside the cell membrane (figure 5.1). Amebas may be testate (shelled) or naked. When tests (shells) are present, they can be made of many different materials, such as sand grains, calcium, or silica. Tests may be simple coverings with only one opening or elaborate and beautiful structures pierced with many holes through which the cell protrudes.

For many years, Protozoa was considered a single phylum, but phylogenetic studies show that protozoa do not form a monophyletic group. Analyses of sequences of bases in genes, primarily those encoding the small subunit of

figure 5.1

Three body forms of the eukaryotic cell. **A,** a ciliate; **B,** a flagellate (*Giardia lamblia*); **C,** an ameba.

ribosomal RNA, but also those encoding several proteins, have revolutionized our concepts of phylogenetic affinities. The origin of the first eukaryote (see prologue to this chapter) was followed by diversification into the many unicellular and multicellular taxa observed today (figure 5.2). In this chapter we briefly describe 16 major eukaryotic clades (figure 5.1) from the more than 30 such clades currently recognized. Only the ciliated body type diagnoses a clade. The ameboid and flagellated forms appear to have evolved independently several times, so these forms occur in many different taxonomic groups. Some traditional group names, such as Foraminifera, are still in use, but other names have been newly created. For example, the taxon Opisthokonta was formed when choanoflagellates, fungi, and animals were found to form a clade. A description of the major clades that contain protozoans begins on page 103, following a description of the features of a eukaryotic cell.

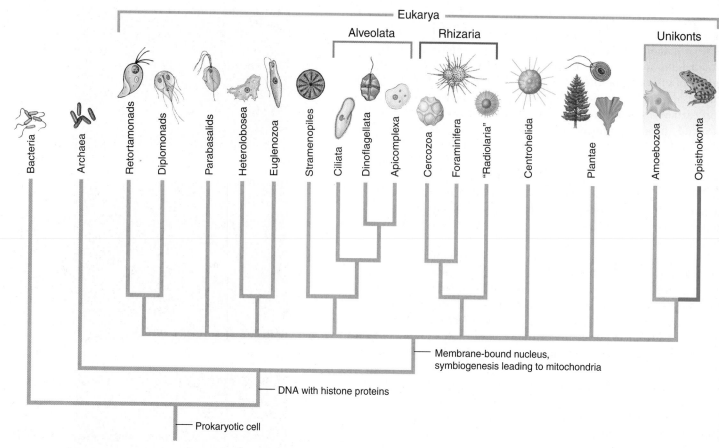

figure 5.2

Cladogram showing two major prokaryotic branches and diversification of 16 eukaryotic lineages. The order of branching remains to be determined for most clades. The multicellular animals, choanoflagellates, and fungi are nested within the Unikont clade in a subgroup called Opisthokonta.

■ Form and Function

Protozoa are unicellular, but they are not simple. They are functionally complete organisms with many complicated microanatomical structures. Particular organelles may perform as skeletons, sensory structures, conducting mechanisms, and feeding structures. Special features of a particular organelle can often be used as diagnostic characters for protozoan clades, as is the case with the shape of the mitochondrial internal membranes called cristae.

Nucleus

The nucleus is a membrane-bound structure whose interior communicates with the cytoplasm by small pores. Within the nucleus, the genetic material (DNA) is borne on chromosomes. Also within the nucleus, one or more **nucleoli** are often present (figure 5.3). Several nuclei, including a large macronucleus and one or more smaller micronuclei, occur in ciliates (see figure 5.16). The persistence of nucleoli during mitosis is one of the diagnostic characters for the Euglenozoa clade.

Mitochondria

A **mitochondrion** is an organelle used in energy acquisition where oxygen serves as the terminal electron acceptor. It contains DNA. The internal membranes of a mitochondrion, called cristae, vary in form, being flat, tubular, discoid (disc-shaped), or branched. The form of the cristae is a diagnostic character of some protozoan clades, such as Ramacristates and Euglenozoa. Cells without mitochondria may have **hydrogenosomes,** which produce molecular hydrogen when oxygen is absent and are assumed to have evolved from mitochondria. **Kinetoplasts** are also assumed to be mitochondrial derivatives, but they work in association with a kinetosome, an organelle at the base of a flagellum (see p. 98).

Golgi Apparatus

The Golgi apparatus is part of the secretory system of the endoplasmic reticulum. Golgi bodies are also called **dictyosomes** in protozoan literature. **Parabasal bodies** are similar structures with potentially similar functions.

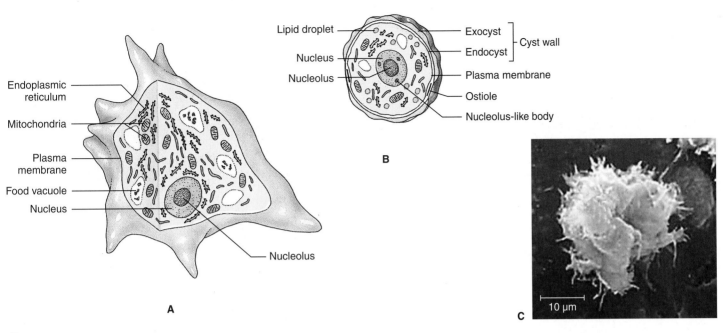

figure 5.3

Structure of *Acanthamoeba palestinensis*, an ameboid protozoan. **A,** Active, feeding form. **B,** Cyst. **C,** *Acanthamoeba castellanii* kills cells of the human cornea; it is spread by contact lenses that have not been correctly disinfected.

Plastids

Plastids are organelles containing photosynthetic pigments. The original addition of a plastid to eukaryotic cells occurred when a cyanobacterium was engulfed and not digested. **Chloroplasts** (see figure 5.12) contain different versions of chlorophylls (*a, b,* or *c*), but other kinds of plastids contain other pigments. For example, red algal plastids contain phycobilins. Particular pigments shared among protozoan lineages may indicate shared ancestry, but plastids could also have been gained by secondary endosymbiosis rather than direct inheritance from a common ancestor (see this chapter's prologue).

Extrusomes

The general term *extrusomes* refers to membrane-bound organelles that a protozoan uses to extrude something from the cell. The wide variety of structures extruded suggests that extrusomes are not all homologous. The ciliate **trichocyst** (see p. 106) is an extrusome.

Locomotor Organelles

Protozoa move chiefly by cilia and flagella and by pseudopodial movement. These mechanisms are extremely important in the biology of higher animals as well. For example, many small metazoans use cilia not only for locomotion but also to create water currents for their feeding and respiration. Ciliary movement is also vital to many species in such functions as ingestion of food, reproduction, excretion, and osmoregulation (as in flame cells, see p. 159).

Cilia and Flagella

No sharp morphological distinction exists between cilia and flagella, and some investigators have preferred to call them both *undulipodia* (L. dim. of *unda,* a wave, + Gr. *podos,* a foot). However, their actions are somewhat different. A cilium propels water parallel to the surface to which the cilium is attached, whereas a flagellum propels water parallel to the main axis of the flagellum. Cilia are typically present in large numbers and are relatively short, whereas flagella are longer and less numerous. Each flagellum or cilium contains nine pairs of longitudinal microtubules arranged in a circle around a central pair (figure 5.4), and this is true for most of the motile flagella and cilia in the animal kingdom, with a few notable exceptions. This "9 + 2" tube of microtubules in a flagellum or cilium is its **axoneme;** an axoneme is covered by a membrane continuous with the plasma membrane covering the rest of the organism. At about the point where an axoneme enters the cell proper, the central pair of microtubules ends at a small plate within the circle of nine pairs. Also at about that point, another microtubule joins each of the nine pairs, so that these form a short tube extending from the base of the flagellum into the cell. The tube consists of nine *triplets* of microtubules and is

B

figure 5.4

Internal structure of a flagellum. **A,** The axoneme is composed of nine pairs of microtubules plus a central pair, and it is enclosed within the plasma membrane. The central pair ends at about the level of the cell surface in a basal plate (axosome). The peripheral microtubules continue inward for a short distance to compose two of each of the triplets in the kinetosome (at level *y* in **A**). **B,** Electron micrograph of a section through several cilia, corresponding to section at *x* in **A** (× 133,000).

called the **kinetosome** (or **basal body**). Kinetosomes are exactly the same in structure as the **centrioles** that organize mitotic spindles during cell division. Centrioles of some flagellates may give rise to kinetosomes, or kinetosomes may function as centrioles. All typical flagella and cilia have a kinetosome at their base, whether borne by a protozoan or metazoan cell.

Description of the axoneme as "9 + 2" is traditional, but it is also misleading because there is only a single pair of microtubules in the center. If we were consistent, we would have to describe the axoneme as "9 + 1." A more precise way to describe the nine sets of doublets and pair of singlets is "9(2) + 2(1)"; the kinetosome is then 9(3) + 0.

The current explanation for ciliary and flagellar movement is the *sliding-microtubule hypothesis.* Movement is powered by the release of chemical bond energy in ATP. Two little arms are visible in electron micrographs on each of the pairs of peripheral tubules in the axoneme (level *x* in figure 5.4), and these bear the enzyme adenosine triphosphatase (ATPase), which cleaves the ATP. When bond energy in ATP is released, the arms "walk along" one of the microtubules in the adjacent pair, causing it to slide relative to the other tubule. Shear resistance, causing the axoneme to bend when the microtubules slide past each other, is provided by "spokes" from one of the tubules in each doublet projecting toward the central pair of microtubules. These spokes are also visible in electron micrographs.

CHARACTERISTICS
of Protozoa

1. **Unicellular;** some colonial, and some with multicellular stages in their life cycles
2. **Mostly microscopic,** although some are large enough to be seen with the unaided eye
3. All symmetries represented in the group; shape variable or constant (oval, spherical, or other)
4. **No germ layer present**
5. No organs or tissues, but **specialized organelles** are found; nucleus single or multiple
6. Free-living, mutualism, commensalism, parasitism all represented in the groups
7. Locomotion by **pseudopodia, flagella, cilia,** and direct cell movements; some sessile
8. Some have a **simple endoskeleton** or **exoskeleton,** but most are naked
9. **Nutrition of all types:** autotrophic (manufacturing own nutrients by photosynthesis), heterotrophic (depending on other plants or animals for food), saprozoic (using nutrients dissolved in the surrounding medium)
10. Aquatic or terrestrial habitat; free-living or symbiotic mode of life
11. Reproduce **asexually** by fission, budding, and cysts and **sexually** by conjugation or by syngamy (union of male and female gametes to form a zygote)
12. The simplest example of **division of labor between cells** is seen in certain protozoa that have both somatic and reproductive zooids (individuals) in the colony

Pseudopodia

Pseudopodia are temporary protrusions of cytoplasm that function in locomotion or ingestion. They are the chief means of locomotion for amebas, but they can also be formed by a variety of other protozoa, as well as by ameboid cells of many invertebrates. In fact, much defense against disease in the human body depends on ameboid white blood cells, and ameboid cells in many other animals, vertebrate and invertebrate, play similar roles.

In protozoa, pseudopodia exist in several forms. The most familiar are lobopodia (figure 5.5; see figure 5.8), which are rather large, blunt extensions of the cell body containing central, granular **endoplasm** and a peripheral, nongranular layer called **ectoplasm.** The endoplasm is in the sol state (fluid), and the ectoplasm is in the gel state of a colloid (jellylike semisolid). The pseudopodia in amebas take many forms (figure 5.6). **Filipodia** are thin extensions, usually branching and containing only ectoplasm,

A   **B**   **C**

figure 5.5

Large blunt pseudopodia, called lobopodia, are used to engulf food items in a process called phagocytosis. **A, B,** the ameba extends a lobopodium toward a *Pandorina* colony. **C,** the ameba surrounds the *Pandorina* before engulfing it by phagocytosis.

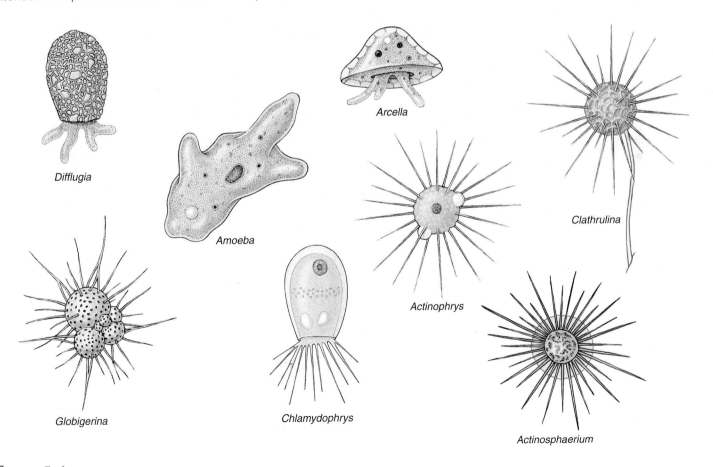

figure 5.6

The amebas are a very diverse group distributed across many clades: *Arcella, Difflugia,* and *Amoeba* have lobopodia and are amoebozoans; *Globigerina* has a testate body with net-like reticulopodia and is a foraminiferan; *Chlamydophrys* and *Clathrulina* are cercozoans, whereas *Actinosphaerium* and *Actinophrys* are stramenopiles. Axopodia are present in *Actinophrys, Actinosphaerium,* and *Clathrulina.*

that occur in amebas such as *Chlamydophrys*. In contrast to filipodia, **reticulopodia** repeatedly rejoin to form a net-like mesh (although some protozoologists consider the distinction between filipodia and reticulopodia artificial). **Axopodia** are long, thin pseudopodia supported by axial rods of microtubules. The microtubules are arranged in a definite spiral or geometric array, depending on the species, and constitute the axoneme of the axopod (figure 5.7). Axopodia can be extended or retracted, apparently by addition or removal of microtubular material. Cytoplasm can flow along the axonemes, toward the body on one side and in the reverse direction on the other.

How pseudopodia work has long interested zoologists, but we have only recently gained some insight into the phenomenon. When a typical lobopodium begins to form, an extension of ectoplasm called a **hyaline cap** appears (figure 5.8). As endoplasmic material flows into the hyaline cap, it fountains out to the periphery and changes from the sol (fluid) to the gel (semisolid) state of the colloid; it thereby becomes ectoplasm. Thus, the ectoplasm is a tube through which the endoplasm flows as the pseudopodium extends. On the trailing side of the organism, ectoplasm becomes endoplasm. At some point, the pseudopodium becomes anchored to the substrate, and the cell is drawn forward.

The current hypothesis of pseudopodial movement involves the participation of actin, myosin, and other components. As endoplasm fountains out in the hyaline cap, actin subunits become polymerized into microfilaments, which in turn become cross-linked to each other by actin-binding protein (ABP) to form a gel; the endoplasm becomes ectoplasm. At the "posterior," the ABP releases the microfilaments, which return to the sol state of endoplasm. Before the microfilaments are disassembled into actin subunits, they interact with myosin, contracting and causing the endoplasm to flow in the direction of the pseudopodium by hydrostatic pressure.

Colloidal systems are permanent suspensions of finely divided particles that do not precipitate, such as milk, blood, starch, soap, ink, and gelatin. Colloids in living systems are commonly proteins, lipids, and polysaccharides suspended in the watery fluid of cells (cytoplasm). Such systems may undergo sol-gel transformations, depending on whether the fluid or particulate components become continuous. In the sol state of cytoplasm, solids are suspended in a liquid, and in the semisolid gel state, liquid is suspended in a solid.

Nutrition and Digestion

Protozoa can be categorized broadly into two groups: autotrophs, which synthesize their own organic constituents from inorganic substrates, and heterotrophs, which must have organic molecules synthesized by other organisms.

Axopodium

Actinosphaerium

figure 5.7

Right, Diagram of axopodium to show orientation of (*left*) electron micrograph of axopodium (from *Actinosphaerium nucleofilum*) in cross section. The axoneme of the axopodium is composed of an array of microtubules, which may vary from three to many in number, depending on the species. Some species can extend or retract their axopodia quite rapidly (x99,000).

Within heterotrophs, those that ingest visible particles of food (**phagotrophs,** or *holozoic feeders*) contrast with those ingesting food in a soluble form (**osmotrophs,** or *saprozoic feeders*).

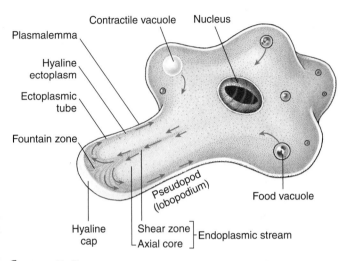

figure 5.8

Ameba in active locomotion. Arrows indicate direction of streaming protoplasm. The first sign of a new pseudopodium (in this case, a lobopodium) is thickening of the ectoplasm to form a clear hyaline cap, into which the fluid endoplasm flows. As endoplasm reaches the forward tip, it fountains out and is converted into ectoplasm, forming a stiff outer tube that lengthens as the forward flow continues. Posteriorly the ectoplasm is converted into fluid endoplasm, replenishing the flow. A substratum is necessary for ameboid movement.

Holozoic nutrition implies **phagocytosis** (figure 5.9; see figure 5.5), in which there is an infolding or invagination of the plasma membrane around the food particle. As the invagination extends farther into the cell, it is pinched off at the surface. A food particle is thus contained in an intracellular, membrane-bound vesicle, the **food vacuole** or **phagosome.** Lysosomes, small vesicles containing digestive enzymes, fuse with the phagosome and pour their contents into it, where digestion begins. As digested products are absorbed across the vacuolar membrane, the phagosome becomes smaller. Any undigestible material may be released to the outside by exocytosis, the vacuole again fusing with the plasma membrane. In most ciliates, many flagellates, and many apicomplexans, the site of phagocytosis is a definite mouth structure, the **cytostome** (see figure 5.16). In amebas, phagocytosis can occur at almost any point by envelopment of the particle with pseudopodia. Many ciliates have a characteristic structure for expulsion of waste matter, called a **cytoproct** (see figure 5.16) or **cytopyge,** occupying a characteristic location.

Excretion and Osmoregulation

Water balance, or osmoregulation, is a function of the one or more **contractile vacuoles** (see figures 5.8, 5.12, and 5.16) possessed by most protozoa, particularly freshwater forms, which live in a hypoosmotic environment. These vacuoles are often absent in marine or parasitic protozoa, which live in a nearly isosmotic medium.

Contractile vacuoles are usually located in the ectoplasm and act as pumps to remove excess water from the cytoplasm. A recent hypothesis suggests that proton pumps on the vacuolar surface and on tubules radiating

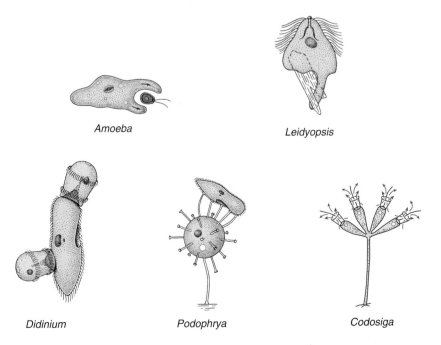

figure 5.9

Types of phagocytosis among protozoa. *Amoeba* surrounds a small flagellate with pseudopodia. *Leidyopsis*, a flagellate living in the intestine of termites, forms pseudopodia and ingests wood chips. *Didinium*, a ciliate, feeds only on *Paramecium*, which it swallows through a temporary cytostome in its anterior end. Sometimes more than one *Didinium* feeds on the same *Paramecium*. *Podophrya* is a suctorian ciliophoran. Its tentacles attach to its prey and suck prey cytoplasm into its body, where the cytoplasm is pinched off to form food vacuoles. *Codosiga*, a sessile Choanoflagellate with a collar of microvilli, feeds on particles suspended in the water drawn through its collar by the beat of its flagellum. The particles are moved to the cell body and ingested (surrounded by small pseudopodia).

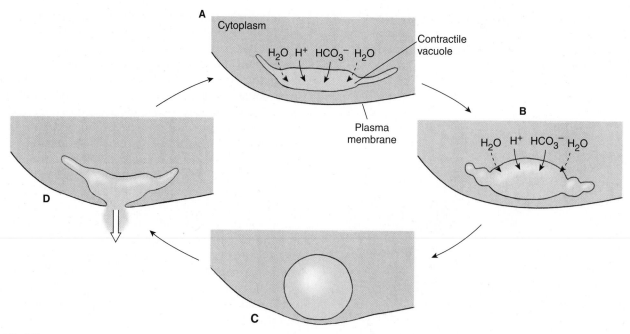

figure 5.10

Proposed mechanism for operation of contractile vacuoles composed of a system of cisternae and tubules. **A, B,** Proton pumps in their membranes transport H^+ and cotransport HCO_3^- into the vacuoles. Water passively diffuses inward to maintain an osmotic pressure equal to that in the cytoplasm. **C,** Once the vacuole has filled, **D,** its membrane fuses with the cell's plasma membrane, expelling water, H^+, and HCO_3^-. Protons and bicarbonate ions are replaced readily by action of carbonic anhydrase on carbon dioxide and water.

from it actively transport H^+ and cotransport bicarbonate (HCO_3^-) (figure 5.10), which are osmotically active particles. According to this hypothesis, as these particles accumulate within a vacuole, water is drawn into the vacuole. Fluid within the vacuole remains isosmotic to the cytoplasm. Then, as the vacuole finally joins its membrane to the surface membrane and empties its contents to the outside, it expels water, H^+, and HCO_3^-. These ions can be replaced readily by action of the enzyme carbonic anhydrase on CO_2 and H_2O. Carbonic anhydrase occurs in the cytoplasm.

Nitrogenous wastes from metabolism apparently diffuse through the plasma membrane, but some are emptied also by contractile vacuoles.

Reproduction

Asexual Reproduction: Fission

The cell multiplication process that produces genetically identical individuals in protozoa is called **fission.** The most common type is **binary fission,** which yields two essentially identical individuals (figure 5.11; see figure 5.17). When a progeny cell is considerably smaller than the parent and then grows to adult size, the process is called **budding.** This process occurs in some ciliates. In **multiple fission,** several nuclear divisions precede division of the cytoplasm (cytokinesis) so that numerous individuals are produced almost simultaneously (see figure 5.21B). Multiple fission, or **schizogony,** is common among the Apicomplexa and some amebas. When multiple fission leads to spore or sporozoite formation, it is called **sporogony.**

Sexual Processes

Although all protozoa reproduce asexually, and some are apparently exclusively asexual, the widespread occurrence of sexual reproduction among protozoa testifies to the value of genetic recombination. Sexual processes may precede certain phases of asexual reproduction, but embryonic development does not occur; protozoa do not have embryos. The essential features of sexual processes include a reduction division of the chromosome number to half (diploid number to haploid number), development of sex cells (gametes) or at least gamete nuclei, and usually fusion of the gamete nuclei (see p. 108).

The gamete nuclei, or pronuclei, which fuse in fertilization to restore the diploid number of chromosomes, are usually borne in special gamete cells. Gametes may be similar in appearance or unlike.

In animals, meiosis usually occurs during or just before gamete formation (called **gametic meiosis**), as seen in the Ciliophora and some flagellated and ameboid groups. However, in other flagellated groups and in the Apicomplexa, the first divisions *after* fertilization (zygote formation) are meiotic (*zygotic meiosis*) (see figure 5.21A), and all individuals produced asexually (mitotically) in the life cycle up to the next zygote are haploid. In some amebas (foraminiferans), haploid and diploid generations alternate (**intermediary meiosis**), a widespread phenomenon among plants.

Fertilization of an individual gamete by another is called **syngamy,** but some sexual phenomena in protozoa do not involve syngamy. Examples are **autogamy,** in which gametic

A *Arcella*

B *Euglypha*

C *Trypanosoma*   **D** *Euglena*

figure 5.11

Binary fission in some amebas and Euglenozoa. **A,** The two nuclei of *Arcella* divide as some of its cytoplasm is extruded and begins to secrete a new shell for the daughter cell. **B,** The shell of another ameba, *Euglypha*, is constructed of secreted platelets. Secretion of the platelets for the daughter cell begins before the cytoplasm enters it. As these form the shell of the daughter cell, the nucleus divides. **C,** *Trypanosoma* has a kinetoplast (part of the mitochondrion) near the kinetosome of its flagellum close to its posterior end in the stage shown. All of these parts must replicate before the cell divides. **D,** Division of *Euglena*.

nuclei arise by meiosis and fuse to form a zygote within the same organism that produced them, and **conjugation,** in which an exchange of gametic nuclei occurs between paired organisms (conjugants). Conjugation is described further in the discussion of *Paramecium*.

Encystment and Excystment

Separated as they are from their external environment only by their delicate external plasma membrane, protozoa are surprisingly successful in habitats frequently subjected to extreme conditions, such as heat, cold, or dryness. Survival under harsh conditions surely is related to an ability to form cysts, dormant forms that have resistant external coverings and undergo a more or less complete shutdown of metabolic machinery. Cyst formation is also important to many parasitic forms that must survive a harsh environment between hosts (see figure 5.3). However, some parasites do not form cysts, apparently depending on direct transfer from one host to another. Reproductive phases such as fission, budding, and syngamy may occur in cysts of some species. Encystment has not been found in *Paramecium*, and it is rare or absent in marine forms.

Cysts of some soil-inhabiting and freshwater protozoa have amazing durability. For instance, cysts of the soil ciliate *Colpoda* can survive 7 days in liquid air and 3 hours at 100°C. Survival of *Colpoda* cysts in dried soil has been shown for up to 38 years, and those of a certain small flagellate (*Podo*) can survive up to 49

years! Not all cysts are so sturdy, however. Those of *Entamoeba histolytica* tolerate gastric acidity but not desiccation, temperature above 50°C, or sunlight.

The conditions stimulating encystment are incompletely understood, although in some cases cyst formation is cyclic, occurring at a certain stage in the life cycle. In most free-living forms, adverse environmental change favors encystment. Such conditions may include food deficiency, desiccation, increased environmental osmotic pressure, decreased oxygen concentration, or change in pH or temperature.

During encystment, a number of organelles, such as cilia or flagella, are resorbed, and the Golgi apparatus secretes cyst wall material, which is carried to the surface in vesicles and extruded.

Although the exact stimulus for excystation (escape from cysts) is usually unknown, a return of favorable conditions initiates excystment in those protozoa in which the cysts are a resistant stage. In parasitic forms, an excystment stimulus may be more specific, requiring conditions similar to those found in the host.

■ Protozoan Taxa

An understanding of protozoan classification can be quite useful if the diagnostic features of each clade are easily recognized. For example, the presence of branching, netlike pseudopodia in a shelled ameba identifies the organism as belonging to Foraminifera (see figure 5.2). We provide identifying morphological features for those clades discussed, if such features are present.

Euglenozoa

Euglenozoa is generally recognized as a monophyletic group, based on the shared persistence of the nucleoli during mitosis, and the presence of discoid mitochondrial cristae. Cristae of this form occur also in a clade of amebas (Heterolobosea), so these groups are shown as sister taxa in figure 5.2. Members of Euglenozoa have a series of longitudinal microtubules just beneath the plasma membrane that help stiffen the membrane into a **pellicle** (figure 5.12). There are two subgroups: Euglenida and Kinetoplasta.

Euglenids have chloroplasts with chlorophyll *b*. These chloroplasts are surrounded by a double membrane and are likely to have arisen via secondary endosymbiosis. *Euglena* is a representative of this group. It has a light-sensitive **stigma**, or eyespot, which is a shallow pigment cup that allows light from only one direction to strike a light-sensitive receptor (figure 5.12).

Kinetoplastans are all parasitic in plants and animals. They have a unique organelle called a **kinetoplast**, which is part of their mitochondrion and is composed of a large disc of DNA. Various species of *Trypanosoma* (Gr. *trypanon*, auger, + *soma*, body) (figure 5.13) cause serious diseases in humans and other animals. Two subspecies of *Trypanosoma brucei* (*T. b. rhodesiense* and *T. b. gambiense*) cause clinically distinct forms of African sleeping sickness in humans. Another subspecies, *T. b. brucei*, along with several other trypanosomes, causes a similar disease in domestic animals, which makes agriculture very difficult in large areas of Africa. *Trypanosoma cruzi* causes American trypanosomiasis, or Chagas disease, a very serious disease of humans in South and Central America. The parasite is spread by biting bugs.

Heterolobosea

Heteroloboseans are naked amebas whose pseudopodia form abruptly, in what is sometimes called an "eruptive" manner. The life cycle of many heteroloboseans includes both amebic and flagellated stages, so group members are sometimes called amoeboflagellates or schizopyrenids. In *Naegleria gruberi,* the amebic stage feeds on bacteria, but once all local food is exhausted, the ameba completely alters its cytoskeleton, transforming into a flagellated cell within 90 minutes. The flagellated stage is better able to seek distant food sources than the amebic stage. Most heteroloboseans feed on bacteria and are harmless, but *Naegleria fowleri* causes a few deaths each year. It lives in hot pools and causes primary amebic meningoencephalitis in humans when water containing amebas is inhaled. The amebas enter through the nasal passages and migrate along olfactory nerves to the brain, where tissue is destroyed.

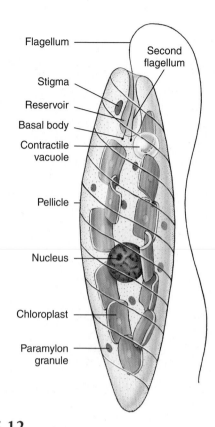

figure 5.12

Eugleng. Features shown are a combination of those visible in living and stained preparations.

Retortamonads and the Diplomonads

Retortamonads and diplomonads both lack mitochondria, causing biologists to wonder whether they branched from the main eukaryotic lineage before the mitochondrial symbiosis. However, recent work has revealed mitochondrial genes in the cell nucleus[1], so it is much more likely that the absence of mitochondria is a secondary loss rather than primary absence.

Retortamonads include commensal and parasitic unicells, such as *Chilomastix* and *Retortamonas. Giardia* (see figure 5.13), a representative diplomonad, is a common intestinal parasite of humans and other animals. It is often asymptomatic but sometimes causes a discomforting, but not fatal, diarrhea. Cysts are passed in the feces, and new hosts are infected by ingestion of cysts, often in contaminated water.

Parabasalids

The parabasalid clade contains some members of the former phylum Axostylata, characterized by an **axostyle**, a stiffening

[1] Roger, A. J. 1999. Reconstructing early events in eukaryotic evolution. Amer. Nat. **154** (supplement): S146–163.

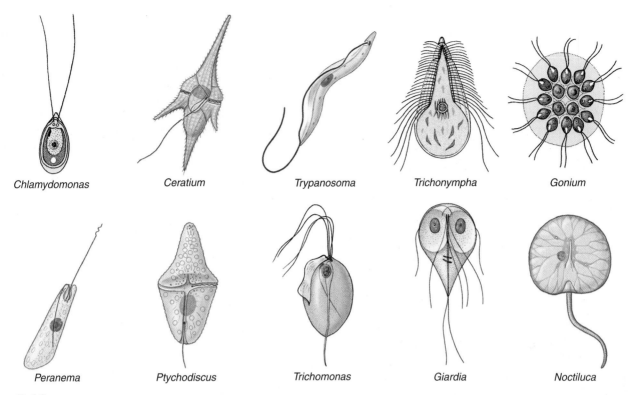

figure 5.13

Representatives of several protozoan taxa having flagella: Diplomonads (*Giardia*); Parabasalids (*Trichomonas, Trichonympha*); Plantae (*Chlamydomonas, Gonium*); Euglenozoa (*Peranema, Trypanosoma*); Dinoflagellata (*Ceratium, Ptychodiscus, Noctiluca*).

rod composed of microtubules, that extends along the longitudinal axis of their body. Parabasalids possess a modified region of the Golgi apparatus called a **parabasal body** (see p. 96) connected to a kinetosome. Mitochondria are absent, but members of order Trichomonadida possess hydrogenosomes (see p. 96).

Some trichomonads (for example, *Trichonympha* and *Trichomonas* in figure 5.13) are of medical or veterinary importance. *Trichomonas vaginalis* infects the urogenital tract of humans and is sexually transmitted. It produces no symptoms in males but is the most common cause of vaginitis in females.

Stramenopiles

Members of the clade Stramenopiles have tubular mitochondrial cristae. They may have flagellated cells, but stramenopiles are heterokont (Gr. *hetero*, different, + *kontos*, pole) flagellates. They have two different flagella, both inserted at the cell anterior, instead of the posterior as in opisthokonts (Gr. *opisth*, posterior). In heterokonts, the forward-directed flagellum is long and hairy, whereas the other is short and smooth, and trails behind the cell. This clade is sometimes called Heterokonta; the name stramenopile (L. *stramen*, straw, + *pile*, hair) refers to three-part tubular hairs covering the flagellum. This clade contains brown algae, yellow algae, and diatoms, all plantlike forms collecting energy with plastids, but

animal-like forms are also present. The opalinids, a group of animal parasites once thought to be modified ciliates, as well as plant pathogens such as slime nets and oomycetes, are among the organisms in this clade. Some amebas with axopodia, such as *Actinosphaerium* (figure 5.6) and *Actinophrys*, are recent additions.

Alveolata

Based on molecular and morphological evidence, the next three taxa (Ciliophora, Dinoflagellata, Apicomplexa) are often grouped together in Alveolata. Alveolates possess **alveoli** or related structures, which are membrane-bound sacs that lie beneath the plasma membrane and serve structural functions or produce pellicles (in ciliates) or thecal plates (in dinoflagellates).

Ciliophora

Ciliates are a large and interesting group, represented in all types of freshwater and marine habitats. They are the most structurally complex and diversely specialized of all protozoa. Most are free-living, although some are commensal or parasitic. They are usually solitary and motile, but some are colonial and others are sessile. There is a great diversity of shape and size. In general, ciliates are larger than most other protozoa, ranging from 10 μm to 3 mm long. At some stage, all have cilia that beat in a coordinated rhythmic manner, although the arrangement of the cilia varies. In suctorians, the young possess cilia and are free-swimming, but the adults grow an attachment stalk, become sessile, and lose their cilia.

The pellicle of some ciliates consists of only a plasma membrane, whereas other species form a thickened armor. Cilia are short and usually arranged in longitudinal or diagonal rows. Cilia may cover the surface of the animal or be restricted to the oral region or to certain bands. In some forms, cilia are fused into a sheet called an **undulating membrane** or into smaller **membranelles;** both structures propel food into the cytopharynx (gullet). Other forms may have fused cilia forming stiffened tufts called **cirri,** which are often used in locomotion by the creeping ciliates, such as *Euplotes* in figure 5.14.

The kinetosomes and a structural system of fibrils just beneath the pellicle form the **infraciliature** in ciliates (figure 5.15). The infraciliature apparently does not coordinate ciliary beat, as formerly hypothesized. Ciliary movement seems to be coordinated by waves of depolarization of the plasma membrane moving down the animal, similar to conduction of a nerve impulse.

Some ciliates have **trichocysts** or **toxicysts** in their ectoplasm between the bases of the cilia (figures 5.15 and 5.16). Upon mechanical or chemical stimulation, these small bodies explosively expel a long, threadlike structure. The mechanism of expulsion is unknown. The function of trichocysts is probably defensive. When a paramecium is attacked by a predatory *Didinium,* it expels its trichocysts but to no avail. Toxicysts, however, release a poison that paralyzes the prey of carnivorous ciliates. Toxicysts are structurally quite distinct from trichocysts. Many dinoflagellates have structures very similar to trichocysts.

figure 5.14

Some representative ciliates. *Euplotes* have stiff cirri used for walking on the surface of an object or an air bubble. Contractile *myonemes* in the ectoplasm of *Stentor* and in the stalks of *Vorticella* allow great expansion and contraction. Note the macronuclei—long and curved in *Euplotes* and *Vorticella,* and shaped like a string of beads in *Stentor.*

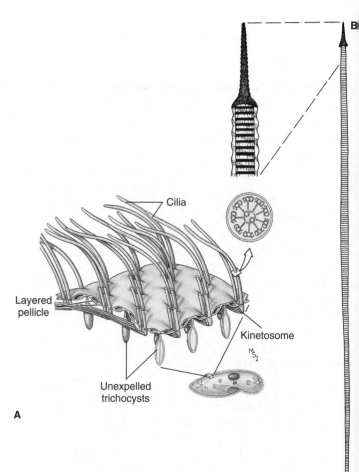

figure 5.15

Infraciliature and associated structures in ciliates. **A,** Structure of the pellicle and its relation to the infraciliature system. **B,** Expelled trichocyst.

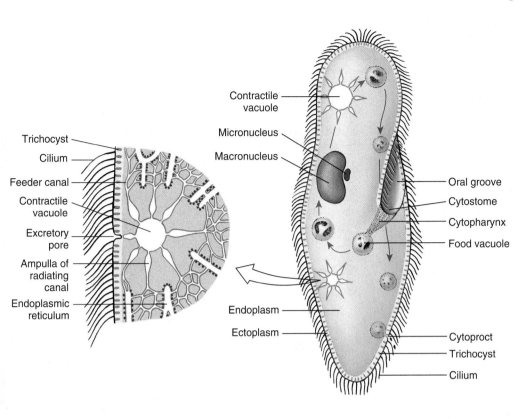

Trichocyst
Cilium
Feeder canal
Contractile vacuole
Excretory pore
Ampulla of radiating canal
Endoplasmic reticulum

Contractile vacuole
Micronucleus
Macronucleus
Oral groove
Cytostome
Cytopharynx
Food vacuole
Endoplasm
Ectoplasm
Cytoproct
Trichocyst
Cilium

figure 5.16

Left, Enlarged section of a contractile vacuole of *Paramecium.* Water is collected by the endoplasmic reticulum, emptied into feeder canals, and then emptied into the vesicle. The vesicle contracts to empty its contents to the outside, thus serving as an osmoregulatory organelle. *Right, Paramecium,* showing gullet (cytopharynx), food vacuoles, macronucleus and micronucleus.

Most ciliates are holozoic, possessing a cytostome (mouth) that in some forms is a simple opening and in others is connected to a gullet or ciliated groove. The mouth in some is strengthened by stiff, rodlike structures for swallowing larger prey; in others, such as paramecia, ciliary water currents carry microscopic food particles toward the mouth. *Didinium* has a proboscis for engulfing the paramecia on which it feeds (see figure 5.9). Suctorians paralyze their prey and then ingest their contents through tubelike tentacles by a complex feeding mechanism that apparently combines phagocytosis with a sliding filament action of microtubules in the tentacles. An example is *Podophrya* in figure 5.9.

Contractile vacuoles occur in all freshwater and some marine ciliates; the number varies from one to many among the different species. In most ciliates, the vacuole is fed by one or more collecting canals (figure 5.16). The vacuoles occupy a fixed position, and each discharges through a more-or-less permanent pore.

Ciliates are always multinucleate, possessing at least one **macronucleus** and one **micronucleus**, but varying from one to many of either type. The macronuclei are apparently responsible for metabolic, synthetic, and developmental functions. Macronuclei are varied in shape among the different species (see figures 5.14 and 5.16). Micronuclei participate in sexual reproduction and give rise to macronuclei after exchange of micronuclear material between individuals. Micronuclei divide mitotically, and macronuclei divide amitotically.

Reproduction and Life Cycles

Life cycles of ciliates usually involve both asexual binary fission (figure 5.17) and a type of sexual reproduction called **conjugation** (figure 5.18). Conjugation is a temporary union

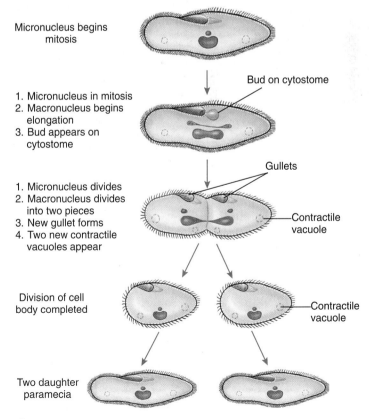

Micronucleus begins mitosis

1. Micronucleus in mitosis
2. Macronucleus begins elongation
3. Bud appears on cytostome

Bud on cytostome

1. Micronucleus divides
2. Macronucleus divides into two pieces
3. New gullet forms
4. Two new contractile vacuoles appear

Gullets
Contractile vacuole

Division of cell body completed

Contractile vacuole

Two daughter paramecia

figure 5.17

Binary fission in a ciliophoran (*Paramecium*). Division is transverse, across rows of cilia.

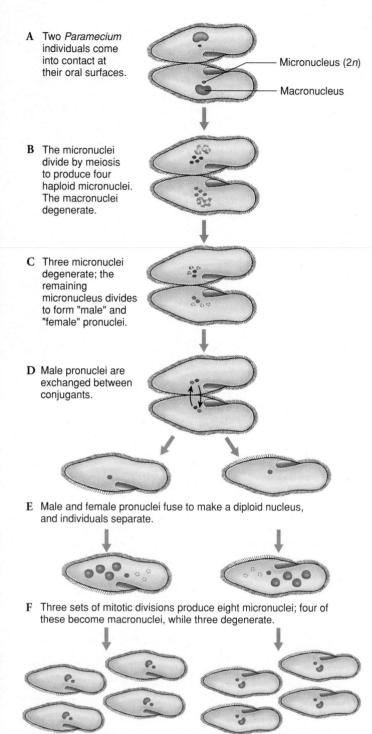

A Two *Paramecium* individuals come into contact at their oral surfaces.

Micronucleus (2*n*)

Macronucleus

B The micronuclei divide by meiosis to produce four haploid micronuclei. The macronuclei degenerate.

C Three micronuclei degenerate; the remaining micronucleus divides to form "male" and "female" pronuclei.

D Male pronuclei are exchanged between conjugants.

E Male and female pronuclei fuse to make a diploid nucleus, and individuals separate.

F Three sets of mitotic divisions produce eight micronuclei; four of these become macronuclei, while three degenerate.

G The remaining micronucleus divides twice, as does the cell, producing four daughter cells.

figure 5.18

Conjugation in *Paramecium*.

of two individuals for the purpose of exchanging chromosomal material. During union, the macronucleus of each cell degenerates, and the micronucleus of each individual undergoes meiosis, giving rise to four haploid micronuclei, three of which degenerate (figure 5.18A and B). The remaining micronucleus then divides into two haploid pronuclei (figure 5.18C), one of which is exchanged for a pronucleus of the conjugant partner. When the exchanged pronucleus unites with the pronucleus of the partner, the diploid number of chromosomes is restored. The two partners, each with fused pronuclei (now comparable to a zygote), separate, and each divides twice by mitosis, thereby giving rise to four daughter paramecia each (figure 5.18D–G).

Conjugation always involves two individuals of different mating types. This prevents inbreeding. Most ciliate species are divided into several varieties, each variety containing two *mating types*. Mating occurs only between the differing mating types within each variety.

Dinoflagellata

Many dinoflagellates are photoautotrophic, with chromatophores bearing chlorophylls *a* and *c* but not *b*. Ecologically, some species are among the most important primary producers in marine environments. They commonly have two flagella, one equatorial and one longitudinal, each borne at least partially in grooves on the body (for example, see *Ceratium*, *Ptychodiscus*, and *Noctiluca* in figure 5.13). The body may be naked or covered by cellulose plates or valves. Most dinoflagellates have brown or yellow chromatophores, although some are colorless. Many species can ingest prey through a mouth region between the plates near the posterior area of the body. *Noctiluca* (see figure 5.13), a colorless dinoflagellate, is a voracious predator and has a long, motile tentacle, near the base of which its single, short flagellum emerges. *Noctiluca* is one of many marine organisms that can produce light (bioluminescence).

Zooxanthellae are dinoflagellates that live in mutualistic association in tissues of certain invertebrates, including other protozoa, sea anemones, horny and stony corals, and clams. The association with stony corals is of ecological and economic importance because only corals with symbiotic zooxanthellae can form coral reefs (see Chapter 7).

Dinoflagellates can damage other organisms, as when they produce a "red tide" (figure 5.19). Although this name originally applied to situations in which the organisms reproduced in such profusion (producing a "bloom") that the water turned red from their color, any instance of a bloom producing detectable levels of toxic substances is now called a red tide. The water may be red, brown, yellow, or not remarkably colored at all, and the phenomenon has nothing to do with tides! The toxic substances are apparently not harmful to the organisms that produce them, but they may be highly poisonous to fishes, other marine life, and humans. Red tides have caused considerable economic losses to the shellfish industry.

Pfiesteria piscicida is one of several related dinoflagellate species that may affect fish in brackish waters along the Atlantic Coast, south of North Carolina. Much of the time, *Pfiesteria* feeds on algae and bacteria, but something in the excreta of large schools of fishes causes it to release a powerful,

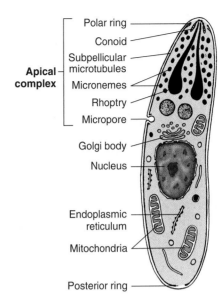

figure 5.20

Diagram of an apicomplexan sporozoite or merozoite at the electron microscopic level, illustrating the apical complex. The polar ring, conoid, micronemes, rhoptries, subpellicular microtubules, and micropore (cytostome) are all considered components of the apical complex.

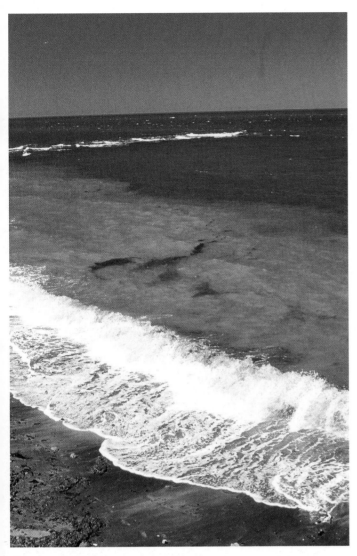

figure 5.19

"Red tide" makes a striking appearance on the coastline. The water appears colored because it contains a high concentration of dinoflagellates.

short-lived toxin. The toxin may stun or kill fishes, often creating skin lesions. *Pfiesteria* has flagellated and ameboid forms among its more than 20 body types; some forms feed on fish tissues and blood. Although it does not have chloroplasts, it may sequester chloroplasts from its algal prey and gain energy from them over the short term. This fascinating group of species was discovered in 1988.

Apicomplexa

All apicomplexans are endoparasites, and they find hosts in many animal phyla. The life cycle usually includes both asexual and sexual reproduction, and there is sometimes an invertebrate intermediate host. Asexual reproduction occurs by multiple fission, called schizogony or sporogony depending on the product (see figure 5.21). At some point in the life cycle, most apicomplexans develop a *spore (oocyst)*, which is infective for the next host and is often protected by a resistant coat. The presence of a certain combination of organelles, the **apical complex,** distinguishes this phylum (figure 5.20). The apical complex usually occurs only in certain life-cycle stages of the organisms—for example, **merozoites** and **sporozoites** (figure 5.21). Some of an apicomplexan's structures, especially the **rhoptries** and **micronemes,** apparently aid in penetrating the host's cells or tissues.

Locomotor organelles are not as obvious in this group as they are in other protozoa. Pseudopodia occur in some intracellular stages, and gametes of some species are flagellated. Tiny contractile fibrils can form waves of contraction across the body surfaces to propel the organism through a liquid medium.

Apicomplexan parasites belong to two groups: **coccidians** (Coccidia [Gr. *kokkos,* kernel or berry]), which infect both vertebrates and invertebrates, and **gregarines,** which live mainly in the digestive tract and body cavity of certain invertebrates. We discuss two examples of medical importance from class Coccidea: *Plasmodium* and *Toxoplasma.*

Plasmodium is the sporozoan parasite that causes **malaria.** The vectors (carriers) of the parasites are female *Anopheles* mosquitos, and the life cycle is depicted in figure 5.21. Malaria is one of the most important diseases in the world: 41% of the earth's people live in malarial regions. Global estimates of deaths caused by malaria range from 700,000 to over 2 million per year, with 75% being African children. Although these statistics represent a decline from the situation 45 years ago, some resurgence is caused by increased resistance of mosquitos to insecticides, rising

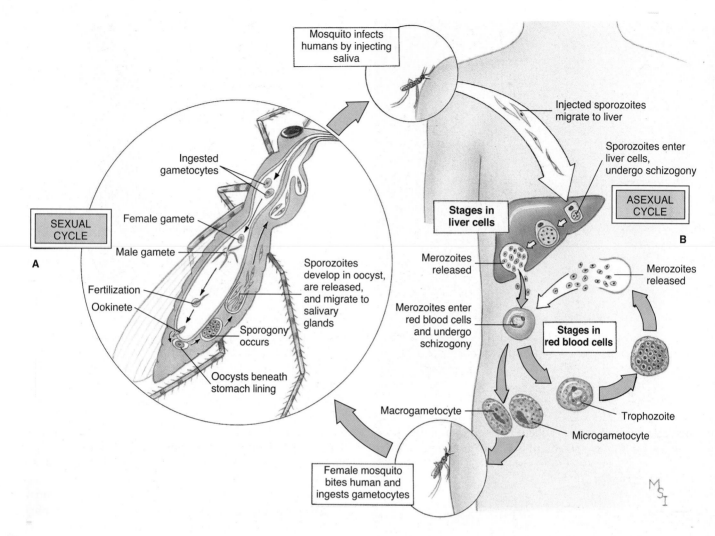

figure 5.21

Life cycle of *Plasmodium vivax*, one of the protozoa (class Coccidea) that causes malaria in humans. **A,** A sexual cycle produces sporozoites in the body of the mosquito. **B,** Sporozoites infect humans and reproduce asexually, first in liver cells, and then in red blood cells. Malaria is spread by the *Anopheles* mosquito, which ingests gametocytes along with human blood. Then, when biting another victim, the mosquito leaves sporozoites in the new wound.

numbers of *Plasmodium* strains that are resistant to drugs, and socioeconomic conditions and civil strife that interfere with control of malaria in tropical countries.

Although malaria has been recognized as a disease and a scourge of humanity since antiquity, its protozoan cause was discovered only 100 years ago by Charles Louis Alphonse Laveran, a French army physician. At that time, the mode of transmission was still mysterious, and "bad air" (hence the name malaria) was a popular suspect. Ronald Ross, an English physician in the Indian Medical Service, determined some years later that the malarial organism was carried by *Anopheles* mosquitos. Because Ross knew nothing about mosquitos or their normal parasites, his efforts were long frustrated by studying the wrong mosquitos and parasites. That he persisted is a tribute to his determination. The momentous discovery earned Ross the Nobel Prize in 1902 and knighthood in 1911.

Toxoplasma (Gr. *toxo*, a bow,+ *plasma*, molded) is a common parasite in the intestinal tissues of cats. But this parasite can produce extraintestinal stages in a wide variety of other animals as well—for example, rodents, cattle, and humans. The extraintestinal forms do not produce gametes and oocysts, but they can initiate the intestinal cycle in a cat if the cat eats infected prey. In humans, *Toxoplasma* causes few or no ill effects except in AIDS patients or in women infected during pregnancy. Prenatal infection greatly increases the chances of a birth defect in the baby; about 2% of all mental retardation in the United States is a result of congenital toxoplasmosis. Humans can become infected by eating insufficiently cooked beef, pork, or lamb or by accidentally ingesting oocysts from the feces of cats. For these reasons, pregnant women should not eat raw meat or empty cats' litterboxes.

Some 16% or more of adults in the United States are infected with *Toxoplasma gondii* but have no symptoms because the parasite is held in check by the immune system. However, *T. gondii* is one of the most important opportunistic infections in AIDS patients. The latent infection is activated in 5% to 15% of AIDS patients, often in the brain, with serious consequences. Another coccidian, *Cryptosporidium parvum*, was first reported in humans in 1976. We now recognize it as a major cause of diarrheal disease worldwide, especially in children in tropical countries. Waterborne outbreaks have occurred in the United States, and the diarrhea can be life-threatening in immunocompromised patients (such as those with AIDS). The latest coccidian pathogen to emerge has been *Cyclospora cayetanensis*. U.S. infection rates for 2005 were about 0.2 cases per 100,000 persons, with diarrhea being the most common symptom of infection. Infection usually occurs by ingestion of contaminated food or water.

Cercozoa

Cercozoans do not share a common body plan, but form a clade in molecular phylogenetic studies. There are flagellated and ameboid members. The ameboid members of the group may be naked or testate. The testate amebas include *Euglypha*, which makes a test from collected particles (see figure 5.11); *Clathrulina*, which has a siliceous capsule (see figure 5.6); and some organisms with an amorphous silica skeleton with magnesium, calcium, and copper added. The skeleton of the latter group contains hollow spines and is known from microfossils dating back to the Cambrian.

Foraminifera

The foraminiferans (L. *foramen*, hole, + *fero*, bear) are mostly marine forms that secrete complex, many-chambered tests of calcium carbonate (figure 5.22). They usually add sand grains to the secreted material, using great selectivity in choosing colors. Slender pseudopodia extend through openings in the test and then run together to form a protoplasmic net (reticulum) in which they ensnare their prey. They are beautiful little creatures with many pseudopodia radiating out from a central test.

The foraminiferan test is a very useful structure because it is distinctive enough to identify particular species and because it preserves very well in ocean sediments, lasting long after the ameba that made it has died. Add to this the fact that different planktonic foraminiferan species thrive only at certain temperatures, and one has the basis to reconstruct paleo-climates by reading history in deep-sea sediments. Such sediments may be 600 to 3600 m deep (approximately 2000 to 12,000 feet) and contain as many as 50,000 foraminiferans per gram. Changing percentages of different species from oldest to youngest sediments tell the story of temperature change in oceanic surface waters.

Radiolaria

Radiolarians (L. *radiolus*, small ray) are marine forms, mostly living in plankton, that have intricate and beautiful siliceous skeletons (figure 5.23). A central capsule that separates the

A **B**

figure 5.22

A, Air-dried foraminiferan, showing spines extending from the test. **B,** A test of the foraminiferan *Vertebralina striata*. Foraminiferans are ameboid marine protozoa that secrete a calcareous, many-chambered test in which to live, and then extrude protoplasm through pores to form a layer over the outside. The animal begins with one chamber, and as it grows, it secretes a succession of new and larger chambers, continuing this process throughout life. Many foraminiferans are planktonic, and when they die, their shells are added to sediments on the sea floor.

figure 5.23

Types of radiolarian tests. In his study of these beautiful forms collected on the famous *Challenger* expedition of 1872 to 1876, Ernst Haeckel worked out our present concepts of symmetry.

inner and outer cytoplasm is perforated to allow cytoplasmic continuity. Around the capsule is a frothy mass of cytoplasm from which axopodia arise. Sticky axopodia capture prey items, which are carried by the streaming protoplasm through the capsule. The ectoplasm on one side of the axial rod moves outward toward the axial tip, while that on the other side moves inward toward the test. Solitary radiolarians feed on bacteria, micro-algae, and micro-flagellates, whereas colonial radiolarians collect nutrients from symbiotic algae. With the exception of one subgroup, which is now placed in Cercozoa (p. 111), organisms described as radiolarians before the advent of molecular phylogenetics are still considered radiolarians.

Radiolarians are among the oldest known protozoa, and they and foraminiferans have left excellent fossil records. For millions of years, tests of dead protozoa have been dropping to the sea floor. Many limestone and chalk deposits on land were formed by foraminiferans when the land was covered by sea. In contrast, the silica skeletons of radiolarians decay to an ooze that lithifies to form chert.

Centrohelida

Prior to the widespread use of molecular methods in phylogenetics, the Heliozoa was a large group of testate amebas forming axopodia. All species formerly placed in Heliozoa, except members of the order Centrohelida, have been moved to other

taxa, hence the name change. Centrohelids are amebas with flattened mitochondrial cristae. The structure of their axopodia is distinctive: microtubules (visible in figure 5.7) within the axoneme are arranged in hexagons or triangles in this group. In most centrohelids, the axopodia extend through a coat of variously shaped silica scales, but this coat may be absent and is replaced in some taxa by a mucous layer. Most centrohelids live in freshwater, but a few clades have colonized brackish or marine habitats. These lovely unicells are predators. Some former heliozoans are now in Stramenopiles (e.g., *Actinophrys*) or Cercozoa (e.g., *Clathrulina*).

Plantae

The clade Plantae contains glaucophytes, red algae, green algae, bryophytes, and vascular plants, among others. Within Plantae, members of the clade Viridiplantae have chlorophylls *a* and *b*. As a subset of Plantae, members of the non-monophyletic phylum Chlorophyta included unicellular, colonial, and multicellular green algae. The volvocine green algae include unicellular *Chlamydomonas* (see figure 5.13), colonial *Gonium* (see figure 5.13), and multicellular *Volvox carteri* (figure 5.24). As discussed in Chapter 3 (page 60), both colonial and multicellur organisms are embedded in an extracellular matrix (ECM). However, in colonial forms, all cells can reproduce sexually, whereas this capacity is limited to cells specialized for that

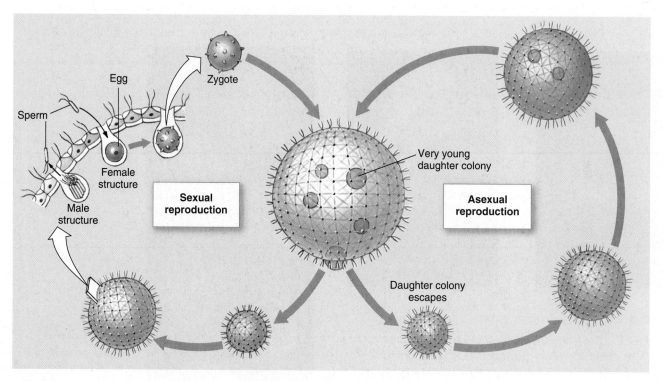

figure 5.24

Life cycle of *Volvox carteri*. Asexual reproduction occurs in spring and summer when specialized diploid reproductive cells divide to form young colonies that remain in the mother colony until they grow large enough to escape. Sexual reproduction occurs largely in autumn when haploid sex cells develop. The zygote may encyst and so survive the winter, developing into a mature asexual colony in the spring. In some species, the colonies have separate sexes; in others, both eggs and sperm are produced in the same colony.

purpose in multicellular organisms. In *Volvox carteri*, cells remain connected to each other after cell division so that a hollow ball of cells is formed, but some cells are somatic and others are reproductive. Reproductive cells develop into eggs or into bundles of sperm. A fertilized egg or zygote undergoes repeated divisions to produce a small colony, which is released from the parental colony in spring. A number of asexual generations may follow before sexual reproduction occurs again.

The persistence of unicellular, colonial, and multicellular forms within one lineage permits biologists to make inferences about steps in the evolution of multicellularity. In the volvocine algae it appears that the ECM evolved through modifications of the inner and outer plant cell walls. A diverse array of cell wall proteins is important for ECM function, but these proteins are very different from the ECM proteins of animals or other multicellur forms.[2]

Amoebozoa

Amoebozoans include naked and testate amebas, as well as amebas with flagellated stages in the life cycle. Amoebozoans typically have branched tubular mitochondrial cristae, but this feature is not unique to the group. Ameboid forms include the fascinating plasmodial and cellular slime molds (e.g., *Dictyostelium*), testate amebas with lobose pseudopodia such as *Arcella* (figures 5.6 and 5.11), and naked lobose amebas such as *Chaos carolinense, Amoeba proteus*, or members of genus *Acanthamoeba* (figure 5.3). *Chaos*

---

[2] Abedin, M., and N. King. 2010. Diverse evolutionary paths to cell adhesion. Trends in Cell Biology **20**: 734–742.

carolinense or *A. proteus* are sometimes used in biology laboratories, but *Acanthamoeba castellani* (figure 5.3) has gained notoriety because of its impact on human health. *Acanthamoeba castellani* kills cells of the human cornea and is spread by contact lenses that are not properly disinfected. Amoebozoa also houses entozoic amebas—those that live inside humans or other animals. *Entamoeba histolytica* causes amebic dysentery in infected people.

Opisthokonta

Opisthokonta is a clade characterized by a combination of flattened mitochondrial cristae and one posterior flagellum on flagellated cells, if such cells are present. Opisthokonta contains animals and fungi as well as microsporidians, choanoflagellates, and some little-known unicells (figure 5.25). The little-known forms include ichthyosporeans (animal parasites sometimes called DRIPs; one type (*Ichthyophonus*) causes skin lesions in fishes), nucleariid amebas, corallochytreans, and ministeriid amebas. Microsporidians are intracellular parasites now recognized as specialized fungi; some cause disease in humans. Choanoflagellates are solitary or colonial protozoans currently placed as the sister group to animals (metazoans).

Choanoflagellates possess genes involved in cell-cell signaling and in cell adhesion, such as tyrosine kinase and cadherin genes, that are homologous to those of metazoans. Choanoflagellate cells are distinctive; they are ovoid in shape and at one end a collar of microvilli surrounds a flagellum. Beating of the flagellum moves water through the collar where food particles carried by the water are collected. Choanoflagellate cells are very similar to the choanocytes of sponges (see Chapter 6). In

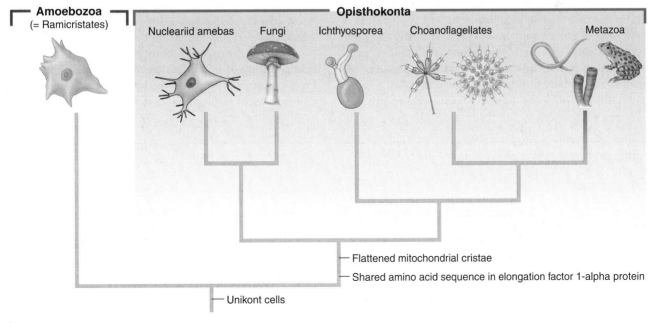

figure 5.25

One hypothesis regarding relationships among some members of Opisthokonta: Choanoflagellates are shown as the sister taxon to Metazoa. Choanoflagellates shown are *Codonosiga* on the left and *Proterospongia* on the right.

fact, such cells may be an ancestral characteristic of metazoans because in addition to sponges, they occur in certain corals and in some echinoderms. If they are ancestral, they have been lost in the evolutionary ancestry of most animal phyla.

Phylogeny and Adaptive Diversification

Phylogeny

The phylogeny of eukaryotes is important to evolutionary biologists and to those with more practical goals. One reason to find phylogenetic homes for disease-causing organisms is that treatments effective against particular pathogens often work on related organisms. For example, there were no effective treatments against microsporidians until it was discovered that microsporidians were highly specialized fungi and thus antifungal drugs could kill them.

There has been progress in determining relationships among protozoan taxa. The clade Alveolata is well-supported because its members share a morphological feature, the presence of alveoli. Most research places Alveolata as the sister taxon to the Stramenopiles (see figure 5.2). The supergroup Rhizaria unites Cercozoa, Foraminifera, "Radiolaria" and some other taxa into a clade whose many members are testate amebas, some with rhizopodia. Rhizaria also includes a clade of testate amebas in genus *Gromia. Gromia sphaerica* made headlines when it was discovered rolling across the seafloor off the Bahamas at a depth of 2000 feet. This ameba is several centimeters in diameter, so the size is striking, but the really exciting part is that its track on the sea floor is very like fossil tracks attributed to bilaterally symmetrical organisms. Paleontologists are now reevaluating some trace fossil evidence, wondering about the role of giant amebas.

The union of Stramenopiles, Alveolata, and Rhizaria into a supergroup called SAR has been proposed. Another proposed supergroup is the Unikonts, which unites Amoebozoa and Opisthokonta on the grounds that when flagellated cells are present, there is only one flagellum. Some recent phylogenies based on molecular data place the root of the eukaryote tree between the Unikonts and all other unicellular taxa, which are grouped as "Bikonts" because they have two flagella when flagella are present. It remains to be seen whether this dichotomy will be supported by new research.

New evidence has strengthened the hypothesis that eukaryotes originated through symbiosis and that origins of mitochondria and plastids represent separate endosymbiogenic events. The ancestral form of a mitochondrion was an alphaproteobacterium. The common ancestor of green plants and green algae acquired its chlorophyll-bearing plastids by symbiogenesis with a cyanobacterium. These plastids have two membranes, one apparently derived from the ancestral cyanobacterium, and the other from the host vacuole containing it. Some investigators propose that in the Euglenozoa, plastids were added by secondary symbiogenesis. Euglenozoa apparently acquired their plastids after divergence from ancestors of Kinetoplasta. Among alveolates, many dinoflagellates are photoautotrophs, maintaining their chloroplasts, but some have lost them. Apicomplexans have a small, circular, plastidlike DNA, localized to a discrete organelle surrounded by four membranes, and evidently a relict plastid inherited from their common ancestor with dinoflagellates. Ancestors of ciliophorans either lost their plastid symbionts or diverged from their common ancestor with dinoflagellates before the secondary symbiogenesis occurred.

Adaptive Diversification

In the previous pages, we have described some of the wide range of adaptations of protozoa. Amebas range from bottom-dwelling, naked species to planktonic forms with beautiful, intricate tests such as the foraminiferans and radiolarians. There are many symbiotic species of amebas. Flagellated forms likewise show adaptations for a similarly wide range of habitats, with the added variation of photosynthetic ability in many groups.

Within a single-cell body plan, the division of labor and specialization of organelles are carried furthest by ciliates, the most complex of all protozoa. Apicomplexa have adopted specializations for intracellular parasitism.

Taxonomy of Protozoan Phyla

Phylum Retortamonada (re-tor´ta-mo´ nad-a) (L. *retorqueo,* to twist back, + *monas,* single, unit). Mitochondria and Golgi bodies lacking; three anterior and one recurrent (running toward posterior) flagellum lying in a groove; intestinal parasites or free-living in anoxic environments. This phylum is now divided into two clades with two genera in clade Retortamonads.

Class Diplomonadea (di´plo-mon-a´de-a). (Gr. *diploos,* double, + L. *monas,* unit). One or two karyomastigonts (group of kinetosomes with a nucleus); individual mastigonts with one to four flagella; mitotic spindle within nucleus; cysts present; free-living or parasitic. Nine genera constitute the clade Diplomonads.

Order Diplomonadida (di´plo-mon-a´di-da). Two karyomastigonts, each with four flagella, one recurrent; have a variety of microtubular bands. Example: *Giardia.*

Phylum Parabasala (par´a-ba-sa´la) (Gr. *para,* beside, + *basis,* base; refers to parabasal body = Golgi). Very large Golgi bodies associated with karyomastigont; up to thousands of flagella. *Trichomonas* and two other forms constitute clade Parabasalids.

Order Trichomonadida (tri´ko-mon-a´di-da) (Gr. *trichos,* hair, + *monas,* unit). Typically at least some kinetosomes associated with rootlet filaments characteristic of trichomonads;

parabasal body present; division spindle extranuclear; hydrogenosomes present; no sexual reproduction; true cysts rare; all parasitic. Examples: *Dientamoeba, Trichomonas.*

Phylum Euglenozoa (ū-gle-nə-zōˊə) (Gr. *eu-*, good, true, + *glēnē*, cavity, socket, + *zōon*, animal). Cortical microtubules; flagella often with paraxial rod (rodlike structure accompanying axoneme in flagellum); mitochondria with discoid cristae; nucleoli persist during mitosis. This phylum is equivalent to the clade Euglenozoa.

Subphylum Euglenida (u-glenˊi-da). Have pellicular microtubules that stiffen pellicle.

Class Euglenoidea (u-glen-oydˊe-a) (Gr. *eu-*, good, true, + *glēnē*, cavity, socket, + *-ōideos,* form of, type of). Two heterokont flagella (flagella with different structures) arising from apical reservoir; some species with light-sensitive stigma and chloroplasts. Example: *Euglena.*

Subphylum Kinetoplasta (kī-nētˊə-plast-ə) (Gr. *kinētos,* to move, + *plastos,* molded, formed). Have a unique mitochondrion containing a large disc of DNA; paraxial rod.

Class Trypanosomatidea (try-panˊo-som-a-tidˊe-a) (Gr. *trypanon,* a borer, + *sōma,* the body). One or two flagella arising from pocket; flagella typically with paraxial rod that parallels axoneme; single mitochondrion (nonfunctional in some forms) extending length of body as tube, hoop, or network of branching tubes, usually with single conspicuous DNA-containing kinetoplast located near flagellar kinetosomes; Golgi body typically in region of flagellar pocket, not connected to kinetosomes and flagella; all parasitic. Examples: *Leishmania, Trypanosoma.*

Phylum Heterolobosea (hetˊə-rō-lō bōˊ sē-ə) (Gr. *heteros,* different, + *lobos,* lobe). Naked amebas with eruptive pseudopodia; a typical life cycle includes both amebic and flagellated stages. Group members sometimes called amoeboflagellates or schizopyrenids. Examples: *Naegleria fowleri, Naegleria gruberi.*

Phylum Stramenopiles (stra-menˊō-pīles) (L. *stramen,* straw, + *pile,* hair). Flagellates with two different flagella, one long and one short; forward flagellum is covered with tubular three-part hairs; mitochondria with tubular cristae. Free-living and parasitic plant-like and animal-like forms. Examples: *Phytophthora infestans, Actinosphaerium, Actinophrys.*

Phylum Apicomplexa (apˊi-com-pleksˊa) (L. *apex,* tip or summit, + *complex,* twisted around). Characteristic set of organelles (apical complex) associated with anterior end present in some developmental stages; cilia and flagella absent except for flagellated microgametes in some groups; cysts often present; all parasitic. This phylum is within clade Alveolata.

Class Gregarinea (gre-ga-rinˊe-a) (L. *gregarius,* belonging to a herd or flock). Mature gamonts (individuals that produce gametes) large, extracellular; gametes usually alike in shape and size; zygotes forming oocysts within gametocysts; parasites of digestive tract of body cavity of invertebrates, life cycle usually with one host. Examples: *Monocystis, Gregarina.*

Class Coccidea (kok-sidˊe-a) (Gr. *kokkos,* kernel, grain). Mature gamonts small, typically intracellular; life cycle typically with merogony, gametogony, and sporogony; most species infect vertebrates. Examples: *Cryptosporidium, Cyclospora, Eimeria, Toxoplasma, Plasmodium, Babesia.*

Phylum Ciliophora (sil-ē-ofˊor-a) (L. *cilium,* eyelash, + Gr. *phora,* bearing). Cilia or ciliary organelles in at least one stage of life cycle; two types of nuclei, with rare exceptions; binary fission across rows of cilia; budding and multiple fission; sexuality involving conjugation and autogamy; nutrition heterotrophic; contractile vacuole typically present; most species free-living, but many commensal, some parasitic. (This is a very large group, now divided into three superclasses and numerous classes and orders. Classes are separated on technical characteristics of ciliary patterns, especially around the cytostome, development of the cytostome, and other characteristics.) Examples: *Paramecium, Colpoda, Tetrahymena, Balantidium, Stentor, Blepharisma, Epidinium, Euplotes, Vorticella, Carchesium, Trichodina, Podophrya, Ephelota.* This phylum is within clade Alveolata.

Phylum Dinoflagellata (dī-nō-flaˊjə-lata) (Gr. *dinos,* whirling, + *flagellum,* little whip). Typically have two flagella, one transverse and one trailing; body usually grooved transversely and longitudinally, each groove containing a flagellum; chromoplasts usually yellow or dark brown, occasionally green or blue-green, bearing chlorophylls *a* and *c;* nucleus unique among eukaryotes in having chromosomes that lack or have low levels of histones; mitosis intranuclear; body form sometimes of spherical unicells, colonies, or simple filaments; sexual reproduction present; members free-living, planktonic, parasitic, or mutualistic. Examples: *Zooxanthella, Ceratium, Noctiluca, Ptychodiscus.* This phylum is within clade Alveolata.

Phylum Cercozoa (ser-kō-zoˊa) (Gr. *kerkos,* tail, + *zōon,* animal). A diverse group of unicells, heterogeneous in lifestyle and morphology; monophyly suppored by molecular data. Most are free-living, some parasitic. Examples: *Euglypha, Clathrulina.*

Phylum Foraminifera (forˊa-min-ifˊer-a) (L. *foramin,* hole, + *fero,* to bear). Shelled amebas bearing slender pseudopodia that extend through many openings in the test, forming a net that ensnares prey. Examples: *Vertebralina, Globigerina.*

Phylum Radiolaria (raˊdē-a-laˊrē-a) (L. *radiolus,* small sunbeam). Most are amebas with a well-developed internal skeleton of strontium sulfate or silica, forming beautiful tests. Axopodia are present. Examples: *Tetrapyle, Pterocorys.*

Phylum Centrohelida (cen-trō-helˊi-da) (Gr. *kentron,* center of a circle, + helios, the sun). Amebas with flattened mitochondrial cristae; axoneme of axopodia with microtubules arranged in hexagons or triangles; in most, axopodia extend through a coat of variously shaped silica scales. Mostly freshwater, some marine. Examples: *Acanthocystis, Pterocystis, Heterophrys.*

Phylum Plantae (planˊ tē) (L. *planto,* to set, plant). A photosynthetic group whose chloroplast is derived from the primary endosymbiotic event with a cyanobacterium. Examples: *Chlamydomonas, Volvox, Zea mays.*

Phylum Amoebozoa (uh-mēˊ bō-zōˊ a) (Gr. *amoibē,* to change, + *zōon,* animal). Naked and shelled amebas, many with flagellated stages in life cycle; mitochondria, when present, have tubular and branched cristae. Free-living and parasitic. Examples: *Entamoeba, Dictyostelium, Chaos.*

■ Summary

Animal-like, single-celled organisms were formerly assigned to the phylum Protozoa, but we now recognize that the "phylum" was composed of numerous clades. We use the term *protozoa* informally for all of these highly diverse organisms. Protozoa demonstrate the great adaptive potential of the basic body plan: a single eukaryotic cell. They occupy a vast array of niches and habitats.

Ciliary or flagellar movement is important in both protozoa and metazoa. The most widely accepted mechanism to account for such movement is the sliding-microtubule hypothesis. Pseudopodial or ameboid movement is a locomotory and food-gathering mechanism in protozoa and plays a vital role as a defense mechanism in metazoa. It is accomplished by microfilaments moving past each other, and it requires expenditure of energy from ATP.

Various protozoa feed by holozoic or saprozoic means. The excess water that enters their bodies is expelled by contractile vacuoles. Respiration and waste elimination are through the body surface. Protozoa reproduce asexually by binary fission, multiple fission, and budding; sexual processes are common.

Members of several taxa have photoautotrophic species, including Plantae, Euglenozoa, and Dinoflagellata. Some of these are very important planktonic organisms. Euglenozoa also includes many nonphotosynthetic species, and some of these cause serious diseases in humans, such as African sleeping sickness and Chagas disease. Apicomplexa are all parasitic, including *Plasmodium,* which causes malaria. Ciliophora move by means of cilia or ciliary organelles. They are a large and diverse group, and many are complex in structure. Amebas move by pseudopodia and are now assigned to Heterolobosea, Cercozoa, Foraminifera, Radiolaria, Centrohelida and Amoebozoa. Foraminiferans and radiolarians have an extensive fossil record due to preservation of their tests.

■ Review Questions

1. Explain why a protozoan may be very complex, even though it is composed of only one cell.
2. Distinguish among the following protozoan phyla: Euglenozoa, Apicomplexa, Ciliophora, Dinoflagellata.
3. Explain the transitions of endoplasm and ectoplasm in ameboid movement. What is a current hypothesis regarding the role of actin in ameboid movement?
4. Distinguish lobopodia, filipodia, reticulopodia, and axopodia.
5. Contrast the structure of an axoneme with that of a kinetosome.
6. What is the sliding-microtubule hypothesis?
7. Explain how protozoa eat, digest their food, osmoregulate, and respire.
8. Distinguish the following: sexual and asexual reproduction; binary fission, budding, and multiple fission.
9. What is the survival value of encystment?
10. Contrast and give examples of autotrophic and heterotrophic protozoa.
11. Describe the structure and habitat of a foraminiferan.
12. Outline the general life cycle of malarial organisms. How do you account for the resurgence of malaria in recent years?
13. What is the public health importance of *Toxoplasma,* and how do humans become infected with it? What is the public health importance of *Cryptosporidium* and *Cyclospora?*
14. Define the following with reference to ciliates: macronucleus, micronucleus, pellicle, undulating membrane, cirri, infraciliature, trichocysts, toxicysts, conjugation.
15. Distinguish primary endosymbiosis from secondary endosymbiosis.
16. Explain how surface temperatures in ancient oceans may be reconstructed.

For Further Thought Mark the distribution of photoautotrophic taxa on figure 5.2. Recent research suggests that members of Ciliophora contain genes of algal origin. From a "tree-thinking" perspective, how would you explain the origin of these genes?

■ Selected References

See also general references on page 432.

Abedin, M., and N. King. 2010. Diverse evolutionary paths to cell adhesion. Trends in Cell Biology **20**: 734–742. *A concise article describing the evolution of multicellularity in volvocine algae, animals, and a slime mold,* Dictyostelium.

Baldauf, S. L. 2008. An overview of the phylogeny and diversity of eukaryotes. J. Syst. Evol. **46**: 263–273. *A nice summary of the recent status of eukaryote phylogeny.*

Baldauf, S. L., A. J. Roger, I. Senk-Siefert, and W. F. Doolittle. 2000. A kingdom-level phylogeny of eukaryotes based on combined protein data. Science **290**: 972–977. *The authors contend that combined sequence data for genes encoding several proteins indicate there are 15 kingdoms of organisms.*

Bonner, J. T. 1983. Chemical signals of social amoebae. Sci. Am. **248**: 114–120 (April). *Social amebas of two different species secrete different chemical compounds that act as aggregation signals. The evolution of the aggregation signals in these protozoa may provide clues to the origin of the diverse chemical signals (neurotransmitters and hormones) in more complex organisms.*

Burkholder, J. M. 2002. *Pfiesteria:* the toxic *Pfiesteria* complex. In G. Bitton (ed.), Encyclopedia of environmental microbiology pp. 2431–2447. New York, Wiley

Publishers. *A nice summary of recent work on habitat and life cycles of Pfiesteria, including its effects on fish, shellfish, and humans.*

Cavalier-Smith, T. 1999. Principles of protein and lipid targeting in secondary symbiogenesis: euglenoid, dinoflagellate, and sporozoan plastid origins and the eukaryote family tree. J. Euk. Microbiol. **46:** 347–366. *Many organisms are the products of secondary symbiogenesis (a eukaryote is consumed by another eukaryote, both products of primary symbiogenesis, and symbiont becomes an organelle), but tertiary symbiogenesis also has occurred (product of secondary symbiogenesis itself becomes a symbiont … and organelle).*

Lee, J. J., G. F. Leedale, and P. Bradbury (eds). 2000. An illustrated guide to the protozoa, ed. 2, 1432 pp., 2 vols. Lawrence, Kansas, Society of Protozoologists. *This long-awaited guide appeared in 2002. It is an essential reference for students of protozoa.*

Patterson, D. J. 1999. The diversity of eukaryotes. Amer. Nat. **154** (supplement): S96–S124.

Patterson provides morphological descriptions and synapomorphies for many clades containing protozoans.

Pawlowski, J., and F. Burki. 2009. Untangling the phylogeny of amoeboid protists. J. Eukaryot. Microbiol. **56**:16–25. *Many amebas have found new phylogenetic homes as shown on detailed trees included here.*

Pawlowski, J., and A. J. Gooday. 2008. Precambrian biota: protistan origin of trace fossils? Curr. Biol. **19**:R28–R30. *A discussion of the impact of the giant rolling ameba on paleobiology.*

Roberts, L. S., and J. J. Janovy, Jr. 2008. Foundations of parasitology, ed. 8. Dubuque, Iowa, McGraw-Hill Higher Education. *Up-to-date and readable information on parasitic protozoa.*

Roger, A. J. 1999. Reconstructing early events in eukaryotic evolution. Amer. Nat. **154** (supplement): S146–S163. *Methods in determining whether absence of mitochondria is primary or due to a secondary loss are discussed here.*

Sachs, J. L. 2008. Resolving the first steps to multicellularity. Trends Ecol.

Evol. **23**:245–248. *An interesting summary of work on Volvox evolution covering morphological change and selection pressures.*

Sebé-Pedros, A., A. J. Roger, F. B. Lang, N. King, and I Ruiz-Trillo. 2010. Ancient origin of the integrin-mediated adhesion and signaling machinery. Proceedings of the National Academy of Sciences **107**:10142–10147. *Cells of multicellular organisms must interact with the extracellular matrix (ECM). In animals, these interactions are mediated by integrins. This research shows that parts of the integrin protein complex were present before the split between fungi and animals, and some parts predate the Amoebozoa. Ancient genes were modified for new functions in the evolution of multicellularity.*

Steenkamp, E. T., J. Wright, and S. L. Baldauf. 2006. The protistan origins of animals and fungi. Mol. Biol. Evol. **23**:93–106. *Opisthokonta is a well-supported clade whose members share a short amino acid sequence in elongation factor 1-alpha.*

■ Custom Website

The *Animal Diversity* Online Learning Center is a great place to check your understanding of chapter material. Visit www.mhhe.com/hickmanad6e for access to key terms, quizzes, and more! Further enhance your knowledge with Web links to chapter-related material.

Explore live links for these topics:

Classification and Phylogeny of Animals
Protistan Groups
Phylum Sarcomastigophora
Phylum Apicomplexa
Phylum Ciliophora
Other Protozoan Phyla

Parasitic Protists
Free-living Protists
Dinoflagellates
Pfisteria
Red Tides
Adaptations for Planktonic Life

6

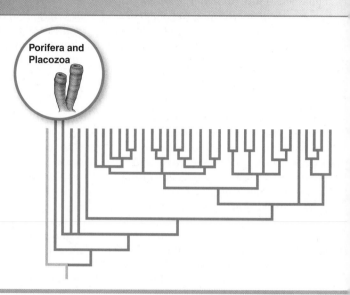

Porifera and
Placozoa

Sponges
Phylum Porifera

The Advent of Multicellularity

Sponges are the simplest multicellular animals. Because the cell is the elementary unit of life, organisms larger than unicellular protozoa arose as aggregates of such building units. Nature has experimented with producing larger organisms without cellular differentiation—certain large, single-celled marine algae, for example—but such examples are rare. Multicellularity has many advantages over simply increasing the mass of a single cell. Because metabolic exchange occurs at cell surfaces, dividing a mass into smaller units greatly increases the surface area available for metabolic activities. It is impossible to maintain a workable surface-to-mass ratio by simply increasing the size of a single-celled organism. Thus, multicellularity is a highly efficient path toward increasing body size that has evolved many times.

Strangely, while sponges are multicellular, their organization is quite distinct from that of other metazoans. A sponge body is an assemblage of cells embedded in an extracellular matrix and supported by a skeleton of minute, needlelike spicules and protein. Because sponges neither look nor behave like other animals, it is understandable that zoologists did not completely accept them as animals until well into the nineteenth century. Nonetheless, molecular evidence reveals that sponges are animals.

Metazoans are sometimes divided into two grades: Parazoa and Eumetazoa. Parazoa, whose name means "beside-animals," comprises the sponges in phylum Porifera and tiny, platelike animals in phylum Placozoa. Placozoans are marine organisms, 2 to 3 mm in size, whose simple bodies are made of dorsal and ventral epithelia with a syncytial middle layer. A syncytial layer consists of a single plasma membrane with multiple nuclei inside it. A syncytium sometimes forms when cells fuse; the syncytium in glass sponges (see p. 125) forms this way. We discuss sponges in this chapter, but omit further mention of placozoans due to space limitations.

A Caribbean demosponge, *Aplysina fistularis*.

ponges belong to phylum Porifera (po-rif´-er-a) (L. *porus,* pore, + *fera,* bearing). Sponges bear myriads of tiny pores and canals that constitute a filter-feeding system adequate for their inactive lifestyle. They are sessile animals that use water currents carried through their unique canal systems to bring them food and oxygen and to carry away their body wastes. Their bodies are masses of cells embedded in a gelatinous extracellular matrix and stiffened by a skeleton of minute **spicules** of calcium carbonate or silica and collagen (see p. 123). Most sponges have no organs or true tissues, and even their cells show a certain degree of independence. For example, dissociated sponge cells can re-form a

new sponge body. As sessile animals with limited body movement, they lack a nervous system or sense organs and have only the simplest contractile elements.

Sponges vary in size from a few millimeters to the great loggerhead sponges, which may reach 2 m or more across. Many sponge species are brightly colored because of pigments in their dermal cells. Red, yellow, orange, green, and purple sponges are not uncommon. However, their color fades quickly when sponges are removed from water. Some sponges, including the simplest, are radially symmetrical, but many are quite irregular in shape. Some stand erect, some are branched or lobed, and others are low, even encrusting, in form (figure 6.1). Some bore holes into shells or rocks.

Sponges are an ancient group, with an abundant fossil record extending back to the early Cambrian period and even, according to some claims, the Precambrian. Living poriferans traditionally have been assigned to three classes: Calcarea (having calcareous spicules), Demospongiae (having a skeleton of siliceous spicules or spongin or both), and Hexactinellida (having six-rayed siliceous spicules) (figure 6.2). A fourth class (Homoscleromorpha) was recently created from a subset of the Demospongiae. Homoscleromorphs are thin encrusting sponges that have simple spicules or lack spicules entirely.

■ Ecological Relationships

Most of the 5000 or more sponge species are marine, although some 150 species occupy fresh water. Marine sponges abound in all seas and at all depths, and a few even occupy brackish water. Although their embryos are free-swimming, adults are always attached, usually to rocks, shells, corals, or

figure 6.1

Some growth habits and forms of sponges.

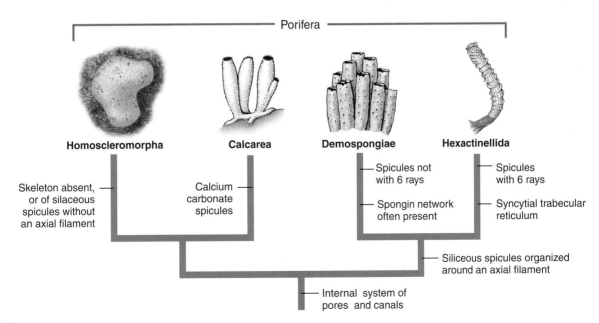

figure 6.2

Cladogram depicting evolutionary relationships among the four classes of sponges with living representatives.

figure 6.3

This orange demosponge, *Mycale laevis,* often grows beneath platelike colonies of the stony coral, *Montastrea annularis.* The large oscula of the leuconoid sponge are seen at the edges of the plates. Unlike some other sponges, *Mycale* does not burrow into the coral skeleton and may actually protect the coral from invasion by more destructive species. Pinkish radioles of a Christmas tree worm, *Spirobranchus giganteus* (phylum Annelida, class Polychaeta), also project from the coral colony. An unidentified reddish sponge is seen farther to the right of and below the Christmas tree worm.

other submerged objects (figure 6.3). Some bottom-dwelling forms even grow on sand or mud. Their growth patterns often depend on shape of the substratum, direction and speed of water currents, and availability of space, so that the same species may differ markedly in appearance under different environmental conditions. Sponges in calm waters may grow taller and straighter than those in rapidly moving waters.

Many animals (crabs, nudibranchs, mites, bryozoans, and fish) live as commensals or parasites in or on sponges. Larger sponges particularly tend to harbor a large variety of invertebrate commensals. On the other hand, sponges grow on many other living animals, such as molluscs, barnacles, brachiopods, corals, and hydroids. One sponge has been described that preys on shrimp. Some crabs attach pieces of sponge to their carapace for camouflage and protection, since most predators seem to find sponges distasteful. Certainly one reason for the success of sponges as a group is that they have few enemies. Because of a sponge's elaborate skeletal framework and often noxious odor, most potential predators find sampling a sponge about as pleasant as eating a mouthful of glass splinters embedded in evil-smelling gristle. Some reef fishes, however, do graze on shallow-water sponges.

Sponges, and the microorganisms that inhabit them, produce a wide variety of bioactive chemicals. One extract from a marine sponge appears effective against leishmaniasis, a disease caused by an apicomplexan parasite (see Chapter 5, p. 109), and another shows promise for treating herpetic infections. Many bacteria isolated from marine taxa also have antimicrobial or antiviral effects: for example, some inhibit *Staphylococcus aureus* infections and others are active against *Escherichia coli,* some strains of which may cause food poisoning. These and other results have increased interest in sponge culturing as a source of valuable pharmaceuticals.

■ Form and Function

Sponges are sessile. Their only body openings are pores, usually many tiny ones called **ostia** for incoming water, and one to a few large ones called **oscula** (sing., **osculum**) that serve as water outlets. These openings are connected by a system of canals, some of which are lined with peculiar flagellated collar cells called **choanocytes,** whose flagella maintain a current of environmental water through the canals. Water enters the canals through a multitude of tiny incurrent pores (**dermal ostia**) and leaves by way of one or more large oscula. Choanocytes not only keep the water moving but also trap and phagocytize food particles from the water. Collapse of the canals is prevented by the skeleton, which, depending on the species, may be composed of needlelike calcareous or siliceous spicules, a meshwork of organic spongin fibers, or a combination of the two.

CHARACTERISTICS
of Phylum Porifera

1. Multicellular, body an aggregation of several types of cells differentiated for various functions, some of which are organized into **incipient tissues** of a low level of integration. However, the pinacoderm form a true tissue epithelium in homoscleromorph sponges.
2. Body with pores (ostia), canals, and chambers that form a unique system of **water currents** on which sponges depend for food and oxygen
3. Mostly marine; all aquatic
4. Radial symmetry or none
5. Outer surface of flat pinacocytes; most interior surfaces lined with flagellated collar cells (choanocytes) that create water currents; a gelatinous protein ECM called mesohyl contains amebocytes of various types and skeletal elements
6. Skeletal structure of fibrillar collagen (a protein) and calcareous or siliceous crystalline spicules, often combined with variously modified collagen (spongin); type IV collagen, characteristic of other metazoans, occurs only in homoscleromorph sponges
7. No organs or true tissues; digestion intracellular; excretion and respiration by diffusion
8. Reactions to stimuli apparently local and independent in cellular sponges, but electrical signals occur in syncytial glass sponges; nervous system probably absent
9. All adults sessile and attached to substratum
10. Asexual reproduction by buds or gemmules and sexual reproduction by eggs and sperm; free-swimming flagellated larvae in most

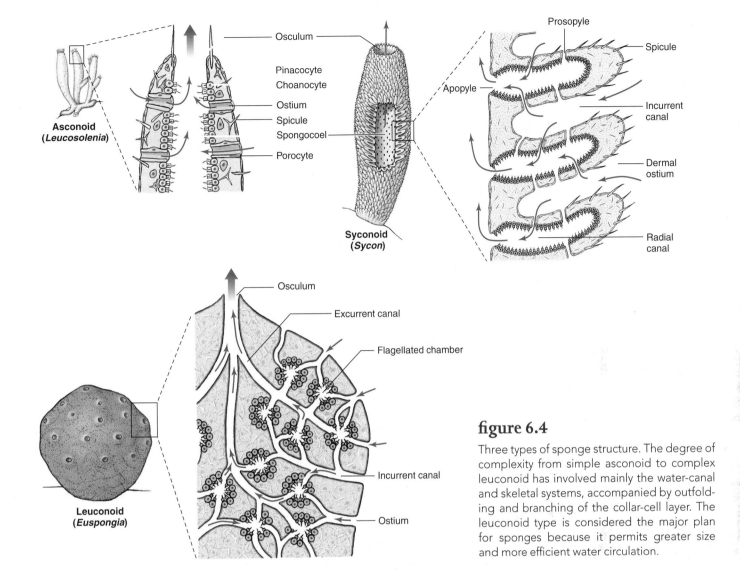

figure 6.4

Three types of sponge structure. The degree of complexity from simple asconoid to complex leuconoid has involved mainly the water-canal and skeletal systems, accompanied by outfolding and branching of the collar-cell layer. The leuconoid type is considered the major plan for sponges because it permits greater size and more efficient water circulation.

Types of Canal Systems

Most sponges have one of three types of canal systems—asconoid, syconoid, or leuconoid (figure 6.4).

Asconoids—Flagellated Spongocoels

Asconoid sponges have the simplest organization. They are small and tube-shaped. Water enters through microscopic dermal pores into a large cavity called a **spongocoel,** which is lined with choanocytes. Choanocyte flagella pull water through the pores and expel it through a single large osculum (figure 6.4). *Leucosolenia* (Gr. *leukos,* white, + *solen,* pipe) is an asconoid type of sponge. Its slender, tubular individuals grow in groups attached by a common **stolon,** or stem, to objects in shallow seawater. *Clathrina* (L. *clathri,* latticework) is an asconoid with bright yellow, intertwined tubes (figure 6.5). All asconoids are in the Calcarea.

figure 6.5

Clathrina canariensis (class Calcarea) is common in caves and under ledges on Caribbean reefs.

Syconoids—Flagellated Canals

Syconoid sponges look somewhat like larger editions of asconoids, from which they were derived. They have a tubular body and a single osculum, but instead of a simple choanocyte layer lining the spongocoel, as in asconoids, this layer in syconoids is folded back and forth to make canals. The choanocytes line certain folds called **radial canals** (see figure 6.4). Water, entering the body through dermal pores, moves first to incurrent canals and then into radial canals via small lateral openings called prosopyles (figure 6.6). From the radial canals, filtered water moves through apopyles into the spongocoel, finally exiting by the osculum. The spongocoel in syconoids is lined with epithelial-type cells rather than with the choanocytes found in asconoids. Syconoids are in the Calcarea. *Sycon* (Gr. *sykon*, a fig) is a commonly studied example of the syconoid type of sponge (figure 6.6).

Leuconoids—Flagellated Chambers

Leuconoid organization is the most complex of the sponge types and permits an increase in sponge size. Most leuconoids form large masses with numerous oscula (see figure 6.3).

Clusters of flagellated chambers are filled from incurrent canals and discharge water into excurrent canals that eventually lead to the osculum (see figure 6.4). There is no spongocoel. Most sponges are of the leuconoid type, which occurs in most Calcarea and in all other classes.

The three types of canal systems—asconoid, syconoid, and leuconoid—demonstrate an increase in the complexity and efficiency of the water-pumping system, but they do not imply an evolutionary or developmental sequence. The leuconoid grade of construction has evolved independently many times in sponges. Possession of a leuconoid plan is of clear adaptive value; it increases the proportion of flagellated surfaces compared with the volume, thus providing more collar cells to meet food demands.

Types of Cells

Sponge cells are loosely arranged in a gelatinous matrix called **mesohyl** (figure 6.7). The mesohyl is the "connective tissue" of the sponges; in it are found various ameboid cells, fibrils, and skeletal elements. Several types of cells occur in sponges.

Pinacocytes

The nearest approach to a true tissue in sponges is arrangement of the **pinacocyte** cells (figure 6.7) of the external epithelium. These thin, flat, epithelial-type cells cover the exterior surface and some interior surfaces. Some are T-shaped, with their cell bodies extending into the mesohyl. Pinacocytes are somewhat contractile and help to regulate the surface area

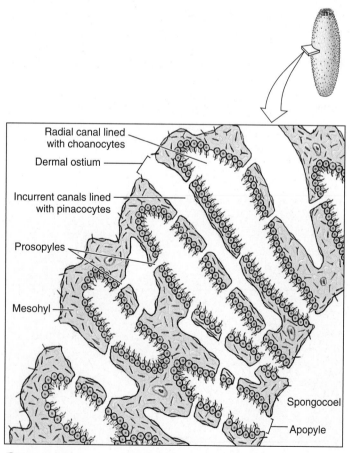

figure 6.6

Cross section through the body wall of the sponge *Sycon*, showing the canal system.

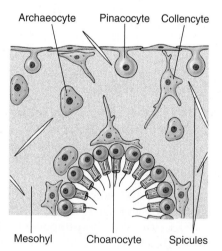

figure 6.7

Small section through the body wall of a sponge, showing four types of sponge cells. Pinacocytes are protective and contractile; choanocytes create water currents and engulf food particles; archaeocytes have a variety of functions, including phagocytosis of food particles and differentiation into other cell types; collencytes secrete collagen.

of the sponge. Some pinacocytes are modified as contractile **myocytes,** which are usually arranged in circular bands around the oscula or pores, where they help to regulate the rate of water flow. Myocytes contain microfilaments similar to those found in muscle cells of other animals.

Although a pinacocyte layer is a grouping of cells specialized for a particular function, it is not considered a true tissue epithelium because the cells do not rest on a basal (basement) membrane or ECM. However, because it is tissuelike, it is sometimes called an "incipient tissue." It is interesting that a basement membrane is present in the newly created sponge class Homoscleromorpha, so this group has a true tissue epithelium.

Porocytes

Tubular cells that pierce the body wall of asconoid sponges, through which water flows, are called **porocytes** (see figure 6.4).

Choanocytes

Choanocytes, which line flagellated canals and chambers, are ovoid cells with one end embedded in mesohyl and the other exposed. The exposed end bears a flagellum surrounded by a collar (figures 6.7 and 6.8). Electron microscopy shows that the collar is composed of adjacent microvilli, connected to each other by delicate microfibrils, so that the collar forms a fine filtering device for straining food particles from the water (figure 6.8B and C). The beat of a flagellum pulls water through the sievelike collar and forces it out through the open top of the collar. Particles too large to enter the collar become trapped in secreted mucus and slide down the collar to the base, where they are phagocytized by the cell body. Larger particles have already been screened out by the small size of the dermal pores and prosopyles. Food engulfed by the cells is passed to a neighboring archaeocyte for digestion.

Archaeocytes

Archaeocytes are ameboid cells that move through the mesohyl (see figure 6.7) and perform a number of functions. They can phagocytize particles at the external epithelium and receive particles for digestion from choanocytes. Archaeocytes apparently can differentiate into any of the other types of more specialized cells in the sponge. Some, called **sclerocytes,** secrete spicules. Others, called **spongocytes,** secrete the spongin fibers of the skeleton, and **collencytes** secrete fibrillar collagen.

Types of Skeletons

The skeleton gives support to a sponge, preventing collapse of canals and chambers. The major structural protein in the animal kingdom is collagen, and fibrils of collagen occur throughout the extracellular matrix of all sponges. In addition, various Demospongiae secrete a form of collagen traditionally called spongin (figure 6.9A). Demospongiae also secrete siliceous spicules, and calcareous sponges secrete spicules composed mostly of crystalline calcium carbonate that have one, three, or four rays (figure 6.9A). Glass sponges (Hexactinellida) have siliceous spicules with six rays arranged in three planes at right

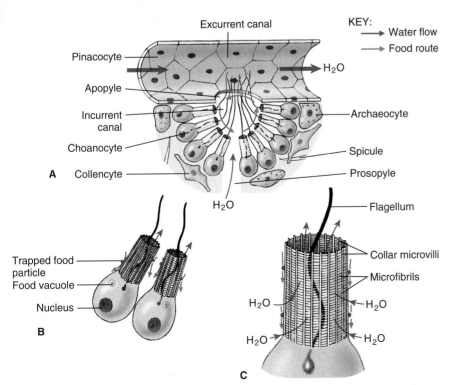

figure 6.8

Food trapping by sponge cells. **A,** Cutaway section of canals showing cellular structure and direction of water flow. **B,** Two choanocytes. **C,** Structure of the collar. Small red arrows indicate movement of food particles. Blue arrows indicate the direction of water flow.

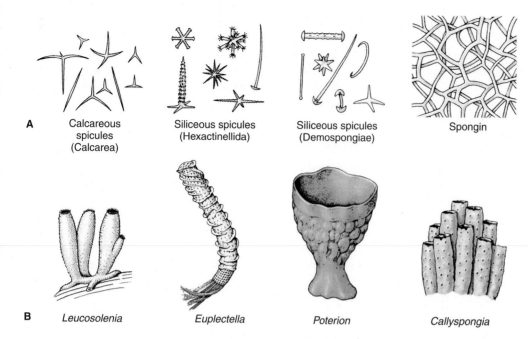

figure 6.9

A, The many types of spicules in sponges provide amazing diversity, beauty, and complexity of form. **B,** Some examples of sponge body forms possessing the support structure shown above in panel A.

A Calcareous spicules (Calcarea)   Siliceous spicules (Hexactinellida)   Siliceous spicules (Demospongiae)   Spongin

B *Leucosolenia*   *Euplectella*   *Poterion*   *Callyspongia*

angles to each other. There are many variations in the shape of spicules, and these structural variations are of taxonomic importance. Sponge bodies possessing each type of support in figure 6.9A are shown in figure 6.9B.

Sponges harbor microalgae and cyanobacteria on the body surface and deep inside the body. The presence of photosynthetic organisms inside the sponge led some to propose that spicules were able to transmit light into the body. The fiber-optic properties of siliceous spicules have now been confirmed. This has sparked interest among materials scientists and engineers in the enzymatic machinery needed to form silica nanoparticles and to fuse these particles into spicules inside and outside sponge cells. In many cases, the exterior simplicity of a sponge masks chemical and functional sophistication.

Sponge Physiology

Sponges feed primarily on particles suspended in water pumped through their canal systems. They consume detritus particles, planktonic organisms, and bacteria in sizes ranging from 50 μm (average diameter of ostia) to 0.1 μm (width of spaces between the microvilli of the choanocyte collar). Pinacocytes may phagocytize particles at the surface, but most larger particles are consumed in the canals by archaeocytes that move close to the lining of the canals. The smallest particles, accounting for about 80% of the particulate organic carbon, are phagocytized by choanocytes. Digestion is entirely **intracellular** (occurs within cells), a chore performed by the archaeocytes.

Sponges consume a significant portion of their nutrients in the form of organic matter dissolved in water circulating through the system. Such material is apparently ingested by a process similar to phagocytosis.

Sponges have no respiratory or excretory organs; these functions are performed by diffusion. Contractile vacuoles occur in the archaeocytes and choanocytes of freshwater sponges.

All of a sponge's life activities depend on a current of water flowing through its body. A sponge pumps a remarkable amount of water. Some large sponges can filter 1500 liters of water a day. At least some sponges can crawl (move laterally over their supporting substratum) at speeds of up to 4 mm per day. This ability may give them an advantage over more sessile encrusting organisms in competition for space.

Reproduction and Development

All sponges can reproduce both sexually and asexually. In sexual reproduction, most sponges are **monoecious** (having both male and female sex cells in one individual). Sperm arise from transformation of choanocytes. In Calcarea and at least some Demospongiae, oocytes also develop from choanocytes; in other demosponges, oocytes apparently develop from archaeocytes. Sperm are released into the water by one individual and taken into the canal system of another. There, choanocytes phagocytize them, transform into carrier cells, and then carry the sperm through the mesohyl to the oocytes.

Other sponges are oviparous and expel both oocytes and sperm into the water. Ova are fertilized by motile sperm (without carrier cells) in the mesohyl. There, the zygotes develop into flagellated larvae, which break loose and are carried away by water currents. The free-swimming larva of most sponges is a solid-bodied **parenchymula** (figure 6.10). The outwardly directed, flagellated cells on the larval surface migrate to the interior after the larva settles and become choanocytes in the flagellated chambers.

The loose organization of sponges is ideally suited for regeneration of injured and lost parts, and for asexual reproduction.

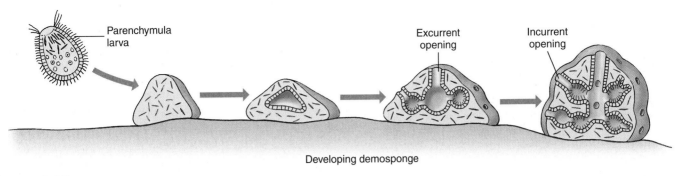

figure 6.10

Development of a demosponge.

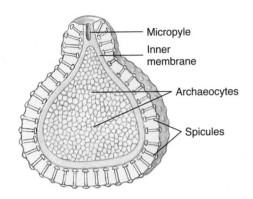

figure 6.11

Section through a gemmule of a freshwater sponge (Spongillidae). Gemmules are a mechanism for surviving the harsh conditions of winter. Upon the return of favorable conditions, the archaeocytes exit through the micropyle to form a new sponge. The archaeocytes of the gemmule give rise to all cell types of the new sponge.

Sponges reproduce asexually by fragmentation and by forming external buds that detach or remain to form colonies. In addition to external buds, which all sponges can form, freshwater sponges and some marine sponges reproduce asexually by regularly forming internal buds called **gemmules** (figure 6.11). These dormant masses of encapsulated archaeocytes form during unfavorable conditions. They can survive periods of drought and freezing and more than three months without oxygen. Later, with the return of favorable conditions for growth, archaeocytes in the gemmules escape and develop into new sponges.

■ Brief Survey of Sponges

Class Calcarea (Calcispongiae)

Calcarea are calcareous sponges, so called because their spicules are composed of calcium carbonate. Their spicules are straight monaxons or have three or four rays (see figure 6.9A). The sponges tend to be small—10 cm or less in height—and

tubular or vase shaped. They may be asconoid, syconoid, or leuconoid in structure. Although many are drab, some are bright yellow, red, green, or lavender. *Leucosolenia, Clathrina* (see figure 6.5), and *Sycon* are common examples.

Class Hexactinellida (Hyalospongiae)

Hexactinellids are nearly all deep-sea forms. Most are radially symmetrical and range in length from 7 to 10 cm to more than 1 m. One distinguishing feature, reflected in the Hexactinellida, is the skeleton of six-rayed siliceous spicules bound together in an exquisite glasslike latticework (see figure 6.9A).

The tissue structure of hexactinellids differs so dramatically from that of other sponges that some scientists advocate placing them in a subphylum separate from other sponges. The body is syncytial—there are many nuclei inside a single very large cell. This single, continuous syncytial tissue is called a **trabecular reticulum.** A 1-m-diameter glass sponge makes the largest syncytium known within Metazoa.

The trabecular reticulum is bilayered and can be sheetlike or tubular. Between the layers of the sheet, or inside the tubes, is a thin collagenous mesohyl in which cells, such as archeocytes or choanoblasts, occur (figure 6.12). Choanoblasts and other cells are connected to each other, and to the trabecular reticulum, by cytoplasmic bridges. Choanoblasts are unusual cells that make two or more flagellated outgrowths called **collar bodies.** The flagellum on a collar body beats to drive water flow in the same way that it would on a choanocyte.

An assemblage of collar bodies forms a flagellated chamber. Here, the trabecular reticulum branches to become two distinct bilayered sheets: a primary reticulum and a thin secondary reticulum that lacks mesohyl. The two sheets make a sandwich around the center of a collar body (figure 6.12). Collar bodies extend through openings in both sheets, but the openings surround the collar bodies tightly. There is a space between the two sheets. To collect food, the incurrent water is directed to the primary reticulum, where pores enter the space between the primary and secondary reticular sheets. Water entering this space must leave by moving through the mesh of microvilli on collar bodies; water cannot go anywhere else because it is

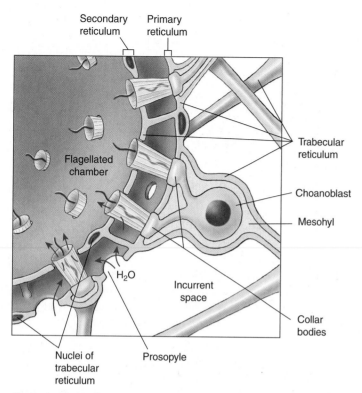

Secondary reticulum

Primary reticulum

Flagellated chamber

Trabecular reticulum

Choanoblast

Mesohyl

H₂O

Incurrent space

Collar bodies

Nuclei of trabecular reticulum

Prosopyle

figure 6.12

Diagram of part of a flagellated chamber of hexactinellids. The primary and secondary reticula are branches of the trabecular reticulum, which is syncytial. Cell bodies of the choanoblasts and their processes are borne by the primary reticulum and are embedded in a thin, collagenous mesohyl. Processes of the choanoblasts end in collar bodies, whose collars extend upward through the secondary reticulum. Flagellar action propels water (arrows) to be filtered through the mesh of collar microvilli.

blocked by the secondary reticulum. Food particles captured on microvilli are shared throughout the syncytium.

The latticelike network of spicules in many glass sponges is of exquisite beauty, as seen, for example, in *Euplectella* (NL. from Gr. *euplektos,* well-plaited), a classic example of Hexactinellida (see figure 6.9B).

Class Demospongiae

Demospongiae comprise approximately 80% of all sponge species, including most larger sponges. Their skeletons may be composed of siliceous spicules, spongin fibers, or both (see figure 6.9A). All members of the class are leuconoid, and all are marine except one family, the freshwater Spongillidae. Freshwater sponges occur widely in well-oxygenated ponds and streams, where they encrust plant stems and old pieces of submerged wood. They resemble a bit of wrinkled scum, pitted with pores, and are brownish or greenish in color. Freshwater sponges die and disintegrate in late autumn, leaving gemmules to survive the winter.

Marine Demospongiae vary in both color and shape. Some are encrusting; some are tall and fingerlike; and others are shaped like fans, vases, cushions, or balls (figure 6.13). Some sponges bore into and excavate molluscan shells and coral skeletons. Loggerhead sponges may grow several meters in diameter. So-called bath sponges belong to the group called horny sponges, which have only spongin skeletons. They can be cultured by cutting out pieces of the individual sponges, fastening them to a weight, and dropping them into the proper water conditions. It takes many years for a horny sponge to grow to market size. Most commercial "sponges" now on the market are synthetic, but the harvest and use of bath sponges persist.

A

B

C

figure 6.13

Marine Demospongiae on Caribbean coral reefs. **A,** *Pseudoceratina crassa* is a colorful sponge growing at moderate depths. **B,** *Ectyoplasia ferox* is irregular in shape, and its oscula form small, volcano-like cones. It is toxic and may cause skin irritation if touched. **C,** *Monanchora unguifera* with commensal brittle star, *Ophiothrix suensoni* (phylum Echinodermata, class Ophiuroidea).

Taxonomy of Phylum Porifera

Class Calcarea (cal-cáre̅-ə) (L. *calcis*, lime, + Gr. *spongos*, sponge) **(Calcispongiae).** Have spicules of calcium carbonate that often form a fringe around the osculum; spicules needle-shaped or three- or four-rayed; all three types of canal systems (asconoid, syconoid, leuconoid) represented; all marine. Examples: *Sycon, Leucosolenia, Clathrina.*

Class Hexactinellida (hex-ak-tin-el´i-da) (Gr. *hex,* six, + *aktis,* ray) **(Hyalospongiae).** Have six-rayed, siliceous spicules extending at right angles from a central point; spicules often united to form a network; body often cylindrical or funnel-shaped. Flagellated chambers in simple syconoid or leuconoid arrangement. Habitat mostly deep water; all marine. Examples: Venus' flower basket *(Euplectella), Hyalonema.*

Class Demospongiae (de-mo-spuńje̅) (tolerated misspelling of Gr. *desmos,* chain, tie, bond, + *spongos,* sponge). Have skeleton of siliceous spicules that are not six-rayed, or spongin, or both. Leuconoid-type canal systems. One family found in fresh water; all others marine. Examples: *Thenea, Cliona, Spongilla, Myenia,* and all bath sponges.

Class Homoscleromorpha (ho̅-mo̅-skle̅´-ro̅-mor-fə) (Gr. *homos,* same, + *skleros,* hard, + *morphe,* form). Previously a subgroup of Demospongiae; spicules may be absent as in *Oscarella;* if present, spicules are small, simple in shape, and do not form around an axial filament; pinacoderm with a distinct basement membrane. Examples: *Oscarella, Corticium.*

Class Homoscleromorpha

Homoscleromorphs are marine sponges that occur in a range of colors, but live in cryptic habitats, so they are often overlooked. They are more common in nearshore habitats, but they do occur in deep water. Sponges in this class were formerly placed in Class Demospongiae, but were separated because they possess unique features such as a pinacoderm layer with a true basement membrane or ECM (see p. 60). The cells of this layer also differ from those in cells of other sponge classes because adherens cell junctions join them together. Thus, the pinacoderm is a true tissue in this group. The class is divided into two clades, one whose members lack spicules entirely, and the other with spicules that do not form around a central longitudinal (axial) filament. Representative genera are *Plakina, Oscarella,* and *Corticium.*

Phylogeny and Adaptive Diversification

Phylogeny

Sponges originated before the Cambrian period. Two groups of calcareous spongelike organisms occupied early Paleozoic reefs. The Devonian period saw rapid development of many glass sponges. Sponges are the sister taxon to a group comprising all animal phyla, as shown on the cladogram inside the front cover. The simple body plans of most sponges, aside from hexactinellids, might suggest that sponges share few features with other animals, but this is not true. To form a multicellular body, cells must adhere to each other in a stable fashion. The proteins used in cell adhesion and cell signaling in sponges are homologous to those in other metazoans; in fact many of these occur in choanoflagellates, evolving before the last common ancestor of all animals. The animal epithelium is a unique structure where cells adhere to each other in a single layer attached to a basement membrane that contains collagen type IV. Adherens junctions in which cadherin molecules function as adhesives[1] connect cells in an epithelium to each other. A true epithelium was presumed absent in sponges until the homoscleromorphs were examined in detail. We know now that epithelia are present in at least some sponges and are elaborated in other animals. Sponge development includes the characteristic metazoan blastula stage, and some sponges actually develop to the two-layered gastrula stage before reorganizing their bodies into asymmetrical adults. It is possible that the sessile lifestyle of sponges favored a deceptively simple body in most species and that a closer look at sponges will reveal more features typical of metazoans.

Adaptive Diversification

Porifera are a highly successful group that includes several thousand species in a variety of marine and freshwater habitats. Their diversification centers largely on their unique water-current system and its various degrees of complexity. The leuconoid body plan, with its many flagellated chambers, has a relatively large surface area for food capture and gas exchange when compared with the asconoid and syconoid plans; this may account for the large size of leuconoid sponges.

One very novel way of feeding has evolved within a family of sponges, called cladorhizids, inhabiting nutrient-poor deep-water caves. Each sponge has a fine coating of tiny, hook-like spicules over its highly branched body. The spicule layer entangles the appendages of tiny crustaceans swimming near the sponge surface. Later, filaments of the sponge body grow over the prey, enveloping and digesting them. These animals

[1]Abedin, M., and N. King. 2010. Diverse evolutionary paths to cell adhesion. Trends in Cell Biology **20:** 734–742.

are carnivores, not suspension feeders; they lack choanocytes and internal canals, but have silicious spicules like typical members of class Demospongiae. In addition to capturing prey, some augment their diets with nutrients obtained from symbiotic methanotrophic bacteria. To colonize such a nutrient-poor habitat initially, the ancestors of this group must have had at least one alternative feeding system, either carnivory or chemoautotrophy, already in place. Presumably, after the alternative method of food capture was in use, the choanocytes and internal canals were no longer formed.

■ Summary

Sponges (phylum Porifera) are an abundant marine group with some freshwater representatives. They have various specialized cells; these cells are not organized into tissues or organs in most sponges, but the pinacoderm forms a true tissue in one class, Homoscleromorpha. Sponges depend on the flagellar beat of their choanocytes to circulate water through their bodies for gathering food and exchanging respiratory gases. They are supported by secreted skeletons of fibrillar collagen, collagen in the form of large fibers or filaments (spongin), calcareous or siliceous spicules, or a combination of spicules and spongin in most species. A newly erected sponge class, Homoscleromorpha, is interesting because only its members have type IV collagen characteristic of other animals.

Most sponges are monoecious but produce sperm and oocytes at different times. Embryogenesis is unusual because flagellated cells migrate from the embryo surface to the interior of the larva. The larva is typically a free-swimming parenchymula. Sponges have great regenerative capabilities and may reproduce asexually by budding, fragmentation, or gemmules (internal buds).

Sponges are an ancient group present before the Cambrian. Their adaptive diversification is centered on elaboration of the water circulation and filter-feeding system, with the exception of a unique group of cave-dwelling, carnivorous sponges (cladorhizids) that do not filter-feed.

■ Review Questions

1. Give six characteristics of sponges.
2. Briefly describe asconoid, syconoid, and leuconoid body types in sponges.
3. What sponge body type is most efficient and makes possible the largest body size?
4. Define the following: ostia, osculum, spongocoel, mesohyl.
5. Define the following: pinacocytes, choanocytes, archaeocytes, sclerocytes, collencytes.
6. What material occurs in the skeleton of all sponges?
7. Describe the skeleton of each of the four classes of sponges.
8. Describe how sponges feed, respire, and excrete.
9. What is a gemmule?
10. Why are gemmules likely to occur in sponges in areas with harsh winters?
11. Describe how gametes are produced and how fertilization occurs in most sponges.
12. What is the largest class of sponges, and what is its body type?
13. What is the definition of a true tissue and what structures meet this definition in sponges?

For Further Thought Mutualisms are interactions between two species where both participants benefit. How might the relationship between a sponge with glassy spicules and micro-algae or cyanobacteria be a mutualism?

■ Selected References

See also general references on page 432.
Abedin, M., and N. King. 2010. Diverse evolutionary paths to cell adhesion. Trends in Cell Biology **20**:734–742. *A concise article describing the evolution of multicellularity in volvocine algae, animals, and a slime mold,* Dictyostelium.
Dohrmann, M., D. Janussen, J. Reitner, A. G. Collins, and G. Worheide. 2008. Phylogeny and evolution of glass spooges (Porifera, Hexactinellida). Syst. Biol. **57**:388–405. *Homoscleromorpha*

forms a clade within the monophyletic sponges.
Ereskovsky, A. V., C. Borchiellini, E. Gazave, J. Ivanisevic, P. Lapé´bie, T. Perez, E. Renard, and J. Vacelet. 2009. The homoscleromorph sponge *Oscarella lobularis*, a promising sponge model in evolutionary and developmental biology. BioEssays **31**:89–97. *A good descriptive article that covers many aspects of the biology of these interesting animals; photographs are very nice.*

Hooper, J. N. A., and R. W. M. van Soest (eds.). 2002. Systema Porifera: a guide to the classification of sponges. New York, Kluwer Academic/Plenum. *A large and comprehensive work on sponge systematics and biology.*
Leys, S. P., and A. E. Ereskovsky. 2006. Embryogenesis and larval differentiation in sponges. Can. J. Zool. **84**: 262–287. *A review of sponge development with clearly explained terms and excellent photomicrographs.*

Reiswig, H. M., and T. L. Miller. 1998. Freshwater sponge gemmules survive months of anoxia. Invert. Biol. **117**:1–8. *Hatchability of gemmules kept in the absence of oxygen was equal to controls, but they would not hatch unless oxygen was present.*

Sperling, E. A., J. M. Robinson, D. Pisani, and K. J. Peterson. 2010. Where's the glass? Biomarkers, molecular clocks, and microRNAs suggest a 200-Myr missing Precambrian fossil record of siliceous sponges. Geobiology **8**:24–36. *A chemical found only in demosponges indicates a Precambrian origin of sponges with glass spicules, but no fossil spicules are found.*

Vacelet, J., and N. Boury-Esnault. 1995. Carnivorous sponges. Nature **373**:333–335. *A fascinating article on a novel feeding method. Later work demonstrates that symbiotic methanotrophic bacteria produce a second source of nutrition for the sponges.*

Wyeth, R. C. 1999. Video and electron microscopy of particle feeding in sandwich cultures of the hexactinellid sponge, *Rhabdocalyptus dawsoni.* Invert. Biol. **118**:236–242. *Phagocytosis is not by choanoblasts but by trabecular reticulum, especially primary reticulum.*

■ Custom Website

The *Animal Diversity* Online Learning Center is a great place to check your understanding of chapter material. Visit www.mhhe.com/hickmanad6e for access to key terms, quizzes, and more! Further enhance your knowledge with Web links to chapter-related material.

Explore live links for these topics:

Classification and Phylogeny of Animals

Phylum Porifera

Radiate Animals
Cnidarians and Ctenophores

A Fearsome Tiny Weapon

Although members of phylum Cnidaria are more highly organized than sponges (described in Chapter 6), they are still relatively simple animals. Most are sessile. Most of those that are unattached, such as jellyfish, can swim only feebly. Indeed, we might easily get the false impression that cnidarians provide easy meals for other animals. However, many cnidarians are very effective predators that kill and eat much more highly organized, swift, and intelligent animals. They manage these feats because they possess tentacles that bristle with tiny, remarkably sophisticated weapons called nematocysts.

As a nematocyst forms within a cell, it is endowed with potential energy to power its discharge. A nematocyst is like a factory-manufactured gun that rolls off the assembly line cocked and ready with a bullet in its chamber. Like the cocked gun, the completed nematocyst requires only a small stimulus to make it fire. Rather than a bullet, a tiny thread bursts from the nematocyst. Achieving a velocity of 2 m/sec and an acceleration of 40,000 × gravity, it instantly penetrates its prey and injects a paralyzing toxin. A small animal unlucky enough to brush against one of the tentacles is suddenly speared with hundreds or even thousands of nematocysts and quickly immobilized. Some nematocyst threads can penetrate human skin, causing sensations ranging from minor irritation to great pain, or even death, depending on the species. A nematocyst is a fearsome, but wondrous, tiny weapon.

Tentacles of a Caribbean sea anemone, *Condylactis gigantea*.

P hyla Cnidaria and Ctenophora exhibit primary radial or biradial symmetry. Radial symmetry, in which body parts are arranged concentrically around an oral-aboral axis, is particularly suitable for sessile or sedentary animals and for free-floating animals because they approach their environment (or it approaches them) from all sides equally. Biradial symmetry is basically a type of radial symmetry in which only two planes through the oral-aboral axis divide the animal into mirror images because the body contains a paired structure. All other eumetazoans have a primary bilateral symmetry; they are bilateral or were derived from an ancestor that was bilateral.

Neither phylum has advanced generally beyond the tissue level of organization, although a few organs occur. In general, ctenophores are structurally more complex than cnidarians. Members of both groups are sometimes considered diploblastic because their adult epidermis and gastrodermis develop from two embryonic layers (ectoderm and endoderm, respectively). Diploblastic organisms cannot produce true muscles, which come from mesoderm, so there is debate about the nature of muscles in ctenophores.

■ Phylum Cnidaria

Phylum Cnidaria (nī-dar´-ēə) (Gr. *knidē,* nettle, + L. *aria* [pl. suffix]; like or connected with) is a group of more than 9000 species. It takes its name from cells called **cnidocytes,** which contain the stinging organelles (cnidae) characteristic of the phylum. Cnidae come in several types, including the common **nematocysts.** Nematocysts are formed only by cnidarians.

Cnidarians are an ancient group with the longest fossil history of any animal phylum, reaching back more than 700 million years. Although their organization has a structural and functional simplicity not found in other animals, they form a significant proportion of the biomass in some locations. They are widespread in marine habitats, and some occupy fresh water. Although cnidarians are mostly sessile or, at best, fairly slow moving or slow swimming, they are quite efficient predators of organisms that are much swifter and more complex.

The phylum Cnidaria includes some of nature's strangest and loveliest creatures: branching, plantlike hydroids; flowerlike sea anemones; jellyfishes; and those architects of the ocean floor, gorgonian corals (sea whips, sea fans, and others) and stony corals, whose thousands of years of calcareous house-building have produced great reefs and coral islands. Algae frequently live as mutualists in the tissues of cnidarians, notably in some freshwater hydras and in reef-building corals. The presence of algae in reef-building corals limits coral reefs to relatively shallow, clear water where sunlight is sufficient for the photosynthetic requirements of the algae. These corals are an essential component of coral reefs, and reefs are extremely important habitats in tropical waters. Coral reefs are discussed further later in the chapter (see p. 147).

Four classes of Cnidaria were traditionally recognized: Hydrozoa (the most variable class, including hydroids, fire corals, Portuguese man-of-war, and others), Scyphozoa ("true" jellyfishes), Cubozoa (cube jellyfishes), and Anthozoa (the largest

CHARACTERISTICS
of Phylum Cnidaria

1. **Cnidocytes** present, typically housing stinging organelles called **nematocysts**
2. Entirely aquatic, some in freshwater, but most marine
3. **Radial symmetry** or biradial symmetry around a longitudinal axis with oral and aboral ends; no definite head
4. Two types of individuals: **polyps** and **medusae**
5. Adult body two-layered (**diploblastic**) with epidermis and gastrodermis derived from embryonic ectoderm and endoderm, respectively
6. Mesoglea, an extracellular matrix ("jelly") lies between body layers; amount of mesoglea varies; some have mesoglea with cells and connective tissue from ectoderm
7. Incomplete gut called **gastrovascular cavity;** often branched or divided with septa
8. Extracellular digestion in gastrovascular cavity and intracellular digestion in gastrodermal cells
9. Extensible tentacles usually encircle mouth or oral region
10. Muscular contractions via epitheliomuscular cells, which form an outer layer of longitudinal fibers at base of epidermis and an inner layer of circular fibers at base of gastrodermis; modifications of plan in hydrozoan medusae (independent ectodermal muscle fibers) and other complex cnidarians
11. Sense organs include well-developed statocysts (organs of balance) and ocelli (photosensitive organs); complex eyes in members of Cubozoa
12. Nerve net with symmetrical and asymmetrical synapses; diffuse conduction; two nerve rings in hydrozoan medusae
13. Asexual reproduction by budding in polyps forms clones and colonies; some colonies exhibit **polymorphism***
14. Sexual reproduction by gametes in all medusae and some polyps; monoecious or dioecious; holoblastic indeterminate cleavage; planula larval form
15. No excretory or respiratory system
16. No coelomic cavity

---

* Note that polymorphism here refers to more than one structural form of individual within a species, as contrasted with the use of the word in genetics (see p. 27), where it refers to different allelic forms of a gene in a population.

class, including sea anemones, stony corals, soft corals, and others) (figure 7.1). A fifth class, Staurozoa, has been proposed for a group of odd animals whose bodies resemble polyps with a medusa-like region on the top (see p. 142).

Ecological Relationships

Cnidarians occur most abundantly in shallow marine habitats, especially in warm temperatures and tropical regions. There are some freshwater species, but no terrestrial species. Colonial hydroids are usually found attached to mollusc

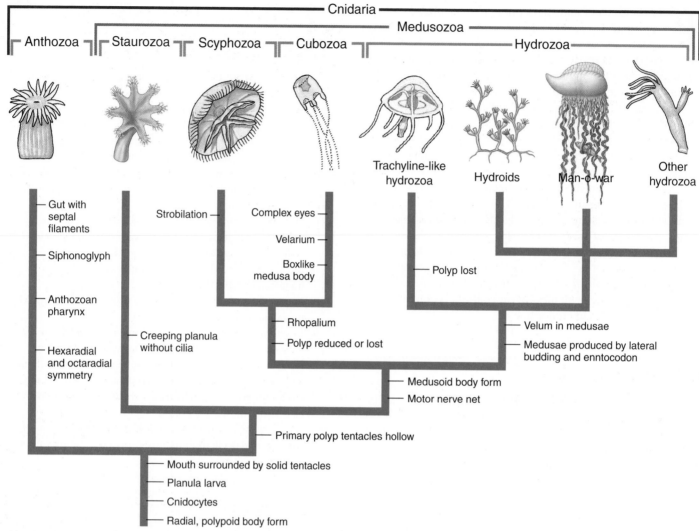

figure 7.1

Cladogram showing hypothetical relationships of cnidarian classes with some shared derived characters (synapomorphies) indicated. Relationships are according to data of Collins et al. (2006, Syst. Biol. **55**:97–115). Synapomorphies are adapted from Brusca and Brusca (2003, Invertebrates ed.2. Sunderland, Massachusetts, Sinauer Associates, Inc.).

shells, rocks, wharves, and other animals in shallow coastal water, but some species occur at great depths. Floating and free-swimming medusae occur in open seas and lakes, often far from shore. Colonies such as the Portuguese man-of-war and *Velella* (L. *velum,* veil, + *ellus,* dim. suffix) have floats or sails by which the wind carries them.

Some ctenophores, molluscs, and flatworms eat hydroids bearing nematocysts and use these stinging structures for their own defense. Other animals, such as some molluscs and fishes, feed on cnidarians, but cnidarians rarely serve as food for humans. Planktonic medusae may be of some importance as food for commercially valuable fishes, but the reverse is also true—young fish fall prey to cnidarians.

Cnidarians sometimes live symbiotically with other animals, often as commensals on the shell or other surface of their host. Certain hydroids (figure 7.1) and sea anemones commonly live on snail shells inhabited by hermit crabs (figure 7.2), giving the crabs some protection from predators.

Although many cnidarians have little economic importance, reef-building corals are an important exception. Fish and other animals associated with reefs provide substantial amounts of food for humans, and reefs are of economic value as tourist attractions. Precious coral is used for jewelry and ornaments, and coral rock for buildings.

Form and Function

Dimorphism and Polymorphism in Cnidarians

One of the most interesting—and sometimes puzzling—aspects of phylum Cnidaria is the dimorphism and often polymorphism displayed by many of its members. All cnidarian forms fit into one of two morphological types (dimorphism): a **polyp,** or hydroid form, which has a sedentary or sessile lifestyle, and a **medusa,** or jellyfish form, which has a floating or free-swimming existence (figure 7.3).

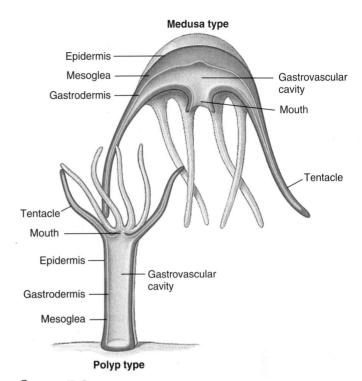

figure 7.3

Comparison between the polyp and medusa types of individuals.

A

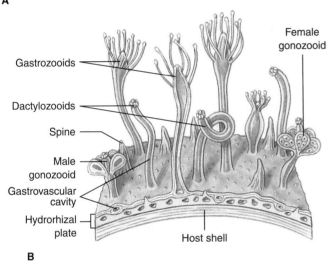

B

figure 7.2

A, A hermit crab with its cnidarian mutuals. The shell is blanketed with polyps of the hydrozoan *Hydractinia Symbiolongicarpus*. The crab gets some protection from predation by the presence of cnidarians, and the cnidarians get a free ride and bits of food from their host's meals. **B,** Portion of a colony of *Hydractinia*, showing the types of zooids and the stolon (hydrorhiza) from which they grow. Note that male and female gonozooids do not naturally occur on the same colony.

Most polyps have tubular bodies. A mouth surrounded by tentacles defines the oral end of the body. The mouth leads into a blind gut or gastrovascular cavity (figure 7.3). The aboral end of the polyp is usually attached to a substratum by a pedal disc or other device.

Polyps may reproduce asexually by budding, fission, or laceration of the pedal disc. In **budding,** a knob of tissue forms on the side of an existing polyp and develops a functional mouth and tentacles (see figure 7.8). A bud that detaches from the polyp that made it is a clone. Clones can also be formed by **fission,** in which one-half of a polyp pulls away from the other, or by laceration of the pedal disc, in which tissue torn from the pedal disc develops into tiny new polyps (see p. 145). Polyps that do not bud are solitary; others form clones or colonies. The distinction between clones and colonies is blurred when a colony fragments.

When buds stay attached to the polyp that made them, a colony forms, and food may be shared through a common gastrovascular cavity (see figure 7.2). A shared gastrovascular cavity permits some polyps (often called zooids) to specialize for particular functions, such as feeding (gastrozooids), defense (dactylozooids), or making sexually reproducing stages (gonozooids). Gastrozooids and dactylozooids do not collect food, but are fed through the shared gastrovascular cavity. A colony that contains several body forms is polymorphic; such colonies are common in class Hydrozoa.

The name "medusa" was suggested by a fancied resemblance to the Gorgon Medusa, a mythological creature with snaky tresses that turned to stone any who gazed upon her.

Medusae are usually free-swimming and have bell-shaped or umbrella-shaped bodies and tetramerous symmetry (body parts arranged in fours). The mouth is usually centered on the concave side, and tentacles extend from the rim of the umbrella.

Superficially, polyps and medusae seem very different, but each has retained the saclike body plan basic to the phylum (figure 7.3). A medusa is essentially an unattached polyp with the tubular portion widened and flattened into the bell shape.

Both a polyp and a medusa possess two tissue layers with an acellular mesoglea between them, but the jellylike layer of mesoglea is much thicker in a medusa, forming the bulk of the animal and making it more buoyant. Because of this mass of mesoglea ("jelly"), medusae are commonly called jellyfishes.

Locomotion

Colonial polyps are permanently attached, but hydras can move freely across a substrate by gliding on their pedal disc, aided by mucous secretions. Sea anemones can move similarly on their pedal discs. Hydras can also use a "measuring worm" movement, looping along a surface by bending over and attaching their tentacles to the substratum. They may even turn handsprings or detach and, by forming a gas bubble on the pedal disc, float to the surface.

Most medusae can move freely, and they swim by contracting the bell, which expels water from the concave, oral side. The muscular contractions are antagonized by the compressed mesoglea and elastic fibers within it. Usually, medusae contract several times and move generally upward, then sink slowly. Cubozoan medusae, however, can swim strongly.

Life Cycles

In a cnidarian life cycle, polyps and medusae play different roles. The exact life cycle varies among the classes, but in general a zygote develops into a free-swimming **planula** larva. The planula settles and metamorphoses into a polyp. A polyp may make other polyps asexually, but polyps in classes Hydrozoa and Scyphozoa eventually make medusae. Although medusae are made asexually from the polyp body, they reproduce sexually. Medusae develop into either male or female individuals (dioecious) and produce gametes. Fertilization typically occurs in open water, although some species brood their zygotes until the planula stage.

Sea anemones and corals (class Anthozoa) are all polyps; hence, both sexual and asexual reproduction occur in the polyp phase. The true jellyfishes (class Scyphozoa) have a conspicuous medusoid form, although most also have a polypoid larval stage. Colonial hydroids of class Hydrozoa typically have life histories that feature both a polyp stage and a free-swimming medusa stage. A species that has both an attached polyp and a floating medusa within its life history can take advantage of the feeding and distribution possibilities of both pelagic (open-water) and benthic (bottom) environments.

Feeding and Digestion

Cnidarians prey on a variety of organisms of appropriate size; larger species are usually capable of killing and eating larger prey. The mouth opens into the **gastrovascular cavity,** which communicates with cavities in the tentacles. The mouth may be surrounded by an elevated manubrium or by elongated oral lobes in the medusa stage. Normally, prey organisms are drawn into the gastrovascular cavity into which gland cells discharge enzymes that initiate *extracellular digestion*. In addition, nutritive-muscular cells phagocytize many food particles for *intracellular digestion*. Ameboid cells can carry undigested particles to the gastrovascular cavity, where they are eventually expelled with other indigestible matter.

Body Wall

In a cnidarian polyp, such as *Hydra,* the body wall surrounding the gastrovascular cavity consists of an outer **epidermis** (ectodermal) and an inner **gastrodermis** (endodermal) with **mesoglea** between them (see figure 7.4). Each layer and the cells it contains are discussed briefly.

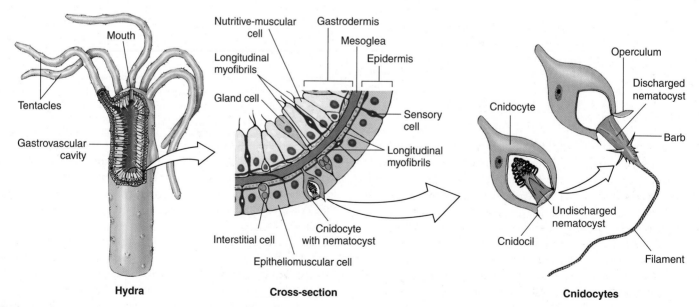

Hydra   **Cross-section**   **Cnidocytes**

figure 7.4

A Cnidarian polyp, illustrated by *Hydra*, (*left*), has specialized cells in the body wall (*center*). The epidermal layer of the body wall contains specialized cnidocytes called nematocysts. Nematocysts (*right*) arise from epidermal interstitial cells and are used to sting prey.

Mesoglea   Mesoglea lies between the epidermis and the gastrodermis and adheres to both layers (see figure 7.3). It is gelatinous, or jellylike, and has no fibers or cellular elements in hydrozoan polyps. It is thicker in medusae and has elastic fibers; in scyphozoan medusae, it has ameboid cells. The mesoglea of anthozoans contains ameboid cells and epitheliomuscular cells (see below).

Gastrodermis   The gastrodermis, a layer of cells lining the gastrovascular cavity, consists chiefly of large, ciliated, columnar epithelial cells with irregular flat bases (figure 7.4, *center*). Cells of the gastrodermis include nutritive-muscular, interstitial, and gland cells and, in classes other than Hydrozoa, cnidocytes. Gonads are gastrodermal in most cnidarians.

Nutritive-muscular cells are usually tall columnar cells that have laterally extended bases containing myofibrils. In hydrozoans, the myofibrils run at right angles to the body or tentacle axis and so form a circular muscle layer. However, this muscle layer is very weak, and longitudinal extension of the body and tentacles occurs mostly by increasing the volume of water in the gastrovascular cavity. Water is brought into the cavity through the mouth by the beating of cilia on the nutritive-muscular cells in hydrozoans or by ciliated cells in the pharynx of anthozoans. Thus, water in the gastrovascular cavity serves as a **hydrostatic skeleton.** The two cilia on the free end of each cell also serve to circulate food and fluids in the digestive cavity. The cells often contain large numbers of food vacuoles. Gastrodermal cells of green hydras (*Chlorohydra* [Gr. *chlōros,* green, + *hydra,* a mythical nine-headed monster slain by Hercules]) bear green algae (*zoochlorellae*), but in marine cnidarians they bear a type of dinoflagellate (see p. 108) (*zooxanthellae*). Both are examples of mutualism, with the algae furnishing organic compounds to their cnidarian hosts.

Interstitial cells scattered among the bases of the nutritive cells can transform into other cell types. Gland cells are tall cells that secrete digestive enzymes.

Epidermis   The epidermal layer contains epitheliomuscular, interstitial, gland, cnidocyte, and sensory and nerve cells. Gonads are epidermal in Hydrozoa.

> **Epitheliomuscular cells** form most of the epidermis (figure 7.5) and serve both for covering and for muscular contraction. The bases of most of these cells are extended parallel to the tentacle or body axis and contain myofibrils, thus forming a layer of longitudinal muscle next to the mesoglea. Contraction of these fibrils shortens the body or tentacles.
>
> **Interstitial cells** are undifferentiated stem cells found among the bases of the epitheliomuscular cells. Differentiation of interstitial cells produces cnidoblasts, sex cells, buds, nerve cells, and others, but generally not epitheliomuscular cells (which reproduce themselves).

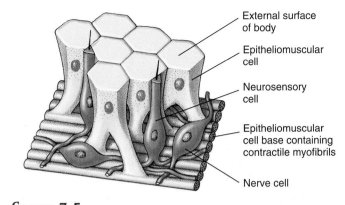

External surface of body
Epitheliomuscular cell
Neurosensory cell
Epitheliomuscular cell base containing contractile myofibrils
Nerve cell

figure 7.5

Epitheliomuscular and nerve cells in hydra.

> **Gland cells** are particularly abundant around the mouth and in the pedal disc of hydra. They secrete mucus or adhesive material.
>
> **Cnidocytes** (containing cnidae) occur throughout the epidermis (figure 7.4 *right*). They may be between the epitheliomuscular cells or housed in invaginations of these cells, and they are most abundant on the tentacles.
>
> **Sensory cells** are scattered among the other epidermal cells, especially around the mouth and tentacles. The free end of each sensory cell bears a flagellum, which is the sensory receptor for chemical and tactile stimuli. The other end branches into fine processes, which synapse with nerve cells.
>
> **Nerve cells** of the epidermis are often multipolar (have many processes), although in more highly organized cnidarians, the cells may be bipolar (with two processes). Their processes (axons) form synapses with sensory cells and other nerve cells, and junctions with epitheliomuscular cells and cnidocytes. Both one-way and two-way synapses with other nerve cells are present.

Cnidae   Over 20 different types of cnidae have been described in cnidarians, and they are important taxonomic characters. **Cnidae** are tiny capsules contained within cnidocytes. They are composed of material similar to chitin and contain a coiled tubular "thread" or filament, which is a continuation of the narrowed end of the capsule (figure 7.6). This end of the capsule is covered by a little lid, or **operculum.** The inside of the undischarged thread may bear tiny barbs, or spines, as in the most common cnida, the **nematocyst.** There are three functional types of cnidae in hydras: those that penetrate prey and inject venom (penetrants, see figure 7.4); those that recoil and entangle prey (volvents, see figure 7.6); and those that secrete an adhesive substance used in locomotion and attachment (glutinants).

A cnida is enclosed in the cell that produced it. (During its development, a cnidocyte is properly called a **cnidoblast.**) Except in Anthozoa, cnidocytes are equipped with a triggerlike **cnidocil,** which is a modified cilium. Anthozoan cnidocytes

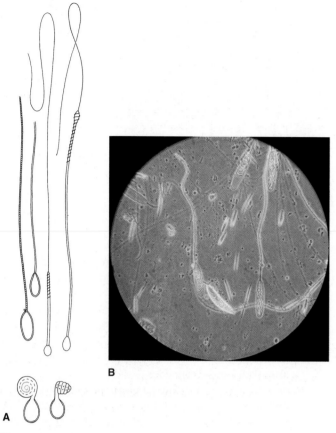

B

A

figure 7.6

A, Several types of cnidae shown after discharge. At bottom are two views of a type that does not impale prey but recoils like a spring, catching any small part of the prey in the path of the recoiling thread. **B,** Fired and unfired cnidae from *Corynactis californica*.

Note the distinction between osmotic and hydrostatic pressure. A nematocyst is never actually required to contain 140 atmospheres of hydrostatic pressure within itself; such a hydrostatic pressure would doubtless cause it to explode. As water rushes in during discharge, osmotic pressure falls rapidly, while hydrostatic pressure increases rapidly.

The nematocysts of most cnidarians are not harmful to humans and are a nuisance at worst. However, stings of the Portuguese man-of-war (see figure 7.13) and certain jellyfish are quite painful and sometimes dangerous. The venom of box jellyfishes (Class Cubozoa, p. 142) may be lethal.

Nerve Net   The nerve net of the cnidarians is one of the best examples of a diffuse nervous network in the animal kingdom. This plexus of nerve cells occurs both at the base of the epidermis and at the base of the gastrodermis, forming two interconnected nerve nets. Nerve processes (axons) end on other nerve cells at synapses or at junctions with sensory cells or effector organs (nematocysts or epitheliomuscular cells). Nerve impulses are transmitted from one cell to another by release of a neurotransmitter from small vesicles on one side of the synapse or junction. One-way transmission between nerve cells in higher animals is ensured because the vesicles are located on only one side of the synapse. However, cnidarian nerve nets are peculiar in that many of the synapses have vesicles of neurotransmitters on both sides, allowing transmission across the synapse in either direction. Another peculiarity of cnidarian nerves is the absence of any sheathing material (myelin) on the axons.

Cnidarians have no concentrated grouping of nerve cells to suggest a "central nervous system." However, nerves are grouped in the "ring nerves" of hydrozoan medusae and in the marginal sense organs of scyphozoan medusae. In some cnidarians, the nerve nets form two or more systems: Scyphozoa has a fast conducting system to coordinate swimming movements and a slower one to coordinate movements of tentacles.

Note that there is little adaptive value for a radially symmetrical animal in having a central nervous system with a brain. The environment approaches from all sides equally, allowing no contro over the direction of approach to a prey organism.

have a somewhat different ciliary mechanoreceptor. In some sea anemones and perhaps other cnidarians, small organic molecules from a prey "tune" the mechanoreceptors, sensitizing them to the frequency of vibration caused by the prey swimming. Tactile stimulation causes a nematocyst to discharge. Cnidocytes are borne in invaginations of ectodermal cells and, in some forms, in gastrodermal cells, and they are especially abundant on the tentacles. When a nematocyst has discharged, its cnidocyte is absorbed and a new one replaces it.

The mechanism of nematocyst discharge is remarkable. Present evidence indicates that discharge is caused by a combination of tensional forces generated during nematocyst formation and also by an astonishingly high osmotic pressure within the nematocyst: 140 atmospheres. When stimulated to discharge, the permeability of the nematocyst changes, and the high internal osmotic pressure causes water to rush into the capsule. The operculum opens, and the rapidly increasing *hydrostatic pressure* within the capsule forces the thread outward with great force, the thread turning inside out as it goes. At the everting end of the thread, the barbs flick to the outside like tiny switchblades. This minute but awesome weapon then injects venom when it penetrates prey.

Nerve cells of the net have synapses with slender sensory cells that receive external stimuli, and the nerve cells have junctions with epitheliomuscular cells and nematocysts. Together with the contractile fibers of epitheliomuscular cells, the sensory cell and nerve-net combination is often considered a cellular neuromuscular system, an important landmark in the evolution of nervous systems. The nerve net arose early in metazoan evolution, and it has never been completely lost phylogenetically. Annelids, such as the earthworm,

Radiate Animals **137**

have it in their digestive systems. In the human digestive system, it occurs as nerve plexuses in the musculature controlling the rhythmic peristaltic movements of the stomach and intestine.

Class Hydrozoa

Most Hydrozoa are marine and colonial in form, and the typical life cycle includes both an asexual polyp and a sexual medusa stage. Some, however, such as the freshwater hydras, have no medusa stage. Certain marine hydroids do not have free medusae (figure 7.7), whereas some hydrozoans occur only as medusae and have no polyp stage.

Hydras, although not typical hydrozoans, have become favorites as an introduction to Cnidaria because of their size and ready availability. Combining study of a hydra with that of a representative colonial marine hydroid such as *Obelia* (Gr. *obelias,* round cake) gives an excellent idea of class Hydrozoa.

Hydra: A Freshwater Hydrozoan The common freshwater hydra (figure 7.8) is a solitary polyp and one of the few cnidarians found in fresh water. Its normal habitat is the underside of aquatic leaves and lily pads in cool, clean freshwater pools and streams. The hydra family occurs throughout the world, with 16 species in North America.

Over 230 years ago, Abraham Trembley was astonished to discover that isolated sections of the stalk of hydra could regenerate and each become a complete animal. Since then, over 2000 investigations of hydra have been published, and the organism has become a classic model for the study of morphological differentiation. The mechanisms governing morphogenesis have great practical importance, and the simplicity of hydra lends itself to these investigations. Substances controlling development (morphogens), such as those determining which end of a cut stalk will develop a mouth and tentacles, can function at extremely low concentrations (10^{-10}M).

The body of a hydra can extend to a length of 25 to 30 mm or can contract to a tiny, jellylike mass. It is a cylindrical tube with the lower (aboral) end drawn out into a slender stalk, ending in a pedal disc for attachment. This pedal disc has gland cells that enable a hydra to adhere to a substratum and also to secrete a gas bubble for floating. In the center of the disc may be an excretory pore. The mouth, located on a conical elevation called the **hypostome,** is encircled by six to ten hollow tentacles that, like the body, can greatly extend when the animal is hungry. The mouth opens into the gastrovascular cavity, which communicates with the cavities in the tentacles.

Hydras feed on a variety of small crustaceans, insect larvae, and annelid worms. The hydra awaits its prey with

figure 7.7

In some hydroids, such as this *Tubularia crocea,* medusae are reduced to gonadal tissue and do not detach. These reduced medusae are known as gonophores.

figure 7.8

Hydra with developing bud on the right and ovary on the left.

figure 7.9

Hydra catches an unwary water flea with the nematocysts of its tentacles. This hydra already contains one water flea eaten previously.

tentacles extended (figure 7.9). A food organism that brushes against its tentacles may find itself harpooned by scores of nematocysts that render it helpless, even though it may be larger than the hydra. The tentacles move the prey toward the mouth, which slowly widens. Well lubricated with mucous secretions, the mouth glides over and around the prey, totally engulfing it.

The chemical activator that actually causes the mouth to open is the reduced form of *glutathione,* which occurs to some extent in all living cells. The prey releases glutathione through the wounds made by the nematocysts, but not all animals release enough of the chemical to activate a feeding response by a hydra. This explains how a hydra distinguishes between *Daphnia,* which it relishes, and some other forms that it refuses. If glutathione is added to water containing hydras, each hydra will go through the motions of feeding, even though no prey is present.

In asexual reproduction, buds appear as outpocketings of the body wall and develop into young hydras that eventually detach from the parent. In sexual reproduction, temporary gonads (see figure 7.8) usually appear in autumn, stimulated by lower temperatures and perhaps also by reduced aeration of stagnant waters. Testes or ovaries, when present, appear as rounded projections on the surface of the body (see figure 7.8). Eggs in the ovary usually mature one at a time and are fertilized by sperm shed into the water. A cyst forms around the embryo before it breaks loose from the parent, enabling it to survive the winter. Young hydras hatch in spring when the weather is favorable.

Hydroid Colonies   Far more representative of class Hydrozoa than hydras are those hydroids that have a medusa stage in their life cycle. *Obelia* is often used in laboratory exercises for beginning students to illustrate the hydroid type (figure 7.10).

A typical hydroid has a base, a stalk, and one or more terminal polyps (zooids). The base by which colonial hydroids attach to the substratum is a rootlike stolon, which gives rise to one or more stalks. The living cellular part of the stalks secretes a nonliving chitinous sheath. Attached to the ends of the branches of the stalks are individual zooids. Most zooids are feeding polyps called hydranths, or *gastrozooids.* They may be tubular, bottle shaped, or vaselike, but all have a terminal mouth and a circlet of tentacles. In some forms, such as *Obelia,* the chitinous sheath continues as a protective cup around the polyp into which the polyp can withdraw for protection (figure 7.10). In others, the polyp is naked. Dactylozooids are polyps specialized for defense in some species (see figure 7.2B).

Hydranths, much like hydras, capture and ingest prey, such as tiny crustaceans, worms, and larvae, thus providing nutrition for the entire colony. After partial digestion in a hydranth, the digestive broth passes into the common gastrovascular cavity where intracellular digestion occurs.

Circulation within the gastrovascular cavity is a function of the ciliated gastrodermis, but rhythmic contractions and pulsations of the body, which occur in many hydroids, also aid circulation.

In contrast to hydras, new individuals that bud do not detach from the parent; thus the size of the colony increases. New polyps may be hydranths or reproductive polyps called **gonangia.** Medusae are produced by budding within the gonangia. Young medusae leave the colony as free-swimming individuals that mature and produce gametes (eggs and sperm) (figure 7.10). In some species, medusae remain attached to the colony and shed their gametes there. In other species, medusae never develop, gametes being shed by male and female gonophores. Development of a zygote produces a ciliated planula larva that swims for a time. Then

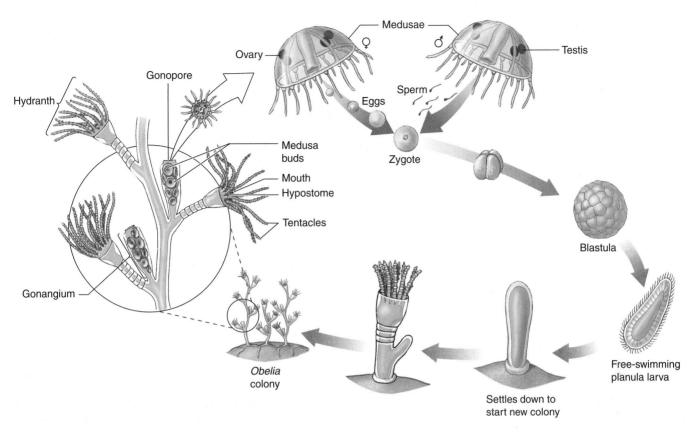

figure 7.10

Life cycle of *Obelia*, showing alternation of polyp (asexual) and medusa (sexual) stages. *Obelia* is a calyptoblastic hydroid, meaning that a nonliving covering protects its polyps as well as its stem.

figure 7.11

The medusa of *Polyorchis penicillatus* is thumb-sized and commonly encountered on the Pacific Coast of North America, although the polyp stage has never been identified.

it settles onto a substratum, becoming a minute polyp that gives rise, by asexual budding, to the hydroid colony, thus completing the life cycle.

Hydroid medusae are usually smaller than their scyphozoan counterparts, ranging from 2 or 3 mm to several centimeters in diameter (figure 7.11). The margin of the bell projects inward as a shelflike **velum,** which partly closes the open side of the bell and functions in swimming (figure 7.12). Muscular pulsations that alternately fill and empty the bell propel the animal forward, aboral side first, with a sort of "jet propulsion." Tentacles attached to the bell margin are richly supplied with nematocysts.

The mouth opening at the end of a suspended **manubrium** leads to a stomach and four radial canals that connect with a ring canal around the margin. This in turn connects with the hollow tentacles. Thus the coelenteron is continuous from mouth to tentacles, and the entire system is lined with gastrodermis. Nutrition is similar to that of hydranths.

The nerve net is usually concentrated into two nerve rings at the base of the velum. The bell margin is liberally supplied with sensory cells. It usually also bears two kinds of specialized sense organs: **statocysts,** which are small organs of equilibrium (figure 7.12), and **ocelli,** which are light-sensitive organs.

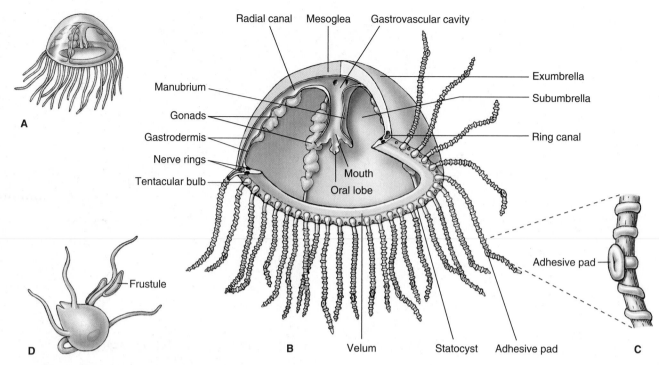

figure 7.12

Structure of *Gonionemus*. **A,** Medusa with typical tetramerous arrangement. **B,** Cutaway view showing morphology. **C,** Portion of a tentacle with its adhesive pad and ridges of nematocysts. **D,** Tiny polyp, or hydroid stage, that develops from the planula larva. It can produce more polyps by budding (frustules) or produce medusa buds.

figure 7.13

A Portuguese man-of-war colony, *Physalia physalis* (order Siphonophora, class Hydrozoa). Colonies often drift onto southern ocean beaches, where they are a hazard to bathers. Each colony of medusa and polyp types is integrated to act as one individual. As many as a thousand zooids may be found in one colony. The nematocysts secrete a powerful neurotoxin.

Other Hydrozoans Siphonophores are hydrozoans that form floating colonies and include *Physalia* (Gr. *physallis,* bladder), the Portuguese man-of-war (figure 7.13). These colonies include several types of modified medusae and polyps. *Physalia* has a rainbow-hued float, probably a modified polyp, which carries it at the mercy of winds and currents. It contains an air sac filled with secreted gas and acts as a carrier for the generations of individuals that bud from it and hang suspended in the water. There are several types of individuals, including feeding polyps, reproductive polyps, long stinging tentacles, and so-called jelly polyps. Many swimmers have experienced the uncomfortable sting that these colonial floaters can inflict. The pain, along with the panic of the swimmer, can increase the danger of drowning.

Other hydrozoans secrete massive calcareous skeletons that resemble true corals (figure 7.14). They are sometimes called **hydrocorals.**

Class Scyphozoa

Class Scyphozoa (si-fō-zo´ə) (Gr. *skyphos,* cup) includes most of the larger jellyfishes, or "cup animals." A few, such as *Cyanea* (Gr. *kyanos,* dark-blue substance), may attain a bell diameter exceeding 2 m and tentacles 60 to 70 m long (figure 7.15). Most scyphozoans, however, range from 2 to 40 cm in diameter. Most float in the open sea, some even at depths of 3000 m. Their coloring ranges from colorless to striking orange and pink hues.

The medusae of scyphozoans, unlike those of hydrozoans, have no velum. The bells of different species vary in depth from a shallow saucer shape to a deep helmet or goblet shape, and many have a scalloped margin, each notch bearing a sense organ called a **rhopalium** and a pair of lobelike projections called lappets. *Aurelia* (L. *aurum,* gold) has eight such notches

A

B

figure 7.15

Giant jellyfish, *Cyanea capillata* (order Semeaeostomeae, class Scyphozoa). A North Atlantic species of *Cyanea* reaches a bell diameter exceeding 2 m. Many fishermen know it as "sea blubber."

figure 7.14

These hydrozoans form calcareous skeletons that resemble true coral. **A,** *Stylaster roseus* (order Stylasterina) occurs commonly in caves and crevices in coral reefs. These fragile colonies branch in only a single plane and may be white, pink, purple, red, or red with white tips. **B,** Species of *Millepora* (order Milleporina) form branching or platelike colonies and often grow over the skeletons of gorgonians. They have a generous supply of powerful nematocysts that produce a burning sensation on human skin, justly earning them the common name fire coral.

(figures 7.16 and 7.17); others may have four or sixteen. Each rhopalium bears a statocyst for balance, two sensory pits containing concentrations of sensory cells, and sometimes an ocellus (simple photoreceptor). The mesoglea is thick and contains cells as well as fibers.

The mouth is centered on the subumbrellar side. The manubrium is usually drawn out into four frilly **oral lobes** used in food capture and ingestion. Marginal tentacles may be many or few; they may be short, as in *Aurelia,* or long, as in *Cyanea.* The tentacles, the manubrium, and often the entire body surface of scyphozoans are well supplied with nematocysts. Scyphozoans feed on small organisms, from protozoa to fishes. Capture of some prey involves stinging and manipulation with tentacles and oral lobes, but methods vary. *Aurelia* also feeds on small planktonic animals caught in mucus of the umbrella surface, carried to "food pockets" on the umbrella margin by cilia, and picked from the pockets by the oral lobes, whose cilia carry the food to the gastrovascular cavity. Cilia on the gastrodermis keep a current of water moving to bring food and oxygen into the gastrovascular cavity and to remove wastes.

Internally, four gastric pouches containing nematocysts connect a complex system of **radial canals** that branch from the pouches to the ring canal (see figure 7.16) forming the gastrovascular cavity, through which nutrients circulate.

The sexes are separate in scyphozoans; their gonads are located in the gastric pouches. Fertilization is internal, with sperm carried by ciliary currents into the gastric pouch of the female. Zygotes may develop in seawater or may be brooded in folds of the oral arms. The ciliated planula larva attaches and develops into a **scyphistoma,** a hydralike form (see figure 7.16) that may bud to make other polyps. By a process called **strobilation,** the scyphistoma of *Aurelia* forms a series of saucerlike buds (**ephyrae**) and becomes a **strobila** (see figure 7.16). When the ephyrae break loose, they grow into mature jellyfish.

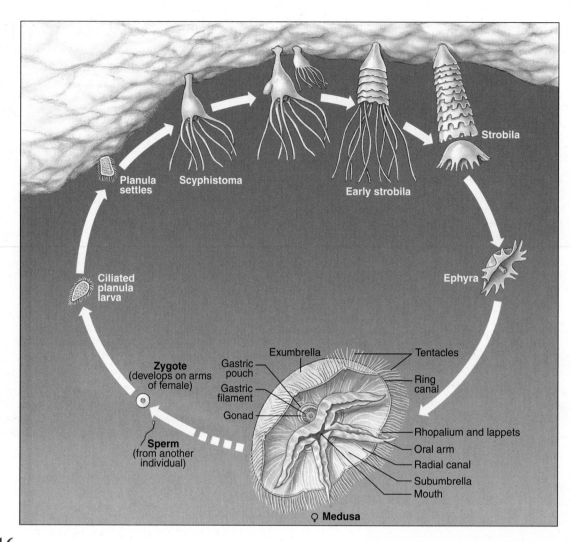

figure 7.16

Life cycle of *Aurelia*, a marine scyphozoan medusa.

Class Staurozoa

Animals in class Staurozoa are commonly called stauromedu-sans and were previously considered unusual scyphozoans, even though their life cycle does not include a medusa phase. The solitary polyp body is stalked (figure 7.18) and uses an adhesive disc to attach to seaweeds and other objects on the sea floor. The top of the polyp resembles a medusa. It has eight extensions ("arms"), ending in tentacle clusters, surrounding the mouth. Polyps reproduce sexually. The nonswimming planula develops directly into a new polyp.

Class Cubozoa

In Cubozoa the medusa is the predominant form (figure 7.19); the polyp is inconspicuous and in most cases unknown. In transverse section, medusa bells are almost square. A tentacle or group of tentacles occurs at each corner of the square at the umbrella margin. The base of each tentacle is differentiated

figure 7.17

Moon jellyfish *Aurelia aurita* (class Scyphozoa) is worldwide in distribution. It feeds on planktonic organisms caught in mucus on its umbrella.

figure 7.18

Thaumatoscyphus hexaradiatus in class Staurozoa is an unusual cnidarian. The oral end of the animal resembles a medusa.

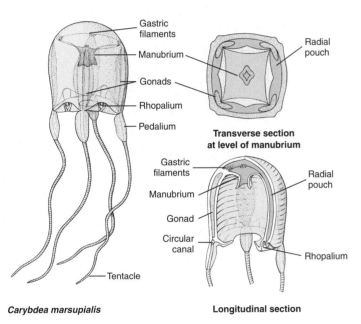

Carybdea marsupialis

figure 7.19

Carybdea, a cubozoan medusa.

into a flattened, tough blade called a **pedalium** (figure 7.19). Rhopalia each house six eyes, of three different types, in addition to other sense organs. Some eyes are image-forming. The umbrella margin is not scalloped, and the subumbrella edge turns inward to form a **velarium.** The velarium functions as a velum does in hydrozoan medusae, increasing swimming efficiency, but it differs structurally. Cubomedusae are strong swimmers and voracious predators, feeding mostly on fish. Members of at least one species exhibit complex mating behaviors, including twined tentacles as the male passes a spermatophore to the female.

Chironex fleckeri (Gr. *cheir*, hand, + *nexis*, swimming) is a large cubomedusa called the sea wasp. Its stings are dangerous and sometimes fatal. Most fatalities have been reported from tropical Australian waters, usually following massive stings. Witnesses have described victims as covered with "yards and yards of sticky wet string." Stings are very painful, and death, if it is to occur, ensues within minutes. If the victim does not die within 20 minutes after stinging, complete recovery is likely.

Class Anthozoa

Anthozoans, or "flower animals," are polyps with a flower-like appearance (figure 7.20). There is no medusa stage. Anthozoa are all marine and occur in both deep and shallow water and in polar seas as well as tropical seas. They vary greatly in size and may be solitary or colonial. Many are supported by skeletons.

Class Anthozoa has three subclasses: **Zoantharia** (or **Hexacorallia**), composed of the sea anemones, hard corals, and others; **Ceriantipatharia,** which includes only tube anemones and thorny corals; and **Octocorallia** (or **Alcyonaria**), containing soft and horny corals, such as sea fans, sea pens, sea pansies, and others. Zoantharians and ceriantipatharians have a **hexamerous** body plan (based on six or multiples of six) or polymerous symmetry and have simple tubular tentacles arranged in one or more circlets on the oral disc. Octocorallians are **octomerous** (built on a plan of eight) and always have eight pinnate (featherlike) tentacles arranged around the margin of the oral disc (figure 7.21).

The gastrovascular cavity is large and partitioned by septa, or mesenteries, that are inward extensions of the body wall. Where one septum extends into the gastrovascular cavity from the body wall, another extends from the diametrically opposite side; thus, these septa are *coupled*. In Zoantharia, septa are not only coupled, but also *paired* on each side (figure 7.22). The muscular arrangement varies among different groups, but the body wall usually has circular muscles, while the septa have longitudinal and transverse muscles.

Subclass Ceriantipatharia has been created from the Ceriantharia and Antipatharia, formerly considered orders of Zoantharia. Ceriantharians are tube anemones that live in soft sediments, buried to the level of the oral disc. Antipatharians are thorny or black corals. They are colonial and have a hard axial skeleton. Both of these groups are small in numbers of species and limited to warmer waters of the sea.

Anthozoans tend to exhibit biradial symmetry in the septal arrangement and in the shape of the mouth and pharynx. They have no special organs for respiration or excretion.

Sea Anemones Sea anemone polyps (subclass Zoantharia, order Actiniaria) are larger and heavier than hydrozoan polyps

figure 7.20

Sea anemones (order Actiniaria, subclass Zoantharia) are the familiar and colorful "flower animals" of tide pools, rocks, and pilings of the intertidal zone. Most, however, are subtidal, their beauty seldom revealed to human eyes. These are rose anemones, *Tealia piscivora*.

A

B

figure 7.21

A, Orange sea pen *Ptilosarcus gurneyi* (order Pennatulacea, subclass Octocorallia, class Anthozoa). Sea pens are colonial forms that inhabit soft sediment. The base of the fleshy body of the primary polyp is buried in sediment. It gives rise to numerous secondary, branching polyps. **B,** Close-up of a gorgonian. The pinnate tentacles characteristic of the subclass Octocorallia are apparent.

(see figures 7.20 and 7.21). Most range from 5 mm or less to 100 mm in diameter, and from 5 mm to 200 mm long, but some grow much larger. Some are quite colorful. Anemones occur in coastal areas all over the world, especially in warmer waters, and they attach by means of their pedal discs to shells, rocks, timber, or whatever submerged substrata they can find. Some burrow in mud or sand.

Sea anemones are cylindrical, with a crown of tentacles arranged in one or more circles around the mouth on the flat **oral disc** (figure 7.22). The slit-shaped mouth leads into a **pharynx.** At one or both ends of the mouth is a ciliated groove called a **siphonoglyph,** which extends into the pharynx. Siphonoglyphs create water currents directed into the pharynx. Cilia elsewhere on the pharynx direct water outward. Currents thus created bring oxygen and remove wastes. They also help to maintain an internal fluid pressure, providing a hydrostatic skeleton that supports opposing muscles.

The pharynx leads into a large gastrovascular cavity divided into radial chambers by pairs of septa that extend vertically from the body wall toward the pharynx (figure 7.22). These chambers communicate with each other and are open below the pharynx. In many anemones, the lower ends of the septal edges are prolonged into threads called **acontia,** which are also provided with nematocysts and gland cells, that can be protruded through the mouth or through pores in the body wall to overcome prey or to provide defense. The pores can rapidly discharge water from the body when the animal is endangered and contracts to a small size.

Anemones form some interesting mutualistic relationships with other organisms. Many anemones house unicellular algae in their tissues (as do reef-building corals), from which they derive some nutrients. Some hermit crabs place anemones on the snail shells in which the crabs live, gaining some protection from predators by the presence of the anemone, while the anemone dines on particles of food dropped by the crab. Anemone fishes (figure 7.23) of the tropical Indo-Pacific form associations with large anemones. A property of the skin mucus of this fish prevents the anemone's nematocysts from discharging. By studying the mucus, biologist Amit Lotan was able to develop a protective lotion for swimmers.

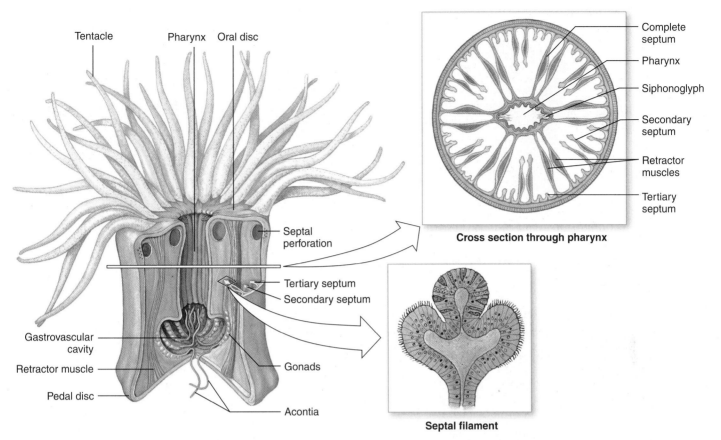

Cross section through pharynx

Septal filament

figure 7.22

Structure of the sea anemone. The free edges of the septa and the acontia threads are equipped with nematocysts to complete the paralyzation of prey begun by the tentacles.

figure 7.23

Orangefin anemone fish *(Amphiprion chrysopterus)* nestles in the tentacles of its sea anemone host. Anemone fishes do not elicit stings from their hosts but may lure unsuspecting other fish to become meals for the anemone.

Sea anemones are carnivorous, feeding on fish or almost any live animal of suitable size. Some species live on minute forms caught by ciliary currents.

The sexes are separate in some sea anemones (dioecious), and other anemones are monoecious (hermaphroditic). Gonads are arranged on the margins of the septa. Fertilization is external in some species, whereas in others the sperm enter the gastrovascular cavity to fertilize eggs. The zygote develops into a ciliated larva that settles to become a polyp. Asexual reproduction commonly occurs by pedal laceration or by transverse fission.

Zoantharian Corals   Zoantharian corals belong to order Scleractinia of subclass Zoantharia, sometimes called the stony (or true) corals. Stony corals look like miniature sea anemones that live in calcareous cups they themselves have secreted (figures 7.24 and 7.25). Like that of anemones, a coral polyp's gastrovascular cavity is subdivided by septa arranged in multiples of six (hexamerous), and its hollow tentacles surround the mouth, but there is no siphonoglyph.

Instead of a pedal disc, the epidermis at the base of the column secretes a limy skeletal cup, including sclerosepta, which project upward into the polyp between its true septa (figure 7.25). Living polyps can retract into the safety of their cup when not feeding. Since the skeleton is secreted below

A

B

C

figure 7.24

Zoantherian corals. **A,** Cup coral, *Tubastrea* sp. The polyps form clumps resembling groups of sea anemones. Although often found on coral reefs, *Tubastrea* is not a reef-building coral (ahermatypic) and has no symbiotic zooxanthellae in its tissues. **B,** The polyps of *Montastrea cavernosa* are tightly withdrawn in the daytime but open to feed at night, as in **C** (order Scleractinia, subclass Zoantharia).

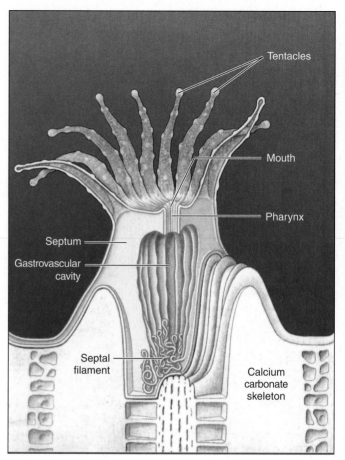

figure 7.25

Polyp of a zoantharian coral (order Scleractinia), showing calcareous cup (exoskeleton), gastrovascular cavity, septa, and septal filaments.

the living tissue rather than within it, the calcareous material is an exoskeleton. In many colonial corals, the skeleton may become massive, accumulating over many years, with the living coral forming a sheet of tissue over the surface (figure 7.26). The gastrovascular cavities of the polyps all connect through this sheet of tissue.

Octocorallian Corals   Octocorals include soft corals, sea pansies, sea pens, and sea fans and other gorgonian corals (horny corals). They have strict octomerous symmetry, with eight pinnate tentacles and eight unpaired, complete septa (figure 7.27). Almost all are colonial, and the gastrovascular cavities of polyps in colonies communicate through a system of gastrodermal tubes called **solenia.** The tubes run through an extensive mesoglea in most octocorals, and epidermis covers the surface of the colony.

The graceful beauty of octocorals—in hues of yellow, red, orange, and purple—helps to create the "submarine gardens" of coral reefs (figure 7.28). Octocoral colonies may be thick encrusting mats or ribbons, tall forms resembling an antique quill pen, or whiplike clusters of long vertical branches. Soft

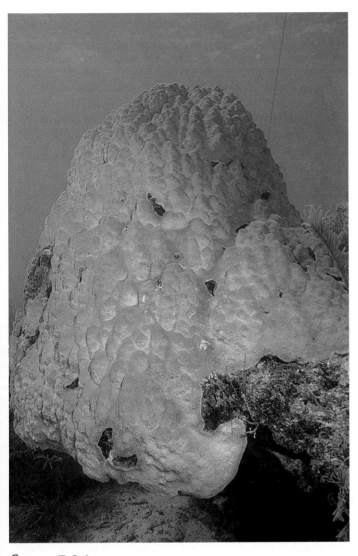

figure 7.26

Boulder star coral, *Montastrea annularis* (subclass Zoantharia, class Anthozoa). Colonies can grow up to 10 feet (3 m) high.

corals have fleshy bodies with calcareous spicules in the mesoglea, but the skeletons of other octocorals typically combine fused or separate spicules with a stiffened yet flexible protein called gorgonin. Gorgonin has chemical similarities to keratin and to collagen; it is responsible for structural support in horny corals, including gorgonians. Because skeletal structures are secreted within the mesoglea, most octocorals have an endoskeleton.

Coral Reefs Coral reefs are among the most productive of all ecosystems, and their diversity of life forms is rivaled only by tropical rain forests. Coral reefs are large formations of calcium carbonate (limestone) in shallow tropical seas laid down by living organisms over thousands of years; living plants and animals are confined to the top layer of reefs, where they add more calcium

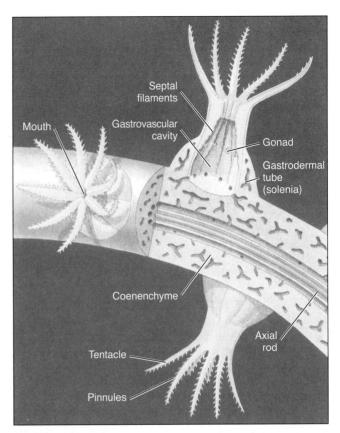

figure 7.27

Polyps of an octocorallian coral. Note the eight pinnate tentacles, coenenchyme, and solenia. These corals have an endoskeleton of limy spicules, often with a hard structural protein, which may form an axial rod.

figure 7.28

A soft coral, *Dendronephthya* sp. (order Alcyonacea, subclass Octocorallia, class Anthozoa), on a Pacific coral reef. The showy hues of this soft coral vary from pink and yellow to bright red and contribute much color to Indo-Pacific reefs.

Taxonomy of Phylum Cnidaria

Class Hydrozoa (hī-drō-zō´ə) (Gr. *hydra*, water serpent, + *zōon*, animal). Solitary or colonial; asexual polyps and sexual medusae, although one body type may be suppressed; hydranths with no mesenteries; medusae (when present) with a velum; both freshwater and marine. Examples: *Hydra, Obelia, Physalia, Tubularia.*

Class Scyphozoa (si-fō-zō´ə) (Gr. *skyphos*, cup, + *zōon*, animal). Solitary; polyp stage sometimes reduced or absent; bell-shaped medusae without velum; gelatinous mesoglea much enlarged; margin of bell or umbrella typically has eight notches containing sensory structures; all marine. Examples: *Aurelia, Cassiopeia, Cyanea, Rhizostoma.*

Class Staurozoa (stō-ro-zōá) (Gr. *strauros*, a cross, + *zōon*, animal). Solitary; polyps only; medusa absent; polyp surface extended into eight clusters of tentacles surrounding mouth; attachment via adhesive disc; all marine. Examples: *Haliclystis, Lucernaria.*

Class Cubozoa (kū´ bō-zō´ə) (Gr. *kybos*, a cube, + *zōon*, animal). Solitary; polyp stage reduced; bell-shaped medusae are square in cross section, with tentacle or group of tentacles hanging from a bladelike pedalium at each corner of the umbrella; margin of umbrella entire, without velum but with velarium; all marine. Examples: *Tripedalia, Carybdea, Chironex, Chiropsalmus.*

Class Anthozoa (an-thō-zō´ə) (Gr. *anthos*, flower, + *zōon*, animal). All polyps; no medusae; solitary or colonial; enteron subdivided by mesenteries or septa bearing nematocysts; gonads endodermal; all marine.

Subclass Zoantharia (zo´an-tha´re-ə) (N.L. from Gr. *zōon*, animal, + *anthos*, flower, + L. *aria*, like or connected with) (**Hexacorallia**). Have simple unbranched tentacles; mesenteries in pairs, in multiples of six; include sea anemones, hard corals, and others. Examples: *Metridium, Anthopleura, Tealia, Astrangia, Acropora, Montastrea, Tubastrea.*

Subclass Ceriantipatharia (se´re-ant-ip´a-tha´re-ə) (N.L. combination of Ceriantharia and Antipatharia). Have simple, unbranched tentacles; mesenteries unpaired, initially six; include tube anemones and black or thorny corals. Examples: *Cerianthus, Antipathes, Stichopathes.*

Subclass Octocorallia (ok´to-ko-ral´e-ə) (L. *octo*, eight, + Gr. *korallion*, coral) (**Alcyonaria**). Have eight pinnate tentacles; eight complete, unpaired mesenteries; include soft and horny corals. Examples: *Tubipora, Alcyonium, Gorgonia, Plexaura, Renilla, Ptilosarcus.*

---

carbonate to that deposited by their predecessors. The most important organisms that take dissolved calcium and carbonate ions from seawater and precipitate it as limestone to form reefs are *reef-building corals* and *coralline algae.* Reef-building corals have mutualistic algae (zooxanthellae) living in their tissues. Coralline algae include several types of red algae, and they may be encrusting or form upright, branching growths. Not only do they contribute to the total mass of calcium carbonate, but their deposits help hold the reef together. Some octocorals and hydrozoans (especially *Millepora* [L. *mille*, a thousand, + *porus*, pore] spp., the "fire coral," see figure 7.14B) contribute in some measure to the calcareous material, and an enormous variety of other organisms contribute small amounts. However, reef-building corals are essential to the formation of large reefs, since such reefs do not occur where these corals cannot live.

Because zooxanthellae are vital to reef-building corals and water absorbs light, reef-building corals rarely live below a depth of 30 m (100 feet). Interestingly, some deposits of coral-reef limestone, particularly around Pacific islands and atolls, reach great thickness—even thousands of feet. Clearly, the corals and other organisms could not have grown from the bottom in the abyssal blackness of the deep sea and reached shallow water where light could penetrate. Charles Darwin was the first to realize that such reefs began their growth in *shallow* water around volcanic islands; then, as the islands slowly sank beneath the sea, growth of the reefs kept pace with the rate of sinking, thus explaining the deep deposits.

The distribution of coral reefs in the world is limited to locations that offer optimal conditions for their zooxanthellae. Because they require warmth, light, and the salinity of undiluted seawater, coral reefs are limited to shallow waters between 30° N and 30° S latitude and excluded from areas with upwelling of cold water or where river outflows produce low salinity and high turbidity. Photosynthesis and fixation of carbon dioxide by the zooxanthellae furnish food molecules for the coral. Zooxanthellae recycle phosphorus and nitrogenous waste compounds that otherwise would be lost, and they enhance the ability of the coral to deposit calcium carbonate.

Despite their great intrinsic and economic value, coral reefs in many areas are threatened by a variety of factors, mostly of human origin. These include overenrichment with nutrients (from sewage and runoff of agricultural fertilizer used on nearby land) and overfishing of herbivorous fishes, both of which contribute to overgrowth of multicellular algae. Agricultural pesticides, sediment from tilled fields and dredging, and oil spills contribute to reef degradation. When such environmental stresses do not kill corals directly, they may make the organisms more susceptible to the numerous coral diseases observed in recent years.

Coral reefs are apparently suffering from the effects of global warming. The highly beneficial symbiosis between corals and zooxanthellae is threatened by **coral bleaching** (figure 7.29), where corals become white and brittle after expelling their zooxanthellae. The loss of zooxanthellae correlates with global warming and the resultant increase in ocean temperature. As the water warms, heat damages part of the photosynthetic mechanism in the zooxanthellae, leading to the production of harmful oxidants. The oxidants diffuse into coral tissues, destroying the finely tuned mutualism. Zooxanthellae die or are expelled in what appears to be an immune response from the corals. An initial reduction in the numbers of symbionts worsens the

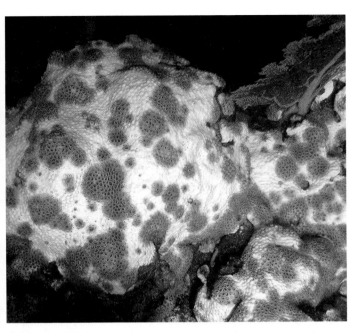

figure 7.29

A comparison of healthy and bleached polyps within a colony of the zoanthid *Palythoa caribbaeorum* in La Parguera, Puerto Rico. *Photo courtesy of Dr. Ernesto Weil, Department of Marine Sciences, University of Puerto Rico.*

problem because the strongly reflective coral skeleton brings more light into the damaged photosynthetic pathway. There are at least eight separate taxa of symbionts that vary in thermal tolerance, but none appears able to survive continued warming or to prevent bleaching. Bleaching occurred prior to global warming, but never at the intensity or on the scale seen now. Bleaching levels in 2002 on the Great Barrier Reef in Australia were the worst in recorded history, with 60% of the entire reef showing bleaching; in some areas bleaching was visible in 90% of the coral present. Caribbean reefs also showed 90% bleaching, which was accompanied by death of half of the affected corals. Furthermore, higher atmospheric concentrations of carbon dioxide (from burning hydrocarbon fuels) tend to acidify ocean water, which makes precipitation of $CaCO_3$ by corals more difficult metabolically.

■ Phylum Ctenophora

Ctenophora (te-nof´o-rə) (Gr. *kteis, ktenos,* comb, + *phora,* pl. of bearing) contains about 150 species. All are marine forms occurring in all seas but especially in warm waters. They take their name from eight rows of comblike plates, or ctenes, which they bear for locomotion. Common names for ctenophores are "sea walnuts" and "comb jellies." Ctenophores, along with cnidarians, represent the only two phyla having primary radial symmetry, in contrast to other metazoans, which have primary bilateral symmetry. In both phyla, the oral/aboral axis is used as a reference point because there is no concentration of sensory or nerve cells to form a head. Like cnidarians, ctenophores have not advanced beyond the tissue level of organization. They have no definite organ systems in a strict sense.

Ctenophores may make adhesive cells called **colloblasts,** but they do not make nematocysts. However, one ctenophore species (*Haeckelia rubra,* named after Ernst Haeckel, nineteenth-century German zoologist) carries undischarged nematocysts on certain regions of its tentacles and lacks colloblasts. These nematocysts apparently are appropriated from cnidarian medusae on which *H. rubra* feeds.

Except for a few creeping and sessile forms, ctenophores are free-swimming (figure 7.30). Although they are feeble swimmers and are more common in surface waters, ctenophores are sometimes found at considerable depths. Highly modified forms, such as *Cestum* (L. *cestus,* girdle), use sinuous body movements, as well as their comb plates, in locomotion (figure 7.31).

The fragile, transparent bodies of ctenophores are easily seen at night when they emit light (luminesce). Ctenophora has a greater proportion of luminescent species than any other phylum.

Form and Function

Pleurobrachia (Gr. *pleuron,* side, + L. *brachia,* arms), is a representative tentaculate ctenophore (figure 7.32). Its surface has eight longitudinal rows of transverse plates bearing long, fused cilia called **comb plates** or **ctenes.** The beating of the cilia in each row starts at the aboral end and proceeds along the rows to the oral end, thus propelling the animal forward. All rows beat in unison. A reversal of the wave direction drives the animal backward. Ctenophores may be the largest animals that swim exclusively by cilia.

A *Pleurobrachia*   **B** *Mnemiopsis*

figure 7.30

A, Comb jelly *Pleurobrachia* sp. (order Cydippida, class Tentaculata). Its fragile beauty is especially evident at night when its comb rows luminesce. **B,** *Mnemiopsis* sp. (order Lobata, class Tentaculata).

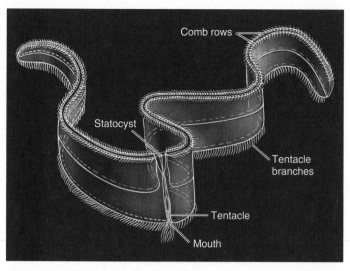

figure 7.31

Venus' girdle (*Cestum* sp.), a highly modified ctenophore. It may reach a length of 5 feet but is usually much smaller.

Two long tentacles carried in a pair of tentacle sheaths (figure 7.32) can stretch to a length of perhaps 15 cm. The surface of the tentacles bears specialized glue cells called colloblasts, which secrete a sticky substance that facilitates catching small prey. When covered with food, the tentacles contract and food is wiped onto the mouth. Digestion occurs as food passes through a pharynx, a gastrovascular cavity, and a system of gastrovascular canals. Rapid digestion occurs in the pharynx; then, partly digested food circulates through the rest of the system, where digestion is completed intracellularly. Residues are regurgitated or expelled through small pores (anal pores) in the aboral end.

Lobate ctenophores, such as *Mnemiopsis* (figure 7.30B), do not have tentacles for prey capture. Instead they use cilia lining the lobes around the mouth to ingest large volumes of water from which prey are captured. The invasive *Mnemiopsis leidyi* is an extremely efficient generalist predator able to capture food items ranging in size from 50 μ to more than 3 mm. Prey are unable to detect the gentle influx of water in

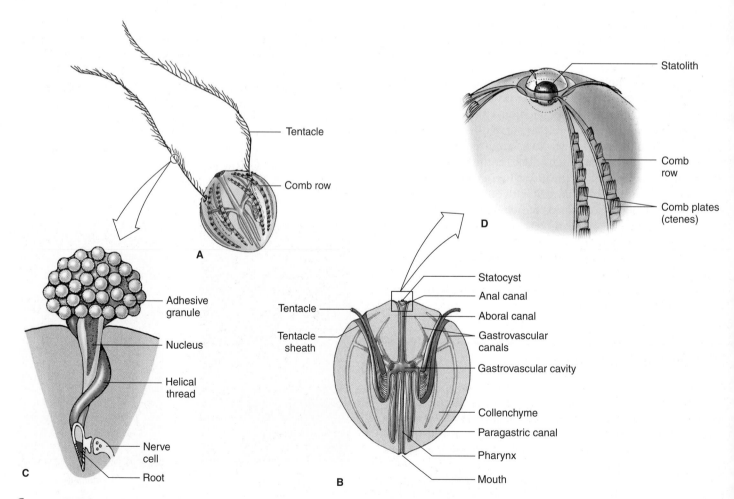

figure 7.32

Comb jelly *Pleurobrachia*, a ctenophore. **A,** External view. **B,** Hemisection. **C,** Colloblast, an adhesive cell characteristic of ctenophores. **D,** Portion of comb rows showing comb plates, each composed of transverse rows of long, fused cilia.

which they are carried into the ctenophore, making *M. leidyi* a serious threat to coastal plankton communities (see boxed reading).

A nerve net system similar to that of the cnidarians includes a subepidermal plexus concentrated under each comb plate.

The sense organ at the aboral pole is a **statocyst,** or organ of equilibrium. The statocyst is also concerned with the beating of the comb rows but does not trigger their beat. Other sensory cells are abundant in the epidermis.

All ctenophores are monoecious, bearing both an ovary and a testis. Gametes are shed into the water, except in a few species that brood their eggs, and there is a free-swimming larva.

Phylogeny and Adaptive Diversification

Phylogeny

Although the origin of the cnidarians and ctenophores is obscure, one hypothesis suggests that the radiate phyla arose from a radially symmetrical, planula-like ancestor with a sessile or free-floating habitat. A planula larva in which an invagination formed to become the gastrovascular cavity would correspond roughly to a cnidarian with an ectoderm and an endoderm.

A central question in cnidarian evolution is this: Which came first, the polyp or the medusa? Some researchers hypothesize that the life cycles of a specialized group of hydrozoans suggest that the medusa came first. In trachyline hydrozoans, the planula develops to a second larva called an actinula and then directly into a medusa without an intervening polyp stage. If the ancestral cnidarian had this life cycle, polyps would have been added to the life cycle later, and anthozoans would have lost the medusa stage.

The alternative hypothesis is that the anthozoan life cycle was ancestral and the medusoid form was added to the life cycle later. This would suggest that hydrozoans, scyphozoans, and cubozoans shared a common ancestor that had added the medusa to the life cycle. The medusa would have been lost in the lineage leading to *Hydra*. Molecular evidence currently supports this hypothesis, and some researchers group the three classes with medusae into the Medusozoa (see figure 7.1).

The evolutionary branching order for cnidarians and ctenophores is not yet known (see cladogram inside front cover). Similarities between cnidarians and ctenophores are mostly of a general nature and do not seem to indicate a close relationship. Ctenophores share a derived cleavage pattern that separates them from all other phyla. Unlike cnidarians, ctenophores may possess muscle tissue derived from the endoderm.

Adaptive Diversification

In Cnidaria, polyp and medusa have similar body plans. They are efficient predators, many feeding on prey quite large in relation to themselves. Some are adapted for feeding on small particles. There are many colonial species, with some colonies growing to great size among corals, and others, such as siphonophores, showing astonishing polymorphism and specialization of individuals within the colony.

Ctenophores have retained the basic arrangement of eight comb plates, with or without tentacles, and biradial symmetry in most. However, some, like Venus' girdle, have a modified body plan.

Summary

Phyla Cnidaria and Ctenophora have a primary radial symmetry; radial symmetry serves sessile or free-floating organisms well because environmental stimuli come from all directions equally. Cnidaria are surprisingly efficient predators because they possess stinging organelles called cnidae. Both phyla are essentially diploblastic, with a body wall composed of epidermis and gastrodermis and a mesoglea between these cell layers. The digestive-respiratory (gastrovascular) cavity has a mouth and no anus, but anal pores occur in ctenophores. Cnidarians are at the tissue level of organization. Many species have two basic body types (polyp and medusa), and in many hydrozoans and scyphozoans, the life cycle involves both an asexually reproducing polyp and a sexually reproducing medusa.

That unique organelle, the cnida, is produced by a cnidoblast (which becomes the cnidocyte) and is coiled within a capsule. When discharged, some types of cnidae, called nematocysts, penetrate prey and inject poison. Discharge is effected by a change in the permeability of the capsule and an increase in internal hydrostatic pressure because of high osmotic pressure within the capsule.

Most hydrozoans are colonial and marine, but the solitary freshwater hydras are commonly demonstrated in class laboratories. They have a typical polyp form, are not colonial, and have no medusa stage. Most marine hydrozoans are in the form of a branching colony of many polyps (hydranths). Most colonies are sessile, but a few drift or swim in the sea. Hydrozoan medusae may be free-swimming or remain attached to the colony.

Scyphozoans are typical jellyfishes, in which the medusa is the dominant body form, and most have an inconspicuous polyp stage. Staurozoans are odd, polyp-like animals whose oral end resembles a medusa. Cubozoans are predominantly medusoid. They include the dangerous sea wasps.

Anthozoans are all marine and are polypoid; there is no medusa stage. The most important subclasses are Zoantharia (having hexamerous or polymerous symmetry) and Octocorallia (having octomerous symmetry). The largest zoantharian orders contain sea anemones, which are solitary and do not have a skeleton, and stony corals, which are mostly colonial and secrete a calcareous exoskeleton. Stony corals are the critical component in coral reefs, which are habitats of great beauty, productivity, and ecological and economic value. Octocorallia contain the soft and horny corals, many of which are important and beautiful components of coral reefs.

Ctenophora are biradial and swim by means of eight comb rows of fused cilia (ctenes). Colloblasts, with which they capture small prey, are characteristic of the phylum.

Cnidaria and Ctenophora are perhaps derived from an ancestor that resembled the planula larva of the cnidarians. Despite their relatively simple level of organization, cnidarians are an important phylum economically, environmentally, and biologically.

■ Review Questions

1. Explain the utility of radial symmetry for sessile and free-floating animals.
2. What characteristics of phylum Cnidaria are most important in distinguishing it from other phyla?
3. Name and distinguish the classes in phylum Cnidaria.
4. Distinguish between polyp and medusa forms.
5. Explain the mechanism of nematocyst discharge. How can a hydrostatic pressure of one atmosphere be maintained within a nematocyst until it receives an expulsion stimulus?
6. What is an unusual feature of the nervous system of cnidarians?
7. Diagram a hydra, and label the main body parts.
8. Name and give the functions of the main cell types in the epidermis and the gastrodermis of hydra.
9. What stimulates feeding behavior in hydras?
10. Give an example of a highly polymorphic, floating, colonial hydrozoan.
11. Distinguish the following from each other: statocyst and rhopalium; scyphomedusae and hydromedusae; scyphistoma, strobila, and ephyrae; velum, velarium, and pedalium; polyp forms in Zoantharia and in Octocorallia.
12. Define the following with regard to sea anemones: siphonoglyph; primary septa or mesenteries; incomplete septa; septal filaments; acontia threads; pedal laceration.
13. Describe three specific symbiotic interactions of anemones with other organisms.
14. Contrast the skeletons of zoantharian and octocorallian corals.
15. Why are coral reefs generally restricted to shallow marine waters?
16. Specifically, what kinds of organisms are most important in deposition of calcium carbonate on coral reefs?
17. How do zooxanthellae contribute to the welfare of reef-building corals?
18. What characteristics of Ctenophora do you consider most important in distinguishing it from other phyla?
19. How do ctenophores swim, and how do they obtain food?
20. What arguments favor the ideas that the first cnidarian was either a medusa or a polyp?

For Further Thought What would be the best way to make non-biologists aware of the ecological and economic costs of global warming as it affects coral reefs?

■ Selected References

See also general references on page 432.

Bentlage, B., P. Cartwright, A. A. Yanagihara, C. Lewis, G. S. Richards, and A. G. Collins. 2010. Evolution of box jellyfish (Cnidaria: Cubozoa), a group of highly toxic invertebrates. Proc. R. Soc. B. **277**:493–501. Fascinating discussion of the unique features, such as mating behavior, in cubozoans.

Coates, M. M. 2003. Visual ecology and functional morphology of Cubozoa (Cnidaria). Integr. Comp. Biol. **43**:542–548. *Collected information on the three different kinds of eyes in cubozoans.*

Colin, S. P., J. H. Costello, L. J. Hansson, J. Titelman, and J. O. Dabiri. 2010. Stealth predation and

the predatory success of the invasive ctenophore. *Mnemiopsis leidyi*. Proceedings of the National Academy of Sciences **107**:17223–17227. *Low-velocity feeding currents generated by cilia of this ctenophore are not detected by prey in time to produce an escape response.*

Collins, A. G., P. Schuchert, A. C. Marques, T. Jankowski, M. Medina, and B. Schierwater. 2006. Medusozoan phylogeny and character evolution clarified by new large and small subunit rDNA and an assessment of the utility of phylogenetic mixture models. Syst. Biol. **55**:97–115. *Authors produce a working cladogram*

of major cnidarian taxa, including class Staurozoa.

Côté, I. M., and J. D. Reynolds, eds. 2006. Coral reef conservation. Conservation Biology **13**. Cambridge, UK, Cambridge University Press. *A modern discussion of biological and social aspects of reef conservation.*

Ferrari, A., and A. Ferrari. 2006. A diver's guide to reef life. Oxford, Mississippi, Nautilus Publishing. *A very recent guide with excellent photographs of reef invertebrates and fishes.*

Humann, P., and N. DeLoach. 2002. Reef creature identification. Florida, Caribbean, Bahamas. Jacksonville, Florida, ed. 2. New

World Publications, Inc. *This is the best field guide available for identification of "noncoral" cnidarians of the Caribbean.*

Humann, P., and N. DeLoach. 2002. Reef coral identification. Florida, Caribbean, Bahamas. Jacksonville, Florida, ed. 2. New World Publications, Inc. *Superb color photographs and accurate identifications make this by far the best field guide* to corals now available; includes sea grasses and some algae.

Weis, V. M. 2008. Cellular mechanisms of cnidarian bleaching: stress causes the collapse of symbiosis. J. Exp. Biol. **211**: 3059-3066. *Clear description of our current understanding of the mechanisms underlying coral bleaching.*

Widder, E. A. 2010. Bicluminescence in the ocean: origins of biological, chemical, and ecological diversity. Science **328**: 704-708. *The chemical basis and potential function for light production across a wide range of invertebrates, including medusae and ctenophores, is described.*

■ Custom Website

The *Animal Diversity* Online Learning Center is a great place to check your understanding of chapter material. Visit www.mhhe.com/hickmanad6e for access to key terms, quizzes, and more! Further enhance your knowledge with Web links to chapter-related material.

Explore live links for these topics:

Classification and Phylogeny of Animals
Radiate Animals
Phylum Cnidaria
Class Hydrozoa
Class Scyphozoa

Class Anthozoa
Coral Reefs
Coral Bleaching
Class Cubozoa
Phylum Ctenophora

8

Acoelomorpha, Platyzoa, and Mesozoa

Flatworms, Gastrotrichs, Gnathiferans, and Mesozoans

Acoelomorpha
Platyhelminthes
Gastrotricha
Gnathostomulida
Micrognathozoa
Rotifera
Acanthocephala
Mesozoa

Getting Ahead

For animals that spend their lives sitting and waiting, as do most members of the two radiate phyla Cnidaria and Ctenophora, radial symmetry is ideal. One side of the animal is just as important as any other for snaring prey coming from any direction. But if an animal is active in seeking food, shelter, home sites, and reproductive mates, it requires a different set of strategies and a new body organization. An elongated body form with head (anterior) and tail (posterior) ends serves active, directional movement. In addition, one side of the body faces up (dorsal), and the other side, specialized for locomotion, faces down (ventral). What results is a bilaterally symmetrical animal in which the body could be divided along only one plane of symmetry to yield two halves that are mirror images of each other. Furthermore, because it is better to perceive where one is going than where one has been, sense organs and centers for nervous control have come to be located at the anterior end. This process is called cephalization. Thus, cephalization and primary bilateral symmetry occur together in almost all triploblastic animals.

Thysanozoon nigropapillosum, a marine turbellarian (order Polycladida).

The term "worm" has been applied loosely to elongated, bilateral invertebrate animals without appendages. At one time, zoologists considered worms (Vermes) a taxon that included a highly diverse assortment of forms. This unnatural assemblage has been reclassified into various phyla. By tradition, however, zoologists still call various groups of these animals flatworms, ribbon worms, roundworms, segmented worms, and others. In this chapter we introduce two phyla of flatworms and six other phyla that lack a mesodermally lined body cavity. All but one phylum contains bilaterally symmetrical, **triploblastic** animals. Beyond bilateral symmetry and the presence of three tissue layers however, these animals differ in body plan.

Members of two phyla that we discuss here have **acoelomate** (Gr. *a*, not, + *koilōma*, cavity) bodies (see p. 62); in other words, they lack a coelom. Readers may recall that diploblasts also lack a coelom, but are not called acoelomate; the term applies only to animals possessing mesoderm.

Typical acoelomates have only one internal space, the digestive cavity (figure 8.1). The region between the epidermis and the digestive cavity lining is filled with a cellular, mesodermally derived **parenchyma**. Parenchyma is a form of packing tissue containing more cells and fibers and less extracellular matrix (ECM) than does the mesoglea of cnidarians. Organs are another derivative of mesoderm that increases internal complexity in triploblasts. We see some internal complexity in members of the acoelomate phyla Acoelomorpha and Platyhelminthes.

A **pseudocoelomate** (Gr. *pseudo*, false, + *koilōma*, cavity) body has an internal cavity, the pseudocoelom, surrounding the gut, but the cavity is not completely lined with mesoderm as it is in a coelomate animal (figure 8.1). The mesodermal layer lies only on the outside of the pseudocoelom; the inner lining of the cavity is formed by the endodermally derived gut lining. A pseudocoelom is an embryonic blastocoel that persists throughout development, so some biologists call animals with this body plan blastocoelomates rather than pseudocoelomates.

A pseudocoelom may be filled with fluid or with a gelatinous matrix containing some mesenchymal cells (derived from mesoderm). It may provide space for expansion of digestive, excretory, and reproductive systems or for storage of waste products, but when filled with fluid, it often functions as a hydrostatic support or permits circulation of body fluids. Many pseudocoelomate animals are quite small and lack true circulatory systems.

Animals sharing a particular body plan do not necessarily form a monophyletic group. Pseudocoelomate groups include rotifers and acanthocephalans, among others discussed in later chapters, whereas acoelomate taxa include acoelomorphs, platyhelminths, gastrotrichs, and gnathostomulids. Members of the Acoelomorpha are hypothesized to have branched from a lineage ancestral to all other bilateral animals before that lineage split to form protostomes and deuterostomes. Examine the cladogram inside the front cover to see this branching pattern. Notice that the other phyla discussed in this chapter belong to Protostomia and are placed in the subgroup Lophotrochozoa (see p. 67).

■ Phylum Acoelomorpha

Acoelomorpha (Gr. *a*, without, + *koilos*, hollow, + *morphe*, form) is a relatively new phylum containing worms that previously belonged to two orders within class Turbellaria in phylum Platyhelminthes (see p. 156). These worms are less than 5 mm long and have a cellular, ciliated epidermis. Some have a saclike gut without an anus, and others entirely lack a digestive cavity. In gutless worms, food particles enter through the mouth and move into a tissue mass derived from endoderm; sometimes, a temporary digestive cavity forms around the food.

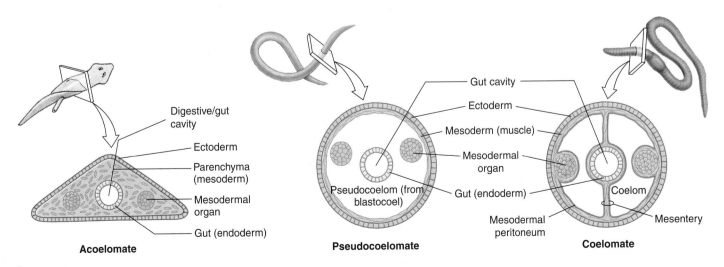

Acoelomate

Digestive/gut cavity
Ectoderm
Parenchyma (mesoderm)
Mesodermal organ
Gut (endoderm)

Pseudocoelomate

Gut cavity
Ectoderm
Mesoderm (muscle)
Mesodermal organ
Pseudocoelom (from blastocoel)
Gut (endoderm)

Coelomate

Coelom
Mesentery
Mesodermal peritoneum

figure 8.1

Acoelomate, pseudocoelomate, and coelomate body plans for comparison. Notice that peritoneum and mesenteries are present only in coelomate animals.

Acoelomorphs have a distinct anteroposterior axis, but their diffuse set of anterior nerve cells lacks the ganglia typical of a "true" brain. Acoelomorphs have a radial arrangement of nerves in the body, instead of the ladderlike pattern (see p. 160) seen in flatworms of phylum Platyhelminthes. Potential synapomorphies include special cilia and a frontal organ.

Acoelomorphs typically inhabit marine sediments, although a few forms are pelagic. Some species live in brackish water. Most acoelomorphs are free-living, but some are symbiotic and others parasitic. The group contains approximately 350 species. Space limitations preclude further discussion of these animals.

■ Clade Platyzoa

Platyzoa is a group of lophotrochozoan protostome phyla. The evolutionary relationships among lophotrochozoans are still under debate because different combinations of molecular and morphological characters yield different cladograms. When gene sequences for certain taxa are not available, these taxa are excluded from the study. However, recent efforts to broaden the number and type of characters as well as the number of taxa in a phylogenetic analysis (see p. 85) have enhanced our understanding of the protostomes. The Platyzoa emerged as a clade in several phylogenies, but readers should be aware that contrasting relationships have been proposed for the phyla forming Platyzoa. As described here, Platyzoa contains Platyhelminthes, Gastrotricha, and four phyla in Gnathifera (figure 8.2).

■ Phylum Platyhelminthes

Platyhelminthes (Gr. *platys,* flat, + *helmins,* worm) range in length from a millimeter or less to many meters (in some tapeworms). Their typically flattened bodies may be slender, broadly leaflike, or long and ribbonlike. There are four classes within Platyhelminthes. Class Turbellaria contains nonparasitic

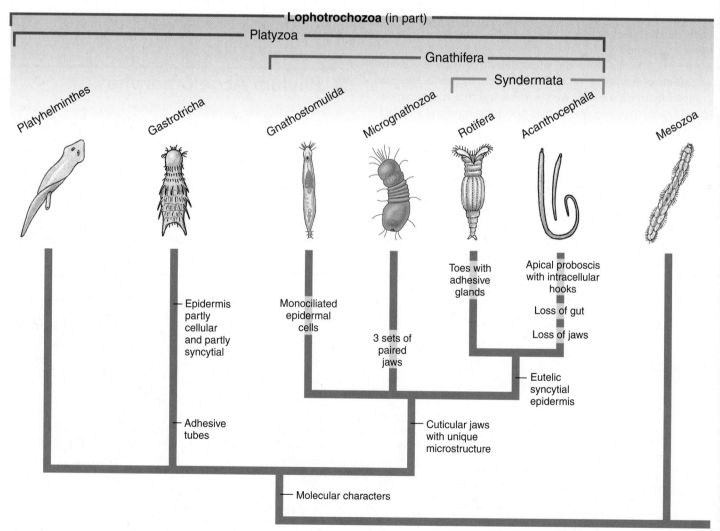

figure 8.2

Hypothetical relationships among members of Platyzoa. Characters are modified subsets of those in Kristensen (2002) and Brusca and Brusca (2003). See references p. 174 and p. 432 for citations.

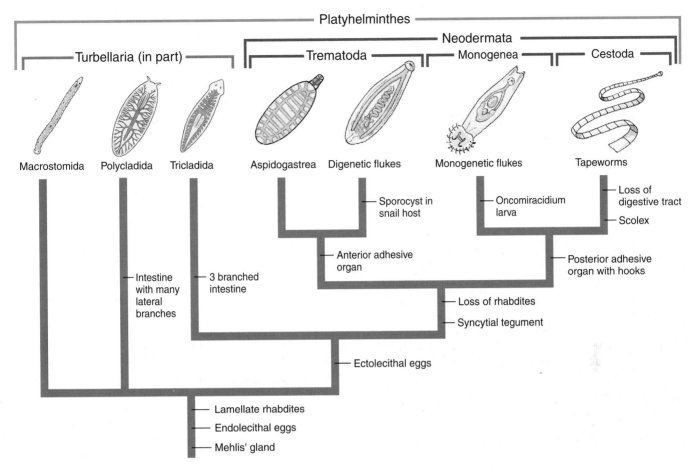

figure 8.3

Hypothetical relationships among Platyhelminthes. The traditionally accepted class Turbellaria is paraphyletic. Some turbellarians have ecto-lecithal development and, together with the Trematoda, Monogenea, and Cestoda, form a clade and a sister group of the endolecithal turbellarians. For the sake of simplicity, the synapomorphies of those turbellarians and of the Aspidogastrea, as well as many others given by Brooks (1989), are omitted.

Source: Modified from D. R. Brooks. The phylogeny of the Cercomeria (Platyhelminthes: Rhabdocoela) and general evolutionary principles. Journal of Parasitology **75**:*606–616, 1989.*

forms, whereas members of classes Trematoda, Monogenea, and Cestoda are parasitic (figure 8.3). Although members of the parasitic classes all possess a syncytial tegument, there is no single feature yet discovered that diagnoses phylum Platyhelminthes.

Ecological Relationships

Free-living flatworms are found exclusively in class Turbellaria. A few turbellarians are symbiotic (commensals or parasites), but most are adapted as bottom-dwellers in marine or fresh water or live in moist places on land. Many, especially larger species, occur on the undersides of stones and other hard objects in freshwater streams or in littoral zones of the ocean.

Relatively few members of class Turbellaria inhabit fresh water, but freshwater planarians, such as *Dugesia* (order Tricladida), are used extensively in introductory laboratory courses. Planarians and some others frequent streams and

spring pools; others prefer the flowing water of mountain streams. Some species occur in moderately hot springs. Terrestrial turbellarians live in fairly moist places such as under stones and logs or on moist vegetation (figure 8.4).

All members of classes Monogenea and Trematoda (flukes) and class Cestoda (tapeworms) are parasitic. Most Monogenea are ectoparasites, but all trematodes and cestodes are endoparasitic. Many species have indirect life cycles with more than one host; the first host is often an invertebrate, and the final host is usually a vertebrate. Humans serve as hosts for some species. Certain larval stages may be free-living.

Form and Function

Tegument, Muscles

Most turbellarians have a ciliated cellular epidermis resting on a basement membrane. The epidermis typically contains rod-shaped **rhabdites,** composed of fused vesicles from the Golgi apparatus, that swell to form a protective mucous sheath around the body when discharged with water; single-celled mucous glands may also be present (figure 8.5). Many orders of turbellarians also possess an interesting locomotory system that allows them quickly to attach and detach from surfaces. In such cases, the epidermis contains **dual-gland adhesive organs** consisting of three cell types: viscid and releasing gland cells and anchor cells (figure 8.6). Secretions of the viscid gland cells apparently fasten microvilli of the anchor cells to the substrate, and secretions of the releasing gland cells provide a quick, chemical detaching mechanism.

figure 8.4

Terrestrial turbellarian (order Tricladida) from the Amazon Basin, Peru.

In contrast to the ciliated cellular epidermis of most turbellarians, all members of the parasitic classes possess a nonciliated body covering called a **syncytial tegument.** The term *syncytial* means that many nuclei occur within a single cell membrane. It might seem that a completely new body covering appeared in parasites, but two atypical forms of the epidermis similar to tegument occur within turbellarians. A few turbellarian species have a syncytial epidermis, and at least one species has a syncytial *insunk* epidermis where cell bodies (containing the nuclei) lie beneath the basement membrane of the epidermis and communicate with the distal (surface) cytoplasm through cytoplasmic channels. The term "insunk" is a misnomer because the cell bodies below the basement membrane send extensions upward to the epidermis, rather than the other way around. The extensions then fuse to form the syncytial covering, much as they do in the parasitic classes.

Adults of all members of Trematoda, Monogenea, and Cestoda share a syncytial covering that entirely lacks cilia and is designated a **tegument** (figure 8.7). This distinctive tegumental plan is the basis for uniting trematodes, monogeneans, and cestodes in a taxon called Neodermata. It is a peculiar epidermal arrangement and may be related to adaptations for parasitism in ways that are still unclear.

In the body wall below the basement membrane of flatworms are layers of muscle fibers that run circularly, longitudinally, and diagonally. A meshwork of parenchyma cells, developed from mesoderm, fills the spaces between muscles and visceral organs. Parenchyma cells in some, perhaps all, flatworms are not a separate cell type, but are noncontractile portions of muscle cells.

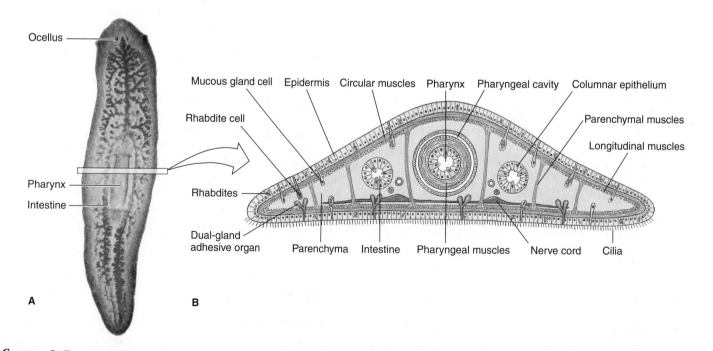

figure 8.5

Cross section through the pharyngeal region of a common planarian, showing relationships of body structures.

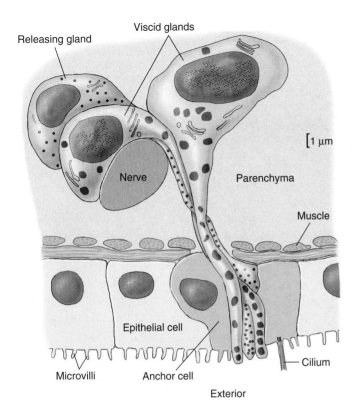

figure 8.6

Reconstruction of a dual-gland adhesive organ of a turbellarian, *Haplopharynx* sp. Two viscid glands and one releasing gland lie beneath the body wall. The anchor cell lies within the epidermis, and one of the viscid glands and the releasing gland contact a nerve.

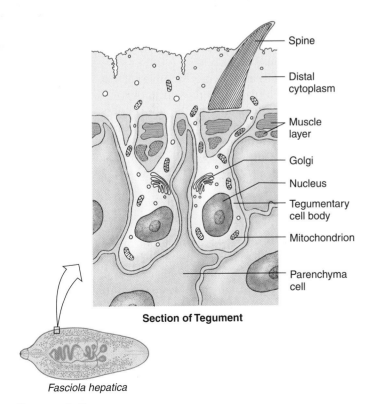

figure 8.7

Structure of the tegument of a trematode, *Fasciola hepatica*.

Nutrition and Digestion

The digestive system of most platyhelminthes includes a mouth, pharynx, and intestine. In turbellarians the muscular **pharynx** opens posteriorly just inside the mouth, through which it can extend (figure 8.8). The mouth is usually at the anterior end in flukes, and the pharynx is not protrusible. The intestine may be simple or branched.

Intestinal secretions contain proteolytic enzymes for some extracellular digestion. Food is sucked into the intestine, where cells of the gastrodermis often phagocytize it and complete **intracellular digestion.** Undigested food is egested through the pharynx. Tapeworms lack a digestive system and must absorb all of their nutrients as small molecules (predigested by the host) directly through their tegument.

Excretion and Osmoregulation

Except in some turbellarians, the osmoregulatory system consists of canals with tubules that end in **flame cells (protonephridia)** (figure 8.8A). Each flame cell surrounds a small space into which

CHARACTERISTICS
of Phylum Platyhelminthes

1. No single diagnostic feature
2. Marine, freshwater, and moist terrestrial habitats
3. Turbellarian flatworms are mostly free-living; classes Monogenea, Trematoda, and Cestoda are entirely parasitic
4. **Bilateral symmetry;** definite polarity of anterior and posterior ends; **body flattened dorsoventrally**
5. Adult body three-layered **(triploblastic)**
6. Body acoelomate
7. Epidermis may be cellular or syncytial (ciliated in some); **rhabdites** in epidermis of most Turbellaria; epidermis a syncytial **tegument** in Monogenea, Trematoda, Cestoda, and some Turbellaria
8. Gut incomplete; may be branched; absent in cestodes
9. Muscular system primarily a sheath form of mesodermal origin; layers of circular, longitudinal, and sometimes oblique fibers beneath the epidermis
10. Nervous system consisting of a **pair of anterior ganglia** with **longitudinal nerve cords** connected by transverse nerves and located in the mesenchyme in most forms
11. Sense organs include statocysts (organs of balance) and ocelli, auricles, and rheoreceptors
12. Asexual reproduction by fragmentation and other methods as part of complex parasitic life cycles
13. Most forms monoecious; reproductive system complex, usually with well-developed gonads, ducts, and accessory organs; internal fertilization; development direct in free-swimming forms and those with single hosts; complicated life cycle often involving several hosts in many internal parasites
14. Excretory system comprises two lateral canals with branches bearing **flame cells (protonephridia);** lacking in some forms
15. Respiratory, circulatory, and skeletal systems lacking; lymph channels with free cells in some trematodes

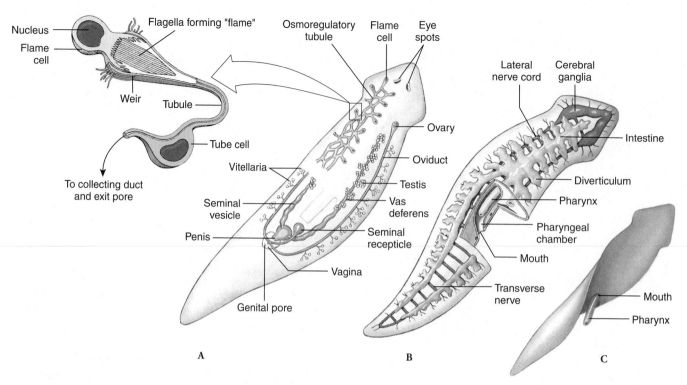

figure 8.8

Structure of a planarian. **A,** Reproductive and excretory systems, shown in part. Inset at left is enlargement of a flame cell. **B,** Digestive tract and ladder-type nervous system. The pharynx is shown in the resting position. **C,** Pharynx extended through ventral mouth.

a tuft of flagella projects. In some turbellarians and in the other classes of flatworms, the protonephridia form a **weir** (Old English *wer,* a fence placed in a stream to catch fish). In a weir, the rim of the cup formed by the flame cell bears fingerlike projections that interdigitate with similar projections of a tubule cell. The beat of the flagella (resembling a flickering flame) provides a negative pressure to draw fluid through the weir into the space (lumen) enclosed by the tubule cell. The lumen continues into collecting ducts that finally open to the outside by pores. The wall of the duct beyond the flame cell commonly bears folds or microvilli that probably function in reabsorption of certain ions or molecules. It is likely that this system is osmoregulatory in most forms; it is reduced or absent in marine turbellarians, which do not have to expel excess water.

Metabolic wastes are largely removed by diffusion through the body wall.

Nervous System and Sense Organs

Flatworms are cephalized but vary in the complexity of their nervous system. The simplest system, found in some turbellarians, is a subepidermal nerve plexus resembling the nerve net of the cnidarians. Other flatworms have, in addition to a nerve plexus, one to five pairs of longitudinal nerve cords (figure 8.8B) lying under the muscle layer.

Connecting nerves form a "ladder-type" pattern. Their brain is a mass of ganglion cells arising anteriorly from the nerve cords. The neurons are organized into sensory, motor, and association types—an important advance in the evolution of nervous systems.

Tactile cells and chemoreceptive cells are abundant over the body, and in planarians they form definitive organs on the **auricles** (the earlike lobes on the sides of the head). Some also have **statocysts** for equilibrium and **rheoreceptors** for sensing water current direction. **Ocelli,** or light-sensitive eyespots, are common in turbellarians (figure 8.8A), monogeneans, and larval trematodes.

Reproduction

Many flatworms reproduce both asexually and sexually. Many freshwater turbellarians can reproduce asexually by fission, merely constricting behind the pharynx and separating into two animals, each of which regenerates the missing parts. In forms such as *Stenostomum* and *Microstomum,* individuals do not separate at once but remain attached, forming chains of zooids (figure 8.9B and C). Flukes reproduce asexually in their snail intermediate host (see p. 162), and some tapeworms, such as *Echinococcus,* can bud off thousands of juveniles in their intermediate host.

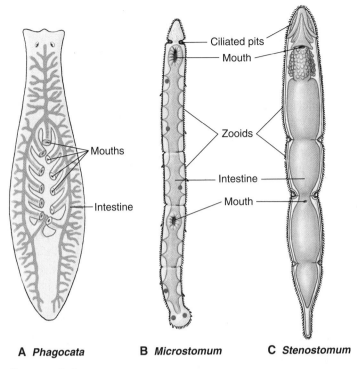

A *Phagocata*　　**B** *Microstomum*　　**C** *Stenostomum*

figure 8.9

Some small freshwater turbellarians. A, *Phagocata* has numerous mouths and pharynges. B, C, Incomplete fission yields a series of temporarily attached zooids.

Most flatworms are monoecious (hermaphroditic) and practice cross-fertilization. Fertilization is internal via a penis or cirrus. In some turbellarians, yolk for nutrition of the developing embryo is contained within the egg cell itself (**endolecithal**), just as it is normally in other phyla of animals. Possession of endolecithal eggs is considered ancestral for flatworms. Other turbellarians, as well as all trematodes, monogeneans, and cestodes, share a derived condition in which female gametes contain little or no yolk, and yolk is contributed by cells released from separate organs (yolk glands). Usually yolk cells surround the zygote within an eggshell (**ectolecithal**).

Development following formation of the zygote may be direct or indirect. In some flatworms, an embryo becomes a larva, which may be ciliated or not, according to the group.

Class Turbellaria

Turbellarians are mostly free-living worms that range in length from 5 mm to nearly 50 cm. Their mouth is on the ventral side and leads into a gut with a variable form. The orders within class Turbellaria are distinguished by the form of the gut (present or absent; simple or branched; pattern of branching) and the pharynx (simple; folded; bulbous). Except for order Polycladida

(Gr. *poly,* many, + *klados,* branch), turbellarians with endolecithal eggs have a simple gut and a simple pharynx. Polyclads have a folded pharynx and a gut with many branches. Polyclads include many marine forms of moderate to large size (3 to more than 40 mm) (see p. 154), and a highly branched intestine is correlated with larger size in turbellarians. Members of order Tricladida (Gr. *treis,* three, + *klados,* branch), which are ectolecithal and include freshwater planaria, have a three-branched intestine.

Turbellarians are typically creeping forms that combine muscular with ciliary movements to achieve locomotion. Very small planaria swim by means of their cilia. Others move by gliding, with the head slightly raised, over a slime track secreted by the marginal adhesive glands. Beating of epidermal cilia in the slime track moves the animal forward, while rhythmic muscular waves pass backward from the head. Large polyclads and terrestrial turbellarians crawl by muscular undulations, much in the manner of a snail.

Some turbellarians have simple life cycles without a distinct larval stage. In some freshwater planarians, egg capsules are attached by little stalks to the undersides of stones or plants, and embryos emerge as juveniles that resemble miniature adults. Marine turbellarians have a ciliated larva that is very similar to the trochophore (see p. 191) of some annelids, molluscs, and members of other phyla.

As traditionally recognized, turbellarians form a paraphyletic group. Several synapomorphies, such as insunk epidermis and ectolecithal development, show that some turbellarians are phylogenetically closer to Trematoda, Monogenea, and Cestoda than they are to other turbellarians. Ectolecithal turbellarians therefore appear to form a clade with trematodes, monogeneans, and cestodes to the exclusion of endolecithal turbellarians (see figure 8.3). Endolecithal turbellarians also are paraphyletic; the presence of a dual-gland adhesive system in some endolecithal turbellarians indicates a clade with ectolecithal flatworms to the exclusion of other endolecithal turbellarian lineages. Turbellaria therefore describes an artificial group, and the term is used here only because it is prevalent in zoological literature.

Class Trematoda

Trematodes are all parasitic flukes, and as adults they are almost all endoparasites of vertebrates. They are chiefly leaflike in form and are structurally similar in many respects to more complex Turbellaria. A major difference is the tegument (described earlier; see figure 8.7).

Other structural adaptations for parasitism are apparent: various penetration glands or glands that produce cyst material; organs for attachment, such as suckers and hooks; and increased reproductive capacity. Otherwise, trematodes share several characteristics with turbellarians, such as a well-developed gut tube (but with the mouth at the anterior, or cephalic, end) and similar reproductive, excretory, and nervous systems, as well as a musculature and parenchyma that differ only slightly from those of turbellarians. Sense organs are poorly developed.

Of the three subclasses of Trematoda, two are small and poorly known groups, but Digenea (Gr. *dis,* double, + *genos,* descent) is a large group that includes many species of medical and economic importance.

Subclass Digenea

With rare exceptions, digenetic trematodes have a complex life cycle, the first host **(intermediate host)** being a mollusc and the final host **(definitive host)** being a vertebrate. A definitive host is one in which the parasite reproduces sexually. In some species, a second, and sometimes even a third, intermediate host intervenes. The group has many species, and they can inhabit diverse sites in their hosts: all parts of the digestive tract, respiratory tract, circulatory system, urinary tract, and reproductive tract.

Among the world's most amazing phenomena are digenean life cycles. Although cycles of different species vary widely in detail, typical stages include adult, egg (shelled embryo), miracidium, sporocyst, redia, cercaria, and metacercaria (figure 8.10). The shelled embryo or larva usually passes from the definitive host in excreta and must reach water to develop further. There, it hatches to a free-swimming, ciliated larva, the **miracidium.** The miracidium penetrates the tissues of a snail, where it transforms into a sacklike **sporocyst.** Sporocysts reproduce asexually to yield either more sporocysts or a number of **rediae.** Rediae, in turn, reproduce asexually to produce more rediae or to produce **cercariae.** In this way, a single zygote can give rise to an enormous number of progeny. Cercariae emerge from the snail and penetrate a second intermediate host or encyst on vegetation or other objects to become **metacercariae,** which

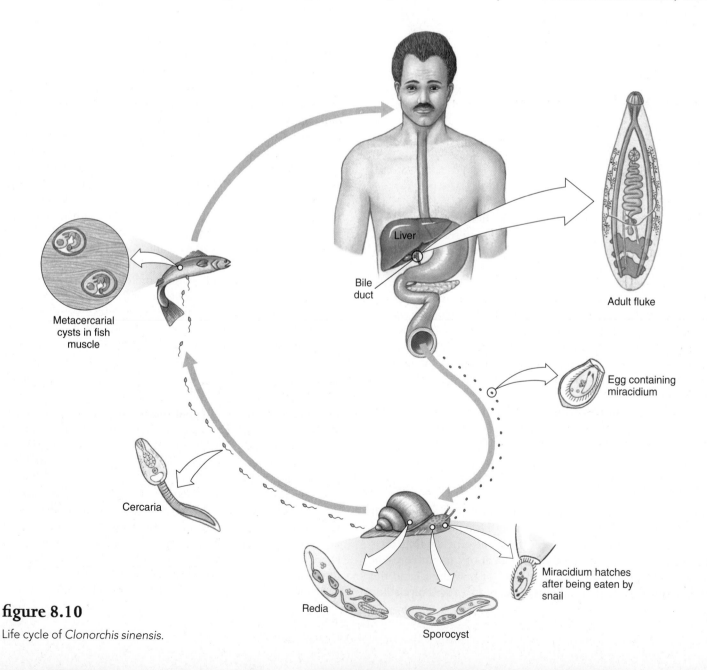

figure 8.10

Life cycle of *Clonorchis sinensis.*

Table 8.1   Examples of Digenetic Flukes Infecting Humans

Common and scientific names	Means of infection; distribution and prevalence in humans
Blood flukes (*Schistosoma* spp.); three widely prevalent species, others reported	Cercariae in water penetrate skin; 200 million people infected with one or more species
S. mansoni	Africa, South and Central America
S. haematobium	Africa
S. japonicum	Eastern Asia
Chinese liver flukes (*Clonorchis sinensis*)	Eating metacercariae in raw fish; about 30 million cases in Eastern Asia
Lung flukes (*Paragonimus* spp.); seven species, most prevalent is *P. westermani*	Eating metacercariae in raw freshwater crabs, crayfish; Asia and Oceania, sub-Saharan Africa, South and Central America; several million cases in Asia
Intestinal fluke (*Fasciolopsis buski*)	Eating metacercariae on aquatic vegetation; 10 million cases in Eastern Asia
Sheep liver fluke (*Fasciola hepatica*)	Eating metacercariae on aquatic vegetation; widely prevalent in sheep and cattle, occasional in humans

are juvenile flukes. Adults grow from the metacercaria when that stage is eaten by a definitive host.

Some of the most serious parasites of humans and domestic animals belong to the Digenea (table 8.1).

Clonorchis (Gr. *clon,* branch, + *orchis,* testis) (figure 8.10) is the most important liver fluke of humans and is common in many regions of the Orient, especially China, southern Asia, and Japan. Cats, dogs, and pigs are also often infected. Adult flukes live in the bile passages, and shelled miracidia are passed in feces. If ingested by certain freshwater snails, sporocyst and redia stages develop, and free-swimming cercariae emerge. Cercariae that manage to find a suitable fish encyst in the skin or muscles as metacercariae. When the fish is eaten raw, the

juveniles migrate up the bile duct to mature and may survive there for 15 to 30 years. The effect of the flukes on humans depends mainly on the extent of infection. A heavy infection may cause pronounced cirrhosis of the liver and death.

Schistosomiasis, an infection with blood flukes of genus *Schistosoma* (Gr. *schistos,* divided, + *soma,* body) (figure 8.11A), ranks as one of the major infectious diseases in the world, with 200 million people infected. The disease is widely prevalent over much of Africa and parts of South America, the West Indies, the Middle East, and the Far East. It is spread when shelled miracidia shed in human feces and urine get into water containing host snails (figure 8.11B). Cercariae that contact human skin penetrate the skin to enter blood vessels, which they follow to

A

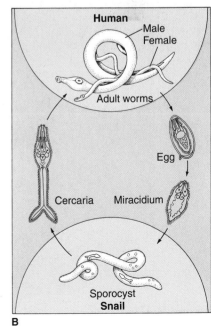

B

figure 8.11

A, Adult male and female *Schistosoma japonicum* in copulation. Blood flukes differ from most other flukes in being dioecious. Males are broader and heavier and have a large, ventral gynecophoric canal, posterior to the ventral sucker. The gynecophoric canal embraces the long, slender female (darkly stained individual) during insemination and oviposition. **B,** Life cycle of *Schistosoma* spp.

certain favorite regions depending on the type of fluke. *Schistosoma mansoni* lives in venules draining the large intestine; *S. japonicum* localizes more in venules draining the small intestine; and *S. haematobium* lives in venules draining the urinary bladder. Blood flukes differ from most other flukes in having separate males and females. In each species, many eggs released by female worms do not find their way out of the body but lodge in the liver or other organs, where they are sources of chronic inflammation. The inch-long adults may live for years in a human host, and their eggs cause such disturbances as severe dysentery, anemia, liver enlargement, bladder inflammation, and brain damage.

Cercariae of several genera whose normal hosts are birds often enter the skin of human bathers in their search for a suitable bird host, causing a skin irritation called "swimmer's itch" (figure 8.12). In this case, humans are a dead end in the fluke's life cycle because the fluke cannot develop further in the human body.

Class Monogenea

Monogenetic flukes were traditionally an order of Trematoda, but they are now a separate class. Cladistic analysis places them as the sister taxon to Cestoda. Monogeneans are mostly external parasites that clamp onto the gills and external surfaces of fish using a hooked attachment organ called an **opisthaptor** (figure 8.13). A few occur in the urinary bladders of frogs and turtles, and one has reportedly been removed from the eye of a hippopotamus. Although widespread and common, monogeneans seem to cause little damage to their hosts under natural conditions. However, like numerous other fish pathogens, they become a serious threat when their hosts are crowded together, as in fish farming.

The life cycles of monogeneans are simple, having a single host, as suggested by the name of the group, which means "single descent." The egg hatches a ciliated larva that attaches to a host, sometimes following a free-swimming phase.

Class Cestoda

Cestoda, or tapeworms, differ in many respects from the preceding classes: they usually have long, flat bodies composed of a **scolex,** for attachment to the host, followed by many reproductive units or **proglottids** (figure 8.14). The scolex, or holdfast, is usually provided with suckers or suckerlike organs and often with hooks or spiny tentacles as well. Tapeworms entirely lack a digestive system, but they have well-developed muscles, and their excretory and nervous systems are somewhat similar to those of other flatworms. They have no special sense organs, but sensory endings in their tegument are modified cilia (figure 8.15).

figure 8.12

Human abdomen, showing schistosome dermatitis caused by penetration of schistosome cercariae, which are unable to complete development in humans. Sensitization to allergenic substances released by cercariae causes rash and itching.

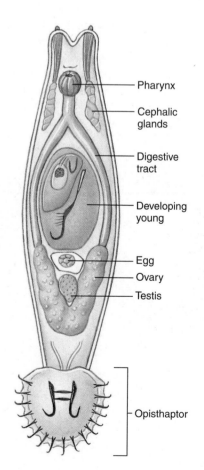

figure 8.13

Gyrodactylus sp. (class Monogenea), ventral view.

Taxonomy of Phylum Platyhelminthes

Class Turbellaria (tur′bel-lar′e-ə) (L. *turbellae* [pl.], stir, bustle, + *-aria*, like or connected with). **Turbellarians.** Usually free-living forms with soft, flattened bodies; covered with ciliated epidermis containing secreting cells and rodlike bodies (rhabdites); mouth usually on ventral surface sometimes near center of body; no body cavity other than the blind gut; mostly hermaphroditic; some undergo asexual fission. A paraphyletic taxon. Examples: *Dugesia* (planaria), *Microstomum, Planocera.*

Class Trematoda (trem′a-to′də) (Gr. *trēma*, a hole, + *eidos*, form). **Digenetic flukes.** Body covered with a syncytial tegument without cilia; leaflike or cylindrical in shape; usually have oral and ventral suckers, no hooks; gut usually has two main branches; mostly monoecious; life cycle complex, with first host a mollusc, final host usually a vertebrate; parasitic in all classes of vertebrates. Examples: *Fasciola, Clonorchis, Schistosoma.*

Class Monogenea (mon′o-gen′e-ə) (Gr. *mono*, single, + *genos*, descent). **Monogenetic flukes.** Body covered with a syncytial tegument without cilia; body usually leaflike to cylindrical in shape; posterior attachment organ with hooks, suckers, or clamps, usually in combination; monoecious; life cycle simple, with single host and usually with free-swimming, ciliated larva; all parasitic, mostly on skin or gills of fish. Examples: *Dactylogyrus, Polystoma, Gyrodactylus.*

Class Cestoda (ses-to′də) (Gr. *kestos*, girdle, + *eidos*, form). **Tapeworms.** Body covered with nonciliated, syncytial tegument; general form of body tapelike; have scolex with suckers or hooks, sometimes both, for attachment; body usually divided into series of proglottids; no gut or digestive organs; usually monoecious; parasitic in digestive tract of all classes of vertebrates; life cycle complex, with two or more hosts; first host may be vertebrate or invertebrate. Examples: *Diphyllobothrium, Hymenolepis, Taenia.*

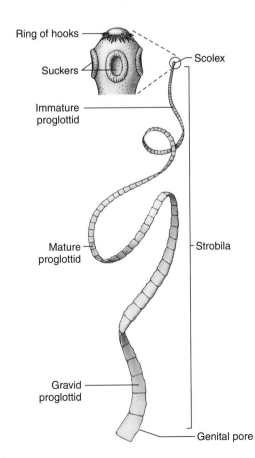

figure 8.14

A tapeworm, showing strobila and scolex. The scolex is the organ of attachment.

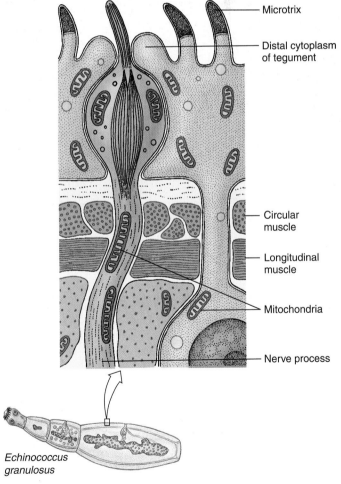

figure 8.15

Schematic drawing of a longitudinal section through a sensory ending in the tegument of *Echinococcus granulosus*, a tapeworm. Note the position of the nerve process.

As in Monogenea and Trematoda, adult cestodes have no external, motile cilia, and the tegument is composed of a distal cytoplasm with sunken cell bodies beneath the superficial layer of muscle (see figure 8.7). In contrast to monogeneans and trematodes, however, their entire surface is covered with minute projections called **microtriches** (sing., **microtrix**) that are similar in certain respects to microvilli of the vertebrate small intestine (figure 8.15). These microtriches greatly amplify the surface area of the tegument, which is a vital adap-

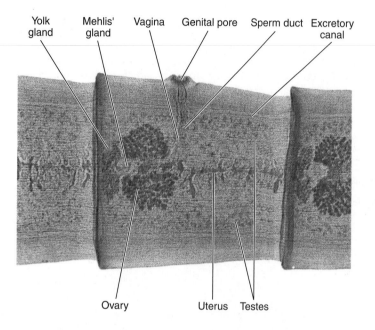

Yolk gland   Mehlis' gland   Vagina   Genital pore   Sperm duct   Excretory canal

Ovary   Uterus   Testes

figure 8.16

Mature proglottid of *Taenia pisiformis*, a dog tapeworm. Portions of two other proglottids are also shown.

tation for a tapeworm, since it must absorb all its nutrients across the tegument.

Tapeworms are nearly all monoecious. The main body of a cestode is a chain of proglottids called a **strobila** (see figure 8.14). Typically, new proglottids form at a **germinative zone** just behind the scolex (proximal to it). As new proglottids differentiate in front of it, each previously formed proglottid moves distally in the strobila, and its gonads mature (figure 8.16). A proglottid is usually fertilized by another proglottid in the same or a different strobila. Shelled embryos form in the uterus of the proglottid, and either they are expelled through a uterine pore, or the entire proglottid detaches from the worm as it reaches the distal end.

About 4000 species of tapeworms are known. With rare exceptions, all cestodes require at least two hosts, and adults are parasitic in the digestive tract of vertebrates. Often one of the intermediate hosts is an invertebrate. Almost all vertebrate species can be infected. Normally, adult tapeworms do little harm to their hosts. Table 8.2 lists the most common tapeworms in humans.

In the beef tapeworm, *Taenia saginata* (Gr. *tainia*, band, ribbon), shelled larvae shed from the human host are ingested by cattle (figure 8.17). The six-hooked larvae (oncospheres) hatch, burrow into blood or lymph vessels, and migrate to skeletal muscle, where they encyst to become "bladder worms" **(cysticerci).** Each of these juveniles develops an invaginated scolex and remains quiescent until the uncooked muscle is eaten by humans. In the new host, the scolex evaginates, attaches to the intestine, and matures in 2 or 3 weeks; then ripe proglottids may be expelled daily for many years. Humans become infected by eating raw or rare infested ("measly") beef. Adult worms may attain a length of 7 m or more, folded back and forth in the host intestine.

The pork tapeworm, *Taenia solium*, uses humans as definitive hosts and pigs as intermediate hosts. Humans can also serve as intermediate hosts by ingesting shelled larvae

Table 8.2   Common Cestodes Infecting Humans

Common and scientific names	Means of infection; prevalence in humans
Beef tapeworm (*Taenia saginata*)	Eating rare beef; most common of large tapeworms in humans
Pork tapeworm (*Taenia solium*)	Eating rare pork; less common than *T. saginata*
Fish tapeworm (*Diphyllobothrium latum*)	Eating rare or poorly cooked fish; fairly common in Great Lakes region of United States and other areas of world where raw fish is eaten
Dog tapeworm (*Dipylidium caninum*)	Unhygienic habits of children (juveniles in flea and louse); moderate frequency
Dwarf tapeworm (*Hymenolepis nana*)	Juveniles in flour beetles; common
Unilocular hydatid (*Echinococcus granulosus*)	Cysts of juveniles in humans; infection by contact with dogs; common wherever humans are in close relationship with dogs and ruminants
Multilocular hydatid (*Echinococcus multilocularis*)	Cysts of juveniles in humans; infection by contact with foxes; less common than unilocular hydatid

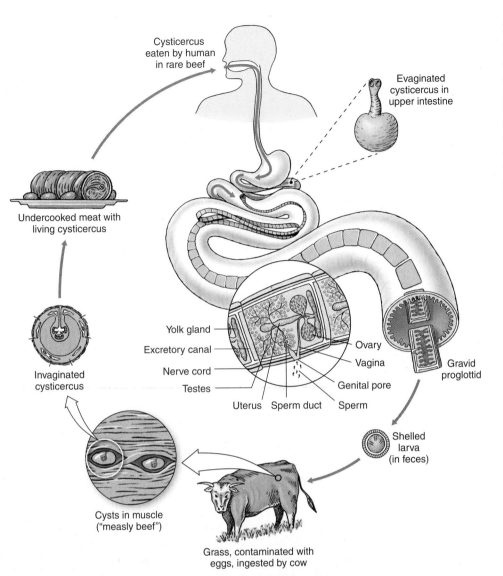

figure 8.17

Life cycle of beef tapeworm, *Taenia saginata.* Ripe proglottids detach in the human intestine, leave the body in feces, crawl onto grass, and are ingested by cattle. Eggs hatch in the cow's intestine, freeing oncospheres, which penetrate muscles and encyst, developing into "bladder worms." A human eats infected rare beef, and a cysticercus is freed in the intestine where it attaches to the intestinal wall, forms a strobila, and matures.

Cysticercus eaten by human in rare beef

Evaginated cysticercus in upper intestine

Undercooked meat with living cysticercus

Yolk gland

Excretory canal

Nerve cord

Testes

Uterus   Sperm duct   Sperm

Ovary

Vagina

Genital pore

Gravid proglottid

Invaginated cysticercus

Cysts in muscle ("measly beef")

Shelled larva (in feces)

Grass, contaminated with eggs, ingested by cow

figure 8.18

Section through the brain of a person who died of cerebral cysticercosis, an infection with cysticerci of *Taenia solium.*

from contaminated hands or food or, in persons harboring an adult worm, by regurgitating segments into the stomach. Cysticerci may encyst in the central nervous system, where great damage may result (figure 8.18).

■ Phylum Gastrotricha

The phylum Gastrotricha (gas-trō-tri′kə) (Gr. *gaster,* belly, + *thrix,* hair) is a small group (about 460 species) of microscopic animals, approximately 65 to 500 µm long (figure 8.19). They are usually bristly or scaly in appearance and flattened on the ventral side; many move by gliding on ventral cilia. Others move in a leechlike fashion by briefly attaching the posterior end by means of adhesive glands. Both marine and freshwater species exist and they are common in lakes, ponds, and seashore sands. They feed on bacteria, diatoms, and small protozoa. Gastrotrichs are hermaphroditic. However, the male system of many freshwater species is nonfunctional, and the female system produces offspring parthenogenetically.

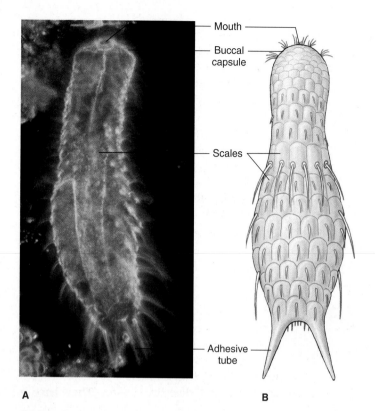

figure 8.19

A, A live *Chaetonotus simrothi,* a common gastrotrich. **B,** External structure of *Chaetonotus.*

Clade Gnathifera

Clade Gnathifera (nath-if′er-ə) comprises four phyla. Members of Gnathostomulida, Micrognathozoa, and Rotifera are tiny, free-living aquatic animals, whereas members of Acanthocephala are wormlike endoparasites of fishes or other vertebrates. Clade Gnathifera ("jaw-bearing") is so named because its members, other than the acanthocephalans, possess small, cuticular jaws with a homologous microstructure. The number of pairs of such jaws varies within the clade, but jaws are absent from acanthocephalans.

Acanthocephala and Rotifera are presumed sister taxa, together forming a clade called Syndermata. Their close relationship first appeared in molecular phylogenies and led morphologists to examine acanthocephalans anew, searching for evidence that these parasites were highly derived rotifers. There is little external similarity between free-swimming rotifers and endoparasitic worms, but members of both groups have a eutelic syncytial epidermis. **Eutely** denotes constancy in the numbers of nuclei present, as illustrated by the constant numbers of nuclei in various organs of one species of rotifer: A researcher, E. Martini (1912), always counted 183 nuclei in the brain, 39 in the stomach, and 172 in the coronal epithelium. Despite the shared syncytial epidermis, the union of two morphologically disparate taxa in clade Syndermata is still controversial.

■ Phylum Gnathostomulida

The first species of Gnathostomulida (nath′ō-sto-mūlid′-a) (Gr. *gnatho,* jaw, + *stoma,* mouth, + L -*ulus,* diminutive) was observed in 1928 in the Baltic, but its description was not published until 1956. Since then, these animals have been found in many parts of the world, including the Atlantic coast of the United States, and over 80 species in 18 genera have been described.

Gnathostomulids are delicate, wormlike animals 0.5 to 1 mm long (figure 8.20). They occupy interstitial spaces of very fine sandy coastal sediments and silt and can endure conditions of very low oxygen.

Gnathostomulids scrape bacteria and fungi from a substratum using a pair of lateral jaws in the pharynx. The pharynx, structurally similar to the rotifer mastax (see p. 170), leads into a blind gut. The body is acoelomate with a poorly developed parenchyma layer. There is no circulatory system, and respiration relies on diffusion. The epidermis is ciliated, but each epidermal cell has only one cilium; this condition is rare among other bilateral animals, although it does occur in some

figure 8.20

A, *Gnathostomula jenneri* (phylum Gnathostomulida) is a tiny member of the interstitial fauna living between grains of sand or mud. Species in this family are among the most commonly encountered jaw worms, found in shallow water and down to depths of several hundred meters. **B,** *Gnathostomula paradoxa* is abundant in sediments near burrows of marine polychaetes in the North Sea. Its ecology is very similar to that of *G. jenneri* from the North American Atlantic coast.

gastrotrichs. Gnathostomulids are simultaneous hermaphrodites. Development proceeds via spiral cleavage.

Phylum Micrognathozoa

The first and only micrognathozoan species, *Limnognathia maerski* (figure 8.21), was collected from Greenland in 1994 but not formally described until 2000. Micrognathozoans are tiny, interstitial animals (living between sand grains) about 142 µm long. The body consists of a two-part head, a thorax, and an abdomen with a short tail. The cellular epidermis has dorsal plates but lacks plates ventrally. These animals move using cilia and also possess a unique ventral ciliary adhesive pad that produces glue. They have a very complicated jaw system of plates and associated teeth; the mouth leads into a relatively simple gut. An anus opens to the outside only periodically. Only animals with female reproductive organs have been found, so reproduction is not well understood.

Phylum Rotifera

Rotifera (rō -tif'e-rə) (L. *rota*, wheel, + *fero*, to bear) derive their name from their characteristic ciliated crown, or **corona**, which, when beating, often gives the impression of rotating wheels. Rotifers range in size from 40 µm to 3 mm in length, but most are between 100 and 500 µm long. Some have beautiful colors, although most are transparent. Some have odd or bizarre shapes, and their shapes are often correlated with their mode of life. Floaters are usually globular and saclike; creepers and swimmers are somewhat elongated and wormlike; and sessile types are commonly vaselike, with a cuticular envelope. Some are colonial.

Rotifers are a cosmopolitan group of about 2000 species, some of which occur throughout the world. Most species are freshwater inhabitants, a few are marine, some are terrestrial, and some are epizoic (living on the surface of other animals) or parasitic. Most often they are benthic, occurring on submerged vegetation in ponds or shallow margins of freshwater lakes.

The body is usually composed of a head, a trunk, and a foot. The head region bears the corona. The corona may form a ciliated funnel, with its upper edges folded into lobes bearing bristles, or the corona may comprise a pair of ciliated discs (figure 8.22). The cilia create currents of water toward the mouth, bringing small planktonic forms for food. The corona may be retractile. The mouth, surrounded by some part of the corona, opens into a modified muscular pharynx called a **mastax,** which is unique to rotifers. The mastax has a set of intricate jaws used for grasping and chewing. The trunk contains visceral organs, and a terminal foot, when present, is segmented and, in some, ringed with joints that can telescope to shorten. The one to four toes secrete a sticky substance from the pedal glands for attachment.

Rotifers have a pair of protonephridial tubules with flame bulbs, which eventually drain into a cloacal bladder that collects excretory and digestive wastes. They have a bilobed "brain" and sense organs that include eyespots, sensory pits, and papillae.

Female rotifers have one or two syncytial ovaries (**germovitellaria**) that produce yolk as well as oocytes. Although rotifers are dioecious, males are unknown in many species; in these, reproduction is entirely parthenogenetic. In parthenogenetic species, females produce only diploid eggs that have not undergone meiosis, cannot be fertilized, and develop only into females. Such eggs are called **amictic eggs.** Other species can produce two kinds of eggs—amictic eggs, which

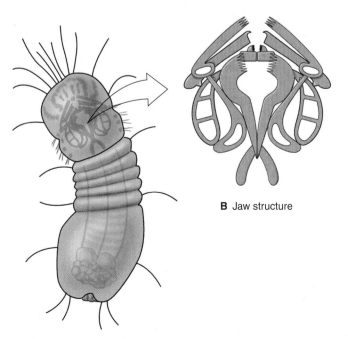

B Jaw structure

A *Limnognathia maerski,* a micrognathozoan

C

figure 8.21

A, *Limnognathia maerski,* a micrognathozoan. **B,** Detail of complex jaws. **C,** A living specimen. This animal was found on moss in a freshwater spring on Disko Island, Greenland. It swims, or crawls, consuming bacteria, blue-green algae, and diatoms.

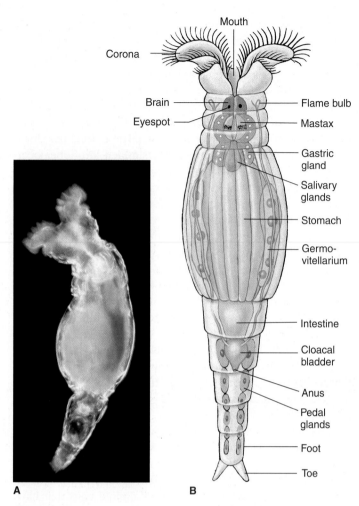

figure 8.22

A, A live *Philodina*, a common rotifer. **B,** Structure of *Philodina*.

develop parthenogenetically into females, and **mictic eggs,** which have undergone meiosis and are haploid. Mictic eggs, if unfertilized, develop quickly and parthenogenetically into males. Males are haploid and make sperm via mitosis. If mictic eggs are fertilized, they secrete a thick shell and become dormant for several months before developing into females. Such dormant eggs can withstand desiccation and other adverse conditions and permit rotifers to inhabit temporary ponds.

Mictic (Gr., *miktos*, mixed, blended) refers to the capacity of haploid eggs to be fertilized (that is, "mixed") with the male's sperm nucleus to form a diploid embryo. Amictic ("without mixing") eggs are already diploid and can develop only parthenogenetically.

■ Phylum Acanthocephala

Members of phylum Acanthocephala (a-kan'tho-sef'a-la) (Gr. *akantha*, spine or thorn, + *kephalē*, head) are commonly

called "spiny-headed worms." The phylum derives its name from one of its most distinctive features, a cylindrical invaginable **proboscis** (figure 8.23) bearing rows of recurved spines, by which the worms attach themselves to the intestine of their host. All acanthocephalans are endoparasitic, living as adults in the intestines of vertebrates.

Over 1100 species of acanthocephalans are known. Most of them parasitize fishes, birds, and mammals, and the phylum is worldwide in distribution. No species is normally a parasite of humans, although rarely humans are infected with species that normally use other hosts.

Various species of acanthocephalans range in length from less than 2 mm to over 1 m, with the females of a species usually larger than the males. In life, the body is normally bilaterally flattened, with numerous transverse wrinkles. These worms are usually cream color, but may be yellowish or brown due to absorption of pigments from intestinal contents.

The body wall is syncytial, and its surface is punctuated by minute crypts 4 to 6 μm deep, which greatly increase the surface area of the tegument. About 80% of the thickness of the tegument is the radial fiber zone, which contains a **lacunar system** of ramifying, fluid-filled canals (figure 8.23). The function of the lacunar system is unclear, but it may help distribute nutrients to the peculiar, tubelike muscles in the body wall of these organisms.

Excretion is across the body wall in most species. One family has a pair of **protonephridia** with flame cells that unite to form a common tube opening into the sperm duct or uterus.

Because acanthocephalans have no digestive tract, they must absorb all nutrients through their tegument. They can absorb various molecules by specific membrane transport mechanisms, and their tegument can perform pinocytosis (cell drinking).

Sexes are separate. Males have a protrusible penis, and at copulation sperm travel up the genital duct and escape into the pseudocoelom of females. Zygotes develop into shelled acanthor larvae. Shelled larvae escape from the vertebrate host in feces, and if eaten by a suitable arthropod, they hatch and work their way into the hemocoel, where they grow to become juvenile acanthocephalans. At, that point, either their development ceases until the arthropod is eaten by a suitable host, or they pass through several transport hosts in which they encyst until eaten.

■ Phylum Mesozoa

Members of the phylum Mesozoa (Gr. *mesos*, in the middle, + *zōon*, animal) are minute, ciliated animals only 0.5 to 7 mm in length. They are highly specialized parasites or symbionts living in marine invertebrates; some live in cephalopod kidneys. Most are composed of only 20 to 30 cells arranged in two layers. The two layers are not homologous to the germ layers of other metazoans, and mesozoan development does not include gastrulation, unlike that of other diploblasts or triploblasts. However, some researchers argue that the simple bodies in modern forms were derived from more complex free-living ancestors.

figure 8.23

A, Intestine of an eider duck filled with a lethal number of acanthocephalan worms, *Polymorphus botulus*. Prior to death, the duck fed on shore crabs, an intermediate host for the endoparasites, due to absence of blue mussels, the preferred prey for eider ducks on Cape Cod. **B,** An adult worm. **C,** An everted proboscis, showing the trait for which spiny-headed worms are named. The proboscis attaches to the intestinal wall; nutrients are absorbed across the tegument. **D,** Diagrammatic longitudinal section of an everted proboscis showing muscles. **E,** Diagrammatic longitudinal section of an inverted proboscis within the pseudocoelom.

Molecular studies indicate that mesozoans possess some genetic and biochemical markers present in triploblastic animals. Their position on the cladogram inside the front cover reflects this evidence. Note that they are not placed within Platyzoa.

Phylogeny and Adaptive Diversification

Phylogeny

Examination of the cladogram inside the front cover shows a large clade called Bilateria that contains almost all the triploblastic animals.

Most current phylogenies place members of phylum Acoelomorpha as the sister taxon to all other Bilateria. Acoelomorphs differ from flatworms in phylum Platyhelminthes in their embryonic cleavage patterns, in the way that their mesoderm forms, and in the structure of the nervous systems.[1]

Within Bilateria, evidence from sequence analysis of ribosomal genes suggests that ancestral protostomes split from ancestral deuterostomes in the Precambrian period. Protostomes later split into two large groups, sometimes called

---

[1] Ruiz-Trillo et al. 2004. Molecular Phylogenetics and Evolution **33**: 321–332.

superphyla, Ecdysozoa and Lophotrochozoa. Evolutionary relationships within Lophotrochozoa are still in flux. We depict a clade called Platyzoa, but not all phylogenies support this grouping. Platyzoa contains Platyhelminthes, Gastrotricha, and Gnathifera. Some phylogenies place Platyhelminthes as the sister taxon and Gastrotricha, but others place Gnathostomulida as the sister taxon of Gastrotricha, so for now we depict a polytomy within Platyzoa.

Flatworms in phylum Platyhelminthes are provisional members of Platyzoa. Within Platyhelminthes, class Turbellaria is clearly paraphyletic, but we are retaining the taxon for the present because thorough cladistic analysis would require introducing many more taxa and characteristics that are beyond the scope of this book and not yet common in zoological literature. For example, ectolecithal turbellarians should be allied with trematodes, monogeneans, and cestodes in the sister group to endolecithal turbellarians. Some ectolecithal turbellarians share a number of other derived characters with trematodes and cestodes and have been placed by Brooks (1989) in a group designated Cercomeria (Gr. *kerkos,* tail, + *meros,* part) (see figure 8.3). Several synapomorphies, including the unique architecture of the tegument, indicate that neodermatans (trematodes, monogeneans, and cestodes) form a monophyletic group, and monophyly of Neodermata is supported by sequence data.

The group Gnathifera emerges in several different phylogenies, but most molecular studies do not include sequences from Micrognathozoa. Within Gnathifera, the sister taxon relationship between Acanthocephala and Rotifera is surprising for reasons already discussed (see p. 168). Both taxa are united as Syndermata, sharing a eutelic syncytial epidermis. Syndermata emerges from phylogenetic studies repeatedly and several recent studies show that acanthocephalans belong within Rotifera. The classification of rotifer taxa is in flux, but a group called bdelloids is of interest.

These rotifers are swimming or creeping forms, most with trochal discs at the top (see *Philodina,* figure 8.22, p. 170). There are no known male bdelloids, and reproduction is by parthenogenesis. Several recent phylogenetic studies place acanthocephalans as the sister taxon to bdelloids, making the Rotifera paraphyletic as presently conceived. Acanthocephala will be subsumed within phylum Rotifera if this result is confirmed. The group name Syndermata would no longer be necessary.

Mesozoans are identified as lophotrochozoan protostomes based on molecular data but are not placed in Platyzoa. Absence of the complex body plan typically associated with protostomes may be related to their lifestyles: they are parasites and endosymbionts.

Adaptive Diversification

The flatworm body plan, with its creeping adaptation, placed a selective advantage on bilateral symmetry and further development of cephalization, ventral and dorsal regions, and caudal differentiation. Acoelomorph flatworms have extremely simple triploblastic bodies; many entirely lack a gut. Free-living worms in phylum Platyhelminthes have an extensible pharynx, a blind gut, and thin, flattened bodies. Because of their body shape and metabolic requirements, early flatworms must have been well predisposed toward parasitism and gave rise to symbiotic descendants in Neodermata. These descendants diversified greatly as parasites, and many flatworms became highly specialized for that mode of existence. Specializations include loss of the head, reduced sensory organs, and loss of the gut in tapeworms.

An ancestral form with complex cuticular jaws may have diversified into the four phyla now placed within Gnathifera. This group includes the acanthocephalans whose jaw loss is assumed to have accompanied the transition to parasitism.

■ Summary

Acoelomorpha and Platyhelminthes, are the simplest phyla that are bilaterally symmetrical, a condition of adaptive value for actively crawling or swimming animals. They are triploblastic and have the organ-system level of organization. They vary in body plan, but lack a coelomic cavity surrounding the gut. Acoelomorphs are presumed to be the sister taxon to a clade comprising all other Bilateria.

The body surface of turbellarians is a cellular epithelium, at least in part ciliated, containing mucous cells and rod-shaped rhabdites that

function together in locomotion. Members of all other classes of flatworms are covered by a nonciliated, syncytial tegument having a vesicular distal cytoplasm and cell bodies beneath superficial muscle layers. Digestion is both extracellular and intracellular in most; cestodes must absorb predigested nutrients across their tegument because they have no digestive tract. Osmoregulation is by flame-cell protonephridia, and removal of metabolic wastes and respiration occur across the body wall. Except for some turbellarians, flatworms have a ladder-type nervous system

with motor, sensory, and association neurons. Most flatworms are hermaphroditic, and asexual reproduction occurs in some groups.

Class Turbellaria is a paraphyletic group with mostly free-living and carnivorous members. Digenetic trematodes have mollusc intermediate hosts and almost always a vertebrate definitive host. The great amount of asexual reproduction that occurs in their intermediate host helps to increase the chance that some offspring will reach a definitive host. Aside from the tegument, digeneans share many basic structural characteristics with the Turbellaria. Digenea includes a number of important parasites of humans and domestic animals. Digenea contrast with Monogenea, which are important ectoparasites of fishes and have a direct life cycle (without intermediate hosts).

Cestodes (tapeworms) generally have a scolex at their anterior end, followed by a long chain of proglottids, each of which contains a complete set of reproductive organs of both sexes. Cestodes live as adults in the digestive tract of vertebrates. They have microvillus-like microtriches on their tegument, which increase its surface area for absorption. The shelled larvae are passed in the feces of their host, and the juveniles develop in a vertebrate or invertebrate intermediate host.

Flatworms and cnidarians may have evolved from a common planuloid ancestor, some of whose descendants became sessile or free floating and radial (cnidarians), while others became creeping and bilateral (flatworms).

Gastrotrichs are tiny aquatic animals. They have ventrally flattened bodies with bristles or scales. They move by cilia or adhesive glands.

Clade Gnathifera contains four phyla whose common ancestor is hypothesized to possess cuticular jaws with a unique microstructure. Included phyla are Gnathostomulida, Micrognathozoa, Rotifera, and Acanthocephala.

Gnathostomulida contains tiny, wormlike animals living among sand grains and silt. The animals do not have an anus.

Micrognathozoa consists of a single species of tiny animals living between sand grains. These animals have three pairs of complex jaws similar to those of rotifers and gnathostomulids.

Phylum Rotifera is composed of small, mostly freshwater species with a ciliated corona that creates currents of water to draw planktonic food toward the mouth. The mouth opens into a muscular pharynx, or mastax, which is equipped with jaws.

Acanthocephalans are all parasitic in the intestine of vertebrates as adults, and their juvenile stages develop in arthropods. They have an anterior, invaginable proboscis armed with spines, which they embed in the intestinal wall of their host. They do not have a digestive tract and so must absorb all nutrients across their tegument.

Mesozoans are wormlike parasites or symbionts of marine invertebrates. Although they have only two body layers, they are presumed to have triploblastic protostome ancestors.

■ Review Questions

1. Why does bilateral symmetry have adaptive value for actively motile animals?
2. How does an acoelomorph flatworm differ from a free-living member of phylum Platyhelminthes?
3. Match the terms in the right column with the classes in the left column:

   _____ Turbellaria   a. Endoparasitic
   _____ Monogenea   b. Free-living and
   _____ Trematoda   commensal
   _____ Cestoda   c. Ectoparasitic

4. Examine the text to discover that there is no unique feature to distinguish Platyhelminthes. How many classes share the syncytial tegument?
5. Distinguish two mechanisms by which flatworms supply yolk for their embryos. Which system is evolutionarily ancestral for flatworms, and which one is derived?
6. How do flatworms digest their food?
7. Briefly describe the osmoregulatory system and the nervous system and sense organs of platyhelminthes.
8. Contrast asexual reproduction in Turbellaria and Trematoda.
9. Contrast the typical life cycle of a monogenean with that of a digenetic trematode.
10. Describe and contrast the tegument of turbellarians and the other classes of platyhelminths. Could this be evidence that the trematodes, monogeneans, and cestodes form a clade within the Platyhelminthes? Why?
11. Answer the following questions with respect to both *Clonorchis* and *Schistosoma:* (a) How do humans become infected? (b) What is the general geographical distribution? (c) What main disease conditions are produced?
12. Define each of the following with reference to cestodes: scolex, microtriches, proglottids, strobila.
13. Why is *Taenia solium* a more dangerous infection than *Taenia saginata?*
14. What are some of the adaptations that permit flukes and tapeworms to have a parasitic lifestyle?
15. What evidence indicates that the traditional class Turbellaria is paraphyletic?
16. Explain the difference between an acoelomate body and one possessing a pseudocoelom.
17. What is the normal size of rotifers; where are they found; and what are their major body features?
18. Explain the difference between mictic and amictic eggs of rotifers.
19. Under what environmental conditions might amictic eggs be more advantageous than mictic eggs?
20. What are the approximate lengths of gastrotrichs, gnathostomulids, and micrognathozoans?
21. What habitat is shared by micrognathozoans and gnathostomulids?
22. What character unites members of clade Gnathifera?
23. What characters are used to unite rotifers and acanthocephalans as members of clade Syndermata?
24. What is eutely?
25. The evolutionary ancestry of acanthocephalans is particularly obscure. Describe some characters of acanthocephalans that challenge the hypothesis that acanthocephalans are derived rotifers.
26. How do acanthocephalans get food?
27. Where do mesozoans live?

For Further Thought How could an endoparasite evolve from a free-living ancestor, and what other ways of living (e.g., symbiont, commensal, ectoparasite) would you predict as part of the transition?

■ Selected References

See also general references on page 432.

Brooks, D. R. 1989. The phylogeny of the Cercomeria (Platyhelminthes: Rhabdocoela) and general evolutionary principles. J. Parasitol. **75**:606–616. *Cladistic analysis of parasitic flatworms.*

Dalton, J. P., P. Skelly, and D. W. Halton. 2004. Role of the tegument and gut in nutrient uptake by parasitic platyhelminthes. Can. J. Zool. **82**:211–232. *The function of the tegument and the evolution of an endoparasitic lifestyle are discussed here.*

Desowitz, R. S. 1981. New Guinea tapeworms and Jewish grandmothers. New York, W. W. Norton & Company. *Accounts of parasites and parasitic diseases of humans. Entertaining and instructive. Recommended for all students.*

Eernisse, D. J., and K. J. Peterson. 2004. The history of animals, pp. 197–208. In J. Cracraft and M. J. Donoghue, eds., Assembling the tree of life. Oxford, UK, Oxford University Press. *A general overview of current metazoan phylogeny with discussion of the major clades and "problem" taxa including acoelomorph flatworms.*

Giribet, G., M. V. Sorenson, P. Funch, R. M. Kristensen, and W. Sterrer. 2004. Investigations into the phylogenetic position of Micrognathozoa using four molecular loci. Cladistics **20**:1–13. *Research supports the placement of micrognathozoans outside any known phylum.*

Kristensen, R. M. 2002. An introduction to Loricifera, Cycliophora, and Micrognathozoa. Integ. and Comp. Biol. **42**:641–651. *A clear and informative description of these little-known animal groups.*

Rieger, R. M., and S. Tyler. 1995. Sister-group relationship of Gnathostomulida and Rotifera-Acanthocephala. Invert. Biol. **114**:186–188. *Evidence that gnathostomulids are the sister group of a clade containing rotifers and acanthocephalans.*

Roberts, L. S., and J. Janovy, Jr. 2005. G. D. Schmidt and L. S. Roberts's foundations of parasitology, ed. 7. Dubuque, Iowa, McGraw-Hill Higher Education. *Up-to-date and readable accounts of parasitic flatworms.*

Ruiz-Trullo, M., M. Riutort, H. M. Fourcade, J. Baguna, and J. L. Boore. 2004. Mitochondrial genome data support the basal position of Acoelomorpha and the polyphyly of the Platyhelminthes. Mol. Phylogenet. Evol. **33**:321–332. *This paper adds to the growing body of evidence that acoelomorphs are the sister taxon to all other bilateral metazoans and not members of Platyhelminthes.*

Schell, S. C. 1985. Handbook of trematodes of North America north of Mexico. Moscow, Idaho, University Press of Idaho. *Good for trematode identification.*

Telford, M. J., A. E. Lockyear, C. Cartwright-Finch, and D. T. J. Littlewood. 2003. Combined large and small subunit ribosomal RNA phylogenies support a basal position of the acoelomate flatworms. Proc. R. Soc. London B **270**:1077–1083. *A review of the molecular evidence that led to removal of the Acoela from Platyhelminthes.*

Wallace, R. L. 2002. Rotifers: exquisite metazoans. Integ. and Comp. Biol. **42**:660–667. *This paper summarizes recent work on rotifers, but assumes basic knowledge of the group.*

Welch, M. D. B. 2000. Evidence from a protein-coding gene that acanthocephalans are rotifers. Invert. Biol. **119**:17–26. *Sequence analysis of a gene coding for a heat-shock protein supports a position of acanthocephalans within Rotifera. Other molecular and morphological evidence is cited that supports this position.*

■ Custom Website

The *Animal Diversity* Online Learning Center is a great place to check your understanding of chapter material. Visit www.mhhe.com/hickmanad6e for access to key terms, quizzes, and more! Further enhance your knowledge with Web links to chapter-related material.

Explore live links for these topics:

Classification and Phylogeny of Animals
Phylum Platyhelminthes
Class Turbellaria
Class Trematoda
Class Cestoda
Class Monogenea

Pseudocoelomate Animals
Phylum Gastrotricha
Phylum Gnathostomulida
Phylum Micrognathozoa
Phylum Rotifera
Phylum Acanthocephala

Polyzoa and Kryptrochozoa

Cycliophora, Entoprocta, Ectoprocta, Brachiopoda, Phoronida, and Nemertea

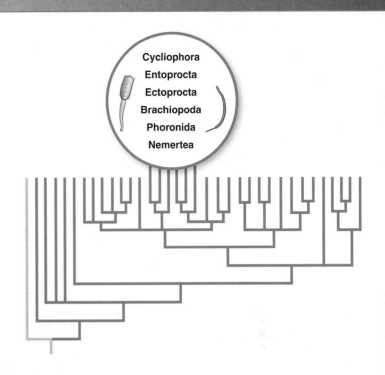

Cycliophora
Entoprocta
Ectoprocta
Brachiopoda
Phoronida
Nemertea

...procts and other animals fouling a boat bottom.

Some Evolutionary Experiments

The early Cambrian period, about 570 million years ago, was the most fertile time in evolutionary history. For 3 billion years before this period, evolution had forged little more than bacteria and blue-green algae. Then, within the space of a few million years, all major phyla, and probably all smaller phyla, became established. This was the Cambrian explosion, the greatest evolutionary "bang" the world has known. In fact, the fossil record suggests that more phyla existed in the Paleozoic era than exist now, but some disappeared during major extinction events that punctuated evolution of life on earth. The greatest of these disruptions was the Permian extinction about 230 million years ago. Thus, evolution has led to many "experimental models." Some of these models failed because they were unable to survive in changing conditions. Others gave rise to abundant and dominant species and individuals that inhabit the world today. Some persisted with small numbers of species, living in very specialized habitats—for example, on the lips of lobsters. The odd shapes and unusual feeding structures of many such animals seem more in the realm of science fiction than reality.

Three phyla—Phoronida, Ectoprocta, and Brachiopoda—possess a crown of ciliated tentacles, called a lophophore, that they use in food capture and respiration. Brachiopods were abundant in the Paleozoic era but began to decline thereafter. Phylum Ectoprocta arose in the Cambrian period, became widespread during the Paleozoic era, and remains a prevalent group today. Phylogenetic evidence indicates that the lophophore evolved more than once.

T he six phyla described in this chapter are lophotrochozoan protostomes. The name "Lophotrochozoa" was developed by merging terms for two features present in many animals in this group: the lophophore and the trochophore. A **lophophore** (Gr. *lophos*, crest or tuft, + *phorein*, to bear) is a crown of tentacles covered with cilia that is borne on a ridge or fold of the body wall. It is an efficient feeding device. The cavity inside the lophophore is part of the coelom and is filled with coelomic fluid. The thin ciliated walls of the lophophore act as a respiratory surface for exchange of gases between environmental water and coelomic fluid. A lophophore can usually be extended for feeding and respiration and withdrawn for protection.

Three animal phyla possess a lophophore: Ectoprocta, Brachiopoda, and Phoronida. Does the common possession of this feature indicate shared ancestry? Most evidence suggests not. Identifying homologous structures requires attention to

detail and a detailed analysis of lophophore structure and function[1] indicates that the lophophore evolved twice, once in Ectoprocta and once in the common ancestor of Brachiopoda and Phoronida. The Ectoprocta are placed in clade with two other taxa that possess ciliated tentacles (Cycliophora and Entoprocta). These three taxa form a clade called Polyzoa (figure 9.1).

Brachiopoda and Phoronida are united in a clade called Brachiozoa (figure 9.1). The sister taxon to Brachiozoa differs among phylogenetic analyses, but we depict Nemertea in this position. Nemerteans are very thin, stretchy, marine worms with an unusual way of capturing prey (see p. 182). The three taxa form a clade called Kryptrochozoa (Gr. *kryptos*, hidden + *trochos*, wheel + *zoa*, animal) that is nested in a larger clade

---

[1] Nielsen, C. 2002. Integ. and Comp. Biol. **42**:685–691.

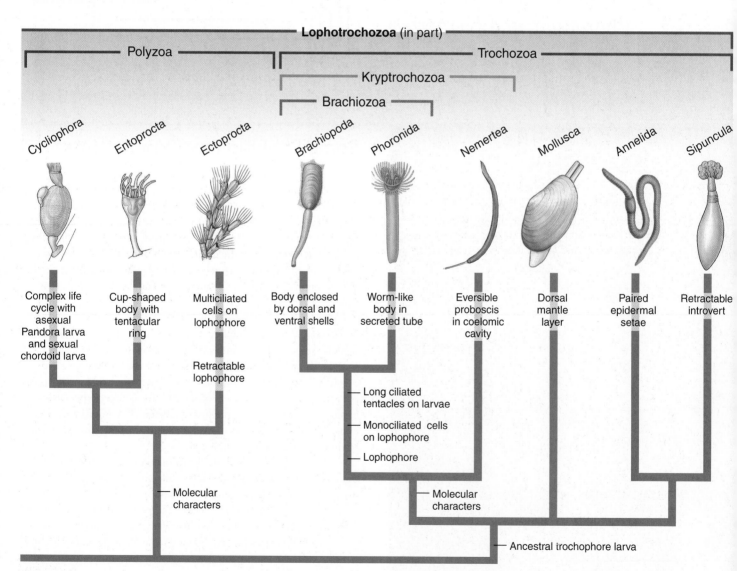

figure 9.1

Hypothetical relationships among polyzoans and kryptrochozoans. Characters are modified subsets of those in Nielsen (2002) and Brusca and Brusca (2003). See references p. 185 and p. 432 for citations.

called Trochozoa (figure 9.1). The remaining members of Trochozoa, Mollusca, Annelida, and Sipuncula, are discussed in Chapters 10 and 11. The term "troch" in the name Trochozoa refers to part of a trochophore larva. A **trochophore** is a free-swimming, feeding, larval stage having a ring of large ciliated cells in front of the mouth (see figure 10.6). This ring of cells is called a prototroch and is used for locomotion primarily, although it may play a role in feeding. The trochophore stage is obvious in the development of molluscs, annelids, and sipunculans, among other taxa. Is there a trochophore in the life cycles of kryptrochozoans? As the name implies, the larval forms in this group may be modified trochophores, carrying a band of cilia or a "hidden" prototroch. Although a detailed discussion of larval forms is beyond the scope of this text, there are undoubtedly biologists who consider some of these larvae well beyond "modified." The developmental patterns, molecular characteristics, and morphologies of the animals covered here are all under intense study, so the evolutionary relationships of lophotrochozoans may change.

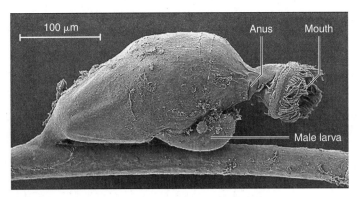

figure 9.2

Symbion pandora, a cycliophoran living on setae on the mouthparts of lobsters. A male larva is attached to the female stage.

■ Clade Polyzoa

The clade Polyzoa unites cycliophorans with entoprocts and ectoprocts, also called bryozoans (see figure 9.1). It is now supported by phylogenetic studies using multiple genes, but a close relationship between Ectoprocta and Entoprocta was proposed on morphological grounds 40 years ago. Cycliophora was not discovered until 1995, so it could not have been included in those early studies, but similarities to entoprocts were noted when the animals were first described. The tiny animals in these three phyla have fascinating body plans and life cycles.

■ Phylum Cycliophora

In December 1995, P. Funch and R. M. Kristensen reported their discovery of some very strange little creatures clinging to the mouthparts of the Norway lobster (*Nephrops norvegicus*). The animals were tiny, only 0.35 mm long and 0.10 mm wide, and did not fit into any known phylum. They were named *Symbion pandora,* the first members of phylum Cycliophora (figure 9.2). Two other species have since been found on other species of lobsters, but they have not been formally described.

Cycliophorans have a very specialized habitat: they live on mouthparts of marine decapod crustaceans in the Northern Hemisphere. They attach to bristles on the mouthparts with an adhesive disc on the end of an acellular stalk. They feed by collecting bacteria, or bits of food dropped from their lobster host, on a ring of compound cilia that surrounds the mouth.

The body plan is relatively simple. The mouth leads into a U-shaped gut ending with an anus that opens outside the ciliated ring (figure 9.2). The body is acoelomate. The epidermis is cellular and surrounded by a cuticle.

The cycliophoran life cycle has complex sexual and asexual phases. Asexual reproduction permits a clone quickly to fill vacant space on lobster mouthparts. As a prelude to sexual reproduction, male larvae are made. Male larvae attach to feeding stages that contain female larvae (figure 9.2). After further asexual reproduction of males, sexual reproduction occurs. Sexually produced larvae swim to new lobster hosts, where they form clones.

■ Phylum Entoprocta

Entoprocta (en´to-prok´ta) (Gr. *entos,* within, + *proktos,* anus) is a small phylum containing less than 100 species of tiny, sessile animals that, superficially, look much like hydroid cnidarians, except that their tentacles are ciliated and tend to roll inward (figure 9.3). Most entoprocts are microscopic, and none is more than 5 mm long. They are all stalked and sessile; some are colonial, and some are solitary. All are ciliary feeders.

With the exception of one genus, all entoprocts are marine forms that are widely distributed from polar regions to the tropics. Most marine species are restricted to coastal and brackish waters and often grow on shells and algae. Some are commensals on marine annelid worms. Freshwater entoprocts (figure 9.3) occur on the undersides of rocks in running water.

The body of an entoproct is cup-shaped, bears a crown or circle of ciliated tentacles, and attaches to a substrate by a stalk in solitary species. In colonial species, the stalks of individuals connect to the stolon of the colony. Both tentacles and stalk are continuations of the body wall.

The gut is U-shaped and ciliated, and both the mouth and anus open within the circle of tentacles. The organism captures food particles in the current created by tentacular cilia and then passes these particles along the tentacles to the mouth. Entoprocts have a pair of protonephridia, but no circulatory or respiratory organs.

Some species are monoecious, some are dioecious, and others seem to be hermaphrodites in which the gonad at first produces sperm and then eggs (a condition described as *protandrous*). Cleavage is modified spiral and mosaic, and a

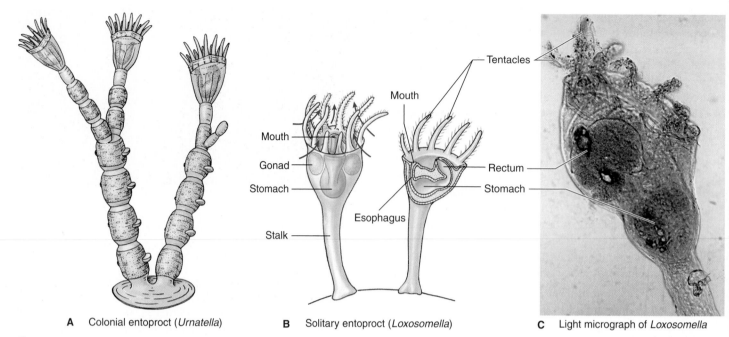

figure 9.3

A, *Urnatella,* a freshwater entoproct, forms small colonies of two or three stalks from a basal plate. **B,** *Loxosomella,* a solitary entoproct. Both solitary and colonial entoprocts can reproduce asexually by budding, as well as sexually. **C,** A live *Loxosomella.*

trochophore-like larva is produced, similar to that of some molluscs and annelids (see p. 191).

Libbie Henrietta Hyman (1888–1969), author of the monumental multi-volume treatise, *The Invertebrates,* was a distinguished American zoologist. She earned all her academic degrees from the University of Chicago, where she received her Ph.D. in 1915 and remained doing research until 1931. She published a laboratory manual for elementary zoology in 1919 and one for comparative vertebrate anatomy in 1922. The latter remained popular for the next 50 years, allowing Hyman to live modestly on proceeds from her books while pursuing her first love, the study of invertebrates. She became a research associate at the American Museum of Natural History and published 135 research papers, chiefly on turbellarians, as well as the treatise on invertebrates for which she is best remembered. Her six-volume treatise covers the protozoan groups through molluscs to gastropods. Thousands of invertebrate zoologists owe Hyman a debt of gratitude.

■ Phylum Ectoprocta

Ectoprocta (ek´to-prok´ta) (Gr. *ektos,* outside, + *proktos,* anus) have long been called bryozoans (Gr. *bryon,* moss, + *zoön,* animal), or moss animals, a term that originally included Entoprocta also.

Of the 4000 or so species of ectoprocts, few are more than 0.5 mm long. All are aquatic, both freshwater and marine, but they largely occur in shallow waters. With very few exceptions, most are colony builders. Ectoprocts, unlike most other phyla

considered in this chapter, were abundant and widespread in the past and remain so today. They have left a rich fossil record since the Ordovician era. Modern marine forms exploit all kinds of firm surfaces, such as shells, rock, large brown algae, mangrove roots, and ship bottoms. As one of the most important groups of fouling organisms on boat hulls, they decrease the efficiency of a hull passing through the water and make periodic scraping of the hull necessary.

Each member of a colony occupies a tiny chamber, called a **zoecium,** which is secreted by its epidermis (figure 9.4A). Each individual, or **zooid,** comprises a feeding **polypide** and a case-forming **cystid.** Polypides include a lophophore, digestive tract, muscles, and nerve centers. Cystids are the body wall of the animal, together with its secreted exoskeleton, which is called a zoecium. A zoecium may, according to the species, be gelatinous, chitinous, or stiffened with calcium and possibly also impregnated with sand. Its shape may be boxlike, vaselike, oval, or tubular.

Some ectoproct colonies form limy encrustations on seaweed, shells, and rocks (figure 9.5); others form fuzzy or shrubby growths or erect, branching colonies that look like seaweed. Some ectoprocts might easily be mistaken for hydroids but can be distinguished under a microscope by the presence of ciliated tentacles (figure 9.6). In some freshwater forms, individuals are borne on finely branching stolons that form delicate tracings on the underside of rocks or plants. Other freshwater ectoprocts are embedded in large masses of gelatinous material. Although zooids are minute, colonies may be several centimeters in diameter; some encrusting colonies may be a meter or more in width, and erect forms may reach 30 cm or more in height.

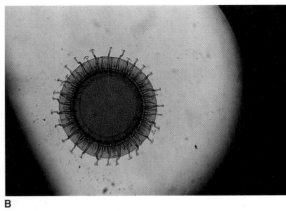

figure 9.4

A, Small portion of freshwater colony of *Plumatella* (phylum Ectoprocta), which grows on the underside of rocks. These tiny individuals disappear into their chitinous zoecia when disturbed. **B,** Statoblast of a freshwater ectoproct, *Cristatella*. Statoblasts are a kind of bud that survives over winter when a colony dies in the autumn. This one is about 1 mm in diameter and bears hooked spines.

figure 9.5

A, A colony of *Membranipora*, a marine encrusting form of Ectoprocta. Each little oblong zoecium is the calcareous home of a tiny ectoproct. **B,** Each zoecium holds one polypide. To feed, each polypide lifts its operculum and extends its lophophore.

A polypide lives a type of jack-in-the-box existence, popping up to feed and then quickly withdrawing into its little chamber, which often has a tiny trapdoor (operculum) that shuts to conceal its inhabitant. To extend the tentacular crown, certain muscles contract, which increases hydrostatic pressure within the body cavity and pushes the lophophore out. Other muscles contract to withdraw the crown to safety with great speed.

Lophophore ridges tend to be circular in marine ectoprocts and U-shaped in freshwater species. When feeding, an animal extends its lophophore, and its tentacles spread out to a funnel shape (figure 9.6). Cilia on the tentacles draw water into the funnel and out between tentacles. Food particles trapped in mucus in the funnel are drawn into the mouth by pumping of the muscular pharynx and by action of cilia in the pharynx. The digestive tract is U-shaped with the mouth opening inside the lophophore ring and the anus opening outside the ring (see figure 9.4).

Respiratory, vascular, and excretory organs are absent. Gas exchange occurs through the body surface, and since

figure 9.6

A, Ciliated lophophore of *Electra pilosa*, a marine ectoproct. **B,** *Plumatella repens*, a freshwater ectoproct. It grows on the undersides of rocks and vegetation in lakes, ponds, and streams.

ectoprocts are small, coelomic fluid is adequate for internal transport. Coelomocytes (cells in coelomic fluid) engulf and store waste materials. A ganglionic mass and a nerve ring surround the pharynx, but no sense organs are present. A septum divides the coelom into an anterior region in the lophophore and a larger posterior area. A third coelomic region is present only in freshwater ectoprocts. Pores in the walls between adjoining zooids permit exchange of materials by way of coelomic fluid.

Feeding individuals dominate most colonies, but polymorphism also occurs. One type of modified zooid resembles a bird beak that snaps at small invading organisms that might foul a colony. Another type has a long bristle that sweeps away foreign particles.

Most ectoprocts are hermaphroditic. Some species shed eggs into seawater, but most brood their eggs, some within the coelom and some externally in a special zoecium in which embryos develop. Cleavage is radial but apparently determinate.

Brooding is often accompanied by degeneration of the lophophore and gut of adults, the remains of which contract into minute dark balls, or **brown bodies.** Later, new internal organs regenerate in old chambers. Brown bodies may remain in the body or be eliminated by the new digestive tract—an unusual kind of excretion.

Freshwater species reproduce both sexually and asexually. Asexual reproduction is by budding or by means of **statoblasts,** which are hard, resistant capsules containing a mass of germinative cells that form during the summer and fall (see figure 9.4B). When a colony dies in late autumn, statoblasts are released, and in spring they can give rise to new polypides and eventually to new colonies.

There are many ways to value biological diversity, but people looking for practical or economic value need go no further than the anti-cancer properties of compounds from a wide range of marine invertebrates. The bryozoan *Bugula neritina*, pictured on page 175, is the source of bryostatin-1, a treatment for tumors and lymphoma. Bryostatin is actually made by a bacterial symbiont of *B. neritina* called *Candidatus endobugula sertula*, but the bacterium cannot be cultured outside the host. Other anti-cancer compounds are derived from sponges (e.g., discodermolide), corals (e.g., eleutherobin analogues), tunicates (e.g., ecteinascidin), clams (e.g., spisulosine), and sea cucumbers (e.g., calcigeroside B). The organisms just named are among the best-known marine invertebrates, so we wonder which unique compounds will be discovered within the animals discussed in this chapter. Known only to specialists, many animals covered here may have untapped potential uses. We know as little about their role in maintaining a healthy ocean (ecosystem services) as we do about their biochemical novelty.

◼ Clade Brachiozoa

The clade Brachiozoa contains brachiopods with phoronids; both molecular characters and morphological features support this pairing (see figure 9.1). Both taxa possess a lophophore, although one group is shelled and the other is wormlike.

◼ Phylum Brachiopoda

Brachiopoda (brak´ē ō-pod ə) (Gr. *brachiōn*, arm, + *pous, podos*, foot), or lamp shells, is an ancient group. Compared with the fewer than 300 species now living, some 30,000 fossil species, which flourished in the Paleozoic and Mesozoic seas, have been described. Brachiopods were once very abundant, but they are now apparently in decline. Some modern forms have changed little from early ones. Genus *Lingula* (L., little tongue) (figure 9.7A) has existed virtually unchanged for over 400 million years. Most modern brachiopod shells range between 5 and 80 mm, but some fossil forms reached 30 cm in length.

Brachiopods are all attached, bottom-dwelling, marine forms that mostly prefer shallow water. Externally, brachiopods resemble bivalve molluscs in having two calcareous shell valves secreted by a mantle. They were, in fact, grouped with molluscs until the middle of the nineteenth century, and their name refers to the arms of their lophophore, which were thought homologous to the mollusc foot. However, brachiopods have **dorsal** and **ventral valves** instead of right and left lateral valves that characterize bivalve molluscs, and also unlike bivalves, most of them are attached to a substrate either directly or by means of a fleshy stalk called a **pedicel** (or pedicle).

In most brachiopods, the ventral (pedicel) valve is slightly larger than the dorsal (brachial) valve, and one end projects in the form of a short, pointed beak that is perforated

where the fleshy stalk passes through (figure 9.7B). In many, the shape of the pedicel valve is that of a classical oil lamp of Greek and Roman times, so that brachiopods came to be called "lamp shells."

There are two classes of brachiopods based on shell structure. The shell valves of Articulata are connected by a hinge with an interlocking tooth-and-socket arrangement (articular process); those of Inarticulata lack the hinge and are held together by muscles only.

The body occupies only the posterior part of the space between the valves (figure 9.8A), and extensions of the body wall form mantle lobes that line and secrete the shell. (Note that the name "mantle" reflects the fact that the brachiopods were once grouped with molluscs; this structure is not homologous to the mollusc mantle. The large, horseshoe-shaped lophophore in the anterior mantle cavity bears long, ciliated tentacles used in respiration and feeding. Ciliary water currents carry food particles between the gaping valves and over the lophophore. Food is caught in mucus on the tentacles and carried in a ciliated food groove along an arm of the lophophore to the mouth (figure 9.8B). The digestive tract is U-shaped with the mouth inside the lophophore ring and the anus outside it (figure 9.8A).

There are only two coelomic cavities in articulates, but inarticulates have three coelomic cavities with the posterior cavity housing the viscera. One or two pairs of nephridia open into the coelom and empty into the mantle cavity. The circulatory system is open, with a contractile heart. A nerve ring with a small dorsal and a larger ventral ganglion is present.

Sexes are separate, and paired gonads discharge gametes through the nephridia. Development of brachiopods is similar in some ways to that of deuterostomes, with radial, mostly equal, holoblastic cleavage, and the coelom forms enterocoely in articulates. Free-swimming larvae of articulates resemble a trochophore.

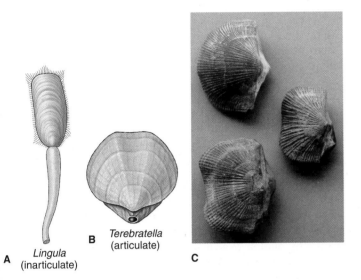

A *Lingula* (inarticulate)
B *Terebratella* (articulate)
C

figure 9.7

Brachiopods. **A,** *Lingula*, an inarticulate brachiopod that normally occupies a burrow. Its contractile pedicel can withdraw the body into the burrow. **B,** An articulate brachiopod, *Terebratella*. Its valves have a tooth-and-socket articulation, and a short pedicel projects through one valve to attach to the substratum (pedicel shown in figure 9.8). **C,** Fossil brachiopods of the genus *Hebertella* from the late Ordovician period, northern Kentucky.

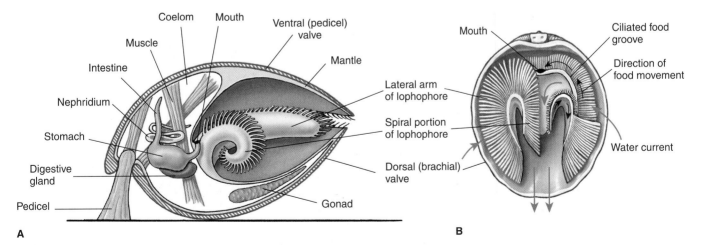

A   **B**

figure 9.8

Brachiopod anatomy. **A,** An articulate brachiopod (longitudinal section). **B,** Feeding and respiratory currents. Blue arrows show water flow over its lophophore; black arrows indicate food movement toward its mouth in the ciliated food groove.

■ Phylum Phoronida

Phylum Phoronida (fo-ron´i-də) (Gr. *phoros*, bearing, + L. *nidus*, nest) comprises approximately 10 species of small wormlike animals that live in benthic substrates under shallow coastal waters, especially in temperate seas. The phylum name refers to their tentacled lophophore. Phoronids range from a few millimeters to 30 cm in length. Each worm secretes a leathery or chitinous tube in which it lies free, but which it never leaves (figure 9.9). Their tubes may be anchored singly or in a tangled mass on rocks, shells, or pilings, or buried in the sand. Tentacles on the lophophore are thrust out for feeding, but if an animal is disturbed, it can withdraw completely into its tube.

Their lophophore has two parallel ridges curved in a horseshoe shape, the bend located ventrally and the mouth lying between the two ridges. Cilia on the tentacles direct a water current toward a groove between the two ridges, which leads toward their mouth. Plankton and detritus caught in this current become entangled in mucus and are carried by cilia to their mouth. As in brachiopoda, the mouth is inside the lophophore ring and the anus is outside.

The coelomic cavity is divided into three parts by mesenteries. Phoronids have a closed system of contractile blood vessels but no heart; their red blood contains hemoglobin. There is a pair of metanephridia. A nerve ring sends nerves to tentacles and the body wall.

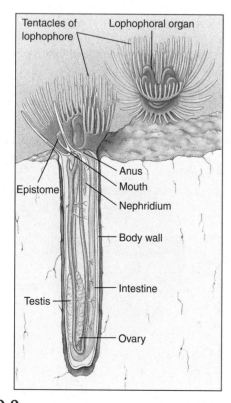

figure 9.9

Internal structure of *Phoronis* (phylum Phoronida), in diagrammatic vertical section.

There are both monoecious (the majority) and dioecious species of Phoronida, and at least one species reproduces asexually. Cleavage combines features of both spiral and radial types.

■ Phylum Nemertea (Rhynchocoela)

Nemerteans are thread- or ribbon-shaped predatory worms, often called ribbon worms. Their name (Gr. *Nemertes,* one of the nereids, unerring one) refers to the unerring aim of the proboscis, a long muscular tube (figures 9.10 and 9.11) that can be thrust out swiftly to grasp the prey. The phylum is also called Rhynchocoela (Gr. *rhynchos,* beak, + *koilos,* hollow), which also refers to the proboscis. There are about 1300 species in the group; nearly all are marine.

Form and Function

Many nemerteans are difficult to examine because they are so long and fragile. *Amphiporus* (Gr. *amphi,* on both sides, + *poros,* pore), one of the smaller genera that ranges from 2 to 10 cm in length, is fairly typical of nemertean structure (see figure 9.11). Its body wall consists of ciliated epidermis and layers of circular and longitudinal muscles. Most movement occurs by gliding over a slime track, although larger species move by muscular contractions.

The mouth is anterior and ventral, and the digestive tract is complete, extending the full length of the body and ending at an anus. The presence of an anus marks a significant advancement over the gastrovascular systems of flatworms and radiates, because ingestion and egestion can occur simultaneously. Cilia move food through the intestine. Digestion is largely extracellular.

figure 9.10

A ribbon worm, *Amphiporus bimaculatus* (phylum Nemertea), is 6 to 10 cm long, but other species range up to several meters. The proboscis of this specimen is partially extended at the top; the head is marked by two brown spots. The animal is photographed on an algal frond.

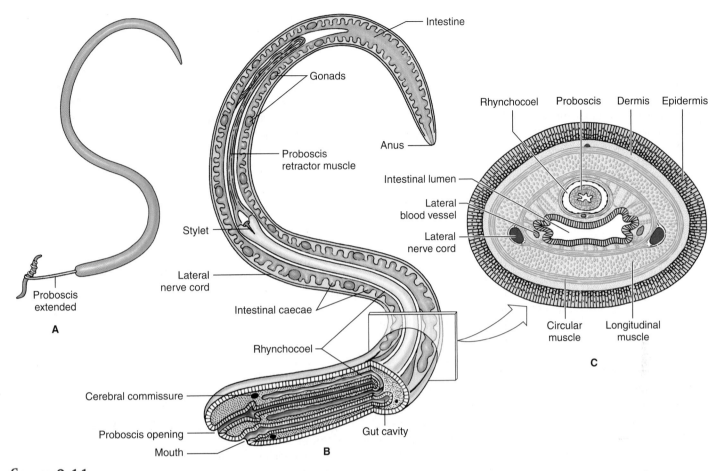

figure 9.11

A, *Amphiporus* with proboscis extended to catch prey. **B,** Internal structure of female ribbon worm (diagrammatic). Dorsal view to show proboscis. **C,** Diagrammatic cross section of female nemertean.

Nemerteans are carnivorous, feeding primarily on annelids and other small invertebrates. They seize their prey with a **proboscis** that lies in an interior cavity of its own, the **rhynchocoel,** above the digestive tract (but not connected with it). The proboscis itself is a long, blind muscular tube that opens at the anterior end at a proboscis pore above the mouth. (In a few nemerteans, the esophagus opens through the proboscis pore rather than through a separate mouth.) Muscular pressure on fluid in the rhynchocoel causes the long tubular proboscis to be everted rapidly through the proboscis pore. Eversion of the proboscis exposes a sharp barb, called a stylet (absent in some nemerteans). The sticky, slime-covered proboscis coils around the prey and stabs it repeatedly with the stylet, while pouring potent neurotoxins on the prey. Then, retracting its proboscis, a nemertean draws the prey near its mouth, through which the esophagus is thrust to engulf the prey.

The rhynchocoel is a cavity entirely contained within mesoderm, so it is a true coelom. However, in all other organisms with a coelom, this fluid-filled cavity surrounds the gut. The nemertean rhynchocoel lies above the gut, rather than around it, and houses the eversible proboscis. It is not known whether the rhynchocoel is an independently derived cavity or some form of modified coelom, but most researchers support the former explanation.

Nemerteans have a true circulatory system, and an irregular flow of blood is maintained by the contractile walls of the vessels. Many flame-bulb protonephridia are closely associated with the circulatory system, so that their function appears to be truly excretory (for disposal of metabolic wastes), in contrast to their apparently osmoregulatory role in Platyhelminthes.

Nemerteans have a pair of nerve ganglia, and one or more pairs of longitudinal nerve cords are connected by transverse nerves.

Some species reproduce asexually by fragmentation and regeneration. Most nemerteans are dioecious.

Phylogeny and Adaptive Diversification

Phylogeny

Since Lophotrochozoa first emerged from phylogenies based on molecular characters, biologists have struggled to understand patterns of morphological change across this diverse group of phyla. The suite of developmental characters associated with spiral cleavage is presumed to be ancestral for the clade. This suite includes spirally cleaving embryos, mosaic cleavage where cell fates are determined by the cytoplasmic factors they contain, and formation of mesoderm from derivatives of a particular cell, called the 4d cell, that is present at about the 64-cell stage of cleavage. Mesoderm is derived from both endoderm (via the 4d cell) and ectoderm in most taxa with spiral cleavage. If some or all of these characters are ancestral, then there must be character loss in several taxa. For example, spiral cleavage does not occur in rotifers, gastrotrichs, bryozoans, phoronids, or brachiopods; nor does mesoderm come from the 4d cell in most of these taxa. The likelihood that these characters were lost over evolutionary time, as opposed to never having been present, cannot be determined until there is a good understanding of the branching pattern within Lophotrochozoa. Phylogenetic analysis has proved challenging because the results vary depending upon which genes are examined. The next few years are likely to yield resolution of this problem as large multigene datasets are developed for most or all taxa.

Placement of Nemertea is a contentious issue. Nemerteans were once allied with Platyhelminthes because they share a ciliated epidermis and flame cell excretory structures, but the presence of a complete digestive tract and an eversible proboscis in a unique coelomic cavity in nemerteans makes a close relationship unlikely. However, the position of the coelomic cavity above the digestive tract, rather than surrounding it, sets nemerteans apart from other coelomate animals.

Adaptive Diversification

The boxlike modular unit of ectoprocts has proved very flexible. Zooids bud to form colonies, but a colony may be encrusting or treelike, to name just two extreme forms. A colony may be soft or calcified, and the shape of the zooids varies greatly, as does the surface ornamentation. Ectoprocts have colonized freshwater habitats, although they are far less widespread in freshwater as compared to marine habitats.

■ Summary

Cycliophorans are very tiny animals living on the setae of the mouthparts of lobsters. They have complex life cycles with sexual and asexual phases.

Entoprocts are small, sessile aquatic animals with a cup-shaped body on a small stalk. A crown of ciliated feeding tentacles encircles both the mouth and anus.

Ectoprocta, Brachiopoda, and Phoronida all bear a lophophore, which is a crown of ciliated tentacles surrounding the mouth but not the anus. They are sessile as adults, have a U-shaped digestive tract, and have a free-swimming larva. The lophophore functions as both a respiratory and a feeding structure, its cilia creating water currents from which food particles are filtered. However, the lophophore is thought to have evolved twice, once in Ectoprocta and once in the common ancestor of brachiopods and phoronids.

Ectoprocts are abundant in marine habitats, and a number of species are common in fresh water. Ectoprocts are colonial, and although each individual is quite small, colonies are commonly several centimeters or more in width or height. Each individual lives in a chamber (zoecium), which is a secreted exoskeleton of chitinous, calcium carbonate, or gelatinous material.

Brachiopods were very abundant in the Paleozoic era but have been declining in numbers and species since the early Mesozoic era. Their bodies and lophophores are covered by a mantle, which secretes a dorsal and a ventral valve (shell). They are usually attached to the substrate directly or by means of a pedicel.

Phoronida are the least common lophophorates, living in tubes mostly in shallow coastal waters. They thrust the lophophore out of the tube for feeding. Phoronids and brachiopods are sister taxa forming a clade called Brachiozoa.

Members of Nemertea have a complete digestive system with an anus and a true circulatory system. They are free-living and mostly marine; they ensnare prey with their long, eversible proboscis.

■ Review Questions

1. Where would you look if you had to find a cycliophoran?
2. How does an entoproct differ from an ectoproct?
3. Examine figure 9.1. How many times did the lophophore evolve? What evidence suggests that lophophorates do not form a clade?
4. Define each of the following for an ectoproct: lophophore, zoecium, zooid, polypide, cystid, brown bodies, and statoblasts.
5. Are lophophorates coelomate, acoelomate or pseudocoelomate?
6. Brachiopods superficially resemble bivalve molluscs. How would you explain the difference to a layperson?
7. Lophophorates are sometimes placed in a phylogenetic position between protostomes and deuterostomes. How would you justify or oppose such placement?
8. In what ways are nemertean worms different from other coelomate taxa?
9. How does a nemertean catch and consume prey?

For Further Thought How would you advocate for more research into the animals discussed in this chapter? Consider economics, practical use, aesthetics, and the role of a diverse and healthy ecosystem in your answer.

■ Selected References

See also general references on page 432.

Cohen, B. L., and A. Weydmann. 2005. Molecular evidence that phoronids are a subtaxon of brachiopods (Brachiopoda : Phoronata) and that genetic divergence of metazoan phyla began long before the early Cambrian. Org. Divers. Evol. **5**:253–273. *Indicates that phoronids arose within Brachiopoda and should no longer be considered a phylum.*

Eernisse, D. J., and K. J. Peterson. 2004. The history of animals, pp. 197–208. In J. Cracraft, and M. J. Donoghue, eds., Assembling the tree of life. New York, Oxford University Press. *A general overview of current metazoan phylogeny with discussion of the major clades and "problem" taxa.*

Hejnol, A. 2010. A twist in time—the evolution of spiral cleavage in the light of animal phylogeny. Integ. Comp. Biol. **50**:695–706. *Interesting review article on the extent to which spiralian taxa are lophotrochozoan taxa.*

Helfenbein, K. G., and J. L. Boore. 2004. The mitochondrial genome of *Phoronis architecta*—Comparisons demonstrate that phoronids are Lophotrochozoan protostomes. Mol. Biol. Evol. **21**:153–157. *Analysis of the mitochondrial DNA sequence shows a gene arrangement very similar to that of a chiton.*

Kristensen, R. M. 2002. An introduction to Loricifera, Cycliophora, and Micrognathozoa. Integ. and Comp. Biol. **42**:641–651. *A clear and informative description of these little-known animal groups.*

Lambert, J. D. 2010. Developmental patterns in spiralian embryos. Current Biology **20**:R72–R77. *A discussion of the spiralian characters as ancestral features for Lophotrochozoa.*

Mayer, A. M. S., and K. R. Gustafson. 2003. Marine pharmacology in 2000: antitumor and cytotoxic compounds. Int. J. Cancer **105**:291–299. *A good summary of results from marine natural products chemistry.*

Nielsen, C. 2002. The phylogenetic position of Entoprocta, Ectoprocta, Phoronida, and Brachiopoda. Integ. and Comp. Biol. **42**:685–691. *Evidence is presented to support the idea that lophophorates do not form a monophyletic group, and that phoronids and brachiopods are deuterostomes.*

Podsiadlowski, L., A. Braband, T. H. Struck, J. von Doehren, and T. Bartolomaeus. 2009. Phylogeny and mitochondrial gene order in Lophotrochozoa in the light of new mitogenomic data from Nemertea. BMC Genomics **10**:364–377. *A subset of lophotrochozoan taxa often emerges as a clade, but the relative relationships of phyla change depending upon the study; mitochondrial gene order data are not helpful.*

Turbeville, J. M. 2002. Progress in nemertean biology: development and phylogeny. Integ. and Comp. Biol. **42**:692–703. *A discussion of nemertean features that suggest shared ancestry with flatworms and other lophotrochozoan phyla.*

Witek, A., H. Herlyn, A. Meyar, L. Boell, G. Bucher, and T. Hankeln. 2008. EST based phylogenomics of Syndermata questions monophyly of Eurotatoria. BMC Evolutionary Biology **8**:345. *New sequence data suggest that acanthocephalans are the sister taxon to bdelloid rotifers.*

■ Custom Website

The *Animal Diversity* Online Learning Center is a great place to check your understanding of chapter material. Visit www.mhhe.com/hickmanad6e for access to key terms, quizzes, and more! Further enhance your knowledge with Web links to chapter-related material.

Explore live links for these topics:

Classification and Phylogeny of Animals
Phylum Cycliophora
Phylum Entoprocta
Phylum Ectoprocta
Phylum Brachiopoda
Phylum Phoronida
Phylum Nemertea

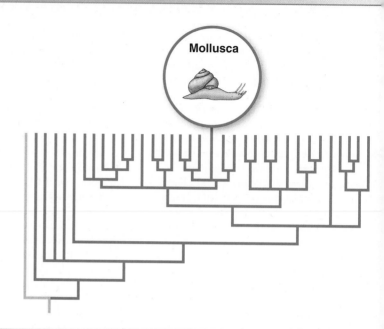

Molluscs

A Significant Space

Long ago in the Precambrian era, the most complex animals populating the seas were acoelomate. The simplest, and probably first, mode of achieving a fluid-filled space within the body was retention of the embryonic blastocoel, as in pseudocoelomates. Organs lay loose in the body cavity.

Some descendants of Precambrian acoelomate organisms evolved a more elegant arrangement: a fluid-filled space *within* the mesoderm, the *coelom*. This space was lined with mesoderm, and organs were suspended by mesodermal membranes, the *mesenteries.* Not only could the coelom serve as an efficient hydrostatic skeleton, with circular and longitudinal body-wall muscles acting as antagonists, but a more stable arrangement of organs with less crowding resulted. Mesenteries provided an ideal location for networks of blood vessels, and the gut tube could become more muscular, more highly specialized, and more diversified without interfering with other organs.

Development of a coelom was a major step in the evolution of larger and more complex forms. All of the major groups discussed in subsequent chapters are coelomates. In some, a large, fluid-filled coelom surrounded by muscles becomes a hydrostatic skeleton capable of rapid shape change and efficient burrowing. However, the heavy molluscan shell prevents large changes in body shape, and the molluscan coelom is relatively small.

Fluted giant clam, *Tridacna squamosa.*

Mollusca (mol-lus´ka) (L. *molluscus,* soft) is among the largest animal phyla after Arthropoda. There are nearly 90,000 named living species and some 70,000 fossil species. Many more molluscs await formal description. The name Mollusca indicates one of their distinctive characteristics, a soft body.

This group includes organisms as disparate as chitons, snails, clams, and octopuses (figure 10.1). They range from fairly simple organisms to some of the most complex invertebrates, and from almost microscopic in size to the giant squid, *Architeuthis harveyi* (Gr. *archi,* primitive, + *teuthis,* squid). The body of this huge species may grow up to 18 m long with tentacles extended and may weigh up to 454 kg (1000 pounds). The shells of some giant clams, *Tridacna gigas* (Gr. *tridaknos,* eaten at three bites), which inhabit the Indo-Pacific coral reefs, reach 1.5 m in length and weigh over 225 kg (see figure 10.23). These are extremes, however, since probably 80% of all molluscs have a maximum shell size of less than 5 cm.

The enormous variety, great beauty, and availability of mollusc shells have made shell collecting a popular pastime. However, many amateur shell collectors, although able to name hundreds of the shells that grace our beaches, know very little about the living animals that created those shells and once occupied them. The largest classes of molluscs are Gastropoda (snails and their relatives), Bivalvia (clams, oysters, and others), Polyplacophora (chitons), and Cephalopoda (squids, octopuses, nautiluses). Monoplacophora, Scaphopoda (tusk shells), Caudofoveata, and Solenogastres are much smaller classes.

The evolutionary relationships among molluscan classes remain controversial. One hypothesis (figure 10.2) recognizes as clade Conchifera all molluscan classes that possess solid shells, although some gastropods have lost this feature. A contrasting hypothesis derived from molecular data groups Monoplacophora and Polyplacophora as a clade called Serialia from their sharing of serially repeated body parts, and it places Caudofoveata closest to the cephalopods. This latter hypothesis requires additional gains and/or losses of the solid shell.

■ Ecological Relationships

Molluscs occupy a great range of habitats: from the tropics to polar seas; at altitudes exceeding 7000 m; in ponds, lakes, and streams; on mudflats; in pounding surf; and in the open ocean, from the surface to abyssal depths. Most live in the sea, and they represent a variety of lifestyles, including bottom-feeders, burrowers, borers, and pelagic forms. The phylum includes some of the most sluggish as well as some of the swiftest and most active invertebrates. It encompasses herbivorous grazers, predaceous carnivores, and ciliary filter feeders.

According to fossil evidence, molluscs originated in the sea, and most have remained there. Much of their evolution occurred along shores, where food was abundant and habitats were varied. Only bivalves and gastropods moved into brackish and freshwater habitats. As filter feeders, bivalves are unable to leave their aquatic surroundings; however, snails (gastropods) invaded land and may have been among the first animals to do so. Terrestrial snails are limited in range by their need for humidity, shelter, and calcium in the soil.

figure 10.1

Molluscs: a diversity of life forms. The basic body plan of this ancient group has become variously adapted for different habitats. **A,** A chiton (*Tonicella lineata*), class Polyplacophora. **B,** A marine snail (*Calliostoma annulata*), class Gastropoda. **C,** A nudibranch (*Chromodoris kuniei*), class Gastropoda. **D,** Pacific giant clam (*Panope abrupta*), with siphons to the left, class Bivalvia. **E,** An octopus (*Octopus briareus*), class Cephalopoda, forages at night on a Caribbean coral reef.

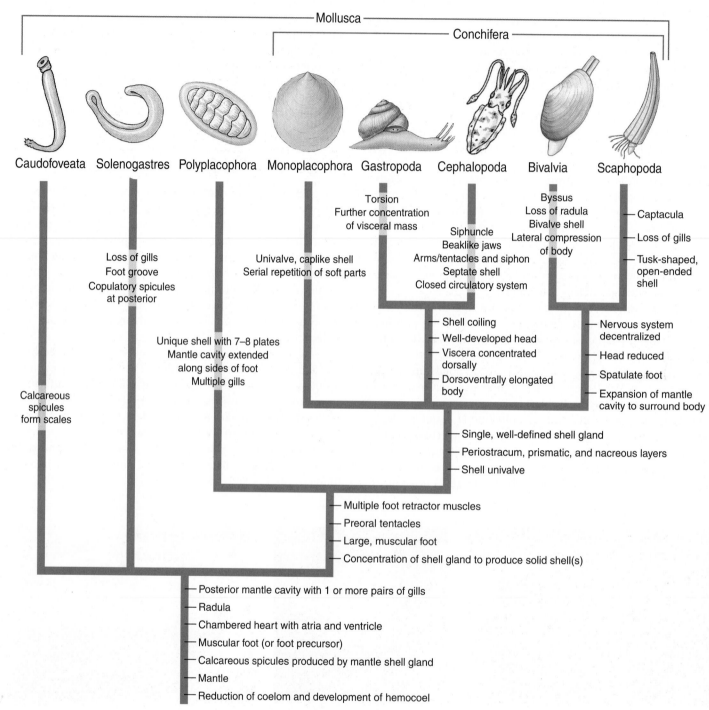

figure 10.2

Cladogram showing hypothetical relationships among classes of Mollusca. Synapomorphies that identify the various clades are shown, although a number of these have been modified or lost in some descendants. For example, the univalve shell (as well as shell coiling) has been reduced or lost in many gastropods and cephalopods, and many gastropods have undergone detorsion. The bivalve shell of Bivalvia was derived from an ancestral univalve shell. The byssus is lost in most adult bivalves but functions in larval attachment in many; therefore, the byssus is considered a synapomorphy of Bivalvia.

Source: Modified from R. C. Brusca and G. J. Brusca, *Invertebrates,* ed. 2. Sinauer Associates, Inc., Sunderland, MA, 2003.

figure 10.3

This representation of a generalized mollusc has been called a "hypothetical ancestral mollusc" (HAM), but the ancestral mollusc may not have had a distinct shell or a well-developed foot.

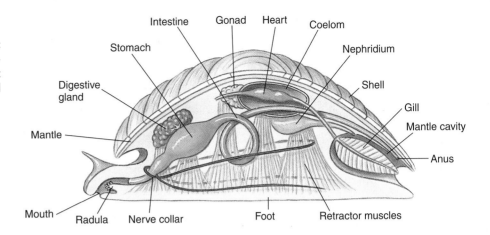

Economic Importance

Many molluscs are economically valuable as food. Pearls, both natural and cultured, are produced in the shells of clams and oysters, mainly in a marine oyster found in eastern Asia (see figure 10.5B).

Some molluscs are destructive. Burrowing shipworms (see figure 10.27), which are bivalves of several species, do great damage to wooden ships and wharves. To prevent the ravages of shipworms, wharves must be either creosoted or built of concrete. Snails and slugs often damage garden and other vegetation. In addition, many snails serve as intermediate hosts for serious parasites, such as the trematodes that cause schistosomiasis (see p. 163). A certain boring snail, the oyster drill, rivals sea stars in destroying oysters.

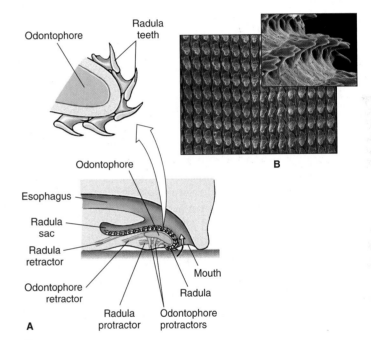

Form and Function

Body Plan

Reduced to its simplest dimensions, a mollusc body consists of a head-foot portion and a **visceral mass** portion. The head-foot region contains feeding, cephalic sensory, and locomotor organs (see mouth and foot in figure 10.3). It depends primarily on muscular action for its function. The visceral mass contains digestive, circulatory, respiratory, and reproductive organs, and it depends primarily on ciliary tracts for its functioning. Two folds of skin, outgrowths of the dorsal body wall, form a protective *mantle*, which encloses a space between itself and the body wall called a mantle cavity. The mantle cavity houses gills or a lung, and in many molluscs the mantle secretes a protective shell over the head-foot and visceral mass. Modifications of structures that form the head-foot and the visceral mass produce the great diversity of molluscan body plans.

Head-Foot

Most molluscs have a well-developed head, which bears a mouth and some specialized sensory organs. Photosensory receptors range from fairly simple to the complex eyes of cephalopods. Tentacles are often present. Within the mouth is

figure 10.4

A, Diagrammatic longitudinal section of gastropod head showing the radula and radula sac. The radula moves back and forth over the odontophore cartilage. As the animal grazes, the mouth opens, the odontophore is thrust forward, the radula gives a strong scrape backward to bring food into the pharynx, and the mouth closes. The sequence is repeated rhythmically. As the radula ribbon wears out anteriorly, it is continually replaced posteriorly. **B,** Radula of a snail prepared for microscopic examination. The inset depicts a magnified side view of several radular teeth.

a structure unique to molluscs, the radula, and usually posterior to the mouth is the chief locomotor organ, or foot.

Radula

The **radula** is a rasping, protrusible, tonguelike organ found in all molluscs except bivalves and some gastropods and solenogasters. It is a ribbonlike membrane on which are mounted rows of tiny chitinous teeth that point backward (figure 10.4).

Protractor and retractor muscles move the radula and its supporting cartilages (**odontophore**) outside and back into the mouth for feeling. A radula may have from a few to as many as 250,000 teeth that, when protruded, can scrape, pierce, tear, or cut particles of food. A radula may carry particles in a continuous stream toward the digestive tract.

Foot

The molluscan foot is variously adapted for locomotion, for attachment to a substrate, or for a combination of functions. The foot is usually a ventral, solelike structure that effects a creeping locomotion through waves of muscular contraction. However, there are many modifications, such as the attachment disc of limpets, the laterally compressed "hatchet foot" of bivalves, or the funnel for jet propulsion in squids and octopuses. Secreted mucus often aids adhesion or provides a slime track for small molluscs that glide on cilia.

Visceral Mass

Mantle and Mantle Cavity

The mantle is a sheath of skin, extending dorsally from the visceral mass, that wraps around each side of the body, protecting the soft parts and creating the mantle cavity between itself and the visceral mass. The outer surface of the mantle secretes the shell.

The mantle cavity plays an enormous role in the life of a mollusc. It usually houses respiratory organs (gills or a lung), which develop from the mantle, and the mantle's own exposed surface also serves gas exchange. Products from the digestive, excretory, and reproductive systems empty into the mantle cavity. In aquatic molluscs, a continuous current of water, kept moving by surface cilia or by muscular pumping, performs multiple functions: it supplies oxygen, and in some forms, food; flushes out wastes; and carries reproductive products out to the environment. In aquatic forms, the mantle is usually equipped with chemoreceptors. In cephalopods (squids and octopuses), the muscular mantle and its cavity create jet propulsion used in locomotion.

Shell

The shell of a mollusc, when present, is secreted and lined by the mantle. A shell typically has three layers (figure 10.5). The **periostracum,** the outer organic layer, is composed of a resistant protein called conchiolin. It protects the underlying calcareous layers from erosion by boring organisms. It is secreted by a fold of the mantle edge, and growth occurs only at the margin of the shell. On the older parts of the shell, the periostracum often wears away. The middle prismatic layer is composed of densely packed prisms of calcium carbonate laid down in a protein matrix. It is secreted by the glandular margin of the mantle, and increase in shell size occurs at the shell margin as the animal grows. The inner **nacre,** or nacreous layer, of the shell is composed of calcium carbonate sheets laid down over a thin protein matrix. This layer is secreted continuously by the mantle surface, so that it becomes thicker during the animal's life.

Freshwater molluscs usually have a thick periostracum that gives some protection against acids produced in the water by decay of leaf litter. The periostracum varies from thick to absent in marine forms. Shell structure varies greatly. Calcium for the shell comes from environmental water or soil or from food. The first shell appears during the larval period and grows continuously throughout life.

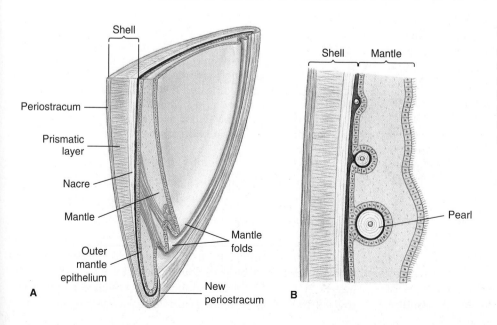

figure 10.5

A, Diagrammatic vertical section of shell and mantle of a bivalve. The outer mantle epithelium secretes the shell; the inner epithelium is usually ciliated. **B,** Formation of pearls between the mantle and shell as nacre accumulates around a parasite or bit of sand under the mantle.

Pearl production is a by-product of a protective device used by a mollusc when a foreign object, such as a grain of sand or a parasite, lodges between the shell and the mantle. The mantle secretes many layers of nacre around the irritating object (figure 10.5). Pearls are cultured by inserting small spheres, usually made from the shells of freshwater clams, in the mantle of a certain species of oyster and by maintaining the oysters in enclosures. The oyster deposits its own nacre around the "seed" in a much shorter time than would be required to form a pearl normally.

Internal Structure and Function

Gas exchange occurs through the body surface, particularly the mantle, and in specialized respiratory organs such as gills or lungs. Most molluscs have an open circulatory system with a pumping heart, blood vessels, and blood sinuses. In an open circulatory system, blood is not entirely contained within blood vessels; rather, it flows through vessels in some parts of the body and enters open sinuses in other parts. An open circulatory system is less efficient at supplying oxygen to all tissues in the body, so it is common in slow-moving animals. Insects are a notable exception, but in these animals oxygen is distributed by the tracheal system, not by the circulatory system. In a closed circulatory system, blood moves to and from tissues within blood vessels. Most members of class Cephalopoda have a closed circulatory system with a heart, vessels, and capillaries.

The digestive tract is complex and highly specialized according to the feeding habits of the various molluscs. Most molluscs have a pair of kidneys (metanephridia), a type of nephridium in which the inner end opens into the coelom; ducts of the kidneys in many forms serve also for discharge of eggs and sperm. The nervous system consists of several pairs of ganglia with connecting nerve cords. Molluscs have various highly specialized sense organs.

Most molluscs are dioecious, although some gastropods are hermaphroditic. Many aquatic molluscs pass through free-swimming **trochophore** (figure 10.6) and **veliger** (figure 10.7) larval stages. A veliger is the free-swimming larva of most marine snails, tusk shells, and bivalves. It develops from a trochophore and has the beginning of a foot, shell, and mantle.

Trochophore larvae (figure 10.6) are minute, translucent, often top-shaped, and they have a prominent circlet of cilia (prototroch) and sometimes one or two accessory circlets. They are the ancestral larval form of molluscs and annelids. A trochophore-like larva occurs also in marine turbellarians, nemerteans, brachiopods, phoronids, sipunculids, and echiurans. Possession of a trochophore or trochophore-like larva supports assigning these phyla to superphylum Lophotrochozoa.

Classes Caudofoveata and Solenogastres

Caudofoveates and solenogasters (figure 10.8) both are worm-like and shell-less, with calcareous scales or spicules in their

figure 10.6

A, Generalized trochophore larva. Molluscs and annelids with ancestral embryonic development have trochophore larvae, as do several other phyla. **B,** Trochophore of a Christmas tree worm, *Spirobranchus spinosus* (Annelida).

integument. Members of both groups have a reduced head and lack nephridia. Caudofoveates have one pair of gills and are dioecious. They burrow in marine sediments, feeding on microorganisms and detritus. Class Caudofoveata is sometimes called Chaetodermomorpha. In contrast to caudofoveates, solenogasters usually have no true gills, and they are hermaphroditic. Solenogasters live freely on the ocean floor and often feed on cnidarians. Class Solenogastres is sometimes called Neomeniomorpha.

CHARACTERISTICS
of Phylum Mollusca

1. Dorsal body wall forms a pair of folds called the **mantle,** which encloses the **mantle cavity,** is modified into **gills** or **lungs,** and secretes the **shell** (shell absent in some); ventral body wall specialized as a muscular **foot,** variously modified but used chiefly for locomotion; radula in mouth
2. Live in marine, freshwater, and terrestrial habitats
3. Free-living or occasionally parasitic
4. Body bilaterally symmetrical (bilateral asymmetry in some); unsegmented; often with definite head
5. Triploblastic body
6. **Coelom** limited mainly to area around heart; and perhaps lumen of gonads, part of kidneys, and occasionally part of the intestine
7. Surface epithelium usually ciliated and bearing mucous glands and sensory nerve endings
8. Complex digestive system; rasping organ **(radula)** usually present; anus usually emptying into mantle cavity; internal and external **ciliary tracts** often of great functional importance
9. Circular, diagonal, and longitudinal muscles in the body wall; mantle and foot highly muscular in some classes (for example, cephalopods and gastropods)
10. Nervous system composed of paired cerebral, pleural, pedal, and visceral ganglia, with nerve cords and subepidermal plexus; ganglia centralized in nerve ring in gastropods and cephalopods
11. Sensory organs of touch, smell, taste, equilibrium, and vision (in some); the highly developed direct **eye** (photosensitive cells in retina face light source) of cephalopods is similar to the indirect eye (photosensitive cells face away from light source) of vertebrates but arises as a skin derivative in contrast to the brain eye of vertebrates
12. No asexual reproduction
13. Both **monoecious** and **dioecious** forms; **spiral cleavage;** ancestral larva a **trochophore,** many with a **veliger** larva, some with direct development
14. One or two kidneys **(metanephridia)** opening into the pericardial cavity and usually emptying into the mantle cavity
15. Gas exchange by **gills, lungs, mantle,** or **body surface**
16. **Open circulatory system** (secondarily closed in cephalopods) of heart (usually three-chambered), blood vessels, and sinuses; respiratory pigments in blood

■ Class Monoplacophora

Until 1952, Monoplacophora (mon-o-pla-kof´-o-ra) were known only from Paleozoic shells. However, in that year, living specimens of *Neopilina* (Gr. *neo,* new, + *pilos,* felt cap) were dredged up from the ocean floor near the west coast of Costa Rica. These molluscs are small and have a low, rounded shell and a creeping foot (figure 10.9). They superficially resemble limpets (see figure 10.20 A), but unlike in most other molluscs, a number of organs, such as metanephridia, gonads, and gills, are

figure 10.7

Veliger of a snail, *Pedicularia,* swimming. The adults are parasitic on corals. The ciliated process (velum) develops from the prototroch of the trochophore (see figure 10.6A).

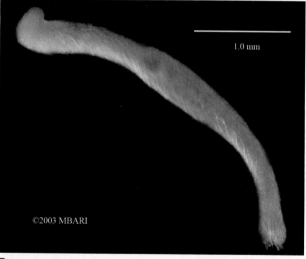

figure 10.8

Spicules are clearly visible on the skin of solenogasters and caudofoveates. **A,** *Neomeniomorpha,* a solenogaster; **B,** *Chaetoderma elegans,* a caudofoveate.

serially repeated. Serial repetition occurs less extensively in chitons. Body structures repeat in each segment of an annelid worm (see p. 213)—are these repeated structures an indication

figure 10.10

Dorsal view of the mossy chiton, *Mopalia muscosa*, on a rock surface. The upper surface of the mantle, or "girdle," is covered with hairs and bristles, an adaptation for defense. Seven of the chiton's eight plates are clearly visible.

figure 10.9

Neopilina, class Monoplacophora. Living specimens range from 3 mm to about 3 cm in length. **A,** Ventral view. **B,** Dorsal view.

that molluscs had a segmented (metameric) ancestor? Some authors consider monoplacophorans truly segmented, but most argue that *Neopilina* shows only pseudometamerism and that molluscs did not have a metameric ancestor. The serialia hypothesis provides another possibility: repeated body parts arose in the common ancestor of monoplacophorans and chitons.

■ Class Polyplacophora: Chitons

Chitons are somewhat flattened and have a convex dorsal surface that bears eight (rarely seven) articulating calcareous plates (or valves), which give them their name (figures 10.10 and 10.11). Polyplacophora means "bearing many plates," in contrast to Monoplacophora, which bear one shell (*mono,* single). The plates overlap posteriorly and are usually dull in color, like the rocks to which chitons cling.

Most chitons are small (2 to 5 cm); the largest rarely exceeds 30 cm. They commonly occur on rocky surfaces in intertidal regions, although some live at great depths. Chitons are stay-at-home organisms, straying only very short distances for feeding. When they feed, a sensory subradular organ protrudes from their mouth to explore for algae or colonial organisms. The radula protrudes to scrape food from the rocks. A chiton can cling tenaciously to rock with its broad flat foot. If detached, it can roll up like an armadillo for protection.

The mantle forms a girdle around the margin of the plates (see figure 10.11), and in some species mantle folds cover part or all of the plates. On each side of the broad ventral foot and lying between the foot and the mantle is a row of gills suspended from the roof of the mantle cavity. With the foot and the mantle margin adhering tightly to the substrate, these grooves become closed chambers, open only at the ends. Water enters the grooves anteriorly, flows across the gills, and leaves posteriorly, thus bringing a continuous supply of oxygen to the gills.

Blood pumped by the three-chambered heart reaches gills by way of an aorta and sinuses. Two kidneys carry waste from the pericardial cavity to the exterior. Two pairs of longitudinal nerve cords connect in the buccal region. Sense organs include shell eyes on the surface of the shell (in some) and a pair of **osphradia** (chemosensory organs for sampling water).

Sexes are separate in most chitons. Eggs are released singly or in strings or masses of jelly; they are shed into the female mantle cavity or into the sea. Sperm shed by males into the excurrent water may enter the gill grooves of females through incurrent openings. Trochophore larvae metamorphose directly into juveniles, without a second larval stage.

■ Class Scaphopoda

Scaphopoda (ska-fo-pod´a), commonly called tusk shells or tooth shells, are sedentary marine molluscs that have a slender tubular shell open at both ends. Inside the shell the mantle is

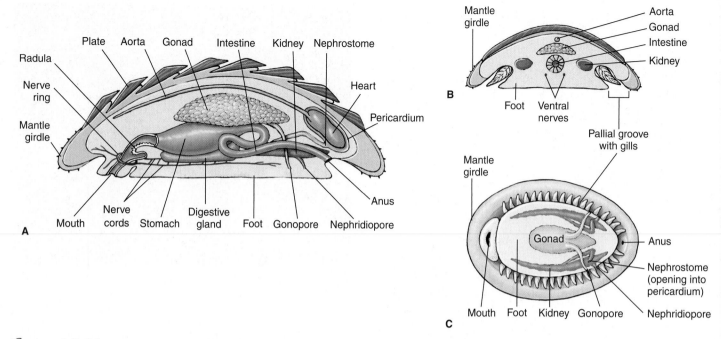

figure 10.11

Anatomy of a chiton (class Polyplacophora). **A,** Midsagittal section. **B,** Transverse section. **C,** Ventral view.

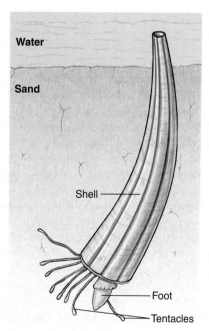

figure 10.12

The tusk shell, *Dentalium*, a scaphopod. It burrows into soft mud or sand and feeds by means of its prehensile tentacles. Respiratory currents of water are drawn in by ciliary action through the small open end of the shell, and then expelled through the same opening by muscular action.

wrapped around the viscera and fused to form a tube. Most scaphopods are 2.5 to 5 cm long, although they range from 4 mm to 25 cm.

The foot, which protrudes through the larger end of the shell, functions in burrowing into mud or sand, always leaving the small end of the shell exposed to water above (figure 10.12). Respiratory water circulates through the mantle cavity both by movements of the foot and by ciliary action. Food is caught on cilia of the scaphopod's foot or on the mucus-covered ciliated knows of its long tentacles. Food is mainly detritus and protozoa from the substrate.

■ Class Gastropoda

Among molluscs, class Gastropoda (gas-tro-pod´-a) (Gr. *gastér,* stomach, + *pous, podos,* foot) is by far the largest and most diverse, containing about 70,000 living and 15,000 fossil species. Its members differ so widely that no single general term in our language can apply to them as a group. They include snails, limpets, slugs, whelks, conchs, periwinkles, sea slugs, sea hares, sea butterflies, and others. They range from marine molluscs with many ancestral characters to air-breathing snails and slugs.

Gastropods are often sluggish, sedentary animals because most of them have heavy shells and slow locomotor organs. When present, the shell is almost always of one piece (univalve) and may be coiled or uncoiled. Some snails have an **operculum,** a hard proteinaceous plate that covers the shell aperture when the body withdraws into the shell. It protects the body and prevents water loss. These animals are basically bilaterally symmetrical, but because of **torsion,** a twisting process that occurs during development, the visceral mass has become asymmetrical. Gastropods develop through a trochophore and a veliger larval stage.

ONTOGENETIC TORSION

Lateral View

Shell — Mouth — Velar lobe — Foot — Anus

Dorsal View

Mouth — Velar lobe — Velar lobe — Anus — Developing mantle cavity

| In **veliger** larva, anus is posterior; no mantle cavity is visible | Shell rotates to adult position; viscera rotates 90°; mantle cavity visible | Anus and mantle continue to rotate; mantle cavity expands | Anus rotated 180°; mantle epithelium reorganizes to enlarge cavity | Adult snail |

figure 10.13

Ontogenetic torsion in a gastropod veliger larva.

Form and Function

Torsion

Gastropod development is highly variable, but in general a trochophore larval stage is followed by a veliger larval stage in which the shell arises. The veliger has two ciliated velar lobes, used in swimming, and the developing foot is visible (see figure 10.7). Initially, the mouth is anterior and the anus posterior, but through the process of **torsion,** the relative positions of body parts change.

Torsion is usually described as a two-step process. In the first step, an asymmetrical foot retractor muscle contracts and pulls the shell and enclosed viscera (containing organs of the body) 90 degrees counterclockwise, relative to the head. This movement brings the anus from the posterior to the right side of the body (figure 10.13). Movement of the shell is independent of visceral movements. The first movements of the shell rotate it between 90 and 180 degrees into a position that persists into adulthood. The mantle cavity develops on the right side of the body near but initially separate from the anus. The anus and mantle cavity usually move farther to the right, and the mantle cavity is remodeled to encompass the anus. In a slower and more variable series of changes, the digestive tract moves both laterally and dorsally so that the anus lies above the head within the mantle cavity (figure 10.13).

After torsion, the anus and mantle cavity open above the mouth and head. The left gill, kidney, and heart atrium are now on the right side, whereas the original right gill, kidney, and heart atrium are on the left, and the nerve cords form a figure eight.

The developmental sequence just described is called *ontogenetic torsion*. It contrasts with *evolutionary torsion*, the series of changes that produced the modern torted gastropod body from the ancestral untorted form. The hypothetical ancestral gastropod was assumed to have a posterior mantle cavity like that of the hypothetical ancestral mollusc (see figure 10.3). It has long been assumed that morphological changes in ontogenetic torsion represent the sequence of evolutionary changes. However, new studies led researchers to hypothesize that the ancestral gastropod had two lateral mantle cavities, much like those in *Neopilina* (see figure 10.9) and in chitons (see figure 10.11). A single mantle cavity over the head may have arisen when the left lateral mantle cavity was lost and the right cavity expanded toward the middle of the body after the first 90 degrees of torsion.

The arrangement that results from torsion creates the possibility of wastes washing back over the gills **(fouling)** and causes us to wonder what evolutionary factors favored such a strange realignment of the body. Several explanations have been proposed, none entirely satisfying. For example, sense

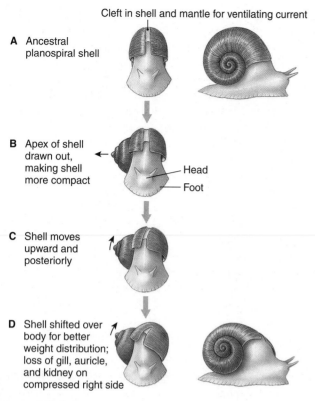

Cleft in shell and mantle for ventilating current

A Ancestral planospiral shell

B Apex of shell drawn out, making shell more compact

Head
Foot

C Shell moves upward and posteriorly

D Shell shifted over body for better weight distribution; loss of gill, auricle, and kidney on compressed right side

figure 10.14

Evolution of a shell in gastropods. **A,** The earliest coiled shells were planospiral, each whorl lying completely outside the preceding whorl. Interestingly, the shell has become planospiral secondarily in some living forms. **B,** Better compactness was achieved by snails in which each whorl lay partially to the side of the preceding whorl. **C, D,** Better weight distribution resulted when the shell was moved upward and posteriorly.

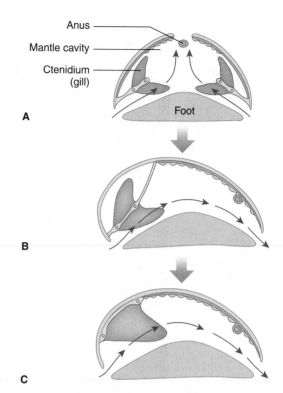

Anus
Mantle cavity
Ctenidium (gill)
Foot

A

B

C

figure 10.15

Evolution of gills in gastropods. The diagrams represent cross sections of the mantle cavity viewed from the posterior. **A,** Primitive condition in prosobranchs with two gills and excurrent water leaving the mantle cavity by a dorsal slit or hole. **B,** Condition after one gill had been lost. **C,** Derived condition in prosobranchs, in which filaments on one side of the remaining gill are lost, and the axis is attached to the mantle wall.

organs of the mantle cavity (osphradia) would better sample water when turned in the direction of travel, and as mentioned already, the head could be withdrawn into the shell. Certainly the consequences of torsion and the resulting need to avoid fouling have been very important in the subsequent evolution of gastropods. We explore these consequences after describing an interacting feature of gastropods—coiling of the shell and visceral mass.

Coiling

Coiling, or spiral winding, of the shell and visceral mass may occur in the larval stage at the same time as torsion, but the fossil record shows that coiling was a separate evolutionary event and originated in gastropods earlier than torsion did. Coiling also occurs in cephalopods. Nevertheless, all living gastropods descend from coiled, torted ancestors, whether or not they now show these characteristics.

Early gastropods had a bilaterally symmetrical shell with all whorls lying in a single plane (figure 10.14A). Such a shell was not very compact, since each whorl had to lie completely outside the preceding one. A few modern species have second-

arily returned to that form. The lack of compactness of the planospiral shell was resolved by a shape in which each succeeding whorl was at the side of the preceding one (figure 10.14B). However, this shape clearly was unbalanced, hanging as it did with much weight over to one side. Later gastropods achieved better weight distribution by shifting the shell upward and posteriorly, with the shell axis oblique to the longitudinal axis of the foot (figure 10.14C and D). However, the weight and bulk of the main body whorl, the largest whorl of the shell, pressed on the right side of the mantle cavity, and apparently interfered with the organs on that side. Accordingly, the gill, auricle, and kidney of the right side have been lost in all except a few living gastropods, leading to bilateral asymmetry.

Adaptations to Avoid Fouling

Although loss of the right gill was probably an adaptation to the mechanics of carrying a coiled shell, that condition made it possible to reduce the effects of fouling (see p. 195). In most modern gastropods, water flows one-way; it is brought into the left side of the mantle cavity and out the right side, carrying with it the wastes from the anus and nephridiopore, which lie

figure 10.16

A, Red abalone, *Haliotus rufescens*. This huge, limpetlike snail is prized as food and extensively marketed. Abalones are strict vegetarians, feeding especially on sea lettuce and kelp. **B,** Moon snail, *Polinices lewisii*. A common inhabitant of West Coast sand flats, the moon snail is a predator of clams and mussels. It uses its radula to drill neat holes in its victim's shell, through which it then extends its proboscis to eat the bivalve's fleshy body.

near the right side (figure 10.15). Some gastropods with the ancestral characteristic of two gills avoid fouling by venting excurrent water through a dorsal cleft above the anus (see figure 10.14), bearing one (keyhole limpet; see figure 10.20A), or a series of holes in the shell (abalone, figure 10.16). Opisthobranchs (nudibranchs and others) have evolved an even more elaborate "twist"; after undergoing torsion as larvae, they develop various degrees of *detorsion* as adults.

Detorsion is a post-veliger change in the orientation of the digestive and nervous systems of opisthobranchs. After detorsion, the anus is posterior rather than anterior, and lies on the right side of the body. Compare the position of the anus in the shelled gastropod after torsion (see figure 10.13) with that of the opisthobranch after torsion and detorsion (figure 10.17). The name suggests that torsion reverses, but in the sea hare (figure 10.17), the mantle complex migrates from the anterior of the body to the posterior right side accompanied by differential growth and folding of the digestive tract. In other taxa, torsion may cease earlier or be followed by different degrees of body reorganization.

Feeding Habits

The feeding habits of gastropods are as varied as their shapes and habitats, but all use some adaptation of the radula. Many gastropods are herbivorous, rasping algae from a substrate. Some herbivores are grazers, some are browsers, and others

figure 10.17

A, A sea hare, *Aplysia dactylomela*, crawls and swims across a coral reef, assisted by large, winglike parapodia, here curled above the body. **B,** When attacked, sea hares squirt a copious protective secretion derived from their red algal food source. Substances from the algae pass from the digestive gland to a "purple gland" where they are modified. Secretions from the purple gland exit the body through the mantle cavity, as do wastes from the anus. The anus and mantle cavity are posterior in sea hares because these animals undergo detorsion during development.

figure 10.18

A, *Conus* extends its long, wormlike proboscis. When the fish attempts to consume this tasty morsel, the *Conus* stings it in the mouth and kills it. **B,** The snail engulfs the fish with its distensible stomach, and then regurgitates the scales and bones some hours later.

planktonic feeders. Abalones (see figure 10.16A) hold seaweed with their foot and break off pieces with their radula. Some snails are scavengers, living on dead and decayed flesh; others are carnivores, tearing prey with their radular teeth. Snails such as oyster borers and moon snails (see figure 10.16B) have an extensible proboscis for drilling holes in the shells of bivalves whose soft parts they find delectable. Some even have a spine for opening shells. Most pulmonates (air-breathing snails) (see figure 10.22) are herbivorous, but some eat earthworms and other snails. After maceration by the radula or by some grinding device, such as the so-called gizzard in sea hares, digestion is usually extracellular in the lumen of the stomach or digestive glands.

Some sessile gastropods, such as slipper shells, are ciliary feeders that use the gill cilia to collect particulate matter, which they roll into a mucous ball and carry to their mouth. Some sea butterflies secrete a mucous net to catch small planktonic forms and then draw the web into their mouth. In ciliary feeders, the stomachs are sorting regions, and most digestion is intracellular in the digestive gland.

Among the most interesting predators are venomous cone shells (figure 10.18), which feed on vertebrates or other invertebrates, depending on the species. When *Conus* senses its prey, a single radular tooth slides into position at the tip of the proboscis. When the proboscis strikes prey, it expels the tooth like a harpoon, and the venom tranquilizes or kills the prey at once. Some species can deliver very painful stings, and the sting of several species is lethal to humans. The venom consists of a series of toxic peptides, and each *Conus* species carries peptides (conotoxins) specific for the neuroreceptors of its preferred prey.

Internal Form and Function

Respiration in most gastropods is performed by a gill (two gills in a few), although some aquatic forms lack gills and depend on the skin. Pulmonates (most freshwater and terrestrial snails) have lost their gill altogether, and the vascularized mantle wall has evolved to function as a lung. The anus and nephridiopore open near the opening of the lung to the outside, which is called the pneumostome (figure 10.19B), and waste is expelled forcibly with air or water from the lung. Freshwater pulmonates must surface to expel a bubble of gas from the lung and to curl the edge of the mantle around the pneumostome to form a siphon for air intake.

Most gastropods have a single nephridium (kidney). The circulatory and nervous systems are well developed (figure 10.19). The nervous system includes three pairs of ganglia connected by nerves. Sense organs include eyes, statocysts, tactile organs, and chemoreceptors.

There are both dioecious and hermaphroditic gastropods. Many forms perform courtship ceremonies. During copulation in hermaphroditic species, an exchange of **spermatophores** (bundles of sperm) avoids self-fertilization. Most land snails lay their eggs in holes in the ground or under logs. Some aquatic gastropods lay their eggs in gelatinous masses; others enclose them in gelatinous capsules or in parchment egg cases. Most marine gastropods go through a free-swimming veliger larval stage during which torsion and coiling occur. Others develop directly into a juvenile within an egg capsule.

Major Groups of Gastropods

Traditional classification of class Gastropoda recognized three subclasses: Prosobranchia, Opisthobranchia, and Pulmonata. Prosobranchia is by far the largest subclass, and its members are almost all marine. Familiar examples of marine prosobranchs are periwinkles, limpets (figure 10.20A), whelks, conchs, abalones (see figure 10.16A), slipper shells, oyster borers, rock shells, and cowries.

Opisthobranchia is an assemblage of marine gastropods including sea slugs, sea hares, nudibranchs, and canoe shells. At

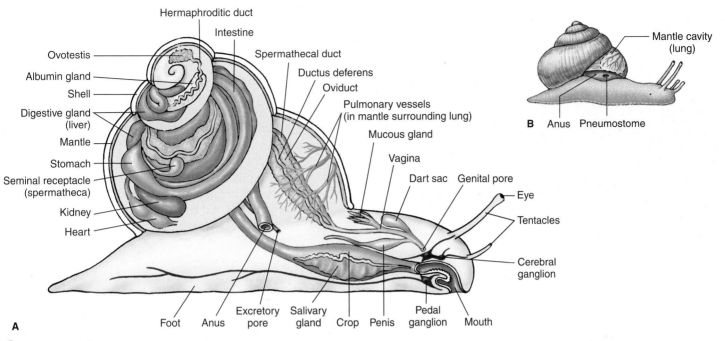

figure 10.19

A, Anatomy of a pulmonate snail. **B,** Position of the mantle cavity serving as a lung. Air enters and exits via the pneumostome.

figure 10.20

A, Keyhole limpet, *Diodora aspera,* a prosobranch gastropod with a hole in the apex through which water leaves the mantle cavity. **B,** Flamingo tongues, *Cyphoma gibbosum,* are showy inhabitants of Caribbean coral reefs, where they are associated with gorgonians. These snails have a smooth, creamy, orange to pink shell that is normally covered by the brightly marked mantle. Here the flexible foot is visible as the snail crawls along a gorgonian coral.

present, 8 to 12 groups of opisthobranchs are recognized. Some have a gill and a shell, although the latter may be vestigial, and some have no shell or true gill. Large sea hares, such as *Aplysia* (see figure 10.17), have large, earlike anterior tentacles, chemosensory rhinophores, and a vestigial shell. Nudibranchs have no shell as adults and rank among the most beautiful and colorful of molluscs (figure 10.21). Having lost the gill, the body surface of some nudibranchs is often increased for gas exchange by small projections (**cerata**) or by a ruffling of the mantle edge.

The third major group, Pulmonata, contains most land and freshwater snails and slugs. Usually lacking gills, their mantle cavity has become a lung, which fills with air by contraction of the mantle floor. Aquatic and a few terrestrial species have one pair of nonretractile tentacles, at the base

of which are eyes; land forms usually have two pairs of tentacles, with the posterior pair bearing eyes (figure 10.22; see figure 10.19). The few nonpulmonate species of gastropods that live in fresh water usually can be distinguished from pulmonates because they have an operculum, which is lacking in pulmonates.

Currently, gastropod taxonomy is in flux, and some workers regard any attempt to present a classification as premature. A phylogeny of the gastropods based on morphological characters divided the class into two groups differing in form of the radula and other features. However, a preliminary molecular analysis did not support this grouping. The molecular data generally support a clade (Euthyneura) combining opisthobranchs and pulmonates, but Opisthobranchia appears to be paraphyletic.

figure 10.21

Phyllidia ocellata, a nudibranch. Like other *Phyllidia* spp., it has a hard body with dense calcareous spicules and bears its gills along the sides, between its mantle and foot.

A

Pneumostome

B

figure 10.22

A, Pulmonate land snail. Note two pairs of tentacles; the second, larger pair bears the eyes. **B,** Banana slug, *Ariolimax columbianus.* Note the pneumostome leading into the lung.

■ Class Bivalvia (Pelecypoda)

Bivalvia (bi-val´ve-a) are also called Pelecypoda (pel-e-sip´o-da) (Gr. *pelekus,* hatchet, + *pous, podus,* foot). They are bivalved (two-shelled) molluscs that include mussels, clams, scallops, oysters, and shipworms and range in size from tiny seed shells 1 to 2 mm in length to the giant South Pacific *Tridacna* (figure 10.23). Most bivalves are sedentary **suspension feeders** that depend on ciliary currents produced by the gills to collect their food. Unlike gastropods, they have no head, no radula, and very little cephalization.

Most bivalves are marine, but many live in brackish water and in streams, ponds, and lakes.

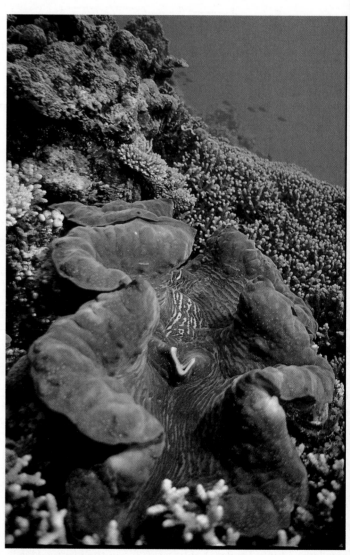

figure 10.23

A clam (*Tridacna gigas*) lies buried in coral rock. The richly colored mantle tissue around the siphons bears enormous numbers of symbiotic dinoflagellates (zooxanthellae) that provide much of the clam's nutrition.

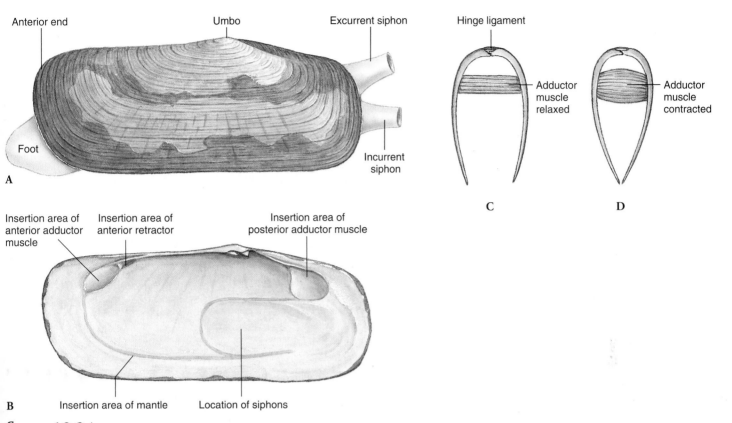

figure 10.24

Tagelus plebius, the stubby razor clam (class Bivalvia). **A,** External view of left valve. **B,** Inside of right shell, showing scars where muscles were attached. The mantle was attached to part of the shell, leaving a pallial line visible at the edge of the mantle insertion area. **C, D,** Sections showing the function of adductor muscles and the hinge ligament. In **C,** the adductor muscle is relaxed, allowing the hinge ligament to pull the valves apart. In **D,** the adductor muscle is contracted, pulling the valves together.

Form and Function

Shell

Bivalves are laterally compressed, and their two shells **(valves)** are held together dorsally by a hinge ligament that causes the valves to gape ventrally. **Adductor muscles** work in opposition to the hinge ligament and draw the valves together (figure 10.24B, C, and D). Projecting above the hinge ligament on each valve is the **umbo** (figure 10.24A), which is the oldest part of the shell. The valves function largely for protection, but those of ship-worms (figure 10.25) have microscopic teeth for rasping wood, and rock borers use spiny valves for boring into rock.

Body and Mantle

The visceral mass is suspended from the dorsal midline, and the muscular foot attaches to the visceral mass anteroventrally (figure 10.25). The gills hang down on each side, each covered by a fold of the mantle. The posterior edges of the mantle folds form dorsal excurrent and ventral incurrent openings (figure 10.24A). In some marine bivalves, part of the mantle is drawn out into extremely long, muscular siphons allowing the clam to burrow into the mud or sand and extend the siphons to the water

above. Cilia on the gills and inner surface of the mantle direct the flow of water over the gills, bringing in food and oxygen (figure 10.25).

Bivalves have a three-chambered heart that pumps blood through the gills and mantle for oxygenation and to the kidneys for waste elimination (figure 10.26). They have three pairs of widely separated ganglia and poorly developed sense organs. A few bivalves have ocelli. The steely blue eyes of some scallops (see figure 10.26), located around the mantle edge, are remarkably complex, each being equipped with a cornea, lens, and retina.

Feeding and Digestion

Most bivalves are suspension feeders. Their respiratory currents bring both oxygen and organic materials to their gills, where ciliary tracts direct them to the tiny gill pores. Gland cells on the gills and labial palps secrete copious amounts of mucus, which entangles food particles suspended in the water entering gill pores. Ciliary tracts move the particle-laden mucus to the mouth (see figure 10.25).

In the stomach, the mucus and food particles are kept whirling by a rotating gelatinous rod, called a **crystalline**

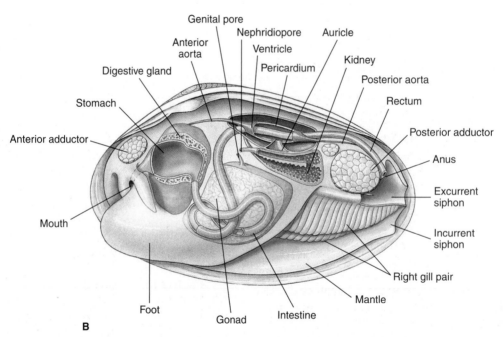

figure 10.25

A, Feeding mechanism of a freshwater clam, with the left valve and mantle removed. Water enters the mantle cavity posteriorly and is drawn forward by ciliary action to the gills and palps. As water enters the tiny openings of the gills, food particles are collected in strings of mucus, which are then carried by cilia to the palps and directed to the mouth. Sand and debris drop into the mantle cavity and are removed by cilia. **B,** Clam anatomy.

style. As the style rotates, enzymes for extracellular digestion dissolve from it in layers. Ciliated ridges of the stomach sort food particles and direct suitable particles to the digestive gland for intracellular digestion.

Shipworms (see figure 10.27) feed on the particles they excavate as they burrow in wood. Symbiotic bacteria occupy a special organ in these bivalves and produce cellulase to digest wood. Other bivalves, such as giant clams gain much of their nutrition from the photosynthetic products of symbiotic algae (zooxanthellae, see p. 108) in their mantle tissue (see figure 10.23).

Locomotion

Most bivalves move by extending their slender muscular foot between the valves (see figure 10.24A). They pump blood into

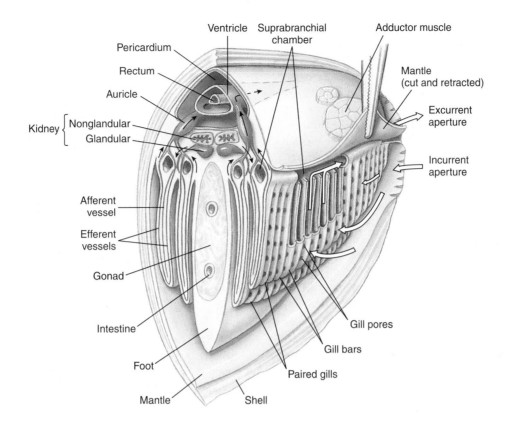

figure 10.26

Section through the heart region of a freshwater clam to show the circulatory and respiratory systems in relation to one another. Respiratory water currents: water is drawn in by cilia, enters gill pores, and then passes up water tubes to suprabranchial chambers and out the excurrent aperture. Blood in the gills exchanges carbon dioxide and oxygen. Blood circulation: the ventricle pumps blood forward to sinuses of their foot and viscera, and posteriorly to the mantle sinuses. Blood returns from the mantle to auricles; it returns from viscera to the kidney, and then goes to the gills, and finally to the auricles.

A

B

figure 10.27

A, Shipworms (*Teredo, Bankia,* and others) are bivalves that burrow in wood, causing great damage to unprotected wooden hulls and piers.
B, The two small, anterior valves, seen at left, are used as rasping organs to extend the burrow.

the foot, causing it to swell and to act as an anchor in mud or sand. Then longitudinal muscles contract to shorten the foot and pull the animal forward. Most bivalves use the foot for burrowing, but a few creep. Some bivalves are sessile: oysters attach their shells to a surface by secreting cement, and mussels (figure 10.28) attach themselves by secreting a number of slender byssal threads. The foot is often reduced in sessile bivalves. A few bivalves, such as scallops (figure 10.29), use their shells for locomotion by clapping the valves together to move in spurts.

Reproduction

Sexes are usually separate, and fertilization is usually external. Marine embryos typically go through three free-swimming larval stages—trochophore, veliger larva, and young spat—before reaching adulthood. In freshwater clams, fertilization is internal, and some gill tubes become temporary brood chambers. There, larvae develop into specialized veligers called **glochidia,** which are discharged with the excurrent flow (figure 10.30). If glochidia come in contact with a passing fish, they hitchhike a ride as parasites in the fish's gills for the next 20 to 70 days before sinking to the bottom to become sedentary adults. In lakes or streams, attachment to a fish prevents the small larvae from being moved out of the lake or farther downstream.

figure 10.28

Mussels, *Mytilus edulis,* occur in northern oceans around the world; they form dense beds in the intertidal zone. A host of marine creatures live protected beneath attached mussels.

Freshwater clams were once abundant and diverse in streams throughout the eastern United States, but they are now easily the most jeopardized group of animals in the country. Of the more than 300 species once present, 24 are extinct, 160 are listed as threatened or endangered, and many more may be listed soon. A combination of causes is responsible, of which decline in water quality is among the most important. Pollution and sedimentation from mining, industry, and agriculture are among the culprits. Habitat destruction due to altered natural water courses and damming is an important factor. Poaching to supply the Japanese cultured pearl industry is partially to blame (see note on p. 191). In addition, the prolific zebra mussels attach in great numbers to native clams, exhausting food supplies (phytoplankton) in the surrounding water.

Zebra mussels, *Dreissena polymorpha,* are a potentially disastrous biological introduction into North America. They were apparently picked up as veligers with ballast water by one or more ships in freshwater ports in northern Europe and then expelled between Lake Huron and Lake Erie in 1986. This 4 cm bivalve spread throughout the Great Lakes by 1990, and by 1994 it was as far south on the Mississippi River as New Orleans, as far north as Duluth, Minnesota, and as far east as the Hudson River in New York. It attaches to any firm surface and filter-feeds on phytoplankton. Large numbers accumulate rapidly. They foul water intake pipes of municipal and industrial plants, impede intake of water for municipal supplies, and have far-reaching effects on the ecosystem. Zebra mussels cost approximately 500 million dollars to control. However, on a slightly brighter note, some evidence suggests that a few species of native bivalves may survive in areas with zebra mussels, although their population sizes are severely reduced.

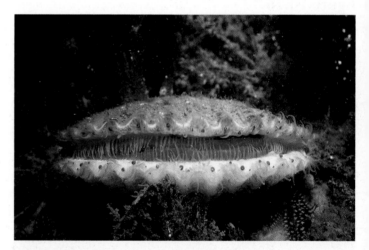

figure 10.29

Representing a group that has evolved from burrowing ancestors, the surface-dwelling bay scallop, *Aequipecten irradians,* has developed sensory tentacles and a series of blue eyes along its mantle edges.

■ Class Cephalopoda

Cephalopoda (sef´-a-lo-pod´a) include squids, octopuses, nautiluses, and cuttlefishes. All are marine, and all are active predators.

Cephalopods (Gr. *kephalē,* head, + *pous, podos,* foot) have an odd body plan (figure 10.31) that develops as the embryonic head and foot become indistinguishable. The ring around the mouth bearing the arms and tentacles develops from the anterior margin of the head, whereas the circle of arms or tentacles itself is derived from the anterior margin of the foot. The foot also forms a **funnel** for expelling water from the mantle cavity.

Cephalopods range in size from 2 to 3 cm up to the giant squid, *Architeuthis*, which is the largest invertebrate known. The squid *Loligo* (L., cuttlefish) is about 30 cm long (figure 10.31A).

Cephalopods are predaceous, feeding chiefly on small fishes, molluscs, crustaceans, and worms. Their arms, used to capture and handle food, have a complex musculature and make delicately controlled movements. They are highly mobile and swiftly seize prey and bring it to the mouth. Octopuses and cuttlefishes have salivary glands that secrete a venom for immobilizing prey. Strong, beaklike jaws grasp prey, and the radula tears off pieces of flesh (figure 10.31A). Digestion is extracellular and occurs in the stomach and cecum.

The enormous giant squid, *Architeuthis*, is very poorly known because there have been few studies of living specimens. The anatomy has been studied in stranded animals, in those captured in nets of fishermen, and in specimens found in stomachs of sperm whales. The mantle length is 5 to 6 m, and the head is up to one meter long. They have the largest eyes in the animal kingdom: up to 25 cm (10 inches) in diameter. They apparently eat fish and other squids, although they have been filmed responding to bait. They are an important food item for sperm whales. Giant squids are thought to live on or near the ocean floor at a depth of 1000 m, but some have been seen swimming at the surface.

Mantle edge

Glochidium

A   **B**

figure 10.30

A, Glochidium, or larval form, for some freshwater clams. When larvae are released from the mother's brood pouch, they may become attached to a fish's gill by clamping their valves closed. They remain as parasites on the fish for several weeks. Their size is approximately 0.3 mm. **B,** Some clams have adaptations that help their glochidia find a host. The mantle edge of this female pocketbook mussel (*Lampsilis ovata*) mimics a small minnow, complete with eye. When a smallmouth bass comes to dine, it gets doused with glochidia.

Form and Function

Shell

Fossil records of cephalopods extend to the Cambrian. The earliest shells were straight cones. Cephalopods without shells or with internal shells (such as octopuses and squids)

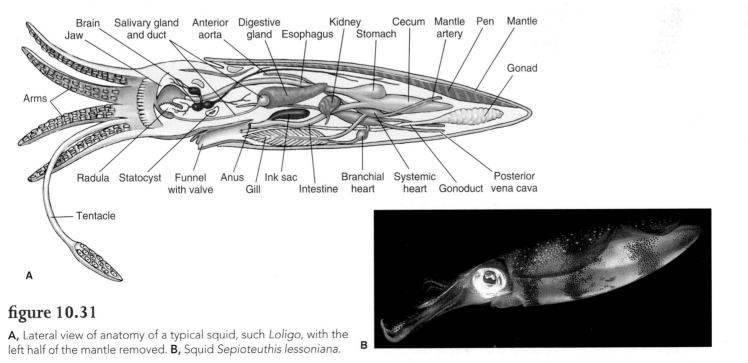

Brain Salivary gland Anterior Digestive Kidney Cecum Mantle Pen Mantle
Jaw and duct aorta gland Esophagus Stomach artery

Gonad

Arms

Radula Statocyst Funnel Anus Ink sac Branchial Systemic Posterior
with valve Gill Intestine heart heart Gonoduct vena cava

Tentacle

A

figure 10.31

A, Lateral view of anatomy of a typical squid, such *Loligo*, with the left half of the mantle removed. **B,** Squid *Sepioteuthis lessoniana*.

B

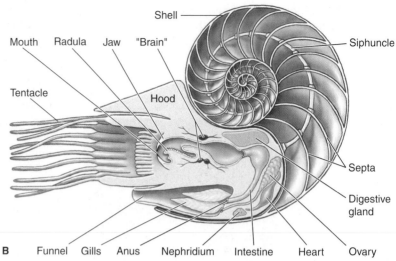

figure 10.32

Nautilus, a cephalopod. **A,** Live *Nautilus*, feeding on a fish. **B,** Longitudinal section, showing the gas chambers of the shell and a diagram of the internal body structure.

figure 10.33

Cuttlefish, *Sepia latimanus*, has an internal shell familiar to keepers of caged birds as "cuttlebone."

probably evolved from a straight-shelled ancestor. Other fossil cephalopods had curved or coiled shells, as in ammonoids and nautiloids. The only remaining members of the once-flourishing nautiloids are the modern *Nautilus* species (Gr. *nautilos*, sailor) (figure 10.32). Ammonoids were widely prevalent in the Mesozoic era but became extinct by the end of the Cretaceous period. Reasons for their extinction remain a mystery. Present evidence suggests that they were gone before the asteroid bombardment at the end of the Cretaceous period (see p. 34).

Although early nautiloid and ammonoid shells were heavy, they were made buoyant by a series of gas chambers, as in of *Nautilus* (figure 10.32B), which enabled the animal to swim while carrying its shell. The shell of *Nautilus*, although coiled, is

quite different from that of a gastropod. Transverse septa divide the shell into internal chambers (figure 10.32B). The living animal inhabits only the last chamber. As it grows, it moves forward, secreting behind it a new septum. The chambers are connected by a cord of living tissue called a **siphuncle,** which extends from the visceral mass. Ammonoids had chambered shells analogous to those of nautiloids, but the septa were more complex.

Cuttlefishes also have a small, coiled or curved shell, but it is entirely enclosed by the mantle (figure 10.33). In squids, most of the shell has disappeared, leaving only a thin, flexible strip called a **pen,** which the mantle encloses. *Octopus* (Gr. *oktos*, eight, + *pous*, *podos*, foot) has no shell.

After *Nautilus* secretes a new septum, the new chamber is filled with fluid similar in ionic composition to that of the *Nautilus's* blood (and of seawater). Fluid removal involves the active secretion of ions into tiny intercellular spaces in the siphuncular epithelium, so that a very high local osmotic pressure is produced, and the water is drawn from the chamber by osmosis. The gas in the chamber is only the respiratory gas from the siphuncle tissue that diffuses into the chamber as the fluid is removed. Thus the gas pressure in the chamber is 1 atmosphere or less because it is in equilibrium with the gases dissolved in the seawater surrounding the *Nautilus*. These dissolved gases are in turn in equilibrium with the air at the surface of the sea, despite the fact that the *Nautilus* may be swimming at 400 m beneath the surface. That the shell can withstand implosion by the surrounding 41 atmospheres (about 600 pounds per square inch), and that the siphuncle can remove water against this pressure, are marvelous feats of natural engineering!

Body and Mantle

In *Nautilus*, the head with its tentacles can be extruded from the opening of the body compartment of the shell (see figure 10.32). Its 60 to 90 or more tentacles have no suckers but adhere to prey by secretions. The tentacles search for, sense, and grasp food. Beneath the head is the funnel. The shell shelters the mantle, mantle cavity, and visceral mass. Two pairs of gills occupy the mantle cavity.

Cephalopods other than nautiloids have only one pair of gills. Octopuses have 8 arms with suckers. Squids and cuttlefishes (decapods) have 10 arms: 8 arms with suckers and a pair of long, retractile tentacles. The thick mantle covering the trunk fits loosely at the neck region, allowing water to be taken into the mantle cavity. When the mantle edges contract closely about the neck, water is forcefully expelled through the funnel. The water current thus created provides oxygen for gills in the mantle cavity, jet power for locomotion, and a means of carrying wastes and sexual products away from the body.

The active habits of cephalopods are reflected in their internal anatomy, particularly their respiratory and circulatory systems. Ciliary propulsion would not circulate enough water over the gills for an active animal, so cephalopods ventilate their gills by muscular action of the mantle wall. They have a closed circulatory system with a network of vessels, and blood flows through the gills via capillaries. **Branchial (gill) hearts**, or *accessory hearts* (see figure 10.31A), increase blood pressure in the gills, which occur at the end of the molluscan circulatory system. This increased blood pressure permits more rapid gas exchange.

Cephalopods have well-developed nervous systems, including the most complex brain among invertebrates (see figure 10.31A). Except for *Nautilus,* which has relatively simple eyes, cephalopods have elaborate eyes with a cornea, lens, chambers, and a retina (figure 10.34)—similar to the camera-type eye of vertebrates.

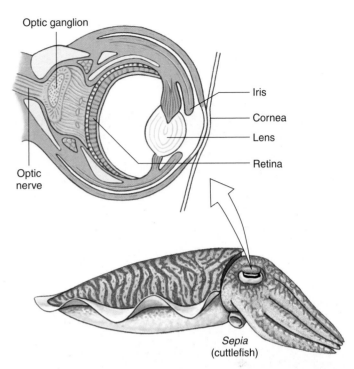

figure 10.34

Eye of a cuttlefish (*Sepia*). The structure of cephalopod eyes shows a high degree of convergent evolution with the eyes of vertebrates.

Color Changes

Special pigment cells called **chromatophores** in the skin of most cephalopods produce color changes by expanding and contracting. They are controlled by the nervous system and perhaps by hormones. Some color changes are protective to match background hues; most are behavioral and associated with alarm or courtship. The degree to which cephalopods are able to match colorful backgrounds made it difficult to believe that they lack color vision. However, in an experiment where background colors were equal in intensity, cephalopods were unable to match colors. This result, coupled with the physiology of their eyes, makes color vision unlikely. Cephalopod photoreceptors are rhabdomere cells, not rods and cones, as in vertebrates. Their ability to see color depends on the number of visual pigments present; almost all cephalopods have only one such pigment and cannot detect color. An exception is the bioluminescent firefly squid, which has three visual pigments, the same number found in humans. It appears that the wonderful color patterns we see on cephalopod skin are visible to other cephalopods as polarized light patterns; all cephalopods detect differences in polarized light. They also use this ability to enhance discrimination of translucent prey and fish with silvery reflective scales in much the same way that fishermen use polarized glasses to reduce glare from water.

Ink Production

Most cephalopods other than nautiloids have an ink sac that empties into the rectum. The sac contains an ink gland that secretes a dark fluid containing the pigment melanin. When the animal is alarmed, it releases a cloud of ink through the anus to form a "smokescreen" to confuse an enemy.

Locomotion

Most cephalopods swim by forcefully expelling water from the mantle cavity through a ventral funnel—a sort of jet-propulsion method. The funnel is mobile and can be pointed forward or backward to control direction; the force of water expulsion determines speed.

Squids and cuttlefishes are excellent swimmers. The squid body is streamlined and built for speed (see figure 10.31B). Cuttlefishes (see figure 10.33) swim more slowly. Both squids and cuttlefishes have lateral fins that can serve as stabilizers, but they are held close to the body for rapid swimming. The gas-filled chambers of *Nautilus* keep the shell upright. Although not as fast as squids, nautiluses move surprisingly well.

Octopus has a rather globular body and no fins (see figure 10.1E). Octopuses can swim backward by spurting jets of water from their funnel, but they are better adapted to crawling over rocks and coral, using the suction discs on their arms to pull or to anchor themselves. Some deep-water octopods have fins and arms webbed like an umbrella; they swim in a

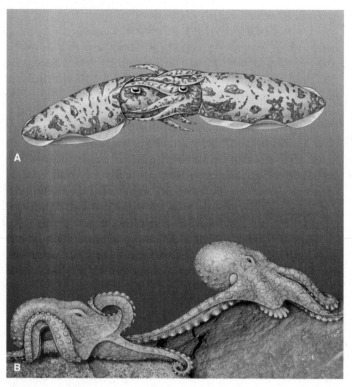

figure 10.35

Copulation in cephalopods. **A,** Mating cuttlefishes. **B,** A male octopus uses its modified arm to deposit spermatophores in the female mantle cavity to fertilize her eggs. Octopuses often tend their eggs during development.

medusa-like fashion. Very large octopuses may flatten their bodies and swim by undulation.

Reproduction

Sexes are separate in cephalopods. Before copulation, males often display changes in skin pigmentation and patterning, apparently directed toward females and against rival males. In the male seminal vesicle, spermatozoa are encased in spermatophores and stored in a sac that opens into the mantle cavity. During copulation, one arm of an adult male plucks a spermatophore from his own mantle cavity and inserts it into the mantle cavity of a female near the oviduct opening (figure 10.35). Eggs are fertilized as they leave the oviduct and are usually attached to stones or other objects to develop. Some octopuses tend their eggs.

▮ Phylogeny and Adaptive Diversification

Phylogeny

Fossils attributed to Mollusca date from the early Cambrian (figure 10.36). Some had shells and others had wormlike bodies without shells. Did the wormlike forms arise first? Two classes of living molluscs have wormlike bodies: Caudofoveata and Solenogastres. According to the Conchifera hypothesis shown in figure 10.2, caudofoveates and solenogasters are the earliest branches on the molluscan tree. For many years, biologists assumed that the wormlike form was ancestral for both classes and that caudofoveates possessed many features of the ancestral mollusc. On this basis the ancestral mollusc was thought to be wormlike with a ventral gliding surface and a dorsal mantle with a chitinous cuticle and calcareous scales. The mantle cavity was assumed to be posterior and to house two gills; the body had a ladder-like nervous system, an open circulatory system with a heart, and a radula used to feed. Caudofoveates have most of these features, but instead of a ventral foot, they have only an oral shield. In solenogasters, the foot is represented by a ventral groove and gills are absent. Thus both solenogasters and caudofoveates are morphologically less complex than other modern gastropods.

However, according to the Serialia hypothesis, only the solenogasters have an ancestral wormlike body, and the wormlike body of caudofoveates is derived from a body plan resembling those of gastropods or cephalopods. Under this hypothesis caudofoveates, cephalopods, scaphopods, and gastropods form a clade; relationships among these four taxa vary depending on the method of tree construction used, but in some trees, caudofoveates are the sister taxon to cephalopods. Thus, the morphology of a caudofoveate must be interpreted in entirely different ways under the two main hypotheses for evolution of the molluscan classes. For example, under the Serialia hypothesis, the foot has been reduced to an oral shield and the shell has been reduced to calcareous scales. At present there is limited statistical support for the Serialia hypothesis, but researchers are trying to improve the data sets.

The two main hypotheses for molluscan evolution also differ in the phylogenetic position of the two taxa with repeated body parts. In chitons, there are typically eight, and occasionally seven, calcareous shell plates along the back of the animal. Plates have their own musculature; the number of pairs of gills ranges from 6 to more than 80. In the monoplacophorans, there are three to six pairs of gills, two pairs of auricles, three to seven pairs of metanephridia, one or two pairs of gonads, and a ladderlike nervous system with ten pairs of pedal nerves. The body is not segmented in chitons or monoplacophorans—there are no septa, so what is the significance of the repeated parts? One possibility is that repeated parts are remnant features of a truly segmented ancestor; another is that they evolved independently in each group.

Under the Conchifera hypothesis, multiple foot retractor muscles are ancestral traits for all shelled molluscs, but multiple gills and other serially repeated features evolved independently in chitons and monoplacophorans (see figure 10.2). In strong contrast, under the Serialia hypothesis, these features arose in a common ancestor of the Serialia clade and

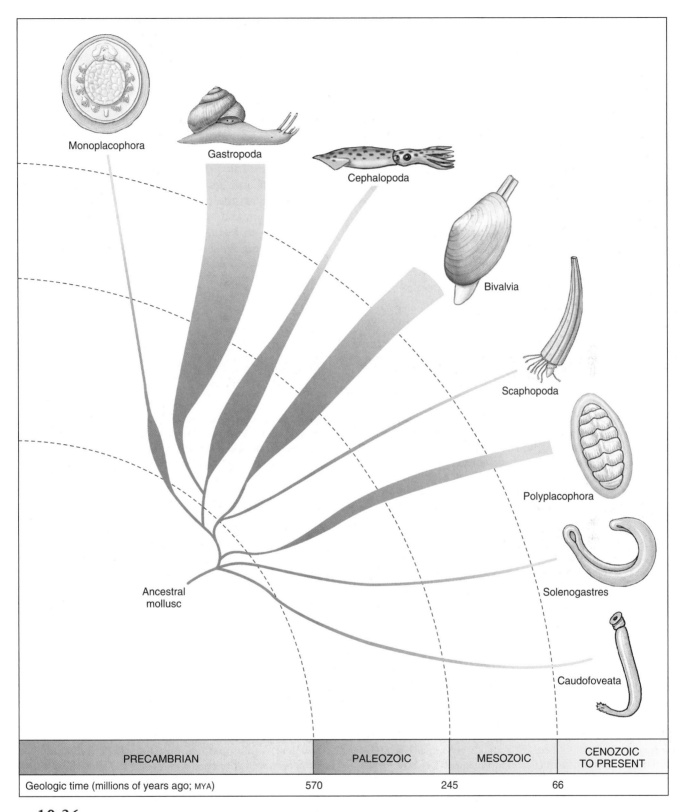

figure 10.36

Classes of Mollusca, showing their derivations and relative abundance.

were inherited by the sister taxa: chitons and monoplacophorans. It is worth noting that some biologists interpret structural and developmental differences between the shells of chitons and those of other molluscs within Conchifera as evidence of independent evolution of the shell. A detailed comparison of the shells of chitons and monoplacophorans might yield interesting results. New phylogenetic analyses are likely to present a better picture of molluscan evolution in the next few years.

Adaptive Diversification

Both gastropods and cephalopods have a greatly expanded visceral mass. In gastropods, the mantle cavity was brought toward the head by torsion, but in cephalopods the mantle cavity was extended ventrally. Evolution of a chambered shell in cephalopods was a very important contribution to their freedom from the substrate and their ability to swim. Elaboration of their respiratory, circulatory, and nervous systems is correlated with their predatory and swimming habits.

Scaphopods and bivalves have an expanded mantle cavity that essentially envelops the body. Adaptations for burrowing characterize these taxa: a spatulate foot and reduction of the head and sense organs.

Most of the diversity among molluscs is related to their adaptation to different habitats and modes of life and to a wide variety of feeding methods, ranging from sedentary filter feeding to active predation. Many adaptations for food gathering appear within the phylum including an enormous variety in radular structure and function particularly among gastropods.

The versatile glandular mantle has evolved a wide variety of forms and functions. Besides secreting the shell and forming the mantle cavity, it is variously modified into gills, lungs, siphons, and apertures, and it sometimes functions in locomotion, in feeding, or in a sensory capacity. The shell, too, has undergone a variety of evolutionary adaptations.

Taxonomy of Phylum Mollusca

Class Caudofoveata (kaw´do-fo-ve-at´a) (L. *cauda,* tail, + *fovea,* small pit): **caudofoveates.** Wormlike; shell, head, and excretory organs absent; radula usually present; mantle with chitinous cuticle and calcareous scales; oral pedal shield near anterior mouth; mantle cavity at posterior end with pair of gills; sexes separate; often united with solenogasters in class Aplacophora. Examples: *Chaetoderma, Limifossor* (see figure 10.8).

Class Solenogastres (so-len´o-gas´trez) (Gr. *solēn,* pipe, + *gastēr,* stomach): **solenogasters.** Wormlike; shell, head, and excretory organs absent; radula usually present; mantle usually covered with scales or spicules; mantle cavity posterior, without true gills, but sometimes with secondary respiratory structures; foot represented by long, narrow, ventral pedal groove; hermaphroditic. Example: *Neomenia* (see figure 10.8).

Class Monoplacophora (mon´o-pla-kof´o-ra) (Gr. *monos,* one, + *plax,* plate, + *phora,* bearing): **monoplacophorans.** Body bilaterally symmetrical with a broad, flat foot; a single limpetlike shell; mantle cavity with five or six pairs of gills; large coelomic cavities; radula present; six pairs of nephridia, two of which are gonoducts; separate sexes. Example: *Neopilina* (see figure 10.9).

Class Polyplacophora (pol´y-pla-kof´o-ra) (Gr. *polys,* many, several, + *plax,* plate, + *phora,* bearing): chitons. Elongated, dorsoventrally flattened body with reduced head; bilaterally symmetrical; radula present; shell of seven or eight dorsal plates; foot broad and flat; gills multiple, along sides of body between foot and mantle edge; sexes usually separate, with a trochophore but no veliger larva. Examples: *Mopalia* (see figure 10.10), *Chaetopleura.*

Class Scaphopoda (ska-fo-pod´a) (Gr. *skaphē,* trough, boat, + *pous, podos,* foot): **tusk shells.** Body enclosed in a one-piece tubular shell open at both ends; conical foot; mouth with radula and tentacles; head absent; mantle for respiration; sexes separate; trochophore larva. Example: *Dentalium* (see figure 10.12).

Class Gastropoda (gas-tro-pod´a) (Gr. *gastēr,* belly, + *pous, podos,* foot): **snails and relatives.** Body asymmetrical; usually in a coiled shell (shell uncoiled or absent in some); head well developed, with radula; foot large and flat; dioecious or monoecious, some with trochophore, typically with veliger, some without larva. Examples: *Busycon, Polinices* (see figure 10.16B), *Physa, Helix, Aplysia* (see figure 10.17).

Class Bivalvia (bi-val´ve-a) (L. *bi,* two, + *valva,* folding door, valve) **(Pelecypoda): bivalves.** Body enclosed in a two-lobed mantle; shell of two lateral valves of variable size and form, with dorsal hinge; head greatly reduced but mouth with labial palps; no radula; no cephalic eyes; gills platelike; foot usually wedge-shaped; sexes usually separate, typically with trochophore and veliger larvae. Examples: *Mytilus* (see figure 10.28), *Venus, Bankia* (see figure 10.27).

Class Cephalopoda (sef-a-lo-pod´-a) (Gr. *kephalē,* head, + *pous, podos,* foot): **squids and octopuses.** Shell often reduced or absent; head well developed with eyes and a radula; head with arms or tentacles; foot modified into a funnel; nervous system of well-developed ganglia, centralized to form a brain; sexes separate, with direct development. Examples: *Loligo, Sepioteuthis* (see figure 10.31), *Octopus* (see figure 10.1E), *Sepia* (see figure 10.33), *Nautilus* (see figure 10.32).

Summary

Mollusca is one of the largest and most diverse phyla, its members ranging in size from very small organisms to the largest invertebrates. Their basic body divisions are the head-foot and the visceral mass, usually covered by a shell. A majority are marine, but some are freshwater, and a few are terrestrial. They occupy a variety of niches; a number are economically important, and a few are medically important as hosts of parasites.

Molluscs are coelomate (have a coelom), although their coelom is limited to the area around the heart. If the ancestors of molluscs used the coelom as an efficient hydrostatic skeleton, reduction of the coelom in molluscs may have been correlated with the evolution of a protective shell. The coelom could not function as a hydrostatic skeleton once a hard shell covered the body.

The mantle and mantle cavity are important characteristics of molluscs. The mantle secretes the shell and overlies a part of the visceral mass to form a cavity housing the gills. The mantle cavity has been modified into a lung in some molluscs. The foot is usually a ventral, solelike, locomotory organ, but it may be variously modified, as in the cephalopods, where it has become a funnel and arms. Most molluscs except bivalves and some solenogasters and gastropods have a radula, which is a protrusible, tonguelike organ with teeth used in feeding. The circulatory system of molluscs is open, with a heart and blood sinuses, except in cephalopods, which have a closed circulatory system. Molluscs usually have a pair of nephridia connecting with the coelom and a complex nervous system with a variety of sense organs. The ancestral larva of molluscs is the trochophore, and most marine molluscs have a derived second larva, the veliger.

Classes Caudofoveata and Solenogastres are small groups of wormlike molluscs with calcareous spines or scales, but no shell. Scaphopoda is a slightly larger class with a tubular shell, open at both ends, and the mantle wrapped around the body.

Class Monoplacophora is a tiny, univalve marine group with many serially repeated body parts. Polyplacophora are more common and diverse marine organisms with shells in the form of a series of seven or eight plates. They are rather sedentary animals with a row of gills along each side of their foot.

Gastropoda is the largest class of molluscs. Gastropods are the only molluscs to exploit terrestrial environments. Their interesting evolutionary history includes torsion, where, early in development, the posterior end rotates to the anterior so that the anus and head are at the same end, as well as coiling, elongation, and spiraling of the visceral mass. Torsion could cause fouling, a release of excreta over the head and in front of the gills. Fouling is avoided in various ways among different gastropods, including bringing water into one side of the mantle cavity and out the other (seen in many gastropods), some degree of detorsion (opisthobranchs and pulmonates), and conversion of the mantle cavity into a lung (pulmonates).

Class Bivalvia includes marine and freshwater forms. Bivalves have two shells joined by a dorsal ligament and held together by an adductor muscle. Most bivalves are suspension feeders, drawing water through their gills by ciliary action.

Members of class Cephalopoda are the most complex molluscs; all are predators, and many can swim rapidly. Their tentacles and arms capture prey by adhesive secretions or by suckers. They swim by expelling a jet of water from their mantle cavity through a funnel, which was derived from the foot.

There is strong embryological and molecular evidence that molluscs are lophotrochozoan protostomes. Evolutionary relationships among the molluscan classes are not yet clear; there are two main hypotheses for these relationships: Conchifera and Serialia.

Review Questions

1. How does a coelom develop embryologically? Why was the evolutionary development of a coelom important?
2. Members of phylum Mollusca are extremely diverse, and yet the phylum clearly constitutes a monophyletic group. What evidence can you cite in support of this statement?
3. How are molluscs important to humans?
4. Distinguish among the following classes of molluscs: Polyplacophora, Gastropoda, Bivalvia, Cephalopoda.
5. Define the following: radula, odontophore, periostracum, prismatic layer, nacreous layer, trochophore, veliger.
6. Briefly describe the habitat and habits of a typical chiton.
7. Define the following with respect to gastropods: operculum, torsion, fouling, bilateral asymmetry.
8. Torsion in gastropods created a selective disadvantage: fouling. Suggest one or more potential selective advantages that could have offset the disadvantage. How have gastropods evolved to avoid fouling?
9. Distinguish between opisthobranchs and pulmonates.
10. Briefly describe how a typical bivalve feeds and how it burrows.
11. What is the function of the siphuncle of *Nautilus*?
12. Describe how cephalopods swim and eat.
13. Cephalopods are actively swimming predators, but they evolved from a slow-moving, probably grazing ancestor. Describe evolutionary modifications of the ancestral plan that make the cephalopod lifestyle possible.
14. Briefly describe differences in the way that the morphology of caudofoveates is interpreted under the Conchifera and Serialia hypotheses.

For Further Thought Provide two potential evolutionary explanations for the presence of serially repeated body parts in monoplacophorans.

■ Selected References

See also general references on page 432.

Abbott, R. T., and P. A. Morris. 2001. A field guide to shells: Atlantic and Gulf Coasts and the West Indies (R. T. Peterson, Ed.). Boston, Houghton Mifflin Company. *A popular handbook for shell collectors.*

Barinaga, M. 1990. Science digests the secrets of voracious killer snails. Science **249:**250-251. *Describes research on the toxins produced by cone snails.*

Bossenbroek, J. M., L. E. Johnson, B. Peters, and D. M. Lodge. 2007. Forecasting the expansion of zebra mussels in the United States. Conservation Biology **21:**800-810. *Mussel dispersal and the likelihood of invasion are discussed, along with potential consequences.*

Colgan, D. J., W. F. Ponder, and P. E. Eggler. 2000. Gastropod evolutionary rates and phylogenetic relationships assessed using partial 28S rDNA and histone H3 sequences. Zool. Sci. **29:**29-63. *Phylogenies constructed using molecular data do not agree well with hypotheses about gastropod evolution developed from morphological studies.*

Giribet, G., A. Okusu, A. R. Lindgren, S. W. Huff, M. Schrödl, and M. Nishiguchi. 2006. Evidence for a clade composed of molluscs with serially repeated structures: monoplacophorans are related to chitons. Proc. Natl. Acad. Sci. USA **103:**7723-7728. *Molluscs with repeated structures are derived taxa and not likely to have inherited repeated structures from a segmented ancestor.*

Gosline, J. M., and M. D. DeMont. 1985. Jet-propelled swimming in squids. Sci. Amer. **252:**96-103 (Jan.). *Mechanics of swimming in squid are analyzed; elasticity of collagen in mantle increases efficiency.*

Hanlon, R. 2007. Cephalopod dynamic camouflage. Curr. Biol. **17:**400-404. *Experiments clearly demonstrate that cephalopods cannot see background colors and that they match backgrounds by matching intensity.*

Holloway, M. 2000. Cuttlefish say it with skin. Nat. Hist. **109** (3):70-76. *Cuttlefish and other cephalopods can change texture and color of their skin with astonishing speed. Fifty-four components of cuttlefish "vocabulary" have been described, including color display, skin texture, and a variety of arm and fin signals.*

Kubodera, T., and K. Mori. 2005. First-ever observations of a live giant squid in the wild. Proc. Biol. Sci. **272:**2583-2586. *Camera recordings of a giant squid feeding on bait.*

Page, L. R. 2003. Gastropod ontogenetic torsion: developmental remnants of an ancient evolutionary change in body plan. J. Exp. Zool. Part B: **297:**11-26. *A torted body is the result of asynchronous movements of the shell and viscera, accompanied by remodeling of mantle epithelium.*

Ponder, W. F., and D. R. Lindberg. 1997. Towards a phylogeny of gastropod molluscs: an analysis using morphological characters. Zool. J. Linn. Soc. **119:**83-265. *Analysis of 117 morphological characters from 40 taxa suggested several new gastropod subgroups.*

Strayer, D. L., and H. M. Malcom. 2007. Effects of zebra mussels (*Dreissena polymorpha*) on native bivalves: the beginning of the end or the end of the beginning? J. N. Am. Benthol. Soc. **26:**111-122. *Some species of native bivalves rebound after population decline due to zebra mussel establishment.*

Ward, P. D. 1998. Coils of time. Discover **19** (3):100-106. *Present Nautilus has apparently existed essentially unchanged for 100 million years, and all other species known were derived from it, including king nautilus (Allonautilus), a recent derivation.*

Wilson, N. G., G. W. Rouse, and G. Giribet. 2010. Assessing the molluscan hypothesis Serialia (Monoplacophora + Polyplacophora) using novel molecular data. Mol. Phyl. Evol. **54:**187-193. *Several pathways for the evolution of molluscan classes are explored.*

■ Custom Website

The *Animal Diversity* Online Learning Center is a great place to check your understanding of chapter material. Visit www.mhhe.com/hickmanad6e for access to key terms, quizzes, and more! Further enhance your knowledge with Web links to chapter-related material.

Explore live links for these topics:

Classification and Phylogeny of Animals
Phylum Mollusca
Primitive Classes of Molluscs

Class Polyplacophora
Class Gastropoda
Class Bivalvia
Class Cephalopoda

Annelids and Allied Taxa

Annelida

Sipuncula

ia sp., a polychaete.

Dividing the Body

Although a spacious, fluid-filled coelom provides an efficient hydrostatic skeleton for burrowing, precise control of body movements is not possible because the force of muscle contraction in one area is carried throughout the body by fluid in an undivided coelom. In contrast, ancestral annelids had distinct coelomic compartments in each body segment. Each fluid-filled compartment could respond individually to local muscle contraction—one segment being long and thin, and another being short and round. The annelid body illustrates *metamerism,* meaning that it is composed of serially repeated units called segments or metameres. Each segment is separated from its neighbors by partitions called septa, and each contains components of most of the organ systems, such as circulatory, nervous, and excretory systems.

The evolutionary advent of metamerism was highly significant because it made possible the development of much greater complexity in structure and function. Metamerism makes possible independent and separate movements by separate segments. Moreover, repetition of body parts gives the organisms a built-in redundancy, as in some human-made systems. This redundancy provides a safety factor: if one segment is damaged, blood vessels can seal it, and others still function. Thus, an injury to one part would not necessarily be fatal.

The evolutionary potential of a metameric body plan is amply demonstrated by the large and diverse phylum Arthropoda, where it may have arisen independently. Metamerism also arose independently in the deuterostome line, which includes the numerous and adaptively diverse vertebrates.

The wormlike animal phyla described in this chapter are coelomate protostomes belonging to Lophotrochozoa. They develop by spiral mosaic cleavage, form mesoderm from derivatives of the 4d cell (see p. 62) make a coelom by schizocoely (see p. 63), and share a trochophore as the ancestral larval form. Two phyla are discussed: Annelida and Sipuncula; echiurans are placed within Annelida.

Members of phylum Annelida are segmented worms living in marine, freshwater, and moist terrestrial habitats. Marine bristle worms, leeches, and the familiar earthworms belong to this group. Annelida also includes pogonophoran and vestimentiferan worms, which were formerly either placed together in phylum Pogonophora or placed in distinct phyla: Pogonophora and Vestimentifera. These deep-ocean worms belong in Siboglinidae within class Polychaeta. Annelida now includes members of former phylum Echiura. Echiurans are not segmented, but many species possess paired epidermal setae, so it is likely that segmentation has been lost in this group.

Worms in phylum Sipuncula are benthic marine animals with unsegmented bodies. We depict sipunculans as the sister taxon Annelida (figure 11.1).

Phylum Annelida

Annelida (an-nel´i-də) (L. *annellus,* little ring, + *-ida,* suffix) consists of worms with paired epidermal setae; almost all are segmented. It is a large phylum, numbering approximately 15,000 species, the most familiar of which are earthworms and freshwater worms (Oligochaeta) and leeches (class Hirudinida). However, the less familiar marine worms (Polychaeta) form approximately two-thirds of the phylum. Some polychaetes are strange, even grotesque, whereas others are graceful and beautiful. They include clam worms, plumed worms, parchment worms, scaleworms, lugworms, and many others. Beardworms, or pogonophorans, have recently been added to this group along with echiurans.

Annelids are a highly developed group in which the nervous system is more centralized and the circulatory system more complex than those of the phyla we have studied so far. Annelids are sometimes called "bristle worms" because, except for leeches, most annelids bear tiny, chitinous bristles called **setae** (L. *seta,* hair or bristle). Short, needlelike setae help anchor segments during locomotion to prevent backward slipping; long, hairlike setae aid aquatic forms in swimming. Because many annelids either are burrowers or live in secreted tubes, stiff setae also help to prevent a worm from being pulled or washed out of its home. Robins know from experience how effective an earthworm's setae are.

Ecological Relationships

Annelids are worldwide in distribution, occurring in the sea, fresh water, and terrestrial soil. Some marine annelids live quietly in tubes or burrow into bottom mud or sand. Some feed on organic matter in the mud through which they burrow; others feed on suspended particles with elaborate ciliary or mucous devices for trapping food. Many are predators, either swimming in the open sea or hiding in crevices of coral or rock except when hunting. Freshwater annelids burrow in mud or sand, live among vegetation, or swim freely. The most familiar annelids are terrestrial earthworms, which move through moist soil. Some leeches are bloodsuckers, and others are carnivores; most of them live in fresh water.

Economic Importance

Much of the economic importance of annelids is indirect, deriving from their ecological roles. Many are members of grazing food chains or detritus food chains, serving as prey for other organisms of more direct interest to humans, such as fishes. Consequently, there is a thriving market for some polychaetes and oligochaetes as fish bait. Earthworm burrows increase the drainage and aeration of soils, and migrating worms help mix the soil and distribute organic matter to deeper layers (see p. 223). Some marine annelids that burrow serve an analogous role;

CHARACTERISTICS
of Phylum Annelida

1. Unique annelid head and paired epidermal setae present (lost in leeches); parapodia present in the ancestral condition
2. Marine, freshwater, and terrestrial
3. Most free-living, some symbiotic, some ectoparasitic
4. Body bilaterally symmetrical, **metameric,** often with a distinct head; metamerism lost in echiurans
5. Triploblastic body
6. Coelom (schizocoel) well developed and divided by septa, except in leeches and echiurans; coelomic fluid functions as hydrostatic skeleton
7. Epithelium secretes outer transparent, moist cuticle
8. Digestive system complete and not segmentally arranged
9. Body wall has outer circular and inner longitudinal muscle layers
10. Nervous system has a double ventral nerve cord and a pair of ganglia with lateral nerves in each segment; brain is a pair of dorsal cerebral ganglia with connectives to ventral nerve cord
11. Sensory system of tactile organs, taste buds, statocysts (in some), photoreceptor cells, and eyes with lenses (in some); specialization of head region into differentiated organs, such as tentacles, palps, and eyespots of polychaetes
12. Asexual reproduction by fission and fragmentation; capable of complete regeneration; asexual reproduction by budding in some
13. Hermaphroditic or separate sexes; larvae, if present, are trochophore type; spiral cleavage and mosaic development
14. Excretory system typically a **pair of nephridia for each segment;** nephridia remove waste from blood as well as from coelom
15. Respiratory gas exchange through skin, **gills,** or **parapodia**
16. **Circulatory system closed** with muscular blood vessels and aortic arches ("hearts") for pumping blood, segmentally arranged; respiratory pigments (hemoglobin, hemerythrin, or chlorocruorin) often present; amebocytes in blood plasma

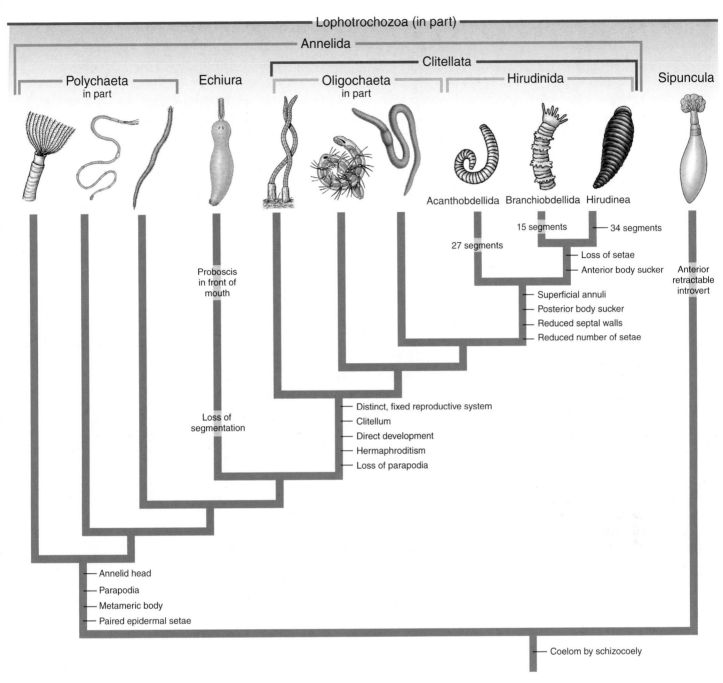

figure 11.1

Cladogram of annelids, showing the appearance of shared derived characters that specify the five monophyletic groups (based on Brusca and Brusca, 2003). Acanthobdellida and Branchiobdellida are two small groups that Brusca and Brusca place with Hirudinea ("true" leeches), in Hirudinida. This clade has several synapomorphies: a tendency toward reduction of septal walls, the appearance of a posterior sucker, and the subdivision of body segments by superficial annuli. However, Oligochaeta has no synapomorphies; it is recognized by retained ancestral characters, and molecular data reveal it to be paraphyletic. Oligochaeta and Hirudinida together form the monophyletic group Clitellata. Polychaeta, like Oligochaeta, is not a monophyletic group.

Source: Modified from R. C. Brusca and G. J. Brusca, Invertebrates, ed. 2. Sinauer Associates, Inc., Sinderland, MA, 2003.

lugworms (*Arenicola*, see figure 11.11) are sometimes called "earthworms of the sea." New medical uses for leeches (see p. 228) have revived the market in blood-sucking leeches and established "leech farms" where these organisms are raised in captivity.

Body Plan

An annelid body typically has a two-part head, composed of a **prostomium** and a **peristomium**, a series of segments (sometimes called metameres), and a terminal **pygidium**

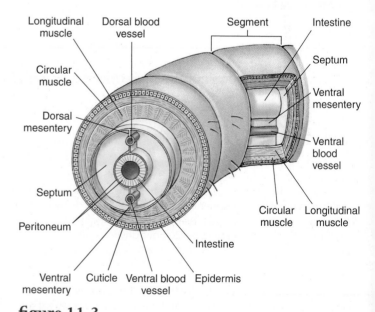

figure 11.2

Nereis virens, an errant polychaete. **A,** Anterior end, with pharynx everted. **B,** External structure. **C,** Posterior end. **D,** Generalized transverse section through region of the intestine.

bearing an anus (figure 11.2). Neither the head nor the pygidium is considered a true segment. During growth, new segments form just in front of the pygidium; thus, the oldest segments are at the anterior end, and the youngest segments are at the posterior end. Each segment typically contains circulatory, respiratory, nervous, and excretory structures, as well as a coelom.

In most annelids, the coelom develops embryonically as a split in the mesoderm on each side of the gut **(schizocoel),** forming a pair of coelomic compartments in each segment. Each compartment is surrounded by **peritoneum** (a layer of mesodermal epithelium), which lines the body wall, forms dorsal and ventral mesenteries (double-membrane partitions that support the gut), and covers all organs (figure 11.3). Where peritonea of adjacent segments meet, **septa** (singular is **septum**) are formed. The body wall surrounding the peritoneum and coelom contains strong circular and longitudinal muscles adapted for swimming, crawling, and burrowing.

Except in leeches, the coelom is filled with fluid and serves as a **hydrostatic skeleton.** Because the volume of fluid in a coelomic compartment is essentially constant, contraction of longitudinal body-wall muscles causes a segment to shorten

figure 11.3

Annelid body plan.

and become larger in diameter, whereas contraction of circular muscles causes it to lengthen and become thinner. The presence of septa means that widening or elongation occurs in restricted areas; crawling motions are effected by *alternating* waves of contraction by longitudinal and circular muscles passing down the body (peristaltic contractions). Longitudinal muscles contract in some segments, allowing those segments to widen and thus to anchor themselves against burrow walls or other substrates. In other segments, circular muscles contract, allowing the segments to elongate and stretch forward. Forces powerful enough for burrowing as well as locomotion can thus be generated. Swimming forms use undulatory rather than peristaltic movements in locomotion.

An annelid body has a thin, outer layer of nonchitinous cuticle surrounding the epidermis (figure 11.3). Paired epidermal setae (see figure 11.2D) are considered ancestral for annelids, although they have been reduced or lost in some. The annelid digestive system is not segmented although it is regionalized: the gut runs the length of the body, perforating each septum (figure 11.3). Longitudinal dorsal and ventral blood vessels follow the same path, as does a ventral nerve cord.

Annelids are divided among three subgroups: Polychaeta, Oligochaeta, and Hirudinida. These groups were once considered classes, but modern phylogenetic methods have demonstrated that only Hirudinida is a monophyletic group, as befits a formally recognized taxon (see figure 11.1). We retain the names Polychaeta and Oligochaeta as descriptive terms awaiting taxonomic revision. Pogonophorans and echiurans are placed within Polychaeta in light of recent morphological and molecular analyses. Polychaeta is a paraphyletic class because ancestors of oligochaetes and leeches arose from within the polychaetes. Oligochaetes and leeches together form a monophyletic group

called Clitellata (see figure 11.1), which is characterized by the presence of a reproductive structure called the **clitellum.** The clitellum is visible in earthworms as a distinctive fat band present about one-third of the way down the body. The clitellum in leeches is visible only during the reproductive season.

Polychaeta

Polychaetes (Gr. *polys,* many, + *chaite,* long hair) are the largest group of annelids, with more than 10,000 described species, mostly marine. Although a majority are from 5 to 10 cm long, some measure less than a millimeter, and others may be as long as 3 m. Some are bright reds and greens; others are dull or iridescent. Some are picturesque, such as "feather duster" worms, or fanworms (figure 11.4).

Polychaetes live under rocks, in coral crevices, or in abandoned shells, or they burrow into mud or sand. Some build their own tubes on submerged objects or in soft substrates on the sea floor; others adopt the tubes or homes of other animals. Some are pelagic, forming part of the planktonic population. They are extremely abundant in some areas; for example, a square meter of mudflat may contain thousands of polychaetes. They play a significant part in marine food chains because they are eaten by fishes, crustaceans, hydroids, and many others.

Polychaetes differ from other annelids in having a well-differentiated head with specialized sense organs; paired, paddlelike appendages **(parapodia)** on most segments; and no clitellum (see p. 224 and figure 11.2). As the name polychaete implies, they have many setae, usually arranged in bundles on the parapodia. They show pronounced differentiation of some body segments and a specialization of sensory organs practically unknown among clitellates (oligochaetes and leeches).

A B

figure 11.4
Tube-dwelling sedentary polychaetes. **A,** Christmas tree worm, *Spirobranchus giganteus,* lives in a calcareous tube. On its head are two whorls of modified tentacles (radioles) used to collect suspended food particles from the surrounding water. Notice the finely branched filters visible on the edge of one radiole. **B,** Sabellid polychaetes, *Bispira brunnea,* live in leathery tubes. Because of their appearance, they are commonly called "feather duster" worms or fanworms.

A polychaete typically has a prostomium, which may or may not be retractile and often bears eyes, antennae, and sensory palps (see figure 11.2). The peristomium surrounds the mouth and may bear setae, palps, or, in predatory forms, chitinous jaws. Suspension feeders may have a tentacular crown that can be opened like a fan or withdrawn into the tube (see figure 11.4).

The trunk is segmented, and most segments bear parapodia, which may have lobes, cirri, setae, and other structures on them (see figure 11.2D). Parapodia are composed of two main parts—a dorsal **notopodium** and a ventral **neuropodium**—either of which may be prominent or reduced in a given species. Parapodia are used for crawling, swimming, or anchoring in tubes. They usually serve as the chief respiratory organs, although some polychaetes also have gills. *Amphitrite,* for example, has three pairs of branched gills and long, extensible tentacles (figure 11.5). *Arenicola,* the lugworm (see figure 11.11), which burrows through sand leaving characteristic castings at the entrance to its burrow, has paired gills on certain segments.

Sense organs are more highly developed in polychaetes than in oligochaetes and include eyes and statocysts. Eyes, when present, may range from simple eyespots to well-developed organs. Usually eyes consist of retinal cups, with rodlike photoreceptor cells lining the cup wall and directed toward the lumen of the cup.

In contrast to clitellates, polychaetes have no permanent sex organs, possess no permanent ducts for their sex cells, and usually have separate sexes. Gonads appear as temporary swellings of the peritoneum and shed their gametes into the coelom. Gametes leave the body through gonoducts, through nephridia, or by rupture of the body wall. Fertilization is external, and development is indirect with a trochophore larva preceding the adult.

Some polychaetes are free-moving pelagic forms, some are active burrowers and crawlers, and some are sedentary, living in tubes or burrows that they rarely (or never) leave. An example of an active, predatory worm is *Nereis* (from Greek mythology, a Nereid, or daughter of Nereus, ancient sea god), the clam worm (see figure 11.2). *Nereis* has a muscular, eversible pharynx equipped with jaws that can be thrust out with surprising speed and dexterity for capturing prey. Scale worms (figure 11.6) often live as commensals with other invertebrates, and fireworms (figure 11.7) feed on gorgonians and stony corals.

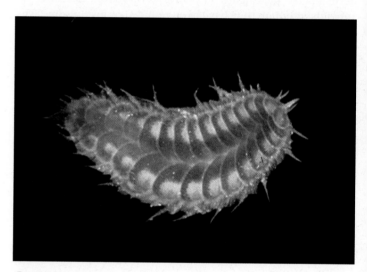

figure 11.6

The scale worm, *Hesperonoe adventor,* normally lives as a commensal in the tubes of *Urechis* (phylum Echiura).

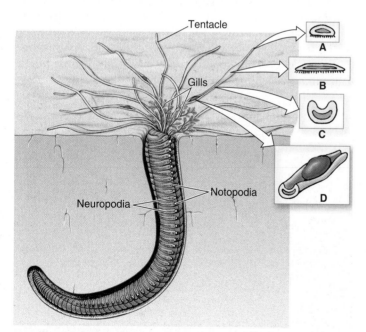

figure 11.5

Amphitrite, which builds its tubes in mud or sand, extends long, grooved tentacles out over the mud to gather bits of organic matter. The smallest particles are moved along food grooves by cilia, and larger particles are carried by peristaltic movement. The plumelike gills are blood red. **A,** Section through the exploratory end of a tentacle. **B,** Section through a tentacle in an area adhering to the substrate. **C,** Section showing a ciliary groove. **D,** A particle being carried toward the mouth.

figure 11.7

A fireworm, *Hermodice carunculata,* feeds on gorgonians and stony corals. Its setae are like tiny glass fibers and serve to repel predators.

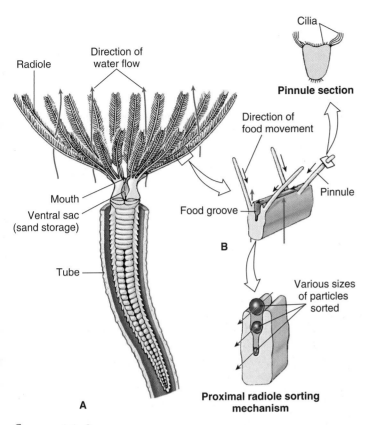

figure 11.8

Sabella, a polychaete suspension feeder. **A,** Anterior view of the crown. Cilia direct small food particles along grooved radioles to the mouth and discard larger particles. Sand grains are directed to storage sacs and later used in tube building. **B,** Distal portion of a radiole, showing ciliary tracts of pinnules and food grooves.

Most sedentary tube and burrow dwellers are particle feeders, using ciliary or mucoid methods of obtaining food. Their principal food source is plankton and detritus. Some, such as *Amphitrite* (from Greek mythology, sea goddess) (see figure 11.5), protrude their heads from the mud and send long, extensible tentacles over the surface. Cilia and mucus on the tentacles entrap particles found on the sea floor and move them toward the mouth, a practice called deposit feeding.

Fanworms, or "feather duster" worms, are beautiful tubeworms, fascinating to watch as they emerge from their secreted tubes and unfurl their lovely tentacular crowns to feed (see figure 11.4). A slight disturbance, sometimes even a passing shadow, causes them to retract quickly into the safety of the homes they have built. Food attracted to the feathery arms, or **radioles,** by ciliary action is trapped in mucus and carried down ciliated food grooves to the mouth (figure 11.8). Particles too large for the food grooves are carried along the margins and discarded. Further sorting may occur near the mouth, where only small particles of food can enter, and sand grains are stored in a sac to be used later in enlarging the tube.

Some polychaetes live most of the year as sexually unripe animals called *atokes,* but during the breeding season, a portion of the body develops into a sexually ripe worm called an *epitoke,* which is swollen with gametes (figure 11.9). For example, palolo worms live in burrows among coral reefs of the South Seas. During the reproductive cycle, their posterior segments become swollen with gametes. During the swarming period, which occurs at the beginning of the last quarter of the October-November moon, these epitokes break off and swim to the surface. Just before sunrise, the sea is literally covered with them, and at sunrise they burst, freeing eggs and sperm for fertilization. The anterior portions of the worms regenerate new posterior sections. A related form swarms in the Atlantic in the third quarter of the June-July moon. Swarming is of great adaptive value because synchronous maturation of all epitokes ensures the maximum number of fertilized eggs. However, it is very hazardous; many types of predators, including local residents who consider the worms a great delicacy, have a feast. In the meantime, the atoke remains safe in its burrow to produce another epitoke at the next cycle!

figure 11.9

Eunice viridis, the Samoan palolo worm. Posterior segments make up the epitokal region, consisting of segments packed with gametes. Each segment has one eyespot on the ventral side. Once a year, the worms swarm, and the epitokes detach, rise to the surface, and discharge their ripe gametes, leaving the water milky. By the next breeding season, epitokes are regenerated.

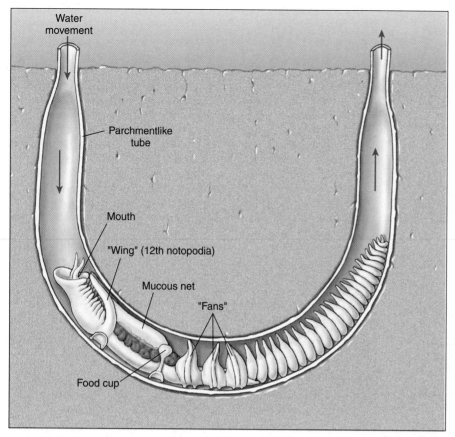

figure 11.10

Chaetopterus, a sedentary polychaete, occupies a U-shaped tube in the sea floor. It pumps water through a parchmentlike tube (of which one-half has been cut away here) with its three pistonlike fans. The fans beat 60 times per minute to keep water currents moving. Winglike notopodia of the twelfth segment continuously secrete a mucous net that strains food particles. As the net fills with food, the food cup rolls it into a ball, and when the ball is large enough (about 3 mm), the food cup bends forward and deposits the ball in a ciliated groove to be carried to the mouth and swallowed.

filtered food particles. The worms ingest the food-laden sand (figure 11.11).

Tube dwellers secrete many types of tubes. Some are parchmentlike; some are firm, calcareous tubes attached to rocks or other surfaces (see figure 11.4A); and some are grains of sand or bits of shell or seaweed cemented together with mucous secretions. Many burrowers in sand and mudflats simply line their burrows with mucus (figure 11.11).

The 2004 discovery of bone-eating marine worms in genus *Osedax* was very exciting. These sIboglinid polychaetes bore into bones of whale carcasses through a "root" system and use endosymbiotic bacteria to digest the bone. The green, branching, vascularized "root" system develops from the posterior part of the egg sac and invades bone marrow. One might wonder if whale falls (carcasses) are sufficiently abundant for this life style to succeed—surprisingly, they are typically only 5 to 15 km apart in coastal regions where the worms occur. Female worms take no chances in finding mates; dwarf males share the tubes of females. How do males and females find each other? Sex determination seems to be environmental, so larvae that land on females become male. Several species of these animals have been found and surprising aspects of their biology appear regularly in the literature.

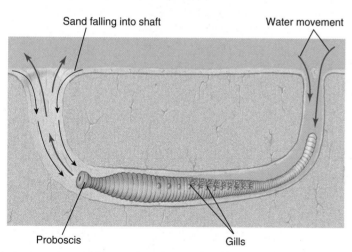

figure 11.11

Arenicola, the lugworm, lives in an L-shaped burrow in intertidal mudflats. It burrows by successive eversions and retractions of its proboscis. By peristaltic movements, it keeps water coming in the open end of the tube and filtering through the sand at the head end. The worm then ingests the food-laden sand.

Some worms, such as *Chaetopterus* (Gr. *chaitē*, long hair, + *pteron*, wing), secrete mucous filters through which they pump water to collect edible particles (figure 11.10). Lugworms, *Arenicola* (L. *arena*, sand, + *colere*, to inhabit), live in L-shaped burrows in which, by peristaltic movements, they cause water to flow. Sand at the front of their burrows collects

Siboglinidae (Pogonophorans)

Members of the former phylum Pogonophora (po-go-nof´e-ra) (Gr. *pōgōn*, beard, + *pherō*, to bear), or beardworms, were entirely unknown before the twentieth century. The first specimens to be described were collected from deep-sea dredgings in 1900 off the coast of Indonesia. They have since been discovered in several seas, including the western Atlantic off the eastern coast of the United States. Some 150 species have been described so far. The taxonomic status of pogonophorans has been controversial for some time, but recent molecular and morphological analyses indicate that these worms are derived polychaetes. Segmentation occurs only in one body region, but the cuticle and setae of pogonophorans are homologous to

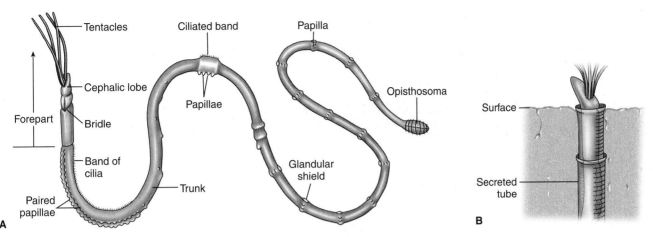

figure 11.12

Diagram of a typical siboglinid. **A,** External features. The body, in life, is much more elongated than shown in this diagram. **B,** Position in its tube.

those of polychaetes. As a clade within Polychaeta, this group is now called Siboglinidae.

Most siboglinids live in soft substrates on the ocean floor, usually at depths of more than 200 m. Their usual length varies from 5 to 85 cm, and their diameter is usually less than a millimeter. They are sessile, and secrete very long, chitinous tubes in which they live, probably extending the anterior end only for feeding.

The siboglinid body has a short forepart, a long, very slender trunk, and a small, segmented opisthosoma (figure 11.12). It is covered with a cuticle and bears setae on the trunk and opisthosoma. A series of coelomic compartments divides the body. The forepart bears from one to many tentacles.

Siboglinids are remarkable in having no mouth or digestive tract, making their mode of nutrition puzzling. They absorb some nutrients, such as glucose, amino acids, and fatty acids dissolved in seawater, through pinnules and microvilli of their tentacles. However, they apparently derive most of their energy from a mutualistic association with chemoautotrophic bacteria. These bacteria oxidize hydrogen sulfide to provide energy for producing organic compounds from carbon dioxide. An expanded region of the midgut fills with endoderm to form an organ called a **trophosome** that bears the bacteria. All traces of a foregut and hindgut are absent in adults.

Among the most amazing animals in the deep-water, Pacific rift communities are giant pogonophorans, *Riftia pachyptila.* Much larger than any pogonophorans previously reported, they measure up to 1.5 m in length and 4 cm in diameter. (Some authors considered them a separate phylum, *Vestimentifera,* but they are now placed within Siboglinidae.) Trophosomes of other siboglinids are confined to the posterior part of the trunk, which is buried in sulfide-rich sediments, but the trophosome of *Riftia* occupies most of its large trunk. It has access to a much larger supply of hydrogen sulfide, enough to nourish its large body, in the effluent of hydrothermal vents.

Siboglinids have a well-developed, closed, blood vascular system. Their photoreceptor cells are very similar to those of other annelids (oligochaetes and leeches). Sexes are separate.

Echiuridae

Members of former phylum Echiura (ek-ē-yur´ə) (Gr. *echis,* viper, + *oura,* tail) are now considered polychaetes in Echiuridae. Echiuridae consists of about 140 species of marine worms that burrow into mud or sand or live in empty snail shells, sand dollar tests, or rocky crevices. They live in all oceans, most commonly in littoral zones of warm waters. They vary in length from a few millimeters to 40 to 50 cm.

Echiurans are cylindrical and somewhat sausage-shaped. Anterior to the mouth is a flattened, extensible proboscis, which cannot be retracted into the trunk. Echiurans are often called "spoon worms" because of the shape of the contracted proboscis in some of them (figure 11.13). The proboscis has a ciliated groove leading to the mouth. There is a complete gut with a posterior anus. While the animal lies buried, its proboscis can extend out over the mud for exploration and deposit feeding (figure 11.14). A different feeding method is used by *Urechis* (Gr. *oura,* tail, + *echis,* viper); it secretes a mucous net in a U-shaped burrow through which it pumps water and strains food particles. *Urechis* is sometimes called the "fat innkeeper" because characteristic species of commensals live with it in its burrow, including species of crab, fish, clam, and polychaete annelid.

Echiurans, with the exception of *Urechis,* have a closed circulatory system with a contractile vessel; most have one to three pairs of nephridia (some have many pairs), and all have a nerve ring and ventral nerve cord. A pair of anal sacs arises from the rectum and opens into the coelom; they are probably respiratory in function and possibly accessory nephridial organs. Early cleavage and trochophore stages are similar to those of other annelids.

Echiurans are not segmented, but they presumably had a segmented ancestor. Many species retain the ancestral paired epidermal setae.

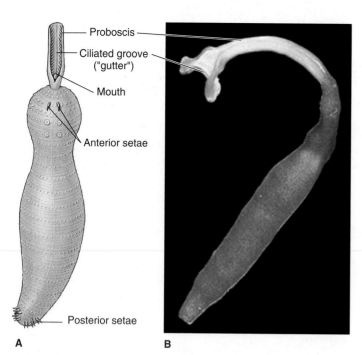

figure 11.13

A, *Echiurus,* an echiuran common on both Atlantic and Pacific coasts of North America. **B,** *Anelassorhynchus,* an echiuran of the tropical Pacific. The shape of their proboscis lends them the common name of "spoon worms."

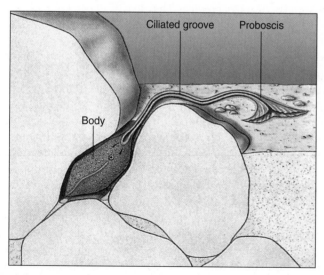

figure 11.14

Bonellia (phylum Echiura) is a detritus feeder. Lying in its burrow, it explores the surface with its long proboscis, which gathers organic particles and carries them along a ciliated groove to its mouth.

In some species of echiurans, sexual dimorphism is pronounced, with females much larger than males. *Bonellia* has an extreme sexual dimorphism, and sex is determined in a very interesting way. At first, free-swimming larvae are sexually undifferentiated. Then those that come into contact with the proboscis of a female become tiny males (1 to 3 mm long) that migrate to the female uterus. About 20 males are usually found in a single female. Larvae that do not contact a female proboscis metamorphose into females. The stimulus for development into males is apparently a pheromone produced by the female proboscis.

Clitellata

As shown in figure 11.1, Clitellata is a monophyletic group comprising oligochaetes and leeches (Hirudinida). Group members share a reproductive structure called a clitellum (see figure 11.15B). Oligochaetes do not form a monophyletic group because leeches arose from an oligochaete ancestor.

Oligochaeta

More than 3000 species of oligochaetes (Gr. *oligos,* few, + *chaitē,* long hair) occur in a great variety of habitats. They include the familiar earthworms as well as many species that live in fresh water. Most are terrestrial or freshwater forms, but some are parasitic, and a few live in marine or brackish water.

Oligochaetes, with few exceptions, bear setae, which may be long or short, straight or curved, blunt or needlelike, or arranged singly or in bundles. They are less numerous in oligochaetes than in polychaetes. Aquatic forms usually have longer setae than do earthworms.

The most familiar oligochaetes are earthworms ("night crawlers"), which burrow in moist, rich soil, emerging at night to feed on surface detritus and vegetation and to breed. In damp, rainy weather they stay near the surface, often with their mouth or anus protruding from their burrow. In very dry weather, they may burrow several feet underground, coil in a slime chamber, and become dormant. *Lumbricus terrestris* (L. *lumbricum,* earthworm), the form commonly studied in school laboratories, is approximately 12 to 30 cm long (figure 11.15). Giant tropical earthworms may have from 150 to 250 or more segments and may grow to 3 to 4 m in length. They usually live in branched, interconnected tunnels.

Form and Function

In earthworms, a fleshy prostomium overhangs the mouth at the anterior end, with the anus on the terminal end (figure 11.15B). In most earthworms, each segment bears four pairs of chitinous setae (figure 11.15C), although in some oligochaetes each segment may have up to 100 or more setae. Each seta is a bristle-like rod set in a sac within the body wall and moved by tiny muscles (figure 11.16). The setae project through small pores in the cuticle to the outside. During locomotion and burrowing, setae anchor parts of the body to prevent slipping. Earthworms move by peristaltic movement. Contraction of circular muscles in the anterior end lengthens the body, thus pushing the anterior end forward where it is anchored by setae; contractions of longitudinal muscles then shorten the body, pulling the posterior end forward. As these waves of contraction pass along the entire body, it gradually moves forward.

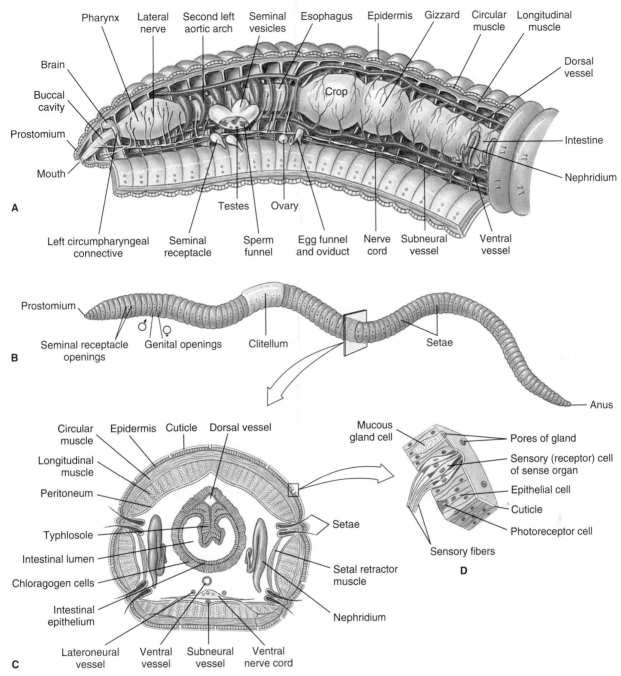

figure 11.15

Earthworm anatomy, as illustrated by *Lumbricus terrestris*. **A,** Internal structure of anterior portion of worm. **B,** External features, lateral view. **C,** Generalized transverse section through region posterior to clitellum. **D,** Portion of epidermis showing sensory, glandular, and epithelial cells.

The food of earthworms is mainly decayed organic matter and bits of vegetation drawn in by the muscular **pharynx** (see figure 11.15A). As in other annelids, the digestive tract is unsegmented and extends the length of the worm. The upper digestive tract has distinct regions, including the crop, where food is stored prior to grinding in the gizzard (figure 11.14A). The intestine has a U-shaped fold, the **typhlosole,** that increases the surface area for nutrient absorption. In the typhlosole and around the intestine are **chloragogen cells**

which synthesize glycogen and fat and can break free to distribute these nutrients through the coelom. Chloragogen cells also serve an excretory function.

Annelids have a double transport system—the coelomic fluid and the circulatory system. Both carry food, wastes, and respiratory gases in varying degrees. The blood is in a closed system of blood vessels, including capillary systems in the tissues. There are five main longitudinal trunks, of which the dorsal blood vessel is the main pumping organ (see figure 11.15A

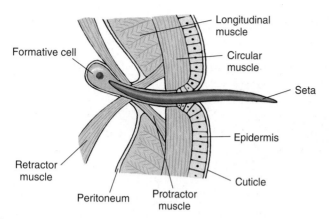

figure 11.16

A seta with its muscle attachments, showing relation to adjacent structures. Setae lost by wear and tear are replaced by new ones, which develop from formative cells.

and C). Their blood contains colorless ameboid cells and a dissolved respiratory pigment, **hemoglobin.** The blood of other annelids may have respiratory pigments other than hemoglobin.

Aristotle called earthworms the "intestines of the soil." Some 22 centuries later, Charles Darwin published his observations in his classic, *The Formation of Vegetable Mould Through the Action of Worms.* He showed how worms enrich soil by bringing subsoil to the surface and mixing it with topsoil. An earthworm can ingest its own weight in soil every 24 hours, and Darwin estimated that from 10 to 18 tons of dry earth per acre pass through their intestines annually, thus bringing up potassium and phosphorus from the subsoil and also adding to the soil nitrogenous products from their own metabolism. Earthworms expose the mold to the air and sift it into small particles. They also drag leaves, twigs, and organic substances into their burrows, closer to the roots of plants. Their activities are important in aerating the soil. Darwin's views were at odds with those of his contemporaries, who thought earthworms were harmful to plants. Later research has amply confirmed Darwin's findings, and many countries now practice earthworm management.

The organs of excretion are **nephridia,** a pair of which occurs in each segment except the first three and the last one. Each one occupies parts of two successive segments (figure 11.17). A ciliated funnel, called a *nephrostome,* lies just anterior to an intersegmental septum and leads by a small ciliated tubule through the septum into the segment behind, where it connects with the main part of the nephridium. This part of the nephridium contains several loops of increasing size, which finally terminate in a bladderlike structure leading to an aperture, or nephridiopore, which opens to the outside near the ventral row of setae. Cilia draw fluid from the coelom into the nephrostome and tubule. In the tubule, water and salts are resorbed, forming a dilute urine that discharges to the outside through the nephridiopore.

The nervous system in oligochaetes is typical of all annelids. A pair of cerebral ganglia, which form the animal's brain, lies above the pharynx joined to a ventral nerve cord by a pair of connectives around the pharynx (see figure 11.15A).

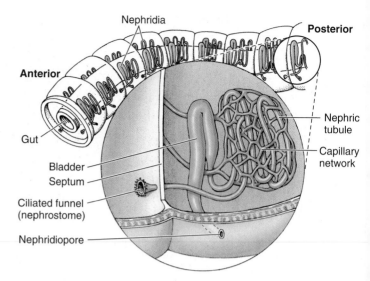

figure 11.17

Nephridium of an earthworm. Wastes are drawn into the ciliated nephrostome in one segment, then passed through loops of the nephridium, and expelled through the nephridiopore of the next segment.

The ventral nerve cord bears a pair of ganglia in each segment, giving off segmental nerves containing both sensory and motor fibers. For rapid escape movements, most annelids have one or more very large axons commonly called giant axons, or giant fibers, in the ventral nerve cord. Speed of conduction in these giant nerve fibers is much greater than that in small axons.

Earthworms are hermaphroditic and exchange sperm during copulation, which usually occurs at night. When mating, worms extend their anterior ends from their burrows and bring their ventral surfaces together (figure 11.18). They are held together by mucus secreted by each worm's **clitellum** and by special ventral setae, which penetrate each other's bodies in the regions of contact. Sperm are discharged and travel to the seminal receptacles of the other worm in its seminal grooves. After sperm exchange, the worms separate. Each worm secretes around its clitellum first a mucous tube and then a tough, chitinlike band that forms a **cocoon** (figure 11.18). The cocoon slides forward along the body. As it moves, eggs from the oviducts, albumin from skin glands, and sperm from the mate (stored in the seminal receptacles) are poured into it. Fertilization of eggs now occurs within the cocoon. When the cocoon slides off the head end of the worm, its ends close, producing a lemon-shaped body. Embryonic development occurs within the cocoon, and the form that hatches from the egg is a young worm similar to the adult. It does not develop a clitellum until it is sexually mature.

In the dorsal median giant axon of *Lumbricus,* which is 90 to 160 μm in diameter, speed of conduction has been estimated at 20 to 45 m/second, several times faster than in ordinary neurons of this species. This is also much faster than in polychaete giant axons, probably because the giant axons in earthworms are enclosed in myelinated sheaths. Speed of conduction may be altered by changes in temperature.

figure 11.18

Earthworm copulation and formation of egg cocoons. **A,** Mutual insemination; sperm from genital pore (segment 15) pass along seminal grooves to seminal receptacles (segments 9 and 10) of each mate. **B, C,** After the worms separate, the clitellum secretes first a mucous tube and then a tough band that forms a cocoon. The developing cocoon passes forward to receive eggs from oviducts and sperm from seminal receptacles. **D,** As the cocoon slips off over the anterior end, its ends close and seal. **E,** Cocoon is deposited near burrow entrance. **F,** Young worms emerge in 2 to 3 weeks. **G,** Two earthworms in copulation. Their anterior ends point in opposite directions as their ventral surfaces are held together by mucous bands secreted by the clitella.

Freshwater Oligochaetes

Freshwater oligochaetes are usually smaller than earthworms and have more conspicuous setae. They are also more mobile than earthworms and tend to have better-developed sense organs. They are generally benthic forms that creep across a substrate or burrow into soft mud. Both freshwater and marine oligochaetes provide an important food source for fishes. A few are ectoparasitic.

Some aquatic forms have gills. In some, gills are long, slender projections from the body surface. Others have ciliated anal gills (figure 11.19D), which they extend from their tubes and use to keep water moving. Most forms respire through their skin as do earthworms.

The chief foods are algae and detritus, which worms may gather by extending a mucus-coated pharynx. Burrowers swallow mud and digest the organic material. Some, such as

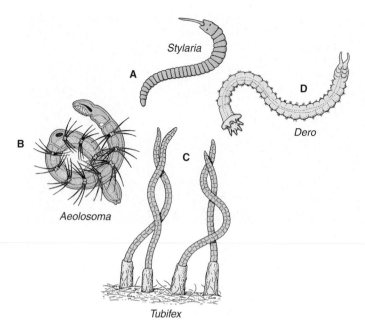

figure 11.19

Some freshwater oligochaetes. **A,** *Stylaria* has its prostomium drawn out into a long snout. **B,** *Aeolosoma* uses cilia around its mouth to sweep in food particles, and it buds off new individuals asexually. **C,** *Tubifex* lives head down in long tubes. **D,** *Dero* has ciliated anal gills.

Aeolosoma, are ciliary feeders that use currents produced by cilia at the anterior end of the body to sweep food particles into their mouth (figure 11.19B).

Class Hirudinida

Leeches, numbering over 500 species, occur predominantly in freshwater habitats, but a few are marine, and some have even adapted to terrestrial life in moist, warm areas. Most leeches are between 2 and 6 cm in length, but some are smaller, and some reach 20 cm or more (figure 11.20). They are found in a variety of patterns and colors—black, brown, red, or olive green. They are usually flattened dorsoventrally.

Leeches are highly modified descendants of a group of oligochaetes (see figure 11.1). It is possible to infer some steps in leech evolution from the morphology of members of Acanthobdellida and Branchiobdellida.

One genus of leech, *Acanthobdella,* has some characteristics of leeches and some of oligochaetes; it is sometimes separated from other leeches into a special class, Acanthobdellida, that characteristically has 27 segments, setae on the first five segments, and no anterior sucker. Another ancestral character of *Acanthobdella* is its five anterior coelomic compartments separated by septa; septa have disappeared in all other leeches. Branchiobdellida is composed of small annelids that are parasitic or commensal on crayfish and show similarities to both oligochaetes and leeches. They have 14 or 15 segments and bear a head sucker.

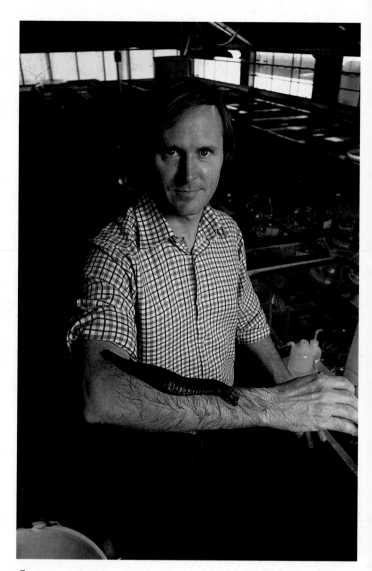

figure 11.20

The world's largest leech, *Haementeria ghilianii,* on the arm of Dr. Roy K. Sawyer, who found it in French Guiana, South America.

Form and Function

Members of Hirudinea, sometimes called "true leeches," have a fixed number of segments, usually 34, and typically have both an anterior and a posterior sucker (figure 11.21). They have no parapodia and lack septa. Their coelom has become filled with connective tissue and muscle, substantially reducing its effectiveness as a hydrostatic skeleton.

Many leeches live as carnivores on small invertebrates; some are temporary parasites, sucking blood from vertebrates; and some are permanent parasites, never leaving their host. Most leeches have a muscular, protrusible proboscis or a muscular pharynx with three jaws armed with teeth. They feed on body juices of their prey, penetrating its surface with their proboscis or jaws and sucking fluids with their powerful, muscular pharynx. Bloodsucking leeches secrete an anticoagulant in their saliva. Predatory leeches feed frequently, but those that feed on the blood of vertebrates consume large meals (up to

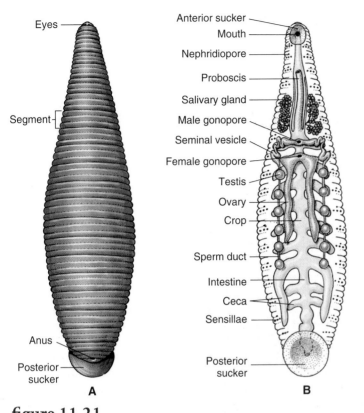

figure 11.21

Structure of a leech, *Placobdella*. **A,** External appearance, dorsal view. **B,** Internal structure, ventral view.

Labels for figure A:
Eyes
Segment
Anus
Posterior sucker
A

Labels for figure B:
Anterior sucker
Mouth
Nephridiopore
Proboscis
Salivary gland
Male gonopore
Seminal vesicle
Female gonopore
Testis
Ovary
Crop
Sperm duct
Intestine
Ceca
Sensillae
Posterior sucker
B

several times their body weight) and digest their food slowly. The slow digestion of their meals results from an absence of amylases, lipase, or endopeptidases in their gut secretions. In fact, they apparently depend mostly on bacteria in their gut for digestion of a blood meal.

Leeches are highly sensitive to stimuli associated with the presence of a prey or host. They are attracted by—and will attempt to attach to—an object smeared with appropriate host substances, such as fish scales, oil secretions, or sweat. Those that feed on the blood of mammals are attracted by warmth; terrestrial haemadipsid leeches of the tropics will converge on a person standing in one place.

Leeches are hermaphroditic but practice cross-fertilization during copulation. A penis or hypodermic impregnation transfers sperm to the mating partner. Leeches have a clitellum, but it is evident only during the breeding season. After copulation, the clitellum secretes a cocoon that receives eggs and sperm. Cocoons are buried underwater in mud, attached to submerged objects, or placed in damp soil (in the case of terrestrial species). Development is similar to that of oligochaetes.

figure 11.22

Hirudo medicinalis feeding on blood from a human arm.

Taxonomy of Phylum Annelida

Classification of annelids is based primarily on the presence or absence of parapodia, setae, and other morphological features. Because both oligochaetes and hirudineans (leeches) bear a clitellum, these two groups are often placed under the heading Clitellata (cli-tel-láta), whose members are called clitellates.

Polychaeta (pol´e-ke´ta) (Gr. *polys*, many, + *chaitē*, long hair) Mostly marine; head distinct and bearing eyes and tentacles; most segments have parapodia (lateral appendages) bearing tufts of many setae; clitellum absent; sexes usually separate; gonads transitory; asexual budding in some; trochophore larva usually present. Examples: *Nereis, Aphrodita, Glycera, Arenicola, Chaetopterus, Amphitrite, Riftia, Bonellia.*

Oligochaeta (ol´i-go-ke´ta) (Gr. *oligos*, few, + *chaitē*, long hair). Body with conspicuous segmentation; number of segments variable; setae few per segment; no parapodia; head not distinct; coelom spacious and usually divided by intersegmental septa; hermaphroditic; development direct, no larva; chiefly terrestrial and freshwater. Examples: *Lumbricus, Stylaria, Aeolosoma, Tubifex.*

Class Hirudinida (hir´u-din´i-da) (L. *hirudo*, leech, + *ida*, pl. suffix): leeches. Body has fixed number of segments (normally 34; 15 or 27 in some groups) with many annuli; oral and posterior suckers usually present; clitellum present; no parapodia; setae absent (except in Acanthobdellida); coelom closely packed with connective tissue and muscle; development direct; hermaphroditic; terrestrial, freshwater, and marine. Examples: *Hirudo, Placobdella, Macrobdella.*

For centuries, "medicinal leeches" (*Hirudo medicinalis*; figure 11.22) were used for blood letting because of the mistaken idea that body disorders and fevers were caused by an excess of blood. A 10 to 12 cm long leech can extend to a much greater length when distended with blood, and the amount of blood it can suck is considerable. Leech collecting and leech culture in ponds were practiced in Europe on a commercial scale during the nineteenth century. Wordsworth's poem "The Leech-Gatherer" was based on this use of leeches.

Leeches are once again being used medicinally. When fingers or toes are severed, microsurgeons can often reconnect arteries but not the more delicate veins. Leeches are used to relieve congestion until veins can grow back into a healing digit.

Phylum Sipuncula

Phylum Sipuncula (sī-punʹkū lə) (L. *sipunculus*, little siphon) consists of about 250 species of benthic marine worms, predominantly littoral or sublittoral. Sometimes called "peanut worms," they live sedentary lives in burrows in mud or sand (figure 11.23), occupy borrowed snail shells, or live in coral crevices or among vegetation. Some species construct their own rock burrows by chemical and perhaps mechanical means. Most species are restricted to tropical zones. Some are tiny, slender worms, but most range from 15 to 30 cm in length.

Sipunculans are not metameric, nor do they possess setae. Their head forms an **introvert,** which is crowned by ciliated tentacles surrounding the mouth (figure 11.23). They are largely deposit feeders, extending their introvert and tentacles from their burrow to explore and to feed. They have a cerebral ganglion, nerve cord, and pair of nephridia. They have a U-shaped gut lying within a spacious coelom. The coelomic

fluid contains red blood cells bearing a respiratory pigment called hemerythrin.

Sipunculan larvae are trochophores (see figure 10.6). Sexes are separate in most cases, and gonads develop seasonally.

Phylogeny and Adaptive Diversification

Phylogeny

Two worm groups, previously considered separate phyla, are evolutionarily derived polychaetes. Phylogenetic analysis clearly supports inclusion of the pogonophorans and echiurans within Annelida. All pogonophorans and some echiurans possess the paired epidermal setae characteristic of annelids. Segments, another annelid character, are visible in parts of the pogonophoran body, but have been completely lost in echiurans. However, there are some serially repeated structures in echiuran larvae, such as nerve cord ganglia and mucus glands, and adults of some species have multiple nephridia.

Sipunculans or peanut worms are a third phylum (Sipuncula) that some workers include within Annelida. Other possible placements for Sipuncula are as the sister taxon to Annelida or as the sister taxon to Mollusca. Two new phylogenetic analyses shed light on this problem. One study examined mitochondrial gene order and the other recorded the presence or absence of unique microRNAs in various annelids and sipunculans. In both studies, Sipuncula was placed outside Annelida with its most likely position as the sister taxon to Annelida. The microRNA study also included molluscs, but there were no shared microRNAs to support a sister-taxon relationship between Sipuncula and Mollusca. Thus, we depict Sipuncula as the sister taxon to Annelida.

The annelid fossil record begins in the early Cambrian with forms that are clearly polychaetes. Parapodia with upper and lower clusters of setae like those found in modern forms (see figure 11.2D) are visible. The fossil record supports the ancestral position of polychaetes and the derivation of the clitellates. From recent cladograms, we infer that the clitellates arose within the polychaetes, perhaps from mud-dwelling worms where parapodia and setae are reduced. Thus, polychaetes are not a monophyletic group and a taxonomic revision is needed to define new annelid classes.

Clitellates form a monophyletic group in all analyses; the group is diagnosed by presence of a clitellum. The clitellum in oligochaetes is always visible, but that in leeches is visible only during the reproductive season. Oligochaetes do not form a monophyletic group because leeches (hirudinideans) arose from within them. Oligochaetes await taxonomic revision. Class Hirudinida (leeches) is monophyletic and is composed of two small clades and a large clade, Hirudinea, comprising the "true leeches."

A recent reconstruction of annelid phylogeny identified *Chaetopterus* (see figure 11.10) and sipunculans as basal annelids, and divided polychaetes and clitellates into two clades: Errantia and Sedentaria. Errantia comprised active predatory worms such as *Nereis* (see figure 11.2) and *Hermodice* (see

figure 11.23

A, *Themiste*, a sipunculan, has extended its introvert. **B,** *Phascolosoma*, another sipunculan, has the introvert tucked into the trunk.

figure 11.7) among others, whereas tube-dwellers and in-faunal forms such as siboglinids, echiurans, *Amphitrite* (see figure 11.5), and *Arenicola* (see figure 11.11) were in Sedentaria. Clitellates were also placed in Sedentaria. It will be very interesting to see whether these results are confirmed by other analyses, because errantiate and sedentary worms were separated on morphological grounds many years ago.

The origins of a metameric (segmented) body are puzzling. True metamerism occurs in annelids, arthropods, and chordates. The placement of the annelids and arthropods in Protostomia and of the chordates in Deuterostomia makes it unlikely that segmentation is homologous among these three taxa because such an ancient origin for segmentation would require that many taxa later lost segmentation. However, a recent hypothesis suggests that ancient bilaterian gene regulatory networks involved in determining the anterior-posterior body axis, such as Notch and Wnt, were secondarily recruited to the development of body segments. In the three segmental lineages, this recruitment occurred independently and somewhat differently, but common genetic elements were used.

Within the protostomes, annelids are placed in clade Lophotrochozoa, whereas arthropods are in clade Ecdysozoa. In both clades, most phyla are not segmented, again making it unlikely that members of these two phyla inherited a segmented body plan from a common ancestor because many evolutionary losses are required. However, some research does support the idea that the common ancestor of all protostomes was segmented.

The selective advantage of a segmented body for annelids appears to lie in the efficiency of burrowing made possible by shape change in individual coelomic compartments of the hydrostatic skeleton. However, this explanation cannot be extended to the arthropods because, as Chapter 13 describes, the rigid exoskeleton of the arthropods prohibits shape change among segments, and the coelom is small in comparison to that of annelids. Clearly, there is much to learn about metamerism.

Adaptive Diversification

Annelids, particularly polychaetes, are very diverse in morphology and lifestyle. A basic adaptive feature in the evolution of annelids is their septal arrangement, producing fluid-filled coelomic compartments. Fluid pressure in these compartments is used as a hydrostatic skeleton in precise movements such as burrowing and swimming. Powerful circular and longitudinal muscles can flex, shorten, and lengthen the body. The basic body structure, particularly that of polychaetes, lends itself to great modification. As marine worms, polychaetes have a wide range of habitats in an environment that is not physically or physiologically demanding. Unlike earthworms, whose environment imposes strict physical and physiological demands, polychaetes have achieved a wide range of adaptive features. In polychaetes, the parapodia have been adapted in many ways and for a variety of functions, chiefly locomotion and respiration.

Feeding adaptations show great variation, from the sucking pharynx of oligochaetes and the chitinous jaws of carnivorous polychaetes to the specialized tentacles and radioles of particle feeders. The evolution of a trophosome to house the chemoautotrophic bacteria that provide nutrients to siboglinids is an adaptation to deep-sea life.

In leeches, many adaptations, such as suckers, cutting jaws, pumping pharynx, distensible gut, and the secretion of anticoagulants, are related to their predatory and blood-sucking habits.

■ Summary

Phylum Annelida is a large, cosmopolitan group containing marine polychaetes, earthworms and freshwater oligochaetes, and leeches. Certainly the most important structural innovation underlying diversification of this group is metamerism, division of the body into a series of similar segments, each containing a repeated arrangement of many organs and systems. The coelom is also highly developed in annelids, and this, together with the septal arrangement of fluid-filled compartments and a well-developed body-wall musculature, forms an effective hydrostatic skeleton for precise burrowing and swimming movements. Further metameric specialization occurs in arthropods (Chapter 13).

Polychaeta is a large paraphyletic group containing predominantly marine annelids. On each segment they have many setae, which are borne on paired parapodia. Parapodia show a wide variety of adaptations among polychaetes, including specialization for swimming, respiration, crawling, maintaining position in a burrow, pumping water through a burrow, and accessory feeding. Some polychaetes are mostly predaceous and have an eversible pharynx with jaws. Other polychaetes rarely leave the burrows or tubes in which they live. Members of this group exhibit several styles of deposit and filter feeding. Siboglinids (pogonophorans) were recently added to this class. These so-called beardworms have neither mouths nor guts; instead, they derive their nutrition from a mutualistic association with bacteria. Echiurans are burrowing marine worms recently placed within Polychaeta. Most are deposit feeders, with a proboscis anterior to their mouth. Some species bear epidermal setae. They lack segmentation. Polychaetes are dioecious, and have a primitive reproductive system, no clitellum, external fertilization, and a trochophore larva.

Clitellata is a monophyletic group comprising oligochaetes and the monophyletic Hirudinida (leeches). The presence of a reproductive structure called a clitellum defines members of this group. Oligochaeta, a paraphyletic taxon, contains earthworms and many freshwater forms; they have a small number of setae per segment (compared to Polychaeta) and no parapodia. They have a closed circulatory system, with a dorsal blood vessel serving as the main pumping organ. Paired nephridia occur in most segments. Earthworms contain the typical annelid nervous system: dorsal cerebral ganglia connected to a double, ventral nerve cord with segmental ganglia running the length of the worm. Oligochaetes are hermaphroditic and practice cross-fertilization. The clitellum plays an important role in reproduction; it secretes mucus to surround the worms during copulation, and it also secretes a cocoon to receive eggs and sperm and to serve as the site of fertilization and embryo development. A small, juvenile worm hatches from the cocoon.

Leeches (class Hirudinida) are mostly freshwater, although a few are marine and a few are terrestrial. They feed mostly on fluids; many are predators, some are temporary parasites, and a few are permanent parasites. The hermaphroditic leeches reproduce in a fashion similar to that of oligochaetes, with cross-fertilization and cocoon formation by the clitellum.

Sipunculans are small, burrowing marine worms with an eversible introvert at their anterior end. The introvert bears tentacles used for deposit feeding. Sipunculans are not segmented.

Embryological evidence places annelids with molluscs as lophotrochozoan protostomes. Sipunculans are currently considered the sister taxon to annelids.

The evolution of a segmented body plan is a very active research area.

■ Review Questions

1. What characteristics of phylum Annelida distinguish it from other phyla?
2. How are members of Clitellata distinguished from polychaetes?
3. Describe the annelid body plan, including the body wall, segments, coelom and its compartments, and coelomic lining.
4. Explain how the hydrostatic skeleton of annelids helps them burrow. How is burrowing efficiency increased by metamerism?
5. Describe three ways that various polychaetes obtain food.
6. Define each of the following: prostomium, peristomium, radioles, parapodium, setae.
7. Explain the functions of each of the following in earthworms: crop, gizzard, typhlosole, chloragogen cells.
8. Compare the main features of each of the following in each class of annelids: circulatory system, nervous system, excretory system.
9. Describe the functions of the clitellum and the cocoon.
10. How are freshwater oligochaetes generally different from earthworms?
11. Describe how leeches obtain food.
12. What are the main differences in reproduction and development among the three groups of annelids?
13. What is the largest siboglinid known? Where is it found, and how is it nourished?
14. What features of echiurans are shared with other annelids?
15. Where does a sipunculan live, and how does it collect food?
16. What is the significance of a trochophore larva for those phyla that possess it?

For Further Thought Review the extent of segmentation in polychaetes, leeches, siboglinids, echiurans, and sipunculans. Which morphological features must change to produce an unsegmented body cavity? Is there evidence of transitional stages for these features?

■ Selected References

See also general references on page 432.

Chipman, A. D. 2010. Parallel evolution of segmentation by co-option of ancestral gene regulatory networks. Bioessays **32**:60–70. *Elements of ancient genetic regulatory networks that determine body axes are independently recruited into the developmental pathways for segmentation in three bilaterian lineages.*

Cutler, E. B. 1995. The Sipuncula. Their systematics, biology, and evolution. Ithaca, New York, Cornell University Press. *The author tried to "bring together everything known about" sipunculans.*

Fischer, A., and U. Fischer. 1995. On the lifestyle and life-cycle of the luminescent polychaete *Odontosyllis enopla* (Annelida: Polychaeta). Invert. Biol. **114**:236–247. *If epitokes of this species survive their spawning swarm, they can return to a benthic existence.*

Halanych, K. M., T. D. Dahlgren, and D. McHugh. 2002. Unsegmented annelids? Possible origins of four lophotrochozoan worm taxa. Integ. and Comp. Biol. **42**:678–684. *A nice summary of morphological and molecular studies on classification of pogonophorans, echiurans, myzostomids, and sipunculans.*

McClintock, J. 2001. Blood suckers. Discover **22**:56–61 (Dec.). *Describes medical uses for leeches.*

Mirsky, S. 2000. When good hippos go bad. Sci. Am. **282**:28 (Jan.). Placobdelloides jaegerskioeldi *is a parasitic leech that breeds only in the rectum of hippopotamuses.*

Mwinyi, A., A. Meyer, C. Bleidorn, B. Lieb, T. Bartolomaeus, and L. Podsiadlowski. 2009. Mitochondrial genome sequence and gene order of *Sipunculus nudus* give additional support for an inclusion of Sipuncula in Annelida. BMC Genomics **10**:27. *Although the title suggests that sipunculans belong with annelids, data support a sister taxon relationship with Annelida.*

Pernet, B. 2000. A scaleworm's setal snorkel. Invert. Biol. **119**:147–151. Sthenelais berkeleyi *is an apparently rare but large (20 cm) polychaete that buries its body in sediment and communicates with the water above just by its anterior end. Ciliary movement on parapodia pumps water into the burrow for ventilation. The worm remains immobile for long periods, except when prey comes near; it then rapidly everts its pharynx to capture prey.*

Prpic, N. 2008. Parasegmental appendage homology in annelids and arthropods and the homology of parapodia and arthropodia. Frontiers in Zoology **5**:17. *Presents* the hypothesis that the ancestral protostome had both parasegments and parasegmental appendages.

Rouse, G. W., S. K. Gofferedi, and R. C. Vrijenhoek. 2004. *Osedax*: Bone-eating marine worms with dwarf males. Science **305**:668–671. *Fascinating worms that feed on carcasses of whales.*

Seaver, E. C. 2003. Segmentation: mono- or polyphyletic. Int. J. Dev. Biol. **47**:583–595. *Preliminary comparisons of the segmentation process in annelids, arthropods, and chordates suggest that annelids and arthropods do not share mechanisms of segmentation, but vertebrates and arthropods may share some mechanisms.*

Sperling, E. A., J. Vinther, V. N. Moy, B. M. Wheeler, M. Se´mon, D. E. G. Briggs, and K. J. Peterson. 2009. MicroRNAs resolve an apparent conflict between annelid systematics and their fossil record. Proc. R. Soc. B. **276**:4315–4322. *Polychaetes with parapodia represent the ancestral condition for annelids; sipunculans are the annelid sister taxon.*

Struck, T. H., C. Paul, N. Hill, S. Hartmann, C. Hösel, M. Kubes, B. Lieb, A. Meyers, R. Tiedemann, G. Purschke, and C. Bleidorn. 2011. Phylogenomic analyses unravel annelid evolution. Nature **471**:95–98. *A new analysis of annelid evolution reconstructs two old groups, errantiate and sedentary worms, as clades.*

■ Custom Website

The *Animal Diversity* Online Learning Center is a great place to check your understanding of chapter material. Visit www.mhhe.com/hickmanad6e for access to key terms, quizzes, and more! Further enhance your knowledge with Web links to chapter-related material.

Explore live links for these topics:

Classification and Phylogeny of Animals
Phylum Annelida
Class Polychaeta
Class Oligochaeta

Class Hirudinea
Phylum Pogonophora
Phylum Echiura
Phylum Sipuncula

Smaller Ecdysozoans

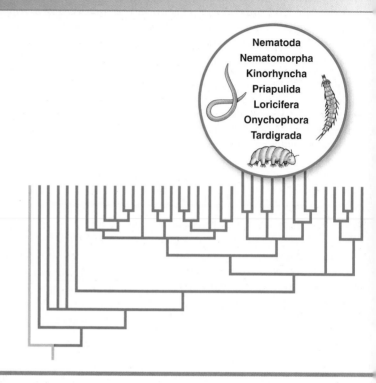

Nematoda
Nematomorpha
Kinorhyncha
Priapulida
Loricifera
Onychophora
Tardigrada

A World of Nematodes

Without any doubt, nematodes are the most important smaller ecdysozoan animals in terms of both numbers and their impact on humans. Nematodes are abundant over most of the world, and yet most people are only occasionally aware of them as parasites of humans or of pets. We are not aware of the billions of these worms in the soil, in ocean and freshwater habitats, in plants, and in all kinds of other animals. Their dramatic abundance moved N.A. Cobb to write in 1914:

> If all the matter in the universe except the nematodes were swept away, our world would still be dimly recognizable, and if, as disembodied spirits, we could then investigate it, we should find its mountains, hills, vales, rivers, lakes and oceans represented by a thin film of nematodes. The location of towns would be decipherable, since for every massing of human beings there would be a corresponding massing of certain nematodes. Trees would still stand in ghostly rows representing our streets and highways. The location of the various plants and animals would still be decipherable, and, had we sufficient knowledge, in many cases even their species could be determined by an examination of their erstwhile nematode parasites.[1]

Male *Trichinella spiralis*, a nematode.

[1]N. A. Cobb. 1914. *Yearbook of the United States Department of Agriculture, 1914.*

rotostome animals include flatworms, roundworms, molluscs, annelids, and arthropods, among many other taxa (see cladogram on inside front cover). Many protostomes, such as annelids, roundworms, and arthropods, possess a **cuticle,** a nonliving external layer secreted by the epidermis. A firm cuticle surrounding the body wall, like that present in roundworms and arthropods, restricts growth. As the body increases in size, the cuticle is molted and the outer layer shed, a process called **ecdysis.**

Protostome phyla are divided between two large clades: Lophotrochozoa and Ecdysozoa. Ecdysozoa (figure 12.1) comprises those taxa that molt the cuticle as they grow. Where it has been studied, molting is regulated by the hormone **ecdysone;** biologists assume that a homologous set of biochemical steps regulates molting among all ecdysozoans. Ecdysozoan taxa, other than loriciferans, were first united as a clade in phylogenies based on molecular characters.

As with lophotrochozoan phyla, ecdysozoans do not share a common body plan. Members of Nematoda, Nematomorpha, and Kinorhyncha have pseudocoelomate bodies.

Members of Priapulida have not been carefully studied, but they are assumed to be pseudocoelomate. The pseudocoelom is used as a hydrostatic skeleton in nematodes, kinorhynchs, and perhaps in priapulids. Within Loricifera, species apparently vary in body plan: some are described as pseudocoelomate and others appear acoelomate. Members of clade Panarthropoda have coelomate bodies, but their coeloms are small. Panarthropoda is an enormous group of animals containing three phyla: Onychophora, Tardigrada, and Arthropoda.

Arthropoda, the largest phylum in terms of numbers of described species, forms the subject of Chapter 13. This chapter describes all other ecdysozoan phyla.

■ Phylum Nematoda: Roundworms

Nematodes (Gr., *nematos,* thread) occur in nearly every conceivable kind of ecological niche. Approximately 12,000 species have been named, but Libbie Hyman estimated that if all

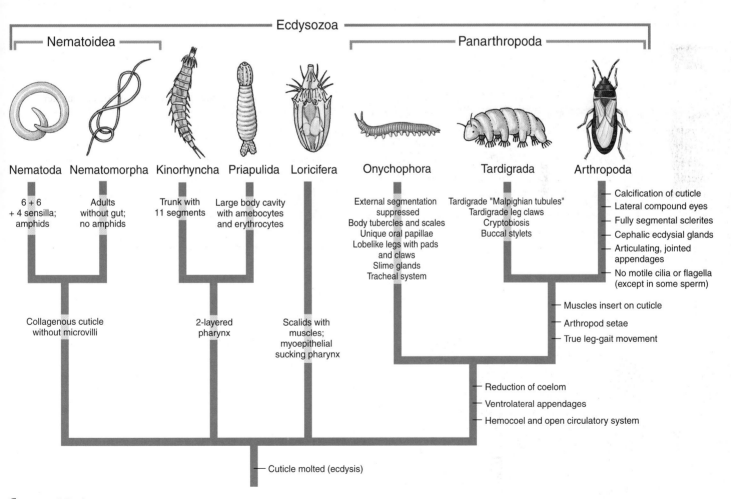

figure 12.1

Cladogram depicting one hypothesis for relationships among ecdysozoan phyla. Characters shown are subsets of those in Nielsen (1995), Neuhaus and Higgins (2002), and Brusca and Brusca (2003); the nematode character "6 + 6 + 4 sensilla" refers to the anterior rings of sensory papillae.

species were known, the number would be nearer 500,000.[2] They live in the sea, in fresh water, and in soil, from polar regions to tropics, and from mountaintops to the depths of seas. Good topsoil may contain billions of nematodes per acre. Nematodes also parasitize virtually every type of animal and many plants. The effects of nematode infestation on crops, domestic animals, and humans make this phylum one of the most important of all parasitic animal groups.

Form and Function

Most nematodes are under 5 cm long, and many are microscopic, but some parasitic nematodes are over a meter in length.

Use of the pseudocoelom as a hydrostatic skeleton is a conspicuous feature of nematodes, and we can best understand much of their functional morphology in the context of the high **hydrostatic pressure** within the pseudocoelom.

The outer body covering is a relatively thick, noncellular cuticle, secreted by the underlying epidermis (**hypodermis**). A cuticle is of great functional importance, serving to contain the high hydrostatic pressure exerted by fluid in the pseudocoelom. Layers of the cuticle are primarily composed of **collagen,** a structural protein present throughout the Metazoa and abundant in vertebrate connective tissue. Crisscrossing fibers form three of the layers, conferring some longitudinal elasticity on the worm but severely limiting its capacity for lateral expansion.

Beneath the hypodermis is a layer of longitudinal muscles. The muscles are arranged in four bands, or quadrants, separated by four epidermal cords that project inward to the pseudocoelom (figure 12.2). There are no circular muscles in the body wall. Another unusual feature of body-wall muscles in nematodes is that the muscles extend processes to synapse with nerve cords, rather than nerves extending an axon to synapse with muscle.

The fluid-filled pseudocoelom, in which internal organs lie, constitutes a hydrostatic skeleton. Many invertebrates use hydrostatic skeletons, which lend support by transmitting the force of muscle contraction to the enclosed, noncompressible fluid. Normally, muscles are arranged antagonistically so that, as movement is effected in one direction by contraction of one group of muscles, movement in the opposite direction is effected by contraction of an antagonistic set of muscles. However, nematodes do not have circular body-wall muscles to antagonize the longitudinal muscles; therefore, the cuticle must serve that function. As muscles on one side of the body contract, they compress the cuticle on that side, and the force of

figure 12.2

A, Structure of a nematode as illustrated by an *Ascaris* female. *Ascaris* has two ovaries and uteri, which open to the outside by a common genital pore. **B,** Cross section. **C,** Relationship of muscle cells to hypodermis and hypodermal cord.

[2]Hyman, L. H. 1951. The invertebrates, vol. III. Acanthocephala, Aschelminthes, and Entoprocta, the pseudocoelomate Bilateria. New York, McGraw-Hill Book Company.

CHARACTERISTICS
of Phylum Nematoda

1. Unique sensory **amphids** or **phasmids**
2. Marine, freshwater, and terrestrial
3. Free-living and parasitic
4. Body bilaterally symmetrical
5. Triploblastic body
6. **Pseudocoelom** functions as a hydrostatic skeleton
7. Nonliving, flexible cuticle is molted
8. Digestive system complete; muscular pharynx is triradiate in cross-section
9. Body wall has longitudinal muscles only
10. Ring of nerve tissue with ganglia around pharynx, dorsal and ventral nerve cords
11. Sensory system includes papillae and setae for touch; anterior amphids and posterior phasmids are likely chemoreceptors
12. Parthenogenesis occurs rarely
13. Sexes separate in most; fertilization is internal, but sperm are ameboid, not flagellated; cleavage pattern unique
14. Excretory system may include gland cells and canals, opens via excretory pore
15. No respiratory or circulatory systems

the contraction is transmitted (by the fluid in the pseudocoelom) to the other side of the nematode, stretching the cuticle on that side. This compression and stretching of the cuticle serve to antagonize the muscle and are the forces that return the body to resting position when the muscles relax; this action produces the characteristic thrashing motion of nematode movement. An increase in the efficiency of this system can be achieved only by an increase in hydrostatic pressure. Consequently, the hydrostatic pressure in a nematode's pseudocoelom is much higher than is usually found in animals that have both hydrostatic skeletons and antagonistic muscle groups.

The gut tube of a nematode comprises a mouth, a muscular **pharynx**, a long, nonmuscular intestine, a short rectum, and a terminal anus (figure 12.2). The cylindrical pharynx has radial muscles that insert on the cuticular lining of its lumen and on a basement membrane at its periphery. At rest, the lumen is closed. When muscles in the anterior of the pharynx contract, they open its lumen and suck food inside. Relaxation of muscles anterior to the food mass closes the pharyngeal lumen and forces the food posteriorly toward the intestine. The intestine is only one cell layer thick and has no muscles. Food matter moves posteriorly in the intestine by body movements and as additional food is passed into the intestine from the pharynx. Defecation is accomplished by muscles that simply pull the anus open, and pseudocoelomic pressure surrounding the rectum provides expulsive force.

Adults of many parasitic nematodes have an anaerobic energy metabolism; a Krebs cycle and cytochrome system characteristic of aerobic metabolism are absent. They derive energy through glycolysis and some additional electron-transport sequences. Interestingly, some free-living nematodes and free-living

stages of parasitic nematodes are obligate aerobes (require oxygen) and have a Krebs cycle and a cytochrome system.

A ring of nerve tissue and ganglia around the pharynx gives rise to small nerves to the anterior end and to two nerve cords, one dorsal and one ventral. Some sensory organs around the lips and around the posterior end are rather elaborate.

In 1963, Sydney Brenner initiated some extremely fruitful research by studying a free-living nematode, *Caenorhabditis elegans*. This small worm has now become one of the most important experimental models in biology. The origin and lineage of all the cells in its body (959) have been traced from zygote to adult, and the complete "wiring diagram" of its nervous system is known—all of the neurons and all of the connections between them. The nematode's genome has been completely mapped, and scientists have sequenced its entire genome of 3 million bases containing 19,820 genes. Many basic discoveries about gene function, such as how genes encode the proteins essential for programmed cell death, have been and will be made using *C. elegans*.

Most nematodes are dioecious. Males are smaller than females, and their posterior end usually bears a pair of copulatory spicules (figure 12.3). Fertilization is internal, and shelled zygotes or embryos are stored in the uterus until deposition. After embryonic development, a juvenile worm hatches from an egg. There are four juvenile stages, each separated by a shedding of the cuticle. Many parasitic nematodes have free-living juvenile stages, and others require an intermediate host to complete their life cycles.

The copulatory spicules of male nematodes are not true intromittent organs, since they do not conduct sperm, but are another adaptation to high internal hydrostatic pressure. Spicules must hold the vulva of a female open while ejaculatory muscles in the male reproductive tract overcome the hydrostatic pressure in the female and rapidly inject sperm into her reproductive tract. Furthermore, nematode spermatozoa are unique in the animal kingdom in having no flagellum or acrosome. Within the female reproductive tract, sperm become ameboid and move by pseudopods.

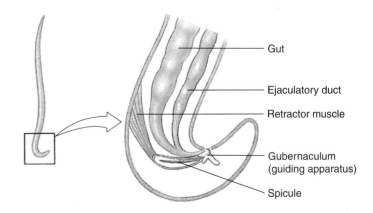

figure 12.3
Posterior end of a male nematode.

Representative Nematode Parasites

Nearly all vertebrates and many invertebrates are parasitized by nematodes, a number of which are very important pathogens of humans and domestic animals. A few nematodes are common in humans in North America (table 12.1), but they and many others usually abound in tropical countries. We have space to mention only a few.

Ascaris lumbricoides: Large Roundworm of Humans

The large human roundworm, *Ascaris lumbricoides* (Gr. *askaris,* intestinal worm), is one of the most common worm parasites of humans (figure 12.4A). Recent surveys have shown a prevalence of up to 25% in some areas of the southeastern United States, and 1.27 billion people worldwide likely have this worm. *Ascaris suum,* a parasite of pigs, is morphologically similar and was long considered the same species (figure 12.4B). Females of both species are up to 30 cm in length and can produce 200,000 eggs a day. Adults occupy their host's small intestine, and eggs leave the host's body in feces. These worms are extremely resistant to adverse conditions, other than direct sunlight and high temperatures, and can survive for months or years in soil.

A

B

figure 12.4

A, Intestinal roundworm *Ascaris lumbricoides,* male and female. Male *(top)* is smaller and has a characteristic sharp kink in the end of its tail. The females of this large nematode may be over 30 cm long. **B,** Intestine of a pig, nearly completely blocked by *Ascaris suum.* Such heavy infections are also fairly common with *A. lumbricoides* in humans.

Table 12.1 Common Parasitic Nematodes of Humans in North America

Common and scientific names	Means of infection; prevalence in humans
Hookworm (*Ancylostoma duodenale* and *Necator americanus*)	Contact with soil infested with juveniles that burrow into skin; common in southern states
Pinworm (*Enterobius vermicularis*)	Inhalation of dust containing ova and by contamination with fingers; most common worm parasite in United States
Intestinal roundworm (*Ascaris lumbricoides*)	Ingestion of ova containing embryos in contaminated food; common in rural areas of Appalachia and southeastern states
Trichina worm (*Trichinella spiralis*)	Ingestion of infected muscle; occasional in humans throughout North America
Whipworm (*Trichuris trichiura*)	Acquired through unhygienic habits or ingestion of contaminated food; usually common wherever *Ascaris* is found

Other ascarids are common in wild and domestic animals. Species of *Toxocara,* for example, infect dogs and cats. Their life cycle is generally similar to that of *Ascaris,* but juveniles often do not complete their tissue migration in adult dogs, remaining in the host's body in a stage of arrested development. Pregnancy in the female dog, however, stimulates juveniles to wander, and they infect the embryos in the uterus. Puppies are then born with worms. These ascarids also survive in humans but do not complete their development, leading to an occasionally serious condition in children called visceral larva migrans. This condition is a good reason for pet owners to practice hygienic disposal of canine wastes!

When humans eat uncooked vegetables contaminated with shelled juveniles, or when children put soiled fingers or toys in their mouths, the consumed juveniles hatch in the host's intestine. They penetrate the intestinal wall and travel in the blood through the heart to the lungs, where they enter alveoli. At this stage, they may cause a serious pneumonia. From the alveoli, juveniles make their way up the bronchi, trachea, and pharynx to be swallowed, eventually reaching the intestine again, where they grow to maturity. In the intestine, worms cause abdominal symptoms and allergic reactions, and in large numbers they may block the intestine.

Hookworms

Hookworms are so named because their anterior end curves dorsally, suggesting a hook. The most common species is *Necator americanus* (L. *necator,* killer), whose females are up to 11 mm long. Males can reach 9 mm in length. Large plates in the mouth (figure 12.5) cut into the intestinal mucosa of the host, where they suck blood and pump it through their intestine, partially digesting it and absorbing the nutrients. They suck much more blood than they need for food, and heavy infections cause anemia in the patient. Hookworm disease in children may retard mental and physical growth and cause a general loss of energy.

Shelled embryos leave the host's body in feces, and juveniles hatch in the soil, where they live on bacteria. When human skin contacts infested soil, the juveniles burrow through the skin to the blood, reaching the lungs and finally the intestine in a manner similar to that described for *Ascaris.*

Trichina Worm

Trichinella spiralis (Gr. *trichinos,* of hair, + *-ella,* diminutive) is one of the species of tiny nematodes responsible for the potentially lethal disease trichinosis. Adult worms burrow in the mucosa of the host's small intestine, where females produce living young. Juveniles penetrate blood vessels and are carried throughout the body, where they occupy almost any tissue or body space. Eventually, they penetrate skeletal muscle cells, becoming one of the largest known intracellular parasites. Juveniles cause astonishing redirection of gene expression in their host cell, which loses its striations and becomes a **nurse cell** that nourishes the worm (figure 12.6). When meat containing live juveniles is swallowed, the worms are liberated into the host's intestine, where they mature.

Trichinella spp. can infect a wide variety of mammals in addition to humans, including pigs, rats, cats, and dogs. Pigs become infected by eating pork scraps infested with juveniles or by eating infected rats. *Trichinella spiralis* has four sibling species, which differ in geographic distribution, infectivity to different host species, and freezing resistance. Heavy infections may cause death, but lighter infections are much more common—about 25 cases of trichinosis are reported per year in the United States.

Pinworms

Pinworms, *Enterobius vermicularis* (Gr. *enteron,* intestine, + *bios,* life), cause relatively little disease, but they are the most common worm parasite in the United States, occurring in an estimated 30% of children and 16% of adults. Adult parasites (figure 12.7A) occupy the large intestine and

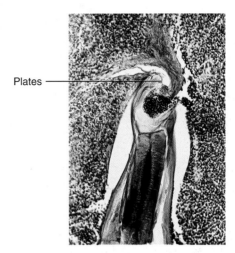

Plates

figure 12.5

Section through anterior end of hookworm attached to dog intestine. Note the cutting plates of the mouth pinching off bits of mucosa from which the thick muscular pharynx sucks blood. Esophageal glands secrete an anticoagulant to prevent blood clotting.

Collagenous nurse cell capsule

Juvenile worm

Enlarged muscle cell nucleus

figure 12.6

Section of a human muscle infected with the trichina worm, *Trichinella spiralis.* Juveniles lie within muscle cells that the worms have induced to transform into nurse cells (commonly called cysts). An inflammatory reaction is evident around the nurse cells. Juveniles may live 10 to 20 years, and nurse cells eventually may calcify.

cecum. Females reach about 12 mm in length and migrate to the anal region at night to lay their eggs (figure 12.7B). Scratching the resultant itch effectively contaminates hands and bedclothes. Eggs develop rapidly and become infective within 6 hours at body temperature. After they are swallowed, they hatch in the duodenum, and the worms mature in the large intestine.

figure 12.7

Pinworms, *Enterobius vermicularis*. **A,** Female worm from human large intestine (slightly flattened in preparation), magnified about 10 times. **B,** Group of shelled juveniles of pinworms, which are usually discharged at night around the anus of the host, who may get fingernails and clothing contaminated by scratching during sleep. This may be the most common and widespread of all human worm parasites.

Most intestinal roundworm infections are usually diagnosed by examining a small bit of feces under a microscope and finding characteristic shelled embryos or juveniles. However, pinworm eggs are not often found in feces because females deposit them on the skin around the anus. For pinworms, the "Scotch tape method" is more effective. The sticky side of cellulose tape is applied around the anus to collect the shelled embryos, and then the tape is placed on a glass slide and examined under a microscope. Several drugs are effective against this parasite, but all members of a family should be treated at the same time, since the worm easily spreads through a household.

Members of this order of nematodes have **haplodiploidy,** a characteristic shared with a few other animal groups, notably many hymenopteran insects (see p. 278). Males are haploid and are produced parthenogenetically; females are diploid and arise from fertilized eggs.

Filarial Worms

At least eight species of filarial nematodes infect humans, and some cause major diseases. Some 120 million people in tropical countries are infected with *Wuchereria bancrofti* (named for Otto Wucherer) or *Brugia malayi* (named for S. L. Brug), placing these species among the scourges of humanity. The worms infect the lymphatic system, and females may be as long as 100 mm.

figure 12.8

Elephantiasis of leg caused by adult filarial worms of *Wuchereria bancrofti*, which live in lymph passages and block the flow of lymph. A mosquito ingests tiny juveniles, called microfilariae, with its blood meal. Microfilariae then develop to the infective stage and can be transmitted by the mosquito to a new host.

Disease symptoms include inflammation and obstruction of the lymphatic system. Female worms release tiny living microfilariae into the blood and lymph of the host. These microfilariae are then ingested by mosquitos as the insects feed, and they develop in mosquitos to the infective stage. When the mosquito again feeds on a human, the microfilariae escape from the mosquito and penetrate the wound made by the mosquito bite.

Long and repeated exposure to filarial worms can lead to elephantiasis, a condition marked by excessive growth of connective tissue and enormous swelling of affected parts, such as the scrotum, legs, arms, and more rarely, the vulva and breasts (figure 12.8).

figure 12.9

Dirofilaria immitis in a dog's heart. This nematode is a major menace to the health of dogs in North America. Adults live in the heart, and juveniles circulate in the blood, where they are ingested and transmitted by mosquitos.

Another filarial worm causes river blindness (onchocerciasis) and is carried by blackflies. It infects more than 37 million people in parts of Africa, the Mideast, Central America, and South America.

The most common filarial worm in the United States is probably the dog heartworm, *Dirofilaria immitis* (figure 12.9). Carried by mosquitos, it also can infect other canids, cats, ferrets, sea lions, and occasionally humans. Along the Atlantic and Gulf Coast states and northward along the Mississippi River throughout the midwestern states, prevalence in dogs reaches 45%. It occurs in other states at a lower prevalence. This worm causes a very serious disease among dogs, and no responsible owner should fail to provide "heartworm pills" for a dog during mosquito season.

Phylum Nematomorpha

The popular name for the Nematomorpha (nem´a-to-mor´fa) (Gr. *nema, nematos,* thread, + *morphē,* form) is "horsehair worms," based on an old superstition that the worms arise from horsehairs that happen to fall into water, and indeed the worms do resemble hairs from a horse's tail. They range from 0.5 to 3 mm in diameter and may reach 1 m in length. They were long included in phylum Nematoda because both groups have a cuticle, epidermal cords, longitudinal muscles only, and a similar nervous system pattern. Several recent studies indicate that nematomorphs are the sister taxon to nematodes; the two groups are united in clade Nematoidea.

figure 12.10

A horsehair worm, or nematomorph, *Paragordius tricuspidatus,* emerges from the body of a European cricket, *Nemobius sylvestris.* Nematomorphs are very long and very thin. Their pharynx is usually a solid cord of cells and appears nonfunctional. *Paragordius,* whose pharynx opens through to the intestine, is unusual in this respect and also in the possession of a photosensory organ ("eye").

About 320 species of horsehair worms have been named. Worldwide in distribution, horsehair worms are free-living as adults and parasitic in arthropods as juveniles (figure 12.10). Early larval forms of some species have a striking resemblance to priapulids. Adults have a vestigial digestive tract and absorb organic molecules through the vestigial gut and body wall, much as do juveniles. They can live almost anywhere in wet or moist surroundings if oxygen is adequate.

Life cycles of nematomorphs are poorly known. In the cosmopolitan genus, *Gordius* (named for an ancient king who tied an intricate knot), juveniles may encyst on vegetation likely to be eaten by a grasshopper or other arthropod. Gordiid larval stages also have hooks or stylets that may be used to bore into a host, perhaps via the integument or the gut lining. In other cases, the gordiid may infect the host via its drinking water. Larvae encyst in the host; in some cases, it seems that development continues after the first host is eaten by a second host. In the marine nematomorph, *Nectonema* (Gr. *nektos,* swimming, + *nema,* a thread), juveniles occur in hermit crabs and other crabs.

After several months in the hemocoel (p. 241) of an arthropod host, juveniles complete a single molt and emerge into water as mature adults. If the host is a terrestrial insect, the parasite somehow stimulates the host to seek water. Worms do not emerge from the host unless water is nearby.

Phylum Kinorhyncha

There are only 179 described species of Kinorhyncha (Gr. *kineo,* to move, + *rhynchos,* beak or snout). These tiny marine worms, usually less than 1 mm long, live in mud or sandy mud. They have no external cilia, but are covered with a cuticle that is molted during growth. The body is divided into a head, a neck, and a trunk with 11 external segments (figure 12.11). Kinorhynchs burrow into mud by extending the head, anchoring it by its recurved spines (**scalids**), and drawing their body forward until their head is retracted. They feed on organic sediment in mud, and some feed on diatoms. Kinorhynchs are dioecious.

Phylum Priapulida

Priapulida (pri´a-pyu´li-da) (Gr. *priapos,* phallus, + *-ida,* pl. suffix) is a small group (only 16 species) of marine worms found chiefly in colder water in both hemispheres. Their cylindrical bodies are rarely more than 12 to 15 cm long. Most are burrowing predaceous animals that usually orient themselves upright in mud or sand with their mouth at the surface.

Long considered pseudocoelomate, priapulids were judged coelomate when nuclei were found in membranes lining the body cavity, the membranes thus representing a peritoneum. However, electron microscopy showed that the nuclei of their muscle cells were peripheral, and that the muscles secreted an extracellular membrane. The muscle nuclei and extracellular membrane gave the appearance of an epithelial lining.

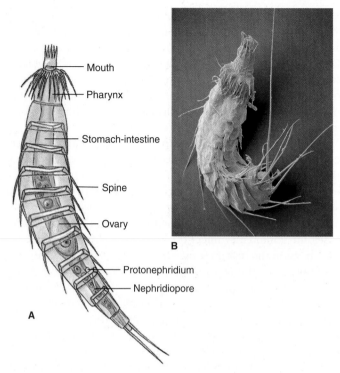

figure 12.11

A, *Echinoderes,* a kinorhynch, is a minute marine worm. Segmentation is superficial. The head with its circle of spines is retractile. **B,** Colored scanning electron micrograph (SEM) of the kinorhynch *Antigomonas* sp.

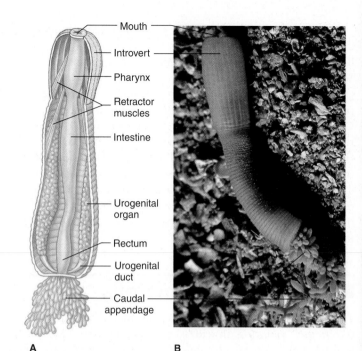

figure 12.12

A, Major internal structures of *Priapulus.* **B,** *Priapulus caudatus* from Lurefjord, Norway.

The body includes a retractable introvert, a trunk, and usually one or two caudal appendages (figure 12.12). The eversible introvert usually ends with rows of curved spines around the mouth; it is used to sample the surroundings as well as to capture small, soft-bodied prey. Priapulids are not metameric. A chitinous cuticle, molted periodically, covers the body.

Priapulids have no circulatory system, but coelomocytes in their body fluids contain the respiratory pigment hemerythrin. There are a nerve ring and ventral cord with nerves as well as a protonephridial tubule that serves also as a gonoduct. The anus and urogenital pores open at the end of the trunk.

Sexes are separate, fertilization is external, and embryogenesis is only poorly known.

■ Phylum Loricifera

Loricifera (L. *lorica,* corselet, + *fero,* to bear) is a very recently described phylum of animals (1983). Its members occupy spaces between grains of marine gravel, to which they cling tightly. Although loriciferans were described from specimens collected off the coast of France, they are apparently widely

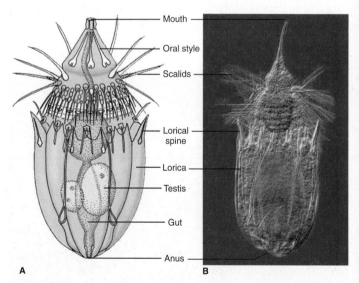

figure 12.13

A, Dorsal view of adult loriciferan, *Nanaloricus mysticus,* showing internal features. **B,** Live animal, 0.3 mm.

distributed. About 100 species have been found. They have oral styles and scalids rather similar to those of the kinorhynchs, and the entire forepart of the body can be retracted into the circular lorica (figure 12.13). Loriciferans are dioecious, and their life cycles contain multiple stages. Larvae molt as they grow.

Clade Panarthropoda

Panarthropoda contains Arthropoda and two allied phyla, Onychophora and Tardigrada (see figure 12.1). In onychophorans and arthropods, a coelom develops by schizocoely, but coelom formation has been described as enterocoelic in tardigrades. In all three phyla, the main coelomic cavity later fuses with the blastocoel to form a new cavity called a **hemocoel,** or mixocoel. The hemocoel is lined by an extracellular matrix, not the mesodermal peritoneum that originally lined the coelom. Blood from the open circulatory system enters the hemocoel and surrounds the internal organs. There is no separation of extracellular fluid into blood plasma and lymph, so the combined fluid is often called **hemolymph.** A muscular heart is present, but tubular blood vessels occur in only one part of the body; blood enters and leaves the hemocoel through blood vessels. Small coelomic cavities surround a few organs in other parts of the body.

Phylum Onychophora

Members of phylum Onychophora (on-i-kof´o-ra) (Gr. *onyx,* claw, + *pherō,* to bear) are called "velvet worms" or "walking worms." About 70 species of these caterpillar-like animals, 0.5 to 15 cm long, occupy rain forests and other tropical and semitropical leafy habitats.

The fossil record of onychophorans describes a 500-million-year history. They were originally marine animals and were probably far more common than they are now. Extant species are all terrestrial and are extremely retiring, being active only at night or when the air is nearly saturated with moisture. Most are predators, feeding on insects, snails, and worms that they entangle in slime extruded from slime glands. Once immobilized by the slime, prey are digested by salivary enzymes and only the liquid portion is consumed.

Onychophorans are covered by a soft cuticle that contains chitin and protein. Their head (figure 12.14A) bears a pair of flexible antennae with annelid-like eyes at the base. Their wormlike bodies are carried on 14 to 43 pairs of stumpy, unjointed legs (figure 12.14B), each ending with a flexible pad and two claws.

Onychophorans are air breathers, using a **tracheal system** that connects with pores scattered over their body. This tracheal system, although similar to that of arthropods, probably evolved independently. Other arthropod-like characteristics include an open circulatory system with a tubular heart, a hemocoel for a body cavity, a large brain, and a cuticle that is molted. However, unlike arthropods, velvet worms molt the cuticle in patches. Annelid-like characteristics include segmentally arranged nephridia, a muscular body wall, and pigment-cup ocelli.

Nearly all onychophorans are dioecious. In some species, there is a placental attachment between mother and young, and young are born as juveniles (viviparous); others have young that develop in the uterus without attachment (ovoviviparous). Two Australian genera are oviparous and lay shell-covered eggs in moist places.

Phylum Tardigrada

Tardigrada (tar-di-grā´ da) (L. *tardus,* slow, + *gradus,* step), or "water bears," are minute forms usually less than a millimeter in length. About 900 species have been described. Some live in freshwater and marine habitats, but most species live on land, occupying a film of water surrounding mosses and lichens.

The tardigrade body bears eight short, unjointed legs, each with claws (figure 12.15). Unable to swim, the animal creeps awkwardly, clinging to the substrate with its claws. There is a wide range of body shapes. A pair of sharp stylets and a sucking pharynx are adaptations for piercing

figure 12.15

Scanning electron micrograph of *Echiniscus maucci,* phylum Tardigrada. This species is 300 to 500 µm long. Unable to swim, it clings to moss or water plants with its claws, and when water is scarce, it enters a state of suspended animation and "sleeps away" the drought.

Antenna

Oral papilla

Oral lobes

First leg

A Ventral view of head   B

figure 12.14

Peripatus, a caterpillar-like onychophoran, has characteristics in common with both annelids and arthropods. **A,** Ventral view of head showing the region around the mouth. **B,** *Peripatus* in its natural habitat.

and sucking plant cells or small prey, such as nematodes and rotifers.

A body covering of nonchitinous cuticle is molted several times during the life cycle. As in arthropods, muscle fibers are attached to the cuticular exoskeleton, and the body cavity is a hemocoel.

The nervous system is annelid in type and surprisingly complex. Some species have a pair of eyespots. Circulatory and respiratory organs are lacking.

One of the most intriguing features of terrestrial tardigrades is their capacity to enter a state of suspended animation, called cryptobiosis, during which metabolism is virtually imperceptible. Under gradual drying conditions, they reduce the water content of their body from 85% to only 3%, movement ceases, and the body becomes barrel shaped. In a cryptobiotic state, tardigrades can withstand harsh environmental conditions—temperature extremes, ionizing radiation, oxygen deficiency, etc.—and many survive for years. Activity resumes when moisture is again available.

Tardigrades may reproduce by parthenogenesis or sexually. Females may deposit their eggs in the old cuticle as they molt or attach them to a substrate. Embryonic formation of the coelom is enterocoelous, a deuterostome characteristic. Nevertheless, molecular data and numerous arthropod-like characteristics strongly suggest phylogenetic affinity with Arthropoda (see figure 12.1).

Phylogeny and Adaptive Diversification

Phylogeny

Evolutionary relationships among ecdysozoans are not well understood. Members of this clade do not share a common cleavage pattern. Cleavage in nematodes and nematomorphs is described as unique, or not obviously spiral or radial. Cleavage in priapulids is somewhat similar to radial cleavage. Cleavage has not been studied in kinorhynchs, loriciferans, and tardigrades. In onychophoran eggs containing large amounts of yolk, the cytoplasm does not cleave, but nuclei do divide. Development is similar to that in arthropods. In onychophoran eggs with little yolk, the cleavage pattern varies, appearing spiral in some taxa and radial in others.

Phylogenetic relationships are not completely resolved for all ecdysozoans, but roundworms, phylum Nematoda, are often united with horsehair worms, phylum Nematomorpha, in clade Nematoidea (see figure 12.1) where the sister taxa share a collagenous cuticle and unpaired dorsal and ventral nerve cords. However, a very recent phylogenetic study proposed a sister taxon relationship for nematomorphs and loriciferans.

Phylum Kinorhyncha is shown as the sister taxon to phylum Priapulida based on the shared two-layered pharynx. Kinorhynchs have mouthparts (oral styles on a noninversible mouth cone) similar to those of loriciferans, but loriciferans also share some morphological features with larval nematomorphs and with priapulids. Some workers construct clade Scalidophora to contain kinorhynchs, priapulids, and loriciferans, but more work is needed on these animals before a branch order is specified.

Clade Panarthropoda unites three phyla whose evolutionary association has been clear for a long time. Velvet worms, phylum Onychophora, are shown as the sister taxon to a clade composed of arthropods and tardigrades, but some recent phylogenies support a sister taxon relationship between onychophorans and tardigrades, with this pair as the sister taxon of arthropods.

Onychophoran characteristics shared with arthropods include a tubular heart and hemocoel with open circulatory system, presence of tracheae (probably not homologous), absence of ectodermal cilia, and a large brain.

Tardigrades have some similarities to rotifers, particularly in their mode of reproduction and their cryptobiotic tendencies. Several important synapomorphies suggest grouping them with arthropods (see figure 12.1), and DNA sequence analysis also supports alignment with arthropods in Ecdysozoa. Tardigrades and arthropods share two morphological features: arthropod-type setae and muscles that insert on the cuticle (see figure 12.1).

Reconstructing the evolutionary history of life is a fascinating pursuit, but biologists lack developmental and morphological information for many taxa, as is clear from the comments presented here. Many of the less well-known taxa are very small animals living in obscure habitats—for example, the spaces between sand grains—but until all character states can be described for these animals, our knowledge of both ecdysozoan and metazoan phylogeny will remain incomplete.

Adaptive Diversification

Certainly the most impressive adaptive diversification in this group of phyla is by nematodes. They are by far the most numerous in terms of both individuals and species, and they have adapted to almost every habitat available to animal life. Their basic pseudocoelomate body plan, with the cuticle, hydrostatic skeleton, and longitudinal muscles, has proved generalized and plastic enough to adapt to an enormous variety of physical conditions. Free-living lines gave rise to parasitic forms on at least several occasions, and virtually all potential hosts have been exploited. All types of life cycles occur: from simple and direct to complex, with intermediate hosts; from normal dioecious reproduction to parthenogenesis, hermaphroditism, and alternation of free-living and parasitic generations. A major factor contributing to the evolutionary opportunism of the nematodes is their extraordinary capacity to survive suboptimal conditions—for example, developmental arrests in many free-living and animal parasitic species and the ability to undergo cryptobiosis (survival in harsh conditions by assuming a very low metabolic rate) in many free-living and plant-parasitic species.

Summary

Phyla covered in this chapter are in the ecdysozoan branch of Protostomia. These taxa have an external cuticle that is molted periodically to accomodate growth.

With as many as 500,000 species, nematodes form by far the largest and most important of these phyla. They are more or less cylindrical, tapering at the ends, and covered with a tough, secreted cuticle. Their body-wall muscles are longitudinal only, and to function well in locomotion, such an arrangement must enclose a volume of fluid in the pseudocoelom at high hydrostatic pressure. This fact of nematode life has a profound effect on most of their other physiological functions—for example, ingestion of food, egestion of feces, excretion, copulation, and others. Most nematodes are dioecious, and there are four juvenile stages, each separated by a molt of the cuticle. Nematodes have undergone much greater adaptive diversification than any other pseudocoelomate phylum.

Almost all invertebrate and vertebrate animals and many plants host nematode parasites, and many other nematodes are free-living in soil and aquatic habitats. The life cycles of parasitic nematodes vary. Some are free-living during part of their life cycle; some undergo a tissue migration in their host; and some have an intermediate host. Some parasitic nematodes cause severe diseases in humans and other animals.

Nematomorpha, or horsehair worms, are related to the nematodes and have parasitic juvenile stages in arthropods, followed by a free-living, aquatic, nonfeeding adult stage.

Kinorhyncha and Loricifera are small phyla of tiny, aquatic pseudocoelomates. Kinorhynchs anchor and then pull themselves by the spines on their head. Loriciferans can withdraw their bodies into the lorica. Priapulids are marine burrowing worms.

Clade Panarthropoda contains onychophorans, tardigrades, and arthropods. They have open circulatory systems with a hemocoel. Onychophora are caterpillar-like predatory animals living in humid, mostly tropical habitats. They are segmented and crawl by means of a series of unjointed, clawed appendages. Tardigrades are minute animals, living in the water film that surrounds mosses and lichens. They have eight unjointed legs and a nonchitinous cuticle. They can undergo cryptobiosis, withstanding adverse conditions for long periods.

Review Questions

1. A skeleton is a supportive structure. Explain how a hydrostatic skeleton supports an animal.
2. What feature of body-wall muscles in nematodes requires a high hydrostatic pressure in the pseudocoelomic fluid for efficient function?
3. Explain how high pseudocoelomic pressure affects feeding and defecation in nematodes. How could ameboid sperm be an adaptation to high hydrostatic pressure in the pseudocoelom?
4. Explain the interaction of cuticle, body-wall muscles, and pseudocoelomic fluid in the locomotion of nematodes.
5. Outline the life cycle of each of the following: *Ascaris lumbricoides*, hookworm, *Enterobius vermicularis*, *Trichinella spiralis, Wuchereria bancrofti*.
6. Where in a human body are adults of each species in question 5 found?
7. Where are juveniles and adults of nematomorphs found?
8. What are the approximate lengths of loriciferans, priapulids, and kinorhynchs? Where are they found?
9. Give the main distinguishing characteristics of Onychophora and Tardigrada.
10. What do the members of each of the aforementioned groups eat, and where do they live?
11. Some investigators regard Onychophora as a "missing link" between Annelida and Arthropoda. Give evidence for and against this hypothesis.
12. What is the survival value of cryptobiosis in tardigrades?
13. How is a hemocoel different from a true coelom?
14. In what sense is a hemocoel part of the circulatory system?
15. Describe the two major protostome clades, and give a defining feature for each.

For Further Thought The comments of N. A. Cobb quoted in the opening essay for this chapter make clear just how successful are nematodes. Do any characteristics of nematodes explain their much greater abundance than the other ecdysozoan phyla covered in this chapter?

Selected References

See also general references on page 432.
Aguinaldo, A. M. A., J. M. Turbeville, L. S. Linford, M. C. Rivera, J. J. F. R. Garey, R. A. Raff, and J. A. Lake. 1997. Evidence for a clade of nematodes, arthropods and other moulting animals. Nature **387**:489–493. *DNA sequence analysis to support a superphylum Ecdysozoa.*

Despommier, D. D. 1990. *Trichinella spiralis:* the worm that would be virus. Parasitol. Today **6**:193–196. *Juveniles of* Trichinella *are among the largest of all intracellular parasites.*
Dopazo, H., and J. Dopazo. 2005. Genome-scale evidence of the nematode-arthropod clade. Genome Biol. **6**:R41. *Phylogenetic*

trees constructed under the hypothesis that a coelomate body is homologous are compared with those assuming that ecdysis is a homologous trait.
Halanych, K. M., and Y. Passamaneck. 2001. A brief review of metazoan phylogeny and future prospects in Hox-research1. Am. Zool. **41**:629–639. *A good review of the*

arguments for and against the lophotro-chozoa and ecdysozoa hypotheses.

Holterman, M., A. van der Wurff, S. van den Elsen, H. van Megen, T. Bongers, O. Holovachov, J. Bakker, and J. Helder. 2006. Phylum-wide analysis of SSU rDNA reveals deep phylogenetic relationships among nematodes and accelerated evolution toward crown clades. Mol. Biol. Evol. **23:**1792–1800. *A new nematode phylogeny shows that plant-parasitic species arose from fungivorous ancestors in three lineages.*

Kristensen, R. M. 2002. An introduction to Loricifera, Cycliophora, and Micrognathozoa. Integ. and Comp. Biol. **42:**641–651. *A clear and informative description of these little-known animal groups.*

Lee, D. L. (ed.). 2002. The Biology of Nematodes. Taylor and Francis, Inc. London, UK. *The best general reference currently available.*

Neuhaus, B., and R. P. Higgins. 2002. Ultrastructure, biology, and phylogenetic relationships of Kinorhyncha. Integ. and Comp. Biol. **42:**619–632. *A detailed summary of the biology and morphology of kinorhynchs.*

Nielsen, C. 1995. Animal evolution: interrelationships of the living phyla. Oxford University Press, New York. *The author proposes homologous features for many lesser-known taxa.*

Rota-Stabeli, O., E. Kayal, D. Gleeson, J. Daubs, and J. L. Boore. 2010. Ecdysozoan mitogenomics: evidence for a common origin of the legged invertebrates, the Panarthropoda. Genome Biol. Evol. **2:**425–440. *Some support for a sister-taxon relationship between onychophorans and tardigrates, as well as for the Mandibulate hypothesis for arthropod phylogeny.*

Sorensen, M. V., M. B. Hebsgaard, I. Heiner, H. Glenner, E. Willerslev, and R. M. Kristensen. 2008. New data from an enigmatic phylum: evidence from molecular sequence data supports a sister-group relationship between Loricifera and Nematomorpha. J. Zool. Syst. Evol. Res. **46:**231–239. *The similarities between larval nematomorphs and loriciferans are emphasized.*

Taylor, M. J., and A. Hoerauf. 1999. *Wolbachia* bacteria of filarial nematodes. Parasitol. Today **15:**437–442. *All filarial parasites of humans have endosymbiotic Wolbachia, and most filarial nematodes of all kinds are infected. Nematodes can be "cured" by treatment with the antibiotic tetracycline. If cured, they cannot reproduce. Bacteria are passed vertically from females to offspring.*

■ Custom Website

The *Animal Diversity* Online Learning Center is a great place to check your understanding of chapter material. Visit www.mhhe.com/hickmanad6e for access to key terms, quizzes, and more! Further enhance your knowledge with Web links to chapter-related material.

Explore live links for these topics:

Classification and Phylogeny of Animals
Phylum Kinorhyncha
Phylum Loricifera
Phylum Priapulida
Phylum Nematoda

Human Diseases Caused by Nematodes
Caenorhabditis elegans
Phylum Nematomorpha
Phylum Onychophora
Phylum Tardigrada

Arthropods

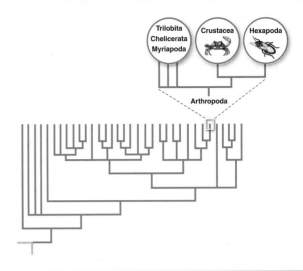

Trilobita Chelicerata Myriapoda | Crustacea | Hexapoda

Arthropoda

...tted spiny lobster, *Panulirus guttatus*.

A Winning Combination

Humans suffer staggering economic losses due to insects. Locust outbreaks in Africa seem a thing of the past to many, but this is far from true. Locust populations fluctuate between quiet phases, during which they cover only 16 million square kilometers in 30 African countries, and plague phases, when they cover 29 million square kilometers of land in 60 countries. A swarm of locusts, *Schistocerca gregaria*, contains 40 to 80 million insects per square kilometer. In peak phases, they cover 20% of the earth's land surface and affect the livelihood of one-tenth of the earth's population. The last plague phase was 1986–1989, but the Food and Agriculture Organization (FAO) of the United Nations monitors and maps population sizes continuously to respond quickly to outbreaks (http://www.fao.org/ag/locusts/en/info/info/faq/index.html).

In the western United States and Canada, an outbreak of mountain pine beetles in the 1980s and 1990s killed pines on huge acreages, and the 1973 to 1985 outbreak of spruce budworm in fir/spruce forests killed millions of conifer trees. Since its introduction in the 1920s, a fungus that causes Dutch elm disease, mainly transmitted by European bark beetles, has virtually obliterated American elm trees in North America. Since 2004, another alien invader, the emerald ash borer, a beetle, threatens the continent's ash trees.

These examples remind us of our ceaseless struggle with the dominant group of animals on earth today: insects. Insect species far outnumber all other species of animals combined. Some scientists have estimated that there are 200 million insects for every single human alive today! Insects have an unmatched ability to adapt to all land environments and virtually all climates. Many have exploited freshwater and shoreline habitats, and have evolved extraordinary abilities to survive adverse environmental conditions.

How can we explain the enormous success of these creatures? Insects, like other members of phylum Arthropoda, have a combination of valuable structural and physiological adaptations, including a versatile exoskeleton, metamerism, an efficient respiratory system, and highly developed sensory organs. In addition, many have a waterproof cuticle and an extraordinary ability to survive adverse environmental conditions. We describe these adaptations and others in this chapter.

245

Phylum Arthropoda (ar-thro-pod´a) (Gr. *arthron,* joint, + *podos,* foot) embraces spiders, scorpions, ticks, mites, crustaceans, millipedes, centipedes, insects, and some smaller groups. In addition its rich fossil record extends back to the mid-Cambrian period (figure 13.1).

Arthropods are coelomate protostomes with well-developed organ systems, and their cuticular exoskeleton containing chitin and tanned protein is a prominent characteristic. Like annelids, they are conspicuously segmented; their ancestral body pattern is a linear series of similar segments, each with a pair of jointed appendages. However, unlike annelids, arthropods often have various segments combined or fused into functional groups, called **tagmata** (sing., **tagma**). The head and thorax are two such tagmata. Appendages, too, are frequently differentiated and specialized for walking, swimming, flying, or eating. Arthropods also differ from annelids in having a small coelom.

Few arthropods exceed 60 cm in length, and most are far smaller. The largest is a Japanese crab (*Macrocheira kaempferi*) with a span of approximately 3.7 m; the smallest is a parasitic mite less than 0.1 mm long.

Arthropods are usually active, energetic animals. Whether judged by their great range of morphologies, their wide ecological distribution, or their vast numbers of species, they are the most abundant and diverse of all animals.

Although arthropods compete with humans for food supplies and spread serious diseases, they are essential pollinators of many food plants, and they also serve as food, yield drugs and dyes, and create such products as silk, honey, and beeswax.

■ Ecological Relationships

In diversity of ecological distribution, arthropods have no rivals. Arthropods occur in all types of environments, from low ocean depths to very high altitudes and from the tropics far into both the north and south polar regions. Some species are adapted for life on land or in fresh, brackish, and marine waters; others live in or on plants and other animals. Most species use flight to move among their favored habitats.

Although all feeding types—carnivorous, omnivorous, and herbivorous—occur in this vast group, most are herbivorous. Most aquatic arthropods depend on algae for their nourishment, and most land forms live chiefly on plants. Many are parasites.

Why Are Arthropods So Diverse and Abundant?

The following structural and physiological patterns have helped arthropods achieve their amazing degree of diversity:

1. **Versatile exoskeleton.** Arthropods have an exoskeleton, called the cuticle, that is highly protective without sacrificing mobility. The cuticle is an outer covering secreted by underlying epidermis.

   The cuticle consists of an outer, relatively thin epicuticle and an inner and thicker procuticle (figure 13.2). Both the epicuticle and the procuticle are composed of several layers. The epicuticle is made of protein and often lipids. The protein is stabilized and hardened by a chemical process called tanning, adding further protection. The procuticle is divided into exocuticle, which is secreted before a molt, and endocuticle, which is secreted after molting. Both layers of the procuticle contain **chitin** bound with protein. Chitin is a tough, resistant, nitrogenous polysaccharide that is insoluble in water, alkalies, and weak acids. Thus, the procuticle is not only flexible and lightweight but also affords protection, particularly against dehydration. In most crustaceans, some areas of the procuticle are also impregnated with calcium salts, which reduce flexibility. In the hard shells of lobsters and crabs, for instance, this calcification is extreme.

   The cuticle may be soft and permeable or may form a veritable coat of armor. Between body segments and between segments of appendages, it is thin and flexible, permitting free movement of joints. In crustaceans and

A   **B**

figure 13.1

Fossils of early arthropods. **A,** Trilobite fossils, dorsal view. These animals were abundant in the mid-Cambrian period. **B,** Eurypterid fossil; eurypterids flourished in northern seas from the Ordovician to the Permian period.

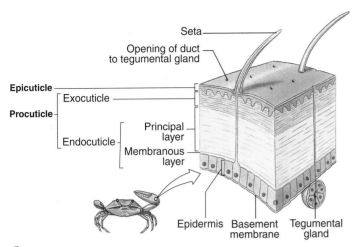

figure 13.2
Structure of crustacean cuticle.

insects, the cuticle forms ingrowths for muscle attachment. It may also line the foregut and hindgut, line and support the tracheae, and be adapted for a variety of other purposes.

The nonexpansible cuticular exoskeleton imposes important conditions on growth. To increase body size, an arthropod must shed its outer covering at intervals and produce a larger one—a process called **ecdysis,** or **molting.** Arthropods molt from four to seven times before reaching adulthood, and some continue to molt after that. Much of an arthropod's physiology centers on molting, particularly in young animals—preparation, molting itself, and then all of the processes that must be completed in the postmolt period. More details of the molting process are given for crustaceans (see p. 257) and for insects (see p. 274).

An exoskeleton is also relatively heavy and becomes proportionately heavier with increasing size. The weight of the exoskeleton tends to limit ultimate body size.

2. **Segmentation and appendages for more efficient locomotion.** Typically, each segment has a pair of jointed appendages, but this arrangement is often modified, with both segments and appendages specialized for adaptive functions. Limb segments are essentially hollow levers that are moved by muscles, most of which are striated for rapid action. The jointed appendages are equipped with sensory hairs and are variously modified for sensory functions, food handling, and swift, efficient walking or swimming.

3. **Air piped directly to cells.** Most land arthropods have a highly efficient tracheal system of air tubes, which delivers oxygen directly to tissues and cells and makes a high metabolic rate possible. Tracheae open to the outside through openings called spiracles. Aquatic arthropods breathe mainly by gills.

4. Highly **developed sensory organs.** Sensory organs show great variety, from compound (mosaic) eyes to senses of touch, smell, hearing, balance, and chemical reception. Arthropods are keenly alert to environmental stimuli.

CHARACTERISTICS
of Phylum Arthropoda

1. **Jointed appendages;** ancestrally, one pair to each segment, but number often reduced; appendages often modified for specialized functions
2. Live in marine, freshwater, and terrestrial habitats; many capable of flight
3. Free-living and parasitic taxa
4. Bilateral symmetry; **segmented body** divided into functional groups called **tagmata:** head and trunk; head, thorax, and abdomen; or cephalothorax and abdomen; definite head
5. Triploblastic body
6. **Small coelom** in adult; body cavity consists mostly of hemocoel (spaces in the tissues) filled with blood
7. **Cuticular exoskeleton;** contains protein, lipid, chitin, and often calcium carbonate secreted by underlying epidermis and shed (molted) at intervals; although **chitin** occurs in a few groups other than arthropods, its use is better developed in arthropods
8. **Complete digestive system;** mouthparts modified from ancestral appendages and adapted for different methods of feeding; gut tube shows great specialization by having, in various arthropods, chitinous teeth, compartments, and gastric ossicles
9. **Complex muscular system,** with exoskeleton for attachment, **striated muscles** for rapid actions, smooth muscles for visceral organs; no cilia
10. **Nervous system** similar to that of annelids, with dorsal brain connected by a ring around the gullet to a double nerve chain of ventral ganglia; fused ganglia in some species
11. Well-developed sensory organs; behavioral patterns much more complex than those of most invertebrates, with more **social organization**
12. Parthenogenesis in some taxa
13. **Sexes usually separate,** with paired reproductive organs and ducts; usually internal fertilization; oviparous, ovoviviparous, or viviparous; often with **metamorphosis**
14. Paired excretory glands called **coxal, antennal,** or **maxillary glands** present in some; others have excretory organs called **Malpighian tubules**
15. Respiration by **body surface, gills, tracheae** (air tubes), or **book lungs**
16. **Open circulatory system,** with dorsal **contractile heart,** arteries, and hemocoel (blood sinuses)

5. **Complex behavior patterns.** Arthropods exceed most other invertebrates in the complexity and organization of their activities. Innate (unlearned) behavior unquestionably controls much of what they do, but many arthropods also demonstrate learned behaviors.

6. **Trophic breadth through metamorphosis.** Many arthropods pass through metamorphic changes, including a larval form quite different from the adult structure. Larval forms are often adapted for eating a different kind of food than adults do, permitting a single species to exploit diverse resources.

Arthropoda is so astonishingly diverse that zoologists have argued over whether all its members belong within a single phylum. Current evidence supports Arthropoda as a monophyletic group. Formerly, arthropods were divided among four subphyla: the extinct Trilobita and the three extant taxa Chelicerata, Crustacea, and Uniramia. The subphylum Uniramia was diagnosed by the shared presence of **uniramous** appendages, with a single branch (figure 13.3, *top*), and included insects and myriapods (centipedes, millipedes, and a few less well-known forms). Uniramia is not supported by molecular data, so we divide uniramians into two subphyla: Myriapoda and Hexapoda. Hexapoda contains class Insecta and one other very small class. Arthropod subphyla themselves are relatively easy to distinguish based on the number of tagmata and the types of appendages present, as outlined next.

Relationships among the subphyla are controversial. One hypothesis proposes that all arthropods possessing a particular mouthpart, called a **mandible,** form a single clade,

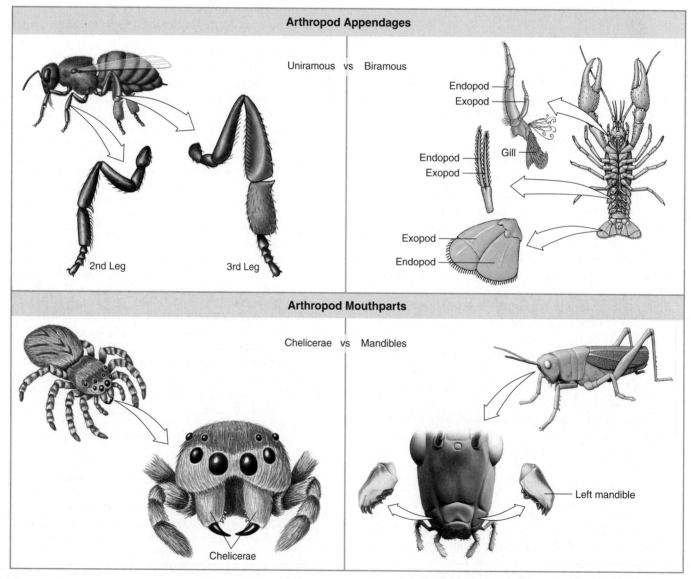

figure 13.3

Two important arthropod characters: Appendages may be uniramous (honey bee leg) or biramous (lobster limb); mouthparts may include chelicerae (spider) or mandibles (grasshopper). Note that the presence or absence of gills is unrelated to appendage form.

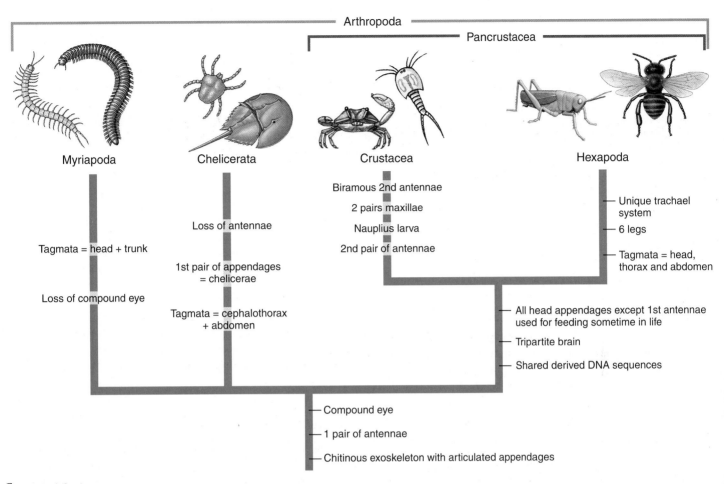

figure 13.4

Cladogram of arthropods showing probable relationships of the four extant subphyla. Only a few synapomorphies are included here. Crustaceans and hexapods are shown as sister taxa, but no branching order for Myriapoda, Chelicerata, or Pancrustacea is specified.

Mandibulata. This clade includes members of Myriapoda, Hexapoda, and Crustacea. Arthropods that do not have mandibles possess **chelicerae** (see figure 13.3, *bottom*), as exemplified by spiders. Thus, according to the "mandibulate hypothesis," myriapods, hexapods, and crustaceans are more closely related to each other than any of them are to chelicerates. Critics of the mandibulate hypothesis argue that the mandibles in each group are so different from each other that they could not be homologous. The mandibles of crustaceans are multijointed, with chewing or biting surfaces on the mandible bases (called a gnathobasic mandible), whereas those of myriapods and hexapods have a single joint with the biting surface on the distal edge (called an entire-limb mandible). There are also some differences in the muscles controlling the two types. Proponents of the mandibulate hypothesis respond that the 550-million-year history of the mandibulates makes possible the evolution of diverse mandibles from an ancestral type.

An alternative to the "mandibulate hypothesis" is the "myriochelata hypothesis" under which myriapods and chelicerates are sister taxa. In this latter hypothesis, the clade comprising myriapods and chelicerates together is the sister taxon of a group containing crustaceans and hexapods.

We assume that subphylum Trilobita is the sister taxon to a group containing all other known arthropods. We depict subphylum Crustacea as the sister taxon of subphylum Hexapoda, but do not specify a branching order for subphylum Myriapoda, subphylum Chelicerata, or the combined branch with hexapods and crustaceans (figure 13.4). Evidence that hexapods and crustaceans form a clade emerged from several phylogenetic studies using molecular characters; these studies prompted a reevaluation of the morphological characters in members of both taxa. We unite subphylum Crustacea with subphylum Hexapoda in clade Pancrustacea.

■ Subphylum Trilobita

Trilobites (see figure 13.1A) probably had their beginnings a million or more years before the Cambrian period. They have been extinct some 250 million years, but were abundant during the Cambrian and Ordovician periods. Their name denotes the trilobed shape of the body, caused by a pair of longitudinal grooves. They were largely bottom-dwellers, probably scavengers. Most of them could roll up like pill bugs.

Subphylum Chelicerata

Chelicerate arthropods are a very ancient group that includes eurypterids (extinct), horseshoe crabs, spiders, ticks and mites, scorpions, sea spiders, and others. They are characterized by the presence of two tagmata and six pairs of cephalothoracic appendages that include a pair of **chelicerae** (see figure 13.3), a pair of **pedipalps,** and **four pairs of walking legs.** They have no mandibles and no antennae. Most chelicerates suck liquid food from their prey.

Class Merostomata

Subclass Eurypterida

Eurypterids, or giant water scorpions (see figure 13.1B), lived 200 to 500 million years ago, and some were perhaps the largest of all arthropods, reaching a length of 3 m. They had some resemblances to marine horseshoe crabs (figure 13.5) and to scorpions, their terrestrial counterparts.

Subclass Xiphosurida: Horseshoe Crabs

Xiphosurids are an ancient marine group that dates from the Cambrian period. Only three genera (five species) survive today. *Limulus* (L. *limus,* sidelong, askew) (figure 13.5), which lives in shallow water along the North American Atlantic coast,

appears in the fossil record practically unchanged in external morphology back to the Triassic period. Horseshoe crabs have an unsegmented, horseshoe-shaped **carapace** (hard dorsal shield) and a broad abdomen, which has a long, spinelike **telson,** or tailpiece. On some abdominal appendages, **book gills** (flat, leaflike gills) are exposed. Horseshoe crabs can swim awkwardly by means of their abdominal plates and can walk on their walking legs. They feed at night on worms and small molluscs and are harmless to humans.

Class Pycnogonida: Sea Spiders

Pycnogonids, commonly called sea spiders, are much more common than most of us realize. They move on four pairs of long, thin walking legs, drinking juices from hydroids and soft-bodied animals with their large suctorial proboscis (figure 13.6). Their odd appearance is enhanced by a much reduced abdomen attached to an elongated cephalothorax. The small abdomen is completely occupied by the digestive tract, so the gonads extend into the legs. Males often use a pair of legs called **ovigers** to carry the egg masses. Most pycnogonids are only a few millimeters long; they are common in all oceans.

Class Arachnida

Arachnids (Gr. *arachnē,* spider) are numerous and diverse, with over 80,000 species described so far. They include spiders, scorpions, pseudoscorpions, whip scorpions, ticks, mites, harvestmen (daddy longlegs), and others. The arachnid tagmata are a cephalothorax and an abdomen.

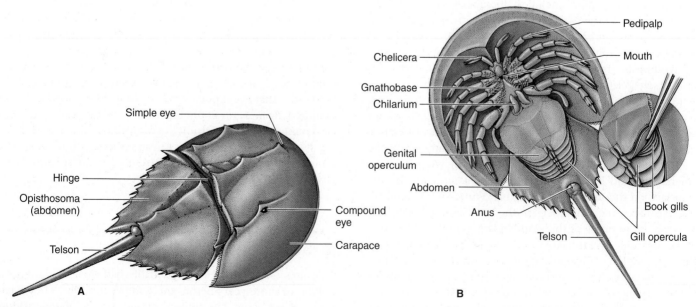

figure 13.5

A, Dorsal view of horseshoe crab *Limulus* (class Merostomata). Individuals grow to 0.5 m in length. **B,** Ventral view of a female.

Order Araneae: Spiders

Spiders are a large group of 40,000 recognized species, distributed all over the world. The cephalothorax and abdomen show no external segmentation, and the tagmata are joined by a narrow, waistlike pedicel (figure 13.7).

All spiders are predaceous and feed largely on insects (figure 13.8). Their chelicerae function as fangs and bear ducts from their venom glands, with which they effectively dispatch their prey. Some spiders chase their prey, others ambush them, and many trap them in a net of silk. After a spider seizes its prey with its chelicerae and injects venom, it liquefies the prey's tissues with a digestive fluid and sucks the resulting broth into the stomach. Spiders with teeth at the bases of their chelicerae

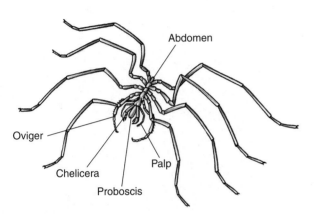

figure 13.6

A pycnogonid, *Nymphon* sp. In this genus, all anterior appendages (chelicerae, palps, and ovigers) are present in both sexes, although ovigers are often not present in females of other genera.

crush or chew prey, and enzymes from their mouth further aid digestion. Many spiders provision their young with previously captured prey.

Spiders breathe by **book lungs** or **tracheae** or both. Book lungs, which are unique to spiders, consist of many parallel air pockets extending into a blood-filled chamber (see figure 13.7C). Air enters the chamber through a slit in the body wall. Tracheae are a system of air tubes that carry air directly to tissues from openings called **spiracles.** The tracheae of spiders are similar to those of insects (see p. 272), but much less extensive.

Spiders and insects have a unique excretory system of **Malpighian tubules** (see figure 13.7C) that work in conjunction with specialized rectal glands. Potassium and other solutes and waste materials are secreted into the tubules, which drain the fluid, or "urine," into the intestine. Rectal glands reabsorb most of the potassium and water, leaving behind wastes such as uric acid. Through this cycling of water and potassium, species living in dry environments conserve body fluids, producing a nearly dry mixture of urine and feces. Many spiders also have coxal glands, which are modified nephridia that open at the coxa, or base, of the first and third walking legs.

Spiders usually have eight simple eyes, each provided with a lens, optic rods, and a retina (see figure 13.7B). Chiefly, the eyes perceive moving objects, but some, such as those of the hunting and jumping spiders, may form images. Because vision is usually poor, a spider's awareness of its environment depends especially on its hairlike sensory setae. Every seta on its surface is useful in communicating some information about the spider's surroundings, air currents, or changing tensions in its web. By sensing vibrations of its web, a spider can judge the size and activity of its entangled prey or receive a message tapped on a silk thread by a prospective mate.

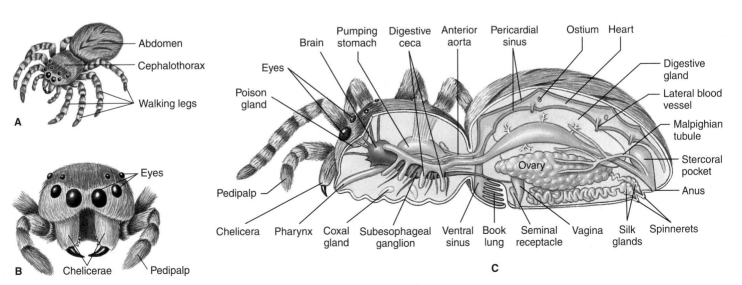

figure 13.7

A, External anatomy of a jumping spider. **B,** Anterior view of head. **C,** Internal anatomy of a female spider.

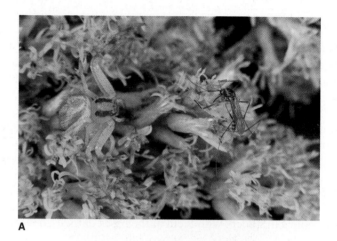

A

figure 13.8

A grasshopper, snared and helpless in the web of a golden garden spider (*Argiope aurantia*), is wrapped in silk while still alive. If the spider is not hungry, its prize is saved for a later meal.

Web-Spinning Habits   The ability to spin silk is important in the life of a spider, and in some other arachnids. Two or three pairs of spinnerets containing hundreds of microscopic tubes connect to special abdominal silk glands (see figure 13.7C). A protein secretion emitted as a liquid hardens on contact with air to form a silk thread. Spiders' silk threads are stronger than steel threads of the same diameter and are probably second in tensional strength only to fused quartz fibers. The threads can stretch one-fifth of their length before breaking.

The spider web used for trapping insects is familiar to most people. The webs of some species consist of merely a few strands of silk radiating out from a spider's burrow or place of retreat. Other species spin beautiful, geometric orb webs. Besides making webs, spiders use silk threads to line their nests; form sperm webs or egg sacs; build draglines; make bridge lines, warning threads, molting threads, attachment discs, or nursery webs; or wrap their prey securely (figure 13.8). Not all spiders spin webs for traps. Some, such as crab spiders, hide from their prey (figure 13.9A), whereas wolf spiders, jumping spiders (figure 13.9B), and fisher spiders (figure 13.10) simply chase and catch their prey.

Reproduction   A courtship ritual usually precedes mating. Before mating, a male spins a small web, deposits a drop of sperm on it, and then lifts the package and stores it in special cavities of his pedipalps (second pair of appendages). When he mates, he inserts a pedipalp into a female's genital opening. Sperm are stored in a female's seminal receptacle, sometimes for weeks or months, until eggs are ready. A female lays her fertilized eggs in a silken cocoon, which she may carry or may attach to a web or plant. A cocoon may contain hundreds of eggs, which hatch in approximately 2 weeks. The young usually remain in their egg sac for a few weeks and molt once before leaving it. Several molts occur before adulthood.

B

figure 13.9

A, A camouflaged crab spider (on the left), *Misumenoides* sp., awaits its insect prey. Its coloration matches the petals among which it lies, thus deceiving insects that visit the flowers in search of pollen or nectar. **B,** A jumping spider, *Eris aurantius*. This species has excellent vision and stalks an insect until it is close enough to leap with unerring precision, fixing its chelicerae into its prey.

Are Spiders Dangerous?   It is truly amazing that such small and helpless creatures as spiders have generated so much unreasoned fear in humans. Spiders are timid creatures that, rather than being enemies to humans, are allies in our continuing conflict with insects. The venom spiders produce to kill prey is usually harmless to humans. Even the most venomous spiders bite only when threatened or when defending their eggs or young. American tarantulas (figure 13.11), despite their fearsome appearance, are not dangerous. They rarely bite, and their bite is not considered serious.

Two spider genera in the United States can inflict severe or even fatal bites: *Latrodectus* (L. *latro*, robber, + *dektes*, biter) and *Loxosceles* (Gr. *loxos*, crooked, + *skelos*, leg). The most important species are *Latrodectus mactans*, or the black widows, and *Loxosceles reclusa*, or the brown recluse. Black widows are moderate to small in size and shiny black, with a bright orange or red "hourglass" on the underside of their abdomen (figure 13.12A). Their venom is

figure 13.10

A fisher spider, *Dolomedes triton*, feeds on a minnow. This handsome spider feeds mostly on aquatic and terrestrial insects but occasionally captures small fishes and tadpoles. It pulls its paralyzed victim from the water, pumps in digestive enzymes, and then sucks out the predigested contents.

A

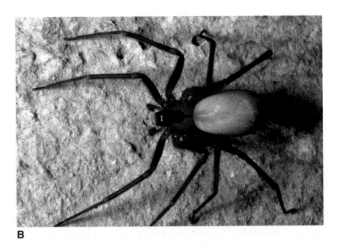

B

figure 13.12

A, A black widow spider, *Latrodectus mactans*, next to her egg sac. Note the orange "hourglass" on the ventral side of her abdomen. **B,** The brown recluse spider, *Loxosceles reclusa*, is a small venomous spider. Note the small, violin-shaped marking on its cephalothorax. The venom is hemolytic and dangerous.

figure 13.11

A tarantula, *Brachypelma vagans*.

neurotoxic, acting on the nervous system. About four or five of each 1000 bites reported are fatal.

Brown recluse spiders, which are smaller than black widows, are brown, and bear a violin-shaped dorsal stripe on their cephalothorax (figure 13.12B). Their venom is hemolytic rather than neurotoxic, destroying the tissues and skin surrounding a bite. Their bite can be mild to serious and occasionally fatal.

Dangerous spiders in other parts of the world include funnel-web spiders (*Atrax robustus*) in Australia and certain ctenid spiders in South America, such as *Phoneutria fera*. In contrast to most spiders, these are quite aggressive.

Order Scorpionida: Scorpions

Although scorpions are more common in tropical and subtropical regions, some occur in temperate zones. Scorpions are generally reclusive, hiding in burrows or under objects by day and feeding at night. They feed largely on insects and spiders, which they seize with chelate pedipalps and rip with jawlike chelicerae.

A scorpion's body consists of a rather short cephalothorax, which bears appendages and from one to six pairs of eyes, and a clearly segmented abdomen without appendages. The abdomen is divided into a broader preabdomen and a taillike postabdomen, which ends in a stinging apparatus used to inject venom. The venom of most species is not harmful to humans, although that of certain species of *Androctonus* in Africa and *Centruroides* in Mexico, Arizona, and New Mexico can be fatal.

Scorpions bear well-developed young, which their mother carries on her back until after their first molt (figure 13.13A).

A   **B**   **C**

figure 13.13

A, An emperor scorpion (order Scorpionida), *Paninus imperator*, with young, which stay with their mother until their first molt. **B,** A camel spider. **C,** A harvestman (order Opiliones). Harvestmen run rapidly on their stiltlike legs. They are especially noticeable during the harvesting season; hence the common name.

A   **B** SEM of house dust mite, *Dermatophagoides farinae*   **C** Follicle mite, *Demodex follicuorum*

figure 13.14

Ticks and mites are arachnids that often affect humans. **A,** The western blacklegged tick, *Ixodes pacificus* (order Acari), is a vector for the bacterium *Borrelia burgdorferi*, which causes Lyme disease. **B,** Dust mites, *Dermatophagoides farinae*, often trigger allergic reactions. **C,** The human follicle mite, *Demodex follicuorum*, is tiny (100 to 400 μm) and lives in follicles, particularly around the nose and eyes. Its prevalence ranges from about 20% in persons 20 years of age or younger to nearly 100% in the aged.

Order Solpugida: Sun or Camel Spiders

Solpugids, sometimes also called solfugids and by such common names as sun, camel, or wind spiders, are nonvenomous arachnids that shred prey with their large chelicerae (figure 13.13B). They are often less than 1 cm long, but some species approach 15 cm. They are common in tropical and subtropical deserts in America, the Middle East, Africa, and Asia.

Order Opiliones: Harvestmen

Harvestmen, often called "daddy longlegs," are common in the United States and other parts of the world (figure 13.13C). They are easily distinguished from spiders by a broad joining of their abdomen and cephalothorax without the constriction of a pedicel, and by the external segmentation of their abdomen.

They have four pairs of long, spindly legs and, without apparent ill effect, can cast off one or more legs if they are grasped by a predator (or human hand). The ends of their chelicerae are pincerlike, and they feed much more as scavengers than do most arachnids.

Order Acari: Ticks and Mites

Acarines differ from all other arachnids in having their cephalothorax and abdomen completely fused, with no external division or segmentation (figure 13.14A). Their mouthparts are on a little anterior projection, or **capitulum.** They occur almost everywhere—in both fresh and salt water, on vegetation, on the ground, and parasitic on vertebrates and invertebrates. Over 40,000 species have been described, many of which are important to humans, but this is probably only a fraction of the species that exist.

Many species of mites are entirely free-living. *Dermatophagoides farinae* (Gr. *dermatos*, skin, + *phago*, to eat, + *eidos*, likeness of form) (figure 13.14B) and related species are denizens of house dust all over the world, sometimes causing allergies and dermatoses. Some mites are marine, but most aquatic species live in fresh water. They have long, hairlike setae on their legs for swimming, and their larvae may be parasitic on aquatic invertebrates. Their sheer abundance gives acarines ecological importance, but many acarines have direct effects on our food supply and health as well. Spider mites (family Tetranychidae) are serious agricultural pests on fruit trees, cotton, clover, and many other plants. Larvae of genus *Trombicula* are called chiggers or redbugs. They feed on dermal tissues of terrestrial vertebrates, including humans, and cause an irritating dermatitis; some species of chiggers transmit a disease called Asiatic scrub typhus. Hair-follicle mites, *Demodex* (figure 13.14C), are apparently nonpathogenic in humans; they infect most of us although we are unaware of them. Other species of *Demodex* and other genera of mites cause mange in domestic animals.

The inflamed welt and intense itching that follow a chigger bite are not caused by a chigger burrowing into the skin, as is popularly believed. Rather, the chigger bites through skin with its chelicerae and injects a salivary secretion containing powerful enzymes that liquefy skin cells. Human skin responds defensively by forming a hardened tube that the larva uses as a sort of drinking straw and through which it gorges itself with host cells and fluid. Scratching usually removes the chigger but leaves the tube, which remains a source of irritation for several days.

Ticks are usually larger than mites. They pierce the skin of vertebrates and suck blood until their bodies become enormously distended; then they drop off and digest their meal. After molting, they are ready for another meal. In addition to disease conditions that they themselves cause, ticks are among the world's premier disease vectors, ranking second only to mosquitos. They carry a greater variety of infectious agents than any other arthropod; such agents include protozoan, rickettsial, viral, bacterial, and fungal organisms. Species of *Ixodes* carry the most common arthropod-borne infection in the United States, Lyme disease (see accompanying note). Species of *Dermacentor* and other ticks transmit Rocky Mountain spotted fever, a misnamed disease since most cases occur in the eastern United States. *Dermacentor* also transmits tularemia and agents of several other diseases. Texas cattle fever, also called red-water fever, is caused by a protozoan parasite transmitted by the cattle tick *Boophilus annulatus*.

In the 1970s, people in the town of Lyme, Connecticut, experienced an epidemic of arthritis-like symptoms. Subsequently called Lyme disease, it is caused by a bacterium and carried by ticks of the genus *Ixodes*. Thousands of cases are reported each year in Europe and North America, and other cases have been reported from Japan, Australia, and South Africa. Many people bitten by infected ticks recover spontaneously or do not suffer any ill effects. Others, if not treated at an early stage, develop a chronic, disabling disease. Lyme disease is now the leading arthropod-borne disease in the United States.

Subphylum Myriapoda

The term myriapod (Gr. *myrias*, a myriad, + *podos*, foot) denotes several classes that share a pattern of two tagmata—head and trunk—with paired appendages on most or all trunk segments. Myriapods include Chilopoda (centipedes), Diplopoda (millipedes), Pauropoda (pauropods), and Symphyla (symphylans). We describe only the chilopods and diplopods.

The head of myriapods has **one pair of antennae**. It also has **mandibles** (see figure 13.3, *bottom*) and two pairs of **maxillae** (one pair of maxillae in millipedes). The legs are all **uniramous**.

Respiratory exchange occurs through body surface and tracheal systems. In some species, aquatic juvenile stages may have gills.

Class Chilopoda: Centipedes

Centipedes are active predators with a preference for moist places such as under logs or stones, where they feed largely on earthworms and insects. Their bodies are somewhat flattened dorsoventrally, and they may contain from a few to 177 segments (figure 13.15). Each segment, except the one behind the head and the last two, bears one pair of appendages. Those of the first body segment (maxillipeds) are modified to form venom claws, which centipedes use to kill their prey. Most species are harmless to humans.

The centipede head bears a pair of eyes, each consisting of a group of ocelli (simple eyes). Respiration is by tracheal tubes with a pair of spiracles in each segment. Sexes are separate, and all species are oviparous. Young are similar to adults. Familiar genera are the common house centipedes *Scutigera*, with 15 pairs of legs, and *Scolopendra* (figure 13.15), with 21 pairs of legs.

Class Diplopoda: Millipedes

Diplopods, or "double-footed" arthropods, are commonly called millipedes, which means "thousand feet," although they do not literally have a thousand legs (figure 13.16). Their cylindrical bodies contain 25 to 100 segments. The four thoracic segments bear only one pair of legs each, but the abdominal segments each have two pairs, a condition that may have evolved from fusion of segments. Two pairs of spiracles occur on each abdominal segment, each opening into an air chamber that gives rise to tracheal tubes.

Millipedes are less active than centipedes and are generally herbivorous, living on decayed plant and animal matter and

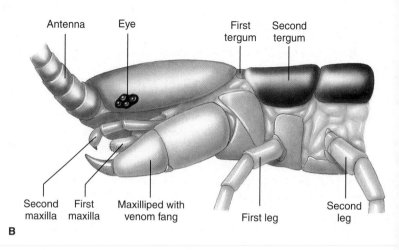

figure 13.15

A, A centipede, *Scolopendra* (class Chilopoda), from the Amazon Basin, Peru. Most segments have one pair of appendages each. The first segment bears a pair of venom claws, which in some species can inflict serious wounds. Centipedes are carnivorous. **B,** Head of a centipede.

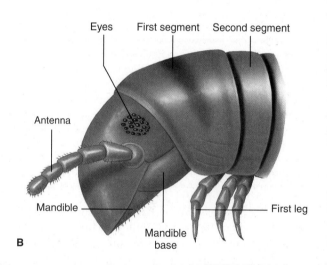

figure 13.16

A, A pair of mating millipedes from South Africa. **B,** Head of a millipede.

sometimes living plants. They prefer dark, moist places under stones and logs. Females lay eggs in a nest and guard them carefully. Larval forms have only one pair of legs per segment.

■ Subphylum Crustacea

The 67,000 or more species of Crustacea (L. *crusta,* shell) include lobsters, crayfishes, shrimp, crabs, water fleas, copepods, and barnacles. Crustaceans compose the only arthropod subphylum that is primarily aquatic; they are mainly marine, but many freshwater and a few terrestrial species are known. Most are free-living, but some are sessile, commensal, or parasitic. Crustaceans are often very important components of aquatic ecosystems, and several have considerable economic importance.

Crustaceans are the only arthropods with **two pairs of antennae** (figure 13.17). In addition to antennae and **mandibles,** they have **two pairs of maxillae** on the head, followed by a pair of appendages on each body segment (although appendages on some segments are absent in some groups). All appendages, except perhaps the first antennae (antennules), are primitively **biramous** (see figure 13.3, *top*), and at least some appendages of all present-day adults show that condition. Organs specialized for respiration, if present, are in the form of gills. Malpighian tubules are absent.

Ancestral crustaceans typically had 60 segments or more, but most modern forms have between 16 and 20 segments and increased tagmatization. The major tagmata are head, thorax, and abdomen, but these are not homologous throughout the

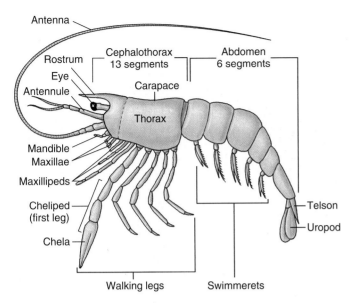

figure 13.17

Archetypical plan of Malacostraca. Note that maxillae and maxillipeds have been separated diagrammatically to illustrate the general plan. Typically, in living animals, only the third maxilliped is visible externally. In order Decapoda, the carapace covers the cephalothorax, as shown here. The head region is beneath the carapace to the left of the dashed line.

subphylum (or even within some classes) because of varying degrees of segment fusion—for example, in the cephalothorax.

In many crustaceans, the dorsal cuticle of the head extends posteriorly and around the sides of the animal to cover or fuse with some or all of the thoracic and abdominal segments. This covering is called a **carapace.** In some groups, the carapace forms clamshell-like valves that cover most or all of the body. In decapods (including lobsters, shrimp, crabs, and others), the carapace covers the entire cephalothorax but not the abdomen. Crustacea has recently expanded to include the wormlike pentastomids; these parasites previously were considered a phylum of animals, but are now placed in Crustacea, near the fish lice, subclass Branchiura.

Form and Function

Appendages

Some modifications of crustacean appendages are illustrated by those of crayfishes and lobsters (class Malacostraca, order Decapoda, see p. 264). Swimmerets, or abdominal appendages, retain the ancestral biramous condition. Such an appendage consists of inner and outer branches, called the **endopod** and **exopod,** which are attached to one or more basal segments collectively called a protopod (figure 13.18).

There are many modifications of this plan. In the ancestral character state for crustaceans, all trunk appendages are rather similar in structure and adapted for swimming. The evolutionary trend, shown in crayfishes, has been toward a reduced number of appendages and an increased variety of modifications

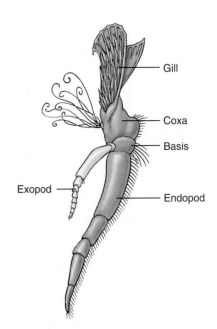

figure 13.18

Parts of a biramous crustacean appendage (third maxilliped of a crayfish).

that fit the appendages for many functions. Some appendages are foliaceous (flat and leaflike), as are maxillae; some are biramous, as are swimmerets, maxillipeds, uropods, and antennae; and others have lost one branch and are secondarily uniramous, as are walking legs.

In crayfishes, the first three pairs of thoracic appendages, called **maxillipeds,** serve along with the two pairs of maxillae as food handlers; the other five pairs of appendages are lengthened and strengthened for walking and defense (figure 13.19). The first pair of walking legs, called **chelipeds,** is enlarged with a strong claw, or **chela,** for defense. Abdominal swimmerets serve not only for locomotion; in males the first pair, called gonopods, are modified for copulation, and in females they all serve as a nursery for attached eggs and young. The last pair of appendages, called **uropods,** are wide and serve as paddles for swift backward movements; with the telson, they form a protective device for eggs or young on the swimmerets.

Zoologists have not adopted a uniform terminology to describe crustacean appendages. At least two systems are in wide use. For example, alternative terms to those used in this book are exopodite, endopodite, basipodite, and coxopodite (see figure 13.18). The first and second pairs of antennae may be called antennules and antennae (see figure 13.17), and the first and second maxillae are often called maxillules and maxillae. A rose by any other name….

Ecdysis

Ecdysis (Gr. *ekdysis*, to strip off) permits growth despite a restrictive exoskeleton in crustaceans and other arthropods. This periodic shedding of old cuticle and formation of a larger

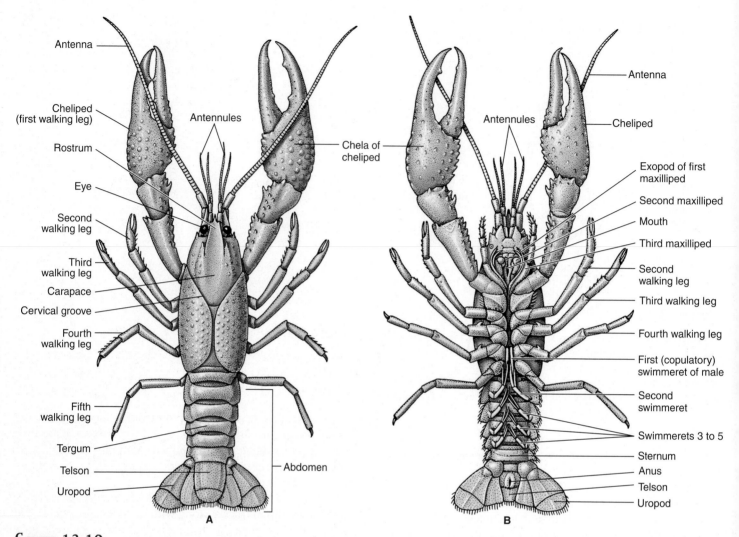

figure 13.19

External structure of crayfishes. **A,** Dorsal view. **B,** Ventral view.

new one occurs most frequently during larval stages and less often as an animal reaches adulthood. Although actual shedding of the cuticle is periodic, the molting process and preparations for it, involving storage of reserves and changes in the integument, are a continuous process lasting most of an animal's life.

During each premolt period, the old cuticle becomes thinner as inorganic salts are withdrawn from it and stored in tissues. Other reserves, both organic and inorganic, also accumulate and are stored. The underlying epidermis begins to grow by cell division; it secretes first a new inner layer of epicuticle and then enzymes that digest away the inner layers of old endocuticle (figure 13.20). Gradually, a new cuticle forms inside the degenerating old one. Finally, actual ecdysis occurs as the old cuticle ruptures, usually along the middorsal line, and the animal backs out (figure 13.21). The animal swells with air or water to stretch the new, larger cuticle to its full size. During the postmolt period, the cuticle thickens, its outer layer hardens by tanning, and its inner layer uses salvaged inorganic salts and other constituents for strengthening. Usually an animal is

very reclusive during its postmolt period when its defenseless condition makes it particularly vulnerable to predation.

That ecdysis is under hormonal control has been demonstrated in both crustaceans and insects, but the process is often initiated by a stimulus perceived by the central nervous system. In decapods, the stimulus decreases production of a molt-inhibiting hormone from neurosecretory cells in the **X-organ** of the eyestalk. The sinus gland, also in the eyestalk, releases the hormone. When the level of molt-inhibiting hormone drops, **Y-organs** near the mandibles produce **molting hormone.** This hormone initiates processes leading to premolt. Y-organs are homologous to the prothoracic glands of insects, which produce ecdysone.

Neurosecretory cells are modified nerve cells that secrete hormones. They are widespread in invertebrates and also occur in vertebrates. Cells in the vertebrate hypothalamus and in the posterior pituitary are good examples.

Intermolt condition

- Epicuticle
- Exocuticle
- Endocuticle
- Epidermis

STEP 1:
In the pre-molt stage, the old procuticle separates from the epidermis, which secretes a new epicuticle.

- New epicuticle

STEP 2:
Still in the pre-molt stage, new exocuticle is secreted as molting fluid dissolves the old endocuticle. Solution products are reabsorbed.

- Dissolving endocuticle
- New exocuticle

STEP 3:
At ecdysis, the old epicuticle and exocuticle are discarded.

- Discarded old epicuticle and exocuticle

STEP 4:
In postecdysis, new cuticle is stretched and unfolded, and endocuticle is secreted.

- New endocuticle

figure 13.20

Cuticle secretion and reabsorption in ecdysis.

Other Endocrine Functions

Pigments in special branched cells (**chromatophores**) in the epidermis produce body color in crustaceans. Chromatophores change color by concentrating pigment granules in the center of each cell, which causes a lightening effect, or by dispersing pigment throughout each cell, which causes darkening. Neurosecretory cells in the eyestalk control pigment behavior. Neurosecretory hormones also control pigment in the eyes for light and dark adaptation, and other neurosecretory hormones control rate and amplitude of heartbeat.

Androgenic glands, which are not neurosecretory, occur in male malacostracans, and their secretion stimulates the expression of male sexual characteristics. If androgenic glands are artificially implanted in a female, her ovaries transform to testes and begin to produce sperm, and her appendages begin to acquire male characteristics at the next molt.

Feeding Habits

Feeding habits and adaptations for feeding vary greatly among crustaceans. Many forms can shift from one type of feeding to another depending on environment and food availability, but fundamentally all use the same set of mouthparts. Maxillipeds hold and manipulate food; mandibles and maxillae shred food and place it in the mouth. In predators, the walking legs, particularly chelipeds, serve in food capture.

Many crustaceans, both large and small, are predatory, and some have interesting adaptations for killing prey. One shrimplike form, *Lygiosquilla,* has on one of its walking legs a specialized digit that can be drawn into a groove and released suddenly to pierce passing prey. Pistol shrimp, *Alpheus,* have one enormously enlarged chela that can be cocked like the hammer of a gun and snapped shut at great speed, forming a cavitation bubble that implodes with a snap sufficient to stun its prey.

The food of crustaceans ranges from plankton, detritus, and bacteria, used by **suspension feeders;** to larvae, worms, crustaceans, snails, and fishes, used by predators; to dead animal and plant matter, used by scavengers. Suspension feeders, such as fairy shrimps, water fleas, and barnacles, use their legs, which bear a thick fringe of setae, to create water currents that sweep food particles through the setae, where they are captured. Mud shrimps, *Upogebia,* use long setae on their first two pairs of thoracic appendages to strain food material from water circulated through their burrow by movements of their swimmerets.

Crayfishes have a two-part stomach. The first contains a gastric mill in which food, already shredded by the mandibles, is ground further by three calcareous teeth into particles fine enough to pass through a filter of setae in the second part of the stomach; food particles then pass into the intestine for chemical digestion.

Respiration, Excretion, and Circulation

The gills of crustaceans vary in shape. They may be treelike, leaflike, or filamentous, and all are provided with blood vessels or sinuses. They are usually attached to appendages and kept ventilated by movement of the appendages through water. The overlapping carapace usually protects the **branchial chambers.** Some smaller crustaceans breathe through their general body surface.

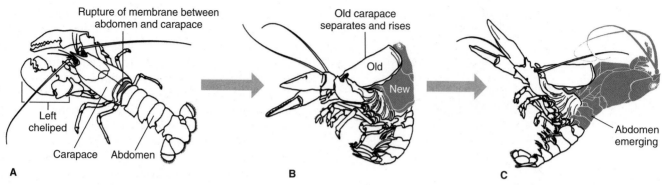

figure 13.21

Molting sequence in a lobster, *Homarus americanus.* **A,** The membrane between the carapace and the abdomen ruptures, and the carapace begins a slow elevation. This step may take up to 2 hours. **B, C,** Head, thorax, and finally the abdomen are withdrawn. This process usually takes no more than 15 minutes. Immediately after ecdysis, chelipeds are desiccated and the body is very soft. The lobster now begins rapid absorption of water so that within 12 hours its body increases about 20% in length and 50% in weight. Water will be replaced by tissue in succeeding weeks.

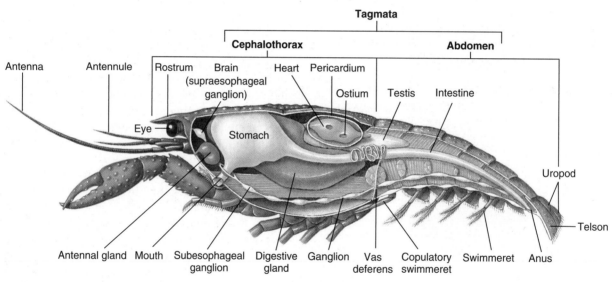

figure 13.22

Internal structure of a male crayfish.

Excretory and osmoregulatory organs in crustaceans are paired glands located in the head, with excretory pores opening at the base of either antennae or maxillae; the glands are thus called **antennal glands** or maxillary glands, respectively (figure 13.22). The antennal glands of decapods are also called **green glands** because of their color in life. They resemble the coxal glands of chelicerates. Waste products are mostly ammonia with some urea and uric acid. Some wastes diffuse through the gills as well as through the excretory glands.

Circulation, as in other arthropods, is an open system consisting of a heart, either compact or tubular, and arteries, which transport blood to different areas of the hemocoel. Some smaller crustaceans lack a heart. An open circulatory system depends less on heartbeats for circulation because movement of organs and limbs circulates blood more effectively in open sinuses than in capillaries. Blood may contain as respiratory pigments either hemocyanin or hemoglobin (hemocyanin in decapods), and it has the property of clotting to prevent loss of blood in minor injuries.

Nervous and Sensory Systems

A cerebral ganglion above the esophagus sends nerves to the anterior sense organs and connects to a subesophageal ganglion by a pair of connectives around the esophagus. A double ventral nerve cord has a ganglion in each segment that sends

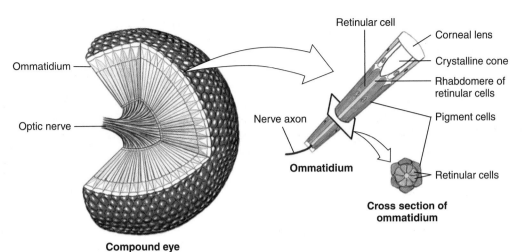

figure 13.23

Compound eye of an insect. *Right,* A single ommatidium is shown enlarged.

nerves to the viscera, appendages, and muscles (figure 13.22). Giant fiber systems are common among crustaceans.

Sensory organs are well developed. There are two types of eyes—a median (or nauplius) eye and compound eyes. A median eye usually consists of a group of three pigment cups containing retinal cells, and it may or may not have a lens. Median eyes occur in nauplius larvae and in some adult forms, and they may be an adult's only eye, as in copepods.

Most crustaceans have compound eyes similar to insect eyes. In crabs and crayfishes, the eyes are on the ends of movable eyestalks (figure 13.22). Compound eyes are precise instruments, different from vertebrate eyes, and yet especially adept at detecting motion; they can analyze polarized light. The convex corneal surface provides a wide visual field, particularly in stalked eyes where the surface may cover an arc of 200 degrees or more.

Compound eyes are composed of many tapering units called **ommatidia** set close together (figure 13.23). Facets, or corneal surfaces, of the ommatidia give the eye surface the appearance of a fine mosaic. Most crustacean eyes are adapted either to bright or to dim light, depending on their diurnal or nocturnal habits, but some are able, by means of screening pigments, to adapt somewhat to both bright and dim light. The number of ommatidia varies from a dozen or two in some small crustaceans to 15,000 or more in a large lobster. By comparison, some insects have approximately 30,000.

Other sensory organs include statocysts, tactile setae on the cuticle of most of the body, and chemosensitive setae, especially on antennae, antennules, and mouthparts.

Reproduction and Life Cycles

Most crustaceans have separate sexes, and numerous specializations for copulation occur among different groups. Almost all barnacles are monoecious but generally practice cross-fertilization. In some ostracods, males are scarce, and reproduction is usually parthenogenetic. Most crustaceans brood their eggs in some manner—branchiopods and barnacles have special brood chambers, copepods have egg sacs attached to the sides of their abdomen (see figure 13.25C), and malacostracans usually carry eggs and young attached to their appendages.

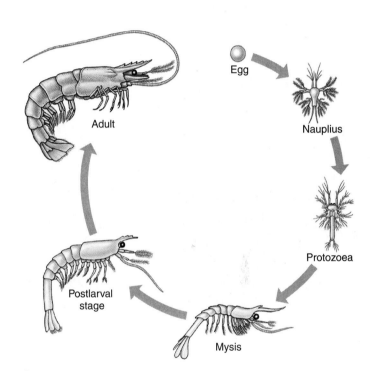

figure 13.24

Life cycle of a Gulf shrimp, *Penaeus*. Penaeids spawn at depths of 40 to 90 m. Young larval forms are planktonic and move inshore to water of lower salinity to develop as juveniles. Older shrimp return to deeper water offshore.

A crayfish hatchling is a tiny juvenile similar in form to the adult and has a complete set of appendages and segments. However, most crustaceans produce larvae that undergo a series of changes, either gradual or abrupt over a series of molts, to assume adult form (metamorphosis). The ancestral larva of crustaceans is a **nauplius** (pl., **nauplii**) (figure 13.24). It has an unsegmented body, a frontal eye, and three pairs of appendages, representing the two pairs of antennae and the mandibles. The developmental stages and postlarvae of different groups of Crustacea are varied and have special names.

Class Branchiopoda

Members of class Branchiopoda (branke-ă´pōd-ə) have several ancestral characteristics. Branchiopods have small first antennae and second maxillae. Their legs are flattened and leaflike (phyllopodia) and are the chief respiratory organs (hence, the name branchiopods). Most branchiopods also use their legs in suspension feeding, and groups other than cladocerans use their legs for locomotion as well.

Four orders are recognized: **Anostraca** (fairy shrimp and brine shrimp), which lack a carapace; **Notostraca** (tadpole shrimp such as *Triops*), whose carapace forms a large dorsal shield covering most trunk segments; **Conchostraca** (clam shrimp such as *Lynceus*), whose carapace is bivalved and usually encloses the entire body; and **Cladocera** (water fleas such as *Daphnia*, figure 13.25A), with a carapace typically covering the entire body but not the head. The most important and diverse order is Cladocera, which often forms a large segment of freshwater zooplankton.

Class Maxillopoda

Class Maxillopoda subsumes a number of crustacean groups once considered classes. The typical body plan consists of five cephalic, six thoracic, and usually four abdominal segments plus a telson, but fewer segments are common. Appendages are usually on the abdomen. The eyes of nauplii (when present) have a unique structure termed a maxillopodan eye.

Members of subclass **Ostracoda** (os-tra´ kōd-ə) (Gr. *ostrakodes*, having a shell) are, like conchostracans, enclosed in a bivalved carapace and resemble tiny clams, 0.25 to 8 mm long (figure 13.25B). Ostracods show considerable fusion of trunk segments, and their thoracic appendages are reduced to two or none. Most ostracods crawl or burrow in marine or freshwater sediments. They scavenge food, feed on detritus, or collect suspended particles from the water.

Subclass **Copepoda** (ko-pe´-pōd-ə) (Gr. *kōpē*, oar, + *podos*, foot) is an important group of Crustacea, second only to Malacostraca in number of species. Copepods are small (usually a few millimeters or less in length), rather elongate, tapering toward the posterior end, lacking a carapace, and retaining a simple, median, nauplius eye in adults (figure 13.25C). They have four pairs of rather flattened, biramous, thoracic swimming appendages, and a fifth, reduced pair. Their abdomen bears no legs. Free-living copepods occur in planktonic and benthic habitats. Planktonic species are essential members of marine and freshwater food webs, often dominating the primary consumer level (herbivore) in aquatic communities. Many symbiotic species are known, and parasitic forms may be so highly modified as adults (and may depart so far from the description just given) that they can hardly be recognized as arthropods.

Subclass **Branchiura** (bran-kē-u´ra) (Gr. *branchia*, gills, + *ura*, tail) is a small group of primarily fish parasites that, despite its name, has no gills (figure 13.26A). Members of this group are usually between 5 and 10 mm long and parasitize marine or freshwater fishes. They typically have a broad, shieldlike carapace, compound eyes, four biramous thoracic appendages for swimming, and a short, unsegmented abdomen. The second maxillae have become modified as suction cups (figure 13.26B).

Subclass **Pentastomida** (pen-ta-stom´i-da) (Gr. *pente*, five, + *stoma*, mouth), or tongue worms, consist of about 130 species of wormlike parasites of the respiratory system of vertebrates. Adult pentastomids live mostly in the lungs of reptiles, but one species, *Linguatula serrata* (Gr. *lingua*, tongue), lives in the nasopharynx of canines and felines (and occasionally humans). Although more common in tropical areas, they also occur in North America, Europe, and Australia.

Adult pentastomida range from 1 to 13 cm in length. Transverse rings give their bodies a segmented appearance (figure 13.26C and D). Their body is covered with a nonchitinous cuticle that is molted periodically during larval stages. The anterior end may bear five short protuberances (hence the name Pentastomida). Four of these bear chitinous claws, and the fifth bears the mouth (figure 13.26E). The digestive system is simple and adapted for sucking blood from the host. The nervous system, similar to that of other arthropods, has paired ganglia along the ventral nerve cord. The only sense organs appear to be papillae. There are no circulatory, excretory, or respiratory organs.

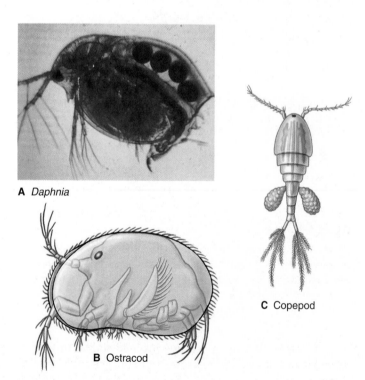

A *Daphnia*

B Ostracod

C Copepod

figure 13.25

A, A water flea, *Daphnia* (order Cladocera), photographed with polarized light. These tiny forms occur in great numbers in northern lakes and are an important component of the food chain leading to fishes. **B,** An ostracod (subclass Ostracoda, class Maxillopoda). **C,** A copepod with attached ovisacs (subclass Copepoda, class Maxillopoda).

A Fish louse

B Anterior, ventral surface of *Argulus nobilis*

figure 13.26

A, Fish lice, in subclass Branchiura, are closely related to pentastomids (**C** and **D**). SEMs of the anterior regions of both animals (**B** and **E**) demonstrate morphological similarities.

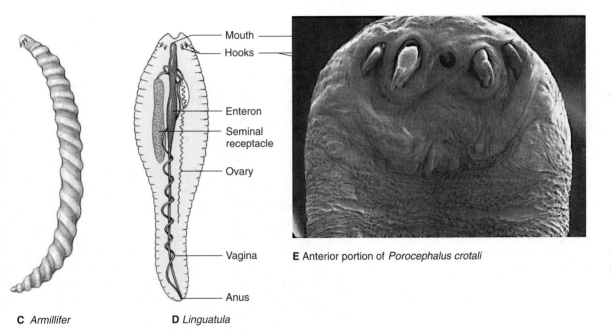

Mouth
Hooks

Enteron
Seminal receptacle

Ovary

Vagina

Anus

C *Armillifer*   **D** *Linguatula*

E Anterior portion of *Porocephalus crotali*

Sexes are separate, and females are usually larger than males. A female may produce several million eggs, which pass up the trachea of the host, are swallowed, and exit with feces. Larvae hatch as oval, tailed creatures with four stumpy legs. Most pentastomid life cycles require an intermediate vertebrate host such as a fish, a reptile, or, rarely, a mammal, that is eaten by the definitive vertebrate host. After ingestion by an intermediate host, larvae penetrate the intestine, migrate randomly in the body, and finally metamorphose into nymphs. After growth and several molts, a nymph finally becomes encapsulated and dormant. When eaten by a final host, a juvenile finds its way to a lung, feeds on blood and tissue, and matures.

Several species have been found encysted in humans, the most common being *Armillifer armillatus* (L. *armilla*, ring, bracelet, + *fero*, to bear) (see figure 13.26C), but usually they cause few symptoms. *L. serrata* is a cause of nasopharyngeal pentastomiasis, or "halzoun," a disease of humans in the Middle East and India.

Subclass **Cirripedia** (sir-i-ped´i-a) (L. *cirrus,* curl of hair, + *pedis,* foot) includes typical barnacles, which are usually enclosed in a shell of calcareous plates, as well as three smaller orders of burrowing or parasitic forms. Barnacles are sessile as adults and may be attached to their substrate directly, as with acorn barnacles (figure 13.27A), or by a stalk, as with gooseneck barnacles (figure 13.27B). Typically, a carapace (mantle) surrounds their body and secretes a shell of calcareous plates. Their head is reduced, they have no abdomen, and their thoracic legs are long, many-jointed cirri with hairlike setae. The cirri are extended through an opening between the calcareous plates to filter from water small particles on which the animal feeds (figure 13.27B).

Barnacles frequently foul ship bottoms by settling and growing there. There may be so many barnacles that the speed of a ship is reduced 30% to 40%, requiring expensive drydocking while the barnacles are removed.

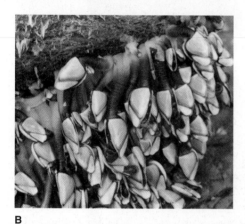

figure 13.27

A, Acorn barnacles, *Semibalanus cariosus* (subclass Cirripedia), are found on rocks along the Pacific Coast of North America. **B,** Common gooseneck barnacles, *Lepas anatifera*. Note the feeding legs, or cirri. Barnacles attach themselves to a variety of firm substrates, including rocks, pilings, and boat bottoms.

Class Malacostraca

Class Malacostraca (mal′a-kos′tra-ka) (Gr. *malakos,* soft, + *ostrakon,* shell) is the largest class of Crustacea and shows great diversity. We mention only 4 of its 12 to 13 orders. The trunk of malacostracans usually has eight thoracic and six abdominal segments, each with a pair of appendages. There are many marine and freshwater species.

Isopoda (i-so′pod-ə) (Gr. *isos,* equal, + *podos,* foot) are commonly dorsoventrally flattened, lack a carapace, and have sessile compound eyes. Their abdominal appendages bear gills. Common land forms are sow bugs or pill bugs (*Porcellio* and *Armadillidium,* figure 13.28A), which live under stones and in other damp places. *Asellus* is common in fresh water, and *Ligia* is abundant on sea beaches and rocky shores. Some isopods are parasites of other crustaceans or of fish (figure 13.29).

Members of **Amphipoda** (am-fi′-pod-ə) (Gr. *amphis,* on both sides, + *podos,* foot) resemble isopods in having sessile compound eyes and no carapace. However, they are usually compressed laterally, and their gills are in the thoracic position, as in other malacostracans. The many marine amphipods (figure 13.30) include beach fleas (*Orchestia*), and there are also numerous freshwater species.

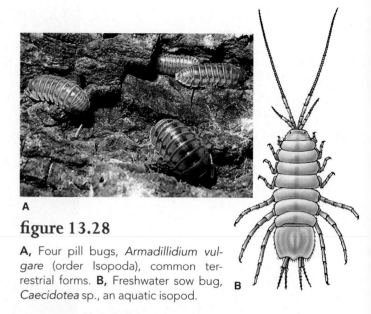

A

figure 13.28

A, Four pill bugs, *Armadillidium vulgare* (order Isopoda), common terrestrial forms. **B,** Freshwater sow bug, *Caecidotea* sp., an aquatic isopod.

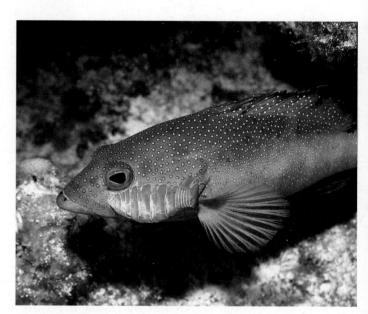

figure 13.29

An isopod parasite (*Anilocra* sp., order Isopoda, class Malacostraca) on a coney (*Cephalopholis fulvus*) inhabiting a Caribbean coral reef.

Euphausiacea (yu-faws′i-a′se-a) (Gr. *eu,* well, + *phausi,* shining bright, + *acea,* L. suffix, pertaining to) is a group of only about 90 species that constitute the important oceanic plankton called "krill." Individuals are about 3 to 6 cm long (figure 13.31) and commonly occur in great oceanic swarms, where they are eaten by baleen whales and many fishes.

Decapoda (deca′-pod-ə) (Gr. *deka,* ten, + *podos,* foot) have five pairs of walking legs of which the first is often modified to form pincers **(chelae)** (see figures 13.17 and 13.19). These are lobsters, crayfishes (see figure 13.19), shrimps (see figure 13.24), and crabs, the largest of the crustaceans (figure 13.32). True crabs differ from others in having a broader carapace and a much reduced abdomen (figure 13.32A and C).

A   **B**   **C**

figure 13.30

Marine amphipods. **A,** Free-swimming amphipod, *Anisogammarus* sp. **B,** Skeleton shrimp, *Caprella* sp., shown on a bryozoan colony, resemble praying mantids. **C,** *Phronima,* a marine pelagic amphipod, occupies the tunic of a salp (subphylum Urochordata, see Chapter 15). Swimming by means of its abdominal swimmerets, which protrude from the opening of the barrel-shaped tunic, the amphipod (order Amphipoda, class Malacostraca) maneuvers to catch its prey. The tunic is not seen.

figure 13.31

Meganyctiphanes, order Euphausiacea, "northern krill."

Familiar examples are fiddler crabs, *Uca,* which burrow in sand just below high-tide level (figure 13.32C); decorator crabs, which cover their carapaces with sponges and sea anemones for camouflage; and spider crabs, such as *Libinia.* Hermit crabs (figure 13.32B) have become adapted to living in snail shells; their abdomen, which lacks a hard exoskeleton, is protected by the snail shell.

figure 13.32

Decapod crustaceans. **A,** A bright orange tropical rock crab, *Grapsus grapsus,* is a conspicuous exception to the rule that most crabs exhibit cryptic coloration. **B,** A hermit crab, *Elassochirus gilli,* which has a soft abdominal exoskeleton, occupies and carries a snail shell into which it can withdraw for protection. **C,** A male fiddler crab, *Uca* sp., uses its enlarged cheliped in territorial displays and in threat and combat. **D,** A red night shrimp, *Rhynchocinetes rigens,* prowls caves and overhangs of coral reefs, but only at night. **E,** Spiny lobster *Panulirus argus* (order Decapoda, class Malacostraca).

A

B

C

D

E

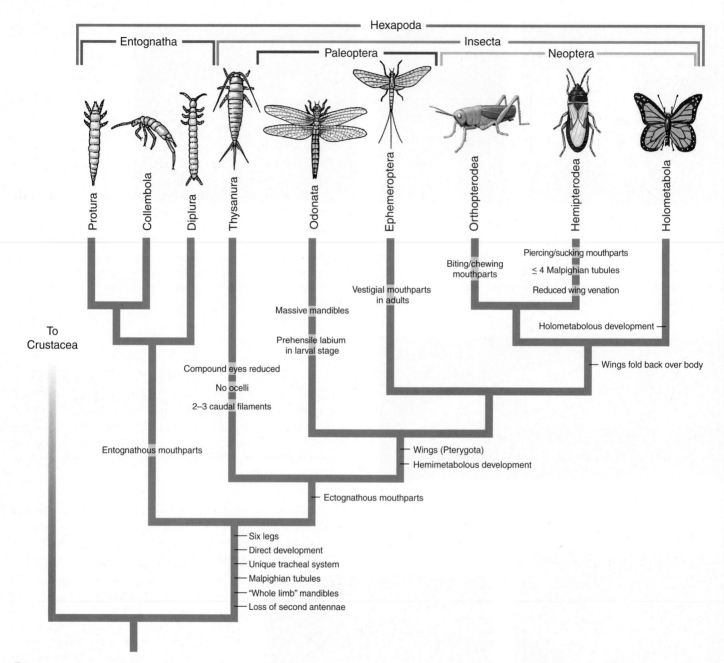

figure 13.33

Cladogram showing hypothetical relationships among hexapods. Many synapomorphies have been omitted. Orders Protura, Collembola, and Diplura are entognathous. These orders, plus Thysanura, originated before the earliest winged ancestors. Orders Odonata and Ephemeroptera form Paleoptera, where wings are outspread. The remaining orders have wings that can fold back over the abdomen (Neoptera). Superorder Orthopterodea includes orders Orthoptera, Blattodea, Phasmatodea, Mantodea, Mantophasmatodea, Isoptera, Plecoptera, Embiidina, and Dermaptera. Hemipterodea includes orders Zoraptera, Psocoptera, Hemiptera, Thysanoptera, and Phthiraptera; and superorder Holometabola encompasses all holometabolous orders.

■ Subphylum Hexapoda

The subphylum Hexapoda is named for **six legs** present in its members. All legs are **uniramous.** Hexapods have **three tagmata**—head, thorax, and abdomen—with appendages on the head and thorax. Abdominal appendages are greatly reduced in size or absent. The subphylum has two classes: Entognatha, a small group whose members have the bases of mouthparts

enclosed within the head capsule, and Insecta, an enormous group whose members have the bases of mouthparts visible outside the head capsule (hence, ectognathous mouthparts).

Winged insects are called pterygotes, and wingless insects are called apterygotes. The wingless order Thysanura forms the living sister taxon to all other insects. Insect wings evolved in a common ancestor of the latter clade (figure 13.33). Thysanurans are called primitively wingless to distinguish them from

figure 13.34

Pie diagram indicating the relative numbers of insect species compared to the rest of the animal kingdom and protozoan groups.

orders whose members do not have wings now, but whose ancestors were winged.

Class Insecta: Insects

Insects are the most numerous and diverse of all groups of arthropods (figure 13.34). There are more species of insects than species in all the other classes of animals combined. The number of insect species named has been estimated at close to 1 million, with thousands, perhaps millions, of other species yet to be discovered and classified.

There are more than 30 orders of insects. Some of the most common ones, such as Diptera and Hymenoptera, are briefly outlined beginning on page 281.

It is difficult to appreciate the great significance of this extensive group and its role in the biological network of animal life. The study of insects (**entomology**) occupies the time and resources of thousands of skilled researchers across the world. The struggle between humans and insect pests seems to be endless, and yet paradoxically, insects are so interwoven into the economy of nature that we would have a difficult time without them.

Insects differ from other arthropods in having **three pairs of legs** and usually **two pairs of wings** on the thoracic region of the body (figure 13.35), although some have one pair of wings or none. In size, insects range from less than 1 mm to 20 cm in length, most being less than 2.5 cm long.

Distribution and Adaptability

Insects have occupied practically all habitats that can support life, but relatively few are marine. They are common in brackish water, in salt marshes, and on sandy beaches. They are abundant in fresh water, soils, forests, and plants, and they even occur in deserts and wastelands, on mountaintops, and as parasites in and on the bodies of plants and animals, including other insects.

A

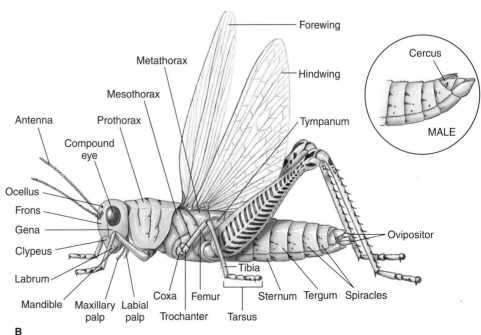

B

figure 13.35

Class Insecta. **A,** A pair of lubber grasshoppers, *Romalea guttata* (order Orthoptera), copulating. The African desert locust mentioned in the chapter prologue (see p. 245) is *Schistocerca gregaria*. **B,** External features of a female grasshopper. The terminal segment of a male with external genitalia is shown in inset.

figure 13.36

A, Praying mantis (order Mantodea), feeding on an insect. **B,** Praying mantis laying eggs.

Their wide distribution is made possible by their powers of flight and their highly adaptable nature. Many insects can easily surmount barriers that are impassable to other animals. Their small size and well-protected eggs allow them to be carried great distances by wind, water, and other animals.

Members of this diverse and vigorous group use all available resources of food and shelter.

Much of insects' success is due to the adaptive qualities of their cuticular exoskeleton, as is true of other arthropods as well. However, the great exploitation of terrestrial environments by insects has been made possible by their array of adaptations for withstanding their rigors. For example, to minimize evaporative water loss, their epicuticle has both a waxy layer and a varnish layer, and they can close their spiracles. Insects extract maximal fluid from food and fecal material, and many can retain water produced in oxidative metabolism. Many can enter a resting stage (diapause) and lie dormant during inhospitable conditions.

External Features

Insect tagmata are **head, thorax,** and **abdomen.** The cuticle of each body segment is typically composed of four plates (**sclerites**), a dorsal notum (**tergum**), a ventral **sternum,** and

a pair of lateral **pleura.** The pleura of abdominal segments are membranous rather than sclerotized (hardened).

The head usually bears a pair of relatively large compound eyes, a pair of antennae, and three ocelli. Mouthparts typically include a **labrum,** a pair each of **mandibles** and **maxillae,** a **labium,** and a tonguelike hypopharynx. The type of mouthparts that an insect possesses determines how it feeds. We discuss some of these modifications in the section titled "Internal Form and Function."

The thorax is composed of three segments: prothorax, mesothorax, and metathorax, each bearing a pair of legs (figure 13.35B). In most insects, the mesothorax and metathorax each bear a pair of wings. The wings consist of a double membrane that contains veins of thicker cuticle, which serve to strengthen the wing. Although these veins vary in their patterns among different species, they often are constant within a species and serve as one means of classification and identification.

The legs of insects are often modified for special purposes. Terrestrial forms have walking legs with terminal pads and claws, as in beetles, for example. These pads may be sticky for walking upside down, as in house flies. The hindlegs of grasshoppers and crickets are adapted for jumping (figure 13.35B). In mole crickets, the first pair of legs is modified for burrowing in the ground. Water bugs and many beetles have paddle-shaped appendages for swimming. For grasping prey, the forelegs of a praying mantis are long and strong (figure 13.36).

Wings and the Flight Mechanism

Insects share the power of flight with birds and flying mammals. However, wings evolved independently in birds, bats, and insects. Insect wings are composed of cuticle and formed by outgrowth from the body wall of the mesothoracic and metathoracic segments.

Most insects have two pairs of wings, but Diptera (true flies, figure 13.37) have only one pair, the hindwings being represented by a pair of small **halteres** (sing., **halter;** balancer) that vibrate and are responsible for equilibrium during flight. Males in order Strepsiptera have only a hind pair of wings and an anterior pair of halteres. Males of scale insects also have one pair of wings but no halteres. Some insects are wingless. Ants and termites, for example, have wings only on males and on females during certain periods; workers (females) are always wingless. Lice, bedbugs, and fleas are always wingless.

Wings may be thin and membranous, as in flies and many others; thick and stiff, as in the forewings of beetles; parchmentlike, as in the forewings of grasshoppers; covered with fine scales, as in butterflies and moths; or hairlike structures, as in caddisflies.

Wing movements are controlled by a complex of thoracic muscles. Direct flight muscles are attached to a part of the wing itself. Indirect flight muscles are not attached to the wing and cause wing movement by altering the shape of the thorax. The wing is hinged at the thoracic tergum and also slightly laterally on a pleural process, which acts as a fulcrum (figure 13.38). In all insects, the upstroke of a wing is effected by contracting indirect muscles that pull the tergum down toward the sternum. Locusts, dragonflies, and cockroaches (figure 13.38A) accomplish the downstroke by contracting direct muscles attached to the wings lateral to the pleural fulcrum. In flies, bees, and midges, all flight muscles are indirect. The downstroke occurs when sternotergal muscles relax and longitudinal muscles of the thorax arch the tergum (figure 13.38B), pulling the tergal articulations upward relative to the pleura. The downstroke in beetles and grasshoppers involves both direct and indirect muscles.

figure 13.37

House fly, *Musca domestica* (order Diptera). House flies can become contaminated with over 100 human pathogens which may be transferred to human and animal food by direct contact, regurgitated food, and feces.

Contraction of flight muscles is governed by two basic types of neural control: synchronous and asynchronous. Larger insects such as dragonflies and butterflies have synchronous muscles, in which a single volley of nerve impulses stimulates a muscle contraction and thus one wing stroke. Asynchronous muscles occur in other insects. Their mechanism of action is complex and depends on storage of potential energy in resilient parts of the thoracic cuticle. As one set of muscles contracts (moving the wing in one direction), the cuticle changes shape. When these muscles relax, the cuticle rebounds, stretching the antagonistic set of muscle passively. These muscles then actively contract (moving the wing in the other direction). Because the muscle contractions are not phase-related to nervous stimulation, only occasional nerve impulses are necessary to keep the muscles responsive to alternating stretch activation. Thus, extremely rapid wing beats are possible. For example, butterflies (with synchronous muscles) may beat as few as four times per second. Insects with asynchronous muscles, such as flies and bees, may vibrate at 100 beats per second or more. Fruit flies, *Drosophila* (Gr. *drosos,* dew, + *philos,* loving), can fly at 300 beats per second, and midges have been clocked at more than 1000 beats per second!

Obviously, flying entails more than simple flapping of wings; a forward thrust is necessary. As the indirect flight muscles alternate rhythmically to raise and lower the wings, the direct flight muscles alter the angle of the wings so that they act as lifting airfoils during both upstroke and downstroke, twisting the leading edge of the wings downward during downstroke and upward during upstroke. This modulation produces a figure-eight movement (see figure 13.38C) that aids in spilling air from the trailing edges of the wings. The quality of the forward thrust depends, of course, on several factors, such as

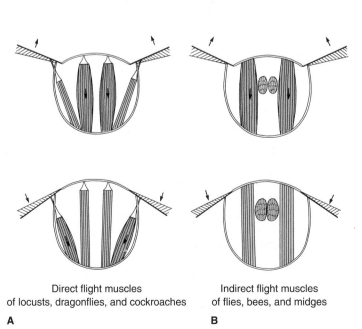

Direct flight muscles of locusts, dragonflies, and cockroaches

Indirect flight muscles of flies, bees, and midges

A

B

figure 13.38

A, Flight muscles of insects such as locusts, dragon flies, and cockroaches, in which upstroke is by indirect muscles and downstroke is by direct muscles. **B,** In insects such as flies, bees, and midges, both upstroke and downstroke are by indirect muscles. **C,** The figure-eight path followed by the wing of a flying insect during the upstroke and downstroke.

C

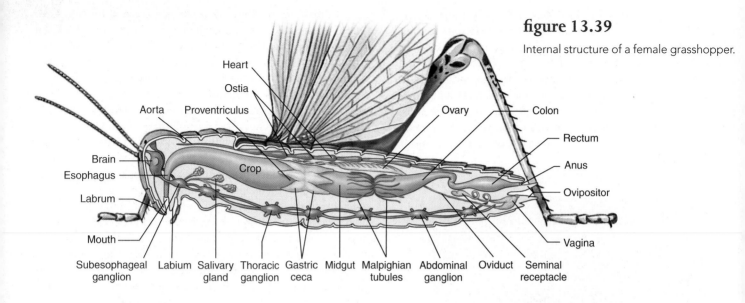

figure 13.39

Internal structure of a female grasshopper.

variations in wing venation, how much the wings are tilted, and how they are tapered.

Flight speeds vary. The fastest flyers usually have narrow, fast-moving wings with a pronounced tilt and a strong figure-eight component. Sphinx moths and horse flies are said to achieve approximately 48 km (30 miles) per hour, and dragon-flies approximately 40 km (25 miles) per hour. Some insects are capable of long, continuous flights. Migrating monarch butter-flies, *Danaus plexippus* (Gr. after Danaus, mythical king of Ara-bia) (see figure 13.47), travel south for hundreds of miles in the fall, flying at a speed of approximately 10 km (6 miles) per hour.

Internal Form and Function

Nutrition   The digestive system (figure 13.39) consists of a foregut (mouth with salivary glands, esophagus, **crop** for storage, and **proventriculus** for grinding), midgut (stomach and gastric ceca), and hindgut (intestine, rectum, and anus). The foregut and hindgut are lined with cuticle, so absorption of food is confined largely to the midgut, although some absorption may occur in all sections. Most insects feed on plant juices and plant tissues, a feeding habit described as **phytophagous.** Some insects feed on specific plants; others, such as grasshoppers, can eat almost any plant. Caterpillars of many moths and butterflies eat the foliage of only certain plants. Certain species of ants and termites cultivate fungus gardens as a source of food.

Many beetles and the larvae of many insects live on dead animals, a feeding habit called **saprophagous.** A number of insects are **predaceous,** catching and eating other insects as well as other types of animals.

Many insects are parasitic as adults or as larvae, and in some cases, both juveniles and adults are parasites. For exam-ple, fleas (figure 13.40) live on the blood of mammals as adults, but their larvae are free-living scavengers. Lice (figures 13.41 and 13.42) are parasitic throughout their life cycle. Many

figure 13.40

Female human flea, *Pulex irritans* (order Siphonaptera).

figure 13.41

Linognathus vituli, the long-nosed cattle louse, attaches to hairs. Five egg cases, also cemented to hairs, are visible. Larvae are emerging from two egg cases.

figure 13.42

The head and body louse of humans, *Pediculus humanus* (order Phthiraptera, suborder Anoplura), feeding.

parasitic insects are themselves parasitized by other insects, a condition called **hyperparasitism.**

Larvae of many varieties of wasps live inside the bodies of spiders or other insects (figure 13.43), consuming their hosts and eventually killing them. Because they always destroy their hosts, they are **parasitoids** (a particular type of parasite); typical parasites normally do not kill their hosts. Parasitoid insects are enormously important in controlling other insect populations.

The feeding habits of insects are determined to some extent by their mouthparts, which are highly specialized for each type of feeding.

Biting and chewing mouthparts, such as those of grasshoppers and many herbivorous insects, are adapted for seizing and crushing food (figure 13.44). The mandibles of chewing insects are strong, toothed plates with edges that can bite or tear while the maxillae hold the food and pass it toward the mouth. Enzymes secreted by the salivary glands add chemical action to the chewing process.

Sucking mouthparts are greatly varied. House flies and fruit flies have no mandibles; their labium is modified into two soft lobes containing many small tubules that absorb liquids with a capillary action, much as do the holes of a commercial sponge (figure 13.44D; see figure 13.37). Horse flies, however, are fitted not only to soak up surface liquids but to bite into skin with their slender, tapering mandibles and then collect blood. Mosquitos combine piercing by means of needlelike stylets and sucking through a food channel (figure 13.44B). In honey bees, the labium forms a flexible and contractile "tongue" covered with many hairs. When a bee plunges its proboscis into nectar, the tip of the tongue bends upward and moves back and forth rapidly. Liquid enters the tube by capillarity and is drawn up the tube continuously by a pumping pharynx. In butterflies and moths, mandibles are usually absent, and maxillae are modified into a long, sucking proboscis (figure 13.44C) for drawing nectar from flowers. At rest, the proboscis is coiled into a flat spiral. In feeding, it extends, and pharyngeal muscles pump fluid.

A

B

figure 13.43

A, Larval stage of the tomato hornworm, *Manduca sexta* (order Lepidoptera). The more than 100 species of North American sphinx moths are strong fliers and mostly nocturnal feeders. Their larvae, called hornworms because of the large fleshy posterior spine, are often pests of tomatoes, tobacco, and other plants. **B,** Hornworm parasitized by a tiny wasp, *Apanteles*, which laid its eggs inside the caterpillar. The wasp larvae have emerged, and their pupae are on the caterpillar's skin. Young wasps emerge in 5 to 10 days, and the caterpillar usually dies.

Circulation   A tubular heart in the pericardial cavity (see figure 13.39) moves **hemolymph,** consisting of plasma and amebocytes, forward through the only blood vessel, a dorsal aorta. The heartbeat is a peristaltic wave. Accessory pulsatory organs help move hemolymph into the wings and legs, and flow is also facilitated by various body movements. Hemolymph apparently has little role in oxygen transport; rather, it distributes substances such as molting hormones and nutrients throughout the body. Oxygen transport is done by the **tracheal system,** an unusual system of tubes that pipe air directly to each cell.

Gas Exchange   Terrestrial animals obtain oxygen from the air as it dissolves across a wet membrane. Maintaining a wet membrane in a terrestrial environment is difficult, so gas exchange typically occurs in an internal cavity. The twin goals of an efficient respiratory system are to permit rapid oxygen–carbon dioxide exchange and to restrict water loss. In insects, this is the function of the tracheal system, an extensive network of thin-walled tubes that branch into every part of the body (figure 13.45). Tracheal trunks open to the outside by paired **spiracles,** usually

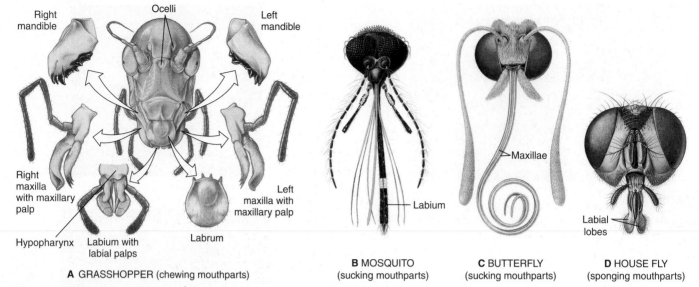

figure 13.44

Four types of insect mouthparts.

figure 13.45

A, Relationship of spiracles, tracheae, taenidia (chitinous bands that strengthen the tracheae), and tracheoles (diagrammatic). **B,** Generalized arrangement of insect tracheal system (diagrammatic). Air sacs and tracheoles not shown. **C,** Spiracles and tracheae of a caterpillar visible through its transparent cuticle.

two on the thorax and seven or eight on the abdomen. A spiracle may be a simple hole in the integument, as in primitively wingless insects, but it more often has a valve or other closing mechanism that decreases water loss. Evolution of such a device must have been very important in enabling insects to move into drier habitats.

Tracheae are composed of a single layer of cells and are lined with cuticle that is shed, along with the outer cuticle, during molts. Spiral thickenings of the cuticle, called **taenidia,** support the tracheae and prevent their collapse. Tracheae branch into smaller tubes, ending in very fine, fluid-filled tubules called **tracheoles** (not lined with cuticle), which branch into a fine network over the cells. Scarcely any living cell is located more than a few micrometers away from a tracheole. In fact, the ends of some tracheoles actually indent the membranes of the cells that they supply, so that they terminate close to mitochondria.

The tracheal system affords an efficient system of transport without the use of oxygen-carrying pigments in hemolymph.

In some very small insects, gas transport occurs entirely by diffusion along a concentration gradient. As oxygen is used, a partial vacuum develops in the tracheae, drawing air inward through the spiracles. Larger or more active insects employ a ventilation device for moving air into and out of the tubes. Usually muscular movements in the abdomen perform the pumping action that draws air in or expels it.

The tracheal system is primarily adapted for breathing air, but many insects (nymphs, larvae, and adults) live in water. In small, soft-bodied aquatic nymphs, gas exchange may occur by diffusion through the body wall, usually into and out of a tracheal network just under the integument. The aquatic nymphs of stone-flies and mayflies are equipped with tracheal gills, which are thin extensions of the body wall containing a rich tracheal supply. The gills of dragonfly nymphs are ridges in the rectum (rectal gills) where gas exchange occurs as water enters and leaves.

Although diving beetles, *Dytiscus* (G. *dytikos*, able to swim), can fly, they spend most of their life in water as excellent swimmers. They use an "artificial gill" in the form of a bubble of air held under the first pair of wings. The bubble is kept stable by a layer of hairs on top of the abdomen and is in contact with the spiracles on the abdomen. Oxygen from the bubble diffuses into the tracheae and is replaced by diffusion of oxygen from the surrounding water. However, nitrogen from the bubble diffuses into the water, slowly decreasing the size of the bubble; therefore, diving beetles must surface every few hours to replace the air. Mosquito larvae are not good swimmers but live just below the surface, putting out short breathing tubes like snorkels to the surface for air. Spreading oil on the water, a favorite method of mosquito control, clogs the tracheae with oil and suffocates the larvae. "Rattailed maggots" of syrphid flies have an extensible tail that can stretch as much as 15 cm to the water surface.

Excretion and Water Balance   Malpighian tubules (see figure 13.39) occur in most insects. As in spiders (see p. 251), Malpighian tubules are very efficient, both as excretory organs and as a means of conserving body fluids—an important factor in the success of terrestrial animals.

Because water requirements vary among different types of insects, this ability to cycle water and salts is very important. Insects living in dry environments may resorb nearly all of the water from the rectum, producing a nearly dry mixture of urine and feces. Leaf-feeding insects take in and excrete large quantities of fluid. Freshwater larvae need to excrete water and conserve salts. Insects that feed on dry grains need to conserve water and to excrete salt.

Nervous System   The insect nervous system in general resembles that of larger crustaceans, with a similar tendency toward fusion of ganglia (see figure 13.39). Some insects have a giant fiber system. There is also a visceral nervous system that corresponds in function with the autonomic nervous system of

vertebrates. Neurosecre parts of the brain have an endocrine fu except for their role in molting and metamorph , little is known of their activity.

Sense Organs   The sensory perceptions of insects are usually keen. Organs receptive to mechanical, auditory, chemical, visual, and other stimuli are well developed and are scattered over the body, but especially on the appendages.

Photoreceptors include both ocelli and compound eyes. Compound eyes are large and constructed of ommatidia, as are those of crustaceans (see p. 261). Apparently, the visual acuity of insect eyes is much lower than that of human eyes, but most flying insects rate much higher than humans in flicker-fusion tests. Flickers of light become fused in human eyes at a frequency of 45 to 55 per second, but in bees and blow flies they do not fuse until 200 to 300 per second. This would be an advantage in analyzing a fast-changing landscape. Most insects have three ocelli on their head, and they also have dermal light receptors on their body surface.

Insects may detect sounds by means of sensitive, hairlike **sensilla** or by relying on tympanic organs sensitive to sonic or ultrasonic sound. Sensilla are modifications in the cuticular surface for reception of sensory stimuli other than light and are supplied with one or more neurons. Tympanic organs, found in grasshoppers (see figure 13.35B), crickets, cicadas, butterflies, and moths, involve a number of sensory cells extending to a thin tympanic membrane that encloses an air space in which vibrations can be detected.

Chemoreceptive sensilla, which are peglike setae, are especially abundant on the antennae, mouthparts, or legs. Mechanical stimuli, such as contact pressure, vibrations, and tension changes in the cuticle, are detected by sensilla or by sensory cells in the epidermis. Insects also sense temperature, humidity, body position (proprioception), and gravity.

Reproduction   Sexes are separate in insects, and fertilization is usually internal. Insects have various means of attracting mates. Female moths emit a chemical (pheromone) that males can detect for a great distance—several miles from females. Fireflies use flashes of light; other insects find each other by means of sounds or color signals and by various kinds of courtship behavior.

At the time of copulation, sperm are usually deposited in the female's vagina (see figure 13.35A). In some orders, sperm are encased in spermatophores that may be transferred at copulation or deposited on the substrate to be collected by a female. A male silverfish deposits a spermatophore on the ground, and then spins signal threads to guide a female to it. During the evolutionary transition from aquatic to terrestrial life, spermatophores were widely used, with copulation evolving much later.

Usually sperm are stored in the seminal receptacle of a female in numbers sufficient to fertilize more than one batch of eggs. Many insects mate only once during their lifetime, and none mates more than a few times. Sperm storage allows fertilization to occur much later.

Insects usually [...] A queen honey bee, for example, may lay more than [...] eggs during her lifetime. On the other hand, some flies are ovoviviparous and bring forth only a single offspring at a time. Forms that make no provision for care of young usually lay many more eggs than those that provide for their young or those that have a very short life cycle.

Most species lay their eggs in a particular habitat to which they are guided by visual, chemical, or other clues. Butterflies and moths lay their eggs on the specific kind of plant on which their caterpillars must feed. A tiger moth may look for a pigweed, a sphinx moth for a tomato or tobacco plant, and a monarch butterfly for a milkweed plant. Insects whose immature stages are aquatic lay their eggs in water. A tiny braconid wasp lays her eggs on a caterpillar of the sphinx moth where they will feed and pupate in tiny white cocoons (see figure 13.43B). An ichneumon wasp, with unerring accuracy, seeks out a certain kind of larva in which her young will live as internal parasites. Her long ovipositor may have to penetrate 1 to 2 cm of wood to deposit her eggs in the larva of a wood wasp or a wood-boring beetle (figure 13.46).

Metamorphosis and Growth

The early development of insects occurs within the eggshell, and hatching young escape in various ways. During postembryonic development, most insects change in form; they undergo **metamorphosis.** A number of molts are necessary during the growth period, and each stage between molts is called an **instar.** Although many animals undergo metamorphosis, insects illustrate it more dramatically than does any other group. The transformation of a caterpillar into a beautiful moth or butterfly is indeed an astonishing morphological change.

Approximately 88% of insects undergo **holometabolous (complete) metamorphosis** (Gr. *holo,* complete, + *metabolē,* change), which separates the physiological processes of growth **(larva)** from those of differentiation **(pupa)** and reproduction **(adult).** Each stage functions efficiently without competition with other stages, because the larvae and adults often live in entirely different surroundings and eat different foods. The wormlike larvae, which usually have chewing mouthparts, have various common names, such as caterpillars, maggots, bagworms, fuzzy worms, and grubs. After a series of instars, a larva forms a case or cocoon around itself and becomes a pupa, or chrysalis, a nonfeeding stage in which many insects pass the winter. When the final molt occurs, a full-grown adult emerges (figure 13.47), pale and with wrinkled wings. In a short time, the wings expand and harden, and the insect flies for the first time. Adults undergo no further molting.

Some insects undergo **hemimetabolous (gradual, or incomplete) metamorphosis** (Gr. *hemi,* half, + *metabolē* change) (figure 13.48). These include bugs, scale insects, lice, and grasshoppers, which have terrestrial young, and mayflies, stoneflies (figure 13.49A), and dragonflies (figure 13.49B), which lay their eggs in water. The young are called **nymphs** or simply juveniles (figure 13.49C), and their wings develop externally as budlike outgrowths in the

figure 13.46

An ichneumon wasp with the end of the abdomen raised to thrust her long ovipositor into wood. She can bore 13 mm or more into wood to find a tunnel made by the larva of a wood wasp or wood-boring beetle. After she lays her eggs in the larva, the wood-boring beetle larva becomes host for the ichneumon larvae. Other ichneumon species attack spiders, moths, flies, crickets, caterpillars, and other insects.

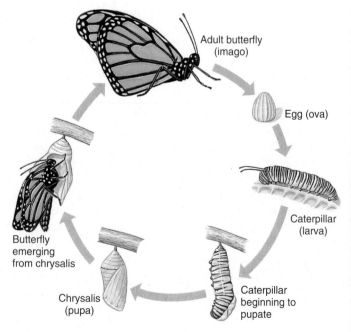

Adult butterfly (imago)

Egg (ova)

Caterpillar (larva)

Caterpillar beginning to pupate

Chrysalis (pupa)

Butterfly emerging from chrysalis

figure 13.47

Holometabolous (complete) metamorphosis in a butterfly, *Danaus plexippus* (order Lepidoptera). Eggs hatch to produce the first of several larval instars. The last larval instar molts to become a pupa. At the pupal molt, an adult emerges.

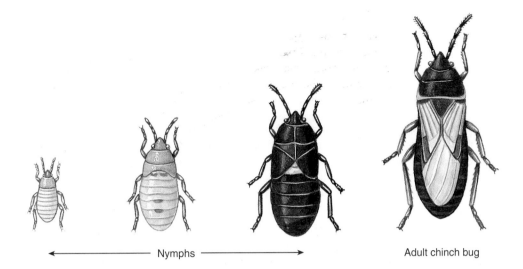

figure 13.48

Life history of a hemime-
tabolous insect.

Egg   ← Nymphs →   Adult chinch bug

A   B   C

figure 13.49

A, A stonefly, *Perla* sp. (order Plecoptera). **B,** A ten-spot dragonfly, *Libellula pulchella* (order Odonata). **C,** Nymph (juvenile) of a dragonfly.
Both stoneflies and dragonflies have aquatic larvae that undergo gradual metamorphosis.

A   B

figure 13.50

A, Ecdysis in a cicada, *Magicicada septendecium* (order Hemiptera, suborder Auchenorrhyncha). The old cuticle splits along a dorsal
midline due to increased blood pressure and air forced into the thorax by muscle contraction. The emerging insect is pale, and its new
cuticle is soft. The wings expand as hemolymph is pumped into veins, and the insect enlarges by taking in air. **B,** An adult cicada.

early instars and increase in size as the animal grows by successive molts and becomes a winged adult (figure 13.50).
Aquatic nymphs have tracheal gills or other modifications
for aquatic life. The stages are egg, nymph (several instars),
and adult.

A few insects, such as silverfish (see figure 13.60) and
springtails, undergo direct development. The young, or juveniles, are similar to the adults except in size and sexual maturation. The stages are egg, juvenile, and adult. Such insects
include the primitively wingless insects.

In common English usage, people often call all insects "bugs," even extending the word to such nonanimals as bacteria, viruses, and glitches in computer programs. Biologically speaking, however, a bug is a member of order Hemiptera, suborder Heteroptera, and nothing else.

Hormones control and regulate metamorphosis in insects. Three major endocrine organs guide development through juvenile instars and the eventual emergence of adults. These organs and the hormones that they produce are the brain (ecdysiotropin), the **ecdysial** or **prothoracic glands** (ecdysone), and the **corpora allata** (juvenile hormone). Hormonal control of molting and metamorphosis is the same in holometabolous and hemimetabolous insects.

Diapause   Many animals can enter a state of dormancy during adverse conditions, and the life cycle of many insects includes a long dormant period during which external climatic conditions are too harsh for normal activity. Most insects enter such a stage facultatively when some environmental factor, such as temperature, becomes unfavorable, and the state continues until conditions again become favorable.

However, some species experience a prolonged arrest of growth that is internally programmed and usually seasonal. This type of dormancy is called **diapause** (di´a-poz) (Gr. *dia*, through, dividing into two parts, + *pausis*, a stopping), and it is an important adaptation to adverse environmental conditions. Diapause is usually triggered by some external signal, such as shortening day length. Diapause always occurs at the end of an active growth stage of the molting cycle so that, when the diapause period is over, the insect is ready for another molt.

Behavior and Communication

Insects' keen sensory perceptions make them extremely responsive to many stimuli. The stimuli may be internal (physiological) or external (environmental), and the responses are governed by both the physiological state of the animal and the pattern of nerve pathways involved. Many responses are simple, such as orientation toward or away from a stimulus as occurs when a moth flies toward light, a cockroach avoids light, or carrion flies approach the odor of dead flesh.

Much insect behavior, however, is not a simple matter of orientation but involves a complex series of responses. A pair of dung beetles chew off a bit of dung, roll it into a ball, and roll the ball laboriously to where they intend to bury it after laying their eggs in it (figure 13.51). A female cicada slits the bark of a twig and then lays an egg in each of the slits. A female potter wasp (*Eumenes*) scoops up clay into pellets, carries them one by one to her building site, and fashions them into dainty, narrow-necked clay pots, into each of which she lays an egg. Then she hunts and paralyzes a number of caterpillars, pokes them into the opening of a pot, and closes the opening with clay. Each egg, in its own protective pot, hatches to find food awaiting.

figure 13.51

Dung beetles, *Canthon pilularis* (order Coleoptera), chew off a bit of dung, roll it into a ball, and then roll it to where they bury it in soil. One beetle pushes while the other pulls. They then lay their eggs in the ball, and the larvae feed on the dung. Dung beetles are black, an inch or less in length, and common in pastures.

Some insects can memorize and perform in sequence tasks involving multiple signals in various sensory areas. Worker honey bees have been trained to walk through mazes that involve five turns in sequence, using such clues as the color of a marker, the distance between two spots, or the angle of a turn. The same is true of ants. Workers of one species of *Formica* learned a six-point maze at a rate only two or three times slower than that of laboratory rats. Foraging trips of ants and bees often wind and loop in a circuitous route, but once the forager has found food, the return trip is relatively direct. One investigator suggested that the continuous series of calculations necessary to figure the angles, directions, distance, and speed of the trip and to convert it into a direct return could involve a stopwatch, a compass, and integral vector calculus. How an insect does it is unknown.

Much insect behavior is "innate," meaning that entire sequences of actions apparently have been programmed. However, more learning is apparently involved than we once thought. A potter wasp, for example, must learn where she has left her pots if she is to return to fill them with caterpillars one at a time. Social insects, which have been studied extensively, are capable of most of the basic forms of learning used by mammals. An exception is insight learning. Apparently insects, when faced with a new problem, cannot reorganize their memories to construct a new response.

Insects communicate with other members of their species by chemical, visual, auditory, and tactile signals. Chemical signals take the form of **pheromones,** which are substances secreted by one individual that affect the behavior or physiological processes of another individual. Examples of pheromones include sex attractants, releasers of certain behavior patterns, trail markers, alarm signals, and territorial markers. Like hormones, pheromones are effective in minute quantities. Social insects, such as bees, ants, wasps, and termites, can recognize a nestmate—or an alien in the nest—by means of identification pheromones. Pheromones determine caste in termites, and to some extent in ants and bees. In fact, pheromones are probably a primary integrating force in populations of social insects. Many insect pheromones have been extracted and chemically identified.

Sound production and sound reception (phonoproduction and phonoreception) in insects have been studied extensively, and although not all insects have a sense of hearing, those that do use sounds as warning devices, to advertise territorial claims, or for courtship. For example, the sounds of crickets and grasshoppers function in both courtship and aggression. Male crickets scrape the modified edges of their forewings together to produce their characteristic chirping. The long, drawn-out sound of male cicadas, a call to attract females, is produced by vibrating membranes in a pair of organs located on the ventral side of the basal abdominal segment.

Insects practice many forms of **tactile communication,** such as tapping, stroking, grasping, and antennae touching, which evoke responses varying from recognition to recruitment and alarm. Certain kinds of flies, springtails, and beetles manufacture their own visual signals in the form of **bioluminescence.** The best-known luminescent beetles are fireflies, or lightning bugs (which are neither flies nor bugs, but beetles), which use a flash of light to locate a prospective mate. Each species has its own characteristic flashing rhythm produced on the ventral side of the last abdominal segments. Females flash an answer to the species-specific pattern to attract males. This interesting "love call" has been adopted by species of *Photuris,* which prey on male fireflies of other species they attract (figure 13.52).

Social Behavior   In terms of being organized into social groups, insects rank very high in the animal kingdom, and cooperation within such complex groups depends heavily on chemical and tactile communication. However, not all social communities are complex. Some community groups are temporary and uncoordinated, such as hibernating associations of carpenter bees or feeding gatherings of aphids (figure 13.53). Some insects are coordinated for only brief periods, such as the tent caterpillars, *Malacosoma,* that join in building a home web and a feeding net. Still, all these are open communities with social behavior.

In true societies of some orders, such as Hymenoptera (honey bees and ants) and Isoptera (termites), a complex social life occurs. Such societies are closed. They involve all stages of the life cycle, communities are usually permanent, all activities are collective, and reciprocal communication occurs. Division of labor is highly efficient. Such a society is essentially a family group in which the mother or perhaps both parents remain with their young, sharing the duties of the group in a cooperative manner. The society usually demonstrates polymorphism, or **caste differentiation.**

Honey bees have one of the most complex social organizations in the insect world. Instead of lasting one season, their organization continues for a more or less indefinite period. As many as 60,000 to 70,000 honey bees may live in a single hive. Of these, there are three castes: a single, sexually mature female, or **queen;** a few hundred drones, which are sexually mature males; and thousands of workers, which are sexually inactive genetic females (figure 13.54).

Workers tend the young, secrete wax with which they build the six-sided cells of the honeycomb, gather nectar from flowers, manufacture honey, collect pollen, and ventilate and guard the hive. One drone, or sometimes more,

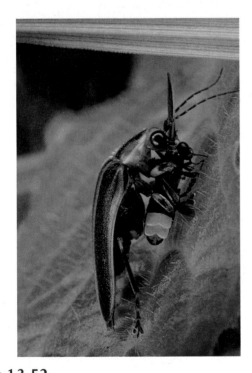

figure 13.52

Firefly femme fatale, *Photuris versicolor,* eating a male *Photinus tanytoxus,* which she has attracted with false mating signals.

figure 13.53

A red carpenter ant, *Camptonotus ferrugineus,* (order Hymenoptera) tending a group of aphids (order Hemiptera, suborder Sternorrhyncha). The aphids feed copiously on plant juices and excrete the excess as a clear liquid rich in carbohydrates ("honeydew"), which ants cherish as a food.

fertilizes the queen during the mating flight, at which time enough sperm are stored in her seminal receptacle to last her lifetime.

Castes are determined partly by fertilization and partly by what is fed to the larvae. Drones develop parthenogenetically from unfertilized eggs (and consequently are haploid); queens and workers develop from fertilized eggs (and thus are diploid; see haplodiploidy, p. 238). Female larvae that will

figure 13.54

Queen bee surrounded by her court. The queen is the only egg layer in the colony. The attendants, attracted by her pheromones, constantly lick her body. As food is transferred from these bees to others, the queen's presence is communicated throughout the colony.

become queens are fed royal jelly, a secretion from the salivary glands of nurse workers. Royal jelly differs from the "worker jelly" fed to ordinary larvae, but the components in it that are essential for queen determination have not yet been identified. Honey and pollen are added to the worker diet about the third day of larval life. Pheromones in "queen substance," which is produced by the queen's mandibular glands, prevent female workers from maturing sexually. Workers produce royal jelly only when the level of "queen substance" pheromone in the colony drops. This change occurs when the queen becomes too old, dies, or is removed. Then workers start enlarging a larval cell and feeding a larva royal jelly that produces a new queen.

Honey bees have evolved an efficient system of communication by which, through certain body movements, their scouts inform workers of the location and quantity of food sources.

Termite colonies contain several castes, consisting of fertile individuals, both males and females, and sterile individuals (figure 13.55). Some fertile individuals may have wings and leave the colony, mate, lose their wings, and start a new colony as king and **queen.** Wingless fertile individuals may, under certain conditions, substitute for the king or queen. Sterile members are wingless and become workers and soldiers. Soldiers, which have large heads and mandibles, defend the colony. As in bees and ants, extrinsic factors cause caste differentiation. Reproductive individuals and soldiers secrete inhibiting pheromones that pass throughout the colony to nymphs via a mutual feeding process called **trophallaxis,** so that they become sterile workers. Workers also produce pheromones, and if the level of "worker substance" or "soldier substance" falls, as might happen after an attack by marauding predators, for example, the next generation produces compensating proportions of the appropriate caste.

Ants also have highly organized societies. Superficially, they resemble termites, but they are quite different (belong to a different order) and can be distinguished easily. In contrast to termites, ants are usually dark in color, are hard bodied, and have a constriction posterior to their first abdominal segment.

figure 13.55

A, Workers of the western subterranean termite, *Reticulitermes hesperus* (order Isoptera). Workers are wingless immature animals that tend the nest and care for the young. **B,** The termite queen becomes a distended egg-laying machine. Here a queen is surrounded by a few workers and soldiers.

A

B

Entomologists describe insect societies by borrowing terms commonly used to describe human societies: queen, king, royal, soldier, worker, and caste. Such terms can be misleading by implying a correspondence between human and insect societies that does not exist. For example, in human societies "queen" suggests a position of political power that has no correspondence to the reproductive role of the "queen" of a bee colony. Confusion of these terms led a famous population geneticist, Ronald Fisher, to argue that human societies could achieve greater stability by emulating insect societies, specifically by concentrating reproduction among members of the upper classes. This argument is now considered an embarrassment of his otherwise highly regarded and influential book, *The Genetical Theory of Natural Selection* (1930).

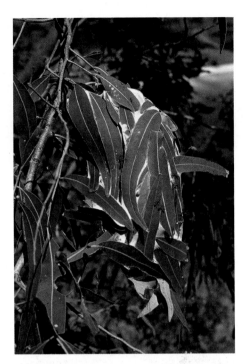

figure 13.56

A weaver ant nest in Australia.

In ant colonies, males die soon after mating, and the queen either starts her own new colony or joins an established colony and does the egg laying. Sterile females are wingless workers and soldiers that do the work of the colony—gather food, care for young, and protect the colony. Many larger colonies may have two or three types of individuals within each caste.

Ants have evolved some striking patterns of "economic" behavior, such as making slaves, farming fungi, herding "ant cows" (aphids or other hemipterans, see figure 13.53), sewing their nests together with silk (figure 13.56), and using tools.

Insects and Human Welfare

Beneficial Insects   Although most of us think of insects primarily as pests, humanity would have great difficulty surviving if all insects were suddenly to disappear. Insects are necessary for cross-fertilization (pollination) of many crops. Bees pollinate over $14 billion worth of food crops per year in the United States alone, and this value does not include pollination of forage crops for livestock or pollination by other insects. In addition, some insects produce useful materials—for example, honey and beeswax from bees, silk from silkworms, and shellac from a wax secreted by lac insects.

Very early in their evolution, insects and flowering plants developed mutual adaptations that have functioned to each other's advantage. Insects exploit flowers for food, and flowers exploit insects for pollination. Each floral petal and sepal arrangement is correlated with the sensory adjustment of certain pollinating insects. Among these mutual adaptations are amazing allurements, traps, specialized structures, and precise timing.

Many predaceous insects, such as tiger beetles, aphid lions, ant lions, praying mantids, and lady beetles, destroy harmful insects (figure 13.57A and B). Some insects control harmful ones by parasitizing them or by laying their eggs where their young, when hatched, may devour the host (figure 13.57C). Dead animals are quickly consumed by maggots hatched from eggs laid on carcasses.

Insects and their larvae serve as an important source of food for many birds, fish, and other animals.

Harmful Insects   Harmful insects include those that eat and destroy plants and fruits; examples are grasshoppers, chinch bugs, corn borers, boll weevils, grain weevils, San Jose scale, and scores of others (figure 13.58). Practically every cultivated crop has several insect pests. Humans expend enormous resources in all agricultural activities, in forestry, and in the food industry to counter insects and the damage they engender. Outbreaks of bark beetles or defoliators, such as spruce budworms and gypsy moths, have generated tremendous economic losses and have had a major impact on the composition of forests in the United States. Gypsy moths, introduced into the United States in 1869 in an ill-advised attempt to breed a better silkworm, have spread throughout the Northeast as far south as Virginia and as far west as Minnesota. In outbreak years, they defoliate oak forests; for example, in 1981, they defoliated 13 million acres in 17 northeastern states.

Lice, blood-sucking flies, warble flies, bot flies, and many others attack humans or domestic animals or both. Malaria (see p. 109), carried by the *Anopheles* mosquito (figure 13.59), is still one of the world's major diseases; mosquitos also

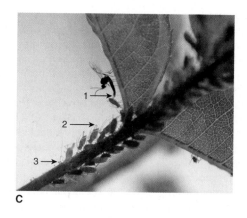

figure 13.57

Some beneficial insects. **A,** A predaceous spined soldier bug (order Hemiptera) feeds on a larva of the Mexican bean beetle. Note the sucking proboscis of the bug. Mexican bean beetles are pests of snap beans and of soybeans. **B,** A lady beetle ("ladybug," order Coleoptera). Adults (and larvae of most species) feed voraciously on plant pests such as mites, aphids, scale insects, and thrips. **C,** A parasitic braconid wasp (*Lysiphlebus testaceipes*, order Hymenoptera) pierces an aphid (arrow 1) to lay her eggs inside. Nearby aphids release alarm pheromones (arrow 2). Aphids, such as *Macrosiphum rosae* shown here, damage plants by feeding on phloem; droplets of phloem from aphid feeding tubes on a rose bush are visible (arrow 3). The green aphid in the lower left is giving birth to live offspring.

figure 13.58

Insect pests. **A,** Japanese beetles, *Popillia japonica* (order Coleoptera), are serious pests of fruit trees and ornamental shrubs. They were introduced into the United States from Japan in 1917. **B,** Mealybugs, *Pseudococcus so.* (order Hemiptera, suborder Sternorrhyncha). Many mealybugs are pests of commercially valuable plants. **C,** Corn ear worms, *Heliothis zea* (order Lepidoptera). An even more serious pest of corn is the infamous corn borer, an import from Europe in 1908 or 1909.

figure 13.59

A mosquito in genus *Aedes* (order Diptera) from South Africa. Members of this genus spread dengue fever and yellow fever, among other illnesses.

transmit yellow fever and lymphatic filariasis (see p. 238). Fleas carry plague, which at times in history has eradicated significant portions of the human population. House flies are vectors of typhoid, as are lice for typhus fever; tsetse flies carry African sleeping sickness (see p. 104); and certain blood-sucking bugs, *Rhodnius* and related genera, transmit Chagas disease (see p. 104).

Tremendous destruction of food, clothing, and property is caused by weevils, cockroaches, ants, clothes moths, termites, and carpet beetles. Not the least of the insect pests are bed bugs, *Cimex,* blood-sucking hemipterous insects originally contracted by humans from bats that shared their caves early in human evolution.

West Nile virus, a disease agent spread by mosquitos, affects mammals and birds throughout the world. First identified in Uganda in 1937, it spread to North America in 1999. Birds serve as a reservoir for the virus: a bird bitten by an infected mosquito plays host to the virus for 1 to 4 days, during which the virus can be picked up by other mosquitos and spread to new hosts. Human response to infection varies, with about 80% showing no symptoms, close to 20% exhibiting flu-like symptoms such as fever and achiness, and less than 1% developing potentially fatal encephalitis or other possibly permanent neurological effects. Preventing mosquito bites is the best way to avoid infection, so studies of mosquito behavior are useful. For example, researchers wonder whether mosquitos that initially feed on infected birds tend to choose another bird for the next bite, or simply feed opportunistically on any available animal. Understanding virus transmission makes possible mathematical models to predict how and when the disease will spread.

Control of Insects   Because all insects are an integral part of the ecological communities to which they belong, their total destruction would probably do more harm than good. Food chains would be disturbed. Some of our favorite birds would disappear. The biological cycles by which dead animal and plant matter disintegrates and returns to enrich the soil would be seriously impeded. The beneficial roles of insects in our environment are often overlooked, and in our zeal to control the pests, we spray the landscape indiscriminately with extremely effective "broad-spectrum" insecticides that eradicate good, as well as harmful, insects. We have also found, to our dismay, that many chemical insecticides persist in the environment and accumulate as residues in the bodies of animals higher in the food chain. Furthermore, many insects have developed resistance to insecticides in common use.

Methods of control other than chemical insecticides have been under intense investigation, experimentation, and development. Economics, concern for the environment, and consumer demand are causing farmers across the United States to use alternative methods to control insect pests, rather than depending strictly on chemicals.

Several types of biological controls have been developed and are under investigation. All of these areas present problems but also show great potential. One method uses bacterial, viral, and fungal pathogens. A bacterium, *Bacillus thuringiensis,* is quite effective in controlling lepidopteran pests (cabbage looper, imported cabbage worm, tomato worm, and gypsy moth). Other strains of *B. thuringiensis* attack insects in other orders, and the species diversity of target insects is being widened by genetic engineering techniques. Genes coding for the toxin produced by *B. thuringiensis* (Bt) have also been introduced into the DNA of the plants themselves, which makes the plants resistant to insect attack, while the Bt is harmless to humans. Genes for Bt and for herbicide resistance have been incorporated into much of the soybeans, corn, cotton, and canola produced in the United States, thus reducing the need for hazardous chemical sprays. However, some insects have now evolved resistance to the toxin. Concerns about the health risks of consuming genetically modified crops have arisen, especially in Europe, but such fears are supported by little scientific evidence.

Taxonomy of Subphylum Hexapoda

Hexapods are divided into orders chiefly by morphology including developmental features. Entomologists are not unanimous on the names of the orders or on the contents of each order. Some tend to combine and others to divide groups. However, the following synopsis of major orders is rather widely accepted.

Class Entognatha

Order Protura (pro-tu´ra) (Gr. *protos,* first, + *oura,* tail). Minute (1 to 1.5 mm); no eyes or antennae; appendages on abdomen as well as thorax; live in soil and dark, humid places; slight, gradual metamorphosis.

Order Diplura (dip-lu´ra) (Gr. *diploos,* double, + *oura,* tail): **japygids.** Usually less than 10 mm; pale, eyeless; have a pair of long terminal filaments or a pair of caudal forceps; live in damp humus or rotting logs; direct development.

Order Collembola (col-lem´bo-la) (Gr. *kolla,* glue, + *embolon,* peg, wedge): **springtails** and **snowfleas.** Small (5 mm or less); no eyes; respiration by trachea or body surface; use a springing organ folded under the abdomen for leaping; abundant in soil; sometimes swarm on pond surface film or on snowbanks in spring; direct development.

Class Insecta

Order Thysanura (thy-sa-nu´ra) (Gr. *thysanos,* tassel, + *oura,* tail): **silverfish** (figure 13.60) and **bristletails.** Small to medium size; large eyes; long antennae; three long terminal cerci; live under stones and leaves and around human habitations; wingless; direct development.

Order Ephemeroptera (e-fem-er-op´ter-a) (Gr. *ephēmeros,* lasting but a day, + *pteron,* wing): **mayflies** (figure 13.61). Wings membranous; forewings larger than hindwings; adult mouthparts vestigial; nymphs aquatic, with lateral tracheal gills; hemimetabolous development.

Order Odonata (o-do-na´ta) (Gr. *odontos,* tooth, + *ata,* characterized by): **dragonflies** (see figure 13.49B) and **damselflies.** Large; membranous wings are long, narrow,

figure 13.60

Silverfish, *Lepisma* (order Thysanura), is often found in homes.

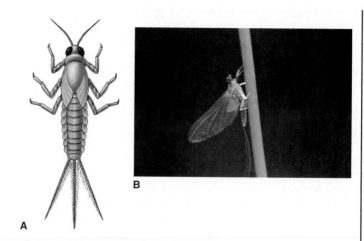

figure 13.61

Mayfly (order Ephemeroptera). **A,** Nymph. **B,** Adult.

net-veined, and similar in size; long and slender body; aquatic nymphs have aquatic gills and prehensile labium for capturing prey; hemimetabolous development.

Order Orthoptera (or-thop´ter-a) (Gr. *orthos,* straight, + *pteron* wing): **grasshoppers** (see figure 13.35), **locusts, crickets,** and **katydids.** Wings, when present, have forewings thickened and hindwings folded like a fan under forewings; chewing mouthparts.

Order Blattodea (blə-´tō-dē ə) (L. *blatta,* cockroach, + Gr. *eidos,* form, + *ea,* characterized by): **cockroaches.** Common insects in tropical areas; often infest houses in northern areas; oval, flattened bodies may exceed 5 cm in length; tarsi with 5 segments; wings typically present, often reduced.

Order Phasmatodea (faz-mə-´tō-dēə) (Gr. *phasma,* apparition, + *eidos* form, + *ea,* characterized by): **walkingsticks** and **leaf insects.** Bodies elongated and sticklike or flattened and laterally expanded; herbivorous, tropical forms may be very large (up to 30 cm).

Order Mantodea (man-´tō-dēə) (Gr. *mantis,* soothsayer, + *eidos,* form, + *ea,* characterized by): **mantids** (see figure 13.36). Bodies elongated with raptorial front legs; predatory; may reach 10 cm in length.

Order Mantophasmatodea (man-´tō-faz-mə-´tō-dēə) (an amalgamation of the order names for praying mantids [Mantodea] and walkingsticks [Phasmatodea]): **gladiators.** Secondarily wingless; chewing mouthparts; resemble a combination of a praying mantis and a walkingstick; nocturnal predators on insects and spiders; described in 2002; rare, found in Africa; 6 to 8 species.

Order Dermaptera (der-map´ter-a) (Gr. *derma,* skin, + *pteron,* wing): **earwigs.** Very short forewings; large and membranous hindwings folded under forewings when at rest; chewing mouthparts; forceps-like cerci.

Order Plecoptera (ple-kop´ter-a) (Gr. *plekein,* to twist, + *pteron,* wing): **stoneflies** (see figure 13.49A). Membranous wings; larger and fanlike hindwings; aquatic nymph with tufts of tracheal gills.

Order Isoptera (i-sop´ter-a) (Gr. *isos,* equal, + *pteron,* wing): **termites** (see figure 13.55). Small; membranous, narrow

wings similar in size with few veins; wings shed at maturity; erroneously called "white ants"; distinguishable from true ants by broad union of thorax and abdomen; complex social organization; hemimetabolous development.

Order Embiidina (em-bē-ə´ di-nə) (Gr. *embios,* lively, + *eidos,* form, + *ina* resembling): **webspinners.** Small; male wings membranous, narrow, and similar in size; wingless females; chewing mouthparts; colonial; make silk-lined channels in tropical soil.

Order Psocoptera (so-cop´ter-a) (Gr. *psoco,* rub away, + *pteron,* wing) **(Corrodentia): psocids, book lice,** and **bark lice.** Body usually small, may be as large as 10 mm; membranous, narrow wings with few veins, usually held rooflike over abdomen when at rest; some wingless species; found in books, bark, bird nests, on foliage.

Order Zoraptera (zo-rap´ter-a) (Gr. *zōros,* pure, + *apteryos,* wingless): **zorapterans.** As large as 2.5 mm; membranous, narrow wings usually shed at maturity; colonial and termite-like.

Order Phthiraptera (thī-rap´ter-a) (Gr. *phteir,* louse, + *apteros,* wingless): **lice.** Wingless ectoparasites adapted for clinging to warm-blooded hosts. **Sucking lice** (see figure 13.42) in former order Anoplura now constitute suborder Anoplura, mouthparts adapted for piercing and sucking, includes head lice, crab lice, and body lice. **Chewing lice** (see figure 13.41) in former order Mallophaga now divided among three suborders.

Order Thysanoptera (thy-sa-nop´ter-a) (Gr. *thysanos,* tassel, + *pteron,* wing): **thrips.** Length 0.5 to 5 mm (a few longer); wings, if present, long, very narrow, with few veins, and fringed with long hairs; sucking mouthparts; destructive plant-eaters, but some feed on insects.

Order Hemiptera (he-mip´ter-a) (Gr. *hemi,* half, + *pteron,* wing). Members have unique mouthparts specialized for piercing and sucking. Hemiptera is divided into three suborders: Heteroptera, Auchenorrhyncha, and Sternorrhyncha. Heteroptera contains **true bugs;** size 2 to 100 mm; wings present or absent; forewings with basal portion thickened and partly sclerotized; apical portion membranous; hindwings membranous; at rest, wings held flat over abdomen; many with odorous scent glands; includes water scorpions, water striders, bedbugs, squash bugs, assassin bugs, chinch bugs, stink bugs, soldier bugs, (see figure 13.57A), plant bugs, lace bugs, and many others. Auchenorrhyncha contains **hoppers** (figure 13.62) and **cicadas** (see figure 13.50); four wings typical if wings are present. Sternorrhyncha contains **whiteflies, psyllids, aphids, mealybugs** (see figure 13.58B), and **scale insects;** four wings typical if wings are present; often have complex life histories; many species are plant pests.

Order Neuroptera (neu-rop´ter-a) (Gr. *neuron,* nerve, + *pteron,* wing): **dobsonflies, ant lions** (figure 13.63), and **lacewings.** Medium to large size; similar, membranous wings with many cross veins; chewing mouthparts; dobsonflies have greatly enlarged mandibles in males and aquatic larvae; ant lion larvae (doodlebugs) make craters in sand to trap ants; holometabolous development.

Order Coleoptera (ko-le-op´ter-a) (Gr. *koleos,* sheath, + *pteron,* wing): **beetles** (see figure 13.58A and B), **fireflies** (see figure 13.52), and **weevils.** The largest order of animals; forewings (elytra) thick, hard, opaque; membranous hindwings folded under forewings at rest; mouthparts for biting and chewing; includes ground beetles, carrion beetles, whirligig beetles, darkling beetles, stag beetles, dung beetles (see figure 13.51), diving beetles, boll weevils, fireflies, lady beetles (ladybugs), others; holometabolous development.

Order Strepsiptera (strep-sip´ter-a) (Gr. *strepsis,* a turning, + *pteron,* wing): **stylops** or **twisted wing parasites.** Females wingless, without eyes or antennae; males have vestigial forewings and fan-shaped hindwings; females and larvae parasites of bees, wasps, and other insects.

Order Mecoptera (me-kop´ter-a) (Gr. *mekos* length, + *pteron,* wing): **scorpionflies.** Small to medium size; wings long, slender, with many veins; at rest, wings held rooflike over back; males have scorpion-like clasping organ at end of abdomen; carnivorous; live in most woodlands.

Order Lepidoptera (lep-i-dop´ter-a) (Gr. *lepidos,* scale, + *pteron,* wing): **butterflies** and **moths** (see figure 13.47 and figure 13.58C (larval form)). Membranous wings covered with overlapping scales, wings coupled at base; mouthpart is a sucking tube, coiled when not in use; larvae (caterpillars) have chewing mandibles for eating plants, stubby prolegs on the abdomen, and silk glands for spinning cocoons; antennae knobbed in butterflies and usually plumed in moths; holometabolous development.

Order Diptera (dip´ter-a) (Gr. *dis,* two, + *pteron,* wing): **true flies.** Single pair of wings, membranous and narrow; hindwings reduced to inconspicuous balancers (halteres); sucking mouthparts or adapted for sponging, lapping, or piercing; legless larvae called maggots or, when aquatic, wigglers; include crane flies, mosquitos (see figure 13.59), moth flies, midges, fruit flies, flesh flies, house flies (see figure 13.37), horse flies, bot flies, blow flies, gnats, and many others; holometabolous development.

Order Trichoptera (tri-kop´ter-a) (Gr. *trichos,* hair, + *pteron,* wing): **caddisflies.** Small, soft-bodied; wings well-veined and hairy, folded rooflike over hairy body; chewing mouthparts; aquatic larvae construct cases of leaves, sand, gravel, bits of shell, or plant matter, bound together with secreted silk or cement; some make silk feeding nets attached to rocks in streams; holometabolous development.

Order Siphonaptera (si-fon-ap´ter-a) (Gr. *siphon,* a siphon, + *apteros,* wingless): **fleas** (see figure 13.40). Small; wingless; bodies laterally compressed; legs adapted for leaping; no eyes; ectoparasitic on birds and mammals; larvae legless and scavengers; holometabolous development.

Order Hymenoptera (hi-men-op´ter-a) (Gr. *hymen,* membrane, + *pteron,* wing): **ants, bees** (see figure 13.54), and **wasps** (see figure 13.46). Very small to large; membranous, narrow wings coupled distally; subordinate hindwings; mouthparts for lapping liquids and biting; ovipositor sometimes modified into stinger, piercer, or saw; both social and solitary species; most larvae legless, blind, and maggot-like; holometabolous development.

figure 13.62

An oak treehopper, *Platycotis vittata* (order Hemiptera, suborder Auchenorrhyncha).

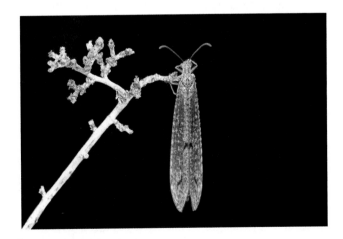

figure 13.63

Adult ant lion (order Neuroptera).

Introduction of natural predators or parasites of the insect pests has had some success. In the United States, vedalia beetles from Australia help control the cottony-cushion scale on citrus plants, and numerous instances of control by using insect parasites have been recorded. However, the introduction of exotic species to control insect pests may have unexpected negative consequences, and should be done with caution.

Another approach to biological control is to interfere with the reproduction or behavior of insect pests by sterilizing males or by using naturally occurring organic compounds that act as hormones or pheromones. Such research, although very promising, is slow because of our limited understanding of insect behavior and the problems of isolating and identifying complex compounds that an insect produces in such minute amounts. Nevertheless, pheromones may play an important role in biological pest control in the future.

A systems approach called integrated pest management is practiced with many crops. This approach involves integrating all possible, practical techniques to contain pest infestations at a tolerable level—for example, cultural techniques (resistant plant varieties, crop rotation, tillage techniques, timing of sowing, planting, or harvesting, and others), use of biological controls, and sparing use of insecticides.

Phylogeny and Adaptive Diversification

Phylogeny

Biologists hypothesize that the ancestral arthropod had a segmented body with one pair of appendages per segment. During evolution, adjacent segments fused to make body regions (tagmata). How many segments contributed to a head in each group of arthropods? A partial answer to this question comes from studies of *Hox* genes that are highly correlated with specific anterior-posterior differentiation of body parts in bilateral animals. *Hox* gene studies indicate that the first five segments, at least, fused to form the head tagma in all four extant subphyla. It is surprising to find the same pattern of fusion in chelicerates as in other subphyla because a head is not immediately obvious in a chelicerate. Spider bodies have two tagmata: prosoma, or cephalothorax, and opisthosoma, or abdomen. Is the head part of the prosoma? *Hox* gene comparisons indicate that the entire prosoma corresponds to the head of other arthropods.

Another controversial area of arthropod biology for which genetic studies have proved helpful involves the evolution and antiquity of uniramous and biramous appendages. Hexapods and myriapods have uniramous appendages, but trilobites and some crustaceans have biramous appendages. If the ancestral appendage were biramous, then the switch to uniramous appendages might have occurred in one lineage whose descendants now carry this trait. Such reasoning led biologists to group hexapods with myriapods, but phylogenies using molecular characters repeatedly placed

hexapods with crustaceans. Is it likely that the uniramous limb evolved more than once? This question would be more easily answered if the genetic basis of limb structure were understood. It is now known that modulation in expression of the *Distal-less* (*Dll*) gene determines the location of distal ends of arthropod limbs. In each primordial (embryonic) biramous appendage, the gene product of *Dll* can be observed in two groups of cells, each of which becomes a branch of the limb. In a uniramous limb primordium, there is only one such group of cells, and in primordia of phyllopodous limbs (as in class Branchiopoda), there are as many groups expressing *Dll* as there are limb branches. Gene expression can be modified within a lineage, so the number of limb branches is not likely to be an homologous character.

The best explanation for the similarities between crustaceans and insects, such as the basic structure of their ommatidia, is that these two taxa form a clade within Arthropoda. We depict crustaceans and hexapods as sister taxa (see figure 13.4), but some evidence suggests that hexapods arose from *within* the crustacean group, making Crustacea paraphyletic unless it includes hexapods. Insects may be the descendants of freshwater branchiopods. Crustacea now includes members of the former phylum Pentastomida as a subclass. The taxon Pancrustacea includes crustaceans and hexapods.

Phylogenetic placement of subphylum Myriapoda is highly controversial. Molecular phylogenies supporting both the "mandibulata" and "myriochelata" hypotheses continue to emerge, with other research showing that the choice of outgroup (see p. 81) has a large influence on phylogeny in the arthropods. Awaiting further research, we show a polytomy (multiple branches coming from a single point) in figure 13.4 to describe the relationships among Myriapoda, Chelicerata, and Pancrustacea.

Evolution within hexapods involved specialization of the first three postcephalic body segments (somites) to become locomotor segments (thorax) and a loss or reduction of appendages on the rest of the body (abdomen). The wingless orders traditionally have been regarded as having the most primitive characteristics. Three wingless orders (Diplura, Collembola, and Protura) have their mandibles and first maxillae located deeply in pouches in the head, a condition called **entognathy.** All other insects are **ectognathous,** including the wingless order Thysanura. Ectognathous insects do not have their mandibles and maxillae in pouches, and they share other synapomorphies. Entognathous and ectognathous insects form sister groups, and Thysanura diverged from a common ancestor of ectognathous insects before the advent of flight, which unites the remaining ectognathous orders (see figure 13.33).

The earliest phylogenetic split among winged insects separates three taxa that differ in their ability to flex their wings (see figure 13.33). Two of these (Odonata and Ephemeroptera) have outspread wings. The other taxon, which has wings that can fold back over the abdomen, branched into three groups: one with hemimetabolous metamorphosis and chewing mouthparts (Orthopterodea); one group with

Taxonomy Phylum Arthropoda

Subphylum Trilobita (trī´lo-bi´ta) (Gr. *tri-*, three, + *lobos*, lobe): **trilobites.** All extinct forms; Cambrian to Permian periods; body divided by two longitudinal furrows into three lobes; distinct head, thorax, and abdomen; biramous appendages.

Subphylum Chelicerata (ke-lis´e-ra´ta) (Gr. *chēle*, claw, + *keratos*, a horn): **eurypterids, horseshoe crabs, spiders,** and **ticks.** First pair of appendages modified to form chelicerae; pair of pedipalps and four pairs of legs; no antennae, no mandibles; cephalothorax and abdomen often with segments fused.

Class Merostomata (mer´o-sto´ma-ta) (Gr. *meros*, thigh, + *stomatos*, mouth): **aquatic chelicerates.** Cephalothorax and abdomen; compound lateral eyes; appendages with gills; sharp telson; **subclasses Eurypterida** (all extinct) and **Xiphosurida,** the horseshoe crabs.

Class Pycnogonida (pik´no-gon´i-da) (Gr. *pyknos*, compact, + *gonia*, knee, angle): **sea spiders.** Small (3 to 4 mm), but some reach 500 mm; body chiefly cephalothorax; tiny abdomen; usually four pairs of long walking legs (some with five or six pairs); one pair of subsidiary legs (ovigers) for egg bearing; mouth on long proboscis; four simple eyes; no respiratory or excretory system. Example: *Pycnogonum.*

Class Arachnida (ar-ack´ni-da) (Gr. *arachnē,* spider): **scorpions, spiders, mites, ticks,** and **harvestmen.** Four pairs of legs; segmented or unsegmented abdomen with or without appendages and generally distinct from cephalothorax; respiration by gills, tracheae, or book lungs; excretion by Malpighian tubules or coxal glands; dorsal bilobed brain connected to ventral ganglionic mass with nerves; simple eyes; sexes separate; chiefly oviparous; no true metamorphosis. Examples: *Argiope, Centruroides.*

Subphylum Crustacea (crus-ta´she-a) (L. *crusta*, shell, + *acea*, group suffix): **crustaceans.** Mostly aquatic, with gills; cephalothorax usually with dorsal carapace; biramous appendages, modified for various functions; head appendages consist of two pairs of antennae, one pair of mandibles, and two pairs of maxillae; sexes usually separate; development primitively with nauplius stage.

Class Branchiopoda (brank´ē-o-pod-ə) (Gr. *branchia*, gills, + *podos*, foot): **branchiopods.** Flattened, leaflike swimming appendages (phyllopodia) with respiratory function. Examples: *Triops, Lynceus, Daphnia.*

Class Maxillopoda (mak´sil-o-pod-ə) (L. *maxilla*, jawbone, + Gr. *podos*, foot): **ostracods, copepods, branchiurans,** and **barnacles.** Five cephalic, six thoracic, and usually four abdominal segments; typically no appendages on abdomen;

unique maxillopodan eye. Examples: *Cypris, Cyclops, Ergasilus, Argulus, Balanus.*

Class Malacostraca (mal´a-kos´tra-ka) (Gr. *malakos*, soft, + *ostrakon*, shell): **shrimps, crayfishes, lobsters, and crabs.** Usually have eight thoracic and six abdominal segments, each with a pair of appendages. Examples: *Armadillidium, Gammarus, Megacytiphanes, Grapsus, Homarus, Panulirus.*

Subphylum Myriapoda (mir-ē-a´pod-ə) (Gr. *myrias*, a myriad, + *podos*, foot): **myriapods.** All appendages uniramous; head appendages consist of one pair of antennae, one pair of mandibles, and one or two pairs of maxillae.

Class Diplopoda (dip´lō-pod-ə) (Gr. *diploos*, double, + *podos*, foot): **millipedes.** Subcylindrical body; head with short antennae and simple eyes; body with variable number of segments; short legs, usually two pairs of legs to a segment; separate sexes. Examples: *Julus, Spirobolus.*

Class Chilopoda (ki´lō-pod-ə) (Gr. *cheilos*, lip, + *podos*, foot): **centipedes.** Dorsoventrally flattened body; variable number of segments, each with one pair of legs; one pair of long antennae; separate sexes. Examples: *Cermatia, Lithobius, Geophilus.*

Class Pauropoda (pāu-rō´-pod-ə) (Gr. *pauros*, small, + *podos*, foot): **pauropods.** Minute (1 to 1.5 mm), cylindrical body consisting of double segments and bearing nine or ten pairs of legs; no eyes. Example: *Pauropus.*

Class Symphyla (sim´fi-la) (Gr. *syn*, together, + *phylon*, tribe): **garden centipedes.** Slender (1 to 8 mm) with long, threadlike antennae; body consists of 15 to 23 segments with 10 to 12 pairs of legs; no eyes. Example: *Scutigerella.*

Subphylum Hexapoda (hek´sa-pod-ə) (Gr. *hex*, six, + *podos*, foot): **hexapods.** Body has distinct head, thorax, and abdomen; pair of antennae; mouthparts modified for different food habits; head composed of six fused segments; thorax has three segments; abdomen has variable number of segments, usually 11 somites; thorax has two pairs of wings (sometimes one pair or none) and three pairs of jointed legs; separate sexes; usually oviparous; gradual or abrupt metamorphosis.

Class Entognatha (en´tog-na-tha) (Gr. *entos*, within, inside + *gnathos*, jaw): **entognaths.** Base of mouthparts lies within head capsule; mandibles have one articulation. Example: *Entomobrya.*

Class Insecta (in-sek´ta) (L. *insectus,* cut into): **insects.** Bases of mouthparts exposed and exiting head capsule; mandibles generally have two regions of articulation. Examples: *Drosophila, Bombus, Anopheles* (insect orders listed on pp. 281–283).

hemimetabolous metamorphosis and usually sucking mouth-parts (Hemipterodea); and a group with holometabolous meta-morphosis (Holometabola).

How did the first arthropod evolve, and to which other phyla are arthropods closely related? Do all segmented phyla share a segmented common ancestor? A long-held hypoth-esis states that both annelids and arthropods originated from a common ancestral lineage of coelomate segmented protostomes from which two or more then diverged: a pro-toannelid lineage with laterally located parapodia and one or more protoarthropod lineage with more ventrally located appendages. However, lineages placement of annelids in superphylum Lophotrochozoa and arthropods in super-phylum Ecdysozoa implies that metamerism arose inde-pendently in the two groups and is a convergent character. Were metameric bodies convergent, the genetic controls and chemical signals used during development of a seg-mented body should differ among phyla. As discussed on p. 229, very preliminary comparisons suggest that annelids and arthropods do not share mechanisms of segmentation, but some elements of regulatory gene pathways are shared. The independent adoption of genetic elements from ancestral gene pathways is one potential explanation for these shared aspects of segmentation (see p. 229).

Adaptive Diversification

The adaptive trend in arthropods has been toward tagmatization of the body by differentiation or fusion of segments, giving rise to such tagmata as head and trunk; head, thorax, and abdomen; or cephalothorax (fused head and thorax) and abdomen. A series of similar appendages, one pair on each trunk segment, is the primitive character state, still retained by some crustaceans and by myriapods. Derived forms include appendages specialized for specific functions and some appendages lost entirely.

Much of the amazing diversity in arthropods seems to have developed because of modification and specialization of their cuticular exoskeleton and their jointed appendages, thus producing a wide variety of locomotor and feeding adaptations. Whether in the area of habitat, feeding adapta-tions, means of locomotion, reproduction, or general mode of living, the adaptive achievements of arthropods are truly remarkable.

■ Summary

Arthropoda is the largest, most abundant and diverse phylum in the world. Arthropods are metameric, coelomate protostomes with well-developed organ systems. Most show marked tagmatization. They are extremely diverse and occur in all habitats capable of supporting life. Perhaps more than any other single fac-tor, the success of arthropods is explained by adaptations made possible by their cuticular exoskeleton. Other important elements in their success are jointed appendages, tracheal respiration, efficient sensory organs, complex behavior, metamorphosis, and flight.

All arthropods must periodically cast off their cuticle (ecdysis) and grow larger before the newly secreted cuticle hardens. Premolt and postmolt periods are hormonally con-trolled, as are several other structures and functions.

Members of subphylum Chelicerata have no antennae, and their main feeding append-ages are chelicerae. In addition, they have a pair of pedipalps (which may be similar to walking legs) and four pairs of walking legs. A great majority of living chelicerates are in class Arachnida: spiders (order Araneae), scor-pions (order Scorpionida), harvestmen (order Opiliones), and ticks and mites (order Acari).

The tagmata of spiders (cephalothorax and abdomen) show no external segmentation and join at a waistlike pedicel. Their chelicerae have venom glands for paralyzing or killing their prey. Spiders can spin silk, which they use for a variety of purposes.

The cephalothorax and abdomen of ticks and mites are completely fused, and the anterior capitulum bears the mouthparts. Ticks and mites are the most numerous arach-nids; some are important disease carriers, and others are serious plant pests.

Crustacea is a large, primarily aquatic subphylum of arthropods. Crustaceans bear two pairs of antennae, mandibles, and two pairs of maxillae on the head. Their append-ages are primitively biramous, and the major tagmata are head, thorax, and abdomen. Many have a carapace and respire with gills.

Many crustaceans are predators, scaven-gers, filter feeders, and parasites. Respiration is through the body surface or by gills, and excretory organs take the form of maxillary or antennal glands. Circulation, as in other arthro-pods, is through an open system of sinuses (hemocoel), and a dorsal, tubular heart is the chief pumping organ. Most crustaceans have compound eyes composed of units called ommatidia.

Members of class Maxillopoda, subclass Copepoda, lack a carapace and abdominal appendages. They are abundant and are some of the most important primary consumers in many freshwater and marine ecosystems.

Most members of subclass Cirripedia (barnacles) are sessile as adults, secrete a shell of calcareous plates, and filter-feed. Subclass Branchiura contains fish lice. Closely related to fish lice are tongue worms; they are para-sitic in the lungs and nasal cavities of verte-brates. These members of former phylum Pentastomida now form subclass Pentas-tomida in class Maxillopoda.

Malacostraca is the largest crustacean class, and the most important orders are Isopoda, Amphipoda, Euphausiacea, and Decapoda. All have both abdominal and thoracic appendages. Decapods include crabs, shrimp, lobster, cray-fish, and others; they have five pairs of walking legs (including chelipeds) on their thorax.

The former subphylum Uniramia grouped arthropods having uniramous appendages, one pair of antennae, a pair of mandibles, and two pairs of maxillae (one pair of maxillae in milli-pedes) on the head. Uniramia has been divided into two subphyla: Myriapoda, in which the animals have head and trunk tagmata, and Hexapoda, in which the tagmata are head, tho-rax, and abdomen.

Hexapoda is the largest subphylum of the world's largest phylum. Insects are easily recog-nized by the combination of their tagmata and their possession of three pairs of thoracic legs.

The diversification and abundance of insects are largely explained by several features that allow them to exploit terrestrial habitats, such as a waterproof cuticle and other mechanisms to minimize water loss and their ability to become dormant during adverse conditions.

Feeding habits vary greatly among insects, and an enormous variety of specialized mouthparts reflects their particular feeding habits. Insects breathe by means of a tracheal system, a system of tubes that open by spiracles on the thorax and abdomen. Excretory organs are Malpighian tubules.

Sexes are separate in insects, and fertilization is usually internal. Almost all insects undergo metamorphosis during development. In holometabolous (complete) metamorphosis, the last larval molt gives rise to a nonfeeding stage (pupa). An adult, usually winged, emerges at the final, pupal molt. In hemimetabolous (gradual) metamorphosis, larval instars are called nymphs, and adults emerge at the last nymphal molt. Both types of metamorphosis are hormonally controlled.

Insects are important to human welfare, particularly because they pollinate food crop plants, control populations of other, harmful insects by predation and parasitism, and serve as food for other animals. Many insects harm human interests because they feed on crop plants, and many carry important diseases affecting humans and domestic animals.

Within Arthropoda, a preponderance of morphological and molecular evidence supports a monophyletic grouping of Crustacea and Hexapoda. These taxa form Pancrustacea. There is still debate over the contention that hexapods evolved from within Crustacea, or are the sister taxon of Crustaceans. The traditional alliance of Crustacea with insects and myriapods in a group called Mandibulata is supported by some morphological evidence, but some molecular data support a clade composed of crustaceans and insects, as the sister taxon to chelicerates plus myriapods (Myriochelata). Wings, hemimetabolous metamorphosis, and holometabolous metamorphosis evolved among ectognathous insects.

Molecular evidence suggests that annelids and arthropods were derived independently from a protostome ancestor and belong to separate superphyla (Lophotrochozoa and Ecdysozoa), indicating that metamerism evolved independently in each superphylum.

■ Review Questions

1. List some characteristics of arthropods that clearly distinguish them from Annelida.
2. Name the subphyla of arthropods, and give a few examples of each.
3. Much of the success of arthropods has been attributed to their cuticle. Why? Describe some other factors that probably contributed to their success.
4. What is a trilobite?
5. What appendages characterize chelicerates?
6. Briefly describe the appearance of each of the following: eurypterids, horseshoe crabs, pycnogonids.
7. Describe the mechanism of each of the following with respect to spiders: feeding, excretion, sensory reception, webspinning, reproduction.
8. Distinguish each of the following orders from each other: Araneae, Scorpionida, Opiliones, Acari.
9. People fear spiders and scorpions, but ticks and mites are far more important medically and economically. Why? Give examples.
10. Which segments and appendages occur on the head of a malacostracan crustacean? What are some other important characteristics of Crustacea?
11. Among the classes of Crustacea, Branchiopoda, Maxillopoda, and Malacostraca are the most important. Distinguish them from each other.
12. Distinguish among subclasses Ostracoda, Copepoda, Branchiura, and Cirripedia in class Maxillopoda.
13. Copepods sometimes are called "insects of the sea" because marine planktonic copepods probably are the most abundant animals in the world. What is their ecological importance?
14. Define each of the following: swimmeret, maxilliped, cheliped, nauplius.
15. Describe molting in Crustacea, including the action of hormones.
16. Explain the mechanism of each of the following with respect to crustaceans: feeding, respiration, excretion, circulation, sensory reception, reproduction.
17. Distinguish the following from each other: Diplopoda, Chilopoda, Insecta.
18. Define each of the following with respect to insects: sclerite, tergum, sternum, labrum, hypopharynx, haltere, instar, diapause.
19. Explain why wings powered by indirect flight muscles can beat much more rapidly than those powered by direct flight muscles.
20. What different modes of feeding do insects use, and how are these reflected in their mouthparts?
21. Describe each of the following with respect to insects: respiration, excretion and water balance, sensory reception, reproduction.
22. Explain the difference between holometabolous and hemimetabolous metamorphosis in insects, including the stages of each.
23. Describe and give an example of each of four ways insects communicate with each other.
24. What castes occur in honey bees and in termites, and what is the function of each? What is trophallaxis?
25. Name several ways in which insects benefit humans and several ways in which they are detrimental.
26. For the past 50 or more years, people have relied on toxic insecticides for control of harmful insects. What problems have arisen as a result? What are the alternatives? Define integrated pest management.
27. What evidence suggests that metamerism evolved independently in Annelida and Arthropoda?
28. We hypothesize that the earliest insects were wingless, making lack of wings the primitive condition. Does winglessness characterize a hexapod class? In what sense is it useful for classification?
29. Distinguish the Mandibulata hypothesis for arthropod evolution from the Myriochelata hypothesis.

For Further Thought Filter feeding is a very common way to collect food from water, but this method is rarely used on land. Why might web-spinning spiders be considered filter feeders?

■ Selected References

See also general references on page 432.

Arnett, R. H., Jr., and M. C. Thomas, eds. 2000. American beetles, vol. 1. Boca Raton, Florida, CRC Press.

Arnett, R. H., Jr., M. C. Thomas, P. E. Skelley, and J. H. Franks, eds. 2002. American beetles, vol. 2. Polyphaga: Scarabaeoidea through Curculionoidea. Boca Raton, Florida, CRC Press. *These two volumes present recent detailed keys to families of American beetles; the second edition of volume 1 (2001) includes additions and corrections.*

Averof, M. 1998. Evolutionary biology: origin of the spider's head. Nature **395**:436–437. *Summary of research on homology of the head across arthropod subphyla.*

Brown, K. 2001. Seeds of concern. Sci. Am. **284**:52–57 (Apr.). *The first of four articles that examine the pros and cons of genetically modified crops, including those with genes for insecticidal toxin.*

Brusca, R. C. 2001. Unraveling the history of arthropod biodiversification. Ann. Missouri Bot. Garden **87**:13–25. *An easy-to-read discussion of changes in our understanding of arthropod evolution.*

Caravas, J., and M. Freidrich. 2010. Of mites and millipedes: recent progress in resolving the base of the arthropod tree. Bioessays **32**:488–495. *A very nice summary of current research on this controversial phylogenetic issue.*

Downs, A. M. R., K. A. Stafford, and G. C. Coles. 1999. Head lice: prevalence in schoolchildren and insecticide resistance. Parasitol. Today **15**:1–4. *This report is mainly concerned with England, but head lice are one of the most common parasites of American schoolchildren.*

Eldridge, B. F., and J. D. Edman (eds). 2000. Medical Entomology. Kluwer Academic Publishers, Dordrecht. *The best reference currently available.*

Glenner, H., P. F. Thomsen, M. B. Hebsgaard, M. V. Sorensen, and E. Willerslev. 2006. The origin of insects. Science **314**:1883–1884. *The origin of insects as terrestrial crustaceans explains a puzzling aspect of the fossil record.*

Gullan, P. J., and P. S. Cranston. 2005. The insects: an outline of entomology, ed. 3. Malden, Massachusetts, Blackwell Publishing. *An easy-to-use text for general entomology with good figures, modern phylogenies, and data on biogeography.*

Johnson, N. F., and C. Triplehorn. 2005. Borror and DeLong's introduction to the study of insects, ed. 7. Belmont, California, Brooks/Cole Publishing Company. *An up-to-date reference text for the study of insects; keys are included, but some require specialized knowledge of morphological characters. See Arnett and Thomas for keys to the American beetle families.*

Lavrov, D. V., W. M. Brown, and J. L. Boore. 2004. Phylogenetic position of the Pentastomida and (pan)crustacean relationships. Proc. Royal Soc. Biol. Sciences B **271**:537–544. *Mitochondrial gene arrangements and sequences indicate that pentastomids are modified crustaceans, and likely related to branchiurans.*

Levine, M. 2002. How insects lose their limbs. Nature **415**:848–849. *A Hox gene product in the abdomen of insects inhibits action of another gene product that is necessary for limb formation.*

Luoma, J. R. 2001. The removable feast. Audubon **103(3)**:48–54. *During May and June, large numbers of horseshoe crabs ascend the shores of U.S. Atlantic states to breed and to lay eggs. Since the 1980s, they have been heavily harvested to be chopped up and used for bait. This practice has led to serious declines in* Limulus *populations, with accompanying declines in populations of migrating shorebirds that feed on* Limulus *eggs.*

Mallatt, J., J. R. Garey, and J. W. Shultz. 2004. Ecdysozoan phylogeny and Bayesian inference: first use of nearly complete 28s and 18s rRNA gene sequences to classify arthropods and their kin. Mol. Phylogenet. Evol. **31**:178–191. *Results indicate that Crustacea is paraphyletic without hexapods, but Pancrustacea is a monophyletic group, that chelicerates and myriapods are sister taxa, and*

that Panarthropoda is a monophyletic group. There was no support for a mandibulate clade.

Milius, S. 2007. Not-so-elementary bee mystery. Science News **172**:56. *A short summary of research into honey bee colony collapse disorder.*

Rota-Stabelli, O., L. Campbell, H. Brinkmann, G. D. Edgecombe, S. J. Longhorn, K. J. Peterson, D. Pisani, H. Philippe, and M. J. Telford. 2011. A congruent solution to arthropod phylogeny: phylogenomics, microRNAs and morphology support monophyletic Mandibulata. Proc. R. Soc. B **278**:298–306. *New sources of phylogenetic data do not favor the Myriochelata hypothesis.*

Telford, S. R. III., and H. K. Goethert. 2004. Emerging tick-borne infections: rediscovered and better characterized, or truly 'new'? Parasitology **129**:S301–S327. *A serious summary of infectious agents carried by ticks.*

Versluis, M., B. Schmitz, A. von der Heydt, and D. Lohse. 2000. How snapping shrimp snap: through cavitating bubbles. Science **289**:2114–2117. *When* Alpheus *snaps the claw on its chela shut, it generates a jet of water with a velocity so high that the pressure decreases below the vapor pressure of water, producing a cavitation bubble. As the water pressure rises again, the bubble collapses with a force sufficient to stun or kill prey.*

Westneat, M. W., O. Betz, R. W. Blob, K. Fezzaa, W. J. Cooper, and W. Lee. 2003. Tracheal respiration in insects visualized with synchrotron X-ray imaging. Science **299**:558–560. *Tracheae compress actively, exhaling air, but expand passively.*

Whiting, R. 2004. Phylogenetic relationships and evolution of insects, pp. 330–344. In J. Cracraft and M. J. Donoghue, eds., Assembling the tree of life. New York, Oxford University Press. *A detailed discussion of current hypotheses for insect evolution, including the suggestion that entognathous hexapods do not form a monophyletic group, unlike ectognathous forms.*

■ Custom Website

The *Animal Diversity* Online Learning Center is a great place to check your understanding of chapter material. Visit www.mhhe.com/hickmanad6e for access to key terms, quizzes, and more! Further enhance your knowledge with Web links to chapter-related material.

Explore live links for these topics:

Classification and Phylogeny of Animals
Phylum Arthropoda
Subphylum Chelicerata

Subphylum Crustacea
Subphylum Myriapoda
Subphylum Hexapoda
Class Insecta

Chaetognaths, Echinoderms, and Hemichordates

Chaetognatha
Xenoturbellida
Echinodermata
Hemichordata

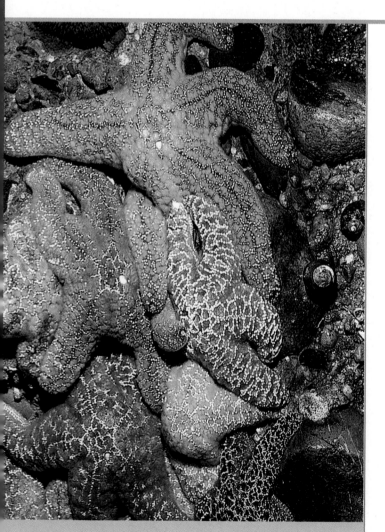

A mass of sea stars (*Pisaster ochraceous*) above the waterline at low tide.

A Design To Puzzle the Zoologist

Libbie Hyman, a distinguished American zoologist, once described echinoderms as a "noble group especially designed to puzzle the zoologist." With a combination of characteristics that should delight the most avid reader of science fiction, echinoderms would seem to confirm Lord Byron's observation that

Tis strange—but true;
for truth is always strange;
Stranger than fiction.

Despite the adaptive value of bilaterality for free-moving animals and the merits of radial symmetry for sessile animals, echinoderms confounded the rules by becoming free-moving but radial. That they evolved from a bilateral ancestor there can be no doubt, for their larvae are bilateral. They undergo a dramatic metamorphosis to a radial adult by a 90° reorientation in body axis. Despite the limitations of radial symmetry, echinoderms are among the most abundant organisms in some marine habitats.

A compartment of the echinoderm coelom has been transformed into a unique water-vascular system that uses hydraulic pressure to operate a multitude of tiny tube feet used in food gathering and locomotion. Dermal ossicles can fuse together to invest echinoderms in armor or can be reduced to microscopic plates. Many echinoderms have miniature jawlike pincers (pedicellariae) scattered on their body surface. These pedicellariae are often stalked and sometimes equipped with venomous glands.

This constellation of characteristics is unique in the animal kingdom. Perhaps strangest of all is that the closest living relatives of chordates, a group that includes humans, are the echinoderms and wormlike hemichordates, and not the more cephalized arthropods.

T riploblastic metazoans are divided into two large clades: Protostomia and Deuterostomia. These clades are diagnosed by a combination of molecular and morphological features, with many morphological differences visible in early development. Classical developmental characters associated with protostomes are spiral mosaic cleavage, formation of the mouth from the embryonic blastopore (protostomy), and formation of a coelom by schizocoely, when a coelom is present (pp. 62–66). Classical deuterostome developmental features include radial regulative cleavage, formation of the mouth from a second opening (deuterostomy), and coelom formation by enterocoely. All deuterostomes except *Xenoturbella* are coelomate.

Members of some phyla possess all the developmental characters in each suite: marine annelids and molluscs are typical protostomes, and echinoderms are typical deuterostomes. However, readers of the preceding chapters know that some taxa have both protostome and deuterostome features. The lophophorates have already been mentioned, and here we include another phylum, Chaetognatha, whose evolutionary position is debated. The chaetognaths are pelagic marine predators commonly called arrow worms. This phylum has been variously placed in Deuterostomia, in Protostomia, or outside the two groups.

We depict Chaetognatha as outside the protostome and deuterostome clades (figure 14.1; see cladogram inside front cover), pending new studies. Chaetognaths are discussed in this chapter, followed by three deuterostome phyla, Xenoturbellida, Echinodermata, and Hemichordata. The latter two are now considered sister taxa in clade Ambulacraria (figure 14.1).

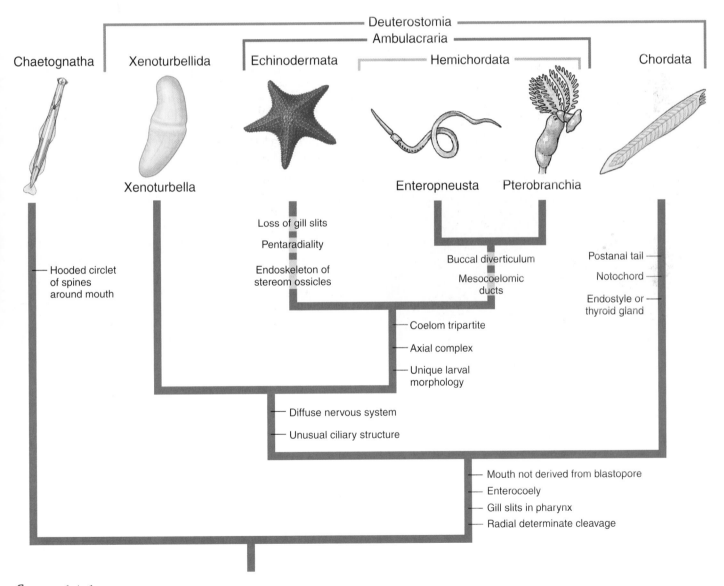

figure 14.1

Cladogram showing hypothetical relationships among deuterostome phyla. Based on this hypothesis, the ancestral deuterostome was a marine form with gill slits. Gill slits apparently were present in stem echinoderms, but lost in modern echinoderms.

Phylum Chaetognatha: Arrow Worms

Chaetognatha (kē-tog´na-thə) (Gr. *chaitē*, long hair, + *gnathos*, jaw) is a group of about 100 species of marine animals adapted for a planktonic existence. Their small, straight bodies resemble miniature torpedoes, or darts, ranging from 1 to 12 cm in length.

Arrow worms usually swim to the surface at night and descend during the day. Much of the time, they drift passively, but they can dart forward in swift spurts, using their caudal fin and longitudinal muscles—a fact that no doubt contributes to their success as planktonic predators. Horizontal fins bordering the trunk function as stabilizers rather than in active swimming.

The body of arrow worms is unsegmented and bears a caudal fin used for swimming (figure 14.2). They have teeth and curved, chitinous spines on the head around the mouth. When an arrow worm captures prey, its teeth and raptorial spines spread apart and then snap shut with startling speed. Arrow worms are voracious feeders, living on other zooplankton, especially copepods, and even small fishes (figure 14.2B).

Their body is covered with a thin cuticle. They have a complete digestive system, a well-developed coelom, eyes and sensory bristles, and a nervous system with a nerve ring containing several ganglia. Respiratory and excretory systems, however, are entirely absent.

Arrow worms are hermaphroditic with either reciprocal fertilization or self-fertilization. Juveniles develop directly without metamorphosis.

Chaetognath embryogenesis suggests deuterostome affinities. The mouth does not arise from the blastopore, and coelom formation usually is considered enterocoelous. Early descriptions of chaetognath cleavage characterize it as radial, complete, and equal, but subsequent studies dispute this description, finding instead that the cleavage planes in four-cell embryos are similar to those of crustaceans and nematodes. Recent phylogenetic studies using nucleotide sequence data place chaetognaths within protostomes as the sister taxon to all other members. Protostome features present in adult animals include a ventral nerve cord. Adult characters and molecular data are at odds with developmental data, so the evolutionary position of chaetognaths remains uncertain.

Phylum Xenoturbellida

It is surprising that a fourth phylum would be added to clade Deuterostomia. The small, yellowish, wormlike ciliated animals in this phylum were first described in 1949, but they have been slow to find a taxonomic home. Xenoturbellida contains a single genus, *Xenoturbella*, with two species. They reach lengths of 3 cm and live in North Sea mud, where they feed on bivalves and bivalve eggs.

The body has two furrows visible externally, which may have sensory functions. Cephalization is absent, but a mouth leads into a blind gut. There are no structural gonads, but sexual reproduction occurs.

Why is *Xenoturbella* placed in Deuterostomia? It has been placed with flatworms (hence the name) or molluscs, but when included in phylogenies using molecular characters, the genus lies within Deuterostomia. Based on some morphological characters, we tentatively place this phylum as the sister taxon of Ambulacraria (see figure 14.1). Descriptions of gametes and larvae are expected soon; these descriptions may shed light on the relationship of *Xenoturbella* to other deuterostomes.

Clade Ambulacraria

Ambulacraria contains two deuterostome phyla: Echinodermata and Hemichordata (see figure 14.1). Echinoderms, including sea stars and sea urchins, are familiar to many people, but hemichordates, including acorn worms and pterobranchs, are much less familiar. Aside from typical deuterostome features, members of Ambulacraria share a three-part (tripartite) coelom, similar larval forms, and an axial complex (a highly specialized metanephridium).

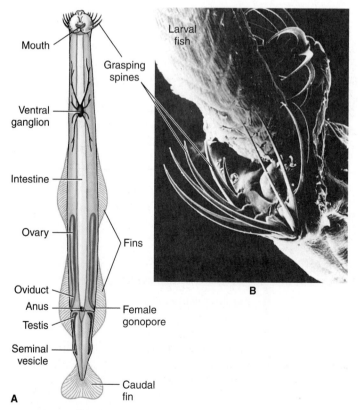

A

Mouth

Grasping spines

Larval fish

Ventral ganglion

Intestine

Ovary

Fins

Oviduct

Anus

Testis

Female gonopore

Seminal vesicle

Caudal fin

B

figure 14.2

Arrow worms. **A,** Internal structure of *Sagitta*. **B,** Scanning electron micrograph of a juvenile arrow worm, *Flaccisagitta hexaptera* (35 mm long), eating a larval fish.

Phylum Echinodermata

Members of Echinodermata (ē-kī´nō-der´mä-tä) (L. *echinatus*, prickly, + Gr. *derma*, skin) are marine forms and include

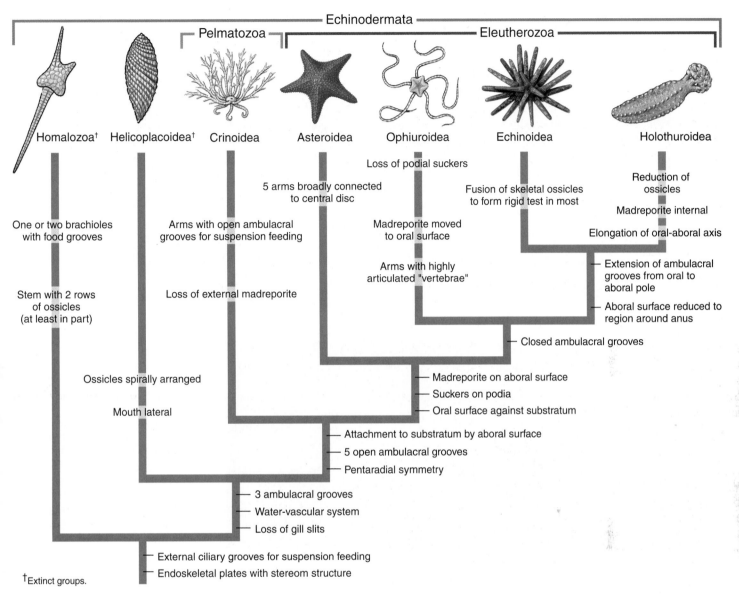

figure 14.3

Cladogram showing hypothetical relationships among echinoderm groups. The extinct Homalozoa (carpoids), which were not radial in symmetry but had stereom endoskeletal plates, represent an early split from all other echinoderms. The extinct helicoplacoids had three ambulacral grooves that wound around their bodies in spiral fashion, and are hypothesized to be the sister group of modern echinoderms. Pentaradial symmetry was an adaptation to sessile existence and is a synapomorphy for modern echinoderms. The cladogram presented here depicts ophiuroids as having arisen separately from asteroids, after evolution of closed ambulacral grooves; possession of five arms would have arisen independently in these groups. An alternate scenario, equally supported, unites Asteroidea and Ophiuroidea in a clade, with five arms being a synapomorphy and independent evolution of closed ambulacral grooves in ophiuroids and in the common ancestor of echinoids and holothuroids.

sea stars (also called starfishes), brittle stars, sea urchins, sea cucumbers, and sea lilies. They have a combination of characteristics found in no other phylum: (1) a calcareous, spiny endoskeleton of plates or ossicles, (2) a water-vascular system, (3) pedicellariae, (4) dermal branchiae, and (5) basic pentaradial symmetry. Despite an excellent fossil record, which extends back to at least the early Cambrian period, the origin and early evolution of echinoderms remains uncertain. It seems clear that they descend from bilateral ancestors because

their larvae are bilateral but become radially symmetrical later in development. Apparently, some of the early echinoderms were sessile and evolved radiality as an adaptation to their sessile existence (figure 14.3).

Ecological Relationships

Echinoderms are all marine; they have no ability to osmoregulate and therefore rarely inhabit brackish waters. Virtually all

figure 14.4

Some sea stars (class Asteroidea) from the Pacific. **A,** Pincushion star, *Culcita novaeguineae*, preys on coral polyps and also eats other small organisms and detritus. **B,** Crown-of-thorns star, *Acanthaster planci*, a major coral predator (see note, p. 297), can have 10–20 arms. **C,** Necklace star, *Fromia monilis*, browses encrusting organisms. **D,** *Choriaster granulatus* scavenges dead animals on shallow Pacific reefs.

benthic as adults, they inhabit all oceans of the world and all depths, from intertidal to abyssal regions.

Some sea stars (figure 14.4) are particle feeders, but many are predators, feeding particularly on sedentary or sessile prey. Brittle stars (see figure 14.12A) are the most active echinoderms, moving by their arms. They may be scavengers, browsers, or deposit or filter feeders. Some are commensal in large sponges. Compared to other echinoderms, sea cucumbers (see figure 14.19) are greatly extended in the oral-aboral axis and are oriented with that axis more or less parallel to their substrate and lying on one side. Most are suspension or deposit feeders. "Regular" sea urchins (see figure 14.15), which are radially symmetrical, prefer hard substrates and feed chiefly on algae. "Irregular" urchins (sand dollars and heart urchins) (see figure 14.16), which are secondarily bilateral, are usually found on sand and feed on detritus. Crinoids (see figures 14.22 and 14.23)

stretch their arms outward and up like a flower's petals, anchor themselves to the substrate, and feed on plankton and suspended particles.

Class Asteroidea: Sea Stars

Sea stars, often called starfishes, demonstrate basic features of echinoderm structure and function. We cover them first, and then comment on major differences among other echinoderm groups.

Sea stars are familiar along shorelines, where large numbers sometimes aggregate on rocks. There are about 1500 living species. They also live on muddy or sandy substrates on the ocean floor and among coral reefs. They are often brightly colored and range in size from a centimeter to about a meter in greatest diameter.

Form and Function

External Features Reflecting their pentaradial symmetry, sea stars typically have five tapering arms (rays), but there may be more (figure 14.4D). These merge gradually with the central disc (figure 14.5). The mouth is on the oral surface; the side opposite the mouth is called the aboral surface.

A band from which **tube feet (podia)** project, called an **ambulacrum** (pl., **ambulacra,** L. *ambulacrum*, an alley), extends from the mouth along the oral sides of each arm to its tip. An ambulacral groove occurs along the middle of each ambulacrum and contains rows of tube feet, which are bordered by movable spines (figure 14.5). Viewed from the oral side, a large radial nerve can be seen in the center of each ambulacral groove (figure 14.6C), between the rows of tube feet. This nerve is very superficially located, covered only by thin epidermis. Under the nerve is an extension of the coelom and the radial canal of the water-vascular system (figure 14.6C). In all other classes of living echinoderms except crinoids, ossicles or other dermal tissue cover these structures; thus, the ambulacral grooves in most asteroids and crinoids (see p. 304) are *open,* and those of the other groups are *closed.*

The aboral surface is usually rough and spiny, although spines of many species are flattened, so that the surface appears smooth. Around the bases of spines in many sea stars are groups of minute, pincerlike **pedicellariae** (ped-i-cel-ar′e-ē), bearing tiny jaws manipulated by muscles (figure 14.7). These jaws keep the body surface free of debris, protect the papulae, and sometimes aid in food capture. **Papulae** (pa-pū-lē) (also called **dermal branchiae** or skin gills) are soft, delicate projections of the coelomic cavity, covered only with epidermis and lined internally with peritoneum; they extend outward through spaces between ossicles (see figure 14.6C). Papulae greatly increase the surface area available for respiratory gas exchange. Also on the aboral side are an inconspicuous **anus** and a circular **madreporite** (see figure 14.5A), a calcareous sieve leading to the water-vascular system.

Endoskeleton and Coelom Beneath the epidermis of sea stars is a mesodermal endoskeleton of small calcareous plates, or **ossicles,** bound together with connective tissue. From these ossicles project spines and tubercles that form the spiny surface. Ossicles are penetrated by a meshwork of spaces, usually filled with fibers and dermal cells. This internal meshwork structure is described as **stereom** (see figure 14.19) and is unique to echinoderms.

Coelomic compartments of larval echinoderms give rise to several structures in adults, one of which is a spacious body **coelom** filled with fluid. Coelomic fluid circulates around the body cavity and into the papulae, propelled by cilia on the peritoneal lining. Exchange of respiratory gases and excretion of nitrogenous waste, principally ammonia, occur by diffusion through the thin walls of papulae and tube feet.

Water-Vascular System The water-vascular system is another coelomic compartment and is unique to echinoderms. It is a set of canals and specialized tube feet that, together with the dermal ossicles, forms a hydraulic system. In sea stars, the primary functions of the water-vascular system are locomotion and food gathering, in addition to respiration and excretion.

Structurally, the water-vascular system opens to the outside through small pores in the madreporite. The madreporite of asteroids is on the aboral surface (see figure 14.5A), and leads into a stone canal, which descends toward a ring canal around the mouth (see figure 14.6B). Radial canals diverge from the ring canal, one into the ambulacral groove of each ray. **Polian vesicles** also attach to the ring canal of most asteroids (but not *Asterias*) and apparently serve as fluid reservoirs for the water-vascular system.

figure 14.5

External anatomy of a sea star, an asteroid. **A,** Aboral view. **B,** Oral view.

A

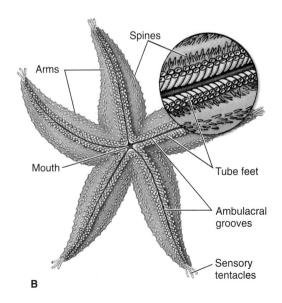

B

figure 14.6

A, Internal anatomy of a sea star. **B**, Water-vascular system. (Polian vesicles are not present in *Asterias*.) **C**, Cross section of an arm at level of gonads, illustrating open ambulacral groove.

A series of small **lateral canals,** each with a one-way valve, connects the radial canal to the cylindrical podia (tube feet), along the sides of the ambulacral groove in each ray. Each podium is a hollow, muscular tube, the inner end of which is a muscular sac, or **ampulla,** which lies within the body coelom (see figure 14.6B), and the outer end of which usually bears a sucker. Podia pass to the outside between ossicles in the ambulacral groove.

The water-vascular system operates hydraulically as an effective locomotor mechanism. Valves in the lateral canals prevent backflow of fluid into the radial canals. Tube feet have in their walls connective tissue that maintains the cylinder at a relatively constant diameter. Contraction of muscles in an ampulla forces fluid into the tube foot, extending it. Conversely, contraction of longitudinal muscles in the tube foot retracts the podium, forcing fluid back into the ampulla. Contraction of muscles in one side of the tube foot bends the organ toward that side. Small muscles at the end of the tube foot can raise the middle of the disclike end, creating a suction-cup effect when the end is applied to a firm substrate. By combining mucous adhesion with suction, a single tube foot can exert a pull equal to 0.25 to 0.3 newtons. Coordinated action of all or many of the tube feet is sufficient to draw a sea star up a vertical surface or over rocks. On a soft surface, such as mud or sand, suckers are ineffective (numerous sand-dwelling species have no suckers), so the tube feet are employed as legs.

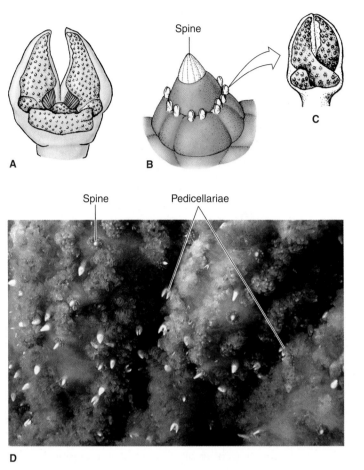

Spine

Spine   Pedicellariae

D

figure 14.7

Pedicellariae of sea stars and sea urchins. **A,** Forceps-type pedicellaria of *Asterias*. **B, C,** Scissors-type pedicellariae of *Asterias;* **B** shows size relative to spine. **D,** Close-up view of the aboral surface of a sea star, *Pycnopodia helianthoides*. Note the large pedicellariae, as well as groups of small pedicellariae around the spines. Many thin-walled papulae can be seen.

Feeding and Digestive System The mouth of a sea star leads into a two-part stomach located in the central disc (see figure 14.6A). The large, lower cardiac stomach can be everted through the mouth during feeding. The smaller upper pyloric stomach connects with pyloric ceca (digestive glands) located in the arms. Digestion is mostly extracellular, occurring in pyloric ceca. A short intestine leads from the stomach to the inconspicuous anus on the aboral side. Some species lack an intestine and anus.

Many sea stars are relatively unselective carnivores, feeding on molluscs, crustaceans, polychaetes, echinoderms, other invertebrates, and rarely small fishes, but some show particular preferences. Some select brittle stars, sea urchins, or sand dollars; swallow prey whole; and later regurgitate undigestible ossicles and spines. Some attack other sea stars, and if small compared to their prey, they may attack and begin eating at the end of one of the prey's arms.

Some asteroids feed heavily on molluscs, and *Asterias* is a significant predator on commercially important mussels and oysters. When feeding on a bivalve, a sea star wraps around its prey, attaching its podia to the valves, and then exerts a steady pull, using its feet in relays. A force of some 12.75 newtons can thus be exerted. In half an hour or so, the adductor muscles of the bivalve fatigue and relax. With a very small gap available, the star inserts its soft everted stomach into the space between the valves, wraps it around the soft parts of the bivalve, and secretes digestive enzymes to begin digestion. After feeding, the sea star draws its stomach back inside its body by contracting the stomach muscles and relaxing the body-wall muscles.

Some sea stars feed on small particles, either entirely or in addition to carnivorous feeding. Tiny plankton or other organic particles coming in contact with a sea star's surface are carried by epidermal cilia to ambulacral grooves and then to the mouth.

Numbers of crown-of-thorns starfish (see figure 14.4D) appear to have increased in recent decades, damaging large areas of coral reef in the Pacific Ocean. Crown-of-thorns stars (*Acanthaster planci* [Gr. *akantha*, thorn, + *asteros*, star]) feed on coral polyps, and their damage to coral reefs is made more severe by recent increases in coral bleaching and other threats to coral reefs. These starfish can be killed by injecting them with sodium bisulfate, a salt that does not damage the coral reefs.

Hemal System The so-called hemal system is composed of tissue strands enclosing unlined channels and is itself enclosed in another coelomic compartment, the perihemal channels. The hemal system is not well developed in asteroids, and in all echinoderms its function is unclear. It has little use in circulation of body fluids; some research suggests it plays a role in distributing digested nutrients.

Nervous and Sensory System The nervous system consists of three subsystems, each formed by a nerve ring and radial nerves placed at different levels in the disc and arms. An epidermal nerve plexus, or nerve net, connects the systems. Sense organs include an eyespot at the tip of each arm and sensory cells scattered over the epidermis.

Reproductive System, Regeneration, and Autotomy Most sea stars have separate sexes. A pair of gonads lies in each interradial space (see figure 14.6A), and fertilization is external.

Echinoderms can regenerate lost parts. Sea star arms can regenerate readily, even if all are lost. Sea stars also exhibit **autotomy** and can detach an injured arm near its base. An arm may take months to regenerate.

If a removed arm contains a part of the central disc (about one-fifth), the arm can regenerate a complete new sea star! In former times, fishermen dispatched sea stars collected

CHARACTERISTICS
of Phylum Echinodermata

1. Unique **water-vascular system** of coelomic origin extends from body surface as a series of tentacle-like projections (**podia**, or **tube feet**) protracted by an increase of fluid pressure within them; opening to exterior (**madreporite** or **hydropore**) usually present
2. Live in marine habitats
3. Free-living taxa
4. Body unsegmented (nonmetameric) with **pentaradial symmetry;** body rounded, cylindrical, or star-shaped, with five or more radiating areas, or **ambulacra,** alternating with interambulacral areas; no head
5. Triploblastic
6. Coelom enterocoelous and extensive, forming perivisceral cavity and cavity of water-vascular system; coelomic fluids circulated by peritoneal cilia
7. **Endoskeleton** of **dermal calcareous ossicles** with **spines** or of calcareous **spicules** in dermis; covered by epidermis (ciliated in most); **pedicellariae** (in some)
8. Digestive system usually complete; axial or coiled; anus absent in ophiuroids
9. Skeletal elements connected by ligaments of moldable collagenous tissue under neural control, ligaments can be "locked" into rigid posture or relaxed to allow free movement; locomotion by **tube feet,** which project from **ambulacra,** by movement of spines, or by movement of arms, which project from the body's central disc
10. Nervous system with circumoral ring and radial nerves; usually two or three systems of networks located at different levels in the body, varying in degree of development according to group
11. **No brain;** few specialized sensory organs; sensory system of tactile and chemoreceptors, podia, terminal tentacles, photoreceptors, and statocysts
12. Autotomy and regeneration of lost parts conspicuous; asexual reproduction by fragmentation in some
13. Sexes separate (except a few hermaphroditic) with large gonads, single in holothuroids but multiple in most; simple ducts, with no elaborate copulatory apparatus or secondary sexual structures; fertilization usually external; eggs brooded in some; development through **free-swimming, bilateral, larval stages** (some with direct development); metamorphosis to radial adult or subadult form; radial cleavage and regulative development
14. **Excretory organs absent**
15. Respiration by **papulae, tube feet, respiratory tree** (holothuroids), and **bursae** (ophiuroids)
16. Blood-vascular system (**hemal system**) much reduced, playing little if any role in circulation, and surrounded by extensions of coelom (**perihemal sinuses**)

figure 14.8

The Pacific sea star, *Echinaster luzonicus*, can reproduce itself by splitting across the disc, and then regenerating missing arms. The one shown here has evidently regenerated six arms from the longer one at top left.

Development   Some species brood their eggs, either under the oral side of the animal or in specialized aboral structures, and development is direct. However, most sea stars produce free-swimming larvae.

Early embryogenesis shows a typical ancestral deuterostome pattern. Gastrulation is by invagination, and the anterior end of the archenteron pinches off to become a coelomic cavity, which expands in a U shape to fill the blastocoel. Each leg of the U, at the posterior end, constricts to become a separate vesicle, and these eventually give rise to the main coelomic compartments of the body (metacoels, called **somatocoels** in echinoderms) (figure 14.9). The anterior portions of the U undergo subdivision to form protocoels and mesocoels (called **axocoels** and **hydrocoels** in echinoderms). The left hydrocoel becomes the water-vascular system, and the left axocoel gives rise to the stone canal and perihemal channels. The right axocoel and hydrocoel disappear. The free-swimming larva, called a **bipinnaria** (figure 14.10A), has cilia arranged in bands. These ciliated tracts become extended into larval arms. The larva grows three adhesive arms and a sucker at its anterior end and is then called a brachiolaria (figure 14.10B). It attaches to the substrate with a temporary stalk and undergoes metamorphosis.

Metamorphosis involves dramatic reorganization of a bilateral larva into a radial juvenile. The anteroposterior axis of the larva is lost. *The larval left side becomes the oral surface, and the larval right side becomes the aboral surface* (see figure 14.9). Correspondingly, the larval mouth and anus disappear, and a new mouth and anus form on what were originally the left and right sides, respectively. As internal reorganization proceeds, short, stubby arms and the first podia appear. The animal then detaches from its stalk and begins life as a young sea star.

from their oyster beds by chopping them in half with a hatchet—a worse than futile activity. Some sea stars reproduce asexually under normal conditions by cleaving the central disc, each part regenerating the rest of the disc and missing arms (figure 14.8).

figure 14.9

Asteroid metamorphosis. The left somatocoel becomes the oral coelom, and the right somatocoel becomes the aboral coelom. The left hydrocoel becomes the water-vascular system, and the left axocoel the stone canal and perihemal channels. The right axocoel and hydrocoel disappear.

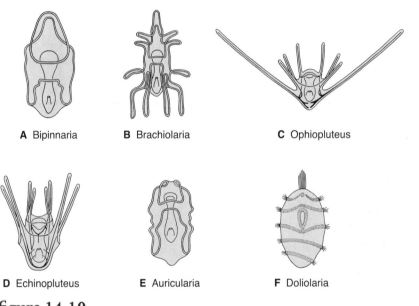

A Bipinnaria   **B** Brachiolaria   **C** Ophiopluteus

D Echinopluteus   **E** Auricularia   **F** Doliolaria

figure 14.10

Larvae of echinoderms. **A,** Bipinnaria of asteroids. **B,** Brachiolaria of asteroids. **C,** Ophiopluteus of ophiuroids. **D,** Echinopluteus of echinoids. **E,** Auricularia of holothuroids. **F,** Doliolaria of crinoids.

Sea Daisies Strange little (less than 1 cm diameter) disc-shaped animals, sometimes called sea daisies (figure 14.11), were discovered in water more than 1000 meters deep off New Zealand. The two species in this group are pentaradial but have no arms, and their tube feet are around the periphery of the disc, rather than along the ambulacral areas. Their water-vascular system includes two concentric ring canals; the outer ring may represent radial canals because podia arise from it. A hydropore, homologous to the madreporite, connects the inner ring canal to the aboral surface. Although they were originally described (1986) as a new class of echinoderms (Concentricycloidea), phylogenetic analysis of rDNA places them within Asteroidea.

Class Ophiuroidea: Brittle Stars and Basket Stars

Brittle stars are the largest group of echinoderms, with over 2000 extant species, and they are probably the most abundant also. They abound in all types of benthic marine habitats, even carpeting the abyssal sea floor in many areas.

Apart from their typical possession of five arms, brittle stars are surprisingly different from asteroids. The arms of brittle stars are slender and sharply set off from the central disc (figure 14.12). They have no pedicellariae or papulae, and their ambulacral grooves are closed and covered with arm ossicles. Their tube feet, which are without suckers, aid in feeding but are of limited use in locomotion. In contrast to asteroids, the madreporite of ophiuroids is located on the oral surface, on one of the oral-shield ossicles (figure 14.13).

Each jointed arm consists of a column of articulated ossicles connected by muscles and covered by plates. Locomotion is by arm movement.

Five movable plates that serve as jaws surround the mouth (figure 14.13). There is no anus. Their skin is leathery, with dermal plates and spines arranged in characteristic patterns.

Visceral organs are confined to the central disc, because the arms are too slender to contain them (figure 14.14). The stomach is saclike, and there is no intestine. Indigestible material is cast out of the mouth.

Five pairs of invaginations called **bursae** open toward the oral surface by bursal slits at the bases of the arms (figure 14.14). Water circulates in and out of these sacs for exchange of gases. On the coelomic wall of each bursa are small **gonads** that discharge their ripe sex cells into the bursa. Gametes pass through the bursal slits into the water for fertilization. Sexes are usually separate; a few ophiuroids are hermaphroditic.

Most species produce a free-swimming larva called an ophiopluteus, and its ciliated bands extend onto delicate, beautiful arms (see figure 14.10C). During metamorphosis to the juvenile stage, there is no temporarily attached phase as occurs in asteroids.

Water-vascular, nervous, and hemal systems are similar to those of sea stars.

Brittle stars tend to be secretive, living on hard substrates where no light penetrates. They are often negatively phototropic and insinuate themselves into small crevices between rocks, becoming more active at night. They are commonly fully exposed in the permanent darkness of deep seas. Ophiuroids consume a variety of small particles, either deposit or suspension feeding. Podia are important in transferring food to the mouth. Some brittle stars extend their arms into the water and catch suspended particles in mucous strands between arm spines. Basket stars (see figure 14.12B) perch themselves on corals, extending their branched arms to capture small plankton.

Regeneration and autotomy are even more pronounced in brittle stars than in sea stars. Many seem very fragile, releasing an arm or even part of the disc at the slightest provocation. Some can reproduce asexually by cleaving the disc; each new individual then regenerates missing parts.

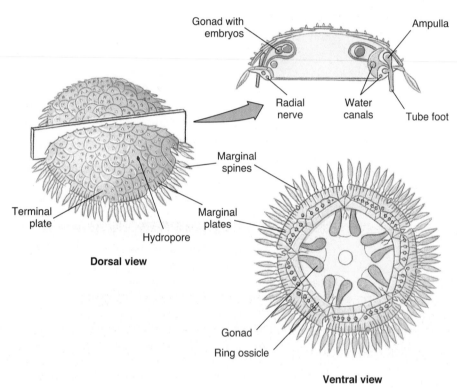

Dorsal view

Terminal plate

Hydropore

Gonad with embryos

Radial nerve

Water canals

Ampulla

Tube foot

Marginal spines

Marginal plates

Gonad

Ring ossicle

Ventral view

figure 14.11

Sea daisies, *Xyloplax* (class Asteroidea), are peculiar little disc-shaped echinoderms. In contrast to all other echinoderms, their podia lie around the margin instead of being distributed along ambulacral areas.

Class Echinoidea: Sea Urchins, Sand Dollars, and Heart Urchins

There are about 950 living species of echinoids, which have a compact body enclosed in an endoskeletal **test,** or shell. Dermal ossicles, which have become closely fitting plates, form the test. Echinoids lack arms, but their tests reflect the typical pentaramous plan of echinoderms in their five ambulacral areas. The most notable modification of the ancestral body plan is that the oral surface, which bears the tube feet and faces the substrate in sea stars, has expanded around to the aboral side, so that ambulacral areas extend upward to an area close to the anus (**periproct**). Most living species of sea urchins are "regular"; they have a hemispherical shape, radial symmetry, and medium to long spines (figure 14.15). Sand dollars and heart urchins (figure 14.16) are "irregular" because members of their orders have become secondarily bilateral; their

A

B

figure 14.12

A, Brittle star, *Ophiothrix oerstedii* (class Ophiuroidea). Brittle stars do not use their tube feet for locomotion but can move rapidly (for an echinoderm) by means of their arms. **B,** Oral view of a giant basket star, *Astrophyton muricatum.* Basket stars extend their many-branched arms to filter-feed, usually at night.

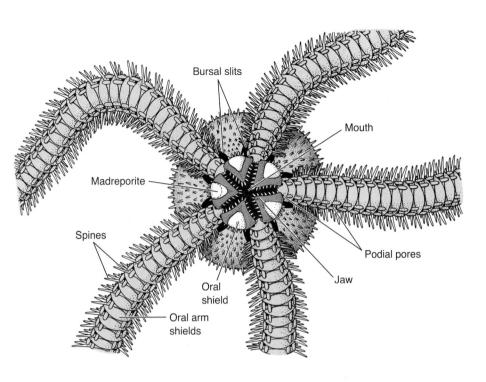

figure 14.13

Oral view of spiny brittle star, *Ophiothrix*.

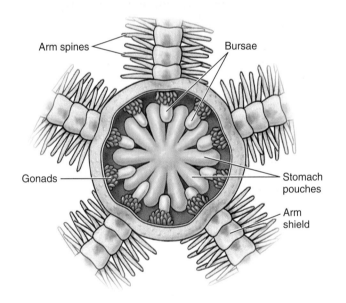

figure 14.14

Ophiuroid with aboral disc wall cut away to show principal internal structures. Bursae are fluid-filled sacs in which water constantly circulates for respiration. They also serve as brood chambers. Only bases of arms are shown.

figure 14.15

Purple sea urchin, *Strongylocentrotus purpuratus* (class Echinoidea), is common along the Pacific Coast of North America where wave action is heavy.

spines are usually very short. Regular urchins move by means of their tube feet, with some assistance from their spines, and irregular urchins move chiefly by their spines. Some echinoids are quite colorful.

Echinoids have wide distribution in all seas, from intertidal regions to deep oceans. Regular urchins often prefer rocky or hard substrates, whereas sand dollars and heart urchins burrow into a sandy substrate.

An echinoid test is a compact skeleton of 10 double rows of plates that bear movable, stiff spines (figure 14.17). The five pairs of ambulacral rows have pores (figure 14.17B) through which long tube feet extend. Spines are moved by small muscles around their bases.

There are several kinds of pedicellariae, the most common of which have three jaws and are mounted on long stalks. Pedicellariae of many species bear venom glands, and their toxin paralyzes small prey.

Five converging teeth surround the mouth of regular urchins and sand dollars. In some sea urchins, branched gills (modified podia) encircle the region around the mouth, although these are of little importance in respiratory gas exchange. An anus, genital pores, and a madreporite are located aborally in the periproct region (figure 14.17). The mouth of sand dollars is located at about the center of the oral side, but their anus has shifted to the margin or even to the oral side of the disc, so that an anteroposterior axis and bilateral symmetry can be recognized. Bilateral symmetry is even more accentuated in heart urchins, with their anus near the posterior on their oral side and their mouth moved away from the oral pole toward the anterior (see figure 14.16).

Inside the test (figure 14.17) are a coiled digestive system and a complex chewing apparatus (in regular urchins and in sand dollars), called **Aristotle's lantern,** to which teeth are attached (figure 14.18). A ciliated siphon connects the esophagus to the intestine and enables water to bypass the stomach to concentrate food for digestion in the intestine. Sea urchins eat algae and other organic material, and sand dollars collect fine particles on ciliated tracts or with podia.

Hemal and nervous systems are similar to those of asteroids. Ambulacral grooves are closed, and radial canals of the water-vascular system run just beneath the test, one in each ambulacral radius (see figure 14.17).

Sexes are separate, and both eggs and sperm are shed into the sea for external fertilization. Echinopluteus larvae (see figure 14.10D) may live a planktonic existence for several months and then metamorphose quickly into young urchins. Sea urchins have been used extensively as models in studies of development.

figure 14.16

An irregular echinoid, *Meoma*, one of the largest heart urchins (test up to 18 cm). *Meoma* occurs in the West Indies and from the Gulf of California to the Galápagos Islands. **A,** Aboral view. **B,** Oral view. Note the curved mouth at the anterior end and the periproct at the posterior end.

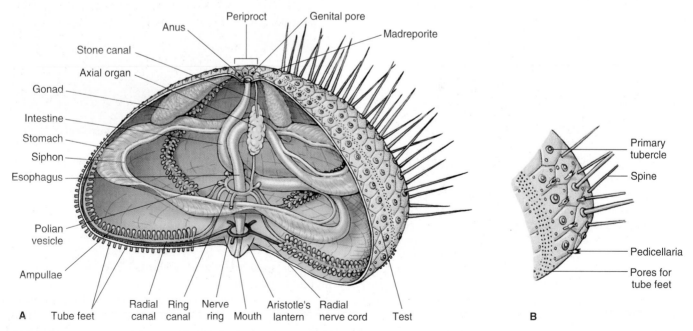

figure 14.17

A, Internal structure of a sea urchin. The water-vascular system is shown in tan. **B,** Detail of a portion of the endoskeleton.

Class Holothuroidea: Sea Cucumbers

In a phylum characterized by odd animals, class Holothuroidea contains members that both structurally and physiologically are among the strangest. There are about 1150 living species of holothuroids. These animals, commonly called sea cucumbers, have a remarkable resemblance to the vegetable after which they are named (figure 14.19). Compared to other echinoderms, sea cucumbers are greatly elongated in their oral-aboral axis, and ossicles are much reduced in most (figure 14.20); consequently, the body wall is usually leathery. Many species crawl along the sea floor, using ventral tube feet and body wall

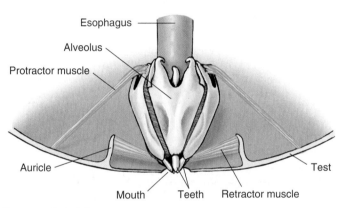

Esophagus

Alveolus

Protractor muscle

Auricle

Mouth   Teeth   Retractor muscle

Test

figure 14.18

Aristotle's lantern, a complex mechanism used by sea urchins for masticating their food. Five pairs of retractor muscles draw the lantern and teeth up into the test; five pairs of protractors push the lantern down and expose the teeth. Other muscles produce a variety of movements. Only major skeletal parts and muscles are shown in this diagram.

muscles. Others live beneath rocks, and some are burrowers. Despite a prominent anterior end, cephalization is absent.

Because sea cucumbers typically lie on one side, their tube feet usually are well developed only in the three strips of ambulacra in contact with the substrate. Tube feet on the dorsal side are reduced and may serve a sensory role (see figure 14.19A). Thus, a secondary bilaterality is present, albeit of different origin from that of irregular urchins. Oral tentacles consist of 10 to 30 retractile, modified tube feet around the mouth. Locomotion in burrowing species, which often lack podia, is primarily accomplished by contraction of well-developed circular and longitudinal muscles in the body wall.

The coelomic cavity is spacious and filled with fluid, serving as a hydrostatic skeleton. A respiratory tree (figure 14.21), composed of two long, many-branched tubes, empties into the **cloaca**. Water is pumped in and out by the muscular cloaca. The respiratory tree serves for both respiration and excretion and is not present in any other group of living echinoderms. Gas exchange also occurs through the skin and tube feet. The water-vascular system is peculiar in that the madreporite lies free in the coelom.

Sexes are separate, but some sea cucumbers are hermaphroditic. Among echinoderms, only sea cucumbers have a single gonad. Fertilization is external, and the free-swimming larva is called an **auricularia** (see figure 14.10E).

Most sedentary species trap suspended food particles in the mucus of their outstretched oral tentacles; others are deposit feeders, collecting fine organic matter in their tentacles. They then stuff the tentacles into their pharynx, one by one, ingesting captured food. As deposit feeders, sea cucumbers are very successful, forming up to 90% of the biomass on the surface of deep-sea floors.

Sea cucumbers have a peculiar power of what appears to be self-mutilation but in reality is a defense mechanism. When disturbed or subjected to unfavorable conditions, they

A

B

figure 14.19

Sea cucumber (class Holothuroidea). **A,** Common along the Pacific Coast of North America, *Parastichopus californicus* grows up to 50 cm in length. Its tube feet on the dorsal side are reduced to papillae and warts. **B,** *Bohadschia argus* expels its Cuvierian tubules, modified parts of its respiratory tree, when it is disturbed. These sticky strands, containing a toxin, discourage potential predators.

discharge **Cuvierian tubules,** which are attached to the posterior part of the respiratory tree (see figure 14.19B). These tubules, which sometimes contain toxins, become long and sticky after expulsion and can entangle a predator. Some species also discharge the digestive tract, respiratory tree, or gonads. All of these structures can be regenerated.

figure 14.20

Ossicles of sea cucumbers are usually microscopic bodies buried in a leathery dermis. They can be extracted from this tissue with commercial bleach and are important taxonomic characteristics. The ossicles shown here, called tables, buttons, and plates, are from the sea cucumber *Holothuria difficilis*. They illustrate the meshwork (stereom) structure observed in ossicles of all echinoderms at some stage in their development. (×250)

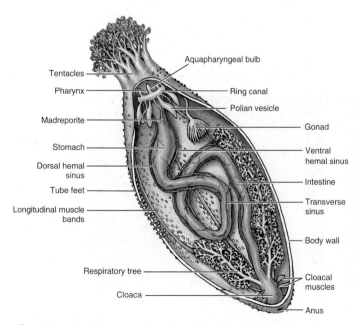

figure 14.21

Internal anatomy of a sea cucumber, *Sclerodactyla*. Hemal system is shown in red.

Class Crinoidea: Sea Lilies and Feather Stars

Crinoids include about 625 living species. As fossil records reveal, crinoids were once far more numerous than they are now. They differ from other echinoderms by being attached to a substrate during a substantial part of their lives. Sea lilies have a flower-shaped body at the tip of an attached stalk (figure 14.22). Feather stars have long, many-branched arms, and adults are free-moving, although they may remain in the same spot for long periods (figure 14.23). They can move by using the cirri to crawl or by sweeping their long, feather arms to swim. Many crinoids are deep-water forms, but feather stars may inhabit shallow waters, especially in Indo-Pacific and West Indian–Caribbean regions, where the largest numbers of species occur.

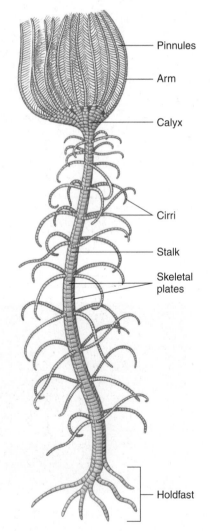

figure 14.22

A sea lily (stalked crinoid). Modern crinoid stalks rarely exceed 60 cm, but fossil forms were as long as 20 m.

figure 14.23

Comantheria briareus, is a feather star that occupies Pacific coral reefs. It extends its arms into the water to catch food particles both during the day and at night.

The crinoid body is covered with a leathery skin containing calcareous plates. Epidermis is poorly developed. Five flexible arms branch to form many more arms, each with many lateral pinnules arranged like barbs on a feather (figure 14.22). Sessile forms have a long, jointed **stalk** attached to the aboral side of the body **(calyx).** This stalk is composed of plates, appears jointed, and may bear jointed, flexible appendages called **cirri.** Cirri at the base of the stalk form a holdfast used to anchor the animal to the substrate. Madreporite, spines, and pedicellariae are absent.

The upper (oral) surface bears a mouth and anus. With the aid of tube feet and mucous strands, crinoids capture small organisms and transfer them to **ambulacral grooves** at the bases of the arms. Their ambulacral grooves, which are open and ciliated, carry food to the mouth. The **water-vascular system** of crinoids follows the basic echinoderm plan. Sense organs are scanty and simple.

Sexes are separate, and gonads are simple. Doliolaria larvae (see figure 14.10F) are free-swimming for a time before they attach and metamorphose.

Taxonomy of Phylum Echinodermata

There are about 7000 living and 20,000 extinct or fossil species of Echinodermata. The traditional classification placed all the free-moving forms that were oriented with their oral side down in subphylum Eleutherozoa, containing most living species. The other subphylum, Pelmatozoa, contained mostly forms with their stems and oral side up; most extinct classes and living Crinoidea belong to this group. Although alternative schemes have strong supporters, cladistic analysis provides evidence that the two traditional subphyla are monophyletic. This listing includes only groups with living members.

Subphylum Pelmatozoa (pel-ma´tō-zō-ə) (Gr. *pelmatos,* a stalk, + *zoōn,* animal). Body in form of cup or calyx, borne on aboral stalk during part or all of life; oral surface directed upward; open ambulacral grooves; madreporite absent; both mouth and anus on oral surface; several fossil classes plus living Crinoidea.

Class Crinoidea (cri-noi´dē-ə) (Gr. *krinon,* lily, + *eidos,* form): **sea lilies** and **feather stars.** Five arms branching at base and bearing pinnules; ciliated ambulacral grooves on oral surface with tentacle-like tube feet for food gathering; spines and pedicellariae absent. Examples: *Antedon, Comantheria.*

Subphylum Eleutherozoa (e-lū´ther-ō-zō´ə) (Gr. *eleutheros,* free, not bound, + *zoōn,* animal). Body form star-shaped, globular, discoidal, or cylindrical; oral surface directed toward substrate or oral-aboral axis parallel to substrate.

Class Asteroidea (as´ter-oi´dē-ə) (Gr. *aster,* star, + *eidos,* form): **sea stars (starfish).** Star-shaped, with arms not sharply demarcated from the central disc; ambulacral grooves open, with tube feet on oral side; tube feet often with suckers; anus and madreporite aboral; pedicellariae present. **Sea daisies,** included in this group, lack arms, have a disc-shaped body, and have a ring of suckerless podia near the body margin. Examples: *Asterias, Pisaster, Xyloplax.*

Class Ophiuroidea (o´fē-ur-oi´dē-ə) (Gr. *ophis,* snake, + *oura,* tail, + *eidos,* form): **brittle stars** and **basket stars.** Star-shaped, with arms sharply demarcated from central disc; ambulacral grooves closed, covered by ossicles; tube feet without suckers and not used for locomotion; pedicellariae absent. Examples: *Ophiothrix, Astrophyton.*

Class Echinoidea (ek´i-noi´dē-ə) (Gr. *echinos,* sea urchin, hedgehog, + *eidos,* form): **sea urchins, sea biscuits,** and **sand dollars.** Globular or disc-shaped, with no arms; compact skeleton or test with closely fitting plates; movable spines; ambulacral grooves closed; tube feet often with suckers; pedicellariae present. Examples: *Arbacia, Strongylocentrotus Meoma.*

Class Holothuroidea (ho´lo-thur-oi´dē-ə) (Gr. *holothourion,* sea cucumber, + *eidos,* form): **sea cucumbers.** Cylindrical, with no arms; spines absent; microscopic ossicles embedded in muscular body wall; anus present; ambulacral grooves closed; tube feet with suckers; oral tentacles (modified tube feet); pedicellariae absent; madreporite internal. Examples: *Sclerodactyla, Parastichopus, Cucumaria.*

■ Phylum Hemichordata

Hemichordata (hem´ē-kor-dä´tə) (Gr. *hemi,* half, + *chorda,* string, cord) are marine animals formerly considered a subphylum of chordates, based on their possession of gill slits and a rudimentary notochord. However, the hemichordate "notochord" is really an evagination of their mouth cavity and not homologous with the chordate notochord, so hemichordates are ranked as a separate phylum. Hemichordates have a three-part coelom typical of deuterostomes.

Hemichordates are wormlike bottom-dwellers, living usually in shallow waters. Some colonial species live in secreted tubes. Most are sedentary or sessile. They are widely distributed, but their secretive habits and fragile bodies make collecting them difficult.

Members of class Enteropneusta (acorn worms) range from 20 mm to 2.5 m in length. Members of class Pterobranchia (pterobranchs) are smaller, usually 1 to 7 mm, not including their stalk. About 75 species of enteropneusts and 30 species of pterobranchs are recognized.

Class Enteropneusta

Enteropneusts, or acorn worms (figure 14.24), are sluggish, wormlike animals that live in burrows or under stones, usually in mud or sand flats of intertidal zones.

The mucus-covered body is divided into a tonguelike **proboscis,** a short collar, and a long trunk (protosome, mesosome, and metasome).

Many acorn worms are **deposit feeders,** extracting the organic components of sediments. Others are **suspension feeders,** capturing plankton and detritus. The muscular proboscis is the active part of the animal, collecting food in mucous strands on its surface. Captured particles are transported by ciliary action to the groove at the edge of the collar and then to the mouth (figure 14.25). The food-laden mucus continues to be directed by cilia along the ventral part of the pharynx and esophagus to the intestine. A row of gill pores extends dorsolaterally on each side of the trunk just behind the collar (figure 14.25). These gill pores open from a series of densely ciliated gill chambers that in turn connect with a series of U-shaped gill slits in the sides of the pharynx. Acorn worms have no gills, but some respiratory gas exchange occurs in the vascular branchial epithelium, as well as on the body surface. The gill slits allow removal of excess water from the pharynx, but they are not used as a filter-feeding structure.

In the posterior end of the proboscis is a small coelomic sac (protocoel) into which extends a buccal diverticulum, a slender, blind pouch of the gut that was formerly considered a notochord. Above this buccal diverticulum is the heart, which receives blood carried forward from a dorsal blood vessel. Blood pumped from the heart flows into a network of blood sinuses called a glomerulus, which has an excretory function, and then flows posteriorly through a ventral blood vessel, passing through extensive sinuses in the body wall and gut. Acorn worms have a ventral nerve cord and a larger dorsal nerve

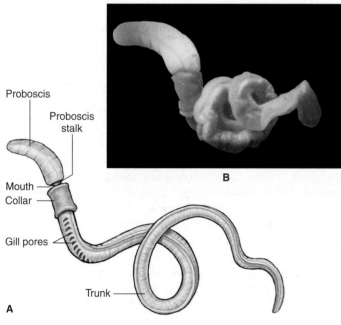

figure 14.24

An acorn worm, *Saccoglossus kovalevskii* (Hemichordata, class Enteropneusta). **A,** External lateral view. **B,** Live animal.

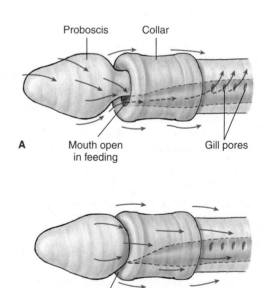

figure 14.25

Food currents of an acorn worm. **A,** Side view with mouth open, showing the direction of currents created by cilia on proboscis and collar. Food particles are directed toward the mouth and digestive tract. Rejected particles move outside of the collar. Water leaves through gill pores. **B,** When the mouth is occluded, all particles are rejected and passed onto the collar. Nonburrowing and some burrowing hemichordates use this feeding method.

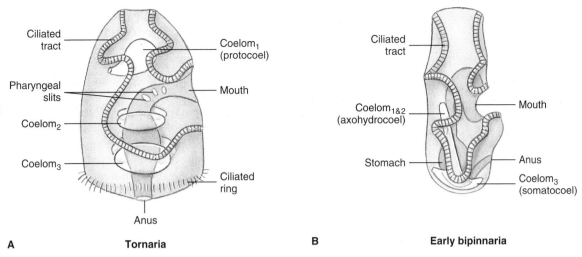

A   **Tornaria**   **B**   **Early bipinnaria**

figure 14.26

Comparison of **A**, a hemichordate tornaria with **B**, an echinoderm bipinnaria.

cord that is hollow in some species. Sexes are separate, and fertilization is external. In some species, a ciliated **tornaria** larva develops that closely resembles the echinoderm bipinnaria (figure 14.26).

Class Pterobranchia

Pterobranchs are small colonial animals. They bear arms with tentacles containing extensions of the coelomic compartment of the mesosome, similar to a lophophore (figure 14.27). Food captured in mucus on the crown of tentacles is transported with ciliated grooves to the mouth. The digestive tract is U-shaped, with the anus near the mouth. A dorsal, hollow nerve cord is absent, but otherwise their nervous system is similar to that of acorn worms. One genus, *Cephalodiscus*, has a single pair of gill slits (figure 14.27). Some species are dioecious, and others are monoecious; asexual reproduction occurs by budding.

Phylogeny

Echinoderms left an extensive fossil record and evolved about 25 anatomically distinct body forms in 20 currently recognized classes. Most of these became extinct by the end of the Paleozoic era, and only five classes survive today. Despite their fossil record, numerous contesting hypotheses on the phylogeny of echinoderms have been proposed. Based on their bilateral larvae, many invertebrate zoologists argue that their ancestors were bilateral and that their coelom had three pairs of spaces (trimeric or tripartite). Some investigators have proposed that radial symmetry arose in a free-moving echinoderm ancestor and that sessile groups were derived several times independently from free-moving ancestors. However, this view does not consider the adaptive significance of radial symmetry for a sessile existence.

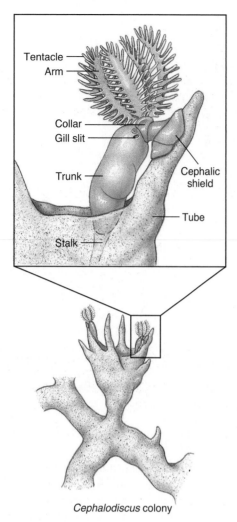

figure 14.27

Cephalodiscus, a pterobranch hemichordate. These tiny (5 to 7 mm) forms live in tubes in which they can move freely. Ciliated tentacles and arms direct currents of food and water toward the mouth.

The more traditional view is that the first echinoderms were sessile, became radial as an adaptation to that existence, and then gave rise to the free-moving groups. Figure 14.3 is consistent with this hypothesis. It depicts endoskeletal plates with stereom structure and the presence of external ciliary grooves for feeding as early echinoderm (or pre-echinoderm) developments. Extinct carpoids (see figure 14.3) had stereom ossicles but were not radially symmetrical, and the status of their water-vascular system, if any, is uncertain. Some investigators regard carpoids as a separate subphylum of echinoderms (Homalozoa). Fossil helicoplacoids (see figure 14.3) show evidence of three true ambulacral grooves, and their mouth was on the side of their body.

Attachment to a substrate by their aboral surface would have selected for radial symmetry, explaining the origin of subphylum Pelmatozoa, whose living members are crinoids. An ancestor that became free-moving and applied its oral surface to the substrate would have given rise to subphylum Eleutherozoa.

Phylogeny within Eleutherozoa is controversial. Most investigators agree that echinoids and holothuroids form a clade, but opinions diverge on the relationship of ophiuroids and asteroids. Figure 14.3 illustrates the view that ophiuroids arose after closure of their ambulacral grooves, but this scheme treats five ambulacral rays (arms) in ophiuroids and asteroids as independently evolved. Alternatively, if ophiuroids and asteroids form a clade, then closed ambulacral grooves must have been lost in the asteroids, or evolved separately in ophiuroids and in the common ancestor of echinoids and holothuroids.

The relationship of hemichordates to other deuterostome phyla has been puzzling. With chordates they share pharyngeal slits. With echinoderms they share a similar ciliated larva, a three-part coelom, and a kind of filtration structure called an axial complex. Phylogenetic analyses of 18S rDNA sequences and *Hox* genes support a close relationship between echinoderms and hemichordates. Nearly all zoologists now agree that echinoderms and hemichordates form a clade, Ambulacraria, which is the sister taxon to chordates (see figure 14.1).

Pharyngeal slits appear to be homologous between chordates and hemichordates. A basal, extinct carpoid echinoderm, *Jaekelocarpus*, has paired openings interpreted as pharyngeal slits, suggesting that pharyngeal slits are ancestral for deuterostomes, and that absence of pharyngeal slits in living echinoderms is derived. Thus, the ancestral deuterostome may have been a small, bilateral, free-moving form with pharyngeal slits.

Adaptive Diversification

Diversification of echinoderms has been limited by their most important characters: radial symmetry, water-vascular system, and dermal endoskeleton. If their ancestors had a brain and specialized sense organs, these were lost in adoption of radial symmetry. Only now are gene expression studies beginning to help researchers identify the anterior-posterior axis in adult echinoderms. The best evidence currently available suggests that the oral surface is anterior and the aboral surface is posterior. Under this hypothesis, the arms represent lateral growth zones.

Echinoderms have diversified within the benthic habitat. There are large numbers of creeping forms with filter-feeding, deposit-feeding, scavenging, and herbivorous habits, and very rare pelagic forms. In this light, the relative success of asteroids as predators is impressive and probably attributable to the extent to which they have exploited the hydraulic mechanism of their tube feet.

Summary

Arrow worms (phylum Chaetognatha) are a small group but an important component of marine plankton. They have a well-developed coelom and are effective predators, catching other zooplankton with the teeth and chitinous spines around their mouth.

Phyla Echinodermata, Xenoturbellida, Chordata, and Hemichordata are deuterostomes. Xenoturbellida houses two species of wormlike deuterostomes with relatively simple bodies, including a blind gut. Echinoderms are an important marine group sharply distinguished from other animals. They have pentaradial symmetry but evolved from bilateral ancestors.

Sea stars (class Asteroidea) usually have five arms, which merge gradually with a central disc. Like other echinoderms, they have no head and few specialized sensory organs. Their mouth is central on the under (oral) side of their body. They have an endoskeleton of dermal ossicles, open ambulacral grooves, respiratory papulae, and in many species, pedicellariae. Their water-vascular system is an elaborate hydraulic system derived embryonically from one of the coelomic cavities. Along the ambulacral areas, branches of the water-vascular system lead to the many tube feet, structures that are important in locomotion, food gathering, respiration, and excretion. Many sea stars are predators, whereas others feed on particulate organic matter. Sexes are separate, and reproductive systems are simple. Bilateral, free-swimming larvae become attached, transform to radial juveniles, and then detach and become motile sea stars.

Brittle stars (class Ophiuroidea) have slender arms that are sharply set off from their central disc; they have no pedicellariae or ampullae, and their ambulacral grooves are closed. Their madreporite is on the oral side. They crawl by means of relatively rapid (for echinoderms) arm movements and use their tube feet to gather food.

Dermal ossicles of sea urchins (class Echinoidea) are fitting plates, and there are no arms. Ambulacral areas are closed and extend around the body toward the aboral pole. Most sea urchins feed on algae, which they remove from the substrate with a skeletal apparatus called Aristotle's lantern. Heart urchins and some sand dollars have returned to adult bilateral symmetry.

Sea cucumbers (class Holothuroidea) have very small dermal ossicles and a soft body wall. They are greatly elongated in the oral-aboral axis and lie on their side. Tube feet around the mouth are modified as oral tentacles, with which they feed. They have an internal respiratory tree and can eject Cuvierian tubules for defense.

Sea lilies and feather stars (class Crinoidea) are the only group of living echinoderms, other than asteroids, with open ambulacral grooves. They are mucociliary particle feeders and lie with their oral side up.

Hemichordates are marine, wormlike, deposit or filter feeders that capture food using mucociliary action. Like chordates, most have paired gill slits. Acorn worms capture food on their proboscis, whereas pterobranchs capture food on their tentacles. Together with echinoderms, they form clade Ambulacraria.

Review Questions

1. How do chaetognaths feed?
2. Molecular sequence data indicate *Xenoturbella* is a deuterostome. Are there corresponding morphological characters?
3. What constellation of characteristics possessed by echinoderms occurs in no other phylum?
4. How do we know that echinoderms were derived from an ancestor with bilateral symmetry?
5. What is an ambulacrum, and what is the difference between open and closed ambulacral grooves?
6. Trace or make a rough copy of figure 14.6B, without labels, and then from memory label the parts of the water-vascular system of a sea star.
7. Name the structures involved in the following functions in sea stars, brittle stars, sea urchins, sea cucumbers, and crinoids, and briefly describe the action of each: respiration, feeding and digestion, reproduction.
8. Match these groups with *all* correct answers in the lettered column.
   _____ Crinoidea
   _____ Ophiuroidea
   _____ Holothuroidea
   _____ Asteroidea
   _____ Echinoidea
   a. Closed ambulacral grooves
   b. Oral surface generally upward
   c. With arms
   d. Without arms
   e. Approximately globular or disc-shaped
   f. Elongated in oral-aboral axis
   g. Have pedicellariae
   h. Madreporite internal
   i. Madreporite on oral plate
9. Define the following: pedicellariae, madreporite, respiratory tree, Aristotle's lantern, papulae, Cuvierian tubules.
10. Give three examples of how echinoderms are important to humans.
11. Distinguish Enteropneusta from Pterobranchia.
12. Members of Hemichordata were once considered chordates. Why?
13. What evidence suggests that hemichordates and echinoderms form a monophyletic group?

For Further Thought The phylogenetic placement of chaetognaths is uncertain. What are the two other possible placements for this group and what would you need to know to decide on the best placement?

■ Selected References

See also general references on page 432.

Baker, A. N., F. W. E. Row, and H. E. S. Clark. 1986. A new class of Echinodermata from New Zealand. Nature **321**:862–864. *Describes the strange sea daisies.*

Ball, E. E., and D. J. Miller. 2006. Phylogeny: the continuing conundrum of chaetognaths. Curr. Biol. **16**:R593–R596. *Nice summary of chaetognath biology and two recent phylogenetic studies in the same issue.*

Bourlat, S. J., T. Juliusdottir, C. J. Lowe, R. Freeman, J. Aronowicz, M. Kirschner, E. S. Lander, M. Thorndyke, H. Nakano, A. B. Kohn, A. Heyland, L. L. Moroz, R. R. Copley, and M. J. Telford. 2006. Deuterostome phylogeny reveals monophyletic chordates and the new phylum Xenoturbellida. Nature **444**:85–88. *The sister taxon to the chordates is Ambulacraria plus* Xenoturbella; *characters uniting clades within Deuterostomia are given.*

Gilbert, S. F. 2010. Developmental biology, ed. 9. Sunderland, Massachusetts, Sinauer Associates. *Any modern text in developmental biology, such as this one, provides a multitude of examples in which studies on echinoderms have contributed (and continue to contribute) to our knowledge of development.*

Hendler, G., J. E. Miller, D. L. Pawson, and P. M. Kier. 1995. Sea stars, sea urchins, and allies: echinoderms of Florida and the Caribbean. Washington, Smithsonian Institution Press. *An excellent field guide for echinoderm identification.*

Hickman, C. P., Jr. 1998. A field guide to sea stars and other echinoderms of Galápagos. Lexington, Virginia, Sugar Spring Press. *Provides descriptions and clear photographs of members of classes Asteroidea, Ophiuroidea, Echinoidea, and Holothuroidea in the Galápagos Islands.*

Hughes, T. P. 1994. Catastrophes, phase shifts and large-scale degradation of a Caribbean coral reef. Science **265**:1547–1551. *Describes the sequence of events, including the die-off of sea urchins, leading to the destruction of the coral reefs around Jamaica.*

Janies, D. 2001. Phylogenetic relationships of extant echinoderm classes. Can. J. Zool. **79**:1232–1250. *This review indicates that* Xyloplax *(sea daisy) is an asteroid and that asteroids and ophiuroids form a monophyletic group.*

Lane, D. J. W. 1996. A crown-of-thorns outbreak in the eastern Indonesian Archipelago, February 1996. Coral Reefs **15**:209–210. *This is the first report of an outbreak of* Acanthaster planci *in Indonesia. It includes a good photograph of an aggregation of these sea stars.*

Smith, A. B., K. J. Peterson, G. Wray, and D. T. J. Littlewood. 2004. From bilateral symmetry to pentaradiality: the phylogeny of hemichordates and echinoderms, pp. 365–383. *In:* Assembling the tree of life. J. Cracraft and M. J. Donoghue (eds.). New York, Oxford University Press. *Discusses evolution of echinoderms and hemichordates and summarizes evidence that they form a monophyletic group.*

Telford, M. J. 2008. Xenoturbellida, the fourth deuterostome phylum and the diet of worms. Genesis **46**:580–586. *A nice summary of the current status of the group.*

Woodley, J. D., P. M. H. Gayle, and N. Judd. 1999. Sea-urchins exert top-down control of macroalgae on Jamaican coral reefs (2). Coral Reefs **18**:193. *In areas where* Tripneustes *(another sea urchin) have invaded fore reefs there is much less macroalgae, and such areas present a better chance that corals can recolonize.* Diadema *recovery has been slow.*

Wray, G. A., and R. A. Raff. 1998. Body builders of the sea. Nat. Hist. **107**:38–47. *Regulatory genes in bilateral animals have taken on new but analogous roles in radial echinoderms.*

■ Custom Website

The *Animal Diversity* Online Learning Center is a great place to check your understanding of chapter material. Visit www.mhhe.com/hickmanad6e for access to key terms, quizzes, and more! Further enhance your knowledge with Web links to chapter-related material.

Explore live links for these topics:

Classification and Phylogeny of Animals
Phylum Echinodermata
Class Asteroidea
Class Ophiuroidea

Class Echinoidea
Class Holothuroidea
Class Crinoidea
Phylum Hemichordata
Phylum Chaetognatha

Vertebrate Beginnings:
The Chordates

Two amphioxus in feeding posture.

It's a Long Way from Amphioxus

Along the southern coasts of North America, half buried in sand on the sea floor, lives a small, fishlike, translucent animal quietly filtering organic particles from seawater. Inconspicuous, of no commercial value, and largely unknown, this creature is nonetheless one of the famous animals of classical zoology. It is amphioxus, an animal that wonderfully exhibits the five distinctive hallmarks of the phylum Chordata: (1) a dorsal hollow nerve cord overlying (2) a supportive notochord; (3) pharyngeal pouches or slits and (4) an endostyle for filter feeding; and (5) a postanal tail for propulsion. Amphioxus is an animal that might have been designed by a zoologist for the classroom. During the nineteenth century, with interest in vertebrate ancestry running high, many zoologists thought amphioxus bore a close resemblance to the direct ancestor of vertebrates. Its exalted position was later acknowledged by Philip Pope in a poem sung to the tune of "Tipperary." It ends with the refrain:

> It's a long way from amphioxus
> It's a long way to us,
> It's a long way from amphioxus
> To the meanest human cuss.
> Well, it's good-bye to fins and gill slits
> And it's welcome lungs and hair,
> It's a long, long way from amphioxus
> But we all came from there.

Amphioxus's place in the sun did not endure. Absence of a head, together with several specialized features, suggests to today's zoologists that amphioxus represents an early departure from vertebrate ancestry. We are a very long way indeed from amphioxus. Nevertheless, amphioxus probably resembles the chordate condition immediately preceding the origin of vertebrates more closely than does any other living animal.

T he animals most familiar to most people belong to the phylum Chordata. Humans are members and share the characteristic from which the phylum derives its name—the **notochord** (Gr. *nōton,* back, + L. *chorda,* cord) (figure 15.1). All members of the phylum possess this structure, either restricted to early development or present throughout life. The notochord is a rodlike, semirigid body of fluid-filled cells enclosed by a fibrous sheath, which extends, in most cases, the length of the body just ventral to the central nervous system. Thus, the notochord is a hydrostatic organ, similar to the hydrostatic skeletons of nematodes (p. 234). Its primary purpose is to stiffen the body, providing skeletal scaffolding for the attachment of swimming muscles.

The structural plan of chordates retains many features that occur in nonchordate invertebrates, such as bilateral symmetry, anteroposterior axis, coelom tube-within-a-tube arrangement, metamerism, and cephalization.

Chordates are deuterostomes and share with other members of that clade several important characteristics: radial cleavage (p. 62), an anus derived from the first embryonic opening (blastopore) and a mouth derived from an opening of secondary origin, and a coelom primitively formed by fusion of enterocoelous pouches (although in most vertebrates coelom formation is schizocoelous, but independently derived, as an accommodation for their large yolks). These uniquely shared characteristics indicate a natural unity among the Deuterostomia.

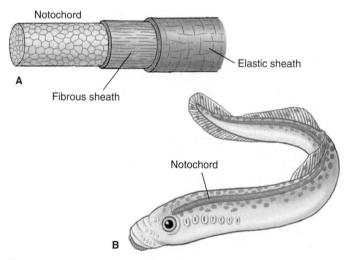

figure 15.1

A, Structure of the notochord and its surrounding sheaths. Cells of the notochord proper are thick walled, pressed together closely, and filled with semifluid. Stiffness is caused mainly by the turgidity of cells with large, fluid-filled vacuoles, enclosed within connective tissue sheaths. This type of endoskeleton is characteristic of all chordates at some stage of the life cycle. The notochord provides longitudinal stiffening of the main body axis, a base for trunk muscles. **B,** In hagfishes and lampreys, the notochord persists throughout life, but in other vertebrates it is largely replaced by vertebrae. In mammals, slight remnants are found in the nuclei pulposi of intervertebral discs.

Traditional and Cladistic Classification of the Chordates

Traditional Linnean classification of chordates (see pp. 324–325) provides a convenient way to indicate the taxa included in each major group. However, in cladistic usage, some of the traditional taxa, such as Agnatha and Reptilia, are no longer recognized. Such taxa do not satisfy the cladistic requirement that only **monophyletic** groups, those groups containing all known descendants of a single common ancestor, are taxonomically valid. The reptiles, for example, are considered a **paraphyletic** grouping because their group does not contain all descendants of their most recent common ancestor. The common ancestor of reptiles as traditionally recognized is also the ancestor of birds. As shown in the cladogram (figure 15.2), reptiles, birds, and mammals compose a monophyletic group called Amniota, so named because all develop from an egg with a membrane called an amnion. Therefore, according to cladistics, "reptiles" can be used only as a term of convenience to refer to amniotes that are not birds or mammals; there are no derived characters that unite the reptiles to the exclusion of birds and mammals. The reasons nonmonophyletic groups are not used in cladistic taxonomy are explained in Chapter 4 (pp. 88–89).

The cladogram of chordates (figure 15.2) and the phylogenetic tree of chordates (figure 15.3) provide different kinds of information. The cladogram shows a nested hierarchy of taxa grouped by their shared derived characters. These characters may be morphological, physiological, embryological, behavioral, chromosomal, or molecular. By contrast, the branches of a phylogenetic tree are intended to represent real lineages that occurred in the evolutionary past. Geological information regarding ages of lineages is added to the information from the cladogram to generate a phylogenetic tree for the same taxa.

In our treatment of chordates, we have retained the traditional Linnean classification (pp. 324–325) because subfields of zoology are organized according to this scheme and because the alternative—thorough revision following cladistic principles—would require extensive change and abandonment of familiar rankings. However, we have tried to use monophyletic taxa as much as possible, because such usage is necessary to reconstruct the evolution of characters in chordates.

Several traditional divisions of phylum Chordata used in Linnean classifications are shown in table 15.1. A fundamental separation is Protochordata from Vertebrata. All vertebrates have a skull enclosing the brain and also are called Craniata. We should note that some cladistic classifications exclude Myxini (hagfishes) from Vertebrata, because they lack vertebrae, but retain hagfishes in Craniata because they have a cranium. The vertebrates (craniates) may be variously subdivided into groups based on shared characteristics. Two such subdivisions shown in table 15.1 are (1) Agnatha, vertebrates lacking jaws (hagfishes and lampreys), and Gnathostomata, vertebrates having jaws (all other vertebrates); and (2) Amniota, vertebrates whose embryos develop within a fluid-filled sac, the amnion (nonavian reptiles, birds, and mammals), and Anamniota, vertebrates lacking this adaptation (fishes and amphibians). The

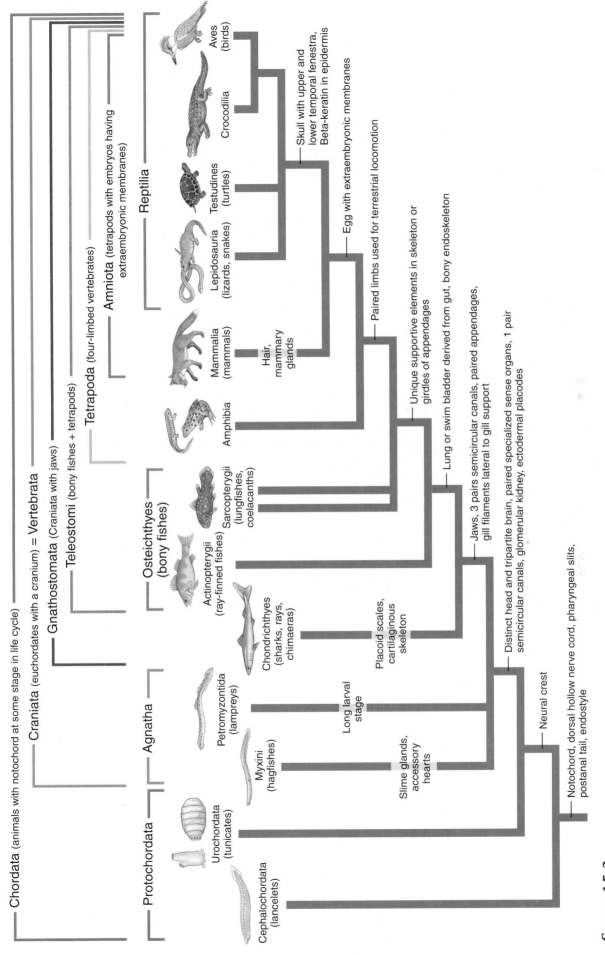

figure 15.2

Cladogram of living members of phylum Chordata, showing probable relationships of the monophyletic groups composing the phylum. Nesting brackets across the top of the cladogram identify monophyletic groupings within the phylum. The term Craniata, although commonly equated with Vertebrata, is preferred by many authorities because it recognizes that some jawless vertebrates have a cranium but no vertebrae. The lower brackets identify the traditional groupings Protochordata, Agnatha, Osteichthyes, and Reptilia. These paraphyletic groups are not recognized in cladistic treatments, but are shown because of widespread use.

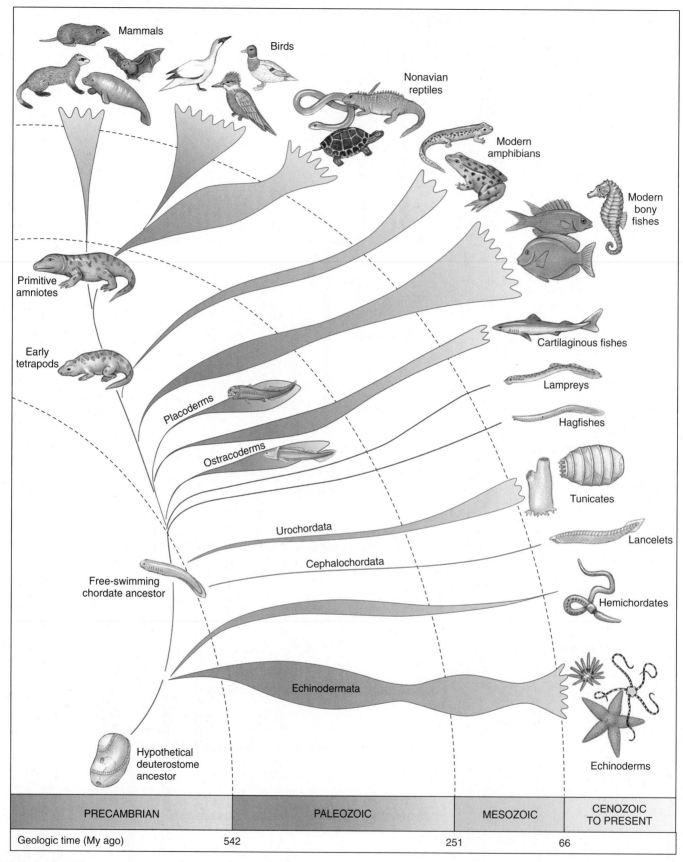

figure 15.3

Phylogenetic tree of the chordates, suggesting their probable origin and relationships. Other schemes have been suggested and are possible. The relative abundance in numbers of species of each group through geological time, as indicated by the fossil record, is indicated here by the bulging and thinning of that group's line of descent.

Table 15.1 Traditional Divisions of the Phylum Chordata

Cephalochordata (lancelets)	Urochordata (tunicates)	Myxini (hagfishes)	Petromyzontida (lampreys)	Chondrichthyes (sharks)	Osteichthyes (bony fishes)	Amphibia (amphibians)	Reptilia (reptiles)	Aves (birds)	Mammalia (mammals)

Chordata

Protochordata / Acraniata — Vertebrata

Craniata

Agnatha — Gnathostomata

Pisces — Tetrapoda

Anamniota — Amniota

CHARACTERISTICS
of Phylum Chordata

1. **Postanal tail; notochord; endostyle or thyroid gland; bone and cartilage** in vertebrates
2. Living in marine, freshwater, and terrestrial habitats; many capable of flight
3. Free-living, but a very few fishes are ectoparasitic
4. Bilateral symmetry; segmented, but segmentation inconspicuous in many
5. Triploblastic
6. **Coelom well developed**
7. Epidermis present in all; dermis in vertebrates; keratinized or bony structures often present in vertebrate integument; glands often diverse and abundant in vertebrates
8. Digestive system complete; muscular gut in vertebrates; **pharyngeal pouches** present early in development, erupting to outside as gill slits in aquatic forms
9. Smooth, skeletal, and cardiac muscle tissue present; segmented myomeres in fishes and amphibians
10. **Nerve cord hollow and dorsal;** distinct, **three-lobed brain** present in vertebrates
11. Protochordates with simple, unpaired photoreceptors and statocysts; vertebrates with well-developed paired sensory organs for vision, chemoreception, hearing, balance, electroreception, and vibration sensitivity
12. Asexual reproduction by parthenogenesis in some fishes, amphibians, and lizards
13. Sexes usually separate; hermaphroditism in sea squirts and some fishes; fertilization internal or external; oviparous or viviparous; distinct larval stage in some; crocodilians, birds, mammals, and some fishes and amphibians with parental care of young
14. Paired, glomerular kidneys and ducts in vertebrates
15. Respiration primarily via gills, lungs, and skin; swim bladder present in many fishes, functioning in buoyancy
16. Closed circulation; **chambered hearts** and red blood cells in vertebrates; distinct aortic arches in all except sea squirts

Gnathostomata in turn can be subdivided into Pisces, jawed vertebrates with appendages, if any, in the form of fins, and Tetrapoda (Gr. *tetras,* four, + *podos,* foot), jawed vertebrates with appendages, if any, in the form of limbs. Note that several of these groupings are paraphyletic (Protochordata, Acraniata, Agnatha, Anamniota, Pisces) and consequently are not accepted in cladistic classifications. Accepted monophyletic taxa are shown at the top of the cladogram in figure 15.2 as a nested hierarchy of increasingly more inclusive groupings.

■ Five Chordate Hallmarks

Five distinctive characteristics that, taken together, set chordates apart from all other phyla are a **notochord,** a dorsal hollow nerve cord, pharyngeal slits or pouches, an **endostyle,** and a postanal tail. These characteristics are always found at some embryonic stage, although they may change or disappear in later stages of the life cycle.

Notochord

The notochord is a flexible rod of fluid-filled cells enclosed in a fibrous sheath extending the length of the body; it is the first part of the endoskeleton to appear in the embryo. Muscles attach to the notochord, and because it can bend laterally without shortening, it permits undulatory movements of the body. In most protochordates and in jawless vertebrates, the notochord persists throughout life, but in all jawed vertebrates it is at least partly replaced by a series of cartilaginous or bony vertebrae.

Notochord

Dorsal Hollow Nerve Cord

In most invertebrate phyla that have a nerve cord, it is ventral to the digestive tract and is solid, but in chordates the single cord is dorsal to the digestive tract and notochord and is a tube (although the hollow center may be nearly obliterated during growth). The anterior end becomes enlarged to form the brain in vertebrates. The hollow cord is produced in the embryo by infolding of ectodermal cells on the dorsal side of the body above the notochord. Among vertebrates, the nerve cord passes through the protective neural arches of the vertebrae, and the brain is surrounded by a bony or cartilaginous cranium.

Brain   Dorsal nerve cord

Pharyngeal Pouches or Slits

Pharyngeal slits are openings that lead from the pharyngeal cavity to the outside. They are formed by the inpocketing of the outside ectoderm (pharyngeal grooves) and the evagination, or outpocketing, of the endodermal lining of the pharynx (pharyngeal pouches). In aquatic chordates, the two pockets break through the pharyngeal cavity where they meet to form the pharyngeal slit. In amniotes, some pockets may not break through the pharyngeal cavity, so only grooves are formed instead of slits. In tetrapod (four-limbed) vertebrates, pharyngeal pouches give rise to several different structures, including the Eustachian tube, middle ear cavity, tonsils, and parathyroid glands.

The perforated pharynx evolved as a filter-feeding apparatus and is used as such in protochordates. Water with suspended food particles is drawn by ciliary action through the mouth and flows out through pharyngeal slits, where food is trapped in mucus. In vertebrates, ciliary action is replaced by muscular pharyngeal contractions that drive water through the pharynx. The addition of a capillary network and thin, gas-permeable walls in the pharyngeal arches led to the development of internal gills, completing the conversion of the pharynx from a filter-feeding apparatus in protochordates to a respiratory organ in aquatic vertebrates.

Pharyngeal slits

Endostyle or Thyroid Gland

Until recently, the endostyle was not recognized as a chordate character. However, it or its derivative, the thyroid gland, occurs in all chordates, and in no other animals. The endostyle, located in the pharyngeal floor, secretes mucus that traps small food particles brought into the pharyngeal cavity. An endostyle is present in protochordates and lamprey larvae. Some cells in the endostyle secrete iodinated proteins. These cells are homologous with the iodinated-hormone-secreting thyroid

gland of adult lampreys and the remainder of vertebrates. In ancestral chordates, an endostyle and a perforated pharynx work together to create an efficient filter-feeding apparatus.

Endostyle

Postanal Tail

A postanal tail, together with somatic musculature and the stiffening notochord, provides the motility that larval tunicates and amphioxus need for their free-swimming existence. As a structure added to the body behind the anus, the tail clearly has evolved specifically for propulsion in water. Its efficiency is later increased in fishes with the addition of fins. A tail is evident in humans only as a vestige (the coccyx, a series of small vertebrae at the end of the spinal column), but most other mammals have a waggable tail as adults.

Muscular, postanal tail

■ Ancestry and Evolution

Since the mid–nineteenth century when Darwin's theory of common descent became the focal point for recognizing relationships among groups of living organisms, zoologists have debated the origin of chordates. Zoologists at first speculated that chordates evolved within the protostome clade (annelids and arthropods) but rejected such ideas when they realized that supposed morphological similarities had no developmental basis. Early in the twentieth century, further theorizing became rooted in developmental patterns of animals, and it became apparent that chordates are deuterostomes. As explained in Chapter 3 (see p. 66 and figure 3.8), Deuterostomia, a grouping that also includes echinoderms and hemichordates, has several important embryological features, as well as shared gene sequences, that clearly separate it from Protostomia and establish its monophyly. Thus, deuterostomes are almost certainly a natural grouping of interrelated animals that have their common origin in ancient Precambrian seas. Somewhat later, at the base of the Cambrian period some 570 million years ago, the first distinctive chordates arose from a lineage related to echinoderms and hemichordates (see figure 15.2). Although more fossil chordates from the Cambrian period (pp. 320–322) have been recently unearthed, reconstruction of the earliest chordates has come primarily from the study of living organisms, namely, the urochordates and cephalochordates, two chordate groups that are not vertebrates.

Most early efforts to determine chordate relationships were based on similarities due to analogy rather than homology (p. 81). Analogous structures perform similar functions but have different origins (such as the wings of birds and butterflies). Homologous structures, on the other hand, share a common origin but can look quite different (at least superficially) and often perform different functions. For example, all vertebrate forelimbs are homologous because they are derived from a pentadactyl limb of the same ancestor, even though they may be modified as differently as a human's arm and a bird's wing. Homologous structures share a genetic heritage; analogous structures do not. Obviously, only homologous similarities reveal common ancestry.

Subphylum Urochordata (Tunicata)

The urochordates ("tail-chordates"), more commonly called tunicates, include about 1600 species. They live in all seas, from near shoreline to great depths. Most are sessile as adults, although some are free-living. The name "tunicate" describes the usually tough, nonliving **tunic,** or test, that surrounds the animal and contains cellulose (figure 15.4). As adults, tunicates are highly specialized chordates, for in most species only the larval form, which resembles a microscopic tadpole, bears all the chordate hallmarks. During adult metamorphosis, the notochord and the tail disappear, while the dorsal nerve cord is reduced to a single ganglion.

Urochordata is divided into three classes—**Ascidiacea** (Gr. *askiolion,* little bag), **Appendicularia** (L. *larva,* ghost), and **Thaliacea** (Gr. *thalia,* luxuriance). Members of Ascidiacea, commonly called ascidians, or sea squirts, are by far the most common and best known. Ascidians may be solitary, colonial, or compound. All but a few ascidian species are sessile, attached to rocks or other hard substances such as pilings or the bottoms of ships. In some areas, they are among the most abundant of intertidal animals.

Solitary or colonial ascidians are usually spherical or cylindrical forms, each bearing its own tunic. Lining the tunic is an inner membrane, the mantle. On the outside are two projections: an **incurrent siphon** and an **excurrent siphon** (figure 15.4). Water enters the incurrent siphon and passes into the pharynx. On the midventral side of the pharynx is a groove, the endostyle, which is ciliated and secretes a sheet of mucus. As the mucous sheet is carried by cilia across the inner surface of the pharynx to the dorsal side, it traps small food particles from water passing through slits in the wall of the pharynx. The mucus with its entrapped food is formed into a rope and passed posteriorly to the esophagus. The water, now largely cleared of food particles, is driven by cilia into the atrial cavity and finally out the excurrent siphon. Nutrients are absorbed in the midgut, and indigestible wastes are discharged from the anus, located near the excurrent siphon.

The circulatory system consists of a ventral heart near the stomach and two large vessels, one on either side of the heart. An odd feature, found in no other chordate, is that the

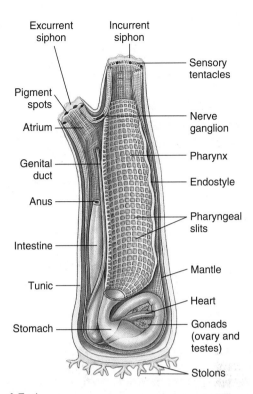

figure 15.4

Structure of a common tunicate, *Ciona.*

heart first drives the blood in one direction for a few beats, then pauses, reverses, and drives the blood in the opposite direction. The nervous system is restricted to a nerve ganglion and a few nerves that lie on the dorsal side of the pharynx. Sea squirts are hermaphroditic. Gametes are carried out of the excurrent siphon into the surrounding water, where fertilization occurs.

Of the five chief characteristics of chordates, adult sea squirts have only two: pharyngeal gill slits and an endostyle. However, the larval forms reveal the secret of their true relationship. The tiny "tadpole" larva (figure 15.5) is an elongate, transparent form with a head and all five chordate characteristics: a notochord, a dorsal hollow nerve cord, a propulsive postanal tail, and a large pharynx with an endostyle and gill slits. The larva does not feed but swims for several hours before fastening vertically by adhesive papillae to some solid object. It then metamorphoses to become a sessile adult.

The remaining two classes of Urochordata—**Appendicularia** and **Thaliacea**—are mostly small, transparent animals of the open sea (figure 15.6). Appendicularians, also called larvaceans, are small, tadpole-shaped forms resembling the larval stage of ascidians. An appendicularian builds and inhabits a hollow, transparent sphere of mucus. Tiny plankton are trapped on feeding filters inside the sphere and ingested. Thaliaceans are barrel- or lemon-shaped forms surrounded by circular muscle bands. They are mostly carried by ocean currents and, as such, form part of the plankton. Many have luminous organs and emit a brilliant light at night.

figure 15.5

Metamorphosis of a solitary ascidian from a free-swimming larval stage.

figure 15.6

Salps. The transparent individuals of this delicate, planktonic species are grouped in a chain. Visible within each individual is an opaque gonad, an opaque gut, and a long, serrated gill bar. Class Thaliacea.

■ Subphylum Cephalochordata

Cephalochordates are lancelets: slender, laterally compressed, translucent animals about 3 to 7 cm in length (figure 15.7) that inhabit sandy sediments of coastal waters around the world. Lancelets originally bore the generic name *Amphioxus* (Gr. *amphi,* both ends, + *oxys,* sharp), later surrendered by priority to *Branchiostoma* (Gr. *branchia,* gills, + *stoma,* mouth). Amphioxus is still used, however, as a convenient common name for all the approximately 29 species in this diminutive subphylum. Five species of amphioxus occur in North American coastal waters.

Amphioxus is especially interesting because it has the five distinctive characteristics of chordates in simple form. Water enters the mouth, driven by cilia in the buccal cavity, and then passes through numerous pharyngeal slits in the pharynx,

where food is trapped in mucus secreted by the endostyle, and is then moved by cilia into the gut. Here, the smallest food particles are separated from the mucus and passed into the **hepatic cecum,** where they are phagocytized and digested intracellularly. As in tunicates, filtered water passes first into an atrium, and then leaves the body by an atriopore (equivalent to the excurrent siphon of tunicates).

The closed circulatory system is complex for so simple a chordate. The flow pattern is remarkably similar to that of fishes, although there is no heart. Blood is pumped forward in the ventral aorta by peristaltic-like contractions of the vessel wall, and then passes upward through branchial arteries (aortic arches) in pharyngeal arches to paired dorsal aortas. From there, blood is distributed to the body tissues by capillaries and then collected in veins, which return it to the ventral aorta. Lacking both erythrocytes and hemoglobin, the blood of a lancelet is thought to circulate nutrients but not respiratory gases. There are no gills specialized for respiration in the pharynx; gas exchange occurs over the surface of the body.

The nervous system is centered around a hollow nerve cord lying above the notochord. Sense organs are simple, including an anterior, unpaired ocellus that functions as a photoreceptor. The anterior end of the nerve cord is not enlarged into the characteristic vertebrate brain.

Sexes are separate. Gametes are released in the atrium, and then pass through the atriopore to the outside, where fertilization occurs. Larvae hatch soon after the eggs are fertilized and gradually assume the shape of adults.

In addition to the five chordate anatomical hallmarks, amphioxus possesses several structural features that resemble the vertebrate plan. Among these are a hepatic cecum, which is similar to the vertebrate liver and pancreas in function, segmented trunk musculature, and the basic circulatory pattern of more advanced chordates.

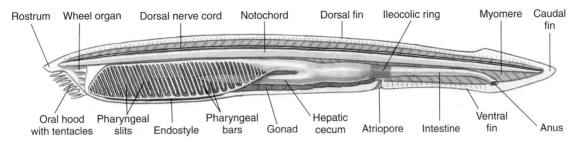

figure 15.7

Amphioxus. This filter-feeding cephalochordate possesses the five distinctive chordate characteristics (notochord, dorsal hollow nerve cord, pharyngeal slits, endostyle, and postanal tail).

■ Subphylum Vertebrata (Craniata)

The third subphylum of chordates is the large and diverse Vertebrata, the subject of Chapters 16 through 20. This monophyletic group shares the basic chordate characteristics with the other two subphyla, but in addition it reveals novel characters that the others do not share. The alternative name of the subphylum, Craniata, more accurately describes the group because all have a cranium (bony or cartilaginous braincase), but some jawless fishes lack vertebrae.

Adaptations That Have Guided Early Vertebrate Evolution

The earliest vertebrates were substantially larger and considerably more active than the protochordates. Modifications of the skeleton and muscles permitted increased speed and mobility. The higher activity level and size of vertebrates also required structures specialized for locating, capturing, and digesting food as well as adaptations designed to support a high metabolic rate.

Musculoskeletal Modifications

The endoskeleton of vertebrates permits almost unlimited body size, with much greater economy of building materials than the exoskeleton of arthropods. Some vertebrates have become the most massive organisms on earth. The endoskeleton forms an excellent jointed scaffolding for attachment of segmented muscles. The segmented body muscles (myomeres) changed from the V-shaped muscles of cephalochordates to the W-shaped muscles of vertebrates. This more complex folding in the myomeres provides powerful control over a long body.

The endoskeleton probably was composed initially of cartilage but later of bone. Cartilage, with its fast growth and flexibility, is ideal for constructing the first skeletal framework of all vertebrate embryos. The endoskeleton of living hagfishes, lampreys, sharks, and their kin, and even that of some "bony" fishes, such as sturgeons, is mostly cartilage. The extracellular matrix of bone tissue is of collagen fibers and the mineral hydroxyapatite, which is mostly formed from calcium and phosphate ions. Bone may have been adaptive in early vertebrates in several ways. Plates of bone in the skin of ostracoderms and other ancient fishes certainly provided protection from predators. In addition, bone is structurally stronger than cartilage, making it ideal for muscle attachment in areas of high mechanical stress. One of the most interesting ideas is that bone's original function was mineral storage and regulation. Phosphorus and calcium are used for many physiological processes and are in particularly high demand in organisms with high metabolic rates.

We should note that most vertebrates possess an extensive exoskeleton, although it is highly modified in terrestrial forms. Some of the earliest fishes, including the ostracoderms and placoderms, were partly covered by a bony dermal armor. This armor is modified as scales in later fishes. Most vertebrates are further protected with keratinized structures derived from the epidermis, such as reptilian scales, hair, and feathers.

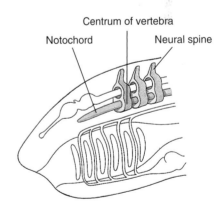

Physiology Upgrade

The vertebrate digestive, respiratory, and circulatory systems are modified to meet an increased metabolic demand. The perforated pharynx evolved as a filter-feeding device in early chordates, using cilia and mucus to move water and to trap small, suspended food particles. In vertebrates, the addition of muscles to the pharynx created a powerful pump for moving water. With the origin of highly vascularized gills, the function of the pharynx shifted primarily to gas exchange. Changes in the gut, including a shift from movement of food by ciliary action to muscular action and the addition of accessory digestive glands, the liver and pancreas, managed the increased amount of food ingested. A ventral, chambered heart as well as the presence of erythrocytes with hemoglobin enhanced transportation of respiratory gases, together with other substances.

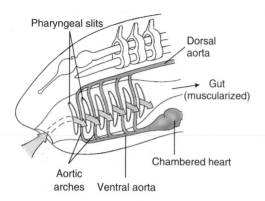

New Head, Brain, and Sensory Systems

When vertebrate ancestors shifted from sessile filter feeding to active predation, new sensory, motor, and integrative controls became essential for locating and capturing larger prey. The anterior end of the nerve cord became enlarged as a tripartite brain (forebrain, midbrain, and hindbrain) and protected by a cartilaginous or bony cranium. Paired special sense organs adapted for distance reception evolved. These included eyes with lenses and inverted retinas; pressure receptors, such as paired inner ears designed for equilibrium and sound reception; chemical receptors, including taste and exquisitely sensitive olfactory organs; lateral-line receptors for detecting water vibrations; and electroreceptors for detecting electrical currents that signal prey.

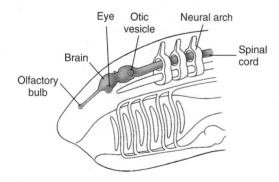

Neural Crest, Ectodermal Placodes, and Hox Genes

Development of the vertebrate head and special sense organs resulted from two embryonic innovations present only in vertebrates: a **neural crest** and ectodermal placodes. The neural crest, a population of ectodermal cells lying along the length of the embryonic neural tube, contributes to the formation of many different structures, including most of the cranium, the pharyngeal skeleton, tooth dentine, Schwann cells, and some

endocrine glands. Ectodermal placodes (Gr. *placo,* plate) are platelike ectodermal thickenings that appear on either side of the neural tube. These give rise to the olfactory epithelium, the lens of the eye, inner-ear epithelium, lateral-line mechanoreceptors, and electroreceptors. Thus, the vertebrate head, with its complex sensory structures located adjacent to the mouth (later equipped with prey-capturing jaws), stemmed from the creation of new embryonic tissues.

Studies of the distribution of **homeobox** (*Hox*) genes that control the body plan of chordate embryos suggest sets of *Hox* genes were duplicated at about the time of the origin of vertebrates. Amphioxus and other invertebrates have only one set of *Hox* genes, whereas most living gnathostomes, including humans, have four sets. These additional copies of *Hox* genes are likely responsible for the evolution of several unique vertebrate characters.

The Search for the Ancestral Vertebrate Stock

Most of the early Paleozoic vertebrate fossils, the jawless ostracoderms (see figure 15.11), share many novel features of organ system development with living vertebrates. These organ systems must have originated in an early vertebrate or invertebrate chordate lineage. Fossil invertebrate chordates are rare and known primarily from two fossil beds—the well-known middle-Cambrian Burgess Shale of Canada (see figure 1.11) and the recently discovered early-Cambrian fossil beds of Chengjiang and Haikou, China. *Pikaia,* a ribbon-shaped, somewhat fishlike creature about 5 cm in length, was discovered in the Burgess Shale (figure 15.8). The presence of V-shaped myomeres and a notochord clearly identifies *Pikaia* as a chordate. The superficial resemblance of *Pikaia* to living amphioxus suggests that it may be an early cephalochordate. *Haikouella lanceolata,* a small, fishlike creature recently discovered in China, possesses several characters that clearly identify it as a chordate, including notochord, pharynx, and dorsal nerve cord, but it also has characters that are characteristic of vertebrates, including pharyngeal muscles, paired eyes, and an enlarged brain (figure 15.9). However, it is not a vertebrate, because the fossils

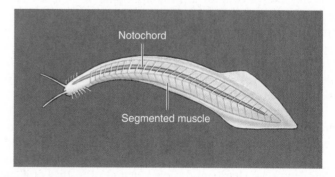

figure 15.8

Pikaia, an early chordate from the Burgess Shale of British Columbia, Canada.

Eye Brain Nerve cord Esophagus Dorsal aorta Notochord Myomere

Endostyle Ventral aorta Pharyngeal bar Atriopore Intestine Anus

figure 15.9

Haikouella, a chordate with several vertebrate features from the early Cambrian shales of Haikou, China. Some zoologists hypothesize that *Haikouella* is the sister group of vertebrates (craniates).

Tadpole larva

METAMORPHOSIS ?

Adult ascidian

Paedomorphic vertebrate ancestor

Early vertebrate

figure 15.10

Garstang's hypothesis of larval evolution (now rejected). More than 550 million years ago, some larvae became paedomorphic, attaining reproductive maturity in the larval body form. These became cephalized, evolving into the first vertebrates.

lack evidence of several diagnostic vertebrate traits, including a cranium and a distinct telencephalon (anterior region of the forebrain). *Haikouella* has been hypothesized to be the sister taxon of vertebrates. Despite recent fossil discoveries of early chordates, many speculations regarding vertebrate ancestry have focused on the living protochordates, in part because we know more about them than we do the fossils.

Garstang's Hypothesis of Chordate Larval Evolution

Chordate evolution has taken two paths, one leading to sedentary urochordates and the other to active, mobile cephalochordates and vertebrates. One hypothesis, proposed in 1928 by Walter Garstang of England, suggested the ancestral chordate was a sedentary filter feeder, like adult ascidian tunicates. Garstang hypothesized the swimming larva failed to metamorphose into an adult, sessile tunicate, instead developing gonads and reproducing in the larval stage. Continued evolution, emphasizing cephalization, would have produced the first vertebrates (figure 15.10).

Garstang called this process **paedomorphosis** (Gr. *pais*, child, + *morphē*, form), a term that describes the evolutionary retention of juvenile or larval traits in the adult body. Paedomorphosis is a well-known phenomenon in several different animal groups (paedomorphosis in amphibians is described on pp. 358–359).

Paedomorphosis, the displacement of ancestral larval or juvenile features into a descendant adult, can be produced by three different evolutionary-developmental processes: neoteny, progenesis, and postdisplacement. In neoteny, the growth rate of body form is slowed so that the animal does not attain the ancestral adult form when it reaches maturity. Progenesis is precocious maturation of gonads in a larval (or juvenile) body that then stops growing and never attains an adult body form. In postdisplacement, the onset of a developmental process is delayed relative to reproductive maturation, so that an ancestral adult form is not attained by the time of reproductive maturity. Neoteny, progenesis, and postdisplacement thus describe different ways in which paedomorphosis can happen. Biologists use the inclusive term paedomorphosis to describe results of these evolutionary-developmental processes.

Recently collected genetic and developmental evidence allows testing of Garstang's hypothesis. Numerous phylogenetic reconstructions, along with fossil evidence, place cephalochordates as sister to a clade consisting of urochordates plus vertebrates, suggesting that cephalochordates and vertebrates retain the ancestral chordate condition and that the sessile ascidians represent a derived condition (see figure 15.2). In addition, incipient neural crest tissue recently has been identified in urochordates, supporting a sister-group relationship with vertebrates. Thus, most zoologists currently reject Garstang's hypothesis and envision the ancestral chordate as a free-swimming creature, perhaps similar to the modern amphioxus.

Although urochordates are now considered the closest living relatives of vertebrates, their sessile body form is evolved in a lineage ancestral only to urochordates and did not occur in vertebrate ancestry. Most zoologists view amphioxus as retaining much of the prevertebrate and ancestral chordate body structure. Cephalochordates share several characters with vertebrates that are absent from tunicates, including segmented musculature and a characteristic circulatory route in the pharynx. Amphioxus is unlike the most recent common ancestor of vertebrates because it lacks the tripartite brain, chambered heart, special sense organs, and other structures inferred to have been present in that ancestor. Despite its phylogenetic distance from vertebrates, amphioxus resembles the prevertebrate condition more than any other living animal.

The Earliest Vertebrates

The earliest known vertebrate fossils, until recently, were armored jawless fishes called **ostracoderms** (os-trak´o-derm) (Gr. *ostrakon*, shell, + *derma*, skin) from late Cambrian and Ordovician deposits. Researchers unearthed from the amazing

Chengjiang deposits a number of 530-million-year-old fossils belonging to two fishlike vertebrates: *Myllokunmingia* and *Haikouichthys*. These fossils push back the origin of vertebrates to at least the early Cambrian period. They show many vertebrate characteristics, including a heart, paired eyes, otic (ear) capsules, and what have been interpreted as rudimentary vertebrae.

The earliest ostracoderms were armored with bone in their dermis and usually lacked the paired fins so important to later fishes for stability (figure 15.11). The swimming movements of one of the early groups, the **heterostracans** (Gr. *heteros*, different, + *ostrakon*, shell), must have been imprecise, although sufficient to propel them along the ocean floor. With fixed, circular or slitlike mouth openings, they may have filtered small food particles from the water or ocean bottom. However, unlike the ciliary filter-feeding protochordates, ostracoderms sucked water into the pharynx by muscular pumping, an important innovation that suggests to some authorities that ostracoderms may have been active predators that fed on soft-bodied animals.

The term "ostracoderm" does not denote a natural evolutionary assemblage, but instead describes several groups of heavily armored, extinct jawless fishes.

During the Devonian period, the heterostracans underwent a major radiation, producing numerous peculiar-looking forms. Without ever evolving paired fins or jaws, these earliest vertebrates flourished for 150 million years until becoming extinct near the end of the Devonian period.

Coexisting with heterostracans throughout much of the Devonian period were **osteostracans** (Gr. *osteon*, bone,

+ *ostrakon*, shell). Osteostracans had paired pectoral fins, an innovation that improved swimming efficiency by controlling yaw, pitch, and roll. A typical osteostracan (figure 15.11) was a small animal, seldom exceeding 30 cm in length. It was covered with a heavy, dermal armor of bone, including a single-piece head shield. Examination of internal features of the braincase reveal a sophisticated nervous system and sense organs, similar to those of modern lampreys.

Another group of ostracoderms, the anaspids (figure 15.11), were more streamlined than other ostracoderms. These and other ostracoderms underwent an impressive diversification in the Silurian and Devonian periods. However, all ostracoderms became extinct by the end of the Devonian period.

For decades, geologists have used strange, microscopic, toothlike fossils called **conodonts** (Gr. *kōnos*, cone, + *odontos*, tooth) to date Paleozoic marine sediments without having any idea what kind of creature originally possessed these elements. The discovery in the early 1980s of fossils of complete conodont animals changed this situation. With their phosphatized toothlike elements, myomeres, and paired eye and otic capsules, conodonts clearly belong to the vertebrate clade (figure 15.12). Although their exact position in this clade is unclear, they are important in understanding the evolution of early vertebrates.

Early Jawed Vertebrates

All jawed vertebrates, whether extinct or living, are collectively called **gnathostomes** ("jaw mouth") in contrast to jawless vertebrates, the **agnathans** ("without jaw"). Gnathostomes

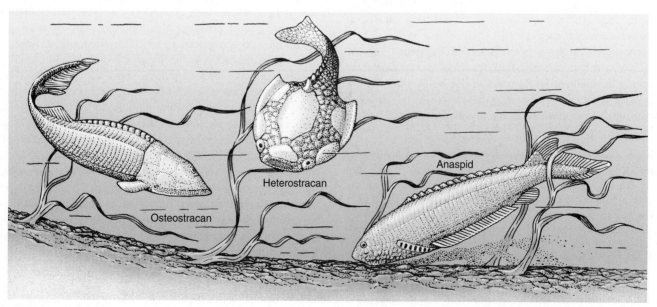

figure 15.11

Three ostracoderms, jawless fishes of Silurian and Devonian times. They are shown as they might have appeared while searching for food on the floor of a Devonian sea. They employed a strong pharyngeal pump to circulate water rather than relying on the much more limiting mode of ciliary action used by their protochordate ancestors (presumably resembling amphioxus in this way).

are a monophyletic group; the presence of jaws is a derived character state shared by all jawed fishes and tetrapods. Agnathans, however, are defined principally by the absence of jaws, a character that is not unique to jawless fishes because jaws are lacking in vertebrate ancestors. Thus, Agnatha is paraphyletic.

The origin of jaws was one of the most important events in vertebrate evolution. The utility of jaws is obvious: they allow predation on large and active forms of food not available to jawless vertebrates. Ample evidence suggests that jaws arose through modifications of the first or second of the serially repeated cartilaginous gill arches. How did this mandibular arch change from a function of gill support and ventilation to one of feeding as jaws? Expansion of this arch and evolution of new, associated muscles may have first assisted gill ventilation, perhaps to meet the increasing metabolic demands of early

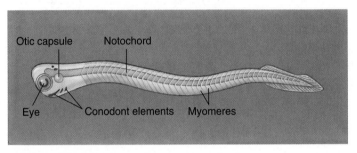

figure 15.12

Restoration of a conodont. Conodonts superficially resembled amphioxus, but they possessed a much greater degree of encephalization (paired eyes and otic [ear] capsules) and bonelike mineralized elements—all indicating that conodonts were probably vertebrates. Conodont elements are thought to be part of a food-handling apparatus.

vertebrates. Once enlarged and equipped with extra muscles, the first pharyngeal arch easily could have been modified to serve as jaws in the manner described in figure 15.13.

An additional feature characteristic of all gnathostomes is the presence of paired pectoral and pelvic appendages in the form of fins or limbs. These likely originated as stabilizers to check the yaw, pitch, and roll generated during active swimming. According to the fin-fold hypothesis, paired fins arose from paired continuous, ventrolateral folds or fin-forming zones. The addition of skeletal supports in the fins enhanced the fins' ability to provide stability during swimming. Evidence for this hypothesis is found in the paired flaps of *Myllokunmingia* and anaspids, and in the multiple paired fins of acanthodians. In one fish lineage, the muscle and skeletal supports in the paired fins became strengthened, adapting them for locomotion on land as limbs. The origin of jaws and paired appendages may be linked to a second *Hox* duplication, near the origin of the gnathostomes. Both jaws and paired fins were major innovations in vertebrate evolution, among the most important reasons for the subsequent major radiations of vertebrates that produced the modern fishes and all tetrapods, including you, the reader of this book.

Among the first jawed vertebrates were the heavily armored **placoderms** (Gr. *plax*, plate, + *derma*, skin). They first appear in the fossil record in the Silurian period (figure 15.14). Placoderms evolved a great variety of forms, some very large (one was 10 m in length!) and grotesque in appearance. They were armored fish covered with diamond-shaped scales or large plates of bone. All became extinct by the end of the Devonian period and appear to have left no descendants. However, **acanthodians** (figure 15.14), a group of early jawed fishes characterized by large, anteriorly set eyes and fins with large spines, are included in a clade that underwent a great evolutionary diversification into the bony fishes that dominate the waters of the world today.

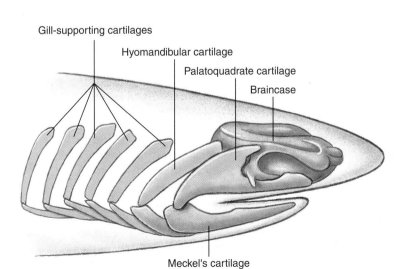

figure 15.13

How vertebrates got their jaws. Resemblance between jaws and gill supports of early fishes, such as this Carboniferous shark, suggests the upper jaw (palatoquadrate) and lower jaw (Meckel's cartilage) evolved from structures that originally functioned as gill supports. The hyomandibular cartilage helped to link the jaws to the braincase in many early gnathostomes. Relics of this transformation are seen during the development of modern sharks.

figure 15.14

Early jawed fishes of the Devonian period, 400 million years ago. Jaws and gill supports from which the jaws evolved develop from neural crest cells, a diagnostic character of vertebrates. Most placoderms were bottom-dwellers that fed on benthic animals, although some were active predators. Acanthodians carried less armor than placoderms and had large, anteriorly placed eyes and prominent spines on paired fins. Most were marine but several species lived in fresh water.

Taxonomy of Living Members of Phylum Chordata

Phylum Chordata

Subphylum Urochordata (yur´ō-kor-dä´tə) (Gr. *oura,* tail, + L. *chorda,* cord, + *ata,* characterized by) **(Tunicata): tunicates.** Notochord and nerve cord in free-swimming larva only; ascidian adults sessile, encased in tunic. About 1600 species.

Subphylum Cephalochordata (sef´ə-lō-kor-dä´tə) (Gr. *kephalē,* head, + L. *chorda,* cord): **lancelets (amphioxus).** Notochord and nerve cord found along entire length of body and persist throughout life; fishlike in form. 29 species.

Subphylum Vertebrata (ver´te-brä´tə) (L. *vertebratus,* backboned). **(Craniata): vertebrates.** Bony or cartilaginous cranium surrounding tripartite brain; well-developed head with paired sense organs, usually with vertebrae; heart with multiple chambers; muscularized digestive tract; paired kidneys.

Superclass Agnatha (ag´na-thə) (Gr. *a,* without, + *gnathos,* jaw): **hagfishes, lampreys.** Without true jaws or paired appendages. A paraphyletic group.

Class Myxini (mik-sē´nē) (Gr. *myxa,* slime): **hagfishes.** Four pairs of tentacles around mouth; buccal funnel absent; 1 to 16 pairs of gill openings; slime glands present; vertebrae absent. About 70 species.

Class Petromyzontida (pet´trō-mī-zon´ti´də) (Gr. *petros,* stone, + *myzon,* sucking): **lampreys.** Buccal funnel with keratinized teeth, nasal sac not connected to pharynx; vertebrae present only as neural arches. 38 species.

Superclass Gnathostomata (na´thō-stō´mä-tə) (Gr. *gnathos,* jaw, + *stoma,* mouth): **jawed fishes, tetrapods.** With jaws and (usually) paired appendages.

Class Chondrichthyes (kon-drik´thē-ēz) (Gr. *chondros,* cartilage, + *ichthys,* a fish): **sharks, skates, rays, chimaeras.** Cartilaginous skeleton, intestine with spiral valve; claspers present in males; no swim bladder. About 970 species.

Class Actinopterygii (ak´ti-nop´te-rij´e´-ī) (Gr. *aktis,* ray, + *pteryx,* fin, wing): **ray-finned fishes.** Skeleton ossified; single gill opening covered by operculum; paired fins supported primarily by dermal rays; appendage musculature within body; swim bladder mainly a hydrostatic organ, if present; atrium and ventricle not divided. About 27,000 species.

Class Sarcopterygii (sär-cop-te-rij´ē-ī) (Gr. *sarkos,* flesh, + *pteryx,* fin, wing): **lobe-finned fishes.** Skeleton ossified; single gill opening covered by operculum; paired fins with sturdy internal skeleton and musculature within appendage; usually with lunglike swim bladder; atrium and ventricle at least partly divided. 8 species. Paraphyletic unless tetrapods are included.

Class Amphibia (am-fib´ē-ə) (Gr. *amphi,* both or double, + *bios,* life): **amphibians.** Ectothermic tetrapods; respiration by lungs, gills, or skin; development through larval stage; skin moist, containing mucous glands and lacking scales. About 6000 species.

Class Reptilia (rep-til´ē-ə) (L. *repere,* to creep): **reptiles.** Ectothermic tetrapods possessing lungs; embryo develops within shelled egg; no larval stage; skin dry, lacking mucous glands, and covered by

epidermal scales. A paraphyletic group unless birds are included. About 8100 species.

Class Aves (ā´vēz) (L. pl. of *avis,* bird): **birds.** Endothermic vertebrates with front limbs modified as wings; body covered with feathers. About 9700 species.

Class Mammalia (ma-mā´lē-ə) (L. *mamma,* breast): **mammals.** Endothermic vertebrates possessing mammary glands; body more or less covered with hair; brain large, with neocortex; three middle ear bones. About 5400 species.

Summary

Phylum Chordata is named for the rodlike notochord that forms a stiffening body axis at least early in the development of every chordate. All chordates share four additional distinctive hallmarks that set them apart from all other phyla: dorsal hollow nerve cord, pharyngeal pouches or slits, endostyle, and postanal tail. Two of the three chordate subphyla are invertebrates and lack a well-developed head. These are Urochordata (tunicates), most of which are sessile as adults, but all of which have a free-swimming larval stage, and Cephalochordata (lancelets), fishlike forms that include the famous amphioxus.

Chordates are deuterostomes, like echinoderms and hemichordates. Most zoologists now consider the chordate ancestor to have been a small, free-swimming, filter-feeding creature.

Subphylum Vertebrata includes the backboned members of the animal kingdom (hagfishes actually lack vertebrae but are included with the Vertebrata because they share numerous homologies with vertebrates). As a group, vertebrates are characterized by a well-developed head and comparatively large size, high degree of motility, and distinctive body plan that permitted their exceptional adaptive diversification. Their most important

distinguishing features are a living endoskeleton that allows continuous growth and provides a sturdy framework for efficient muscle attachment and action, a muscular pharynx with slits and gills (lost or greatly modified in higher vertebrates) with vastly increased respiratory efficiency, a muscularized gut and chambered heart which meets higher metabolic demands, and an advanced nervous system with a distinct brain and paired sense organs. Evolution of jaws and paired appendages likely contributed to the incredible success of one group of vertebrates, the gnathostomes.

Review Questions

1. What characteristics are shared by the three deuterostome phyla that indicate a monophyletic group of interrelated animals?
2. Explain how use of a cladistic classification for vertebrates changes the major important regroupings of traditional vertebrate taxa (refer to figure 15.2). Why are Agnatha and Reptilia, as traditionally recognized, inconsistent with cladistic principles?
3. Name five characteristics shared by all chordates, and explain the function of each.
4. In debating the question of chordate origins, zoologists eventually agreed that chordates must have evolved within the deuterostome assemblage rather than from a protostome group as earlier argued. What embryological evidence supports this view?

5. Offer a description of an adult tunicate that would identify it as a chordate, yet distinguish it from any other chordate group.
6. Both sea squirts (urochordates) and lancelets (cephalochordates) are filter-feeding organisms. Describe the filter-feeding apparatus of a sea squirt, and explain in what ways its mode of feeding is similar to, and different from, that of amphioxus.
7. Explain why it is necessary to know the life history of a tunicate in order to understand why tunicates are chordates.
8. List three groups of adaptations that guided vertebrate evolution, and explain how each has contributed to the success of vertebrates.
9. In 1928, Walter Garstang hypothesized that tunicates resemble the ancestral

stock of the vertebrates. Explain this hypothesis and evaluate its validity based on recent phylogenetic and developmental data.
10. What is the phylogenetic placement of *Haikouella,* and what evidence supports its placement there?
11. Distinguish between ostracoderms and placoderms. What important evolutionary adaptations first appeared in each group? What are conodonts?
12. Explain how zoologists think the vertebrate jaw evolved.

For Further Thought Urochordates, instead of amphioxus, are now considered to be the sister group of vertebrates. Why is amphioxus, instead of urochordates, still considered a better model for the prevertebrate body plan?

■ Selected References

See also general references on page 432.

Ahlberg, P. E. 2001. Major events in early vertebrate evolution. London, Taylor & Francis. *Evolution of vertebrates up to the split of major jawed fish groups, incorporating molecular, fossil, and embryological data. This book makes many important contributions, but some of its conclusions are controversial.*

Cohn, M. J. 2002. Lamprey *Hox* genes and the origin of jaws. Nature **416**:386–387. *Loss of HoxL6 in the mandibular arch apparently facilitated the evolution of jaws.*

Donoghue, P. C. J., P. L. Forey, and R. J. Aldridge. 2000. Conodont affinity and chordate phylogeny. Biol. Rev. **75**:191–251. *In this summary of early chordate evolution, the authors provide evidence that conodonts are vertebrates and that lampreys and hagfishes do not form a monophyletic group.*

Forey, P., and P. Janvier. 1994. Evolution of the early vertebrates. Am. Sci. **82**:554–565. *Summarizes the biology and evolution of many groups of ostracoderms and other ancestral craniates.*

Gee, H. 1996. Before the backbone: views on the origin of the vertebrates. New York, Chapman & Hall. *Outstanding review of the many vertebrate origin hypotheses. Gee links genetic, developmental, and molecular evidence.*

Gould, S. J. 1989. Wonderful life: the Burgess Shale and the nature of history. New York, W.W. Norton & Company. *In this book describing the marvelous Cambrian fossils of the Burgess Shale, Gould "saves the best for last" by inserting an epilogue on Pikaia, one of the earliest chordates.*

Jeffery, W. R., A. G. Strickler, and Y. Yamamoto. 2004. Migratory neural crest-like cells form body pigmentation in a urochordate embryo. Nature **431**:696–699. *Neural-crest-like cells are in sea squirts, supporting a sister-group relationship with vertebrates.*

Long, J. A. 1995. The rise of fishes: 500 million years of evolution. Baltimore, The Johns Hopkins University Press. *An authoritative, liberally illustrated evolutionary history of fishes.*

Maisey, J. G. 1996. Discovering fossil fishes. New York, Henry Holt & Company. *Handsomely illustrated chronology of fish evolution with cladistic analysis of evolutionary relationships.*

Mallatt, J., and J.-Y. Chen. 2003. Fossil sister group of craniates: predicted and found. J. Morph. **258**:1–31. *Reevaluation of Haikouella fossils reveals several features that suggest it is the sister group to craniates.*

Shimeld, S. M., and P. W. H. Holland. 2000. Vertebrate innovations. Proc. Natl. Acad. Sci. **97**:4449–4452. *Focuses on developmental characters, including neural crest, neural placodes, and Hox genes.*

Stokes, M. D., and N. D. Holland. 1998. The lancelet. Am. Sci. **86**(6):552–560. *Describes the historical role of amphioxus in early hypotheses of vertebrate ancestry and summarizes molecular data that rekindled interest in amphioxus.*

■ Custom Website

The *Animal Diversity* Online Learning Center is a great place to check your understanding of chapter material. Visit www.mhhe.com/hickmanad6e for access to key terms, quizzes, and more! Further enhance your knowledge with Web links to chapter-related material.

Explore live links for these topics:

Classification and Phylogeny of Animals
General Chordate References
Subphylum Urochordata
Subphylum Cephalochordata

Subphylum Vertebrata
Systematics and Characteristics of the Craniates
Early Vertebrates

Fishes

Fishes

Chordata

ey snappers, *Lutjanus griseus*, in the Florida Keys.

What Is a Fish?

In common (and especially older) usage, the term "fish" denotes a mixed assortment of water-dwelling animals. We speak of jellyfish, cuttlefish, starfish, crayfish, and shellfish, knowing full well that when we use the word "fish" in such combinations, we are not referring to a true fish. In earlier times, even biologists did not make such a distinction. Sixteenth-century natural historians classified seals, whales, amphibians, crocodiles, and even hippopotamuses, as well as a host of aquatic invertebrates, as fishes. Later biologists were more discriminating, eliminating first the invertebrates and then the amphibians, reptiles, and mammals from the narrowing concept of a fish. Today we recognize a fish as an aquatic vertebrate with gills; appendages, if present, in the form of fins; and usually a skin with scales of dermal origin. Even this modern concept of the term "fish" is used for convenience, not as a taxonomic unit. Fishes do not form a monophyletic group, because the ancestor of land vertebrates (tetrapods) is found within one group of fishes (the sarcopterygians). Thus, fishes can be defined in an evolutionary sense as all vertebrates that are not tetrapods. The world's fishes have diversified to produce an estimated 28,000 living species—more than all other species of vertebrates combined—with adaptations for almost every conceivable aquatic environment. No other animal group threatens their domination of the world's seas, lakes, and streams.

The body of a fish is streamlined for movement through water. Suspended in a medium that is 800 times more dense than air, a trout or pike can remain motionless, varying its neutral buoyancy by adding or removing air from its swim bladder. It may dart forward or at angles, using its fins as brakes and tilting rudders. With excellent organs for salt and water exchange, bony fishes can finely tune their body fluid composition in their freshwater or seawater environment. Their gills are the most effective respiratory devices in the animal kingdom for extracting oxygen from a medium that contains less than 1/20 as much oxygen as air. Fishes have excellent olfactory and visual senses and a unique lateral-line system, which has an exquisite sensitivity to water currents and vibrations. Thus, in mastering the physical problems of their element, early fishes evolved a basic body plan and a set of physiological strategies that both shaped and constrained the evolution of their descendants.

The use of *fishes* as the plural form of *fish* may sound odd to most people accustomed to using *fish* in both the singular and the plural. *Fish* refers to one or more individuals of the same species; *fishes* refers to more than one species.

Ancestry and Relationships of Major Groups of Fishes

Fishes are of early Paleozoic ancestry, having descended from an unknown free-swimming protochordate ancestor (hypotheses concerning chordate and vertebrate origins are discussed in Chapter 15). The earliest fishlike vertebrates were a paraphyletic assemblage of jawless **agnathan** fishes, including ostracoderms (see figure 15.11). One group of ostracoderms gave rise to the jawed **gnathostomes.**

Along with the extinct ostracoderms, the jawless agnathans include the living hagfishes and lampreys, fishes adapted as scavengers or parasites. Although hagfishes have no vertebrae and lampreys have only rudimentary vertebrae, they nevertheless are included within the subphylum Vertebrata because they have a cranium and many other vertebrate homologies. Although hagfishes and lampreys superficially look much alike, they are in fact so different from each other that zoologists have assigned them to separate taxonomic classes.

All remaining living fishes have paired appendages and jaws and are included, along with tetrapods (land vertebrates), in the monophyletic group of gnathostomes. They appear in the fossil record in the late Silurian period with fully formed jaws, and no forms intermediate between agnathans and gnathostomes are known. By the Devonian period, the "age of fishes," several distinct groups of jawed fishes were well represented. One of these, the placoderms (p. 323), became extinct in the subsequent Carboniferous period, leaving no descendants. A second group, cartilaginous fishes of class Chondrichthyes (sharks, rays, and chimaeras), lost the heavy dermal armor of the early jawed fishes and adopted cartilage for the endoskeleton. Most are active predators with sharklike or raylike body forms that changed only slightly over the ages. The other two groups

of gnathostome fishes, **acanthodians** (p. 323) and bony fishes, were abundant and diverse in the Devonian period.

We can recognize two distinct lineages of bony fishes. By far the most diverse are the ray-finned fishes (class Actinopterygii), which radiated to form most modern bony fishes. The other group, lobe-finned fishes (class Sarcopterygii), contains few living species but includes the sister group of the tetrapods. The lobe-finned fishes are represented today by lungfishes and coelacanths—meager remnants of important lineages that flourished in the Devonian period (figure 16.1). A classification of major fish taxa is on p. 348.

Living Jawless Fishes

Living jawless fishes include about 108 species divided between two classes: Myxini (hagfishes) and Petromyzontida (lampreys). Members of each group lack jaws, internal ossification, scales, and paired fins, and both groups share porelike gill openings and an eel-like body form. Based on this morphological similarity, these two groups formerly were united under the name Cyclostomata, a group shown to be paraphyletic when morphological characters were analyzed. Lampreys have many characters, including vertebrae, extrinsic eye muscles, and cerebellum, that are uniquely shared with gnathostomes. In contrast, recent molecular evidence strongly suggests that hagfishes and lampreys form a monophyletic group. If the molecular hypothesis is true, then vertebrae and many other vertebrate features were lost during hagfish evolution. In view of this uncertainty, we show the relationship of hagfishes, lampreys, and gnathostomes as a polytomy (figure 16.2).

Hagfishes: Class Myxini

Hagfishes are an entirely marine group that feeds on annelids, molluscs, crustaceans, and dead or dying fishes. They are not parasitic like lampreys, but are scavengers and predators. There are 70 described species of hagfishes, of which the best known in North America are the Atlantic hagfish, *Myxine glutinosa* (Gr. *myxa,* slime) (figure 16.3), and the Pacific hagfish, *Eptatretus stouti* (N.L., *ept,* Gr. *hepta,* seven, + *tretos,* perforated). Although almost completely blind, hagfishes are quickly attracted to food, especially dead or dying fishes, by their keenly developed senses of smell and touch. A whale that dies and falls to the bottom of the ocean attracts thousands of hagfishes, who feed on the carcass for several years. A hagfish enters a dead or dying animal through an orifice or by digging into the body. Using two toothed, keratinized plates on its tongue that fold together in a pincerlike action, the hagfish rasps bits of flesh from its prey. For extra leverage, the hagfish often ties a knot in its tail, and then passes the knot forward along its body until it is pressed securely against the side of its prey (figure 16.3D).

Unlike other vertebrates, the body fluids of hagfishes are in osmotic equilibrium with seawater, as are those of most marine invertebrates. Hagfishes have several other anatomical and physiological peculiarities, including a low-pressure circulatory system served by three accessory hearts in addition to the main heart positioned behind the gills. Hagfishes also are renowned for their ability to generate enormous quantities of slime.

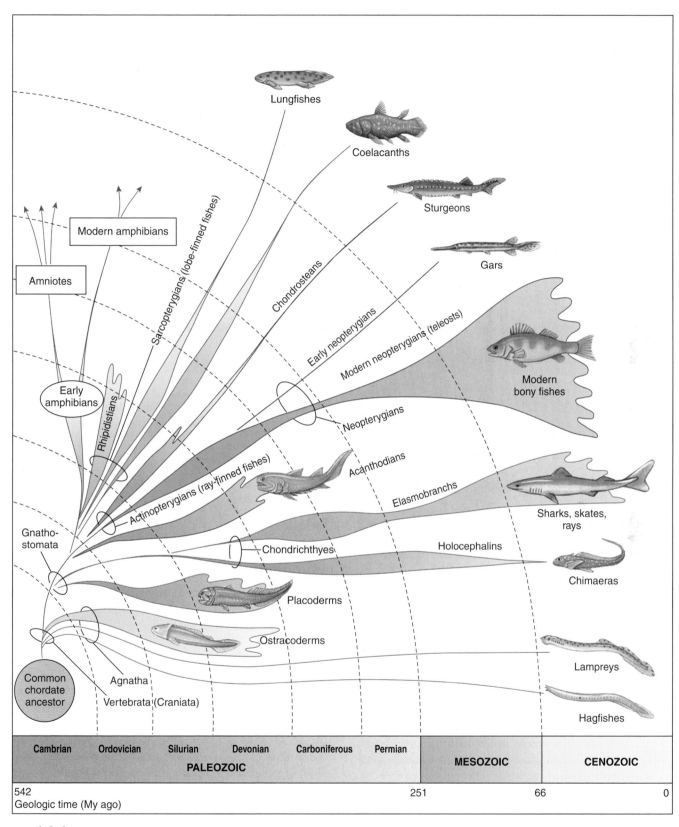

figure 16.1

Graphic representation of the family tree of fishes, showing the evolution of major groups through geological time. Numerous lineages of extinct fishes are not shown. Widened areas in the lines of descent indicate periods of adaptive diversification and the relative number of species in each group. The lobe-finned fishes (sarcopterygians), for example, flourished in the Devonian period, but declined and are today represented by only four surviving genera (lungfishes and coelacanths). Sharks and rays diversified during the Carboniferous period, declined in the Permian, and then diversified again in the Mesozoic era. Johnny-come-latelies in fish evolution are the spectacularly diverse modern fishes, or teleosts, which include most living fishes.

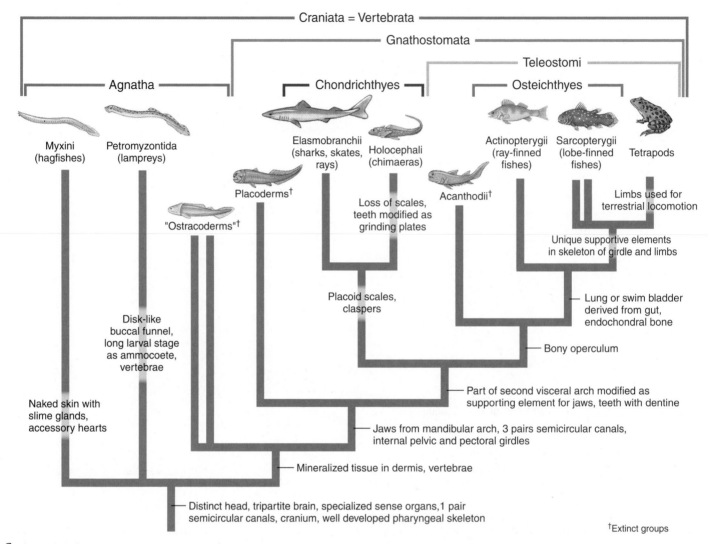

figure 16.2

Cladogram of the fishes, showing the probable relationships of major fish taxa. The groups Agnatha and Osteichthyes, although paraphyletic structural grades considered undesirable in cladistic classification, are conveniently used in systematics because they share broad structural and functional patterns of organization. The polytomy at the base of the cladogram reflects uncertainty concerning the sister group of lampreys: morphological data suggest the sister group is Gnathostomata, but molecular data suggest it is Myxini.

Much of the reproductive biology of hagfishes remains a mystery. Reproduction rarely has been documented; in these cases, females produced small numbers of large, yolky eggs, which might take as long as five months to hatch. There is apparently no larval stage. Almost nothing is known of spawning times, places, and behaviors or age at maturity.

While the strange features of hagfishes fascinate many people, hagfishes have not endeared themselves to commercial fishermen. In earlier days, when commercial fishing was done mainly by gill nets and set lines, hagfishes often bit into the bodies of captured fishes and devoured the contents, leaving behind a useless sack of skin and bones. But as large and efficient trawls came into use, hagfishes ceased to be important pests. The commercial fishing industry "turned the tables" and began targeting hagfishes as a source of leather for golf bags and boots. Fishing pressure has been so intense that some species have greatly declined.

Lampreys: Class Petromyzontida

Of the 38 described species of lampreys distributed around the world, one of the best known is the destructive marine lamprey, *Petromyzon marinus,* which occurs on both sides of the Atlantic Ocean (figure 16.4). The name *Petromyzon* (Gr. *petros,* stone, + *myzon,* sucking) refers to the lamprey's habit of grasping a stone with its mouth to hold its position in a current. There are 20 species of lampreys in North America, of which about half are parasitic; the rest, often called "brook lampreys," are species that never feed after metamorphosis and die soon after spawning.

In North America, all lampreys, marine as well as freshwater forms, spawn in the winter or spring in the shallow gravel of freshwater streams. Males begin building nests and are joined later by females. Using their oral discs to lift stones and pebbles and using vigorous body vibrations to sweep

figure 16.3

Atlantic hagfish, *Myxine glutinosa* (class Myxini). **A,** External anatomy. **B,** Ventral view of head, showing keratinized teeth used to grasp food during feeding. **C,** Sagittal section of head region. (Note retracted position of rasping tongue and internal openings into a row of gill sacs.) **D,** Hagfish knotting, showing how it obtains leverage to tear flesh from prey.

CHARACTERISTICS
of Hagfishes (Myxini) and Lampreys (Petromyzontida)

1. Body slender, eel-like; no paired appendages
2. **Skin naked** (no scales)
3. **Fibrous and cartilaginous skeleton;** notochord persistent; **vertebrae reduced or absent**
4. **Jaws absent;** mouth with keratinized plates (hagfishes) or teeth (lampreys); no distinctive stomach
5. Brain small, but distinct; 10 pairs of cranial nerves
6. Eyes poorly developed (hagfishes) or moderately developed (lampreys); one pair (hagfishes) or two pairs (lampreys) of **semicircular canals**
7. Sexes separate; external fertilization
8. Large yolky eggs and no larval stage in hagfishes; small eggs and long larval stage (**ammocoete**) in lampreys
9. Excretory system of pronephric and mesonephric (hagfishes) or opisthonephric (lampreys) kidneys; kidneys drain via **archinephric duct** to cloaca; **ammonia** main nitrogenous waste
10. Hagfishes with 5–16 pairs of gills; lampreys with 7 pairs of gills
11. Heart with a sinus venosus, atrium, and ventricle; **single circulation;** accessory hearts in hagfishes

figure 16.4

Sea lamprey, *Petromyzon marinus*, feeding on body fluids of a dying fish.

away light debris, they form an oval depression. As a female sheds eggs into the nest, a male fertilizes them. The sticky eggs adhere to pebbles in the nest and soon become covered with sand. Adults die soon after spawning.

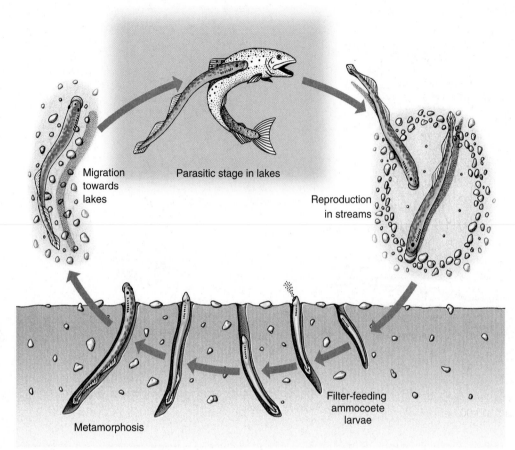

figure 16.5

Life cycle of the "landlocked" form of the sea lamprey *Petromyzon marinus*.

Eggs hatch in about 2 weeks, releasing small larvae called **ammocoetes** (figure 16.5), which stay in the nest until they are approximately 1 cm long; they then burrow into mud or sand and filter-feed on small organisms and fine organic matter. The larvae live from 3 to 7 or more years, and then rapidly metamorphose into adults.

Parasitic lampreys either migrate to the sea, if marine, or remain in fresh water, where they attach themselves by their suckerlike mouth to fish and use their sharp, keratinized teeth to rasp through flesh and suck body fluids. To promote the flow of blood, a lamprey injects an anticoagulant into the wound. When gorged, the lamprey releases its hold but leaves the fish with a large wound that can be fatal. Parasitic freshwater adults live 1 to 2 years before spawning and then die; marine forms live 2 to 3 years.

Nonparasitic lampreys do not feed after emerging as adults, and their digestive tract degenerates to a nonfunctional strand of tissue. Within a few months, they spawn and die.

Invasion of the Great Lakes by the sea lamprey, *Petromyzon marinus*, devastated fisheries. No lampreys were present in the Great Lakes west of Niagara Falls until the Welland Ship Canal was deepened in the 1910s, allowing lampreys to bypass the Falls. Moving first through Lake Erie and then to Lakes Huron, Michigan, and Superior, sea lampreys, combined with the effects of overfishing, caused the total collapse of a multimillion-dollar lake trout fishery in the early 1950s. Other less valuable fish species were attacked and destroyed in turn. After reaching peak abundance in 1951 in Lakes Huron and Michigan and in 1961 in Lake Superior, sea lampreys began to decline, due to depletion of their food supply and expensive control measures (mainly chemical larvicides placed in selected spawning streams). Lake trout, aided by a restocking program, are now recovering, but wounding rates are still high in some lakes.

Cartilaginous Fishes: Class Chondrichthyes

There are about 970 living species in the class Chondrichthyes, an ancient group that appeared in the Devonian period. Although they are a much smaller and less diverse assemblage than bony fishes, their impressive combination of well-developed sense organs, powerful jaws and swimming musculature, and predaceous habits ensures them a secure and lasting niche in the aquatic community. One of their distinctive features is their cartilaginous skeleton. Their skeletons are extensively calcified, but bone is entirely absent throughout the class—a derived feature, because members of Chondrichthyes descended from ancestors having well-developed bone.

Sharks and Rays: Subclass Elasmobranchii

Sharks, which include about 45% of the approximately 937 species in subclass Elasmobranchii, are typically predaceous fishes with five to seven pairs of gill slits and gills and (usually) a spiracle behind each eye. Larger sharks, such as the massive (but harmless) plankton-feeding whale shark, may reach 15 m in length, the largest of all fishes. Dogfish sharks, widely used in zoological laboratories, rarely exceed 1 m.

Although to many people sharks have a sinister appearance and a fearsome reputation, they are at the same time among the most gracefully streamlined of all fishes (figure 16.6). Sharks are heavier than water and will sink if not swimming forward. The asymmetrical **heterocercal tail,** in which the vertebral column turns upward and extends into the dorsal lobe of the tail (see figure 16.14), provides thrust and some lift as it sweeps back and forth in the water, and the broad head and flat pectoral fins act as planes to provide additional lift.

The worldwide shark fishery is experiencing unprecedented pressure, driven by the high price of shark fins for shark-fin soup, an Asian delicacy that may sell for as much as $100.00 per bowl. Coastal shark populations in general have declined so rapidly that "finning" has been outlawed in the United States; other countries, too, are setting quotas to protect threatened shark populations. Even in the Marine Resources Reserve of the Galápagos Islands, one of the world's exceptional wild places, tens of thousands of sharks have been killed illegally for the Asian shark-fin market. Contributing to the threatened collapse of shark fisheries worldwide is the low fecundity of sharks and the long time most sharks require to reach sexual maturity; some species take as long as 35 years.

CHARACTERISTICS
of Class Chondrichthyes

1. Body fusiform or dorsoventrally compressed; caudal fin **heterocercal** (sharks and rays) or diphycercal (chimaeras) (see figure 16.14); **paired pectoral and pelvic fins**
2. **Skin with placoid scales** of dermal origin or **naked**
3. **Skeleton cartilaginous;** notochord present but reduced; **vertebrae** distinct
4. **Jaws present** with **polyphyodont teeth;** stomach large (absent in chimaeras); intestine with **spiral valve** (figure 16.7); liver often large and oil-filled
5. Brain well developed; 10 pairs of cranial nerves
6. Senses of smell, vibration reception (lateral-line), vision and **electroreception** well developed; **three pairs of semicircular canals**
7. Sexes separate; **internal fertilization with claspers**
8. Oviparous or viviparous; embryos of viviparous species nourished by **placenta, yolk sac (ovoviviparity), or cannibalism;** no larval stage
9. Excretory system of **opisthonephric kidneys,** which drain via **archinephric duct** to cloaca; **high concentration of urea and trimethylamine oxide in blood; rectal gland** present
10. **Five to seven pairs of gills** leading to gill slits in rays and sharks or covered by operculum in chimaeras; **no swim bladder or lung**
11. Heart with a sinus venosus, atrium, ventricle, and conus arteriosus; **single circulation**

Sharks are well equipped for their predatory life. Their tough, leathery skin is covered with numerous dermal **placoid scales** (see figure 16.13) that are modified anteriorly to form replaceable rows of teeth in both jaws (figure 16.8). Placoid scales consist of dentine enclosed by an enamel-like substance, like teeth of sharks and other vertebrates. Sharks use their keen

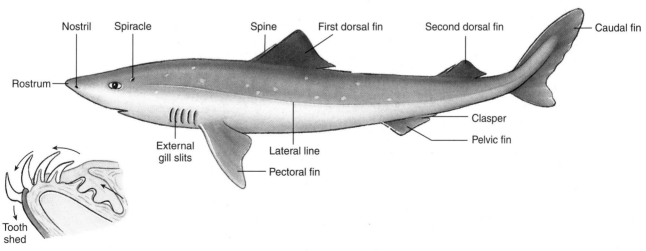

figure 16.6
Male spiny dogfish shark, *Squalus acanthias. Inset:* Section of lower jaw shows new teeth developing inside the jaw. These move forward to replace lost teeth. The rate of replacement varies among species.

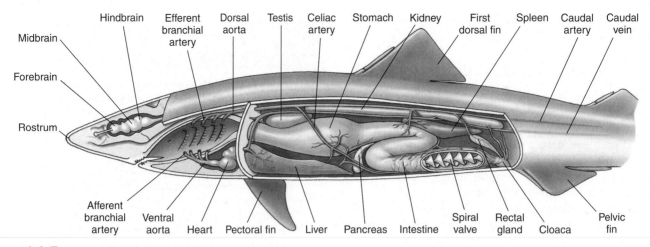

figure 16.7

Internal anatomy of a dogfish shark, *Squalus acanthias*.

figure 16.8

Head of a sand tiger shark, *Carcharias*. Note the series of successional teeth. Also visible below the eye are ampullae of Lorenzini.

sense of smell to guide them to food. They also can locate prey from long distances by sensing low-frequency vibrations with mechanoreceptors in the **lateral-line system.** This system is composed of special receptor organs (**neuromasts**) in interconnected tubes and pores extending along the sides of the body and over the head (figure 16.9). At close range, a shark switches to vision as its primary method of tracking prey. During the final stage of the attack, sharks are guided to their prey by the bioelectric fields that surround all animals. Electroreceptors, the ampullae of Lorenzini, are located on the shark's head.

All chondrichthyans have internal fertilization; sperm is introduced into the female reproductive tract by a modified portion of the male's pelvic fin called a **clasper** (see figure 16.6). Some sharks and all skates are **oviparous,** laying eggs soon after fertilization. Embryos are nourished from the egg yolk for a long period—up to 2 years in one species—before hatching. Many sharks are **ovoviviparous,** retaining developing young in the uterus, where they are nourished by the contents of their yolk sac until birth. Others are **viviparous,** meaning the

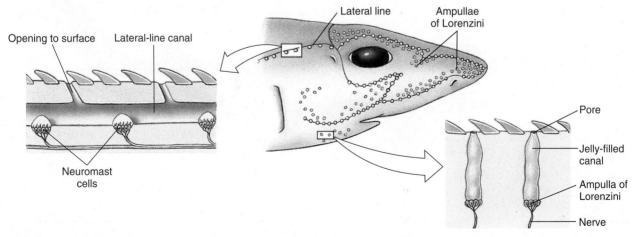

figure 16.9

Sensory canals and receptors in a shark. Ampullae of Lorenzini respond to weak electrical fields. Lateral-line sensors, called neuromast cells, are sensitive to disturbances in the water, enabling a shark to detect nearby objects by reflected waves in the water.

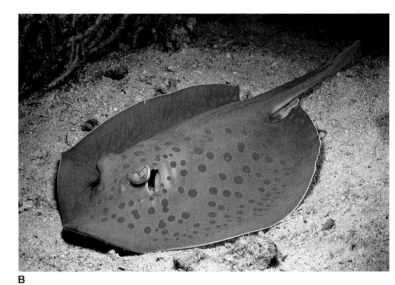

A

B

figure 16.10

Skates and rays are specialized for life on the sea floor. Both **A,** the clearnose skate, *Raja eglanteria,* and **B,** the bluespotted ribbontail ray, *Taeniura lymma,* are flattened dorsoventrally and move by undulations of winglike pectoral fins.

embryos receive nourishment from the mother via a placenta or secretions from the uterus. Regardless of the initial amount of maternal support, all parental care ends once eggs are laid or young are born.

More than half of all elasmobranchs are rays, a group that includes skates, stingrays, electric rays, and manta rays. Rays are distinguished by their dorsoventrally flattened bodies and enlarged pectoral fins, which they move in a wavelike fashion to propel themselves (figure 16.10). Gill openings are on the underside of the head, and the large **spiracles** are on top. Respiratory water enters through these spiracles to prevent clogging of the gills, because the mouth is often buried in sand. Teeth are adapted for crushing prey: molluscs, crustaceans, and an occasional small fish.

Stingrays have a slender, whiplike tail armed with one or more saw-toothed spines that can inflict dangerous wounds. Electric rays have on both sides of their head powerful electrical organs, which produce severe shocks to stun prey.

Chimaeras: Subclass Holocephali

The approximately 33 species of chimaeras (kī-mer´uz; L. monster), also called ratfishes (figure 16.11), are remnants of a group that diverged from the earliest shark lineage, which originated at least 380 million years ago. Fossil chimaeras first appeared in the Mississippian period, reached their zenith in the Cretaceous and early Tertiary periods (120 million to 50 million years ago), and then declined. Anatomically, they have several features that link them to elasmobranchs, but they possess a suite of unique characters, too. Instead of a toothed mouth, their jaws bear large, flat plates. Their food is a mixed diet of molluscs, echinoderms, crustaceans, and fishes. Chimaeras are not commercial species and are seldom caught. Despite their bizarre shape, they are beautifully colored, with a pearly iridescence.

figure 16.11

A spotted ratfish, *Hydrolagus collei,* of the North American west coast. This species is one of the most handsome chimaeras, which tend to be bizarre in appearance.

■ Bony Fishes: The Osteichthyes

In the early to middle Silurian period, a lineage of fishes with bony endoskeletons gave rise to a clade of vertebrates that contains 96% of living fishes and all tetrapods. Fishes of this clade have traditionally been termed "bony fishes" (**Osteichthyes**). Bony fishes and tetrapods are united by endochondral bone (bone that replaces cartilage developmentally), lungs or a swim bladder derived from the gut, and several cranial and dental characters. Because traditional usage of the term Osteichthyes does not describe a monophyletic (natural) group (see figure 16.2), most recent classifications, including the one presented on page 348, do not recognize this term as a valid taxon. Rather, it is a term of convenience to describe

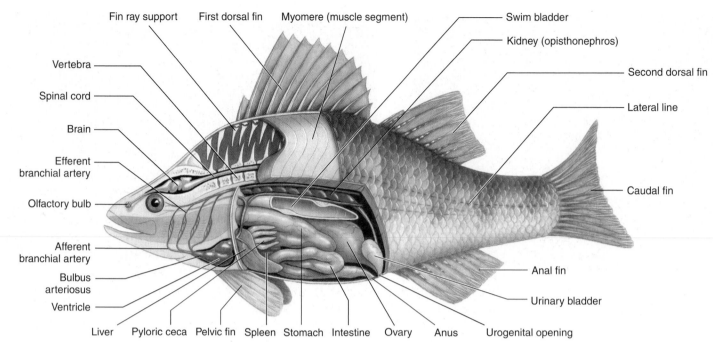

figure 16.12

Anatomy of a yellow perch, *Perca flavescens*, a freshwater teleost fish.

vertebrates with endochondral bone that are conventionally termed "bony fishes."

Fossils of the earliest bony fishes show several structural similarities, including a bony operculum and branchiostegal rays, with acanthodians (see figure 15.14). By the middle Devonian period, bony fishes already had diversified extensively into two major clades, with adaptations that fitted them for every aquatic habitat except the most inhospitable. One of these clades, the ray-finned fishes (class Actinopterygii), includes teleosts (figure 16.12), the most species-rich clade of living vertebrates. A second clade, the lobe-finned fishes (class Sarcopterygii), is represented today by only eight fishlike vertebrates, the lungfishes and coelacanths (see figures 16.19 and 16.20); however, it includes the sister group of land vertebrates (tetrapods).

Several key adaptations contributed to the diversification of bony fishes. They have an **operculum** over the gill composed of bony plates and attached to a series of muscles. This feature increases respiratory efficiency because outward movement of the operculum creates a negative pressure so that water is drawn across the gills, as well as pushed across by the mouth pump. A gas-filled pouch branching from the esophagus provides an additional means of gas exchange in oxygen-poor waters and an efficient means of achieving neutral buoyancy. In fishes that use these pouches primarily for gas exchange, the pouches are called lungs, while in fishes that use these pouches primarily for buoyancy, these pouches are called **swim bladders** (p. 342). Progressive specialization of jaw musculature and skeletal

elements involved in suction feeding is another key feature of bony fish evolution.

Ray-Finned Fishes: Class Actinopterygii

Ray-finned fishes are an enormous assemblage containing all of the bony fishes that we know today—about 27,000 species. The ancestral forms that swam in Devonian waters were small and heavily armored, with thick **ganoid** scales (figure 16.13) and heterocercal caudal fins (figure 16.14).

From those earliest ray-finned fishes arose several clades. Bichirs, in the clade Cladistia, have lungs, heavy ganoid scales, and other characteristics similar to those of their bony-fish ancestors (figure 16.15A). The 16 species of bichirs live in fresh waters of Africa. A second group are the chondrosteans (Gr. *chondros,* cartilage, + *osteon,* bone), represented by 27 species of freshwater and anadromous sturgeons and paddlefishes (figure 16.15B and C).

The third major group to emerge from the early ray-finned stock were **neopterygians** (Gr. *neos,* new, + *pteryx,* fin). Neopterygians appeared in the late Permian period and diversified extensively during the Mesozoic era, when one lineage gave rise to modern bony fishes, the teleosts. The three surviving genera of non-teleost neopterygians are the bowfin *Amia* (Gr., tunalike fish), which inhabits shallow, weedy waters of the Great Lakes and the Mississippi River basin, and the gars *Lepisosteus* (Gr. *lepidos,* scale, + *osteon,* bone) and *Atractosteus* (Gr. *atraktos,* spindle, + *osteon,* bone), both native to eastern and southern North America (figure 16.16). Gars are large,

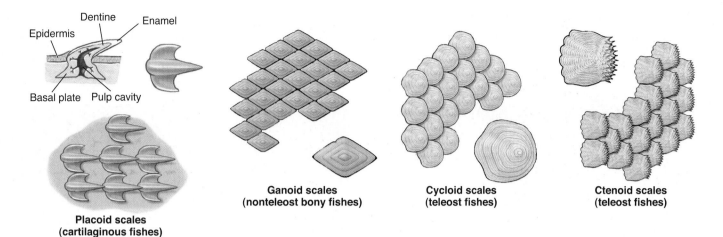

figure 16.13

Types of fish scales. Placoid scales are small, conical, toothlike structures characteristic of Chondrichthyes. Diamond-shaped ganoid scales, present in early bony fishes and living gars, are composed of layers of silvery enamel (ganoin) on the upper surface and bone on the lower. Other bony fishes have either cycloid or ctenoid scales, which are thin and flexible and arranged in overlapping rows.

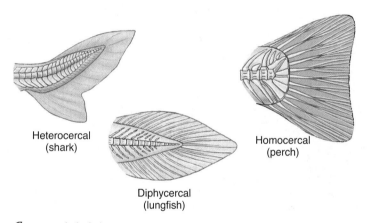

figure 16.14

Types of caudal fins among fishes.

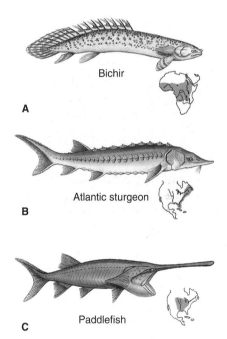

figure 16.15

Nonteleost ray-finned fishes of class Actinopterygii. **A,** Bichir, *Polypterus bichir,* of equatorial West Africa is a nocturnal predator. **B,** Atlantic sturgeon, *Acipenser oxyrhynchus* (now uncommon), inhabits Atlantic coastal rivers. **C,** Paddlefish, *Polyodon spathula,* of the Mississippi River basin reaches 2 m and 80 kg.

ambush predators with elongate bodies and jaws filled with needlelike teeth.

Teleosts (Gr. *teleos,* complete, + *osteon,* bone), the modern bony fishes (see figure 16.12), form the largest clade of neopterygians. Teleost diversity is astounding, with about 27,000 described species representing about 96% of all living fishes and about half of all vertebrates (figure 16.17). They display an array of morphological form and size, occupy nearly every aquatic habitat on Earth, and may even make excursions onto land, as do mudskippers (figure 16.17B).

Several morphological trends in the teleost lineage allowed it to diversify into a truly incredible variety of habitats and forms. The heavy dermal armor of primitive ray-finned fishes was replaced by light, thin, flexible **cycloid** and **ctenoid scales** (see figure 16.13). Increased mobility and speed permitted

by loss of heavy armor improved predator avoidance and feeding efficiency. The symmetrical shape of the **homocercal** tail (see figure 16.14) of most teleosts focused musculature contractions on the tail, allowing greater speed. Various fins were elaborated into forms that permitted precise steering,

figure 16.16

Nonteleost neopterygian fishes. **A,** Bowfin, *Amia calva*. **B,** Longnose gar, *Lepisosteus osseus*. Both species frequent slow-moving streams and swamps of eastern North America, where they can suspend motionless in the water, ready to snatch passing fishes.

figure 16.17

Diversity among teleosts. **A,** Blue marlin, *Makaira nigricans*, one of the largest teleosts. **B,** Mudskippers, *Periopthalmus*, make extensive excursions on land to graze on algae and capture insects; they build nests in which the young hatch and are guarded by the mother. **C,** Protective coloration of the lionfish, *Pterois volitans*, advises caution; the dorsal spines are venomous. **D,** The sucking disc on the sharksucker, *Echeneis naucrates*, is a modification of the dorsal fin.

camouflage, protection, attachment, or social communication (see figure 16.17C and D). Teleost lineages demonstrated an increasingly fine control of gas resorption and secretion in the swim bladder, improving control of buoyancy. Nearly all teleosts capture food by suction feeding, where rapid expansion of the orobranchial cavity creates negative pressure, drawing in water and prey. Changes in jaw suspension allowed teleosts to use this method more effectively than did their ancestors. With so many innovations, teleosts have become the most diverse of fishes.

CHARACTERISTICS
of Class Actinopterygii

1. Caudal fin heterocercal (ancestral condition) or **homocercal** (see figure 16.14); paired pectoral and pelvic fins usually present, supported by **bony rays;** muscles controlling fin movements within trunk
2. Skin with **ganoid (ancestral condition), cycloid, or ctenoid scales** of dermal origin (figure 16.18) or **naked**
3. **Skeleton of bone;** notochord present but reduced; **vertebrae** distinct
4. Jaws present, usually with **enameloid, polyphyodont teeth;** spiral valve present (ancestral condition) or absent
5. Brain well developed, but relatively small; 10 pairs of cranial nerves
6. Development of senses variable; three pairs of semicircular canals
7. Sexes usually separate; many hermaphroditic; some reproduce asexually by parthenogenesis; fertilization usually external, internal in some
8. Oviparous or viviparous; embryos of viviparous specious nourished by placenta or yolk sac (ovoviviparity); larval stage often greatly different from adult
9. Excretory system of opisthonephric kidneys, which drain via archinephric duct to cloaca; ammonia usually main nitrogenous waste
10. **Gills covered by a bony operculum; swim bladder present** usually functioning for buoyancy, sometimes used for respiration
11. Heart with a sinus venosus, atrium, and ventricle; **single circulation;** nucleated red blood cells

Australian lungfish

African lungfishes

South American lungfish

figure 16.19

Lungfishes are lobe-finned fishes of Sarcopterygii. Extinct Paleozoic lungfishes are most similar in morphology to the Australian lungfish, *Neoceratodus forsteri*. The African lungfishes, *Protopterus*, are the best adapted of the three for remaining dormant in mucus-lined cocoons breathing air during prolonged periods of drought.

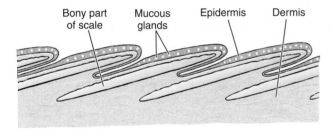

Bony part of scale   Mucous glands   Epidermis   Dermis

figure 16.18

Section through the skin of a bony fish, showing the overlapping scales (yellow). The scales lie in the dermis and are covered by epidermis.

Lobe-Finned Fishes: Class Sarcopterygii

The ancestor of tetrapods is found within a group of extinct sarcopterygian fishes called **rhipidistians,** which included several lineages that flourished in shallow waters in the late Paleozoic era. The evolution of tetrapods from rhipidistians is discussed in Chapter 17.

All early sarcopterygians had lungs as well as gills, and a tail of the **heterocercal** type. However, during the Paleozoic era, the orientation of the vertebral column changed so that the tail became symmetrical and **diphycercal** (see figure 16.14). These fishes had powerful jaws; heavy, enameled scales; and strong, fleshy, paired lobed fins that they may have used like legs to support and possibly move the body over benthic substrates. The sarcopterygian clade today is represented by all tetrapods, but only eight fish species: six species of lungfishes and two species of coelacanths (figures 16.19 and 16.20). Exclusion of tetrapods from clade Sarcopterygii renders it paraphyletic.

Neoceratodus (Gr. *neos,* new, + *keratos,* horn, + *odes,* form), the living Australian lungfish, may attain a length of 1.5 m (figure 16.19) and is similar to early forms. This lungfish, unlike its living relatives, normally relies on gill respiration, and cannot survive long out of water. The South American lungfish, *Lepidosiren* (L. *lepidus,* pretty, + *siren,* mythical mermaid), and the African lungfishes, *Protopterus* (Gr. *protos,* first, + *pteron,*

CHARACTERISTICS
of Sarcopterygian Fishes

1. Caudal fin heterocercal (fossil forms) or **diphycercal** in living forms (see figure 16.14); paired pectoral and pelvic fins usually present, supported by **stout bones and bony rays;** muscles controlling fin movements within fin
2. Skin with dermal scales composed of two layers of bone, a layer of **cosmine** (form of dentin) and a thin layer of **enamel** in fossil species
3. **Skeleton of bone;** vertebrae distinct
4. Jaws present, usually with **polyphyodont teeth covered with true enamel** that typically are crushing plates restricted to the palate; intestine with spiral valve
5. Brain well developed, but relatively small; 10 pairs of cranial nerves
6. Development of senses variable; three pairs of semicircular canals
7. Sexes separate; fertilization external (lungfishes) or internal (coelacanths)
8. Oviparous
9. Excretory system of opisthonephric kidneys, which drain via archinephric duct to cloaca; ammonia and urea main nitrogenous wastes
10. **Gills covered by a bony operculum; swim bladder** present, used primarily for respiration (fat-filled in coelacanths)
11. Heart with a sinus venosus, an atrium, a partly divided ventricle and a conus arteriosus; **pulmonary and systemic circuits incompletely separated;** nucleated red blood cells

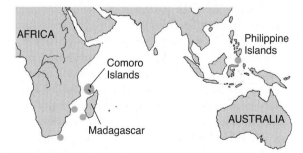

figure 16.20

The coelacanth genus *Latimeria* is a surviving marine relict of a group of lobe-finned fishes that flourished some 350 million years ago.

wing), can live out of water for long periods of time. *Protopterus* lives in African streams and ponds that are baked hard by the hot tropical sun during the dry season. The fish burrows at the approach of the dry season and secretes a copious slime that mixes with mud to form a hard cocoon in which it remains dormant until rain returns.

Coelacanths also arose in the Devonian period, diversified somewhat, and reached their peak of diversity in the Mesozoic era. At the end of that era, they nearly disappeared but left one remarkable surviving genus, *Latimeria* (figure 16.20). Because the last coelacanths were believed to have become extinct 70 million years ago, the scientific world was astonished when a coelacanth was found on a trawl off the coast of South Africa in 1938. An intensive search to locate more specimens was successful off the coast of the Comoro Islands. There, fishermen occasionally catch them at great depths with hand lines, providing specimens for research. This was assumed to be the only population of *Latimeria* until 1998, when the scientific world was again surprised by the capture of a new species of coelacanth in Indonesia, 10,000 km from the Comoros!

The tail of coelacanths is diphycercal (see figure 16.14) but possesses a small lobe between the upper and lower caudal lobes, producing a three-pronged structure (figure 16.20). Their swim bladders are fat-filled and not used for respiration.

Coelacanths are a deep metallic blue with irregular white or brassy flecks, providing camouflage against the dark lava-cave reefs that they inhabit. Young are born fully formed after hatching internally from eggs 9 cm in diameter—the largest among bony fishes.

Structural and Functional Adaptations of Fishes

Locomotion in Water

To the human eye, some fishes appear capable of swimming at extremely high speeds, but our judgment is unconsciously tempered by our own experience that water is a highly resistant medium through which to move. Most fishes, such as a trout or a minnow, can swim maximally about 10 body lengths per second, obviously an impressive performance by human standards. Yet, when these speeds are translated into kilometers per hour, we realize that a 30 cm (1 foot) trout can swim only about 10.4 km (6.5 miles) per hour. As a general rule, the larger the fish, the faster it can swim.

Measuring fish cruising speeds is best done in a "fish wheel," a large, ring-shaped channel filled with water that is turned at a speed equal and opposite to that of the fish. Much more difficult to measure are the sudden bursts of speed that most fish can make to capture prey or to avoid being captured. A hooked bluefin tuna was once "clocked" at 66 km per hour (41 mph); swordfish, sailfish, and marlin may be capable of incredible bursts of speed approaching, or even exceeding, 85 km per hour (53 mph). Such high speeds can be sustained for no more than 1 to 5 seconds.

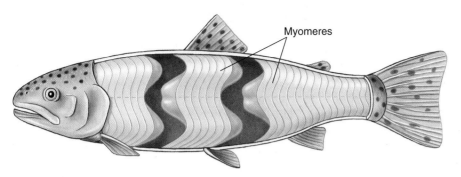

figure 16.21

Trunk musculature of a teleost fish, partly dissected to show the internal arrangement of the muscle bands (myomeres). The myomeres are folded into a complex, nested grouping, an arrangement that favors stronger and more controlled swimming.

The propulsive mechanism of a fish is its trunk and tail musculature. The axial, locomotory musculature is composed of zigzag bands, called **myomeres.** Muscle fibers in each myomere are relatively short and connect the tough connective tissue partitions that separate each myomere from the next. On the surface, myomeres take the shape of a W lying on its side (figure 16.21), but internally the bands are complexly folded and nested so that the pull of each myomere extends over several vertebrae. This arrangement produces more power and finer control of movement, since many myomeres are involved in bending a given segment of the body.

Understanding how fishes swim can be approached by studying the motion of a very flexible fish such as an eel (figure 16.22, *left*). The movement is serpentine, not unlike that of a snake, with waves of contraction moving backward along the body by alternate contraction of the myomeres on either side. The anterior end of the body bends less than the posterior end, so that each undulation increases in amplitude as it travels along the body. While undulations move backward, bending of the body pushes laterally against the water, producing a reactive force that is directed forward, but at an angle. It can be analyzed as having two components: thrust, which is used to overcome drag and propels the fish forward, and lateral force, which tends to make the fish's head "yaw," or deviate from the course in the same direction as the tail. This side-to-side head movement is very obvious in a swimming eel, but many fishes have a large, rigid head with enough surface resistance to minimize yaw.

The movement of an eel is reasonably efficient at low speed, but its body shape generates too much frictional drag for rapid swimming. Fishes that swim rapidly, such as trout, are less flexible and limit body undulations mostly to the caudal region (figure 16.22, *right*). Muscle force generated in the muscle mass is transferred through tendons to the caudal peduncle and caudal fin, where thrust is generated. This form of swimming reaches its highest development in tunas, whose bodies barely flex. Virtually all thrust is derived from powerful beats of the caudal fin (figure 16.23). Many fast oceanic fishes, such as marlin, swordfish, amberjacks, and wahoo, have swept-back caudal fins shaped much like a sickle. Such fins are the aquatic counterpart of the high-aspect wings of the swiftest birds (see p. 398). Swimming is the most economical form of vertebrate

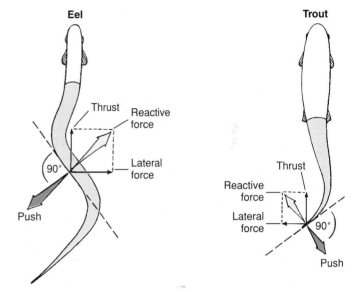

figure 16.22

Movements of swimming fishes, showing the forces developed by an eel-shaped fish and a spindle-shaped fish.

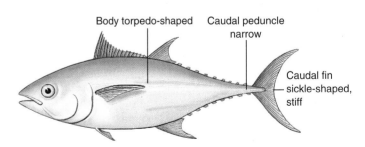

figure 16.23

Bluefin tuna, *Thunnus thynnus,* showing adaptations for fast swimming. Powerful trunk muscles pull on the slender caudal peduncle. Because the body does not bend, all of the thrust comes from beats of the stiff, sickle-shaped caudal fin.

locomotion, largely because aquatic animals are almost perfectly supported by their medium and expend little energy to overcome the force of gravity.

The body temperature of most fishes is the same as their environment, because any heat generated internally is lost quickly into the surrounding water. However, some fishes, such as tunas (figure 16.23) and mako sharks, maintain a high temperature in their swimming muscles and viscera—as much as 10° C warmer than surrounding water. Marlins (see figure 16.17A) and other billfishes elevate the temperature of their brain and retina. Research by F. G. Carey and others explains how these fishes accomplish this kind of thermoregulation, called regional endothermy. Heat is generated as a by-product of various activities, including digestion and swimming, or for billfishes, by a specialized heat-generating organ beneath the brain. This heat is conserved with a **rete mirabile,** a parallel bundle of blood vessels arranged to provide a countercurrent flow of blood. High temperatures apparently promote powerful swimming and enhance digestive and nervous system activity. Fishes with regional endothermy are the fastest in the world.

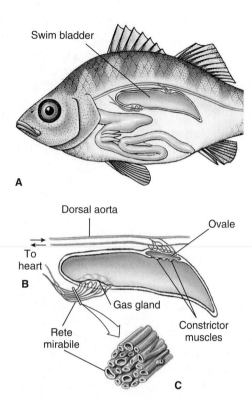

figure 16.24

A, Swim bladder of a teleost fish. The swim bladder lies in the coelom just beneath the vertebral column. **B,** Gas is secreted into the swim bladder by the gas gland. Gas from the blood is moved into the gas gland by the rete mirabile, a complex array of tightly packed capillaries that act as a countercurrent multiplier to increase oxygen concentration. The arrangement of venous and arterial capillaries in the rete is shown in **C.** To release gas during ascent, a muscular valve opens, allowing gas to enter the ovale, from which the gas is removed by diffusion into the blood.

Neutral Buoyancy and the Swim Bladder

All fishes are slightly heavier than water because their skeletons and other tissues contain heavy elements that are present only in trace amounts in natural waters. To keep from sinking, sharks, which lack a swim bladder, must always keep moving forward in the water. The asymmetrical (heterocercal) tail of a shark provides lift as it sweeps through the water, and the broad head and flat pectoral fins (see figure 16.6) act as angled planes to provide additional lift. Sharks also are aided in buoyancy by having very large livers containing a lipid called squalene with a density of only 0.86 grams per milliliter. The liver thus acts like a large sack of buoyant oil that helps compensate for the shark's heavy body.

By far the most efficient device for adjusting buoyancy is a gas-filled space. The **swim bladder** serves this purpose in bony fishes (figure 16.24). It arose from the paired lungs of early Devonian fishes. Swim bladders are present in most pelagic bony fishes but are absent from tunas, most abyssal fishes, and most bottom-dwellers, such as flounders and sculpins. Although their name suggests otherwise, swim bladders are not used to swim.

By adjusting the volume of gas in its swim bladder, a fish can achieve neutral buoyancy and remain suspended indefinitely at any depth with no muscular effort. If a fish swims to a greater depth, the greater pressure exerted by the surrounding water compresses the gas in the swim bladder, so that the fish becomes less buoyant and begins to sink. Gas must be added to the swim bladder to establish a new equilibrium buoyancy. When a fish swims upward, gas in the bladder expands because of the reduced surrounding water pressure, making the fish more buoyant. Unless gas is removed, the fish continues to ascend with increasing speed as the swim bladder continues to expand.

Gases are removed from the swim bladder in two ways. Some fishes (trout, for example) have a pneumatic duct that connects the swim bladder to the esophagus, through which they can expel air. More derived teleosts have lost the pneumatic duct and exchange air in the swim bladder with the blood. Gas is diffused into blood at a highly vascularized region of the swim bladder, the ovale. Gas enters the swim bladder at the gas gland. The gas gland secretes lactic acid, causing oxygen to be released from hemoglobin. A remarkable network of parallel blood capillaries, the **rete mirabile,** supplies the gas gland and acts as a countercurrent multiplier (figure 16.24). The rete allows oxygen to reach high concentrations in the gas gland, permitting it to diffuse into the swim bladder.

The amazing effectiveness of this device is exemplified by a fish living at a depth of 2400 m (8000 feet). To keep the bladder inflated at that depth, the gas inside (mostly oxygen) must have a pressure exceeding 240 atmospheres, which is much greater than the pressure in a fully charged steel gas cylinder. Yet the oxygen pressure in the fish's blood cannot exceed 0.2 atmosphere, in equilibrium with the oxygen pressure in the atmosphere at the sea surface.

Operculum

Water flow

A

Gill rakers   Gill arch   Gill filaments

B

Gill rakers

Gill arch

Filament with lamellae

C

Blood flow

Lamella

Water flow

D   Filament

figure 16.25

Gills of a fish. Muscles attached to the operculum, **A,** pump water over gills and out the gill slit. The bony, protective flap covering the gills (operculum) has been removed, **B,** to reveal the branchial chamber containing the gills. There are four gill arches on each side, each bearing numerous filaments. **C,** A portion of a gill arch shows gill rakers that project forward to strain out food and debris, and gill filaments that project to the rear. **D,** A single gill filament is dissected to show the blood capillaries within the platelike lamellae. Direction of water flow (*blue arrows*) is opposite the direction of blood flow.

Respiration

Fish gills are composed of thin filaments, each covered with a thin epidermal membrane that is folded repeatedly into plate-like **lamellae** (figure 16.25). These lamellae are richly supplied with blood vessels. The gills are located inside a pharyngeal cavity and are covered with an **operculum.** This arrangement protects the delicate gill filaments, streamlines the body, and makes possible a pumping system for moving water through the mouth, across the gills, and out the gill slits. Instead of opercular flaps as in bony fishes, elasmobranchs have a series of gill slits (see figure 16.6) out of which the water flows. In both elasmo-branchs and bony fishes, the branchial mechanism is arranged to pump water continuously and smoothly over the gills, although to an observer it appears that fish breathing is pulsatile.

The direction of water flow is opposite that of blood flow (countercurrent flow), the best arrangement for extracting the greatest possible amount of oxygen from water. Some bony fishes can remove as much as 85% of the dissolved oxygen from water passing over their gills. Very active fishes, such as herring and mackerel, can obtain sufficient water for their high oxygen demands only by swim-ming forward continuously to force water into their open mouth and across their gills. This process is called ram ventilation.

Osmotic Regulation

Fresh water is an extremely dilute medium with a salt concentration (0.001 to 0.005 gram moles per liter [M]) much below that of the blood of freshwater fishes (0.2 to 0.3 M). Water therefore tends to enter their bodies osmotically, and salt is lost by diffusion outward. Although the scaled and mucus-covered body surface is almost totally impermeable to water, water gain and salt loss do occur across thin membranes of the gills. Freshwater fishes are **hyperos-motic regulators** with several defenses against these problems (figure 16.26, *top*). First, excess water is pumped out by the kidneys, which are capable of forming very dilute urine. Second, special salt-absorbing cells located in the gill epithelium actively move salt ions, principally sodium and chloride, from water to the blood. This absorption, together with salt present in the fish's food, replaces diffusive salt loss. These mechanisms are so efficient that a freshwater fish devotes only a small part of its total energy expenditure to maintaining osmotic balance.

Marine fishes are **hypoosmotic reg-ulators** that encounter a completely dif-ferent problem. Having a much lower blood salt concentration (0.3 to 0.4 M) than the seawater around them (about 1 M), they tend to lose water and to gain salt. A marine teleost quite literally risks drying out, much like a desert mammal deprived of water. To compensate for water loss, a marine teleost drinks seawater (figure 16.26, *bottom*). Excess salt accompanying the seawater is eliminated in multiple ways. Major sea salt ions (sodium, chloride, and potassium) are carried by the blood to the gills, where they are expelled by special salt-secretory cells. The remaining sea salt ions, mostly magnesium, sulfate, and calcium, are voided with feces or excreted by the kidney.

Migration

Freshwater Eels

For centuries, naturalists had been puzzled by the life history of freshwater eels, *Anguilla* (an-gwil´ə) (L. eel), a common and

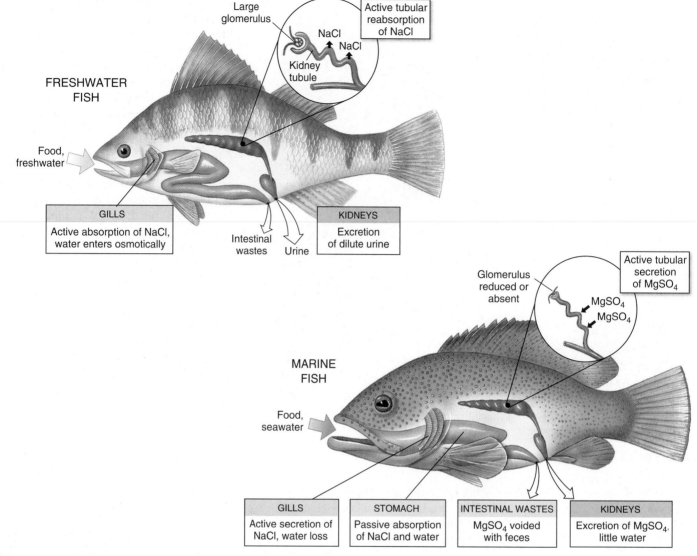

figure 16.26

Osmotic regulation in freshwater and marine bony fishes. *Top:* A freshwater fish maintains osmotic and ionic balance in its dilute environment by actively absorbing sodium chloride (NaCl) across the gills (some salt is gained with food). To flush out excess water that constantly enters the body, the glomerular kidney produces a dilute urine by reabsorbing sodium chloride. *Bottom:* A marine fish must drink seawater to replace water lost osmotically to its salty environment. Sodium chloride and water are absorbed from the stomach. Excess sodium chloride is actively transported outward by the gills. Divalent sea salts, mostly magnesium sulfate ($MgSO_4$), are eliminated with feces and secreted by the tubular kidney.

commercially important species inhabiting coastal streams of the North Atlantic. Eels are **catadromous** (Gr. *kata,* down, + *dromos,* running), meaning they spend most of their lives in fresh water but migrate to the sea to spawn. Each fall, people saw large numbers of eels swimming down rivers toward the sea, but no adults ever returned. Each spring, countless numbers of young eels, called "elvers" (figure 16.27), each about the size of a wooden matchstick, appeared in coastal rivers and began swimming upstream. Beyond the assumption that eels must spawn somewhere at sea, the location of their breeding grounds was completely unknown.

The first clue was provided by two Italian scientists, Grassi and Calandruccio, who in 1896 discovered that elvers were advanced juvenile eels and that true larval eels were tiny,

leaf-shaped, transparent creatures called **leptocephali.** In 1905, Johann Schmidt began a systematic study of eel biology, examining thousands of leptocephali caught in plankton nets from many areas of the Atlantic. By noting where larvae in different stages were captured, Schmidt and his colleagues reconstructed the spawning migration.

When adult eels leave the coastal streams of Europe and North America, they swim to the Sargasso Sea, a vast area of warm oceanic water southeast of Bermuda. Here, at depths of 300 m or more, the eels spawn and die. Minute larvae then begin an incredible journey back to the streams of Europe and North America (figure 16.27). Because the Sargasso Sea is much closer to the American coastline, American eel larvae make their journey in only about 1 year, compared to 3 years

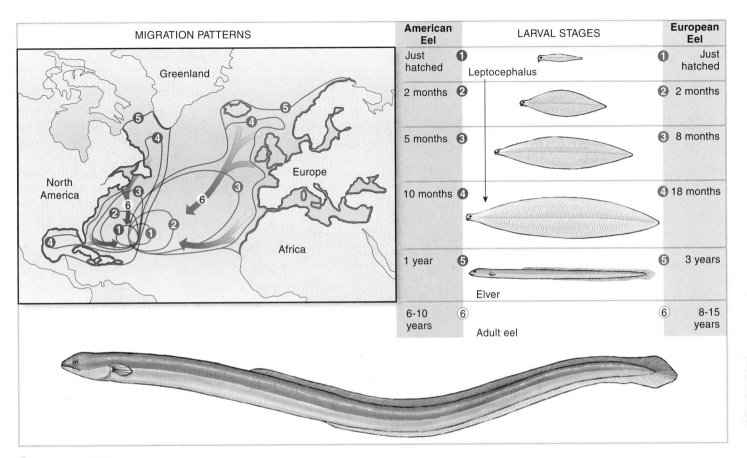

American Eel	LARVAL STAGES		European Eel
1 Just hatched	Leptocephalus	**1**	**1** Just hatched
2 2 months		**2**	**2** 2 months
3 5 months		**3**	**3** 8 months
4 10 months		**4**	**4** 18 months
5 1 year	Elver	**5**	**5** 3 years
6 6-10 years	Adult eel	**6**	**6** 8-15 years

figure 16.27

Life histories of American eels, *Anguilla rostrata*, and European eels, *Anguilla anguilla*. Migration patterns of American eels are shown in blue, while those of European eels are shown in red. Circled numbers refer to stages of development. American eels complete their larval metamorphosis and sea journey in 1 year, while European eels take 3 years to complete their much longer journey.

for European eel larvae. Males typically remain in the brackish water of coastal rivers, while females migrate as far as several hundred kilometers upstream. After 6 to 15 years of growth, females, now 1 m long, return to the ocean to join the smaller males in the journey back to the spawning grounds in the Sargasso Sea.

Enzyme electrophoretic analyses of eel larvae confirm not only the existence of separate European and American species but also Schmidt's hypothesis that the European and American eels spawn in partially overlapping areas of the Sargasso Sea.

Homing Salmon

The life history of salmon is nearly as remarkable as that of freshwater eels and certainly has received far more popular attention. Salmon are **anadromous** (Gr. *anadromous,* running upward), spending their adult lives at sea but returning to fresh water to spawn. Atlantic salmon (*Salmo salar*) can make repeated upstream spawning runs. The seven Pacific salmon (*Oncorhynchus*) species (sockeye, coho, pink, Chinook, chum, steelhead, and Japanese masu) each make a single spawning run (figures 16.28 and 16.29), after which they die.

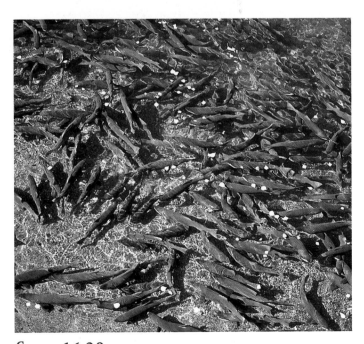

figure 16.28

Migrating Pacific sockeye salmon, *Oncorhynchus nerka.*

Asia

Alaska

Oceanic distribution

Dying salmon

Spawning salmon

Alevins hatching

Adult migrating upstream

Fry

Parr

Smolts

Adults: Feeding for 3–4 years in ocean

figure 16.29

Life history of Pacific salmon, *Oncorhynchus*.

The virtually infallible homing instinct of the Pacific species is legendary. After migrating downstream as a smolt (a juvenile stage; figure 16.29), a sockeye salmon travels many hundreds of kilometers over the Pacific for nearly 4 years, grows to 2 to 5 kg in weight, and then returns almost unerringly to spawn in the headwaters of its parent stream. Some straying does occur and is an important means of increasing gene flow and populating new streams.

Experiments by A. D. Hasler and others show that homing salmon are guided upstream by the characteristic odor of their parent stream. When salmon finally reach the spawning beds of their parents (where they themselves were hatched), they spawn and die. The following spring, newly hatched fry transform into smolts before and during the downstream migration. At this time they are imprinted with the distinctive odor of the stream, which is apparently a mosaic of compounds released by the characteristic vegetation and soil in the watershed of the parent stream. They also seem to imprint on odors of other streams that they pass while migrating downriver and use these odors in reverse sequence as a map during the upriver migration as returning adults.

Salmon runs along the Pacific coast of North America have been devastated by a lethal combination of increased siltation from logging, pollution, and, especially, the presence of more than 50 hydroelectric dams, which obstruct upstream migration of adult salmon and kill downstream migrants as they pass through the dams' power-generating turbines. In addition, the chain of reservoirs behind the dams, which has converted the Columbia and Snake Rivers into a series of lakes, increases the mortality of young salmon migrating downstream by slowing their passage to the sea. The result is that the annual run of wild salmon is today less than 3% of the 10 to 16 million fish that ascended the rivers 150 years ago. While recovery plans have been delayed by the power industry, environmental groups argue that in the long run losing the salmon will be more expensive to the regional economy than making the changes that will allow salmon stocks to recover.

Reproduction

In a group as diverse as fishes, it is no surprise to find extraordinary variations on the basic theme of sexual reproduction. Most fishes favor a simple theme: they are **dioecious,** with external fertilization and external development of their eggs and embryos (oviparity). However, as tropical fish enthusiasts are well aware, the ever-popular ovoviviparous guppies and mollies of home aquaria develop in the ovarian cavity of the mother and then are born (figure 16.30). As described earlier in this chapter (pp. 334–335), some viviparous sharks develop a kind of placental attachment through which the young are nourished during gestation.

Oviparity is the most common mode of reproduction in fishes. Many marine fishes are extraordinarily profligate egg producers. Males and females come together in great schools and, without elaborate courtship behavior, release vast numbers of gametes into the water to drift with currents. A large

figure 16.30

Rainbow seaperch, *Hypsurus caryi*, giving birth. All members of this North Pacific family (Embiotocidae) are ovoviviparous.

figure 16.31

Male banded jawfish, *Opistognathus macrognathus*, orally brooding its eggs. The male retrieves the female's eggs and incubates them until they hatch. During brief periods when the jawfish is feeding, the eggs are left in the burrow.

female cod may release 4 to 6 million eggs at a single spawning. Less than one in a million will survive the numerous perils of the ocean to reach reproductive maturity.

Unlike the minute, semibuoyant, transparent eggs of pelagic marine teleosts, the eggs of many near-shore and bottom-dwelling (benthic) species are larger, yolky, nonbuoyant, and adhesive. Some bury their eggs, many attach them to vegetation, some deposit them in nests, and others even incubate them in their mouths (figure 16.31). Many benthic spawners guard their eggs. Intruders expecting an easy meal of eggs may be met with a vivid and often belligerent display by the guard, which is almost always the male.

Freshwater fishes usually produce nonbuoyant eggs. Those fishes, such as yellow perch, provide no parental care and simply scatter their myriads of eggs among weeds or along the sediment. Freshwater fishes that do provide egg care, such as bullhead catfishes and some darters, produce fewer, larger eggs that have a better chance for survival.

Elaborate preliminaries to mating are the rule for fresh-water fishes. A female Pacific salmon, for example, performs a ritualized mating "dance" with her breeding partner after arriving at the spawning bed in a fast-flowing, gravel-bottomed stream. She then turns on her side and scoops out a depression with her tail. As the eggs are laid by the female, they are fertilized by the male. After the female covers the eggs with gravel, the exhausted fish dies.

Soon after an egg of an oviparous species is fertilized, it absorbs water, and the outer layer hardens. Cleavage follows, and a blastoderm forms astride a relatively large yolk mass. Many fishes hatch as larvae, carrying a semitransparent sac of yolk, which provides their food supply until the mouth and digestive tract have developed and the larvae can feed on their own. After a period of growth, a larva undergoes a metamorphosis, especially dramatic in many marine species, including eels (see figure 16.27). Body shape is refashioned, fin and color patterns change, and the larva becomes a juvenile bearing the unmistakable definitive body form of its species.

Taxonomy of Living Fishes

The following Linnean classification of major fish taxa follows that of Nelson (2006). The probable relationships of these traditional groupings, together with the major extinct groups of fishes, are shown in the cladogram in figure 16.2. Other schemes of classification have been proposed. Because of the difficulty of determining relationships among the numerous living and fossil species, we can appreciate why fish classification has undergone, and will continue to undergo, continuous revision.

Phylum Chordata

Subphylum Vertebrata (Craniata)

Class Myxini (mik-sin´e) (Gr. *myxa*, slime): **hagfishes.** No jaws or paired fins; mouth with four pairs of tentacles; buccal funnel absent; 1–16 pairs of external gill openings; vertebrae absent; slime glands present. Examples: *Myxine, Epaptretus;* about 70 species, marine.

Class Petromyzontida (pet´trō-mī-zon´ti-də) (Gr. *petros*, stone, + *myzon*, sucking): **lampreys.** No jaws or paired fins; mouth surrounded by keratinized teeth but no barbels; buccal funnel present; seven pairs of external gill openings; vertebrae present only as neural arches. Examples: *Petromyzon, Ichthyomyzon, Lampetra;* 38 species, freshwater and anadromous.

Superclass Gnathostomata (na´thō-stō´mä-tə) (Gr. *gnathos*, jaw, + *stoma*, mouth). Jaws present; paired appendages present (secondarily lost in a few forms); three pairs of semicircular canals; notochord partly or completely replaced by centra.

Class Chondrichthyes (kon-drik´the-ēz´) (Gr. *chondros*, cartilage, + *ichthys*, fish): **cartilaginous fishes.** Cartilaginous skeleton; teeth not fused to jaws and usually replaced; no swim bladder; intestine with spiral valve; claspers present in males.

Subclass Elasmobranchii (ē-laz´mō-brank´e-ī) (Gr. *elasmos*, plated, + *branchia*, gills): **sharks, skates,** and **rays.** Placoid scales or derivatives (scutes and spines) usually present; five to seven gill arches and gill slits in separate clefts along pharynx; upper jaw not fused to cranium. Examples: *Squalus, Raja, Sphyrna;* about 937 species, mostly marine.

Subclass Holocephali (hō-lō-sef´ə-lī) (Gr. *holos*, entire, + *kephale*, head): **chimaeras** and **ratfishes.** Scales absent; four gill slits covered by operculum; jaws with tooth plates; accessory clasping organ (tentaculum) in males; upper jaw fused to cranium. Examples: *Chimaera, Hydrolagus;* 33 species, marine.

Class Actinopterygii (ak´ti-nop´te-rij´e-ī) (Gr. *aktis*, ray, + *pteryx*, fin, wing): **ray-finned fishes.** Skeleton ossified; single gill opening covered by operculum; paired fins supported primarily by dermal rays; limb musculature within body; swim bladder mainly for buoyancy, if present; atrium and ventricle not divided; teeth with enameloid covering.

Subclass Cladistia (clə-dis´tē-ə) (Gr. *cladi*, branch): **bichirs.** Rhombic ganoid scales; lungs; spiracle present; dorsal fin consisting of 5–18 finlets. Examples: *Polypterus;* about 16 species, freshwater.

Subclass Chondrostei (kon-dros´tē-ī) (Gr. *chondros*, cartilage, + *osteon*, bone): **paddlefishes, sturgeons.** Skeleton primarily cartilage; caudal fin heterocercal; large scutes or tiny ganoid scales present. Examples: *Polyodon, Acipenser;* 29 species, freshwater and anadromous.

Subclass Neopterygii (nē´op-te-rij´e-ī) (Gr. *neo*, new, + *pteryx*, fin, wing): **gars, bowfin,** and **teleosts.** Skeleton primarily bone; caudal fin usually homocercal; scales cycloid, ctenoid, absent, or rarely, ganoid. Examples: *Amia, Lepisosteus, Anguilla, Oncorhynchus, Perca;* about 27,000 species, nearly all aquatic habitats.

Class Sarcopterygii (sär-cop-te-rij´-e-ī) (Gr. *sarkos*, flesh, + *pteryx*, fin, wing): **lobe-finned fishes.** Skeleton ossified; single gill opening covered by operculum; paired fins with sturdy internal skeleton and musculature within appendage; diphycercal tail; usually with lungs; atrium and ventricle at least partly divided; teeth with enamel covering. Examples: *Latimeria* (coelacanths); *Neoceratodus, Lepidosiren, Protopterus* (lungfishes); 8 species, marine and freshwater. Not monophyletic unless tetrapods are included.

Summary

Fishes are gill-breathing aquatic vertebrates with fins for appendages. They include the oldest vertebrates, having originated from an ancestor in the Cambrian period or possibly earlier. Five classes of living fishes are recognized. The jawless hagfishes (class Myxini) and lampreys (class Petromyzontida) have an eel-like body form without paired fins, a cartilaginous skeleton, a notochord that persists throughout life, and a disclike mouth adapted for sucking or biting. All other vertebrates have jaws, a major development in vertebrate evolution.

Members of class Chondrichthyes (sharks, rays, and chimaeras) have a cartilaginous skeleton, paired fins, well-developed sensory organs, and an active, characteristically predaceous habit. Bony fishes are divided into two classes. Lobe-finned fishes of the class Sarcopterygii, represented today by lungfishes and

coelacanths, form a paraphyletic group if tetrapods are excluded. Terrestrial vertebrates arose from one lineage within this group. The second class is composed of ray-finned fishes (class Actinopterygii), a huge and diverse modern assemblage containing nearly all familiar freshwater and marine fishes.

Modern bony fishes (teleosts) have radiated into approximately 27,000 species that reveal an enormous diversity of adaptations, body form, behavior, and habitat preference. Most fishes swim by undulatory contractions of their body muscles, which generate thrust (propulsive force). Eel-like fishes oscillate the whole body, but in more rapid swimmers the undulations are limited to the caudal region or the caudal fin alone.

Most pelagic bony fishes achieve neutral buoyancy in water using a gas-filled swim

bladder, the most effective gas-secreting device known in the animal kingdom. Gills of fishes, having efficient countercurrent flow between water and blood, facilitate high rates of oxygen exchange. All fishes have highly-developed osmotic and ionic regulation, achieved principally by kidneys and gills.

Many fishes are migratory, and some, such as freshwater eels and anadromous salmon, make remarkable migrations of great length and precision. Most fishes are oviparous, but ovoviviparous and viviparous fishes are not uncommon. Reproductive investment may be in large numbers of eggs with low survival (many marine fishes) or in fewer eggs with greater parental care for better survival (many freshwater fishes).

Review Questions

1. Provide a brief description of fishes, citing characteristics that distinguish them from all other animals.
2. What characteristics distinguish hagfishes and lampreys from all other fishes? How do they differ in morphology from each other?
3. Describe feeding behavior in hagfishes and lampreys. How do they differ?
4. Describe the life cycle of sea lampreys, *Petromyzon marinus,* and the history of their invasion of the Great Lakes.
5. In what ways are sharks well equipped for a predatory life habit?
6. What function does the lateral-line system serve? Where are receptors located?
7. Explain how bony fishes differ from sharks and rays in the following systems or features: skeleton, scales, buoyancy, respiration, and reproduction.
8. Match the ray-finned fishes in the right column with the group to which each belongs in the left column:

    _____ Chondrosteans a. Perch
    _____ Nonteleost   b. Sturgeon
          neopterygians   c. Gar
    _____ Teleosts   d. Salmon
                          e. Paddlefish
                          f. Bowfin

9. Make a cladogram that includes the following groups of fishes: chondrosteans, elasmobranchs, hagfishes, holocephalans, lampreys, lungfishes, teleosts. Add the following synapomorphies to the diagram: claspers, cranium, endochondral bone, fleshy fins, jaws, vertebrae.
10. List four characteristics of teleosts that contributed to their incredible evolutionary diversity.
11. Only eight species of lobe-finned fishes are alive today, remnants of a group that flourished in the Devonian period of the Paleozoic. What morphological characteristics distinguish lobe-finned fishes?
12. Explain how lung fishes are adapted to survive out of water.
13. Describe discoveries of living coelacanths. What is the evolutionary significance of the group to which they belong?
14. Compare the swimming movements of eels with those of trout, and explain why the latter are more efficient for rapid locomotion.
15. Sharks and bony fishes approach or achieve neutral buoyancy in different ways. Describe the methods evolved in each group. Why must a teleost adjust

the gas volume in its swim bladder when it swims upward or downward? How is gas volume adjusted?
16. What is meant by "countercurrent flow" as it applies to fish gills?
17. Compare the osmotic problem and the mechanism of osmotic regulation in freshwater and marine bony fishes.
18. Describe the life cycle of freshwater eels. How does the life cycle of American eels differ from that of European eels?
19. How do adult Pacific salmon find their way back to their parent stream to spawn?
20. What mode of reproduction in fishes is described by each of the following terms: oviparous, ovoviviparous, viviparous?
21. Reproduction in marine pelagic fishes and in benthic freshwater fishes is distinctively different. How and why do they differ?

For Further Thought Regarding osmoregulation, what physiological and behavioral changes would occur as a fish migrates from a freshwater stream to the ocean?

■ Selected References

See also general references on page 432.

Barton, M. 2007. Bond's biology of fishes, ed. 3. Belmont, California, Thomson Brooks/Cole. *A revision of Bond's text that emphasizes anatomy, physiology, and ecology.*

Carey, F. G. 1973. Fishes with warm bodies. Sci. Am. **228**:36–44 (Feb.). *Classic paper about how fishes with regional endothermy keep warm.*

Helfman, G. S., B. B. Collette, D. E. Facey, and B. W. Brown. 2009. The diversity of fishes: biology, evolution, and ecology, ed. 2. Malden, Massachusetts, Wiley-Blackwell. *This information-packed textbook focuses on adaptation and diversity and is particularly strong in evolution and ecology.*

Long, J. A. 1995. The rise of fishes: 500 million years of evolution. Baltimore, The Johns Hopkins University Press. *A lavishly illustrated evolutionary history of fishes.*

Martini, F. H. 1998. Secrets of the slime hag. Sci. Am. **279**:70–75 (Oct.). *Biology of the most ancestral living vertebrate.*

Nelson, J. S. 2006. Fishes of the world, ed. 4. New York, John Wiley & Sons, Inc. *Authoritative classification of all major groups of fishes.*

Paxton, J. R., and W. N. Eschmeyer. 1998. Encyclopedia of fishes, ed. 2. San Diego, Academic Press. *Excellent authoritative reference that focuses on diversity and is spectacularly illustrated.*

Robison, B. H., and K. R. Resenbichler. 2008. *Macropinna microstoma* and the paradox of its tubular eyes. Copeia **2008**:780–784. *A cnidarian-eating fish with tubular eyes located beneath a transparent dome covering its head.*

Springer, V. G., and J. P. Gold. 1989. Sharks in question. Washington, D.C., Smithsonian Institution Press. *Morphology, biology, and diversity of sharks, richly illustrated.*

Weinberg, S. 2000. A fish caught in time: the search for the coelacanth. London, Fourth Estate. *The exciting history of the coelacanth discoveries.*

■ Custom Website

The *Animal Diversity* Online Learning Center is a great place to check your understanding of chapter material. Visit www.mhhe.com/hickmanad6e for access to key terms, quizzes, and more! Further enhance your knowledge with Web links to chapter-related material.

Explore live links for these topics:

Class Myxini
Class Petromyzontida
Subclass Elasmobranchii
Dissection Guides for Elasmobranchs
Vertebrate Laboratory Exercises
Class Osteichthyes

Primitive Bony Fish
Teleosts
Dissection Guides for Teleosts
Fisheries and Conservation Issues Concerning Teleosts
Fisheries
Impact of Humans on the Sea; Harvesting

The Early Tetrapods and Modern Amphibians

Amphibia

Chordata

A frog tadpole undergoing metamorphosis.

From Water to Land in Ontogeny and Phylogeny

The chorus of frogs beside a pond in spring announces the start of a new life cycle. Mating frogs produce masses of eggs, which soon hatch into limbless, gill-breathing tadpole larvae that feed and grow. Then a remarkable transformation occurs. Hindlimbs appear and gradually lengthen. The tail shortens and eventually disappears. Larval teeth and gills are lost. Eyelids develop. The forelegs emerge. In a matter of weeks, the aquatic tadpole has completed its metamorphosis to an adult, land-dwelling frog.

By contrast, the evolutionary transition from water to land occurred not in weeks but over millions of years. A lengthy series of alterations cumulatively fitted vertebrate bodies for life on land. The origin of land vertebrates is no less remarkable, and it is unlikely to happen again because well-established competitors would exclude a transitional form.

Amphibians include the only living vertebrates that have a transition from water to land in both their ontogeny (organismal development) and their phylogeny (evolutionary history). Even after some 350 million years of evolution, amphibians remain quasiterrestrial, hovering between aquatic and land environments. Even the amphibians best adapted for a terrestrial existence cannot stray far from moist conditions. Many, however, have developed ways to keep their eggs out of open water where the larvae would encounter enemies.

L ife on land is a major theme of the remaining vertebrate groups, which form a clade called the superclass Tetrapoda (from the Greek term for "four feet"). Amphibians and amniotes (including nonavian reptiles, birds, and mammals) are the two major extant branches of **tetrapod** phylogeny, which originates in the Devonian period. Many tetrapod lineages have lost one or both of the two pairs of limbs that give the group its name. In this chapter, we review the origins of terrestrial vertebrates and discuss the amphibian branch in detail. We discuss the major amniote groups in Chapters 18 through 20.

Movement onto Land

Movement from water to land is perhaps the most dramatic event in animal evolution, because land is physically hazardous. Life originated in water. Animal bodies are mostly composed of water, and all cellular activities occur in water. Nevertheless, organisms invaded land, carrying their watery composition with them. Vascular plants, pulmonate snails, and tracheate arthropods made this transition much earlier than did vertebrates, and thus generated a source of food for terrestrial vertebrates. Although the invasion of land required modification of almost every organ system, aquatic and terrestrial vertebrates retain many structural and functional similarities. Today, the transition between aquatic and terrestrial vertebrates occurs in many living amphibians that make this transition during their life histories.

When moving from water to land, animals encounter important physical differences in environment: (1) oxygen content, (2) fluid density, (3) temperature regulation, and (4) habitat diversity. Oxygen is at least 20 times more abundant in air, and it diffuses much more rapidly through air than through water. Consequently, terrestrial animals can obtain oxygen far more easily than aquatic ones can, once they possess appropriately adapted respiratory structures, such as lungs. Air, however, has approximately 1000 times less buoyant density and is approximately 50 times less viscous than water is. It therefore provides relatively little support against gravity, requiring terrestrial animals to develop strong limbs and to remodel their skeletons to gain adequate structural support. Air fluctuates in temperature more readily than does water, and terrestrial environments therefore experience harsh and unpredictable cycles of freezing, thawing, drying, and flooding. Terrestrial animals require behavioral and physiological strategies to protect themselves from thermal extremes.

Despite its hazards, the terrestrial environment offers a great variety of habitats, including forests (boreal, temperate, and tropical), grasslands, deserts, mountains, oceanic islands, and polar regions. Animals can find safe shelter for vulnerable eggs and young much more readily in many of these terrestrial habitats than in aquatic ones.

Devonian Origin of Tetrapods

By the Devonian period, beginning about 416 million years ago, bony fishes had diversified to include many freshwater forms. An important combination of characteristics that evolved originally in aquatic habitats now gave its possessors some ability to explore terrestrial habitats. These characteristics included two structures that connected to the pharynx: an air-filled cavity, which could be used as a swim bladder, and paired internal nares (evolutionary origin shown on figure 17.1) serving chemoreception. On land, this combination of structures would draw oxygen-rich air through the nares and into the air-filled cavity, whose surface would permit some respiratory gas exchange with body fluids. The bony elements of paired fins, modified for support and movement on underwater surfaces (evolutionary origin shown on figure 17.2), gained sufficient strength to support the body and to walk on land.

The internal nares, air-filled cavity, and paired limbs of an aquatic tetrapod ancestor therefore were available for modification in later evolution of terrestrial breathing and support. The air-filled cavity illustrates the evolutionary principle of exaptation (p. 27), in which a structure that has evolved by natural selection in an initial utility or role is later recruited or "coopted" in a new role. Note that the air-filled cavities are homologous, with *lung* indicating that the structure's primary role is air breathing, as it is in terrestrial forms, and *swim bladder* denoting that the structure serves primarily in providing buoyancy during swimming in aquatic animals. Zoologists continue to debate the controversial question of whether the lung or swim bladder was the original role of the air-filled cavity.

Freshwater habitats are inherently unstable, being prone to evaporation or depletion of the dissolved oxygen needed to support vertebrate life. It is therefore not surprising that multiple fish groups, given a combination of structures that could be coopted for terrestrial breathing and locomotion, evolved some degree of terrestriality. Mudskippers and lungfish are two familiar examples of evolution of terrestriality by fishes; however, only one such transition, occurring in the early Devonian period, provided the ancestral lineage of all tetrapod vertebrates. This lineage ultimately evolved the characteristic tetrapod adaptations to air breathing, including increased vascularization of the air-filled cavity with a rich capillary network in forming a lung, and a **double circulation** directing deoxygenated blood into the lungs for oxygenation, and oxygenated blood from the lungs to other body tissues.

Tetrapods evolved limbs in an ancestral aquatic habitat during the Devonian period prior to their evolutionary movement onto land. Although fish fins at first appear very different from the jointed limbs of tetrapods, an examination of the bony elements of the paired fins of the lobe-finned fishes reveals their homology with amphibian limbs. In *Eusthenopteron,* a Devonian lobe-fin living approximately 385 million years ago, we recognize an upper arm bone (humerus) and two forearm bones (radius and ulna) as well as other elements that we homologize with the wrist bones of tetrapods (figure 17.2). *Eusthenopteron* could push itself through the bottom mud of pools with its fins, but it could not walk upright because backward and forward movement of the fins was limited to about 20 to 25 degrees. The recently discovered fossil genus *Tiktaalik* which lived approximately 375 million years ago, is morphologically intermediate between lobe-finned fishes and tetrapods. *Tiktaalik* probably inhabited shallow,

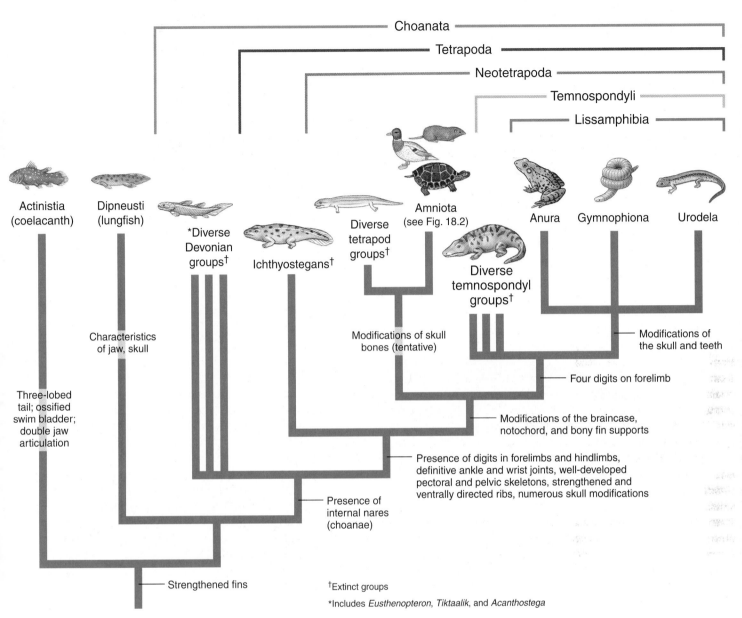

figure 17.1

Tentative cladogram of the Tetrapoda with emphasis on descent of the amphibians. Especially controversial are the relationships of major tetrapod groups (Amniota, Temnospondyli, and diverse early tetrapod groups) and outgroups (Actinistia, Dipneusti, extinct Devonian groups). All aspects of this cladogram are controversial, however, including relationships of the Lissamphibia. Extinct Devonian groups include the fossils *Eusthenopteron*, *Tiktaalik*, and *Acanthostega*.

oxygen-depleted streams or swamps, using its limbs to support the body while placing its snout above water to breathe air. This form also might have traversed land. *Acanthostega,* an early tetrapod that lived approximately 365 million years ago, had well-formed tetrapod limbs with clearly formed digits on both fore- and hindlimbs, but it was clearly an aquatically adapted form whose limbs were too weakly constructed for walking on land. *Ichthyostega,* another fossil genus approximately 365 million years old, had a fully developed shoulder girdle, bulky limb bones, and well-developed muscles; it could pull itself onto land, although it probably did not walk very well. Thus, tetrapods evolved their limbs underwater and only later invaded land.

Evidence points to lobe-finned fishes as the closest relatives of tetrapods; in cladistic terms, they contain the sister group of tetrapods (figure 17.3; see figure 17.1). Both the lobe-finned fishes and early tetrapods such as *Acanthostega* and *Ichthyostega* shared several characteristics of the skull, teeth, and pectoral girdle. *Ichthyostega* (Gr. *ichthys,* fish, + *stegē,* roof or covering, in reference to the roof of the skull, which was shaped like that of a fish) is an early offshoot of tetrapod phylogeny. It possessed several adaptations, in addition to jointed limbs, that equipped it for life on land (see figure 17.2): a stronger backbone and associated muscles that support the body in air, new muscles that elevate the head, strengthened shoulder and hip girdles, a protective rib cage, a more advanced ear

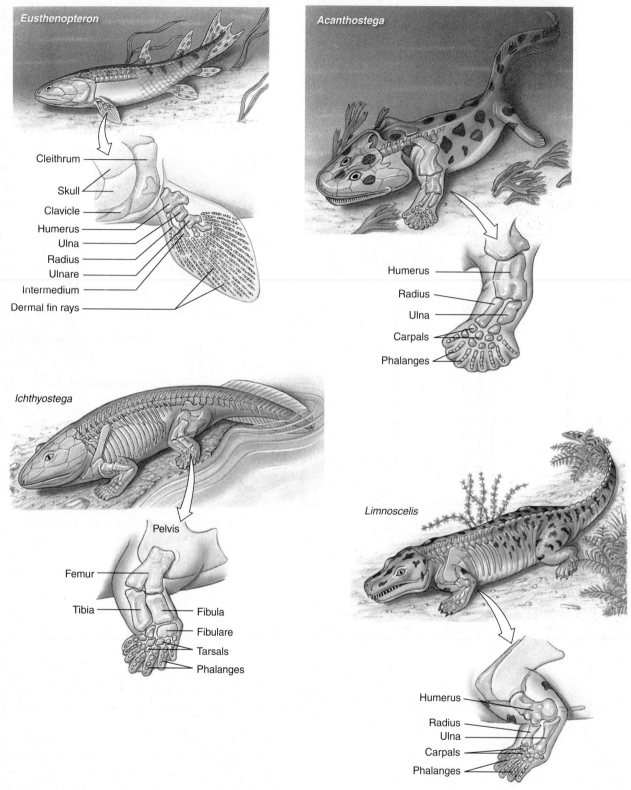

figure 17.2

Evolution of tetrapod limbs. The limbs of tetrapods evolved from the fins of Paleozoic fishes. *Eusthenopteron,* a late Devonian lobe-finned fish, had paired muscular fins supported by bony elements that foreshadowed the bones of tetrapod limbs. The anterior fin contained an upper arm bone (humerus), two forearm bones (radius and ulna), and smaller elements homologous to the wrist bones of tetrapods. As typical of fishes, the pectoral girdle, consisting of the cleithrum, clavicle, and other bones, was firmly attached to the skull. In *Acanthostega,* one of the earliest known Devonian tetrapods (appearing about 365 million years ago), dermal fin rays of the anterior appendage were replaced by eight fully evolved fingers. *Acanthostega* was probably exclusively aquatic because its limbs were too weak for travel on land. *Ichthyostega,* a contemporary of *Acanthostega,* had fully formed tetrapod limbs and must have been able to walk on land. The hindlimb bore seven toes (the number of forelimb digits is unknown). *Limnoscelis,* an anthracosaur of the Carboniferous (about 300 million years ago), had five digits on both fore- and hindlimbs, the basic pentadactyl model that became the tetrapod standard.

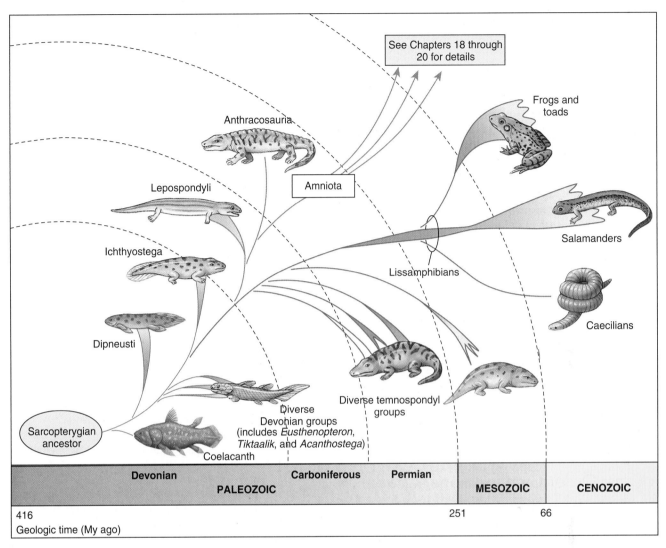

See Chapters 18 through
20 for details

figure 17.3

Early tetrapod evolution and the descent of amphibians. The tetrapods share most recent common ancestry with the extinct Devonian groups shown; of living groups, tetrapods are most closely related to lungfishes. Amphibians share most recent common ancestry with the diverse temnospondyls of the Carboniferous and Permian periods of the Paleozoic era, and the Triassic period of the Mesozoic era.

structure that detects airborne sounds, a foreshortened skull, and a lengthened snout. *Ichthyostega* nonetheless resembled aquatic forms in retaining a tail complete with fin rays and in having opercular (gill-covering) bones.

Bones of *Ichthyostega*, the most thoroughly studied of all early tetrapods, were first discovered on an east Greenland mountainside in 1897 by Swedish scientists looking for three explorers lost two years earlier during an ill-fated attempt to reach the North Pole by hot-air balloon. Later expeditions by Gunnar Säve-Söderberg uncovered skulls of *Ichthyostega*, but Säve-Söderberg died, at age 38, before he could examine the skulls. After Swedish paleontologists returned to the Greenland site, where they found the remainder of *Ichthyostega*'s skeleton, Erik Jarvik, one of Säve-Söderberg's assistants, devoted his life's work to producing the detailed description of *Ichthyostega* available today.

Evolutionary relationships of early tetrapod groups are still controversial. We present a tentative cladogram (see figure 17.1), which almost certainly will be revised as new data are reported. Several extinct lineages plus the **Lissamphibia,** which contains the modern amphibians, are called **temnospondyls** (see figures 17.1 and 17.3). This group generally has only four digits on the forelimb rather than the five digits characteristic of most tetrapods. Lissamphibians diversified during the Carboniferous period to produce ancestors of the three major groups of amphibians alive today: **frogs** (Anura or Salientia), **salamanders** (Caudata or Urodela), and **caecilians** (Apoda or Gymnophiona).

■ Modern Amphibians

The three living amphibian orders comprise more than 6000 species. Most share general adaptations to life on land, including skeletal strengthening. Amphibian larvae and some aquatic adult salamanders use the ancestral lateral-line system for sensory input, but in metamorphosed adults a redesigned olfactory epithelium senses airborne odors and ears detect sounds.

In the ancestral amphibian life history, eggs are aquatic, and they hatch to produce an aquatic larval form that breathes using gills. A metamorphosis follows in which gills are lost. Metamorphosed amphibians use cutaneous respiration on land, and many have lungs, which occur throughout larval life and at metamorphosis become active in breathing air. Gas exchange across the skin also aids air breathing and is the primary means of respiration in many terrestrial salamanders. Many amphibians retain this general pattern, with some important exceptions. Some salamanders do not undergo a complete metamorphosis and retain a permanently aquatic, larval morphology throughout life. Some caecilians, some frogs, and other salamanders live entirely on land and lack the aquatic larval phase completely. Both alternatives are evolutionarily derived conditions.

Even the most terrestrial amphibians remain dependent on very moist environments. Their skin is thin, and it requires moisture for protection against desiccation in air. Amphibians also require moderately cool environments. As ectotherms, their body temperature matches environmental temperatures, greatly restricting where they can live. Cool and wet environments are especially important for reproduction. Eggs must be shed directly into water or onto moist terrestrial surfaces to avoid desiccation.

Caecilians: Order Gymnophiona (Apoda)

The order Gymnophiona (jim´no-fi´o-na) (Gr. *gymnos,* naked, + *ophineos,* of a snake) contains approximately 173 species of elongate, limbless, burrowing creatures commonly called **caecilians** (figure 17.4). They occur in tropical forests of South America (their principal home), Africa, India, and Southeast Asia, and many species are aquatic. They possess a long, slender body, many vertebrae, long ribs, no limbs, and a terminal anus; some have small dermal scales in the skin. Eyes are small, and most species are totally blind as adults. Their food is mostly worms and small invertebrates, which they find underground. Fertilization is internal, and the male has a protrusible copulatory organ. Caecilians usually deposit eggs in moist ground near water. In some species, the female parent carefully guards developing eggs in folds of her body. Some species have aquatic larvae with a tail fin, an open gill slit, and external gills; larval development in other species occurs within the egg. Viviparity is common in some caecilians, with the embryos obtaining nourishment by eating the wall of the oviduct.

Salamanders: Order Urodela (Caudata)

Order Urodela (Gr. *oura,* tail, + *delos,* evident) comprises tailed amphibians, approximately 553 species of salamanders. Salamanders occur in almost all northern temperate regions of the world, and they are abundant and diverse in North America. Salamanders also inhabit tropical areas of Central America and

> ## CHARACTERISTICS
> ### of Modern Amphibians
>
> 1. **Skeleton mostly bony,** with varying numbers of vertebrae; ribs present in some, absent or fused to vertebrae in others
> 2. Body forms vary greatly among species: salamanders usually have distinct head, neck, trunk, and tail; adult frogs have a compressed body with fused head and trunk and no intervening neck; caecilians have an elongated trunk, not strongly demarcated from the head, and a terminal anus
> 3. **Limbs usually four (quadrupedal)** in two pairs with associated shoulder/hip girdle, although some forms have a single pair of limbs and others no limbs; webbed feet often present; no true nails; **forelimb usually has four digits** but sometimes five and sometimes fewer
> 4. Free-living in freshwater or moist terrestrial habitats
> 5. Ectothermic (body temperature dependent upon environmental temperature and not modulated by metabolically generated heat)
> 6. **Skin smooth, moist, and glandular;** integument modified for **cutaneous respiration;** pigment cells (chromatophores) common and of considerable variety; **granular glands** secrete defensive compounds
> 7. Mouth usually large, with small teeth in upper or both jaws; paired **internal nostrils** open into a nasal cavity lined with **olfactory epithelium** at anterior part of mouth cavity and enable breathing (in lung-breathing forms)
> 8. For vision in air, cornea rather than lens is principal refractive surface for bending light; **eyelids** and **lachrymal glands** protect and wash eyes
> 9. Ten pairs of cranial nerves
> 10. Ear with **tympanic membrane** (eardrum) and **stapes** (columella) for transmitting vibrations to inner ear
> 11. Separate sexes; fertilization mostly external in frogs and toads but internal via a spermatophore in most salamanders and caecilians; predominantly oviparous, some ovoviviparous or viviparous; metamorphosis usually present; **moderately yolky eggs** (mesolecithal) **with jellylike membrane coverings**
> 12. Excretory system consists of paired mesonephric or opisthonephric kidneys; urea is main nitrogenous waste
> 13. Respiration by skin and in some forms by gills and/or lungs; presence of gills and lungs varies among species and by developmental stage of some species; aquatic larvae lose gills at metamorphosis if a metamorphosis occurs; many salamanders retain gills and an aquatic existence throughout life
> 14. **Heart with two atria,** one ventricle, a conus arteriosus, and a sinus venosus; **double circulation** through the heart in which pulmonary arteries and veins supply the lungs (when present) and return oxygenated blood to the heart; skin is abundantly supplied with blood vessels

northern South America. Salamanders are typically small; most of the common North American salamanders are less than 15 cm long. Some aquatic forms are considerably longer, and Japanese giant salamanders sometimes exceed 1.5 m in length.

Most salamanders have limbs set at right angles to the trunk, with forelimbs and hindlimbs of approximately equal

A

B

figure 17.4

A, Female caecilian coiled around eggs in a burrow. **B,** Pink-head caecilian (*Herpele multiplicata*), native to western Africa.

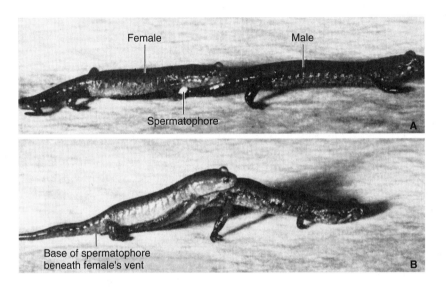

Female

Male

Spermatophore

A

Base of spermatophore beneath female's vent

B

figure 17.5

Courtship and sperm transfer in the pygmy salamander, *Desmognathus wrighti*. After judging the female's receptivity by the presence of her chin on his tail base, the male deposits a spermatophore on the ground, and then moves forward a few paces. **A,** The white mass of the sperm atop a gelatinous base is visible at the level of the female's forelimb. The male moves ahead, the female following until the spermatophore is at the level of her vent. **B,** The female has recovered the sperm mass in her vent, while the male arches his tail, tilting the female upward and presumably facilitating recovery of the sperm mass. The female later uses sperm stored in her body to fertilize eggs internally before laying them.

size. In some aquatic and burrowing forms, limbs are rudimentary or absent.

Salamanders are carnivorous both as larvae and as adults, preying on worms, small arthropods, and small molluscs. Like all amphibians, they are **ectothermic** and have a low metabolic rate.

Breeding Behavior

Some salamanders are either aquatic or terrestrial throughout their life cycle, but the ancestral life cycle is metamorphic, having aquatic larvae and terrestrial adults that live in moist places under stones and rotten logs. Eggs of most salamanders are fertilized internally; a female recovers in her vent (common opening for anus and reproductive system) a packet of sperm (**spermatophore**) deposited by a male on a leaf or stick (figure 17.5). Aquatic species lay eggs in clusters or stringy masses in water. Their eggs hatch to produce an aquatic larva having external gills and a finlike tail. Completely terrestrial species deposit eggs in small, grapelike clusters under logs or in excavations in soft moist earth, and many species guard their eggs (figure 17.6). Terrestrial species undergo **direct development.** They bypass the larval stage and hatch as miniature versions of their parents. A particularly complex life cycle occurs

figure 17.6

Female dusky salamander (*Desmognathus* sp.) attending eggs. Many salamanders care for their eggs, which includes rotating the eggs and protecting them from fungal infections and predation by various arthropods and other salamanders.

in some American newts, whose aquatic larvae metamorphose to form terrestrial juveniles that later metamorphose again to produce secondarily aquatic, breeding adults (figure 17.7).

Respiration

At various stages of their life history, salamanders may have external gills, lungs, both, or neither of these structures. They

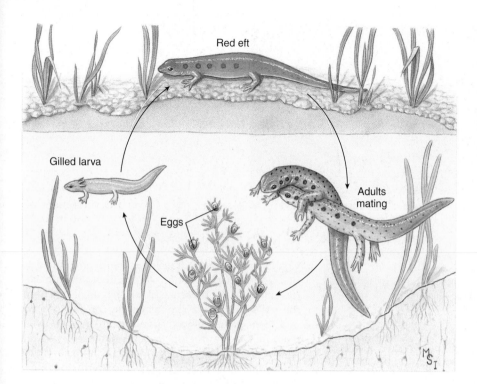

figure 17.7

Life history of the red-spotted newt, *Notophthalmus viridescens* of the family Salamandridae. In many habitats, the aquatic larva metamorphoses into a brightly colored "red eft" stage, which remains on land from one to three years before transforming into a secondarily aquatic adult.

Red eft

Gilled larva

Eggs

Adults mating

also share the general amphibian condition of having extensive vascular nets in their skin for respiratory exchange of oxygen and carbon dioxide with the external environment (called "cutaneous respiration"). Salamanders that have an aquatic larval stage hatch with gills, but lose them later if a metamorphosis occurs. Several diverse lineages of salamanders have evolved permanently aquatic forms that fail to complete metamorphosis and retain their gills and finlike tail throughout life (figure 17.8). Lungs, the most widespread respiratory organ of terrestrial vertebrates, are present from birth in the salamanders that have them, and they become the primary means of respiration following metamorphosis. Amphiumas, while having a completely aquatic life history, nonetheless lose their gills before adulthood and then breathe primarily by lungs, raising their nostrils above the water surface to get air.

In contrast to amphiumas, all species of the large family Plethodontidae (see figures 17.5, 17.6, and 17.9) are lungless, and many of these species are entirely terrestrial. Efficiency of cutaneous respiration is increased by penetration of a capillary network into the epidermis or by thinning of the epidermis over superficial dermal capillaries. Cutaneous respiration is supplemented by air pumped into the mouth, where respiratory gases are exchanged across the vascularized membranes of the buccal (mouth) cavity (buccopharyngeal breathing). Lungless plethodontids probably originated in streams, where lungs would have been disadvantageous by providing excess buoyancy, and where water is so cool and well oxygenated that cutaneous respiration suffices. It is odd that the most completely terrestrial group of salamanders is one that lacks lungs.

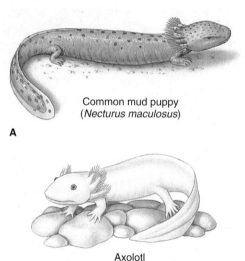

Common mud puppy
(*Necturus maculosus*)

A

Axolotl
(*Ambystoma mexicanum*)

B

figure 17.8

Permanently-gilled, aquatic salamanders. **A,** The mud puppy (*Necturus* sp.) and **B,** the axolotl (*Ambystoma mexicanum*). Some other species of *Ambystoma* are facultatively metamorphic; they may remain permanently gilled or, should their pond habitat evaporate, metamorphose to a terrestrial form that loses its gills and breathes by lungs. The axolotl shown is an albino form commonly used in laboratory experiments but uncommon in natural populations. Evolutionary change from a metamorphosing ancestor to a nonmetamorphosing, permanently gilled descendant illustrates paedomorphosis (juvenile ancestral characters retained by adult descendants).

Paedomorphosis

Whereas most salamanders complete their development by metamorphosis to the adult body form, some species reach sexual maturity while retaining their gills, aquatic lifestyle, and other larval characteristics. This condition illustrates **paedomorphosis** (Gr. "child form"), defined as evolution of an adult form that resembles an ancestral juvenile. Some characteristics of an ancestral adult morphology are consequently eliminated. Examples of such nonmetamorphic, permanently gilled species are mud puppies of the genus *Necturus* (see figure 17.8A), which inhabit the muddy substrates of ponds and lakes, and the **axolotl** of Mexico (see figure 17.8B). These species never metamorphose under any conditions.

Other species of salamanders that reach sexual maturity with larval morphology metamorphose to terrestrial forms under certain environmental conditions. We find good examples in *Ambystoma tigrinum* and related species from North America. They typically inhabit small ponds that can disappear through evaporation in dry weather. When ponds evaporate, the aquatic form metamorphoses to a terrestrial form, losing its gills and developing lungs. It then can traverse land to find new sources of water in which to live and to reproduce.

Paedomorphosis is an important means of evolutionary diversification, even in salamanders that lack an aquatic larval stage. For example, paedomorphosis enables some species of the tropical plethodontid genus *Bolitoglossa* to climb forest vegetation. The highly webbed feet of *Bolitoglossa rufescens* represent paedomorphic evolution in which development of the digits is greatly reduced (figure 17.10B), making the foot an adhesive surface that attaches the animal to banana trees and other smooth leaves and stems. The ancestral structure has greater digital development (figure 17.10A and C) and occurs in related species that live mainly on the ground.

figure 17.9

Longtail salamander, *Eurycea longicauda*, a common plethodontid salamander.

Frogs and Toads: Order Anura (Salientia)

The approximately 5283 species of frogs and toads that form the order Anura (Gr. *an,* without, + *oura,* tail) are for most people the most familiar amphibians. The Anura are an old group, known from the Triassic period, 200–251 million years ago. Frogs and toads occupy a great variety of habitats, despite their aquatic mode of reproduction and water-permeable skin, which prevent them from wandering too far from sources of water, and their ectothermy, which bars them from polar and subarctic habitats. The name of the order, Anura, denotes an obvious group characteristic, the absence of tails in adults (although all pass through a tailed stage during embryonic development and often as tadpole larvae). Frogs and toads are specialized for jumping, as suggested by the alternative order name, Salientia, which means leaping.

figure 17.10

Foot structure of representatives of three different species of the tropical plethodontid salamander genus *Bolitoglossa*. These specimens have been treated chemically to clear the skin and muscles and to stain the bone red/pink and the cartilage blue. The species having the most fully ossified and distinct digits (**A, C**) live primarily on the forest floor. The species having the padlike foot caused by restricted digital growth (**B**) climbs smooth leaves and stems using the foot surface to produce suction or adhesion for attachment. The padlike foot evolved by paedomorphosis; it was derived evolutionarily by truncating development of the body, which prevents full digital development.

A

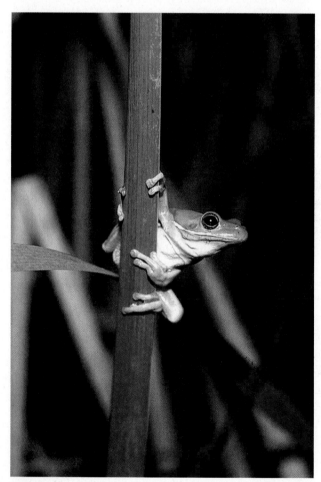

B

figure 17.11
Two common North American frogs. **A,** Bullfrog, *Litho-bates catesbeianus,* largest American frog and mainstay of the frog-leg epicurean market (family Ranidae). **B,** Green tree frog, *Hyla cinerea,* a common inhabitant of swamps of the southeastern United States (family Hylidae). Note the adhesive pads on the feet.

figure 17.12
American toad, *Bufo americanus* of the family Bufonidae. This principally nocturnal yet familiar amphibian feeds on large numbers of insect pests as well as on snails and earthworms. The warty skin contains numerous glands that produce a surprisingly poisonous milky fluid, giving the toad excellent protection from a variety of potential predators.

We see in the appearance and life habit of their larvae further distinctions between Anura and Urodela. Eggs of most frogs hatch into a tadpole having a long, finned tail, both internal and external gills, no legs, specialized mouthparts for herbivorous feeding (some tadpoles and all salamander larvae are carnivorous), and a highly specialized internal anatomy. They bear little resemblance to adult frogs. Metamorphosis of a frog tadpole to an adult frog is thus a striking transformation. The permanently gilled larval condition never occurs in frogs and toads as it does in salamanders.

Frogs and toads are divided into 44 families. The best-known frog families in North America are Ranidae, which contains most of our familiar frogs (figure 17.11A), and Hylidae, the tree frogs (figure 17.11B). True toads, belonging to family Bufonidae, have short legs, stout bodies, and thick skins usually with prominent warts (figure 17.12). However, the term "toad" sometimes denotes thick-skinned, terrestrial members of several other families.

The largest anuran is the West African *Conraua goliath,* which is more than 30 cm long from tip of nose to anus (figure 17.13). This giant eats animals as big as rats and ducks. The smallest frogs recorded are *Eleutherodactylus iberia* and *Psyllophryne didactyla,* measuring less than 1 cm in length; they are also the smallest known tetrapods. These tiny frogs, which can be more than covered by a dime, live respectively in Cuba and in the Brazilian rain forest. The largest American frog is the bullfrog, *Lithobates catesbeianus* (see figure 17.11A), which reaches a head and body length of 20 cm.

Habitats and Distribution

Probably the most familiar frogs are species of the ranid genera *Lithobates, Pelophylax* and *Rana* (Gr. frog), which collectively

figure 17.13

Conraua goliath (family Petropedetidae) of West Africa, the world's largest frog. This specimen weighed 3.3 kg (approximately 7-1/2 pounds).

occupy most temperate and tropical regions of the world. They usually live near water, although some, such as the wood frog, *L. sylvaticus,* spend most of their time on damp forest floors. The larger bullfrogs, *L. catesbeianus,* and the green frogs, *L. clamitans,* are nearly always in or near permanent water or swamps. Leopard frogs, *L. pipiens* and related species, occur in nearly every state and Canadian province and are the most widespread of all North American frogs. The northern leopard frog, *L. pipiens,* is the species most commonly used in biology laboratories and for classical electrophysiological research.

Frog species are often patchy in distribution, being restricted to certain streams or pools and absent or scarce in similar habitats elsewhere. Pickerel frogs (*L. palustris*) are especially noteworthy in this respect. Many populations of frogs worldwide may be declining in numbers and becoming even more geographically fragmented than usual. Causes of decline include high ultraviolet radiation and infections with larval trematodes and fungi.

Most larger frogs are solitary except during breeding season. During breeding periods, most of them, especially males, are very noisy. Each male usually occupies a particular perch near water, where he may remain for hours or even days, trying to attract a female to that spot. At times, frogs are mainly silent, and their presence is not detected until they are disturbed. When they enter the water, they dart swiftly to the bottom of the pool, where they kick the substrate to conceal themselves in a cloud of muddy water. In swimming, they hold the forelimbs near the body and kick backward with their webbed hindlimbs, which propel them forward. When they surface to breathe, only the head and foreparts are exposed, and they conceal themselves behind any available vegetation.

Amphibian populations have declined in various parts of the world, although many species thrive. No single explanation fits all declines, although loss of habitat predominates. In some populations, changes are simply random fluctuations caused by episodic droughts. Frog and toad eggs exposed on the surfaces of ponds are especially sensitive to damaging ultraviolet radiation. Climatic changes that reduce water depth at oviposition sites increase the ultraviolet exposure of embryos and make them more susceptible to fungal infection. Declines in population sometimes are accompanied by an increased incidence of malformed individuals, such as frogs with extra limbs. Malformed limbs are often associated with infection by trematodes (see p. 161).

Declines in some amphibian populations are caused by other amphibians. For example, an exotic frog introduced into southern California thrives in its new American home. African clawed frogs, *Xenopus laevis* (figure 17.14), are voracious, aggressive, primarily aquatic frogs that rapidly displace native frogs and fishes from waterways. This species was introduced into North America in the 1940s when it was used extensively in human pregnancy tests. Some hospitals dumped surplus frogs into nearby streams, where these prolific breeders flourished and are now considered pests. Similar results occurred when giant toads, *Bufo marinus* (to 23 cm in length), were introduced to Queensland, Australia to control agricultural pests. They rapidly spread, producing numerous ecological changes, including displacement of native anurans.

During winter months, most frogs hibernate in the oxygen-rich water of pools and streams. Their life processes reach a very low ebb during hibernation, sustained by diffusion of oxygen across the skin and oxidation of glycogen and fat stored in their bodies during the spring and summer. More terrestrial frogs, such as tree frogs, hibernate in humus of the forest floor. They tolerate low temperatures, and many actually

figure 17.14

African clawed frog, *Xenopus laevis*. The claws, an unusual feature in frogs, are on the hind feet. This frog has been introduced into California, where it is considered a serious pest.

Taxonomy of Class Amphibia

Order Gymnophiona (jim´no-fī´ō-na) (Gr. *gymnos,* naked, + *ophioneos,* of a snake) **(Apoda): caecilians.** Body elongate; limbs and limb girdle absent; mesodermal scales present in skin of some; tail short or absent; 95 to 285 vertebrae; pantropical, 3 families, 33 genera, approximately 173 species.

Order Urodela (ūr´uh-dēl´ə) (Gr. *oura,* tail, + *delos,* evident) **(Caudata): salamanders.** Body with head, trunk, and tail; no scales; usually two pairs of equal limbs; 10 to 60 vertebrae; predominantly Holarctic; 9 living families, 64 genera, approximately 553 species.

Order Anura (uh-nūr´ə) (Gr. *an,* without, + *oura,* tail) **(Salientia): frogs** and **toads.** Head and trunk fused; no tail; no scales; two pairs of limbs; large mouth; lungs; 6 to 10 vertebrae including urostyle (coccyx); cosmopolitan, predominantly tropical; 44 families; 362 genera; approximately 5283 species.

survive prolonged freezing of all extracellular fluid, representing 35% of their body water. Such frost-tolerant frogs prepare for winter by accumulating glucose and glycerol in their body fluids, which protects tissues from ice-crystal formation.

Adult frogs have numerous enemies, including snakes, aquatic birds, turtles, raccoons, and humans; fish prey upon tadpoles, only a few of which survive to maturity. Some adult frogs defend themselves by feigning death. Most anurans can inflate their lungs to make themselves difficult for a predator to swallow. When disturbed along the margin of a pond or brook, a frog often remains quite still. When it senses danger, it jumps, not always into the water where enemies may lurk, but into grassy cover on the bank. When held, a frog might cease its struggles for an instant to put its captor off guard and then suddenly leap, voiding its urine. A frog's best protection is its ability to leap and, in some species, to use poison glands. Many frogs and toads in the tropics and subtropics are aggressive, jumping and biting at predators. Bullfrogs in captivity do not hesitate to snap at tormenters and can inflict painful bites.

Life History

Because frogs and toads are ectothermic, they breed, feed, and grow only during warm seasons, often in a predictable annual cycle. With warming spring temperatures and rainfall, males call vociferously to attract females. After a brief courtship, mated pairs enter the water where the male clasps the female in a process called **amplexus** (figure 17.15), during which eggs are fertilized externally (after leaving the female's body). As the female lays eggs, the male discharges seminal fluid containing sperm over the eggs to fertilize them. After fertilization, the jelly layers absorb water and swell (figure 17.16). Eggs are laid in large masses, often anchored to vegetation, and then abandoned by the parents. The eggs begin development immediately. Within a few days, the embryos have developed into tiny tadpoles visible through the translucent jelly layers surrounding them (see figure 17.16). The embryonic tadpoles grow and develop for several weeks before hatching from their eggs to become free-living. If the eggs were laid in a temporary pond or puddle, tadpoles race against time to complete development before the habitat dries.

At hatching, a tadpole has a distinct head and body with a compressed tail. The mouth is on the ventral side of the head

figure 17.15

A male green frog, *Hyla cinerea,* clasps a larger female during the breeding season in a South Carolina swamp. Clasping (amplexus) lasts until the female deposits her eggs, which are fertilized externally (outside the body). Like most tree frogs, these can make rapid and marked color changes; the male here, normally green, has darkened during amplexus.

and has keratinized jaws for scraping vegetation from objects for food. Behind the mouth is a ventral adhesive disc for clinging to objects. In front of the mouth are two deep pits, which later become nostrils. Swellings occur on each side of the head, and these later become external gills. There are three pairs of external gills, which later transform into internal gills that become covered with a flap of skin (the operculum) on each side. On the right side, the operculum completely fuses with the body wall, but on the left side a small opening, the **spiracle** (L. *spiraculum,* airhole), permits water to exit after entering the mouth and passing the internal gills. The hindlimbs appear first, whereas the forelimbs are hidden for a time by folds of the operculum. The tail is absorbed, the intestine becomes much shorter, the mouth undergoes a transformation into the adult condition, lungs develop, and the gills are absorbed. Leopard frogs usually complete metamorphosis

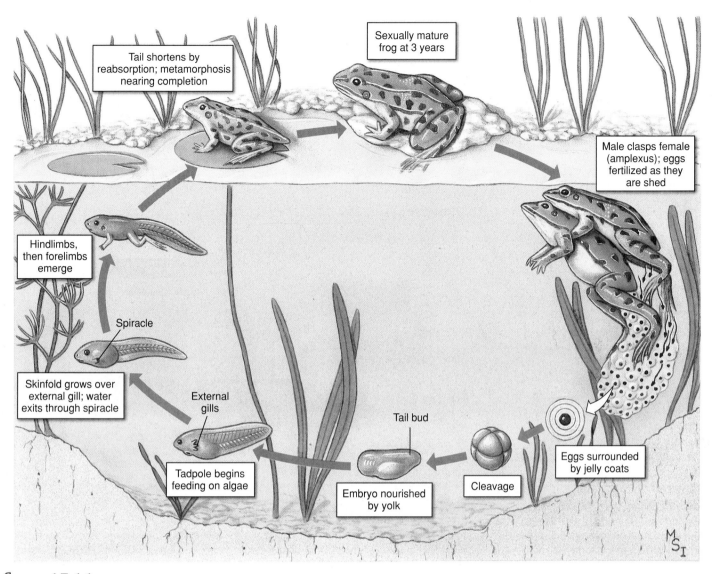

figure 17.16

Life cycle of a leopard frog. Cleavage and embryonic stages occur within the jelly-coated eggs within the egg masses as shown at the right side of the figure. The cleavage and embryonic stages are shown free and disproportionately large to reveal structural detail.

within three months; a bullfrog takes much longer to complete this process.

The life cycle just described is typical of most temperate-zone anurans but is only one of many alternative patterns in tropical anurans. Some remarkable reproductive strategies of tropical anurans are illustrated in figure 17.17. Some species lay their eggs in foam masses that float on the surface of the water; some deposit their eggs on leaves overhanging ponds and streams into which the emerging tadpoles drop; some lay eggs in damp burrows; and others place their eggs in water trapped in tree cavities or in water-filled chambers of some bromeliads (epiphytic plants in the tropical forest canopy).

While most frogs abandon their eggs, some tend their eggs. Marsupial frogs carry their developing eggs in a pouch on the back (figure 17.17A). In Surinam frogs (figure 17.17B), the male and female do backward somersaults during mating, and eggs and sperm slide into the space between the mating pair. The male presses the fertilized eggs into the female's back, which develops a spongy incubating layer that eventually sloughs once young are hatched. Hatchling poison-dart frogs squirm onto the parent's back to be carried for varying lengths of time (figure 17.16C). Tadpoles of Darwin's frog develop into froglets in the protection of their father's vocal pouch (figure 17.7D), whereas Australian gastric-brooding frogs develop in their mother's stomach.

Although most frogs develop through a larval stage (the tadpole), many tropical frogs have evolved direct development. In direct development, no free-living tadpole stage occurs, and the froglet that emerges is a miniature replica of the adult. In the species-rich tropical genus *Eleutherodactylus*, mating occurs on land, and eggs hatch directly into froglets; the aquatic larval stage is eliminated, freeing these frogs from an obligatory association with pools or streams. A Puerto Rican species, *E. jasperi*, has evolved internal fertilization and live birth of offspring.

figure 17.17

Unusual reproductive strategies of anurans. **A,** A female South American pygmy marsupial frog (*Flectonotus pygmaeus*) carries developing larvae in a dorsal pouch. **B,** A female Surinam frog carries eggs embedded in specialized brooding pouches on the dorsum; froglets emerge and swim away when development is complete. **C,** A male poison-dart frog, *Phyllobates bicolor,* carries tadpoles adhering to its back. **D,** Tadpoles of Darwin's frog, *Rhinoderma darwinii,* develop into froglets in their male parent's vocal pouch. When the froglets are ready to emerge, their father opens his mouth to let them escape.

■ Summary

Amphibians are ectothermic, primitively quadrupedal vertebrates that have glandular skin and breathe by lungs, by gills, and/or through their skin. They form one of two major phylogenetic branches of living tetrapods, the other one being the amniotes. Living amphibians comprise three major evolutionary groups. Caecilians (order Gymnophiona) are a small tropical group of limbless, elongate forms. Salamanders (order Urodela, or Caudata) are tailed amphibians that have retained the generalized four-legged body plan of their Paleozoic ancestors. Frogs and toads (order Anura) are the largest group of modern amphibians; adults are specialized for a jumping mode of locomotion on land or in water.

Most amphibians have a biphasic life cycle that begins with an aquatic larva followed by metamorphosis to a terrestrial adult that returns to water to lay eggs. Some frogs, salamanders, and caecilians have evolved direct development that omits the aquatic larval stage, and some caecilians have evolved viviparity. Salamanders are unique among amphibians in having evolved several permanently gilled species that retain a larval morphology throughout life, eliminating the terrestrial phase completely. The permanently gilled condition is obligate in some species, but others metamorphose to a terrestrial form only if the pond habitat dries.

Despite terrestrial life, the adults and eggs of all amphibians require cool, moist environments if not actual pools or streams. Eggs and adult skin have no effective protection against very cold, hot, or dry conditions, greatly restricting the adaptive diversification of amphibians to environments with moderate temperatures and abundant water.

■ Review Questions

1. How did the characteristic differences between aquatic and terrestrial environments influence the early evolution of tetrapods?
2. Describe the different modes of respiration used by amphibians. What paradox do the amphiumas and terrestrial plethodontids present regarding the association of lungs with life on land?
3. Evolution of the tetrapod limb was one of the most important advances in vertebrate history. Describe the inferred sequence of its evolution.
4. Compare the general life-history patterns of salamanders with those of frogs. Which group shows a greater variety of evolutionary changes from the ancestral biphasic amphibian life cycle?
5. Give the literal meaning of the name Gymnophiona. What animals are in

this amphibian order? Describe their appearance and habitats.

6. What are the literal meanings of the order names Urodela and Anura? What major features distinguish members of these two orders from each other?

7. Describe the breeding behavior of a typical woodland salamander.

8. How is paedomorphosis important to evolutionary diversification of salamanders?

9. Briefly describe the reproductive behavior of frogs. In what important ways do frogs and salamanders differ in their reproduction?

For Further Thought On the nineteenth-century notion of a "scale of nature," living amphibians were considered remnants of archaic terrestrial vertebrates largely superseded by "higher" forms including birds and

mammals. Nonetheless, amphibian species are often more abundant and of longer evolutionary duration than avian or mammalian species. In what ways are amphibians unusually well suited for evolutionary persistence?

■ Selected References

See also general references on page 432.

Clack, J. A. 2002. Gaining ground: the origin and evolution of tetrapods. Bloomington, Indiana, Indiana University Press. *An authoritative account of paleontological evidence regarding tetrapod origins.*

Conant, R., and J. T. Collins. 1998. A field guide to reptiles and amphibians: Eastern and Central North America. The Peterson field guide series. Boston, Houghton Mifflin Company. *Updated version of a popular field guide; color illustrations and distribution maps for all* species.

Daeschler, E. B., N. H. Shubin, and F. A. Jenkins, Jr. 2006. A Devonian tetrapod-like fish and the evolution of the tetrapod body plan. Nature **440:**757–763. *Describes* Tiktaalik, *a fossil intermediate between lobefinned fishes and tetrapods.*

Duellman, W. E., and L. R. Trueb. 1994. Biology of amphibians. Baltimore, Johns Hopkins University Press. *Important comprehensive sourcebook of information on amphibians, extensively referenced and illustrated.*

Frost, D. R., T. Grant, J. Faivovich, R. H. Bain, A. Haas, C. F. B. Haddad, R. O. De Sá, A. Channing, M. Wilkinson, S. C. Donnellan, C. J. Raxworthy, J. A. Campbell, B. L. Blotto, P. Moler, R. C. Drewes, R. A. Nussbaum,

J. D. Lynch, D. M. Green, and W. C. Wheeler. 2006. The amphibian tree of life. Bull. Am. Mus. Nat. Hist. **297:**1–370. *A phylogeny for living amphibians derived from a large compilation of morphological and molecular characters.*

Halliday, T. R., and K. Adler, eds. 2002. Firefly encyclopedia of reptiles and amphibians. Toronto, Canada, Firefly Books. *Excellent authoritative reference work with high-quality illustrations.*

Heatwole, H., ed. 1995–2009. Amphibian biology. Baulkham Hills, Australia. Surrey Beatty and Sons. *The first eight volumes of a continuing series covering amphibian biology in great detail.*

Jamieson, B. G. M. (ed.). 2003. Reproductive biology and phylogeny of Anura. Enfield, New Hampshire, Science Publishers, Inc. *Provides detailed coverage of reproductive biology and early evolutionary diversification of frogs and toads.*

Lannoo, M. (ed.). 2005. Amphibian declines: the conservation status of United States species. Berkeley, California, University of California Press. *A survey of conservation status of American amphibians.*

Lewis, S. 1989. Cane toads: an unnatural history. New York, Dolphin/Doubleday. *Based on an amusing and informative film of*

the same title that describes the introduction of cane toads to Queensland, Australia, and the unexpected consequences of their population explosion there. "If Monty Python teamed up with National Geographic, *the result would be* Cane Toads."

Savage, J. M. 2002. The amphibians and reptiles of Costa Rica. Chicago, University of Chicago Press. *Costa Rica hosts diverse caecilians, frogs, and salamanders. The Organization for Tropical Studies offers students a chance to study this amphibian fauna.*

Sever, D. M. (ed.). 2003. Reproductive biology and phylogeny of Urodela (Amphibia). Enfield, New Hampshire, Science Publishers, Inc. *A thorough review of reproductive biology and evolutionary relationships among salamanders.*

Stebbins, R. C., and N. W. Cohen. 1995. A natural history of amphibians. Princeton, New Jersey, Princeton University Press. *World-wide treatment of amphibian biology, emphasizing physiological adaptations, ecology, reproduction, behavior, and a concluding chapter on amphibian declines.*

■ Custom Website

The *Animal Diversity* Online Learning Center is a great place to check your understanding of chapter material. Visit www.mhhe.com/hickmanad6e for access to key terms, quizzes, and more! Further enhance your knowledge with Web links to chapter-related material.

Explore live links for these topics:

Classification and Phylogeny of Animals
Class Amphibia
Order Gymnophiona

Order Urodela
Order Salientia
Dissection Guides for Amphibians
Conservation Issues Concerning Amphibians

18

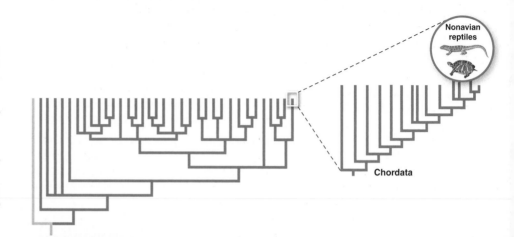

Amniote Origins and Nonavian Reptiles

Chordata

Nonavian reptiles

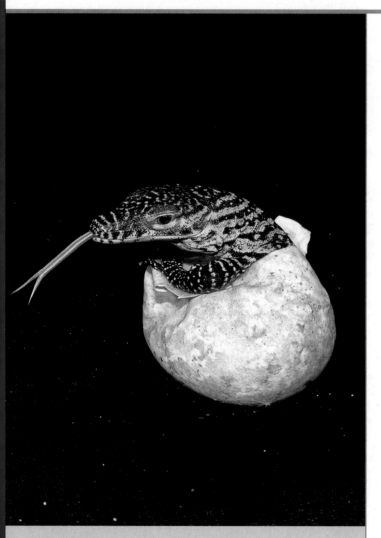

Hatching Komodo dragon, *Varanus komodoensis*.

Enclosing the Pond

Amphibians, with their well-developed limbs, redesigned sensory and respiratory systems, and modifications of the postcranial skeleton for supporting the body in air, have made a notable conquest of land. However, their shell-less eggs, thin, moist skin, and often gilled larvae keep their development hazardously tied to water. An ancestor of a clade containing nonavian reptiles, birds, and mammals evolved an egg better adapted for dry terrestrial conditions. This shelled egg, perhaps more than any other adaptation, unshackled early reptiles from the aquatic environment by freeing the developmental process from dependence on aquatic or very moist terrestrial environments. In fact, the "pond-dwelling" stages were not eliminated but enclosed within a series of extraembryonic membranes that provided complete support for embryonic development. One membrane, the amnion, encloses a fluid-filled cavity, the "pond," which suspends the developing embryo. Another membranous sac, the allantois, serves both as a respiratory surface and as a chamber for storing nitrogenous wastes. Enclosing these membranes is a third membrane, the chorion, through which oxygen and carbon dioxide freely pass. Finally, surrounding and protecting everything is a porous shell.

With the last ties to aquatic reproduction severed, conquest of land by vertebrates was ensured. Paleozoic tetrapods that developed this reproductive pattern and their descendants were called Amniota, named after the innermost of the three extraembryonic membranes, the amnion. Before the end of the Paleozoic era, amniotes diversified into multiple lineages that gave rise to all nonavian reptiles, birds, and mammals.

The paraphyletic class Reptilia (rep-til´ē-ə) (L. *repto,* to creep) includes nearly 8000 species (approximately 340 species in the United States and Canada) occupying a great variety of aquatic and terrestrial habitats, in many of which they are diverse and abundant. Nevertheless, reptiles are perhaps remembered best for what they once were, rather than for what they are now. The "age of reptiles," which lasted for more than 165 million years, saw the diversification of reptilian lineages into a bewildering array of terrestrial and aquatic forms. Among these were herbivorous and carnivorous dinosaurs, many of huge stature and awesome appearance, that dominated animal life on land. Then, during a mass extinction at the end of the Mesozoic era, many reptilian lineages became extinct. Among the lineages to emerge from the Mesozoic extinction are those of today's reptiles. One of these lineages includes the two species of tuataras (*Sphenodon*) of New Zealand, sole survivors of a group that otherwise disappeared 100 million years ago. Their closest living relatives, lizards and snakes, diversified greatly following the Mesozoic era (figure 18.1). In morphology, physiology, and behavior, some reptiles, especially lizards, probably are more similar to the first amniotes than are any other living vertebrates. In the next section, we discuss the origin of amniotes, their diversification into various groups, and their adaptations for life on dry land.

Origin and Early Evolution of Amniotes

As mentioned in the prologue to this chapter, amniotes are a monophyletic group that appeared and diversified in the late Paleozoic era. Most zoologists agree that amniotes are most closely related to anthracosaurs, a group of **anamniotes** (vertebrates lacking an amnion) of the early Carboniferous period. Early diversification of amniotes produced three patterns of holes (fenestrae) in the temporal region of the skull. **Anapsid** (Gr. *an,* without, + *apsis,* arch) skulls have no openings in the temporal area of the skull behind the orbit (opening in the skull for the eye); thus, the temporal region of the skull is completely roofed by dermal bones (figure 18.2). This skull morphology occurs in the earliest amniotes, and in one living group, the turtles, although the anapsid condition in turtles likely evolved secondarily, from ancestors having temporal fenestrae. Two other amniote clades, Diapsida and Synapsida, represent separate evolutionary derivations from the ancestral anapsid condition.

A **diapsid** (Gr. *di,* double, + *apsis,* arch) skull has two temporal openings: one pair located low over the cheeks, and a second pair positioned above the lower pair, in the roof of the skull, and separated from the first by a bony arch (figure 18.2). Diapsid skulls characterize birds and all amniotes traditionally considered "reptiles," except turtles (see figure 18.1). In many living diapsids (lizards, snakes, and birds), one or both of the bony arches and openings have been lost, perhaps to facilitate skull kinesis (see figure 18.9). The earliest diapsids gave rise to five morphologically distinct clades. The lepidosaurs include lizards, snakes, and tuataras. The **archosaurs** include dinosaurs, pterosaurs, crocodilians, and birds. A third, smaller clade, the **sauropterygians,** includes several extinct aquatic groups, the most conspicuous of which are the large, long-necked plesiosaurs (see figure 18.1). **Ichthyosaurs,** represented by extinct, aquatic, dolphinlike forms (see figure 18.1), form a fourth clade of diapsids. Placement of the last clade, the turtles, with diapsids is controversial, although we treat turtles as highly modified members of the clade Diapsida. Turtle skulls lack temporal fenestrae and have been hypothesized to be the only living descendants of parareptiles, an early anapsid group. However, other morphological and genetic evidence published over the past 15 years places turtles within the diapsid clade, suggesting the two pairs of temporal fenestrae characteristic of diapsids were lost early in turtle evolution.

The third skull fenestration condition is **synapsid** (Gr. *syn,* together, + *apsis,* arch), characterized by a single pair of temporal openings located low on the cheeks and bordered by a bony arch (figure 18.2). The synapsid condition occurs in a clade that includes mammals and their extinct relatives, the therapsids and pelycosaurs (see figure 18.1).

What was the functional significance of the temporal openings in early amniotes? In living forms, these openings are occupied by large muscles that elevate (adduct) the lower jaw. Changes in jaw musculature might reflect a shift from suction feeding in aquatic vertebrates (p. 338) to a terrestrial feeding method that required larger muscles to produce more static pressure, for such activities as nipping plant material with the anterior teeth or grinding food with the posterior teeth.

Adaptations of Amniotes

Derived characters of amniotes include an amniotic egg, rib ventilation, and desiccation-resistant skin, in addition to skeletal characteristics of the head, shoulder, and ankle. These and other morphological and physiological characters allowed amniotes to exploit drier habitats than do anamniotes (especially modern amphibians) and to be more active and energetic than anamniotes.

1. **Amniotic egg.** All amniotes have eggs with four extraembryonic membranes, the **amnion, allantois, chorion,** and yolk sac (figure 18.3). The amnion encloses the embryo in fluid, cushioning the embryo and providing an aqueous medium for growth. Metabolic wastes are stored in a sac formed by the allantois. The chorion surrounds the entire contents of the egg, and like the allantois, is highly vascularized. Thus, the chorion and allantois form an efficient respiratory organ for removing carbon dioxide and acquiring oxygen. Most amniotic eggs are surrounded by a mineralized but often flexible shell, although many lizards and snakes and most mammals lack shelled eggs. The shell forms an important mechanical support, and especially for birds, a semipermeable barrier, which allows passage of gases but limits water loss. Like eggs of anamniotes, amniotic eggs have a yolk sac for nutrient storage, although the yolk sac tends to be larger in amniotes. In marsupial and placental mammals, the yolk sac does not store yolk, but it may form a temporary or persistent yolk-sac placenta to transfer nutrients, gases, and wastes between the embryo

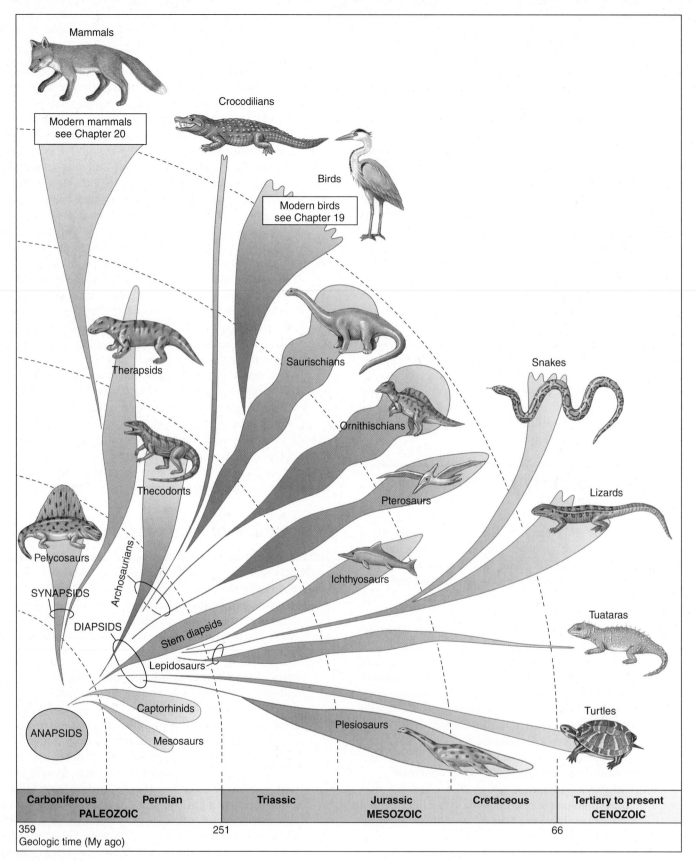

figure 18.1

Evolution of amniotes. The earliest amniotes evolved an amniotic egg, which allowed amniotes to exploit drier habitats than their ancestors. The living amniotes, which include nonavian reptiles, birds, and mammals, evolved from a lineage of small, lizardlike forms that retained the anapsid skull pattern of early, anamniote tetrapods. One lineage that descended from the early amniotes had a synapsid skull pattern and gave rise to mammals. Birds, squamates, and crocodilians have a diapsid skull pattern. Turtles have an anapsid skull, although it is probably independently evolved.

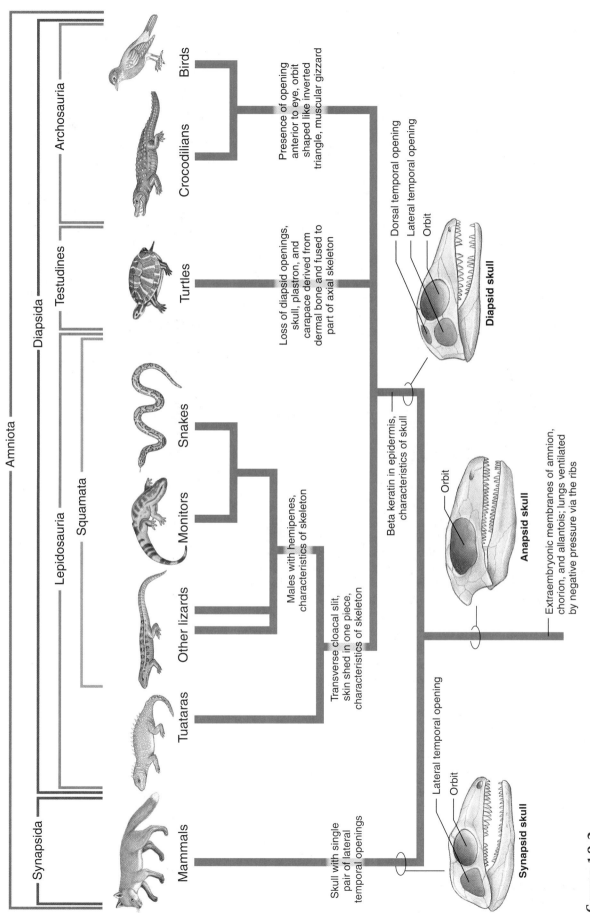

figure 18.2

Cladogram of living Amniota, showing monophyletic groups. Some shared derived characters (synapomorphies) of the groups are given. The skulls represent the ancestral condition of the three groups. Skulls of modern diapsids and synapsids are often highly modified by loss or fusion of skull bones that obscures the ancestral condition. Representative skulls for anapsids are *Nyctiphruetus* of the upper Permian; for diapsids, *Youngina* of the upper Permian; for synapsids, *Aerosaurus*, a pelycosaur of the lower Permian. The relationship of turtles to other reptiles is controversial: some researchers consider them lepidosaurs or direct living descendants of early anapsids.

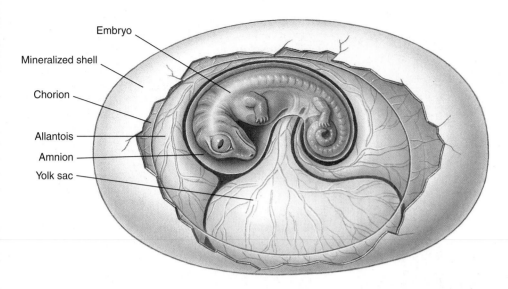

Embryo
Mineralized shell
Chorion
Allantois
Amnion
Yolk sac

figure 18.3

Amniotic egg. The embryo develops within the amnion and is cushioned by amniotic fluid. Food is provided by yolk from the yolk sac, and metabolic wastes are deposited within the allantois. As development proceeds, the allantois fuses with the chorion, a membrane lying against the inner surface of the shell; both membranes are supplied with blood vessels that assist in the exchange of oxygen and carbon dioxide across the porous shell. Because this kind of egg is an enclosed, self-contained system, it is often called a "cleidoic" egg (Gr. *kleidoun*, to lock in).

and mother. For many amniote species, egg or embryonic development occurs in a female's reproductive tract, providing even greater protection from predators and dehydration, and potential for the mother to manage the embryo's nutritional and other physiological needs.

How did the amniotic egg evolve? It is tempting to think of the amniotic egg as *the* land egg. However, many amphibians lay eggs on land and many amniotic eggs, such as those of turtles, must be buried in wet soil or deposited in areas of high humidity. Still, amniotic eggs can be laid in places too dry for any amphibian; clearly, the evolution of amniotic eggs was a major factor in the success of tetrapods on land. Perhaps a more important advantage of the amniotic egg was that it permitted development of a larger, faster-growing embryo. Anamniote eggs are supported mainly by a thick jellylike layer. This jelly layer is inadequate to support large eggs and limits movement of oxygen into the eggs. One hypothesis suggests that a first step in the evolution of the amniotic egg was replacement of the jelly layer with a shell, which provided better support and movement of oxygen. Furthermore, calcium deposited in the shell can be dissolved and absorbed by the growing embryo, to provide one of the raw materials needed for skeleton construction. This hypothesis is supported by physiological studies, which show that embryos of species with the smallest amniotic eggs have a metabolic rate about three times that of embryos of anamniotes with eggs of the same size.

All amniotes lack gilled larvae and have internal fertilization. Shelled amniotic eggs require internal fertilization because sperm cannot penetrate the shell. Internal fertilization in amniotes is accomplished with a copulatory organ, except for tuataras and most birds, which transfer sperm by pressing their cloacas together. The most common copulatory organ in amniotes, a penis derived from the cloacal wall, appears to be another amniote innovation.

2. **Thicker and more waterproof skin.** Amphibians must maintain a thin, moist skin to permit effective gas exchange. However, on land, this skin makes amphibians

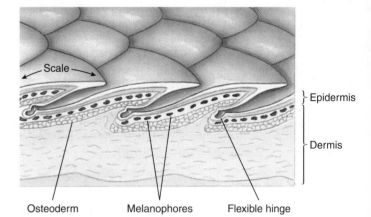

Scale
Epidermis
Dermis
Osteoderm   Melanophores   Flexible hinge

figure 18.4

Section of the skin of a reptile showing overlapping epidermal scales and bony osteoderms in the dermis.

vulnerable to dehydration and physical trauma. In amniotes, a shift away from a respiratory function of the skin is associated with changes in skin morphology. Although the skin varies widely in structure among living amniotes and anamniote tetrapods, amniote skin tends to be much thicker, more keratinized, and less permeable to water. A wide variety of structures composed of keratin, such as scales (figure 18.4), hair, feathers, and claws, project from amniote skin. Keratin gives the skin protection from physical trauma, and hydrophilic lipids in the skin limit water loss through the skin. Unique to clade Reptilia (birds and nonavian reptiles), the epidermis contains a hard form of keratin called beta keratin. The characteristic scales of nonavian reptiles, formed mostly of beta keratin, provide protection against wear in terrestrial environments. The primarily epidermal scales are not homologous to the scales of fishes, which are mostly bony, dermal structures (see figure 16.18).

In crocodilians, scales remain throughout life, growing gradually to replace wear. In other nonavian reptiles, such as lizards and snakes, new keratinized epidermis grows beneath the old, which is then shed at intervals. Turtles add new layers of keratin under the old layers of the platelike scutes, which are modified scales. Crocodiles and many lizards (skinks, for example) possess bony plates called osteoderms (figure 18.4) located in the dermis, beneath the keratinized scales. The dermis has chromatophores, color-bearing cells that give many lizards and snakes their colorful hues. This layer, unfortunately for their bearers, is converted into alligator and snakeskin leather, esteemed for expensive pocketbooks and shoes.

Keratin and lipids limit the skin's ability to exchange respiratory gases—so, unlike most amphibians, few amniotes use their skin as a primary respiratory organ. Amniote gas exchange occurs primarily in the lungs, which have considerably more surface area than do anamniote lungs.

3. **Rib ventilation of the lungs.** Amniote lungs are better developed than those of amphibians; amniote lungs have much more surface area and are ventilated by a different mechanism. These changes reflect both the increased metabolic demands of amniotes and the reduced ability of amniote skin to serve as a gas-exchange organ.

Amphibians, like air-breathing fishes, fill their lungs by *pushing* air from the oral and pharyngeal cavities into the lungs. In contrast, amniotes *draw* air into their lungs (aspiration) by expanding the thoracic cavity using costal (rib) muscles or pulling the liver (with other muscles) posterior. This shift from positive to negative ventilation probably influenced the evolution of amniote limbs. Early tetrapods used their rib muscles to make lateral undulations,

causing a wriggling motion not unlike that of eels (p. 341). Enlargement and mechanical rearrangement of the limbs enabled more efficient locomotion of amniotes, particularly for those with large bodies, while allowing rib muscles to shift to a primary function of lung ventilation. However, we note that limbs are not required for terrestrial locomotion as some amniotes, such as snakes and amphisbaenians, are very successful with no limbs at all! Although lungs are the primary gas exchange organ for nearly all amniotes, other areas of the body can be used. Many aquatic turtles supplement pulmonary respiration with respiration at their pharynx or cloaca, and most gas exchange in sea snakes occurs across their skin.

4. **Stronger jaws.** Most fish jaws are designed for suction and quick closure, but once prey is seized, little static force can be applied. Suction feeding is not possible with terrestrial vertebrates, and the skeleton and muscles of the jaws of early tetrapods became adapted to seize prey. In amniotes, expansion of the jaw musculature, often into temporal openings (p. 367) or notches, provided better mechanical advantage. In contrast to fishes, the tongue of tetrapods is muscular and mobile, functioning to move food in the mouth for mastication and swallowing.

5. **High-pressure cardiovascular systems.** In all amniotes the right atrium, which receives deoxygenated blood from the body, is completely partitioned from the atrium, which receives oxygenated blood from the lungs. Mammals, birds, and crocodilians have two completely separated ventricles as well (figure 18.5); other reptiles have a single ventricle incompletely partitioned into multiple chambers. Even in

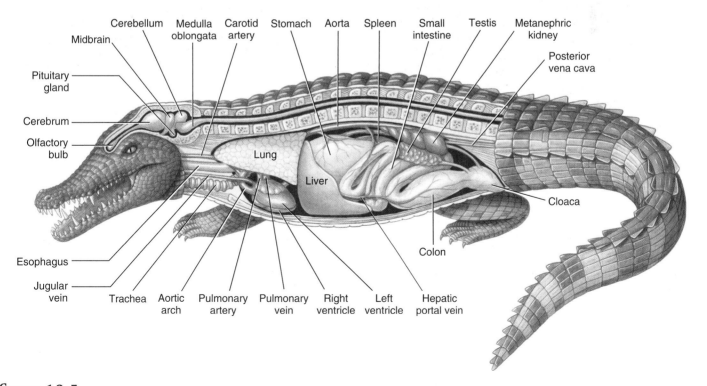

figure 18.5

Internal structure of a male crocodile.

species with incomplete ventricular septa, flow patterns within the heart limit mixing of oxygen-rich blood and oxygen-poor blood; all amniotes therefore have functionally separate circulations. This better separation of circuits provides higher systemic blood pressure in amniotes; fishes and amphibians typically have systemic pressures of 15 to 40 mm Hg, compared with about 80 mm Hg in varanids (a group of large, active lizards). Higher pressure is adaptive for active, terrestrial organisms because of their higher metabolic needs and because the heart must overcome gravity to pump blood "uphill."

Is the incomplete separation of hearts of many nonavian reptiles merely a transitional stage on the evolutionary path to the "advanced" heart of mammals and birds? No, incomplete separation is adaptive for these vertebrates, as it allows blood to bypass the lungs when pulmonary respiration is not occurring (for example, diving or estivation). Thus, during lung-bypass, energy is not expended pumping blood through the capillary beds of the lungs.

6. **Water-conserving nitrogen excretion.** Most amphibians excrete their metabolic waste primarily as ammonia or urea. Ammonia is toxic at relatively low concentrations and must be removed in a dilute solution. Because of the water required for excretion of ammonia, it is not adaptive for vertebrates occupying dry, terrestrial habitats. Mammals excrete their nitrogenous waste as urea, which is concentrated in their kidneys, reducing water loss through excretion. Birds and nonavian reptiles excrete their nitrogenous wastes as concentrated uric acid. Birds and nonavian reptiles have limited abilities to concentrate their urine in the kidneys, so their urinary bladder receives dilute urine. Water (and many salts) are resorbed in the bladder, and "urine" is voided as a semisolid mass of uric acid. Salts are removed in many seabirds and nonavian reptiles by salt glands—located near the nose, eyes, or tongue—which secrete a salty fluid that is strongly hyperosmotic to body fluids.

7. **Expanded brain and sensory organs.** The brain has a relatively large cerebrum and cerebellum in all amniotes, but especially so in birds and mammals. Enlargement of the cerebrum is correlated with integration of sensory information and control of muscles during locomotion. Nonavian reptiles and birds have particularly good vision, and many species display brilliant coloration. Smell is not as developed in many reptiles, but snakes and some lizards use a highly sensitive sense of smell to find prey and mates. In lizards and snakes, olfaction is assisted by well-developed Jacobson's organs, specialized olfactory chambers in the roof of the mouth (p. 376). Some lizards and snakes can detect ultraviolet or infrared light.

Changes in Traditional Classification of Reptiles

With increasing use of cladistic methodology in zoology and its insistence on hierarchical arrangement of monophyletic groups (see p. 88), important changes have been made in the classification of amniotes. As traditionally defined, class Reptilia excludes birds, but includes snakes, lizards, tuataras, crocodilians, and turtles, in addition to several extinct groups, such as dinosaurs, plesiosaurs, pterosaurs, and the "mammal-like reptiles." However, birds and reptiles (excluding "mammal-like reptiles") share several derived characters including certain skull and ankle characteristics as well as the presence of a special type of keratin in the skin called beta keratin, which unites them as a monophyletic group (see figure 18.2). Thus, reptiles as traditionally defined form a **paraphyletic** group because they do not include all descendants of their most recent ancestor.

Crocodilians and birds are sister groups; they are more recently descended from a common ancestor than either is from any other living reptilian lineage. In other words, crocodilians and birds belong to a monophyletic group apart from other reptiles and, according to the rules of cladistics, should be assigned to a clade that separates them from the remaining reptiles. This clade is in fact recognized; it is Archosauria (see figures 18.1 and 18.2), a grouping that also includes the extinct dinosaurs and pterosaurs. Archosaurs, along with their sister

CHARACTERISTICS
of Nonavian Reptiles

1. Two paired limbs, usually with five toes each; limbs vestigial or absent in many; **ectothermic**
2. Body covering of **keratinized epidermal scales** and sometimes bony dermal plates; integument with few glands
3. Skull with **one occipital condyle** (bony bump that connects to the first vertebra); lower jaws of several bones; distinct atlas and axis; usually two sacral vertebrae
4. Teeth **polyphyodont** (replaced many times) or absent (turtles); when present, teeth usually **homodont** (all similar in function and shape), with a single point; **gizzard** in crocodilians
5. Brain moderately well developed with expanded cerebrum; 12 pairs of cranial nerves
6. Eyes with color vision in some; snakes and some lizards with highly developed chemoreception using olfactory epithelia and **Jacobson's organ;** some snakes with **heat-sensitive pit organs;** middle ear with a single bone
7. Usually separate sexes, but some lizards reproduce asexually by parthenogenesis; internal fertilization; copulatory organ a **penis, hemipenes,** or (rarely) absent; sex determined by chromosomes or by environment
8. Fetal membranes of **amnion, chorion, and allantois;** oviparous or viviparous; eggs with leathery or calcareous shells; embryos of viviparous species nourished by **placenta** or **yolk sac** (ovoviviparity); parental care absent, except in crocodilians; no larval stage
9. Excretory system of **metanephric kidneys** and ureters that open into a cloaca; uric acid usually main nitrogenous waste
10. Lungs filled by **aspiration** (negative ventilation); **no gills;** some supplement gas exchange with cloaca, skin, or pharynx
11. Heart with a sinus venosus, two atria, and ventricle incompletely divided into three ventricles; crocodilian heart with sinus venosus, two atria, and two ventricles; **pulmonary and systemic circuits incompletely separated;** nucleated red blood cells

group, the lepidosaurs (tuataras, lizards, and snakes), and turtles, form a monophyletic group that cladists call Reptilia. Here we use Reptilia and reptiles in a cladistic fashion to include living amniote groups traditionally termed "reptiles" together with birds and all extinct groups more closely related to these than to mammals. The term "nonavian reptiles" is used informally to denote a paraphyletic group that includes the living turtles, lizards, snakes, tuataras, and crocodilians, and a number of extinct groups, including plesiosaurs, ichthyosaurs, pterosaurs, and dinosaurs. Nonavian reptiles are the subject of the remainder of this chapter; birds, which form the remainder of clade Reptilia, are discussed in Chapter 19.

Characteristics and Natural History of Reptilian Orders

Turtles: Order Testudines

Turtles appear in the fossil record in the Upper Triassic period some 220 million years ago. Since the Triassic, turtles have survived to the present with little change to their early morphology. The earliest turtles had teeth and a reduced shell, but otherwise were similar to those of today. Lacking teeth, a modern turtle's jaws have tough keratinized plates for gripping food (figure 18.6). They are enclosed in shells consisting of a dorsal **carapace** (Fr., from Sp. *carapacho*, covering) and a ventral **plastron** (Fr., breastplate). The shell is composed of two layers: an outer layer of keratin and an inner layer of bone. Like a medieval coat of armor, the shell offers protection for the head and appendages, which most turtles can retract into it (figure 18.7). Because the ribs are fused to the shell, the turtle cannot expand its chest to breathe. Instead, turtles employ certain abdominal and pectoral muscles as a "diaphragm." They draw air inward by contracting limb flank muscles to make the body cavity larger. To exhale, they draw the shoulder girdle back into the shell, thus compressing the viscera and forcing air out of the lungs.

The terms "turtle," "tortoise," and "terrapin" are applied variously to different members of the turtle order. In North American usage, they are all correctly called turtles. The term "tortoise" is frequently applied to land turtles, especially large forms. British usage of the terms is different: "tortoise" is the inclusive term, whereas "turtle" is applied only to the aquatic members.

Sound perception is poor in turtles, and most turtles are mute. Compensating for poor hearing are good senses of smell and color vision. Turtles are oviparous, and fertilization is internal. All turtles, even marine forms, bury their shelled, amniotic eggs in the ground. An interesting feature of turtle reproduction is that in some turtle families, as in all crocodilians and some lizards, nest temperature determines the sex of the hatchlings. In turtles, low temperatures during incubation produce males, and high temperatures produce females.

Marine turtles, buoyed by their aquatic environment, can reach great size. Leatherbacks are the largest, attaining a length of 2 m and a weight of 725 kg. Green turtles, so named because of their greenish body fat, may exceed 360 kg, although most individuals of this economically valuable and heavily exploited species seldom live long enough to reach anything approaching this size. Some land tortoises weigh several hundred kilograms; examples include the giant tortoises of the Galápagos Islands (figure 18.8) that so intrigued Darwin during his visit there in 1835. Most tortoises move rather slowly; one hour of determined trudging carries a large Galápagos tortoise approximately 300 m. Low metabolism may explain the longevity of turtles, for some live more than 150 years.

figure 18.6

Snapping turtle, *Chelydra serpentina*, showing the absence of teeth. Instead, the jaw edges are covered with a keratinized plate.

figure 18.7

Skeleton and shell of a turtle, showing fusion of vertebrae and ribs with the carapace. The long, flexible neck allows the turtle to withdraw its head into its shell for protection.

figure 18.8

Mating Galápagos tortoises, *Geochelone elaphantopus*. The male has a concave plastron that fits over the highly convex carapace of the female, helping to provide stability during mating. Males utter a roaring sound during mating, the only time they are known to emit vocalizations.

Lizards and Snakes: Order Squamata

Squamates are the most recent and diverse products of diapsid evolution, forming approximately 95% of all known living non-avian reptiles. Lizards appeared in the fossil record in the Jurassic period, but they did not begin their diversification until the Cretaceous period of the Mesozoic era when the dinosaurs were at the climax of their diversity. Snakes appeared during the late Jurassic period, probably evolving from a group of lizards whose descendants include the Gila monster and monitor lizards. Two specializations in particular characterize snakes: (1) extreme elongation of the body with accompanying displacement and rearrangement of internal organs, and (2) skull specializations for swallowing large prey.

Skulls of squamates are modified from the ancestral diapsid condition by loss of dermal bone ventral and posterior to the lower temporal opening. This modification allowed the evolution in most lizards and snakes of a kinetic skull having movable joints (figure 18.9). These joints, located at the posterior of the quadrate and pterygoid bones, palate, and roof of the skull, allow the snout to be tilted. Specialized mobility of the skull enables squamates to seize and to manipulate their prey; it also increases the effective closing force of the jaw musculature. The skull of snakes is even more kinetic than that of lizards. Such exceptional skull mobility is considered a major factor in the diversification of lizards and snakes.

Traditionally, the order Squamata was divided into three suborders: Sauria (lizards), Serpentes (snakes), and Amphisbaenia (amphisbaenians). Amphisbaenians are now considered

figure 18.9

Kinetic diapsid skull of a monitor lizard, *Varanus*, showing joints that allow the snout and upper jaw to move on the rest of the skull. The quadrate bone can move at its dorsal end and ventrally at both the lower jaw and the pterygoid bone. The front part of the braincase is also flexible, allowing the snout to be raised or depressed. Note that the lower temporal opening is very large, with no lower border. The upper temporal opening lies dorsal and medial to the postorbital-squamosal arch and is not visible in this drawing.

modified lizards, and we discuss them with lizards. Snakes form a monophyletic group that evolved within a lizard subgroup (probably the subgroup which includes the monitors and Gila monster), thus making Sauria paraphyletic. Here we use "lizard" and "Sauria" to describe all lizards that are not snakes.

Lizards: Suborder Sauria

Lizards are an extremely diverse group that includes terrestrial, burrowing, aquatic, arboreal, and aerial members. Among the more familiar groups in this varied suborder are geckos (figure 18.10), small, agile, mostly nocturnal forms with adhesive toe pads that enable them to walk upside down and on vertical surfaces;

figure 18.10

The tokay, *Gekko gecko,* of Southeast Asia has a true voice and is named after its strident, repeated *to-kay, to-kay* call.

figure 18.11

A large male marine iguana, *Amblyrhynchus cristatus,* of the Galápagos Islands, feeding underwater on algae. This is the only marine lizard in the world. It has special salt-removing glands in the eye orbits and long claws that enable it to cling to the bottom while feeding on small red and green algae, its principal diet. It may dive to depths exceeding 10 m (33 feet) and remain submerged more than 30 minutes.

figure 18.12

A chameleon snares a dragonfly. After cautiously edging close to its target, the chameleon suddenly lunges forward, anchoring its tail and feet to the branch. A split second later, it launches its sticky-tipped, foot-long tongue to trap the prey. The eyes of this common European chameleon, *Chamaeleo chamaeleon,* are swiveled forward to provide binocular vision and excellent depth perception.

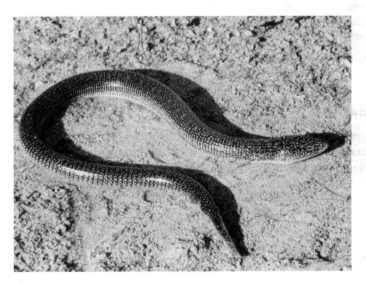

figure 18.13

A glass lizard, *Ophisaurus,* of the southeastern United States. This legless lizard is stiff and has an extremely long, fragile tail that readily fractures when the animal is struck or seized. Most specimens, such as this one, have only a partly regenerated tip to replace a much longer tail previously lost. Glass lizards can be readily distinguished from snakes by the deep, flexible groove running along each side of the body. They feed on worms, insects, spiders, birds' eggs, and small nonavian reptiles.

iguanas, often brightly colored New World lizards with ornamental crests, frills, and throat fans, including the remarkable marine iguana of the Galápagos Islands (figure 18.11); skinks, which have elongate bodies, an armor of tight-fitting osteoderms, and reduced limbs in many species; monitors, large, active predators that include the largest lizard, the Komodo dragon, *Varanus komodoensis* (p. 366); and chameleons, a group of arboreal lizards, mostly of Africa and Madagascar. Chameleons are entertaining creatures that catch insects with a sticky-tipped tongue that can be flicked accurately and rapidly to a distance greater than the length of their body (figure 18.12).

The great majority of lizards have four limbs and relatively short bodies, but in many, the limbs are reduced, and a few, such as glass lizards (figure 18.13), are completely limbless.

Most lizards have movable eyelids, whereas a snake's eyes are permanently covered with a transparent cap. Lizards have keen vision for daylight (retinas rich in both cones and rods), although one group, the nocturnal geckos, has retinas

figure 18.14

A Gila monster, *Heloderma suspectum*, of southwestern United States desert regions. It and its relative, the Mexican beaded lizard, are the only venomous lizards known. These brightly colored lizards feed principally on birds' eggs, nesting birds, mammals, and insects. Unlike venomous snakes, the Gila monster secretes venom from glands in its lower jaw. The chewing bite is painful to humans but seldom fatal.

figure 18.15

An amphisbaenian, or "worm lizard." Amphisbaenians burrow, using their solid skull as a digging tool. The species pictured, *Amphisbaena alba*, is widely distributed in South America.

composed entirely of rods. Most lizards have an external ear opening that snakes lack. However, as with other nonavian reptiles, hearing does not play an important role in the lives of most lizards. Geckos are exceptions because the males are strongly vocal (to announce territory and to discourage the approach of other males), and they must, of course, hear their own vocalizations.

Many lizards occupy the world's hot and arid regions, aided by adaptations that make desert life possible. Lipids in their thick skin minimize water loss. Little water is lost in their urine because they primarily excrete uric acid, as do other groups that are successful in arid habitats (birds, insects, and pulmonate snails). Some, such as Gila monsters of the southwestern United States deserts, store fat in their tails, and then use the fat during droughts to produce energy and metabolic water (figure 18.14).

Lizards, like nearly all nonavian reptiles, are **ectothermic,** adjusting their body temperature by moving among different microclimates. Because cold climates provide few opportunities for ectotherms to raise their body temperature, relatively few nonavian reptile species live in these habitats. Ectotherms use considerably less energy than do endotherms; therefore, nonavian reptiles are successful in ecosystems with low productivity and warm climates, such as tropical deserts, dry, open forests, and grasslands. Thus, ectothermy is not an "inferior" characteristic of nonavian reptiles, but rather is a successful strategy for coping with specific environmental challenges.

The amphisbaenians, or "worm lizards," are lizards highly specialized for a **fossorial** (burrowing) life. Amphisbaenia means "double walk," in reference to their peculiar ability to move backward nearly as effectively as forward. Amphisbaenians have elongate, cylindrical bodies of nearly uniform diameter, and most lack any trace of external limbs (figure 18.15).

Their eyes are usually hidden underneath the skin, and they have no external ear openings. Their skull is solidly built and either conical or spade-shaped, which aids in tunneling through the soil. The skin is formed into numerous, independently moving rings that can grip the soil, creating a movement not unlike that of earthworms. Amphisbaenians have an extensive distribution in South America and tropical Africa; one species occurs in the United States.

Snakes: Suborder Serpentes

Snakes are limbless and usually lack both pectoral and pelvic girdles (the latter persists as a vestige in pythons, boas, and some other snakes). The numerous vertebrae of snakes, shorter and wider than those of most tetrapods, permit quick lateral undulations through grass and over rough terrain. Ribs increase rigidity of the vertebral column, providing more resistance to lateral stresses.

The skull of snakes is more kinetic than that of lizards, enabling snakes to swallow much larger prey. The two halves of the lower jaw of snakes are joined only by muscles and skin, permitting them to spread widely apart (figure 18.16). Snakes differ from lizards in having no movable eyelids and no external ear openings. Most snakes have relatively poor vision, but arboreal snakes possess excellent binocular vision, which is useful for tracking prey through branches where scent trails are difficult to follow (figure 18.17). Snakes' internal ears are mainly sensitive to sounds in a limited range of low frequency (100 to 700 Hz). Snakes are also quite sensitive to vibrations conducted through the ground.

Nevertheless, most snakes employ chemical senses rather than vision or vibration to hunt their prey. In addition to the usual olfactory areas in the nose, which are not well developed, snakes have a pair of pitlike **Jacobson's organs**

The Mesozoic World of Dinosaurs

In 1842, when the English anatomist Richard Owen coined the term *dinosaur* ("fearfully great lizard") to describe fossil Mesozoic reptiles of gigantic size, only three poorly known dinosaur genera were distinguished. New and marvelous fossil discoveries quickly followed, and by 1887 zoologists were able to distinguish two groups of dinosaurs—Saurischia and Ornithischia—based on differences in the structure of the pelvic girdles. The Saurischia ("lizard-hipped") had a simple, three-pronged pelvis with hip bones arranged much as they are in other nonavian reptiles. The Ornithischia ("bird-hipped") had a somewhat more complex pelvis. The ilium and ischium were arranged similarly in ornithischians and saurischians, but the ornithischian pubis was a narrow, rod-shaped bone with anteriorly and posteriorly directed processes lying alongside the ischium. Oddly, while the ornithischian pelvis, as the name suggests, was similar to that of birds, birds are of the saurischian clade.

Dinosaurs and their living relatives, the birds, are archosaurs ("ruling lizards"), a group that includes crocodilians and pterosaurs (refer to the classification of amniotes on p. 381). As traditionally recognized, dinosaurs are a paraphyletic group because they do not include birds, which are descended from the theropod dinosaur lineage.

From among the various archosaurian diversifications of the Triassic there emerged a lineage with limbs drawn under the body which provided an upright posture. This lineage gave rise to the earliest dinosaurs of the Late Triassic period. *Herrerasaurus,* a bipedal dinosaur from Argentina, has one of the most distinctive characteristics of dinosaurs: walking upright on pillarlike legs, rather than on legs splayed outward like those of modern amphibians and nonavian reptiles. This arrangement allowed the legs to support the great weight of the body while providing an efficient and rapid stride.

Two groups of saurischian dinosaurs have been proposed based on differences in feeding habits and locomotion: the carnivorous and bipedal theropods, and the herbivorous and quadrupedal sauropods. *Coelophysis* was an early theropod with a body form typical of all theropods: powerful hindlimbs with three-toed feet; a long, heavy, counterbalancing tail; slender, grasping forelimbs; a flexible neck; and a large head with jaws armed with daggerlike teeth. Large predators such as *Allosaurus,* common during the Jurassic, were replaced by even more massively built carnivores of the Cretaceous, such as *Tyrannosaurus,* which reached a length of 12.5 m (42 ft), stood nearly 6 m high, and weighed more than 7200 kg (8 tons). Not all predatory saurischians were massive; several were swift and nimble, such as *Velociraptor* ("speedy predator") of the Upper Cretaceous.

Herbivorous saurischians, the quadrupedal sauropods, appeared in the Late Triassic. Although early sauropods were small- and medium-sized dinosaurs, those of the Jurassic and Cretaceous attained gigantic proportions and were the largest terrestrial vertebrates ever to have lived. *Brachiosaurus* reached 25 m (82 ft) in length and may have weighed more than 30,000 kg (33 tons). Even larger sauropods have been discovered; *Argentinosaurus* was 40 m (132 ft) long and weighed at least 80,000 kg. With long necks and long front legs, the sauropods were the first vertebrates adapted to feed on trees.

The second group of dinosaurs, the Ornithischia, were all herbivorous. Although more varied in appearance than saurischians, the ornithischians are united by several derived skeletal features that indicate common ancestry. The huge back-plated *Stegosaurus* of the Jurassic is a well-known example of armored ornithischians, which formed two of the five major groups of ornithischians. Even more shielded with bony plates than stegosaurs were the heavily built ankylosaurs, "armored tanks" of the dinosaur world. As the Jurassic gave way to the Cretaceous, several groups of unarmored ornithischians appeared, although many bore impressive horns. The steady increase in ornithiscian diversity in the Cretaceous paralleled a concurrent gradual decline in giant sauropods, which had flourished in the Jurassic. *Triceratops* is representative of horned dinosaurs that were common in the Upper Cretaceous. Even more prominent in the Upper Cretaceous were the hadrosaurs, such as *Parasaurolophus,* which probably lived in large herds. Many hadrosaurs had skulls elaborated with crests that probably functioned as vocal resonators to produce species-specific calls.

Dinosaurs likely provided considerably more complex parental care than most other nonavian reptilians. Because both crocodilians and birds, members of the clade Archosauria in which dinosaurs are contained (see figure 18.1), share complex parental care, it is likely that dinosaurs exhibited similar behavior. Fossil nests of dinosaurs have been discovered for several groups. In one case, a fossil adult of the small theropod *Oviraptor* was found with a nest of eggs. Originally it was hypothesized that the adult was a predator on the eggs (*Oviraptor* means "egg seizer"), but later, an embryo in a similar egg was found and identified as *Oviraptor,* indicating that the adult was probably with its own eggs! Examination of baby *Maiasaura* (a hadrosaur) found in a nest revealed considerable wear on their teeth, suggesting that the babies had remained in the nest and possibly been fed by adults during part of their early life.

Sixty-five million years ago, the last Mesozoic dinosaurs became extinct, leaving birds and crocodilians as the only surviving lineages of archosaurs. The demise of dinosaurs coincided with a large asteroid impact on the Yucatán peninsula that would have produced worldwide environmental upheaval. Although the impact event is generally accepted as the primary cause of the extinctions, other events have been suggested to have contributed to the demise of dinosaurs and other animals. Additional events at the close of Cretaceous, including a massive volcanic eruption in the Deccan Plateau of India and lowering of sea level, also caused substantial changes in landform and climates. We continue to be fascinated by the awe-inspiring, often staggeringly large creatures that dominated the Mesozoic era for 165 million years—an incomprehensibly long period of time. Today, inspired by clues from fossils and footprints from a lost world, scientists continue to piece together the puzzle of how the various dinosaur groups arose, behaved, and diversified.

(continues)

SAURISCHIANS

ORNITHISCHIANS

66
My ago

Titanosaurus
12 m (40 ft)

Parasaurolophus
10 m (33 ft)

Velociraptor
1.8 m (6 ft)

Triceratops
9 m (30 ft)

146
My ago

Brachiosaurus
25 m (82 ft)

Stegosaurus
9 m (30 ft)

Ilium

Ischium

Pubis

Allosaurus
11 m (35 ft)

200
My ago

Coelophysis
3 m (10 ft)

Ilium

Pubis

Ischium

Herrerasaurus 4 m (13 ft)
One of the oldest known
dinosaurs. Has characteristics
of both saurischians and
ornithischians.

251
My ago

CRETACEOUS

JURASSIC

TRIASSIC

figure 18.16

The great mobility of a snake's jaws and cranial bones is evident in this snake, *Dasypeltis*, swallowing an egg.

figure 18.18

A blacktail rattlesnake, *Crotalus molossus*, flicks its tongue to smell its surroundings. Scent particles trapped on the tongue's surface are transferred to Jacobson's organs, olfactory organs in the roof of the mouth. Note the heat-sensitive pit organ between the nostril and eye.

figure 18.17

A parrot snake, *Leptophis ahaetulla*. The slender body of this Central American tree snake is an adaptation for traveling among branches.

figure 18.19

Pit organ of rattlesnake, a pit viper. Cutaway shows the location of a deep membrane that divides the pit into inner and outer chambers. Heat-sensitive nerve endings are concentrated in the membrane.

in the roof of the mouth. These organs are lined with a richly innervated olfactory epithelium. The forked tongue, flicked through the air, collects scent molecules (figure 18.18). The tongue is then drawn past Jacobson's organs and information is then transmitted to the brain, where scents are identified.

Boids (pythons and boas) and pit vipers have special heat-sensitive pit organs on their heads, located between their nostrils and eyes (figure 18.19). They are exceedingly sensitive to radiant energy (long-wave infrared) and can distinguish temperature differences smaller than 0.003°C from a radiating surface. The pits are used to track warm prey and to aim strikes, which are as effective in total darkness as in daylight.

Most snakes capture their prey by grabbing it with their mouth and swallowing it while it is still alive. Swallowing a struggling, kicking, biting animal is dangerous, so most snakes that swallow prey alive specialize on small prey, such as worms, insects, fish, frogs, and, less frequently, small mammals. Many of these snakes, which may be quite fast, locate prey by actively foraging. Snakes that first kill their prey by constriction (figure 18.20) usually specialize on larger, often mammalian, prey. The largest constrictors can kill

figure 18.20

Nonvenomous African house snake, *Boaedon fuluginosus*, constricting a mouse before swallowing it.

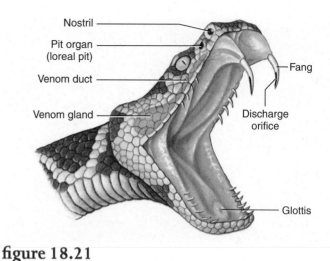

figure 18.21

Head of a rattlesnake showing the venom apparatus. The venom gland, a modified salivary gland, is connected by a duct to the hollow fang.

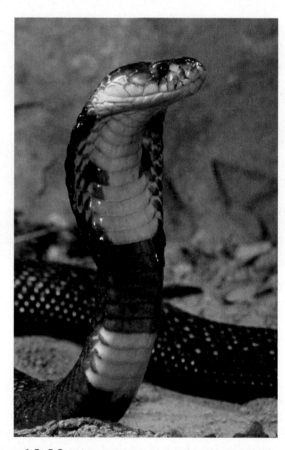

figure 18.22

Monocled cobra (*Naja kaouthia*). Cobras erect the front part of the body when startled and as a threat display. Although a cobra's strike range is limited, all cobras are dangerous because of the extreme toxicity of their venom.

and swallow prey as large as deer, leopards, and crocodilians. However, because muscle rearrangements that permit constricting also reduce speed of travel, most constrictors ambush their prey.

Other snakes kill their prey before swallowing it by injecting it with venom. Less than 20% of all snakes are venomous, although venomous species outnumber nonvenomous species by 4 to 1 in Australia. Venomous snakes are usually divided into five families, based in part on type of fangs.

Vipers (family Viperidae) have large, movable, tubular fangs at the front of their mouth (figure 18.21). The fangs lie in a membranous sheath when the mouth is closed. When a viper strikes, a special muscle and bone lever system erects the fangs as the mouth opens (figure 18.21). Fangs are driven into prey by the thrust, and venom is injected into the wound through a canal in the fangs. A viper immediately releases its prey after the bite and follows it until it is paralyzed or dead. Vipers of the Old World lack facial heat-sensing pits. Snakes of subfamily Crotalinae within family Viperidae are called pit vipers because they possess pit organs on their heads (see figure 18.19). All best-known North American venomous snakes are pit vipers, including rattlesnakes (see figure 18.18), water moccasins, and copperheads. Each year in the United States approximately 7000 bites from pit vipers are reported, causing only about 5 deaths.

A second family of venomous snakes (Elapidae) has short, permanently erect fangs in the front of the mouth and includes cobras (figure 18.22), mambas, coral snakes, kraits, and sea snakes. The large family Colubridae, which contains most familiar (and nonvenomous) snakes, does include a few venomous species, including the rear-fanged African boomslang and the twig snake, both of which have been responsible for some human fatalities.

Even the saliva of harmless snakes possesses limited toxic qualities, which is likely the ancestral state from which highly toxic venom evolved. Although sea snakes and the Australian

Taxonomy of Living Nonavian Reptiles

The following Linnean classification agrees with the genealogical relationships of living nonavian reptiles shown in figure 18.2. Relationships of turtles to other diapsids are controversial.

Subclass Diapsida (dī-a p′si-də) (Gr. *di*, double, + *apsis*, arch): **diapsids.** Amniotes having a skull with two temporal openings.

Order Testudines (tes-tū′di-nēz) (L. *testudo*, tortoise) **(Chelonia): turtles.** Body in a bony case of dorsal carapace and ventral plastron; jaws with keratinized beaks instead of teeth; vertebrae and ribs fused to overlying carapace; temporal openings lost; approximately 300 species.

Superorder Lepidosauria (le-pi-dō-sor′ē-ə) (Gr. *lepidos*, scale, + *sauros*, lizard). Characterized by sprawling posture; no bipedal specializations; diapsid skull often modified by loss of one or both temporal arches; transverse cloacal slit; skin shed in one piece.

Order Squamata (skwä-mä′tə) (L. *squamatus*, scaly, + *ata*, characterized by): **snakes and lizards.** Skin of keratinized epidermal scales or plates, which are shed; quadrate movable; skull kinetic (except in amphisbaenians); vertebrae usually concave in front; paired copulatory organs.

Suborder Lacertilia (la-sur-til′ē-ə) (L. *lacerta*, lizard) **(Sauria): lizards.** Body usually with four limbs but many species lack one or both pairs of limbs; halves of lower jaw fused; eyelids movable; amphisbaenians have eyes hidden beneath skin; external ear present; this paraphyletic suborder contains approximately 4800 species.

Suborder Serpentes (sur-pen′tēz) (L. *serpere*, to creep): **snakes.** Body elongate; limbs, ear openings, and middle ear absent; mandibles joined anteriorly by ligaments; eyelids fused into transparent spectacle; tongue forked and protrusible; left lung reduced or absent; approximately 2900 species.

Order Sphenodonta (sfē′nō-don′tə) (Gr. *sphen*, wedge, + *odontos*, tooth): **tuataras.** Primitive diapsid skull; vertebrae biconcave; quadrate immovable; parietal eye prominent; two extant species in *Sphenodon*.

Superorder Archosauria (ärk′ō-sor′ē-ə) (Gr. *archon*, ruling, + *sauros*, lizard). Lineage tending toward bipedalism; orbit shaped like an upside-down triangle; anteorbital fenestra (opening in the skull anterior to the orbit) and gizzard present; ventricle fully divided; parental care of young.

Order Crocodilia (krok′ə-dil′ē-ə) (L. *crocodilus*, crocodile): **crocodilians.** Skull elongate and massive; nares terminal; secondary palate present; vertebrae usually concave in front; forelimbs usually with five digits; hindlimbs with four digits; quadrate immovable; 23 species of alligators, caimans, crocodiles, and gharials.

---

tiger snake have perhaps the most toxic of snake venoms, several larger snakes are more dangerous. The aggressive king cobra, which may exceed 5.5 m in length, is the largest and one of the most dangerous of all venomous snakes. It is estimated that, worldwide, 50,000–60,000 people die from snakebite each year. Most of the deaths occur in India, Pakistan, Myanmar, and nearby countries where poorly shod people frequently contact venomous snakes and do not get immediate medical attention following a snake bite. The snakes primarily responsible for deaths in these areas are Russell's viper, the saw-scaled viper, and several species of cobras.

Most snakes are **oviparous** (L. *ovum*, egg, + *parere*, to bring forth) species that lay their shelled, elliptical eggs beneath logs, under rocks, or in holes in the ground. The remainder are **viviparous,** giving birth to well-formed young. Many of these, including all American pit vipers except for the tropical bushmaster, are **ovoviviparous,** in which the young receive nourishment only from their yolk sac. Other viviparous snakes, including garter snakes (*Thamnophis*), nourish their young with **placentae,** permitting exchange of nutrients and wastes between the embryonic and maternal bloodstream.

Tuataras: Order Sphenodonta

The order Sphenodonta contains two living species of the genus *Sphenodon* (Gr. *sphenos*, wedge, + *odontos*, tooth) of New Zealand (figure 18.23). Tuataras are sole survivors of the sphenodontid lineage that diversified modestly during the early Mesozoic but declined toward the end of that era. Tuataras were once widespread throughout the two main islands of New Zealand but are now restricted to small islets of Cook Strait and off the northeast coast of North Island. Loss of tuatara populations on the main islands of New Zealand was caused by humans intentionally or accidentally introducing nonnative animals, including cats, dogs, rats, and goats, which preyed upon tuataras and their eggs or destroyed their habitat. Tuataras are particularly vulnerable because they have slow growth and low reproductive rates.

Tuataras are lizardlike forms, measuring 66 cm long or less, that live in burrows they often share with sea birds called petrels. They have one of the slowest reproduction rates of reptiles: they take 10–20 years to become sexually mature and typically produce eggs, which take seven months to hatch, only once every four years. They are slow-growing animals with long lives; one captive male was 112 years old in 2010.

Tuataras have captured the interest of zoologists because of numerous features that are almost identical to those of early Mesozoic diapsids 200 million years old. These features include a diapsid skull with two temporal openings, and a prominent median parietal "third eye."

figure 18.23

Tuatara, *Sphenodon*, a living representative of order Sphenodonta. This "living fossil" has on top of its head a well-developed parietal eye with retina, lens, and nervous connections to the brain. Although covered with scales, this third eye is sensitive to light. The parietal eye may have been an important sense organ in early reptiles.

Crocodilians: Order Crocodilia

Modern crocodilians and birds are the only survivors of the archosaurian lineage that gave rise to the great Mesozoic diversification of dinosaurs and their kin. Crocodilians differ little in structural details from crocodilians of the early Mesozoic. Having remained mostly unchanged for nearly 200 million years, crocodilians face an uncertain future in a world dominated by humans.

All crocodilians have an elongate, robust, well-reinforced skull and massive jaw musculature arranged to provide a wide gape and rapid, powerful closure. Teeth are set in sockets, a type of dentition that was typical of Mesozoic archosaurs as well as the earliest birds. Another adaptation, found in no other vertebrate except mammals, is a complete secondary palate. This feature allows a crocodilian to breathe when its mouth is filled with water or food (or both).

Estuarine crocodiles (*Crocodylus porosus*) of southern Asia and Nile crocodiles (*C. niloticus*) of Africa (figure 18.24A) grow to great size (adults weighing 1000 kg have been reported) and are swift and aggressive. Crocodiles are known to attack animals as large as cattle, deer, and people. Alligators (figure 18.24B) are less aggressive than these crocodiles and far less dangerous to people. In the United States, *Alligator mississipiensis* is the only species of alligator; *Crocodylus acutus*, restricted to extreme southern Florida, is the only species of crocodile.

Crocodilians are oviparous. The female usually lays 20 to 50 eggs in a mass of dead vegetation or buries them in sand, and then guards them. Unlike other nonavian reptiles, crocodilians provide extensive parental care. The mother hears vocalizations

A

B

figure 18.24

Crocodilians. **A,** Nile crocodile, *Crocodylus niloticus*, basking. The lower jaw tooth fits *outside* the slender upper jaw; alligators lack this feature. **B,** American alligator, *Alligator mississipiensis*, an increasingly noticeable resident of rivers, bayous, and swamps of the southeastern United States.

from hatching young and responds by opening the nest to allow the hatchlings to escape. Young are guarded by their mother for two years or more after hatching. Although the young are capable of catching their own food immediately after hatching, they also may feed on small pieces that fall from food the mother is eating. As with many turtles and some lizards, the incubation temperature of the eggs determines the sex ratio of the offspring. However, unlike turtles (p. 373), low nest temperatures produce only females, whereas high nest temperatures produce only males.

Summary

Amniotes diverged from a group of early tetrapods during the late Paleozoic era, about 300 million years ago. Their success as terrestrial vertebrates is attributed to several adaptations, including the amniotic egg. The amniotic egg, with its shell and four extraembryonic membranes—the amnion, allantois, chorion, and yolk sac—permits rapid development of embryos in terrestrial environments. Additional amniote adaptations that permit occupation of dry environments include a thick, water-resistant skin, excretion of urea or uric acid, high-surface-area lungs ventilated by trunk muscles, expanded jaw musculature, and an efficient cardiovascular system.

Before the end of the Paleozoic era, amniotes diversified to form three groups distinguished by skull structure: anapsids, which lack temporal fenestrae; synapsids, which have one pair of temporal fenestrae and diapsids, which have two pairs of temporal fenestrae. Mammals evolved from early synapsids. Early diapsids gave rise to all living nonavian reptiles (tentatively including turtles) and to birds. There are no living descendants of Paleozoic anapsids. One clade of diapsids, the archosaurs, underwent a worldwide diversification during the Mesozoic Era into large and morphologically diverse forms, including dinosaurs.

Turtles (order Testudines), with their distinctive shells, have changed little in anatomy since the Triassic period. Turtles are a small group of long-lived terrestrial, semiaquatic, and aquatic species. They lack teeth, and instead bear keratinized plates on their jaws. All are oviparous, and all, including marine forms, bury their eggs.

Lizards and snakes (order Squamata) represent 95% of all living nonavian reptiles. Both are diversified and successful groups, particularly in hot climates. Lizards are distinguished from snakes by having united lower jaw halves, movable eyelids, and external ear openings. Lizards and snakes are ectothermic, regulating their temperature by moving among different microenvironments. Most lizards and snakes are oviparous, although viviparity is not uncommon, especially in cooler climates. Amphisbaenians are a small group of tropical lizards specialized for burrowing. They have ringed, usually limbless bodies and a solid skull.

Snakes, which evolved from one group of lizards, are characterized by elongate, limbless bodies and a highly kinetic skull that allows them to swallow prey several times larger than their own diameter. Most snakes rely on chemical senses, including Jacobson's organ, to hunt prey, rather than on visual and auditory senses. Two groups of snakes (pit vipers and boids) have unique infrared-sensing organs for tracking prey. Many snakes kill their prey with venom delivered by fangs.

Tuataras of New Zealand (order Sphenodonta) are relicts and the sole survivors of a group that otherwise disappeared 100 million years ago. They bear several features that are almost identical to those of Mesozoic fossil diapsids.

Crocodiles, alligators, caimans, and gharials (order Crocodilia) are the only living nonavian representatives of the archosaurian lineage that gave rise to the extinct dinosaurs and the living birds. Crocodilians have several adaptations for a carnivorous, semiaquatic life, including a massive skull with powerful jaws, and a secondary palate. They have the most complex parental care of any living nonavian reptile.

Review Questions

1. What are the four membranes associated with amniotic eggs? What is the function of each of these membranes?
2. How do the skin and respiratory systems of amniotes differ from those of their early tetrapod ancestors?
3. Amniotes are divided into three groups based on their skull morphologies. What are these three groups, and how do the skulls differ? Which living amniotes, if any, originated from each of these three groups?
4. Why are "reptiles," as traditionally defined, a paraphyletic group? How has cladistic taxonomy revised Reptilia to make it monophyletic?
5. Describe ways in which amniotes are more functionally or structurally suited for terrestriality than were the earliest tetrapods.
6. Describe the principal structural features of turtles that distinguish them from any other nonavian reptilian order.
7. How might nest temperature affect egg development in turtles? In crocodilians?
8. What is meant by a "kinetic" skull, and what benefit does it confer? How are snakes able to eat large prey?
9. In what ways are the special senses of snakes similar to those of lizards, and in what ways have they evolved for specialized feeding strategies?
10. What are amphisbaenians? What morphological adaptations do they have for burrowing?
11. Distinguish ornithiscian and saurischian dinosaurs, based on their hip anatomy. Was a dinosaur's style of parental care more like a lizard's or a crocodilian's?
12. What is the function of Jacobson's organ in snakes and lizards?
13. What is the function of the "pit" in pit vipers?
14. What are the differences in structure or location of fangs in a rattlesnake, a cobra, and an African boomslang?
15. Most lizards and snakes are oviparous, but some are ovoviviparous or have placental viviparity. Compare these methods of reproduction in squamates. In what climate is viviparity more common?
16. Why are tuataras (*Sphenodon*) of special interest to biologists? Why are they rare?
17. From which diapsid lineage have crocodilians descended? What other major fossil and living vertebrate groups belong to this same clade? In what structural and behavioral ways do crocodilians differ from other living nonavian reptiles?
18. How do crocodilians breathe when their mouths are full of food?

For Further Thought How could changes in the environment affect populations of species with temperature-dependent sex determination?

■ Selected References

See also general references on page 432.

Alvarez, W., and F. Asaro. 1990. An extraterrestrial impact. Sci. Am. **263:**78–84 (Oct.). *This article and an accompanying article by V. E. Courtillot, "A volcanic eruption," present opposing interpretations of the cause of the Cretaceous mass extinction that led to the demise of the dinosaurs.*

Cogger, H. G., and R. G. Zweifel, eds. 1998. Encyclopedia of reptiles and amphibians. San Diego, Academic Press. *This comprehensive and lavishly illustrated volume was written by some of the best-known herpetologists in the field.*

Crews, D. 1994. Animal sexuality. Sci. Am. **270:**108–114 (Jan.). *The reproductive strategies of reptiles, including nongenetic sex determination, provide insights into the origins and functions of sexuality.*

Gans, C., and 15 other authors, eds. 1969–2010. Biology of the Reptilia. *This monumental work includes 22 edited volumes that provide a detailed resource about nonavian reptiles.*

Greene, H. W. 1997. Snakes: the evolution of mystery in nature. Berkeley, University of California Press. *Beautiful photographs accompany a well-written volume for the scientist or novice.*

Halliday, T. R., and K. Adler, eds. 2002. Firefly encyclopedia of reptiles and amphibians. Richmond Hill, Ontario, Firefly Books Ltd. *Comprehensive and beautifully illustrated treatment of the reptilian groups with helpful introductory sections on origins and characteristics.*

Lohmann, K. J. 1992. How sea turtles navigate. Sci. Am. **266:**100–106 (Jan.). *Recent evidence suggests that sea turtles use the earth's magnetic field and the direction of ocean waves to navigate back to their natal beaches to nest.*

Norman, D. 1991. Dinosaur! New York, Prentice-Hall. *Highly readable account of the life and evolution of dinosaurs, with fine illustrations.*

Paul, G. 2000. The Scientific American book of dinosaurs. New York, St. Martin's Press. *Essays emphasizing functional morphology, behavior, evolution, and extinction of dinosaurs.*

Pianka, E. R., and L. J. Vitt. 2003. Lizards: windows to the evolution of diversity. Berkeley, University of California Press. *Hundreds of color photographs highlight a discussion of behavior and evolution of lizards.*

Pough, F. H., R. M. Andrews, J. E. Cadle, M. L. Crump, A. H. Savitzky, and K. D. Wells. 2003. Herpetology, ed. 3. Upper Saddle River, New Jersey, Prentice-Hall. *A comprehensive textbook treating diversity, physiology, behavior, ecology, and conservation of reptiles and amphibians.*

Sumida, S. S., and K. L. M. Martin, eds. 1997. Amniote origins: completing the transition to land. San Diego, Academic. *Discusses the diversity, evolution, ecology, and adaptative morphology of early amniotes, with special emphasis on the origin of the amniotic egg.*

Vonk, F. J., and 16 other authors. 2008. Evolutionary origin and development of snake fangs. Nature **454:**630–633. *Despite their incredible morphological variation, all snake fangs are homologous.*

■ Custom Website

The *Animal Diversity* Online Learning Center is a great place to check your understanding of chapter material. Visit www.mhhe.com/hickmanad6e for access to key terms, quizzes, and more! Further enhance your knowledge with Web links to chapter-related material.

Explore live links for these topics:

Class Reptilia
Order Testudines
Marine Turtles
Order Crocodilia
Suborder Lacertilia

Suborder Serpentes
Order Sphenodonta
Superorder Archosauria and Related Mesozoic Reptiles
Dissection Guides for Reptiles
Conservation Issues Concerning Reptiles

Birds

Flock of dunlins, *Calidris alpina,* in flight.

Long Trip to a Summer Home

Some birds, having mastered flight, use this power to make long seasonal migrations. Moving between southern wintering regions and northern summer breeding regions with long days and an abundance of insects provides parents with ample food to rear their young. Predators of birds are not so abundant in the far North, and a brief once-a-year appearance of vulnerable young birds does not encourage buildup of predator populations. Migration also vastly increases the amount of space available for breeding and reduces aggressive territorial behavior. Finally, migration favors homeostasis—the balancing of internal physiological processes—by allowing birds to avoid climatic extremes.

The migratory pageant inspires wonder, and the physiological mechanisms of migration are equally challenging to researchers. What determines the timing of migration, and what induces each bird to store sufficient fuel for the journey? How did the sometimes difficult migratory routes originate, and what cues do birds use in navigation? How does instinct drive the migratory waves in spring and fall, carrying most birds successfully to their northern nests, while countless others fail and die, winnowed by the ever-challenging environment?

O f the vertebrates, birds (class Aves, ā´vēz; L. pl. of *avis,* bird) are the most noticeable, the most melodious, and, many think, the most beautiful. With more than 9700 species distributed nearly worldwide, birds far outnumber any other vertebrate group except fishes. Birds inhabit forests and deserts, mountains and prairies, and all oceans. Four species are known to have visited the North Pole, and one, a skua, was seen at the South Pole. Some birds live in total darkness in caves, finding their way by echolocation, and others dive to depths greater than 45 m to prey on aquatic life.

The single unique feature that distinguishes birds from other living animals is their feathers. If an animal has feathers, it is a bird; if it lacks feathers, it is not a bird. However, we note that feathers were not so diagnostic in the past; some theropod dinosaurs had feathers.

There is great uniformity of structure among birds. Despite approximately 150 million years of evolution, during which they proliferated and adapted to specialized ways of life, we have no difficulty recognizing a living bird as a bird. In addition to feathers, all birds have forelimbs modified into wings (although not always used for flight); all have hindlimbs adapted for walking, swimming, or perching; all have keratinized beaks; and all lay eggs. The reason for this great structural and functional uniformity is that birds evolved into flying machines, which greatly restricts morphological diversity.

A bird's entire anatomy is designed around flight. An airborne life for a large vertebrate is a highly demanding evolutionary challenge. A bird must, of course, have wings for lift and propulsion. Bones must be light and hollow and yet serve as a rigid airframe. The respiratory system must be highly efficient to meet the intense metabolic demands of flight. A bird must have a rapid and efficient digestive system to process an energy-rich diet, and it must have a high-pressure circulatory system. Above all, birds must have a finely tuned nervous system and acute senses, especially superb vision, to handle the complex demands of headfirst, high-velocity flight.

■ Origin and Relationships

Approximately 147 million years ago, a flying animal drowned and settled to the bottom of a shallow marine lagoon in what is now Bavaria, Germany. It was rapidly covered with a fine silt and eventually fossilized. There it remained until discovered in 1861 by a workman splitting slate in a limestone quarry. The fossil was approximately the size of a crow, with a skull not unlike that of modern birds except that the beaklike jaws bore small, bony teeth set in sockets like those of dinosaurs (figure 19.1). The skeleton was decidedly reptilian, with a long bony tail, clawed fingers, and abdominal ribs. It might have been classified as a theropod dinosaur except that it carried an unmistakable imprint of feathers, those marvels of biological engineering that only birds possess. Named *Archaeopteryx lithographica* (är-kē-op´ter-iks lith-ōgraf´i-kə; Gr., meaning "ancient wing inscribed in stone"), the fossil was an especially fortunate discovery because it demonstrated beyond reasonable doubt the phylogenetic relatedness of birds and theropod dinosaurs.

Zoologists had long recognized the similarity of birds and nonavian reptiles because of their many shared morphological, developmental, and physiological homologies. The distinguished English zoologist Thomas Henry Huxley was so impressed with these affinities that he called birds "glorified reptiles" and classified them with a group of dinosaurs called

B

figure 19.1

Archaeopteryx, a 147-million-year-old relative of modern birds. **A,** Cast of the second and most nearly perfect fossil of *Archaeopteryx,* which was discovered in a Bavarian stone quarry. Ten specimens of *Archaeopteryx* have been discovered, the most recent one described in 2007. **B,** Reconstruction of *Archaeopteryx.*

A

CHARACTERISTICS
of Class Aves

1. **Neck elongate and S-shaped;** forelimbs modified as **wings; endothermic**
2. Epidermal covering of **feathers** and **leg scales;** thin epidermis and dermis; no sweat glands; oil gland at base of tail
3. Skull with **one occipital condyle;** many **bones with air cavities;** ribs with strengthening, uncinate processes; tail short, caudal vertebrate reduced to a **pygostyle;** pelvic girdle a **synsacrum; sternum usually large and keeled**
4. **No teeth;** each jaw covered with a keratinized sheath, forming a **beak; gizzard** present
5. Brain well developed, with **large optic lobes and cerebellum;** 12 pairs of cranial nerves
6. Eyes large, with **pecten** (see figure 19.12); middle ear with a single bone
7. Separate sexes; internal fertilization; copulatory organ (penis) only in paleognathids, ducks, geese, and a few others; **females with functional left ovary and oviduct only;** sex determined by chromosomes (female is heterogametic)
8. Fetal membranes of **amnion, chorion, and allantois; oviparous; eggs with much yolk and hard, calcareous shells;** extensive parental care of young
9. Excretory system of **metanephric kidneys** and ureters that open into cloaca; uric acid main nitrogenous waste
10. Lungs of **parabronchi** with **continuous air flow; syrinx** (voice box) present; thin **air sacs** among visceral organs and skeleton
11. Heart with two atria and two ventricles; **separate pulmonary and systemic circuits; persistent right aortic arch;** nucleated red blood cells

theropods (p. 377) that displayed several birdlike characteristics (figures 19.2 and 19.3). Theropod dinosaurs share many derived characters with birds, the most obvious of which is an elongate, mobile, S-shaped neck.

Feathers preceded both birds and flight. Dromeosaurs, a group of theropods that includes *Velociraptor,* share many additional derived characters with birds, including a furcula (fused clavicles) and lunate wrist bones that permit swiveling motions used in flight (figure 19.3). Additional evidence linking birds to dromeosaurs comes from recently described fossils from late Jurassic and early Cretaceous deposits in Liaoning Province, China. These spectacular dromeosaur-like fossils include some with filaments, such as *Sinosauropteryx,* and some with feathers, such as *Protarchaeopteryx* and *Caudipteryx.* These feathered dinosaurs could not fly, however, because they had short forelimbs and symmetrical vaned feathers (the flight feathers of modern flying birds are asymmetrical). Clearly, these filaments and feathers served a different purpose, perhaps providing thermoregulation or crypsis, or used in courtship displays. Later, feathers were exapted (p. 27) for powered flight. Fossils from Spain and Argentina document development of a keeled sternum and alula, loss of teeth, and fusion of bones, characteristics of modern birds

absent in *Archaeopteryx.* Clearly, a phylogenetic approach to classification would group birds with theropod dinosaurs. With this view, dinosaurs are not extinct—they are with us today as birds!

Living birds (Neonithes) are divided into two groups: (1) **Paleognathae** (Gr. *palaios,* ancient, + *gnathos,* jaw), large, flightless, ostrichlike birds and kiwis, often called **ratite** birds, that have a flat sternum with poorly developed pectoral muscles; and (2) **Neognathae** (Gr. *neos,* new, + *gnathos,* jaw), all other birds, nearly all of which are flying birds that have a keeled sternum to which powerful flight muscles insert. There are several flightless neognathus birds, some of which lack a keeled sternum. Flightlessness has appeared independently among many groups of birds; the fossil record reveals flightless wrens, pigeons, parrots, cranes, ducks, auks, and even a flightless owl. Penguins are flightless although they use their wings to "fly" through water (see figure 4.6). Flightlessness has almost always evolved on islands lacking terrestrial predators. Flightless birds living on continents today are the large paleognathids (ostrich, rhea, cassowary, emu), which can run fast enough to escape predators. Ostriches can run 70 km (43 miles) per hour, faster than any other bipedal animal. The evolution of ratites is discussed on p. 19.

The bodies of flightless birds are dramatically redesigned to remove restrictions for flight. The keel of the sternum is lost, and heavy flight muscles (as much as 17% of the body weight of flying birds) disappear. Because body weight is no longer a restriction, flightless birds tend to become large. Several extinct flightless birds were enormous: the giant moas of New Zealand weighed more than 225 kg (500 pounds), and elephantbirds of Madagascar, the largest birds that ever lived, probably weighed nearly 450 kg (about 1000 pounds) and stood nearly 2 m tall.

■ Structural and Functional Adaptations for Flight

Just as an airplane must be designed and built according to rigid aerodynamic specifications, so too must birds meet stringent structural requirements to be airborne. Flight by humans became possible when they developed an internal combustion engine and learned how to reduce the weight-to-power ratio to a critical point. Birds accomplished flight millions of years ago. Unlike airplanes, birds also must feed themselves, convert food into high-energy fuel, escape predators, repair their own injuries, maintain a constant body temperature, and reproduce.

Feathers

Feathers are very lightweight, and yet they possess remarkable toughness and tensile strength. Most bird feathers are contour feathers, vaned feathers that cover and streamline a bird's body. A contour feather consists of a hollow quill, or calamus, emerging from a skin follicle, and a shaft, or rachis,

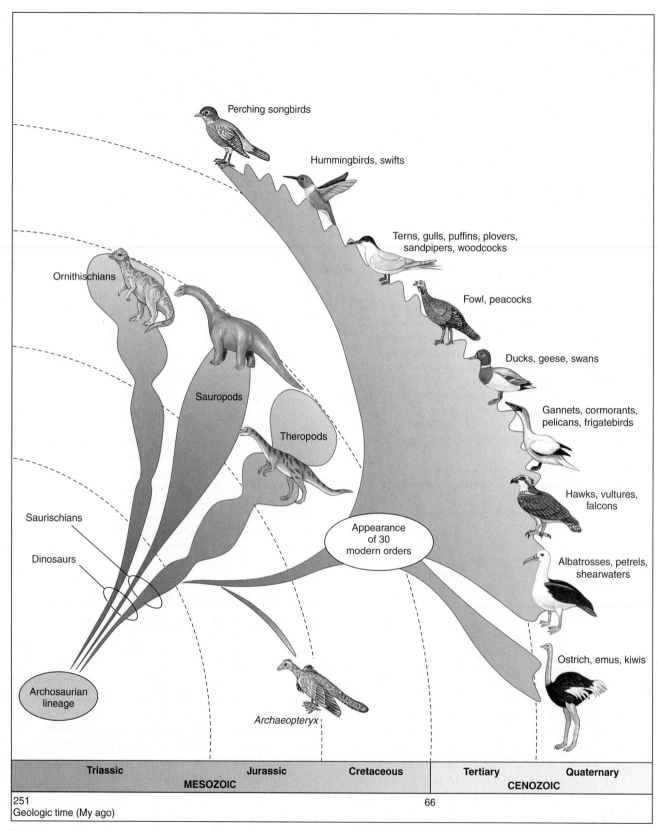

Perching songbirds

Hummingbirds, swifts

Terns, gulls, puffins, plovers, sandpipers, woodcocks

Fowl, peacocks

Ducks, geese, swans

Gannets, cormorants, pelicans, frigatebirds

Hawks, vultures, falcons

Albatrosses, petrels, shearwaters

Ostrich, emus, kiwis

Ornithischians

Sauropods

Theropods

Saurischians

Dinosaurs

Appearance of 30 modern orders

Archosaurian lineage

Archaeopteryx

Triassic	**Jurassic**	**Cretaceous**	**Tertiary**	**Quaternary**
MESOZOIC			**CENOZOIC**	

251
Geologic time (My ago)

66

figure 19.2

Evolution of modern birds. Of 30 living bird orders, 9 of the more important are shown. The earliest known bird, *Archaeopteryx*, lived in the Upper Jurassic period about 147 million years ago. *Archaeopteryx* uniquely shares many specialized aspects of its skeleton with the smaller theropod dinosaurs and is considered to have evolved within the theropod clade. Evolution of modern bird orders occurred rapidly during the Cretaceous and early Tertiary periods.

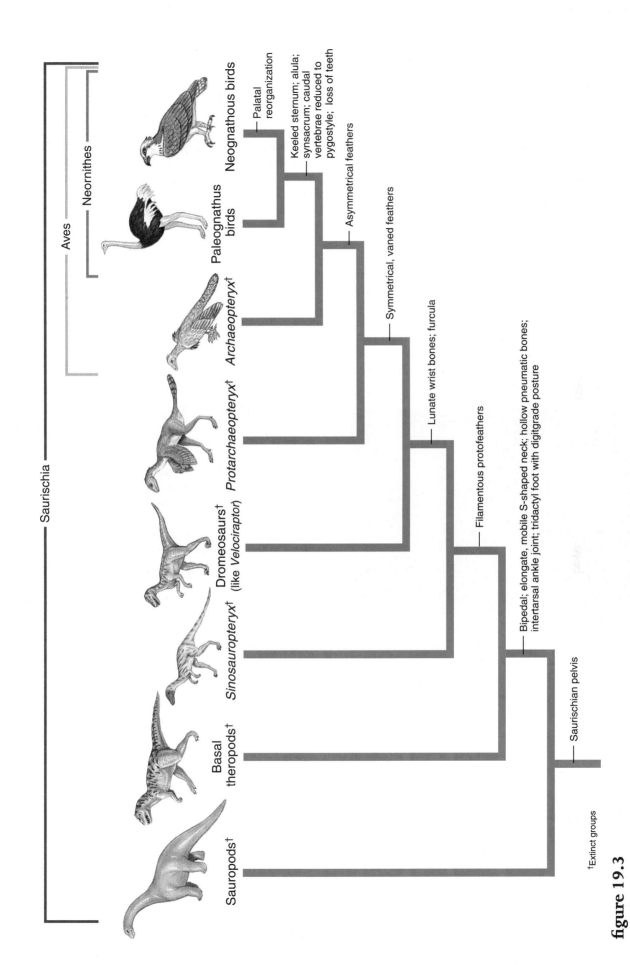

figure 19.3

Cladogram of Saurischia, illustrating the relationship of several taxa to modern birds. Shown are a few of the shared derived characters, mostly related to flight, that were used to construct the genealogy. The ornithischians are the sister group to the saurischians, and all are members of the clade Archosauria (see figures 18.1 and 18.2).

The following labels appear in the cladogram:

Saurischia

Aves
Neornithes

Neognathous birds
Paleognathous birds
Archaeopteryx†
Protarchaeopteryx†
Dromeosaurs† (like *Velociraptor*)
Sinosauropteryx†
Basal theropods†
Sauropods†

Palatal reorganization
Keeled sternum; alula; synsacrum; caudal vertebrae reduced to pygostyle; loss of teeth
Asymmetrical feathers
Symmetrical, vaned feathers
Lunate wrist bones; furcula
Filamentous protofeathers
Bipedal; elongate, mobile S-shaped neck; hollow pneumatic bones; intertarsal ankle joint; tridactyl foot with digitgrade posture
Saurischian pelvis

†Extinct groups

which is a continuation of the quill and bears numerous barbs (figure 19.4). Barbs are arranged in closely parallel fashion and spread diagonally outward from both sides of the central shaft to form a flat, expansive, webbed surface, the vane. There may be several hundred barbs in a vane.

Through a microscope, each barb appears to be a miniature replica of the feather, with numerous parallel filaments called barbules set in each side of the barb and spreading laterally from it. There may be 600 barbules on each side of a barb, adding up to more than 1 million barbules for the feather. Barbules of one barb overlap the barbules of a neighboring barb in a herringbone pattern and are held together with great tenacity by tiny hooks. Should two adjoining barbs become separated—and considerable force is needed to pull the vane apart—they are instantly zipped together again by drawing the feather through the fingertips. Birds do this preening with their bill.

Like a reptile's scale, to which it is homologous, a feather develops from an epidermal elevation overlying a nourishing dermal core. However, rather than flattening like a scale, a feather bud rolls into a cylinder and sinks into the follicle

from which it is growing. During growth, pigments (lipochromes and melanin) are added to epidermal cells. As the feather enlarges and nears the end of its growth, the soft shaft and barbs are transformed into hard structures by deposition of keratin. The protective sheath splits apart, allowing the end of a feather to protrude and barbs to unfold.

When fully grown, a feather, like mammalian hair, is a dead structure. Shedding, or molting, of feathers is a highly orderly process; feathers are discarded gradually, which avoids appearance of bare spots. Flight and tail feathers are lost in exact pairs, one from each side, maintaining balance. Replacements emerge before the next pair is lost, and most birds can continue to fly unimpaired during the molting period; however, many water birds (ducks, geese, loons, and others) lose all their primary feathers at once and are grounded during the molt. Many prepare for molting by moving to isolated bodies of water where they can find food and more easily escape enemies. Nearly all birds molt at least once a year, usually in late summer after nesting season.

Skeleton

A major structural requirement for flight is a light, yet sturdy skeleton (figure 19.5A). Compared with the earliest known bird, *Archaeopteryx* (figure 19.5B), bones of modern birds are phenomenally light, delicate, and laced with air cavities. Such pneumatized bones (figure 19.6) are nevertheless strong. The skeleton of a frigate bird with a 2.1 m (7-foot) wingspan weighs only 114 grams (4 ounces), less than the weight of all its feathers.

As archosaurs, birds evolved from ancestors with diapsid skulls (p. 367). However, skulls of modern birds are so specialized that it is difficult to see any trace of the original diapsid condition. A bird's skull is built lightly and mostly fused into one piece. A pigeon's skull weighs only 0.21% of its body weight; by comparison, a rat's skull weighs 1.25% of its body weight. The braincase and orbits are large in bird skulls to accommodate a bulging brain and large eyes needed for quick motor coordination and superior vision.

In *Archaeopteryx*, both jaws contained teeth set in sockets, an archosaurian characteristic. Modern birds are completely toothless, having instead a keratinous beak molded around the bony jaws. The mandible is a complex of several bones hinged to provide a double-jointed action, which permits the mouth to gape widely. Most birds have kinetic skulls (kinetic skulls of lizards are described on p. 374) with a flexible attachment between upper jaw and skull.

The most distinctive feature of the vertebral column is its rigidity. Most vertebrae, except the **cervicals** (neck vertebrae), are fused together, and along with the pelvic girdle form a stiff but light framework to support the legs and to provide rigidity for flight. To assist in this rigidity, ribs are braced against each other with uncinate processes (see figure 19.5A). Except in flightless birds, the sternum bears a large, thin keel that provides an attachment for powerful flight muscles. Fused clavicles form an elastic **furcula** that apparently stores energy as it

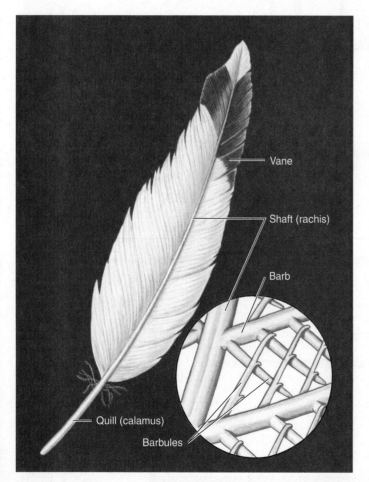

figure 19.4

Contour feather. Inset enlargement of the vane shows the minute hooks on the barbules that cross-link loosely to form a continuous surface of vane.

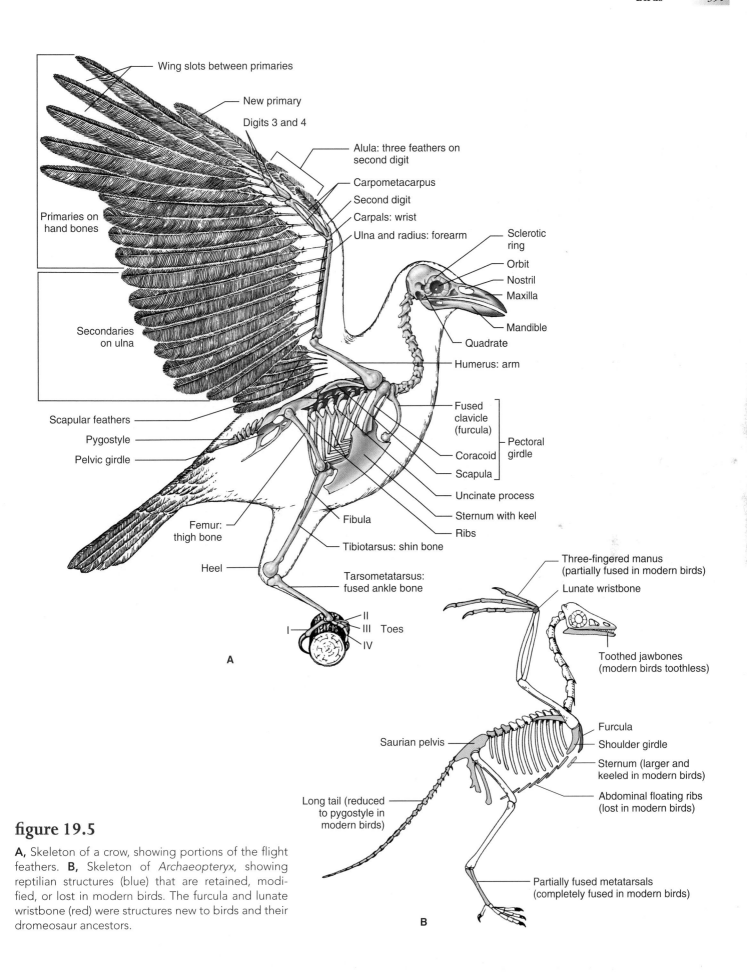

figure 19.5

A, Skeleton of a crow, showing portions of the flight feathers. **B,** Skeleton of *Archaeopteryx*, showing reptilian structures (blue) that are retained, modified, or lost in modern birds. The furcula and lunate wristbone (red) were structures new to birds and their dromeosaur ancestors.

figure 19.6

Hollow wing bone of a songbird, showing stiffening struts and air spaces that replace bone marrow. Such pneumatized bones are remarkably light and strong.

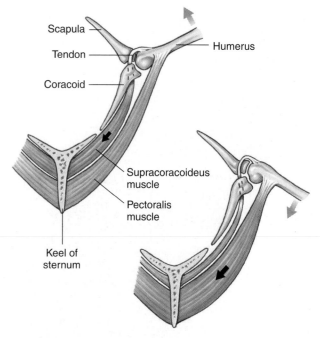

figure 19.7

Flight muscles of a bird are arranged to keep the center of gravity low in the body. Both major flight muscles are anchored to the sternum keel. Contraction of the pectoralis muscle pulls the wing downward. Then, as the pectoralis relaxes, the supracoracoideus muscle contracts and, acting as a pulley system, pulls the wing upward.

flexes during wing beats. The asymmetrical flight feathers and distinct furcula of *Archaeopteryx,* along with the anatomy of its brain and inner ear, suggest it had some flight abilities. However, it would have been a weak flier, because its small sternum, offered relatively little area for flight muscle attachment (see figure 19.5B).

Bones of the forelimbs are highly modified for flight. They are reduced in number, and several are fused together. Despite these alterations for flight, a bird wing is clearly a rearrangement of the basic vertebrate tetrapod limb from which it arose (see figure 17.1), and all the elements—arm, forearm, wrist, and fingers—are represented in modified form (see figure 19.5A).

Muscular System

Locomotor muscles of wings are relatively massive to meet demands of flight. The largest of these is the pectoralis, which depresses the wings in flight. Its antagonist is the supracoracoideus muscle, which raises the wing (figure 19.7). Surprisingly, perhaps, this latter muscle is not located on the backbone (anyone who has been served the back of a chicken knows that it offers little meat) but is positioned under the pectoralis on the breast. It is attached by a tendon to the upper side of the humerus of the wing so that it pulls from below by an ingenious "rope-and-pulley" arrangement. Both the pectoralis and the supracoracoideus are anchored to the keel of the sternum. Positioning the main muscle mass low in the body improves aerodynamic stability.

From the main leg muscle mass in the thigh, thin but strong tendons extend downward through sleevelike sheaths to the toes. Consequently, the feet are nearly devoid of muscles, explaining the thin, delicate appearance of a bird's leg. This

arrangement places the main muscle mass near a bird's center of gravity and at the same time allows great agility to the slender, lightweight feet. Because the feet are composed mostly of bone, tendon, and tough, scaly skin, they are highly resistant to damage from freezing. When a bird perches on a branch, an ingenious toe-locking mechanism (figure 19.8) is activated, which prevents the bird from falling off its perch when asleep. The same mechanism causes the talons of a hawk or owl to sink deeply into its prey as the legs bend under the impact of the strike. The powerful grip of a bird of prey was described by L. Brown.[1]

> When an eagle grips in earnest, one's hand becomes numb, and it is quite impossible to tear it free, or to loosen the grip of the eagle's toes with the other hand. One just has to wait until the bird relents, and while waiting one has ample time to realize that an animal such as a rabbit would be quickly paralyzed, unable to draw breath, and perhaps pierced through and through by the talons in such a clutch.

Digestive System

Beaks of birds are strongly adapted to specialized food habits—from generalized types, such as the strong, pointed beaks of crows and ravens, to highly specialized beaks in

---

[1]Brown, L. 1970, Eagles, New York, Arco Publishing.

figure 19.8

Perching mechanism of a bird. When a bird settles on a branch, tendons automatically tighten, closing the toes around the perch.

flamingos, pelicans, and avocets (figure 19.9). The beak of a woodpecker is a straight, hard, chisel-like device. Anchored to a tree trunk, with its tail serving as a brace, a woodpecker delivers powerful, rapid blows to excavate nest cavities or to expose burrows of wood-boring insects. It then uses its long, flexible, barbed tongue to seek insects in their galleries. A woodpecker's skull is especially thick to absorb shock.

Birds process an energy-rich diet rapidly and thoroughly with efficient digestive equipment. A shrike can digest a mouse in 3 hours, and berries pass completely through the digestive tract of a thrush in just 30 minutes. Although many animal foods find their way into diets of birds, insects constitute by far the largest component. Because birds lack teeth, foods that require grinding are reduced in the gizzard. Many birds have an enlargement (crop) at the lower end of the esophagus, which serves as a storage chamber. In pigeons, doves, and some parrots, the crop not only stores food but also produces a lipid- and protein-rich "milk," composed of sloughed epithelial cells of the crop lining. For a few days after hatching, the young are fed regurgitated crop milk by both parents.

The stomach proper consists of a **proventriculus,** which secretes gastric juice, and a muscular gizzard, a region specialized for grinding food. To assist in grinding food, birds swallow gritty objects or pebbles, which lodge in the gizzard. Certain birds of prey, such as owls, form pellets of indigestible materials, mainly bones and fur, in the proventriculus and eject them through the mouth. At the junction of the small intestine with the colon are paired **ceca;** these are well developed in herbivorous birds in which they serve as fermentation chambers. The terminal part of the digestive system is the **cloaca,** which also receives genital ducts and ureters.

Circulatory System

The general plan of circulation in birds is not greatly different from that of mammals, although it evolved independently. Their four-chambered heart is large with strong ventricular walls; thus birds share with mammals a complete separation of respiratory and systemic circulations. Their heartbeat is extremely fast, and as in mammals an inverse relationship occurs between heart rate and body weight. For example, a turkey has a heart rate at rest of about 93 beats per minute, a chicken has a rate of 250 beats per minute, and a black-capped chickadee has a rate of 500 beats per minute when asleep, which may increase to a phenomenal 1000 beats per minute during exercise. Blood pressure in birds is roughly equivalent to that in mammals of similar size. Birds' blood contains nucleated biconvex erythrocytes. Mammals, the only other endothermic vertebrates, have enucleate biconcave erythrocytes that are somewhat smaller than those of birds.

Respiratory System

The respiratory system of birds differs radically from lungs of nonavian reptiles and mammals and is marvelously adapted for meeting the high metabolic demands of flight. In birds, the finest branches of the bronchi, rather than ending in saclike alveoli as in mammals, are tubelike **parabronchi** through which air flows continuously. Also unique is the extensive system of nine interconnecting air sacs that are located in pairs in the thorax and abdomen and even extend by tiny tubes into the centers of long bones (figure 19.10A). Air sacs are connected to the lungs in such a way that most of the inspired air bypasses the lungs and flows directly into the posterior air sacs, which serve as reservoirs for fresh air. On expiration, this oxygenated air is passed through the lungs and collected in the anterior air sacs. From there, it flows directly to the outside. Thus, it takes two respiratory cycles for a single breath of air to pass through the respiratory system (figure 19.10B). The advantage of such a system is that an almost continuous stream of oxygenated air is passed through a system of richly vascularized parabronchi. It is clearly the most efficient respiratory system of any terrestrial vertebrate.

The remarkable efficiency of a bird's respiratory system is emphasized by bar-headed geese that routinely migrate over the Himalayan mountains and have been sighted flying over Mt. Everest (8848 meters or 29,141 feet) under conditions that are severely hypoxic to humans. The geese reach altitudes of 9000 meters in less than a day, without the acclimatization that is absolutely essential for humans even to approach the upper reaches of Mt. Everest.

Excretory System

Urine is formed in the relatively large, paired metanephric kidneys. As in other vertebrates, urine is formed by glomerular filtration followed by selective modification of filtrate in the

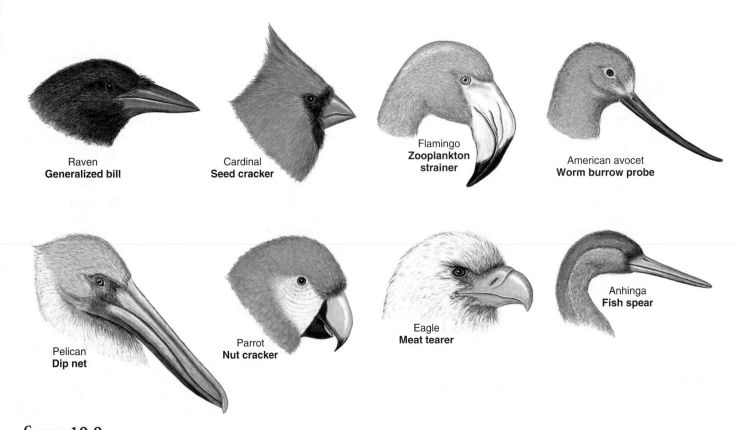

figure 19.9

Some bills of birds showing a variety of adaptations.

A   **B**

figure 19.10

Respiratory system of a bird. **A,** Lungs and air sacs. One side of the bilateral air sac system is shown. **B,** Movement of a single volume of air through a bird's respiratory system. Two full respiratory cycles are required to move air through the system.

tubule. Urine passes by way of ureters to the **cloaca.** There is no urinary bladder.

Like nonavian reptiles, birds excrete their nitrogenous wastes as uric acid. In shelled eggs, all excretory products must remain within the eggshell with the growing embryo. Uric acid crystallizes from solution and can be stored harmlessly in the allantois (p. 367). Because of uric acid's low solubility, a bird can excrete 1 g of uric acid in only 1.5 to 3 ml of water, whereas a mammal can require 60 ml of water to excrete 1 g of urea. Concentration of uric acid occurs almost entirely in the cloaca, where it is combined with fecal matter, and the water is reabsorbed. Bird kidneys are much less efficient than those of mammals in removing salts, especially sodium, potassium, and chloride. The kidneys of most birds concentrate salts only slightly greater than that of blood, but most mammals concentrate salts 4 to 8 times that of blood, and some desert rodents concentrate salts to nearly 25 times that of blood.

To compensate for the weak solute-concentrating ability of the kidneys, some birds, especially marine birds, use extrarenal mechanisms to excrete salts gained from the food they eat and the seawater they drink. **Salt glands,** one located above each eye (figure 19.11), can excrete a highly concentrated solution of sodium chloride, up to twice the concentration of seawater. The salt solution runs out the internal or external nostrils, giving gulls, petrels, and other sea birds a perpetually runny nose.

Nervous and Sensory System

The design of a bird's nervous and sensory system reflects the complex problems of flight and a highly visible existence, in which the bird must gather food, mate, defend territory, incubate and rear young, and correctly distinguish friend from foe. Its brain has well-developed cerebral hemispheres, cerebellum, and optic lobes. The cerebral cortex—chief coordinating center of a mammalian brain—is thin, unfissured, and poorly developed in birds. The core of the cerebrum, the dorsal ventricular ridge, has enlarged into the principal integrative center, controlling such activities as eating, singing,

flying, and complex reproductive behavior. Relatively intelligent birds, such as crows and parrots, have larger cerebral hemispheres than do less intelligent birds, such as chickens and pigeons. The cerebellum, much larger in birds than in nonavian reptiles, coordinates muscle position, equilibrium, and visual information used in movement and balance. The optic lobes, laterally bulging structures of the midbrain, form a visual apparatus comparable to the visual cortex of mammals.

The senses of smell and taste are poor in some birds, but they are relatively well developed in many other birds, such as carnivorous birds, flightless birds, oceanic birds, and waterfowl. Birds have good hearing and superb vision, the keenest in the animal kingdom. The organ of hearing, the **cochlea,** is much shorter than the coiled mammalian cochlea, and yet birds can hear roughly the same range of sound frequencies as humans. Actually, a bird's ear far surpasses that of humans in its capacity to distinguish differences in intensities and to respond to rapid fluctuations in pitch.

A bird's eye resembles that of other vertebrates in gross structure but is relatively larger, less spherical, and almost immobile; instead of turning their eyes, birds turn their heads with their long flexible necks to scan the visual field. The light-sensitive retina (figure 19.12) is generously equipped with rods (for dim light vision) and cones (for color vision). Cones predominate in day birds, and rods are more numerous in nocturnal birds. A distinctive feature of a bird's eye is the **pecten,** a highly vascularized organ attached to the retina and jutting into the vitreous humor (figure 19.12). The pecten is thought to provide nutrients and oxygen to the eye.

The **fovea,** or region of keenest vision on the retina, is in a deep pit (in birds of prey and some others), which requires the bird to focus exactly on the subject. Many birds, moreover, have two sensitive spots (foveae) on the retina (figure 19.12): a central one for sharp monocular views and a posterior one for binocular vision. The visual acuity of a hawk is about eight times that of humans (enabling a hawk to see clearly a

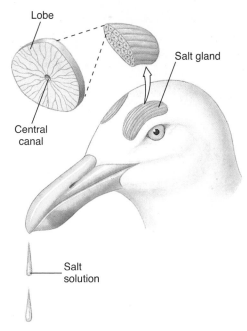

figure 19.11

Salt glands of a marine bird (gull). One salt gland is located above each eye. Each gland consists of several lobes arranged in parallel. One lobe is shown in cross section, much enlarged. Salt is secreted into many radially arranged tubules, and then flows into a central canal that leads into the nose.

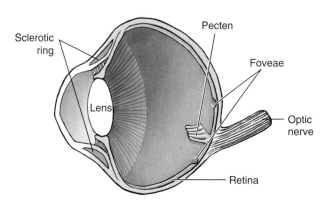

figure 19.12

A hawk's eye has all the structural components of a mammalian eye, plus a pleated structure, the pecten, thought to provide nourishment to the retina. The extraordinarily keen vision of hawks is attributed to the extreme density of cone cells in the foveae: 1.5 million per fovea compared to 0.2 million for humans.

crouching rabbit 2 km away), and an owl's ability to see in dim light is more than 10 times that of a human.

Many birds can see at ultraviolet wavelengths, enabling them to view environmental features inaccessible to humans but accessible to insects (such as flowers with ultraviolet-reflecting "nectar guides" that attract pollinating insects). Several species of ducks, hummingbirds, kingfishers, and passerines (songbirds) can see at near ultraviolet (UV) wavelengths down to 370 nm (human eyes filter out ultraviolet light below 400 nm). For what purpose do birds use their UV sensitivity? Some, such as hummingbirds, may be attracted to nectar-guiding flowers, but for others, the benefit derived from UV sensitivity is unknown.

■ Flight

What prompted evolution of flight, birds' ability to rise free of earthbound concerns, as almost every human has dreamed of doing? The air was a relatively unexploited habitat stocked with flying insects for food. Flight also offered escape from terrestrial predators and opportunity to travel rapidly and widely to establish new breeding areas and to benefit from year-round favorable climate by migrating north and south with the seasons.

Bird Wing as a Lift Device

To fly, a bird must generate lift force greater than its own mass to become airborne and must provide propulsion to move. It uses its wings to do both. A wing is streamlined in cross section, with a slightly concave lower surface (cambered) and with small, tight-fitting feathers where the leading edge meets the air. Air slips smoothly over the wing, creating lift with minimum drag. Some lift is produced by positive pressure against the undersurface of the wing; however, more than two-thirds of the total lift comes from the negative pressure above the wing, where the airstream must travel farther and faster over the convex surface.

The lift-to-drag ratio of an airfoil is determined by the angle of attack (angle of tilt) and by airspeed (figure 19.13A). At high speeds sufficient lift is generated so that the wing is held at a low angle of attack, creating less drag. As speed decreases, lift can be increased by increasing the angle of attack, but drag forces also increase. When the angle of attack becomes too steep, turbulence appears on the upper surface, lift is destroyed, and stalling occurs (figure 19.13B). Stalling can be delayed or prevented by the presence of a wing slot so that a layer of rapidly moving air is directed across the upper wing surface (figure 19.13C). Wing slots are used in aircraft traveling at low speed. In birds, two kinds of wing slots occur: (1) the **alula**, or group of small feathers on the thumb (see figures 19.5A and 19.16), which provides a midwing slot, and (2) gaps which provide a wing-tip slot between the primary feathers. In many songbirds, these together provide

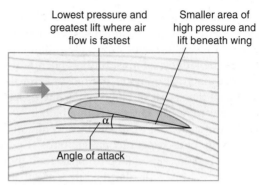

A Air flow around wing

B Stalling at low speed

C Preventing stall with wing slots

D Formation of wing tip vortex

figure 19.13

A, Air patterns formed by the airfoil, or wing, moving from right to left. **B,** At low speed, the angle of attack (α) must increase to maintain lift, but this increases the threat of stalling. **C,** Low-speed stalling can be prevented with wing slots. **D,** Wing tip vortex, a turbulence that tends to develop at high speeds, reduces flight efficiency. The effect is reduced in wings that sweep back and taper to a tip.

figure 19.14

In the normal flapping flight of strong fliers such as ducks, the wings sweep downward and forward fully extended. Thrust is provided by primary feathers at the wing tips. To begin the upbeat, the wing is bent, bringing it upward and backward. The wing then extends, ready for the next downbeat.

figure 19.15

The secret of a hummingbird's ability to change direction instantly, or to hang motionless in the air while sipping nectar from a flower, lies in its wing structure. The wing is nearly rigid, but hinged at the shoulder by a swivel joint and powered by a supracoracoideus muscle (p. 392) that is unusually large for the bird's size. When hovering, the wing moves in a sculling motion. The leading edge of the wing moves forward on the forward stroke, and then swivels nearly 180 degrees at the shoulder to move backward on the backstroke. The effect is to provide lift without propulsion on *both* forward and backward strokes.

stall-preventing slots for nearly the entire outer (and aerodynamically more important) half of the wing.

Flapping Flight

Two forces are required for flapping flight: a vertical *lifting* force to support the bird's weight, and a horizontal *thrusting* force to move the bird forward against resistive forces of drag. Primary feathers at the wing tips provide most of the thrust, while secondary feathers of the highly cambered inner wing, which do not move so far or so fast, act as an airfoil, providing most of the lift. Greatest power is applied on the downstroke. The primary feathers bend upward and twist to a steep angle of attack, biting into the air like a propeller (figure 19.14). The entire wing (and the bird's body) is pulled forward. On the upstroke, the primary feathers bend in the opposite direction so that their upper surfaces twist into a positive angle of attack to produce thrust, just as the lower surfaces did on the downstroke. A powered upstroke is essential for hovering flight, as in hummingbirds (figure 19.15), and is important for fast, steep takeoffs by small birds with elliptical wings.

Basic Forms of Bird Wings

Bird wings vary in size and form because successful exploitation of different habitats has imposed special aerodynamic requirements. Four types of bird wings are easily recognized: elliptical, high-aspect, dynamic soaring, and high-lift (figure 19.16).

Elliptical Wings

Birds such as flycatchers, warblers, sparrows, doves, woodpeckers, and magpies, which must maneuver in forested or brushy habitats, have elliptical wings (figure 19.16A). This type has a low-aspect ratio (ratio of length to average width). Wings of the highly maneuverable British Spitfire fighter plane of World War II fame conformed closely to the outline of a sparrow's wing. Elliptical wings have both an alula and gaps between the primary feathers; this arrangement helps to prevent stalling during sharp turns, low-speed flight, and frequent landing and takeoff. Each separated primary feather behaves as a narrow wing with a high angle of attack, providing high lift at low speed. The high maneuverability of elliptical wings is exemplified by the tiny chickadee, which, if frightened, can change course within 0.03 second.

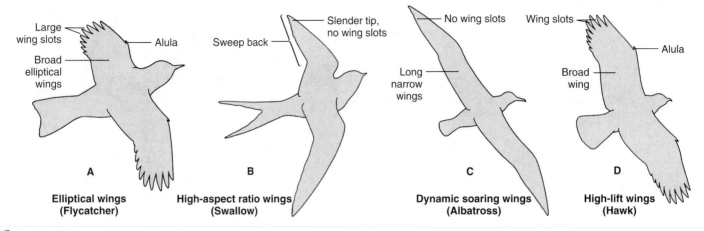

figure 19.16

Four basic forms of bird wings.

High-Aspect Ratio Wings

Birds that feed during flight, such as swallows, hummingbirds, and swifts, or those that make long migrations, such as plovers, sandpipers, terns, and gulls (figure 19.16B), have wings that sweep back and taper to a slender tip. They are rather flat in section, have a high-aspect ratio, and lack wing-tip slotting. Sweepback and wide separation of wing tips reduce "tip vortex" (see figure 19.13D), a drag-creating turbulence that tends to develop at wing tips at faster speeds. This type of wing is aerodynamically efficient for high-speed flight but cannot easily keep a bird airborne at low speeds, except for hummingbirds, which rapidly move their wings in a specialized way to hover (see figure 19.15). The fastest birds, such as sandpipers, clocked at 175 km (109 miles) per hour, belong to this group.

Dynamic Soaring Wings

Oceanic soaring birds, including albatrosses, shearwaters, and gannets (figure 19.16C), also have high-aspect ratio wings, shaped like those of sailplanes. Such long, narrow wings lack slots and are adapted for dynamic soaring. Dynamic soaring can be done only over seas with strong, reliable winds, and it exploits different wind speeds near the ocean surface (slow) and well above the surface (fast). A bird that uses dynamic soaring begins a downwind glide from an elevated position, gaining speed as it descends. Near the surface of the ocean, it turns into the wind and rises into stronger winds. Although its velocity relative to the ocean slows, the strong winds over its wings provide the lift to keep it aloft.

High-Lift Wings

Hawks, vultures, eagles, owls, and ospreys (figure 19.16D)—predators that carry heavy loads—have wings with slotting, alulas, and pronounced camber, all of which promote high lift at low speed. Many of these birds are land soarers, with broad, slotted wings that provide the sensitive response and maneuverability required for static soaring in the capricious air currents over land.

■ Migration and Navigation

We described advantages of migration in the prologue to this chapter. Not all birds migrate, of course, but most North American and European species do, and the biannual journeys of some are truly extraordinary undertakings.

Migration Routes

Most migratory birds have well-established routes trending north and south. Because most birds breed in the Northern Hemisphere, where most of the earth's landmass is concentrated, most birds migrate south for the northern winter and north to nest in the northern summer (figure 19.17). Of the 4000 or more species of migrant birds, most breed in the northern latitudes of the hemisphere. Some, especially certain aquatic species, complete their migratory routes in a very short time. Bar-tailed godwits, *Limosa lapponica,* fly non-stop, 11,000 km from Alaska to New Zealand, relying on large stores of body fat to fuel their nine-day journey. Others, however, make a leisurely trip, often stopping along the way to feed. Some warblers are known to take 50 to 60 days to migrate from their winter quarters in Central America to their summer breeding areas in Canada.

Some species have extremely long-distance migrations. Arctic terns, the greatest globe spanners of all, breed north of the Arctic Circle during the northern summer and then migrate to Antarctic regions for the northern winter. This species is also known to take a circuitous route in migrations from North America, flying to the coastlines of Europe and Africa and then to winter quarters, a trip that may exceed 18,000 km (11,200 miles).

Many small songbirds, such as warblers, vireos, thrushes, flycatchers, and sparrows, and shorebirds, such as sandpipers and plovers, also make great migratory treks (figure 19.17). Migratory birds that nest in Europe or Central Asia spend the northern winter in Africa.

Bobolink

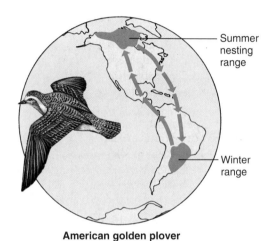

American golden plover

figure 19.17

Migrations of bobolinks, *Dolichonyx oryzivorus*, and American golden plovers, *Pluvialis dominica*. Bobolinks commute 22,500 km (14,000 miles) each year between their nesting sites in North America and their range in Argentina, where they spend the northern winters, a phenomenal feat for such a small bird. Although the breeding range has extended to colonies in western areas, these birds take no shortcuts but adhere to the ancestral seaboard route. American golden plovers fly a loop migration, striking out across the Atlantic in their southward autumnal migration but returning in the spring by way of Central America and the Mississippi Valley because ecological conditions are more favorable at that time.

Stimulus for Migration

Humans have known for centuries that the onset of reproductive cycles in birds is closely related to season. Only relatively recently, however, has it been shown that lengthening days of late winter and early spring stimulate development of gonads and accumulation of fat—both important internal changes that predispose birds to migrate northward. Increasing day length stimulates the anterior lobe of the pituitary into activity. Release of pituitary gonadotropic hormone in turn sets in motion a complex series of physiological and behavioral changes, stimulating gonadal growth, fat deposition, migration, courtship and mating behavior, and care of the young.

Direction Finding in Migration

Numerous experiments suggest most birds navigate chiefly by sight. Birds recognize topographical landmarks and follow familiar migratory routes—a behavior assisted by flock migration, during which the navigational resources and experience of older birds can be pooled. In addition to visual navigation, birds have a variety of orientation cues at their disposal. Birds have a highly accurate sense of time. Numerous studies support a hypothesis that birds can navigate by the earth's magnetic field. In the early 1970s, W. T. Keeton showed that flight bearings of homing pigeons were significantly disturbed by magnets attached to the birds' heads, or by minor fluctuations in the geomagnetic field. Deposits of a magnetic substance called magnetite (Fe_3O_4) have been discovered in the beaks of pigeons. Recent experiments showed that a pigeon could discriminate between the presence and absence of a magnetic anomaly, but not when its upper beak was anesthetized, nor when its trigeminal nerve, which innervates the upper beak, was severed.

Experiments by German ornithologists G. Kramer and E. Sauer and American ornithologist S. Emlen demonstrated convincingly that birds can navigate by celestial cues: the sun by day and the stars by night. Using special circular cages, Kramer concluded that birds maintain compass direction by referring to the sun, regardless of the time of day (figure 19.18). This process is called sun-azimuth navigation (Arabic *azimuth*, compass bearing of the sun). Sauer's and Emlen's ingenious planetarium experiments also strongly suggest that some birds, probably many, are able to navigate by the North Star axis around which the constellations appear to rotate.

Social Behavior and Reproduction

The adage says "birds of a feather flock together," and many birds are indeed highly social creatures. Especially during the breeding season, sea birds gather, often in enormous colonies, to nest and to rear young. Land birds, with some conspicuous exceptions (such as starlings and rooks), tend to be less gregarious than sea birds during breeding and to seek isolation for rearing their brood. Nonetheless, species that covet separation from their kind during breeding may aggregate for migration or feeding. Togetherness offers advantages: mutual protection from enemies, greater ease in finding mates, less opportunity for individual straying during migration, and mass huddling for protection against low night temperatures during migration. Certain species, such as pelicans (figure 19.19), use cooperative behavior to feed. At no time are the highly organized social interactions of birds more evident than during the breeding season, as they establish territorial claims, select mates, build nests, incubate and hatch their eggs, and rear their young.

Reproductive System

During most of the year, testes of male birds are tiny, bean-shaped bodes, but during the breeding season they enlarge greatly, to as much as 300 times their nonbreeding size.

A

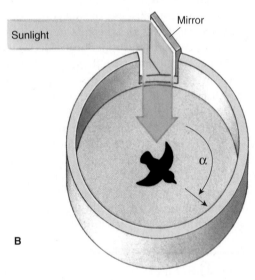

B

figure 19.18

Gustav Kramer's experiments with sun-compass navigation in starlings. **A,** In a windowed, circular cage, the bird fluttered to align itself in the direction it would normally follow if it were free. **B,** When the true angle of the sun is deflected with a mirror, the bird maintains the same relative position to the sun. This shows that these birds use the sun as a compass. The bird navigates correctly throughout the day, changing its orientation to the sun as the sun moves across the sky.

Because males of most species lack a penis, copulation is a matter of bringing cloacal surfaces into contact, usually while the male stands on the back of the female (figure 19.20).

In females of most birds, only the left ovary and oviduct develop (figure 19.21); those on the right dwindle to vestigial structures (loss of one ovary might be an adaptation for reducing weight). Eggs discharged from the ovary enter the oviduct, where fertilization occurs. Several hours later, while eggs are passing down the oviduct, **albumin,** or egg white, from special glands is added to them; farther down the oviduct, shell membrane, shell, and shell pigments are secreted around the egg. Sperm remain alive in the female oviduct for many days after a single mating.

A

B

figure 19.19

Cooperative feeding behavior by white pelicans, *Pelecanus onocrotalus.* **A,** Pelicans form a horseshoe to drive fish together. **B,** Then they plunge simultaneously to scoop fish in their huge bills. These photographs were taken 2 seconds apart. Order Pelecaniformes.

figure 19.20

Copulation in waved albatrosses, *Diomeda irrorata.* In most bird species, males lack a penis. A male passes sperm by standing on the back of a female and pressing his cloaca against that of the female. Order Procellariiformes.

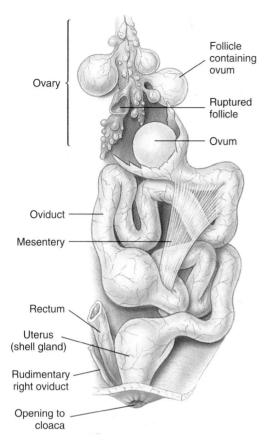

figure 19.21

Reproductive system of a female bird. For most birds, only the left ovary and reproductive tract are functional. The corresponding structures on the right dwindle to vestiges.

Upper-level carnivores, species at the top of the food chain, are vulnerable to decline through biomagnification of toxins. After World War II, DDT was widely used to control insect populations, especially mosquitoes, which transmit malaria, in the United States. Brown pelicans, hawks, eagles, ospreys, and other raptorial birds consumed prey containing DDT, which then became concentrated in the birds' bodies. A primary effect on these birds was thinning of eggshells, possibly because DDT (or its metabolite, DDE) interferes with movement of calcium from blood into the shell glands. The fragile eggs often broke before hatching and these birds were in great decline by the middle of the 20th century. Public outcry led to banning of DDT use in 1972, in part championed by Rachel Carson's *Silent Spring*, warning of the dangers of DDT. Subsequently, most raptor populations rebounded, including bald eagles, leading to their removal from the Endangered Species List. However, DDT is persistent in the environment and continues to affect some birds. In 2010, shell thinning was still a problem in certain California condor populations that feed on sea lions that forage in the highly contaminated Palo Verdes Shelf.

Mating Systems

Two contrasting mating systems are **monogamy,** in which an individual has only one mate, and **polygamy,** in which an individual has more than one mate during a breeding period. Monogamy is rare in most animal groups, but it is common in birds; more than 90% are monogamous. A few bird species, such as swans and geese, choose partners for life and often remain together throughout the year. Seasonal monogamy is more common as the great majority of migrant birds pair only during the breeding season, living independent lives the rest of the year and perhaps choosing a different mate the next breeding season.

One reason monogamy is much more common among birds than among mammals is perhaps because male and female birds are equally adept at most aspects of parental care. Male mammals do not gestate the young and do not lactate; thus they can provide little help in caring for the young. Female and male birds, by contrast, can alternate care of the nest and young, which permits one parent to be at the nest at all times. For some species, a female remains on the nest for months at a time and is fed by the male. This constant attendance to the nest may be particularly important in species that would experience high loss of eggs or young to predators or rival birds if a nest were left unguarded. For many bird species, the high demands on a male to care for the young or his mate preclude the establishment of nests with additional females.

Although most birds have a monogamous mating system (social monogamy), either member of a pair may mate with an individual that is not its partner. Recent DNA analyses have shown that most songbird species frequently are "unfaithful," engaging in extra-pair copulations. As a result, nests of many of these monogamous species contain a sizeable portion (30% or more) of young with fathers other than the attendant male. Why do individuals engage in extra-pair copulations? One possible benefit of extra-pair copulations is increased genetic diversity of the young. Second, extra-pair copulations permit a male to increase the number of sired young, increasing his reproductive success. Third, by mating with an individual of better genetic quality than the present mate, fitness of offspring could be increased. This can be particularly beneficial for females, who cannot easily increase the number of their offspring because the large size of a bird's eggs limits the number of eggs (and offspring) she can produce. Thus, extra-pair copulations allow males to increase the number of their offspring and could allow females to improve the genetic quality of their offspring. Genetic diversity of offspring produced is increased by extra-pair copulation for both males and females.

The most common form of polygamy in birds is **polygyny** ("many females"), in which a male has more than one female mate. In many species of grouse, males gather in a collective display ground, or **lek,** which is divided into individual territories, each vigorously defended by a displaying male (figure 19.22). In a lek, there is nothing of value to the female except the male, and all he can offer are his genes, for only females care for the young. Usually a dominant male and several subordinate males occur in a lek. Competition among males for females is intense, but females appear to choose the dominant male for mating because, presumably, social rank correlates with genetic quality.

figure 19.22

Dominant male sage grouse, *Centrocercus urophasianus*, surrounded by several hens that have been attracted by his "booming" display. Order Galliformes.

Polyandry ("many males"), in which a female mates with several males and the male incubates the eggs, is relatively rare. Females of the spotted sandpiper, a polyandrous shorebird, defend territories and mate with multiple males. Males incubate eggs within the female's territory and provide most parental care. This unusual reproductive strategy may be in response to high predation on spotted sandpiper nests.

Nesting and Care of Young

Most birds build some form of nest in which to rear their young. Some birds simply lay eggs on bare ground or rocks, making no pretense of nest building. Others build elaborate nests, such as the pendant nests constructed by orioles, the delicate lichen-covered mud nests of hummingbirds (figure 19.23) and flycatchers, the chimney-shaped mud nests of cliff swallows, the floating nests of rednecked grebes, and the huge sand and vegetation mounds of Australian brush turkeys. Most birds take considerable pains to conceal their nests from enemies. Nest parasites, such as brown-headed cowbirds and European cuckoos, build no nests at all but simply lay their eggs in the nests of birds smaller than themselves. When their eggs hatch, the foster parents care for the cowbird young, which outcompete the host's own hatchlings.

Newly hatched birds are of two types: **altricial** and **precocial** (figure 19.24). Altricial young, which are naked and unable to walk or see at birth, remain in the nest for a week or more. Precocial young, including quail, fowl, ducks, and most water birds, are covered with down when hatched and can run or swim as soon as their plumage is dry. Young of both types require care from parents for some time after hatching. Parents of altricial species must carry food to their young almost constantly, for young birds may eat more than their weight each day. Some young are not easily categorized as altricial or precocial because they are intermediate in development at birth. For example, gulls and terns are born covered with down and with their eyes open, but are unable to leave the nest for some time.

Although it may seem that precocial chicks have all the advantages, with their greater ability to find food and escape

figure 19.23

Anna's hummingbird, *Calypte anna*, feeding its young in its nest of plant fibers, bound together with spiderwebs and camouflaged with lichens. A female builds the nest, incubates two pea-sized eggs, and rears the young with no assistance from a male. Anna's hummingbird, a common resident of California, is the only hummingbird to overwinter in the United States. Order Apodiformes.

Altricial
One-day-old meadowlark

Precocial
One-day-old ruffed grouse

figure 19.24

Comparison of 1-day-old altricial and precocial young. The altricial meadowlark (*left*) is born nearly naked, blind, and helpless. The precocial ruffed grouse (*right*) is covered with down, alert, strong-legged, and able to feed itself.

predation, altricial chicks have an advantage of their own. Altricial chicks grow faster, perhaps due to the higher growth potential of immature tissue.

■ Humans and Bird Populations

Activities of people may cause spectacular changes in bird distribution. Both starlings (figure 19.25) and house sparrows have been accidentally or deliberately introduced into numerous countries, to become the two most abundant bird species on earth, with the exception of domestic fowl.

Humans also are responsible for the extinction of many bird species. More than 140 species of birds have, since 1681, followed the last dodo to extinction. Many were victims of changes in their habitat or competition with introduced species. Overhunting contributed to the extinction of some species, among them passenger pigeons, which 150 years ago darkened skies over North America in numbers estimated in the billions (figure 19.26).

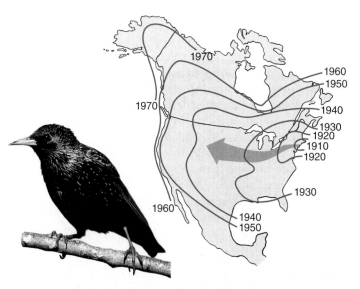

figure 19.25

Colonization of North America by starlings, *Sturnus vulgaris*, after the introduction of 60 birds into Central Park in New York City in 1890. There are now perhaps 200 million starlings in the United States alone, testimony to the great reproductive potential of birds. Starlings are omnivorous, eating mostly insects in spring and summer and shifting to wild fruits in the fall. Order Passeriformes.

Today, game bird hunting is a well-managed renewable resource in the United States and Canada, and while hunters kill millions of game birds each year, none of the 74 bird species legally hunted are endangered. Hunting interests, by acquiring large areas of wetlands for migratory bird refuges and sanctuaries, have contributed to the recovery of both game and nongame birds.

Of particular concern is the recent sharp decline in songbirds in the United States and southern Canada. According to the records of ornithologists and amateur birdwatchers, many songbird species that were abundant as recently as 40 years ago are now scarce. There are several reasons for the decline. Intensification of agriculture, sustained by the use of herbicides, pesticides, and fertilizers, has deprived ground-nesting birds of fields that were formerly left fallow. Excessive fragmentation of forests throughout much of the United States has increased exposure of nests of forest-dwelling species to nest predators, such as blue jays, raccoons, and opossums, and to nest parasites, such as brown-headed cowbirds. House cats also kill millions of small birds every year. From a study of radio-collared farm cats in Wisconsin, researchers estimated that, in that state alone, cats may kill 19 million songbirds in a single year.

The rapid loss of tropical forests—approximately 120,000 square kilometers each year, an area about the size of Louisiana—is depriving some 250 species of songbird migrants of their wintering homes. Of all long-term threats facing songbird populations, tropical deforestation is the most serious and the most intractable to change.

Some birds, such as robins, house sparrows, and starlings, can accommodate various changes, and may even thrive on them, but for most birds the changes are adverse. Unless we take leadership in managing our natural resources wisely, we soon could be facing the "silent spring" that Rachel Carson envisioned in 1962.

figure 19.26

Sport-shooting of passenger pigeons in Louisiana during the nineteenth century before establishment of state and federal hunting regulations. In addition to clearing of the hardwood forests that served as nesting habitats, relentless sport and market hunting eventually dropped the population too low to sustain colonial breeding. The last passenger pigeon died in captivity in 1914.

Lead poisoning of waterfowl is a side effect of hunting and fishing. Before long-delayed federal regulations in 1991 requiring the use of nonlead shot for all inland and coastal waterfowl hunting, shotguns scattered more than 3000 tons of lead each year in the United States alone. When waterfowl eat the pellets (which they mistake for seeds or grist), the pellets are ground and eroded in their gizzards, facilitating absorption of lead into their blood. Lead poisoning paralyzes or weakens birds, leading to death by starvation. Although poisoning of birds from lead shot has declined, lead sinkers used in fishing still poison large amounts of waterfowl. Recently, several states have banned lead sinkers, requiring anglers to use nontoxic alternatives.

Taxonomy of Living Birds of Class Aves

Class Aves contains about 9700 species distributed among 30 orders of living birds. Understanding the relationships of living birds, and consequently placing them in a classification, has been difficult because of the apparent rapid diversification of birds in the Cretaceous and early Tertiary periods. Prior to the study of Sibley and Alquist (1990), which used DNA hybridization, classification had been primarily based on morphological similarity. New attempts at discovering the higher relationships of birds have utilized many kinds of data, but especially mitochondrial DNA (mtDNA) and nuclear DNA sequences. The classification and numbers of living bird species that we present here mostly follow that of Gill (2006), which itself is based on Sibley and Alquist's study and many other, more recent, phylogenetic reconstructions. Nine of the smaller orders are not shown.

Class Aves (L. *avis,* bird)

Superorder Paleognathae (Gr. *palaios,* ancient, + *gnathos,* jaw). Modern birds with primitive archosaurian palate. Ratites, which include ostriches, rheas, cassowaries, emus, and kiwis (with unkeeled sternum), and tinamous (with keeled sternum).

Order Struthioniformes (strū′thē-on-i-for′mēz) (L. *struthio,* ostrich, + *forma,* form): **ostrich.** The ostrich, *Struthio camelus* (figure 19.27), of Africa, is the largest living bird, which reaches 2.4 m tall and 135 kg. The feet have only two toes of unequal size covered with pads, which enable the birds to travel rapidly over sandy ground.

Order Dinornithiformes (din′or-nith′i-for-mēz) (Gr. *din,* terrible, + *ornith,* bird, + form): **kiwis.** Kiwis, about the size of a domestic fowl, are unusual in having only the mere vestige of a wing. This order also includes the extinct, flightless moas, some of which reached 2 m at the shoulder. Three living species, all in New Zealand.

Order Tinamiformes (tin-am′i-for′mēz) (N.L. *Tinamus,* type genus, + form): **tinamous.** Ground-dwelling, grouselike birds of Central and South America. About 47 species.

Superorder Neognathae (Gr. *neos,* new, + *gnathos,* jaw). Modern birds with flexible palate.

Order Anseriformes (an′ser-i-for′mēz) (L. *anser,* goose, + form): swans, **geese,** and **ducks.** Members of this order have broad bills with filtering ridges at their margins, a foot web restricted to the front toes, and a long breastbone with a low keel. About 162 species, worldwide distribution.

figure 19.27

Ostrich, *Struthio camelus* of Africa, the largest of all living birds. Order Struthioniformes.

Order Galliformes (gal′li-for′mēz) (L. *gallus,* cock, + form): **quail, grouse, pheasants, ptarmigans, turkeys,** and **domestic fowl.** Chickenlike ground-nesting herbivores with strong beaks and heavy feet. The bobwhite quail, *Colinus virginianus,* occurs across the eastern half of the United States. The ruffed grouse, *Bonasa umbellus,* occupies the same region, but in woods instead of the open pastures and grain fields the bobwhite frequents. About 290 species, worldwide distribution.

Order Sphenisciformes (sfē-nisʹi-forʹmēz) (Gr. *Sphēniskos,* dim. of *sphen,* wedge, from the shortness of the wings, + form): **penguins.** Web-footed marine swimmers of southern seas from Antarctica north to the Galápagos Islands. Although penguins have a keeled sternum, they use their wings as paddles for swimming rather than for flight. About 17 species.

Order Phoenicopteriformes (fēʹni-cop-terʹi-forʹmēz) (Gr. *phoenico,* reddish-purple, + *pter,* wing, + form): **flamingos** (figure 19.28). Large, colorful, wading birds that use lamellae in their beaks to strain zooplankton from the water. Five species.

Order Procellariiformes (prō-sel-larʹē-i-forʹmēz) (L. *procella,* tempest, + form): **albatrosses, petrels, fulmars,** and **shearwaters.** All are marine birds with hooked beaks and tubular nostrils. In wingspan (more than 3.6 m in some), albatrosses are the largest flying birds. About 112 species, worldwide distribution.

Order Pelecaniformes (pel-e-canʹi-forʹmēz) (Gr. *pelekan,* pelican, + form): **pelicans, cormorants, gannets, boobies,** and **others.** Colonial fish-eaters with throat pouch and all four toes of each foot included within the web. About 65 species, worldwide distribution, especially in the tropics.

Order Ciconiiformes (si-koʹnē-i-forʹmēz) (L. *ciconia,* stork, + form): **herons, bitterns, storks, ibises, spoonbills,** and **vultures.** Long-necked, long-legged, mostly colonial waders and vultures. A familiar eastern North American representative is the great blue heron, *Ardea herodias,* which frequents marshes and ponds. About 116 species, worldwide distribution.

Order Falconiformes (falʹko-ni-forʹmēz) (L. *falco,* falcon, + form): **eagles, hawks, falcons, condors,** and **buzzards.** Diurnal birds of prey. All are strong fliers with keen vision and sharp, curved talons. About 304 species, worldwide distribution.

Order Charadriiformes (kä-ra-drēʹi-forʹmēz) (N.L. *Charadrius,* genus of plovers, + form): **gulls** (figure 19.29), **oyster catchers, plovers, sandpipers, terns, woodcocks, turnstones, lapwings, snipes, avocets, phalaropes, skuas, skimmers, auks,** and **puffins.** All are shorebirds. They are strong fliers and are usually colonial. About 367 species, worldwide distribution.

Order Columbiformes (kō-lumʹbē-i-forʹmēz) (L. *columba,* dove, + form): **pigeons** and **doves.** All have short necks, short legs, and a short, slender bill. The flightless dodo, *Raphus cucullatus,* of the Mauritius Islands became extinct in 1681. About 308 species, worldwide distribution.

Order Psittaciformes (sitʹta-sēʹ-forʹmēz) (L. *psittacus,* parrot, + form): **parrots** and **parakeets.** Birds with hinged and movable upper beak, fleshy tongue. About 364 species, pantropical distribution.

Order Cuculiformes (ku-kuʹli-forʹmēz) (L. *cuculus,* cuckoo, + form): **cuckoos** and **roadrunners.** European cuckoos, *Cuculus canorus,* lay their eggs in nests of smaller birds, which rear the young cuckoos. American cuckoos, black-billed and yellow-billed, usually rear their own young. About 138 species, worldwide distribution.

Order Strigiformes (strijʹi-forʹmēz) (L. *strix,* screech owl, + form): **owls.** Nocturnal predators with large eyes, powerful beaks and feet, and silent flight. About 180 species, worldwide distribution.

Order Caprimulgiformes (kap-ri-mulʹji-forʹmēz) (L. *caprimulgus,* goatsucker, + form): **goatsuckers, nighthawks,** and **whippoorwills.** Night and twilight feeders with small, weak legs and wide mouths fringed with bristles. Whippoorwills, *Antrostomus vociferus,* are common in the woods of the eastern states, and nighthawks, *Chordeiles minor,* are often seen and heard in the evening flying around city buildings. About 118 species, worldwide distribution.

figure 19.28

Greater flamingos, *Phoenicopterus ruber* on an alkaline lake in East Africa. Order Phoenicopteriformes.

figure 19.29

Laughing gulls, *Larus atricilla,* in flight. Order Charadriiformes.

(continued)

(continued)

Order Apodiformes (ə-pōd′ĭ-for′mēz) (Gr. *apous,* footless, + form): **swifts** and **hummingbirds.** Small birds with short legs and rapid wingbeat. The familiar chimney swift, ***Chaetura pelagia,*** fastens its nest in chimneys by means of saliva. A swift in China builds a nest of saliva that Chinese people use for soup making. Most species of hummingbirds occur in the tropics, but there are 14 species in the United States, of which only one, the ruby-throated hummingbird, occurs in the eastern part of the country. About 429 species, worldwide distribution.

Order Coraciiformes (kä-rä′sē′ĭ′for′mēz) (N.L. *coracii* from Gr. *korakias,* a kind of raven, + form): **kingfishers, hornbills,** and **others.** Birds that have strong, prominent bills and nest in cavities. In the eastern half of the United States, belted kingfishers, *Megaceryle alcyon,* are common along most waterways of any size. About 209 species, worldwide distribution.

Order Piciformes (pis′ĭ-for′mēz) (L. *picus,* woodpecker, + form): **woodpeckers, toucans, puffbirds,** and **honeyguides.** Birds having highly specialized bills and two toes extending forward and two backward. All nest in cavities. Many species of woodpeckers inhabit North America; the largest is the pileated woodpecker, usually found in mature forests. About 398 species, worldwide distribution.

Order Passeriformes (pas-ser′ĭ-for′me¯z) (L. *passer,* sparrow, + form): **perching songbirds.** This is the largest order of birds, containing 56 families and 60% of all birds.

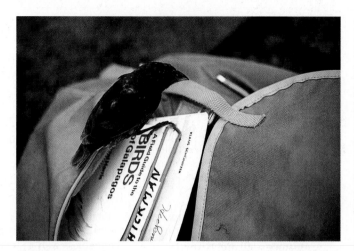

figure 19.30

Ground finch, *Geospiza fuliginosa,* one of the famous Darwin's finches of the Galápagos Islands. Order Passeriformes.

Most have a highly developed syrinx (voice box). Their feet are adapted for perching on thin stems and twigs. The young are altricial. To this order belong many birds with beautiful songs, such as thrushes, warblers, finches (figure 19.30), mockingbirds, meadowlarks, and hosts of others. Others of this order, such as swallows, magpies, starlings, crows, ravens, jays, nuthatches, and creepers, have no songs worthy of the name. About 5750 species, worldwide distribution.

■ Summary

The more than 9700 species of living birds are feathered, egg-laying, endothermic vertebrates whose forelimbs are modified as wings. Birds are in a clade with theropods, a group of small, bipedal dinosaurs. The oldest known fossil bird, *Archaeopteryx* from the Jurassic period of the Mesozoic era, had numerous reptilian characteristics and was almost identical to certain theropod dinosaurs except that it had asymmetrical feathers. It is probably the sister taxon of modern birds.

Adaptations of birds for flight are of two kinds: those reducing body weight and those promoting more power for flight. Feathers, the hallmark of birds, are complex derivatives of reptilian scales that combine lightness with strength, water repellency, and high insulative value. Body weight is further reduced by elimination of some bones, fusion of others (also providing rigidity for flight), and the presence in many bones of hollow, air-filled spaces. The light, keratinized

bill, replacing the heavy jaws and teeth of reptiles, serves as both hand and mouth for birds and is variously adapted for different feeding habits.

Adaptations that provide power for flight include high metabolic rate and body temperature coupled with an energy-rich diet; a highly efficient respiratory system consisting of air sacs arranged to provide a constant, one-way flow of air through the lungs; powerful flight and leg muscles arranged to place muscle weight near the bird's center of gravity; and an efficient, high-pressure circulation.

Birds have keen eyesight, good hearing, and superb coordination for flight. Their metanephric kidneys produce uric acid as the principal nitrogenous waste.

Birds fly by applying the same aerodynamic principles as an airplane and using similar equipment: wings for lift, support, and propulsion; and wing slots for control at low speed. Flightlessness in birds is not uncommon,

and has evolved independently in several bird orders, usually on islands where terrestrial predators are absent; all flightless species are derived from flying ancestors.

Bird migration refers to regular movements between summer nesting places and wintering regions. Spring migration to the north, where more food is available for nestlings, enhances reproductive success. Birds use many cues for finding direction during migration, including an innate sense of direction and the ability to navigate by the sun, stars, or earth's magnetic field.

The highly developed social behavior of birds is manifested in vivid courtship displays, mate selection, territorial behavior, and incubation of eggs and care of the young. Most birds have a monogamous mating system, although extra-pair copulations are common. Young hatch at various levels of development; altricial young are naked and helpless, while precocial young are feathered and able to walk and to feed themselves.

Review Questions

1. Explain the significance of the discovery of *Archaeopteryx*. Why did this fossil demonstrate beyond reasonable doubt that birds are grouped phylogenetically with dinosaurs?
2. The special adaptations of birds contribute to two essentials for flight: more power and less weight. Explain how each of the following contributes to one or both of these two essentials: feathers, skeleton, muscle distribution, digestive system, circulatory system, respiratory system, excretory system, reproductive system.
3. How do marine birds rid themselves of excess salt?
4. In what ways are a bird's ears, eyes, and brain specialized for the demands of flight?
5. Explain how a bird wing is designed to provide lift. What design features help prevent stalling at low flight speeds?
6. Describe four basic forms of bird wings. How does wing shape correlate with flight speed and maneuverability?
7. What advantages does seasonal migration provide for birds?
8. Describe different navigational resources birds may use in long-distance migration.
9. What are some advantages of social aggregation among birds?
10. More than 90% of all bird species are monogamous. Explain why monogamy is much more common among birds than among mammals.
11. Briefly describe examples of polygyny and polyandry among birds.
12. Define altricial and precocial as they relate to birds.
13. Offer some examples of how human activities have affected bird populations.

For Further Thought Reproductive behavior and strategies are better known in birds than in any other vertebrate group. Why?

Selected References

See also general references on page 432.

Ackerman, J. 1998. Dinosaurs take wing. National Geographic **194** (1):74-99. *Beautifully illustrated synopsis of dinosaur-to-bird evolution.*

Bennett, P. M., and I. F. F. Owens. 2002. Evolutionary ecology of birds: life histories, mating systems, and extinction. Oxford, UK, Oxford University Press. *A phylogenetic approach to understanding how natural and sexual selection have led to the incredible diversity of bird mating systems.*

Brooke, M., and T. Birkhead, eds. 1991. The Cambridge encyclopedia of ornithology. New York, Cambridge University Press. *Comprehensive, richly illustrated treatment that includes a survey of all modern bird orders.*

Elphick, J., ed. 1995. The atlas of bird migration: tracing the great journeys of the world's birds. New York, Random House. *Lavishly illustrated collection of maps of birds' breeding and wintering areas, migration routes, and many facts about each bird's migration journey.*

Emlen, S. T. 1975. The stellar-orientation system of a migratory bird. Sci. Am. **233:** 102-111 (Aug.). *Describes fascinating research with indigo buntings, revealing their ability to navigate by the center of celestial rotation at night.*

Feduccia, A. 1996. The origin and evolution of birds. New Haven, Yale University Press. *An updated successor to the author's* The Age of Birds *(1980) but more comprehensive; rich source of information on evolutionary relationships of birds.*

Gill, F. B. 2006. Ornithology, ed. 3. New York, W. B. Freeman and Company. *Popular, comprehensive, and accurate ornithology text.*

Mora, C. V., M. Davison, J. M. Wild, and M. W. Walker. 2004. Magnetoreception and its trigeminal mediation in the homing pigeon. Nature **432:**508-511. *Pigeons can discriminate between the presence and absence of a magnetic anomaly.*

Prum, R. O., and A. H. Brush. 2004. Which came first, the feather or the bird? Sci. Am. **288:**84-93 (March). *Easily readable article which incorporates results of developmental biology with recent fossil finds to understand the evolution of feathers.*

Sibley, C. G., and J. E. Ahlquist. 1990. Phylogeny and classification of birds: a study in molecular evolution. New Haven, Yale University Press. *A comprehensive application of DNA annealing experiments to the problem of resolving avian phylogeny.*

Terborgh, J. 1992. Why American songbirds are vanishing. Sci. Am. **266:**98-104 (May). *The number of songbirds in the United States has been dropping sharply. The author suggests reasons why.*

Videler, J. J. 2006. Avian flight. Oxford, U.K., Oxford Univ. Press. *An up-to-date introduction to all aspects of bird flight and scientific controversies regarding its origin.*

Waldvogel, J. A. 1990. The bird's eye view. Am. Sci. **78:**342-353 (July-Aug.). *Birds possess visual abilities unmatched by humans. So how can we know what they really see?*

Wellnhofer, P. 1990. *Archaeopteryx.* Sci. Am. **262:**70-77 (May). *Description of perhaps the most important fossil ever discovered.*

Custom Website

The *Animal Diversity* Online Learning Center is a great place to check your understanding of chapter material. Visit www.mhhe.com/hickmanad6e for access to key terms, quizzes, and more! Further enhance your knowledge with Web links to chapter-related material.
Explore live links for these topics:

Classification and Phylogeny of Animals
Class Aves

Marine Birds
Dissection Guides for Birds
Conservation Issues Concerning Birds
Movement of Populations

Mammals

Mammalia

Chordata

Juvenile grizzly bear, *Ursus arctos horribilis.*

The Tell-Tale Hair

Hair evolved in a common ancestor of all mammals and has been retained to varying degrees in all species descended from that ancestor. Hair is therefore diagnostic for mammals; except in some pathological conditions, all mammals have hair at some point in their lives, and hair occurs in no other living organisms. Even those living mammals with no obvious hair, such as whales, usually have a few hairs on their bodies. Mammalian hair has undergone numerous adaptive modifications for diverse uses. Mammals use hair for concealment, behavioral signaling, waterproofing, and buoyancy; their hair may serve as sensitive vibrissae on their snouts or as prickly quills. Perhaps the most important use of their hair is thermal insulation, helping to maintain a high, constant body temperature in all climates, and thus support a high level of activity.

Mammals are among the most active animals, exhibiting speed and endurance in aquatic, aerial, and terrestrial habitats. They maintain this activity in nearly all environmental conditions, including cool nights, baking deserts, frigid polar seas, and icy winters. Although hair is perhaps the most obvious feature of mammals, many other evolutionary innovations support their diversification. These adaptions include a set of middle ear bones for transmitting sounds to the inner ear, mammary glands for nourishing newborns, a large brain with a unique covering of the cerebrum (the neocortex), a diaphragm for efficient ventilation of lungs, and highly developed sense of smell. Most mammals have an intrauterine, vascular placenta for feeding the embryo, specialized teeth and jaw musculature for processing food, and an upright gait for rapid and efficient locomotion.

ammals, with their highly developed nervous system and numerous adaptations, occupy almost every environment on earth that supports life. Although not a large group (about 5400 species, compared with more than 9700 species of birds, 28,000 species of fishes, and 1,100,000 species of insects), class Mammalia (ma-ma´lē-ə; L. *mamma*, breast) is among the most biologically differentiated groups in the animal kingdom. Mammals are exceedingly diverse in size, shape, form, and function. They range in size from Kitti's hognosed bat, weighing only 2 g, to blue whales, exceeding 170 metric tons.

Despite their adaptability, and in some instances because of it, mammals have been influenced by humans more than any other group of animals. We have domesticated numerous mammals for food and clothing, as beasts of burden, and as pets. We use millions of mammals each year in biomedical research. We introduce alien mammals into new habitats, occasionally with benign results but more frequently with disastrous effects. Although history provides us with numerous warnings, we continue to overcrop valuable wild stocks of mammals. The whaling industry has threatened itself with total collapse by exterminating its own resource—a classic example of self-destruction, in which competing segments of an industry are intent only on reaping all they can today as though tomorrow's supply were of no concern. In some cases, destruction of a valuable mammalian resource has been deliberate, such as the officially sanctioned (and tragically successful) policy of exterminating bison to drive the Plains Indians into starvation. Although commercial hunting has declined, the ever-increasing human population with accompanying destruction of wild habitats has affected the mammalian fauna. In 2010, 638 species of mammals were listed as "critically endangered" or "endangered" by the International Union for Conservation of Nature and Natural Resources (IUCN), including many bats, cetaceans, cats, and primates.

Because our welfare has been and continues to be closely related to that of other mammals, it is clearly in our interest to preserve the natural environment of which all mammals, ourselves included, are a part. We need to remember that nature can do without humans but humans cannot exist without nature.

Origin and Evolution of Mammals

The evolutionary descent of mammals from their earliest amniote ancestors is perhaps the most fully documented transition in vertebrate history. From the fossil record, we can trace the derivation over 150 million years of endothermic, furry mammals from their small, ectothermic, hairless ancestors. Mammals and their closest fossil relatives have a pair of openings in the temporal region of the skull associated with attachment of mandibular muscles. This condition identifies them as **synapsids,** one of three major groups of amniotes that diversified during the late Paleozoic era (see figure 18.2).

The earliest synapsids included diverse herbivorous and carnivorous forms often collectively called **pelycosaurs** (figures 20.1 and 20.2). Pelycosaurs share a general outward resemblance to lizards, but this resemblance is misleading. Pelycosaurs are not closely related to lizards, which are diapsids, nor are they a monophyletic group. From one group of early carnivorous synapsids arose the **therapsids** (figure 20.2), the only synapsid group to survive beyond the Paleozoic era. Therapsids evolved an efficient erect gait with upright limbs positioned beneath the body rather than sprawled to the side, as in lizards and early pelycosaurs. Because stability was reduced by raising the animal from the ground, the muscular coordination center of the brain, the cerebellum, assumed an expanded role. Changes in the morphology of the skull and jaw-closing muscles increased feeding efficiency. Therapsids diversified into numerous herbivorous and carnivorous forms, but most early forms disappeared during a great extinction at the end of the Permian period. Previously, pelycosaurs and early therapsids were called "mammal-like reptiles," but use of this term is inappropriate, because they are not part of the clade Reptilia (p. 372).

One therapsid group to survive into the Mesozoic era was the **cynodonts** (figures 20.1 and 20.2). Cynodonts evolved several features that supported a high metabolic rate: increased and specialized jaw musculature, permitting a stronger bite; several skeletal changes, supporting greater agility; **heterodont** teeth, permitting better food processing in the mouth; **turbinate bones** in the nasal cavity, aiding retention of body heat (figure 20.3); and a bony, hard (secondary) palate (figure 20.3), enabling an animal to breathe while holding prey in its mouth or chewing food. The secondary palate would be important to subsequent mammalian evolution by permitting the young to breathe while suckling. Loss of lumbar ribs in cynodonts is correlated with the evolution of a **diaphragm** and also may have provided greater dorsoventral flexibility to the spinal column.

The earliest mammals of the late Triassic period were small, mouse- or shrew-sized animals with enlarged crania, redesigned jaws, and a new type of dentition, called **diphyodont,** in which teeth are replaced only once (deciduous and permanent teeth). This condition contrasts with the ancestral amniote pattern of continual tooth replacement throughout life (polyphodont teeth). Two bones, the articular and the quadrate, which previously served as the jaw joint, were reduced in size and relocated to the middle ear, becoming the malleus and incus, respectively. A new jaw joint formed between the dentary and squamosal (temporal) bones; this joint is the defining character for fossil mammals.

The earliest mammals were almost certainly endothermic, although their body temperature would have been lower than that of modern placental mammals. Hair was essential for insulation, and the presence of hair implies that sebaceous and sweat glands must have evolved at this time to condition hair and to facilitate thermoregulation. The fossil record is silent on the appearance of mammary glands, but they must have evolved before the end of the Triassic.

Oddly, early mammals of the Late Triassic, having developed nearly all novel attributes of modern mammals, had to wait

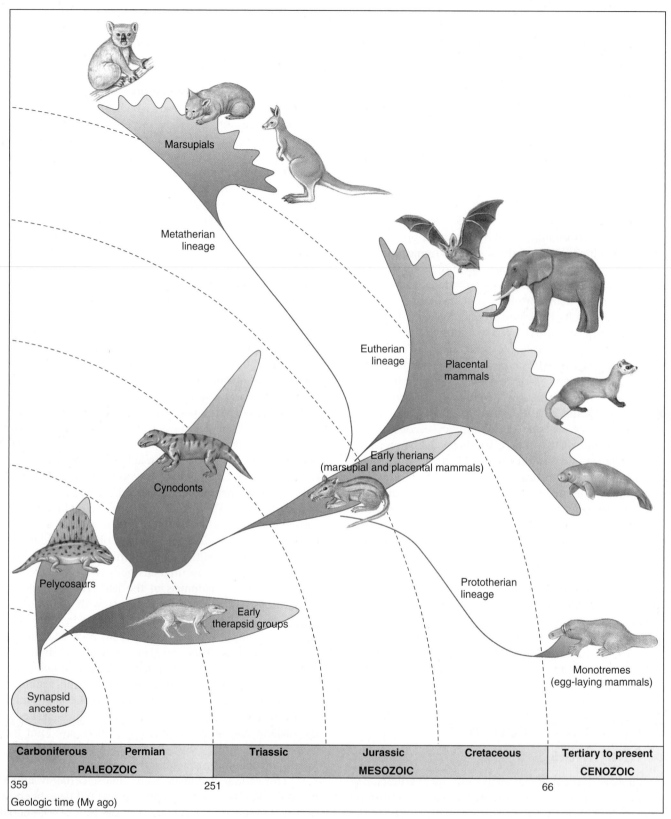

figure 20.1

Evolution of major groups of synapsids. The synapsids, characterized by lateral temporal openings in the skull, began with pelycosaurs, early amniotes of the Permian period. Pelycosaurs diversified extensively and evolved changes in jaws, teeth, and body form that presaged several mammalian characteristics. These trends continued in their successors, the therapsids, especially in cynodonts. One lineage of cynodonts gave rise in the Triassic to early mammals. Fossil evidence, as currently interpreted, indicates that all three groups of living mammals—monotremes, marsupials, and placentals—are derived from the same cynodont lineage. The great diversification of modern placental orders occurred during the Cretaceous and Tertiary periods.

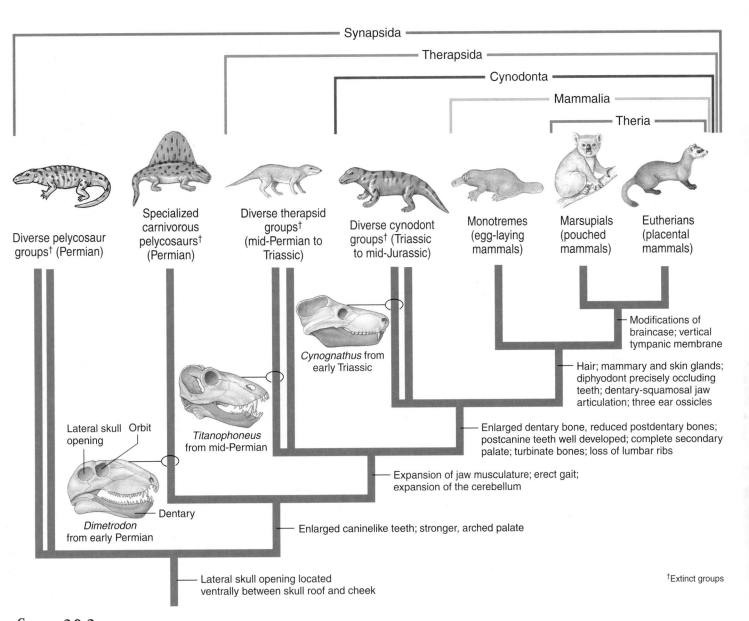

figure 20.2

Abbreviated cladogram of synapsids, emphasizing origins of important characteristics of mammals. The skulls show progressive increase in size of the dentary bone relative to other bones in the lower jaw and increasing heterodonty.

for another 150 million years before they could achieve their great diversity. While dinosaurs became diverse and abundant, all nonmammalian synapsid groups became extinct. Mammals survived mostly as shrewlike, probably nocturnal creatures. Then, in the Cretaceous period, but especially during the Eocene epoch that began about 58 million years ago, modern mammals began to diversify rapidly. The great Cenozoic diversification of mammals is partly attributed to numerous habitats vacated by the extinction of many amniote groups at the end of the Cretaceous. Mammalian diversification was almost certainly promoted by the facts that mammals were agile, endothermic, intelligent, and adaptable, and that they gave birth to living young, which they protected and nourished from their own milk supply, thus dispensing with vulnerable eggs laid in nests.

Class Mammalia includes 29 orders: one order containing **monotremes** (Prototheria), seven orders of **marsupials** (Metatheria), and 21 orders of placentals (Eutheria). A classification is on pp. 427–429.

Structural and Functional Adaptations of Mammals

Integument and Its Derivatives

Mammalian skin and especially its modifications distinguish mammals as a group. In general, skin is thicker in mammals than in other vertebrates, although as in all

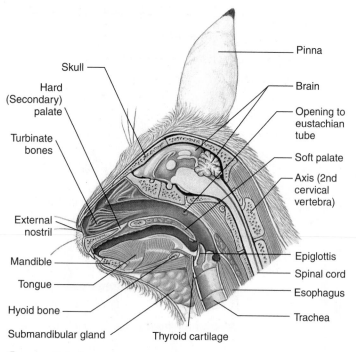

figure 20.3

Sagittal section of the head of a rabbit. The secondary palate, which consists of bony (hard) and nonbony (soft) regions, separates the routes of air (dorsal) and food (ventral).

Labels for figure 20.3: Skull, Hard (Secondary) palate, Turbinate bones, External nostril, Mandible, Tongue, Hyoid bone, Submandibular gland, Thyroid cartilage, Pinna, Brain, Opening to eustachian tube, Soft palate, Axis (2nd cervical vertebra), Epiglottis, Spinal cord, Esophagus, Trachea

figure 20.5

American beaver, *Castor canadensis*, gnawing on an aspen tree. This second largest rodent (the South American capybara is larger) has a heavy waterproof pelage consisting of long, tough guard hairs overlying the thick, silky underhair so valued in the fur trade. Order Rodentia, family Castoridae.

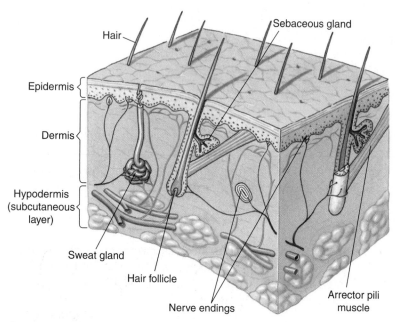

figure 20.4

Structure of human skin (epidermis and dermis) and hypodermis, showing hair and glands.

Labels for figure 20.4: Hair, Epidermis, Dermis, Hypodermis (subcutaneous layer), Sweat gland, Hair follicle, Nerve endings, Sebaceous gland, Arrector pili muscle

contact and use, such as the palms or soles, its outer layers become thick and filled with **keratin,** a fibrous protein that also constitutes nails, claws, hooves, and feathers.

Hair

Hair is especially characteristic of mammals, although humans are not very hairy creatures, and hair in whales is reduced to a few sensory bristles on the snout. A hair grows from a hair follicle that, although epidermal in origin, is sunk into the dermis of the skin (figure 20.4). A hair grows continuously by rapid proliferation of cells in the follicle. As a hair shaft is pushed upward, new cells are carried away from their source of nourishment and die, becoming filled with keratin.

Mammals characteristically have two kinds of hair forming their **pelage** (fur coat): (1) dense and soft underhair for insulation, and (2) coarse and longer guard hair for protection against wear and to provide coloration. Underhair traps a layer of insulating air. In aquatic mammals, such as fur seals, otters, and beavers, underhair is so dense that it is almost impossible to wet. In water, guard hairs become wet and adhere to each other, forming a protective blanket over the underhair (figure 20.5).

When a hair reaches a certain length, it stops growing. Normally it remains in its follicle until a new growth starts, whereupon it falls out. Most mammals experience periodic molts of the entire coat. In humans, hair is shed and replaced throughout life (although replacement is not assured, as evidenced by balding males!).

vertebrates it is composed of **epidermis** and **dermis** (figure 20.4). The epidermis is thinner where it is well protected by hair, but in places that are subject to much

A hair is more than a strand of keratin. It consists of three layers: the medulla or pith in the center of the hair, the cortex with pigment granules next to the medulla, and the outer cuticle composed of overlapping scales. Hair of different mammals shows a considerable range of structure. It may be deficient in cortex, like the brittle hair of deer, or it may be deficient in medulla, like the hollow, air-filled hairs of a wolverine. Hairs of rabbits and some others are scaled to interlock when pressed together. Curly hair, such as that of sheep, grows from curved follicles.

Some mammals, such as foxes and seals, shed their coat every summer. Most mammals have two annual molts, one in spring and one in fall. Summer coats are always much thinner than winter coats and in some mammals may be a different color. Several northern mustelid carnivores, such as weasels, have white winter coats and brown summer coats. It was once hypothesized that the white pelage of arctic animals conserved body heat by reducing radiation loss; in fact, dark and white pelages radiate heat equally well. The white winter pelage of arctic animals is simply camouflage in a land of snow. The snowshoe hare of North America has three annual molts: the white winter coat is replaced by a brownish-gray summer coat, and this is replaced in autumn by a grayer coat, which is soon shed to reveal the white winter coat beneath (figure 20.6).

Most mammals have somber colors that disguise their presence. Often a species is marked with "salt-and-pepper" coloration or a disruptive pattern that helps make it inconspicuous in its natural surroundings. Examples are the spots of leopards and fawns and the stripes of tigers. Skunks advertise their presence with conspicuous warning coloration.

The hair of mammals has become modified to serve many purposes. Bristles of hogs, spines of porcupines and their kin, and vibrissae on the snouts of most mammals are examples. **Vibrissae,** commonly called "whiskers," are really sensory hairs that provide a tactile sense to many mammals. The slightest movement of a vibrissa generates impulses in sensory nerve endings that travel to special sensory areas in the brain. Vibrissae are especially long in nocturnal and burrowing animals.

Porcupines, hedgehogs, echidnas, and a few other mammals have developed an effective and dangerous spiny armor. When cornered, the common North American porcupine turns its back toward its attacker and lashes out with its barbed tail. The lightly attached quills break off at their bases when they enter the skin and, aided by backward-pointing hooks on the tips, work deeply into tissues. Dogs are frequent victims (figure 20.7), but fishers, wolverines, and bobcats can flip the porcupine onto its back to expose vulnerable underparts.

Horns and Antlers

Several kinds of horns or hornlike structures occur in mammals. True horns of the family Bovidae (for example, sheep and cattle) are hollow sheaths of keratinized epidermis

A

B

figure 20.6

Snowshoe hare, *Lepus americanus* in **A,** brown summer coat, and **B,** white winter coat. In winter, extra hair growth on the hind feet broadens the animal's support in snow. Snowshoe hares are common residents of the taiga and are an important food for lynxes, foxes, and other carnivores. Order Lagomorpha, family Leporidae.

figure 20.7

Dogs are frequent victims of a porcupine's impressive quills. Unless removed (usually by a veterinarian), quills continue to work their way deeper into the flesh, causing great distress and sometimes death.

figure 20.8

Annual growth of male deer antlers. **A,** Antlers begin to grow in late spring, stimulated by pituitary gonadotropins. **B,**. Bone grows very rapidly until halted by a rapid rise in testosterone production by the testes. **C,** The skin (velvet) dies and rubs off. **D,** Testosterone levels peak during the fall breeding season. The antlers are shed in January as testosterone levels subside.

that embrace a core of bone arising from the skull. True horns are not shed, usually are not branched (although they may be greatly curved), grow continuously, and occur in both sexes.

Antlers of the deer family Cervidae are branched and composed of solid bone when mature. During their annual spring growth, antlers develop beneath a covering of highly vascular soft skin called velvet (figure 20.8). When growth of antlers is complete just before the breeding season, blood vessels constrict, and the stag tears off the velvet by rubbing its antlers against trees. Antlers are shed after the breeding season. New buds appear a few months later to begin the next set of antlers. For several years, each new pair of antlers is larger and more elaborate than the previous set. Annual growth of antlers places a strain on mineral metabolism because during the growing season an older moose or elk must accumulate 50 or more pounds of calcium salts from its vegetable diet.

Horns of the pronghorn antelope (family Antilocapridae) are similar to the true horns of bovids, except that the keratinized portion is forked and shed annually. Giraffe horns are similar to antlers but retain their integumentary covering and are not shed. Rhinoceros horn consists of hairlike, keratinized filaments that arise from dermal papillae cemented together, but they are not attached to the skull.

Trade in rhino products, especially the horn, has pushed Asian and African rhinos to the brink of extinction. Rhino horn is valued in China as an agent for reducing fever and for treating heart, liver, and skin disease, and in North India as an aphrodisiac. Such supposed medicinal values are totally without pharmacological basis. Until recently the principal use of rhino horns was to fashion handles for ceremonial daggers called jambiyas in the Middle East. Between 1970 and 1997, horns from 22,350 rhinos were imported into northern Yemen alone. Thanks to education efforts, use of rhino horn in Yemen has nearly ceased; the destination of most illegal rhino horn is China and Vietnam. An international ban prohibiting the trade of rhino horn has reduced, but not eliminated, the now illegal trade, and populations continue to suffer.

Glands

Of all vertebrates, mammals have the greatest variety of integumentary glands. Most fall into one of four classes: sweat, scent, sebaceous, or mammary. All are derivatives of epidermis.

CHARACTERISTICS
of Class Mammalia

1. Fleshy external ears **(pinna); endothermic**
2. Body mostly covered with **hair,** but reduced coverage in some; **sweat, scent, and sebaceous glands** present; skin underlain by a thick layer of fat
3. Skull with **two occipital condyles;** lower jaw a **single enlarged bone (dentary);** jaw joint between squamosal and dentary bones; **seven cervical vertebrae** (except in some xenarthrans [edentates] and manatees); **pelvic bones fused**
4. Mouth with **diphyodont teeth;** teeth **heterodont** in most
5. Brain highly developed, especially **cerebral cortex** (a superficial layer of the cerebrum); 12 pairs of cranial nerves
6. Olfactory sense highly developed; middle ear with **three bones** (malleus, incus, and stapes)
7. Separate sexes; internal fertilization; copulatory organ a **penis;** testes usually in a **scrotum;** sex determined by chromosomes (male is heterogametic)
8. Fetal membranes of **amnion, chorion, and allantois;** most **viviparous** with embryos developing in a **uterus** with **placental attachment,** except monotremes, which are **oviparous;** young nourished by **milk from mammary glands**
9. Excretory system of **metanephric kidneys** and ureters that usually open into a bladder; urea main nitrogenous waste
10. Lungs with high surface area from **alveoli** and ventilated by **aspiration;** larynx present; **secondary palate** separates air and food passages (see figure 20.3); muscular **diaphragm** ventilates lungs; convoluted **turbinate bones** in the nasal cavity warm and moisten inspired air
11. Heart with two atria and two ventricles; **separate pulmonary and systemic circuits; persistent left aortic arch;** nonnucleated, biconcave red blood cells

Sweat glands are tubular, highly coiled glands that occur over much of the body surface in most mammals (see figure 20.4). They are absent in other vertebrates. There are two kinds of sweat glands: eccrine and apocrine. **Eccrine glands** secrete a watery fluid that, if evaporated on the skin's surface, draws heat away from the skin and cools it. Eccrine glands occur in hairless regions, especially foot pads, in most mammals, although in horses and most primates they are scattered over the body. **Apocrine glands** are larger than eccrine glands and have longer and more convoluted ducts. Their secretory coil is in the dermis and extends deep into the hypodermis. They always open into a hair follicle or where a hair once was. Apocrine glands develop near sexual puberty and are restricted (in humans) to the axillae (armpits), genital regions, breasts, and external auditory canals. In contrast to the watery secretions of eccrine glands, apocrine secretions are milky fluids, whitish or yellow in color, that dry on the skin to form a film. Apocrine glands are not involved in heat regulation. Their activity is correlated with certain aspects of the reproductive cycle.

Scent glands occur in nearly all mammals. Their location and functions vary greatly. They are used for communication with members of the same species, for marking territorial boundaries, for warning, or for defense. Scent-producing glands are located in orbital, metatarsal, and interdigital regions; at the base of the tail; and in the penis. The most odoriferous of all glands are those of skunks, which open by ducts into the anus; their secretions can be discharged forcefully for 2 to 3 m. During mating season, many mammals produce strong scents for attracting the opposite sex. Humans also are endowed with scent glands. However, we tend to dislike our own scent, a concern that has stimulated a lucrative deodorant industry to produce an endless array of soaps and odor-masking concoctions.

Sebaceous glands are usually associated with hair follicles (see figure 20.4), although some are free and open directly onto the surface. Gland cells are secreted in their entirety and are continuosly renewed by cell division. These cells become distended with a fatty accumulation, then die, and are expelled as a greasy mixture called **sebum** into the hair follicle. Called a "polite fat" because it does not turn rancid, sebum keeps skin and hair pliable and glossy. Most mammals have sebaceous glands over their entire body; in humans, they are most numerous in the scalp and on the face.

Mammary glands, for which mammals are named, occur on all female mammals and in a rudimentary form on all male mammals. In the embryo, they develop by thickening of the epidermis to form a milk line along each side of the abdomen. On certain parts of these lines, the mammae appear while the intervening parts of the ridge disappear. Mammary glands increase in size at maturity, becoming considerably larger during pregnancy and subsequent nursing of young. In human females, adipose tissue begins to accumulate around mammary glands at puberty to form the breast. In most mammals, milk is secreted from mammary glands via nipples, but monotremes lack nipples and simply secrete milk into a depression on the mother's belly where it is lapped by the young.

Food and Feeding

Mammals exploit an enormous variety of food sources; some mammals require highly specialized diets, whereas others are opportunistic feeders that thrive on diversified diets. A mammal's adaptations for finding, capturing, chewing, swallowing, and digesting food all determine its anatomy and habits.

Teeth, perhaps more than any other single physical characteristic, reveal the life habit of a mammal (figure 20.9). All mammals have teeth (with few exceptions), and their modifications are correlated with what a mammal eats.

As mammals evolved during the Mesozoic, major changes occurred in teeth and jaws. Unlike the uniform **homodont** dentition of the first synapsids, mammalian teeth became differentiated to perform specialized functions such as cutting, seizing, gnawing, tearing, grinding, and chewing. Teeth differentiated in this manner are called **heterodont.** Mammalian teeth are differentiated into four types: incisors (I), with simple crowns and sharp edges, used mainly for snipping; canines (C), with long conical crowns, specialized for piercing; premolars (PM) and molars (M), with compressed crowns and one or more cusps, suited for shearing, slicing, crushing, or grinding. The ancestral tooth formula, which expresses the number of each tooth type in one-half of the upper and lower jaw, was I 3/3, C 1/1, PM 4/4, and M 3/3. Shrews, some omnivores, and some carnivores come closest to this ancestral pattern (figure 20.9).

Unlike most other vertebrates, mammals do not continuously replace their teeth throughout their lives. Most mammals grow just two sets of teeth: a temporary set, called **deciduous teeth,** or milk teeth, is replaced by a permanent set when the skull has grown large enough to accommodate a full set. Only incisors, canines, and premolars are deciduous; molars are never replaced and the single permanent set must last a lifetime.

Feeding Specializations

The feeding, or trophic, apparatus of a mammal—teeth and jaws, tongue, and digestive tract—is adapted to the animal's particular feeding habits. Mammals are customarily divided among four basic trophic categories—insectivores, carnivores, omnivores, and herbivores—but many other feeding specializations have evolved in mammals, as in other living organisms, and feeding habits of some mammals defy exact classification. The principal feeding specializations of mammals are shown in figure 20.9.

Insectivorous mammals, such as shrews, moles, anteaters, and most bats, feed on a variety of insects and other small invertebrates. Most insectivorous mammals have teeth with pointed cusps, permitting them to puncture the exoskeleton or skin of their prey. A few insectivores, such as anteaters, completely lack teeth (figure 20.9).

Herbivorous mammals, which feed on grasses and other vegetation, which form two main groups: (1) browsers and grazers, including ungulates (hooved mammals, including horses, deer, antelope, cattle, sheep, and goats), and (2) gnawers, including many rodents, rabbits, and hares. In herbivores, canines are absent or reduced in size, whereas molars and

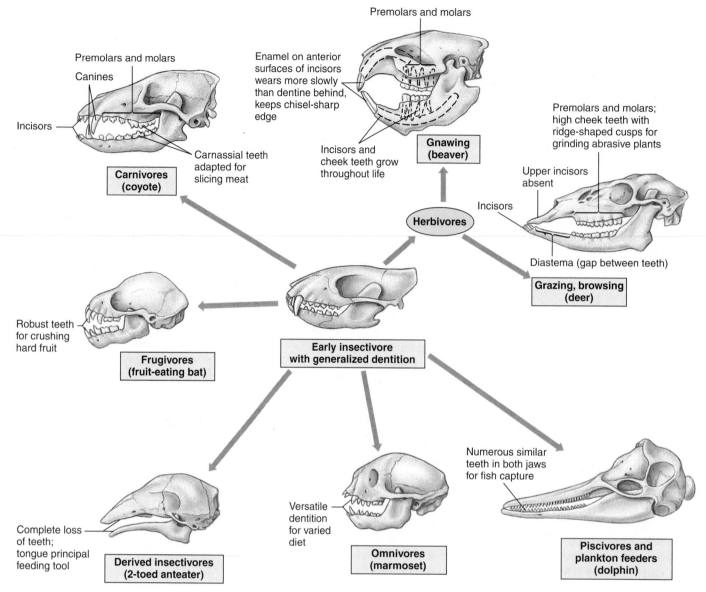

figure 20.9

Feeding specializations of major trophic groups of placental mammals. Early placentals were insectivores; all other types are descended from them.

premolars, which are adapted for grinding, are broad and usually high-crowned. Rodents have chisel-sharp incisors that grow throughout life and must be worn away to keep pace with their continual growth (figure 20.9).

Herbivores have a number of adaptations for processing their fibrous diet of plant food. **Cellulose,** the structural carbohydrate of plants, is composed of long chains of glucose molecules and therefore is a potentially nutritious food resource. However, no vertebrates synthesize cellulose-splitting enzymes **(cellulases).** Instead, herbivorous vertebrates harbor anaerobic bacteria and protozoa that produce cellulase in fermentation, chambers in their gut. Simple carbohydrates, proteins, and lipids produced by the microorganisms can be absorbed by the host animal, and the host can digest the microorganisms as well.

Fermentation in some herbivores, such as rabbits, horses, zebras, elephants, some primates, and many rodents, occurs primarily in the colon and in a spacious side pocket, or diverticulum, called a **cecum** (figure 20.10A). Although some absorption occurs in the colon and cecum, most fermentation occurs posterior to the small intestine, where nutrients are absorbed, so many nutrients are lost in feces. Rabbits and many rodents eat their fecal pellets **(coprophagy),** giving food a second pass through the gut to extract additional nutrients.

Ruminants (cattle, bison, buffalo, goats, antelopes, sheep, deer, giraffes, and okapis) have a huge four-chambered stomach (figure 20.10B). As a ruminant feeds, food passes down the esophagus to the **rumen,** where it is digested by microorganisms and then formed into small balls of cud. The ruminant returns cud to its mouth, where it is deliberately chewed at length

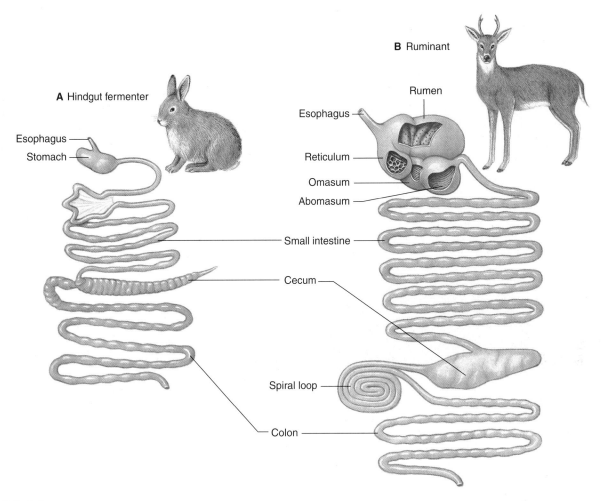

figure 20.10

Digestive tracts of two mammals. **A,** Rabbits are hindgut fermenters; digestion of cellulose by microorganisms occurs in the cecum and colon. **B,** Ruminants are foregut fermenters; digestion of cellulose by microorganisms occurs primarily in the rumen.

to crush the fiber. Swallowed again, the food returns to the rumen, where the cellulolytic bacteria and protozoa continue fermentation. The pulp passes to the **reticulum,** and then to the **omasum,** where water-soluble food and microbial products are absorbed. The remainder proceeds to the **abomasum** ("true" acid stomach) and small intestine, where proteolytic enzymes are secreted and normal digestion occurs. Perhaps because ruminants are particularly good at extracting nutrients from forage, they are the primary large herbivores in ecosystems with low vegetative production, such as tundras and deserts.

Herbivores generally have large, long digestive tracts and must eat a considerable amount of plant food to survive. An African elephant weighing 6 tons must consume 135 to 150 kg (300 to 400 pounds) of rough fodder each day to obtain sufficient nourishment.

Carnivorous mammals, which feed mainly on herbivores, include foxes, dogs, weasels, wolverines, fishers, and cats. Carnivores are well equipped with biting and piercing teeth and powerful clawed limbs for killing their prey. Since their protein diet is more easily digested than the fibrous food of herbivores, their digestive tract is shorter, and the cecum is small or absent. Carnivores organize their feeding into discrete meals rather than feeding continuously (as do most herbivores) and therefore have much more leisure time (figure 20.11).

Note that the terms "insectivore" and "carnivore" have two different uses in mammals: to describe diet and to denote specific taxonomic orders of mammals. For example, not all carnivores belong to the order Carnivora (many marsupials and cetaceans are carnivorous), and not all members of the order Carnivora are carnivorous. Many are opportunistic feeders, and some, such as pandas, are strict vegetarians.

Omnivorous mammals, which use both plants and animals for food, include pigs, raccoons, many rodents, bears, and most primates, including humans. Many carnivorous forms also eat fruits, berries, and grasses when hard pressed. Foxes, which usually feed on mice, small rodents, and birds, eat apples, beechnuts, and corn when their normal food sources are scarce.

Migration

Migration is a more difficult undertaking for mammals than for birds or fishes, because terrestrial locomotion is more energetically expensive than swimming or flying. Not surprisingly, few terrestrial mammals make regular seasonal migrations, instead centering their activities in a defined and limited home range. Nevertheless, some striking terrestrial mammalian migrations occur, especially in northern North America.

figure 20.11

Lionesses, *Panthera leo*, eating a wildebeest. Lacking stamina for a long chase, lions stalk prey and then charge suddenly, surprising their prey. After gorging themselves, lions sleep and rest for periods as long as one week before eating again. Order Carnivora, family Felidae.

The barren-ground caribou of Canada and Alaska undertakes direct and purposeful mass migrations spanning 160 to 1100 km (100 to 700 miles) twice annually (figure 20.12). From winter ranges in boreal forests (taiga), they migrate rapidly in late winter and spring to calving ranges on the barren grounds (tundra). Calves are born in mid-June. As summer progresses, caribou are increasingly harassed by warble and nostril flies that bore into their flesh, by mosquitos that drink their blood (estimated at a liter per caribou each week during the height of the mosquito season), and by wolves that prey on their calves. They move southward in July and August, feeding little along the way. In September, they reach the taiga and feed there almost continuously on low ground vegetation. Mating (rut) occurs in October.

Oceanic seals and whales make the longest mammalian migrations. Gray whales, for example, migrate between Alaska in summer and Baja California, Mexico, in winter, an annual migration of over 18,000 km (11,250 miles). One of the most remarkable migrations is that of northern fur seals, which breed on the Pribilof Islands approximately 300 km (185 miles) off the coast of Alaska and north of the Aleutian Islands. From wintering grounds off southern California, females journey as much as 2800 km (1740 miles) across open ocean, arriving in spring at the Pribilofs, where they congregate in enormous numbers (figure 20.13). Young are born within a few hours or days after the females arrive. Then males, having already arrived and established territories, collect groups of females, which they guard with vigilance during the mating period. After newborn offspring have been nursed for approximately three months, females and juveniles leave for their long migration southward. Males do not follow but remain in the Gulf of Alaska during winter.

figure 20.12

Barren-ground caribou, *Rangifer tarandus*, of Canada and Alaska. **A,** Adult male caribou in autumn pelage and antlers in velvet. **B,** Summer and winter ranges of some major caribou herds in Canada and Alaska. (Other herds not shown occur on Baffin Island and in western and central Alaska.) The principal spring migration routes are indicated by arrows; routes vary considerably from year to year. The same species is known as reindeer in Europe. Order Artiodactyla, family Cervidae.

Although we might expect bats, the only winged mammals, to use their gift of flight to migrate, few of them do. Most spend their winters in hibernation. Four species of American bats that migrate spend their summers in northern or western states and their winters in the southern United States or Mexico.

Flight and Echolocation

Many mammals scamper about in trees with amazing agility; some can glide from tree to tree, and one group, bats, has full flight. Gliding and flying evolved independently in several groups of mammals, including marsupials, rodents, flying lemurs, and bats. Flying squirrels (figure 20.14) actually glide rather than fly, using the gliding skin (patagium) that extends from the sides of their body. Even bats must initiate their flight by gliding from a vertical surface or ledge; a bat resting on a horizontal surface cannot fly upward.

Bats are mostly nocturnal or crepuscular (active at twilight), and thus hold a niche unoccupied by most birds. Their achievement is attributed to two features: flight and capacity to navigate by echolocation. Together, these adaptations enable bats to avoid obstacles in absolute darkness, to locate and to catch insects with precision, and to find their way deep into caves (a habitat largely unexploited by other mammals and birds) where they sleep during daytime hours.

When in flight, a bat emits short pulses 5 to 10 msec in duration in a narrow directed beam from its mouth or nose (figure 20.15). Each pulse is frequency modulated; it is highest at the beginning, up to 100,000 Hz (hertz, cycles per second), and sweeps down to perhaps 30,000 Hz at the end. Sounds at this frequency are ultrasonic to human ears, which have an upper limit of about 20,000 Hz. When bats search for prey, they produce about 10 pulses per second. If prey is detected, the rate increases rapidly up

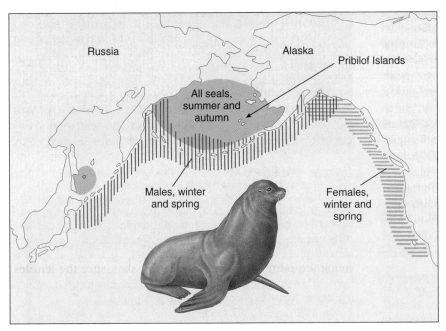

figure 20.13

Annual migrations of northern fur seals, *Callorhinus ursinus,* showing the separate wintering grounds of males and females. Both males and females of the large Pribilof population migrate in early summer to the Pribilof Islands, where females give birth and then mate with males. Order Carnivora, family Otariidae.

figure 20.14

Northern flying squirrel, *Glaucomys sabrinus,* gliding in for a landing. The area of the undersurface is nearly tripled in area when the gliding skin is spread. Glides of 40 to 50 m are possible. Good maneuverability during flight is achieved by adjusting the position of the gliding skin with special muscles. Flying squirrels are nocturnal and have superb night vision. Order Rodentia, family Sciuridae.

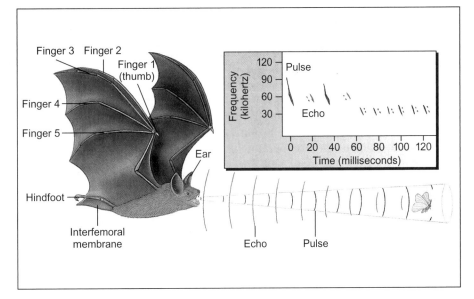

figure 20.15

Echolocation of an insect by a little brown bat, *Myotis lucifugus.* Frequency-modulated pulses are directed in a narrow beam from the bat's mouth. As the bat nears its prey, it emits shorter, lower signals at a faster rate. Order Chiroptera, family Vespertilionidae.

to 200 pulses per second in the final phase of approach and capture. Pulses are spaced so that the echo of each is received before the next pulse is emitted, an adaptation that prevents jamming. Since transmission-to-reception time decreases as a bat approaches an object, the bat can increase pulse frequency to obtain more information about an object. Pulse length is shortened as the bat nears an object. Some nocturnal moths can detect a bat's ultrasonic pulses and avoid predation.

External ears of bats are large, like hearing trumpets, and shaped variously in different species. Less is known about the inner ear of bats, but it obviously receives the ultrasonic sounds emitted. A bat builds a mental image of its surroundings from echo scanning that approaches the visual resolution of a diurnal animal's eyes.

Some bats, including the approximately 170 species of Old World fruit bats, lack echolocation abilities. Even so, most remain nocturnal, using large eyes and olfaction to find their meals of fruits, flowers, and nectar. Vampire bats have razor-sharp incisors used to shave away the epidermis of their prey, exposing underlying capillaries. After infusing an anticoagulant to keep blood flowing, they lap and store their blood meal in a specially modified stomach.

Reproduction

Most mammals have definite mating seasons, usually in winter or spring and timed so that birth and the rearing of young occur at the most favorable time of the year. Many male mammals can copulate at any time, but females are receptive only during a relatively brief period of the **estrous cycle** called **estrus** or heat (figure 20.16).

There are three different patterns of reproduction in mammals, represented by monotremes, marsupials, and placental mammals. **Monotremes** are (oviparous) mammals, such as the duck-billed platypus, which has one breeding season each year. Embryos develop for 10–12 days in the uterus, where they are nourished by yolk supplies deposited prior to ovulation and by secretions from the mother. A thin, leathery shell is secreted around the embryos before the eggs are laid. The platypus lays its eggs in a burrow, where they hatch in a relatively underdeveloped state after about 12 days. After hatching, the young feed on milk produced by the mother's mammary glands. Because monotremes have no nipples, the young lap milk secreted onto the belly fur of the mother.

Marsupials are pouched, viviparous mammals that exhibit a second pattern of reproduction. Although only eutherians are called "placental mammals," marsupials do have a transient type of placenta called a **choriovitteline** or yolk sac placenta. Embryos (blastocysts) of a marsupial are first encapsulated by shell membranes and float freely for several days in uterine fluid. After "hatching" from the shell membranes, embryos of most marsupials do not implant, or "take root," in the uterus as they would in eutherians, but erode shallow depressions in the uterine wall in which they lie and absorb nutrient secretions from the mucosa through the vascularized yolk sac. Gestation (the intrauterine period of development) is brief in

figure 20.16

African lions, *Panthera leo,* mating. Lions breed at any season, although predominantly in spring and summer. During the short period a female is receptive, she may mate repeatedly. Three or four cubs are born after gestation of 100 days. Once the mother introduces cubs into the pride, they are treated with affection by both adult males and females. Cubs go through an 18- to 24-month apprenticeship learning how to hunt and then are frequently driven from the pride to manage themselves. Order Carnivora, family Felidae.

figure 20.17

Opossums, *Didelphis marsupialis,* 15 days old, fastened to nipples in mother's pouch. When born after a gestation period of only 12 days, they are the size of honey bees. They remain attached to the nipples for 50 to 60 days. Order Didelphimorpha, family Didelphidae.

marsupials, and therefore all marsupials give birth to tiny young that are effectively still embryos, both anatomically and physiologically (figure 20.17). However, early birth is followed by a prolonged interval of lactation and parental care (figure 20.18).

The third pattern of reproduction is that of viviparous placental mammals, eutherians. In placentals, the reproductive investment is in prolonged gestation, unlike marsupials in which the reproductive investment is in prolonged lactation (figure 20.18). Like marsupials, most placental embryos initially receive nutrients through a choriovitteline placenta. The choriovitteline placenta may persist, as in mice, but usually is transitory and quickly replaced by a **chorioallantoic placenta,** formed from the chorionic and allantoic membranes. In eutherians, both types of placentae are extensively vascularized. The numerous capillaries of the placenta and the mother's uterus are close, but do not connect, and allow nutrients, respiratory gases, wastes, and other molecules to diffuse between the maternal and embryonic circulation. Length of gestation is longer in placentals than in marsupials, and generally increases with body size. For example, mice have a gestation period of 21 days; rabbits and hares, 30 to 36 days; cats and dogs, 60 days; cattle, 280 days; and elephants, 22 months (the

longest). Important exceptions include baleen whales, the largest mammals, whose gestation period is only 12 months, and small bats no larger than mice, whose gestation period extends 4 to 5 months. The condition of the young at birth also varies. An antelope bears its young well-furred, eyes open, and able to run. Newborn mice, however, are blind, naked, and helpless. We all know how long it takes a human baby to become a toddler. Human growth is slower than that of any other mammal, and this is one of the distinctive attributes that sets us apart from other mammals.

Delayed implantation lengthens the gestation period of many mammals. The blastocyst remains dormant while its implantation in the uterine wall is postponed for periods of a few weeks to several months. For many mammals (for example, bears, seals, weasels, badgers, bats, and many deer), delayed implantation extends gestation so that the young are born at a time of year best for their survival.

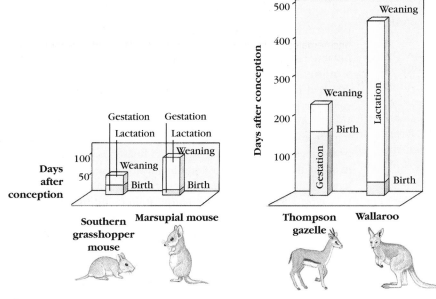

figure 20.18

Comparison of gestation and lactation periods between matched pairs of ecologically similar species of marsupial and placental mammals. The graph shows that marsupials have shorter intervals of gestation and longer intervals of lactation than do similar species of placentals.

figure 20.19

Collared lemming, *Dicrostonyx*, a small rodent of the far north. Populations of lemmings fluctuate widely. Order Rodentia, family Cricetidae.

■ Mammalian Populations

A population of animals includes all members of a species that share a particular space and can potentially interbreed. All mammals (like other organisms) live in ecological communities, each composed of numerous populations of different animal and plant species. Each species is affected by the activities of other species and by the physical environment, especially climate. Thus populations are always changing in size. Populations of small mammals are lowest before breeding season and greatest just after addition of new members.

Irregular fluctuations are commonly produced by variations in climate, such as unusually cold, hot, or dry weather, or by natural catastrophes, such as fires, hailstorms, and hurricanes. These are **density-independent** causes because they affect a population whether it is crowded or dispersed. However, the most spectacular fluctuations are **density-dependent,** associated with population crowding. These extreme limits to growth are discussed in Chapter 2 (see pp. 44–45).

Cycles of abundance are common among many rodent species. Among the best-known examples are mass migrations of Scandinavian and arctic North American lemmings following population peaks. Lemmings (figure 20.19) breed all year, although more in summer than in winter. The gestation period is only 21 days; young born at the beginning of summer are weaned in 14 days and can reproduce by the end of summer. At the peak of their population density, having devastated vegetation by tunneling and grazing, lemmings begin long, mass migrations to find new undamaged habitats

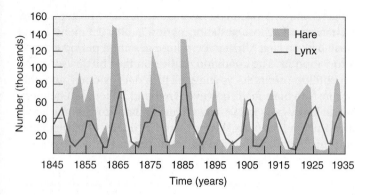

figure 20.20

Changes in population size of snowshoe hare and lynx in Canada as indicated by pelts received by the Hudson's Bay Company. Abundance of lynx (predator) follows that of the hare (prey).

for food and space. They swim across streams and small lakes as they go, but cannot distinguish these from large lakes, rivers, and the sea, in which they drown. Because lemmings are the main diet of many carnivorous mammals and birds, changes in lemming population density affects their predators as well.

Snowshoe hares (see figure 20.6) of North America show 10-year cycles in abundance. The well-known fecundity of rabbits enables them to produce litters of three or four young as many as five times per year. The density may increase to 4000 hares competing for food in each square mile (2.6 square kilometers) of northern forest. Predators (owls, minks, foxes, and especially lynxes) also increase (figure 20.20). Then the population crashes precipitously for reasons that have long puzzled scientists. Hares do not die from lack of food or from an epidemic disease (as once believed). Some data suggest decline of hares is associated with declining winter forage. In addition, hare populations follow patterns of sunspot activity, which may affect vegetative production (p. 49). Whatever the causes, population crashes that follow superabundance, although harsh, permit vegetation to recover, providing survivors with a much better chance for successful breeding.

In his book *The Arctic* (1974. Montreal, Infacor, Ltd.), the Canadian naturalist Fred Bruemmer describes the growth of lemming populations in arctic Canada: "After a population crash one sees few signs of lemmings; there may be only one to every 10 acres. The next year, they are evidently numerous; their runways snake beneath the tundra vegetation, and frequent piles of rice-sized droppings indicate the lemmings fare well. The third year one sees them everywhere. The fourth year, usually the peak year of their cycle, the populations explode. Now more than 150 lemmings may inhabit each acre of land and they honeycomb it with as many as 4000 burrows. Males meet frequently and fight instantly. Males pursue females and mate after a brief but ardent courtship. Everywhere one hears the squeak and chitter of the excited, irritable, crowded animals. At such times they may spill over the land in manic migrations."

■ Human Evolution

Darwin devoted an entire book, *The Descent of Man and Selection in Relation to Sex* (1871), largely to human evolution. The idea that humans share common descent with apes and other animals was repugnant to the Victorian world, which responded with predictable outrage (see figure 1.17). When Darwin's views were first debated, few human fossils had been unearthed, but the current accumulation of fossil evidence has strongly vindicated Darwin's hypothesis that humans descend from other primates. All primates share certain significant characteristics: grasping fingers on all four limbs, flat fingernails instead of claws, and forward-pointing eyes with binocular vision and excellent depth perception.

Primate Diversity

The earliest primate was probably a small, nocturnal animal similar in appearance to a tree shrew. This ancestral primate lineage split into two lineages, one of which gave rise to lemurs and lorises, and the other to tarsiers (figure 20.21), monkeys (figure 20.22), and apes (figure 20.23). Traditionally, the lemurs, lorises, and tarsiers have been called the **prosimians,** a paraphyletic group, and the apes and monkeys have been called **simians** or **anthropoids.** Prosimians and many simians are arboreal (tree-dwelling), which is probably the ancestral lifestyle for both groups. Flexible limbs are essential for active animals moving through trees. Grasping hands and feet, in contrast to the clawed feet of squirrels and other rodents, enable primates to grip limbs, to hang from branches, to seize and to manipulate food, and most significantly, to use tools. Primates have highly developed special senses, especially acute, binocular vision, and proper coordination of limb and finger muscles to assist their active arboreal life. Of course, sense organs are no better than the brain that processes the sensory information. A large cerebral cortex supports timing, judgment of distance, and perception of the environment.

The earliest simian fossils appeared in Africa some 55 million years ago. Many of these primate lineages became day-active rather than nocturnal, making vision the dominant

figure 20.21

A prosimian, the Mindanao tarsier, *Tarsius syrichta carbonarius,* of Mindanao Island in the Philippines. Order Primates, family Tarsiidae.

World monkeys differ from New World monkeys in being more terrestrial and in lacking a grasping tail while having close-set nostrils and better opposable and grasping thumbs.

Apes, which differ from Old World monkeys in having a larger cerebrum and loss of a tail, include gibbons, orangutans, gorillas, chimpanzees, bonobos, and humans. Except for gibbons, all apes are in the family Hominidae and are here called **hominids.** Chimpanzees and bonobos, *Pan,* form the living sister group to modern humans (see figure 4.7). All fossil hominid species that are phylogenetically closer to living humans than to chimpanzees and bonobos are called here humans or **hominins.** The earliest known fossils of apes are from 23-million-year-old rocks in eastern Africa. These early, forest-dwelling hominids subsequently diversified into numerous forms that spread across Africa and Eurasia.

The First Humans and Bipedalism

Trends in the evolution of skeletal differences between humans and other hominids often are associated with changes in diet and posture. Human jaws are less robust and bear smaller canines than those of other hominids, reflecting a more omnivorous diet. Position of the foramen magnum (a hole in the skull through which the spinal cord passes) shifted to directly underneath the braincase in humans, indicative of bipedalism and upright posture. Additional skeletal changes in the human lineage associated with bipedalism include shorter pelvic bones and an S-shaped vertebral column. Bipedalism provided a better view of the landscape and freed the hands for using tools, defense, carrying young, and gathering food.

Genetic evidence suggests that humans diverged from chimpanzees 5.5 to 6.5 million years ago. Fossil evidence of humans from this period is sparse and controversial. In 2001 the desert sands of Chad yielded a remarkably complete skull of a hominid, *Sahelanthropus tchadensis,* dated at about 7 million years ago (figure 20.24). Although its brain was no larger than that of a chimpanzee (320–380 cm^3), its relatively small canine teeth and ventral position of the foramen magnum suggest that the skull could be human.

The earliest well-known human is *Ardipithecus ramidus* from Ethiopia, dated at about 4.4 million years ago (figure 20.24). Until recently this species was known only from teeth, but in 2009 numerous additional fossils were described, including a 45% complete skeleton, named "Ardi." *Ardipithecus* was about 120 cm tall and bipedal, although it retained many ancestral adaptations for arboreal life, including long arms, fingers, and toes. Both *Sahelanthropus* and *Ardipithecus* were probably woodland dwellers, based on associated fossil vertebrates and invertebrates from the same deposits. Earlier hypotheses suggested that bipedalism arose as an adaptation to drying African environments, as woodlands were replaced with open savannas. However, because formation of the savannas of Africa did not occur until about 3 million years ago, it is now clear that bipedalism first arose in woodland-dwelling hominins.

A **B**

figure 20.22

Monkeys. **A,** Geoffroy's spider monkey, *Ateles geoffroyi,* order primates, family Alelidae, is a New World monkey. **B,** The olive baboon, *Papio homadryas,* order Primates, family Cercopithecidae, is an Old World monkey.

figure 20.23

Gorillas, *Gorilla gorilla,* order Primates, family Hominidae, are examples of apes.

special sense, now enhanced by color perception. We recognize three major simian clades: (1) New World monkeys of South America, including howler monkeys, spider monkeys (see figure 20.22A), and tamarins; (2) Old World monkeys of Africa and Asia, including baboons (see figure 20.22B), mandrills, and colobus monkeys; and (3) apes (figure 20.23). Old

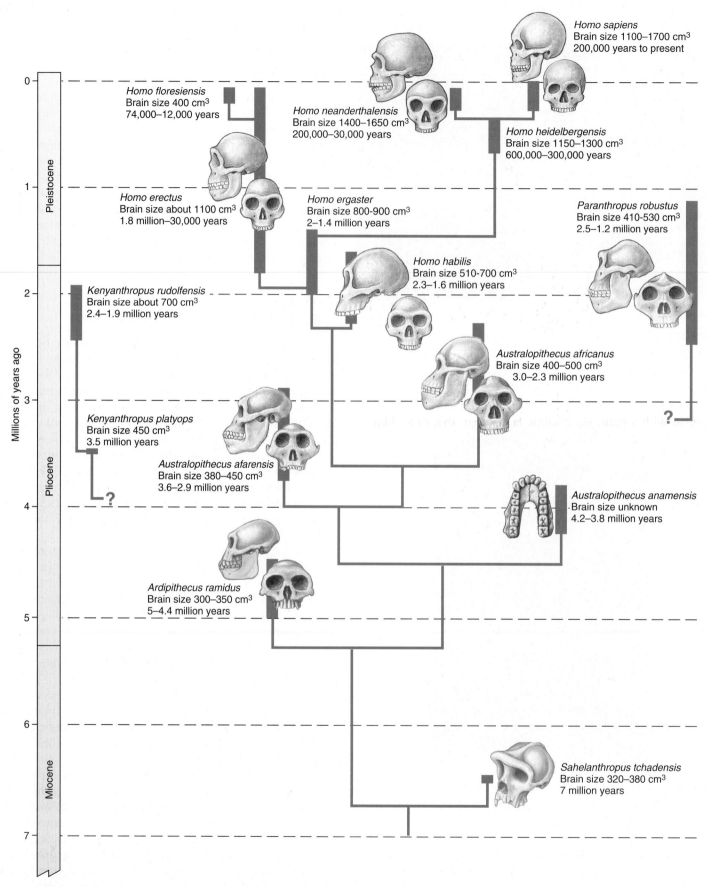

figure 20.24

Evolution of hominins. Shown is one possible phylogenetic reconstruction. Many relationships are controversial. Note that in this reconstruction *Australopithecus* is paraphyletic.

figure 20.25

Lucy (*Australopithecus afarensis*), one of the most nearly complete skeletons of an early human ever found. Lucy is dated at 3.2 million years old. A nearly complete skull of *A. afarensis* was discovered in 1994.

Another celebrated early human fossil is a 40% complete skeleton of a female *Australopithecus afarensis* (figure 20.25). Unearthed in 1974 and named "Lucy" by its discoverer, Donald Johanson, *A. afarensis* was a short, bipedal human with a brain size slightly larger than that of a chimpanzee (figure 20.26). This species was 1 to 1.5 m tall, and its teeth suggest it was primarily a fruit-eater, but also ate meat. The numerous fossils of *A. afarensis* date from 3.7 to 3 million years ago.

An explosion of australopithecine fossil finds during the last three decades documents at least eight species. Most of the early forms are considered gracile australopithecines because of their comparatively light build, especially in skull and teeth (although all were more robust than modern humans). The oldest of these is *Australopithecus anamensis* of Kenya and Ethiopia, which lived from 4.2 to 3.9 million years ago. This species is morophologically intermediate between *Ardipithecus* and *A. afarensis;* some researchers consider it the ancestor or sister group of modern humans and all other australopithecines. The australopithecine most frequently considered to be an immediate ancestor or sister taxon to genus *Homo* is *Australopithecus africanus,* dated at 3.0 to 2.3 million years ago. In 2010 two hominid skeletons discovered from rocks 2 to 1.8 million years ago were described as *Australopithecus sediba* and suggested to be phylogenetically close to *Homo.*

Coexisting with the earliest species of *Homo* and the latest species of gracile *Australopithecus* were at least three robust australopithecines, including *Paranthropus robustus* (see figure 20.24). The "robust" australopithecines, which lived from 2.5 to 1.2 million years ago, were characterized by prominent skull crests, heavy jaws, and large back molars. They had a broad diet, consuming coarse seeds, roots, and nuts in addition to the fruit, soft plants, and invertebrates typical of other early hominins. They form an extinct branch in hominid evolution and not part of our own ancestry.

figure 20.26

A reconstruction of the appearance of Lucy (*right*) compared with a modern human (*left*).

Early *Homo:* Tool-Making and Migration Out of Africa

Although researchers are divided over who the first members of *Homo* were, and indeed how to define the genus *Homo,* most recognize *Homo habilis* as the earliest known species of the genus. *Homo habilis* ("handy man") was similar in form to *Australopithecus,* about 111 to 135 cm tall (3.7-4.5 feet), with long arms and short legs. However, it had a larger brain (500 to 700 cm^3) than did *Australopithecus,* and may have used it in a novel cultural feature the making of stone tools. This capacity to make stone tools may be the defining feature of genus *Homo.* Chimpanzees and bonobos do not make stone tools, and they cannot be taught to make simple stone tools, despite efforts by patient researchers. *Homo habilis* shared the early Pleistocene drying African landscape with many other hominids, including species of *Paranthropus, Kenyanthropus rudolfensis,* and later, *Homo ergaster. Kenyanthropus rudolfensis* (also placed in the genus *Homo* or *Australopithecus*), a slightly more robust species than *H. habilis,* is known from remains formerly assigned to *H. habilis.* Although the expansion of savannas is no longer linked to the origin of bipedalism, perhaps it contributed to the spread of savanna-dwelling *Homo.*

About 1.5 million years ago, *Homo ergaster* appeared, a large hominid standing 150 to 190 cm (5 to 6 feet) tall, with a low but distinct forehead and strong brow ridges. Its brain capacity was about 850 mm^3, intermediate between the brain capacity of *H. habilis* and modern humans (see figure 20.24). *Homo ergaster* used more advanced tools, including hand axes, and controlled and used fire, as indicated by charcoal deposits. *Homo ergaster* migrated out of Africa about 1.7 million years ago and subsequently diversified into several forms, including *Homo erectus. Homo erectus,* a fairly morphologically uniform species, spread throughout southern Europe and in Asia as far east as China and Java, where it survived until about 30,000 years ago. Another amazing hominid find was announced in 2004: *Homo floresiensis,* a species only 1 m tall, from the island of Flores, Indonesia. This species diverged from *H. erectus* and became extinct only about 13,000 years ago.

Modern Humans

Modern humans diverged from African *H. ergaster* about 800,000 years ago. These humans, now assigned to *H. heidelbergensis,* spread throughout Africa, Europe, and the Middle East. The brain of *H. heidelbergensis* (about 1250 cc^3) was larger than those of its ancestors. About 200,000 years ago, when long glacial conditions predominated, *Homo heidelbergensis* was replaced by two hominins, *H. neanderthalensis* in Europe and *H. sapiens* in Africa **Neandertals** (*H. neanderthalensis*) occupied most of Europe and the Middle East. They had a brain size similar to that of modern humans and made more sophisticated stone tools than did *H. erectus.* Their robust, heavily muscled bodies allowed them to survive the cold climates of the Ice Age and to hunt large Pleistocene mammals.

Homo sapiens, as presently defined, arose in Africa about 200,000 years ago. Neandertals and *H. erectus* disappeared about 30,000 years ago, 10,000 years after *H. sapiens* spread into Europe and East Asia. Modern humans were tall people with a culture very different from Neandertals. Implement crafting developed rapidly, and human culture became enriched with aesthetics, artistry, and sophisticated language.

What happened to the Neandertals? Recent evidence demonstrates limited interbreeding with *H. sapiens.* This apparently occurred after *H. sapiens* left Africa, because "Neandertal" genes are found in some living Asians and Europeans, but not in Africans. Direct evidence is lacking, but most anthropologists venture than *H. sapiens,* with its technological and cultural advantages, may have outcompeted other humans.

In closing our discussion of human evolution, it is important to note that human evolution was characterized by the same patterns and processes evident in the evolution of other animal groups, including speciation and extinction. At present there is only one human species alive, an unusual situation when one considers that three to five species of humans have been present during nearly all of the past 4 million years.

Biologically, *Homo sapiens* is a product of the same processes that directed the evolution of every organism since life began. Mutation, isolation, genetic drift, and natural selection have operated for us as they have for other animals. We are nonetheless unique, with a nongenetic cultural evolution that provides constant feedback between past and future experience. Our symbolic languages, capacity for conceptual thought, knowledge of our history, and ability to manipulate our environment emerge from this nongenetic cultural endowment. Finally, we owe much of our cultural and intellectual achievements to our arboreal ancestry, which gave us binocular vision, superb visuotactile discrimination, and manipulative use of our hands. If horses (with one toe instead of five fingers) had human mental capacity, could they have accomplished what humans have?

Taxonomy of Living Mammalian Orders

This classification follows that of Wilson and Reeder (2005). Of the 29 recognized mammalian orders, three smaller marsupial orders and ten smaller placental orders are omitted.

Class Mammalia

Subclass Prototheria (prō´tō-thir´ē-ə) (Gr. *prōtos,* first, + *thēr,* wild animal).

Infraclass Ornithodelphia (or-ni´-thō-del´fē-ə) (Gr. *ornis,* bird, + *delphys,* womb). Monotreme mammals.

Order Monotremata (mon´ō-trē-mä´tä) (Gr. *monos,* single, + *trēma,* hole): **egg-laying (oviparous) mammals: duck-billed platypus,** and **echidnas.** Five species in this order from Australia, Tasmania, and New Guinea. Spiny anteaters, or echidnas, *Tachyglossus* and *Zaglossus,* have a long, narrow snout adapted for feeding on ants, their chief food.

Subclass Theria (thir´ē-a) (Gr. *thēr,* wild animal).

Infraclass Metatheria (met´ə-thir´ē-ə) (Gr. *meta,* after, + *thēr,* wild animal). Marsupial mammals.

Order Didelphimorphia (di´del-fi-mor´fē-ə) (Gr. *di,* two, + *delphi,* uterus, + *morph,* form): **American opossums.** These mammals, like other marsupials, are characterized by an abdominal pouch, or marsupium, in which they rear their young. Most species occur in Central and South America, but one species, the Virginia opossum, is widespread in North America; 87 species.

Order Dasyuormorphia (das-ē-yur´ō-mor´fē-ə) (Gr. *dasy,* hairy, + *uro,* tail, + *morph,* form): **Australian carnivorous mammals.** In addition to a number of larger carnivores, this order includes a number of marsupial "mice," all of which are carnivorous. Confined to Australia, Tasmania, and New Guinea; 71 species.

Order Peramelemorphia (per´ə-mel-e-mor´fē-ə) (Gr. *per,* pouch, + *mel,* badger, + *morph,* form): **bandicoots.** Like placentals, members of this group have a chorioallantoic placenta and a high rate of reproduction for marsupials. Confined to Australia, Tasmania, and New Guinea; 22 species.

Order Diprotodontia (di´prō-tō-don´-tē-ə) (Gr. *di,* two, + *pro,* front, + *odont,* tooth): **koala, wombats, possums, wallabies,** and **kangaroos.** Diverse marsupial group containing some of the largest and most familiar marsupials. Present in Australia, Tasmania, New Guinea, and many islands of the East Indies; 143 species.

Infraclass Eutheria (yu¯-ther´ē-ə) (Gr. *eu,* true, + *thēr,* wild animal). Placental mammals.

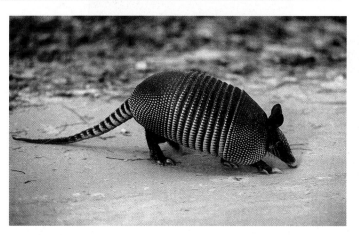

figure 20.27

Nine-banded armadillo, *Dasypus novemcintus.* During the day, this nocturnal species occupies long tunnels, which it digs using powerful, clawed forelimbs. Order Cingulata, family Dasypodidae.

Order Proboscidea (prō´bo-sid´ē-ə) (Gr. *proboskis,* elephant's trunk, from *pro,* before, + *boskein,* to feed): **elephants.** Largest of living land animals, with two upper incisors elongated as tusks and well-developed molar teeth. Three extant species: Indian elephant, with relatively small ears, and two species of African elephants, with large ears.

Order Cingulata (sin´gyū-lätə) (L. *cingul,* belt): **armadillos** (figure 20.27). Insectivorous mammals with small, peglike teeth and beltlike bands of armor. Inhabit South and Central America; the nine-banded armadillo is expanding its range northward in the United States; 21 species.

Order Primates (prī-mā´tēz or prī´māts) (L. *prima,* first): **prosimians, monkeys,** and **apes.** First among animals in brain development with especially large cerebral cortex. Most species are arboreal, with large eyes, binocular vision, grasping hands, and five digits (usually with flat nails) on both forelimbs and hindlimbs; two suborders; 376 species.

Suborder Strepsirhini (strep´sə-rī´-nē) (Gr. *strepsō,* to turn, twist, + *rhinos,* nose): **lemurs, aye-ayes, lorises, pottos,** and **bush babies.** Seven families of arboreal primates concentrated on Madagascar, but with species in Africa, Southeast Asia, and Malay peninsula. All have a wet, naked region surrounding comma-shaped nostrils, long nonprehensile tail, and second toe provided with a claw; 88 species.

Suborder Haplorhini (hap´lō-rī´-nē) (Gr. *haploos,* single, simple, + *rhinos,* nose): **tarsiers, marmosets, New and Old World monkeys, gibbons, gorilla, chimpanzee, orangutan,** and

(continued)

figure 20.28

A pika, *Ochotona princeps*, atop a rockslide in Alaska. This little rat-sized mammal does not hibernate but prepares for winter by storing dried grasses beneath boulders. Order Lagomorpha, family Ochotonidae.

figure 20.29

A shorttail shrew, *Blarina brevicauda*, eating a grasshopper. This tiny but fierce mammal, with a prodigious appetite for insects, mice, snails, and worms, spends most of its time underground and is seldom seen. Shrews resemble the insectivorous ancestors of placental mammals. Order Soricomorpha, family Soricidae.

human. Six families, all except tarsiers, are in the clade Anthropoidea. Haplorhine primates have dry, hairy noses, ringed nostrils, and differences in skull morphology that distinguish them from strepsirhine primates; 288 species.

Order Lagomorpha (lag´ō-mor´fə) (Gr. *lagos*, hare, + *morphē*, form): **rabbits, hares,** and **pikas** (figure 20.28). Dentition resembling that of rodents but with four upper incisors rather than two; 92 species.

Order Rodentia (rō-den´tē-ə) (L. *rodere*, to gnaw): **gnawing mammals: squirrels, rats,** and **woodchucks.**

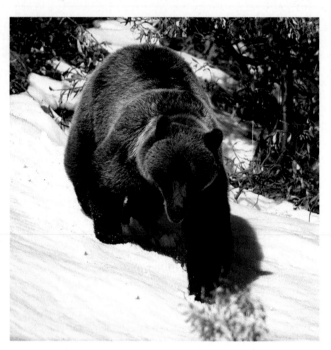

figure 20.30

Grizzly bear, *Ursus arctos horribilis*, of Alaska. Grizzlies, once common in the lower 48 states, are now confined largely to northern wilderness areas. Order Carnivora, family Ursidae.

Most numerous of all mammals both in numbers and species. Characterized by two pairs of chisel-like incisors that grow continually and are adapted for gnawing; 2277 species.

Order Soricomorpha (sor´i-cō-mor´fə) (L. *soric* shrew, + *morph*, form): **shrews** (figure 20.29) and **moles.** Small, sharp-snouted animals, which feed principally on invertebrates; 428 species.

Order Chiroptera (kī-rop´ter-ə) (Gr. *cheir*, hand, + *pteron*, wing): **bats.** Flying mammals with forelimbs modified into wings. Most are nocturnal insect-eaters and navigate by echolocation, but fruit bats of the Old World tropics are diurnal; 1116 species.

Order Carnivora (car-niv´or-ə) (L. *caro*, flesh, + *vorare*, to devour): **flesh-eating mammals: dogs, wolves, cats, bears** (figure 20.30), **weasels, seals, sea lions,** and **walruses.** All except the giant panda have predatory habits, and their teeth are especially adapted for tearing flesh; most have canines for killing prey; worldwide but only seals occur in Australian and Antarctic regions; 286 species.

Order Perissodactyla (pe-ris´sō-dak´til-ə) (Gr. *perissos*, odd, + *dactylos*, toe): **odd-toed hoofed mammals: horses, asses, zebras, tapirs,** and **rhinoceroses.** Mammals with an odd number (one or three) of toes,

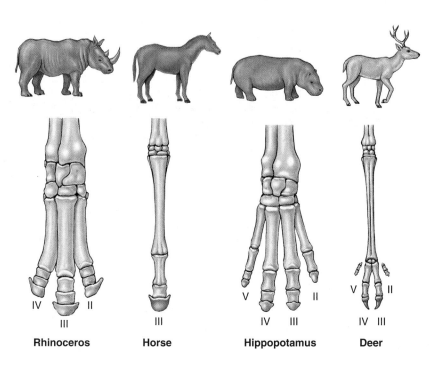

figure 20.31

Odd-toed and even-toed ungulates. Rhinoceroses and horses (order Perissodactyla) are odd-toed; hippopotamuses and deer (order Artiodactyla) are even-toed. Lighter, faster mammals run on only one or two toes.

Rhinoceros   **Horse**   **Hippopotamus**   **Deer**

each with a keratinized hoof (figure 20.31). All are herbivorous. Both Perissodactyla and Artiodactyla often called **ungulates,** or hoofed mammals, with teeth adapted for grinding plants; 17 species.

Order Artiodactyla (är´tē-ō-dak´til-ə) (Gr. *artios,* even, + *daktylos,* toe): **even-toed hoofed mammals: swine, camels, deer and their allies, hippopotamuses, antelopes, cattle, sheep,** and **goats.** Each toe sheathed in a cornified hoof. Most species have two toes, although hippopotamuses and some others have four (figure 20.31). Many, such as cattle, deer, and sheep, have horns or antlers. Most are ruminants; 240 species.

Order Cetacea (se-tā´shē-ə) (L. *cetus,* whale): **whales** (figure 20.32), **dolphins,** and **porpoises.** Anterior limbs of cetaceans are modified into broad flippers; posterior limbs are absent. Nostrils are represented by a single or double blowhole on top of the head. Hairs are limited to a few on the muzzle, skin glands are absent except for mammary glands and those of the eye, and there is no external ear; 84 species.

figure 20.32

Humpback whale, *Megaptera novaeangliae,* breaching. Among the most acrobatic of whales, humpbacks appear to breach to stun fish schools or to communicate information to other herd members. Order Cetacea, family Balaenopteridae.

■ Summary

Mammals are endothermic and homeothermic vertebrates whose bodies are insulated by hair and who nurse their young with milk. The approximately 5400 species of mammals descend from the synapsid lineage of amniotes that arose during the Carboniferous period of the Paleozoic era. Their evolution is traced from pelycosaurs of the Permian period to therapsids of the late Permian and Triassic periods of the Mesozoic era. One group of therapsids, the cynodonts, gave rise during the Triassic to mammals. Mammalian evolution was accompanied by the appearance of many important derived characteristics, among these an enlarged brain with greater sensory integration, high metabolic rate, endothermy, heterodont teeth, and many changes in the skeleton that supported a more active life. Mammals diversified rapidly during the Tertiary period of the Cenozoic era.

Mammals are named for the glandular milk-secreting organs of females (rudimentary in males), a unique adaptation that, combined with prolonged parental care, buffers infants from demands of foraging for themselves and eases the transition to adulthood. Hair, the integumentary outgrowth that covers most mammals, serves variously for mechanical protection, thermal insulation, protective coloration, and waterproofing. Mammalian skin is rich in glands: sweat glands that function in evaporative cooling, scent glands used in social interactions, and sebaceous glands that secrete lubricating skin oil.

All placental mammals have deciduous teeth that are replaced by permanent teeth (diphyodont dentition). Four groups of teeth—incisors, canines, premolars, and molars—may be highly modified in different mammals for specialized feeding tasks, or they may be absent. Insectivores have pointed teeth for piercing the exoskeleton of insects and other small invertebrates. Herbivorous mammals have specialized teeth for grinding cellulose and silica-rich plants and have specialized regions of the gut for harboring microorganisms that digest cellulose. Carnivorous mammals have adaptations, including powerful jaw muscles and teeth, for killing and processing their prey, mainly herbivorous mammals. Omnivores feed on both plant and animal foods and have a variety of tooth types.

Some marine, terrestrial, and aerial mammals migrate; some migrations, such as those of fur seals and caribou, are extensive. Migrations are usually made toward climatic conditions favorable for finding food, mating, or rearing young.

Mammals with true flight, the bats, are mainly nocturnal and thus avoid direct competition with birds. Most employ ultrasonic echolocation to navigate and to feed in darkness.

The only living egg-laying mammals are monotremes of the Australian region. After hatching, young are nourished with their mother's milk. All other mammals are viviparous. Embryos of marsupials have brief gestation periods, are born underdeveloped, and complete their early growth in the mother's pouch, nourished by milk. The remaining mammals are eutherians, mammals that develop a sophisticated placental attachment between mother and embryos through which embryos are nourished for a prolonged period.

Mammal populations fluctuate due to both density-dependent and density-independent causes, and some mammals, particularly rodents, may experience extreme cycles in population density. The unqualified success of mammals as a group reflects their impressive overall adaptability.

Humans are primates, a mammalian group that descends from a shrewlike ancestor. The common ancestor of all modern primates was arboreal and had grasping fingers and forward-facing eyes capable of binocular vision. Primates diversified to form two major groups: (1) lemurs and lorises and (2) tarsiers, monkeys, and apes (including humans). Chimpanzees and bonobos form the living sister group of humans.

Humans appeared in Africa about 6.5 million years ago, and gave rise to several species of australopithecines, which persisted for about 3 million years. *Australopithecines* were shorter and smaller-brained than modern humans, but were bipedal. They gave rise to, and coexisted with, *Homo habilis,* the first maker of stone tools. *Homo ergaster* appeared about 1.8 million years ago and with its close relative, *H. erectus,* spread throughout Africa, Asia, and Europe. They were eventually replaced by modern humans, *Homo sapiens.*

■ Review Questions

1. Describe the evolution of mammals, tracing their synapsid ancestry from early amniote ancestors to true mammals. How would you distinguish pelycosaurs, early therapsids, cynodonts, and mammals?
2. Describe structural and functional adaptations in early amniotes that foreshadowed the mammalian body plan. Which mammalian attributes were especially important to successful diversification of mammals?
3. Hair is hypothesized to have evolved in therapsids as an adaptation for insulation, but modern mammals have adapted hair for several other purposes. Describe these.
4. What is distinctive about each of the following: horns of bovids, antlers of deer, and horns of rhinos? Describe the growth cycle of antlers.
5. Describe the location and principal function(s) of each of the following skin glands: sweat glands, scent glands, sebaceous glands, and mammary glands.
6. Define the terms "diphyodont" and "heterodont," and explain how both terms apply to mammalian dentition.
7. Describe the food habits of insectivores, herbivores, carnivores, and omnivores. Give the common names of some mammals belonging to each group.
8. Most herbivorous mammals depend on cellulose as their main energy source, and yet no mammal synthesizes cellulose-splitting enzymes. How are the digestive tracts of mammals specialized for symbiotic digestion of cellulose?
9. How does fermentation differ between horses and cattle?
10. Describe the annual migrations of barren-ground caribou and fur seals.

11. Explain what is distinctive about the life habit and mode of navigation in bats.

12. Describe and distinguish patterns of reproduction in monotremes, marsupials, and placental mammals. What aspects of mammalian reproduction occur in *all* mammals but in no other vertebrates?

13. What is the difference between density-dependent and density-independent fluctuations in the sizes of mammalian populations?

14. Describe the hare-lynx population cycle, considered a classic example of a prey-predator relationship (see figure 20.20). From your examination of the cycle, formulate a hypothesis to explain the oscillations.

15. What anatomical characteristics distinguish primates from other mammals?

16. What role do the fossils named "Ardi" and "Lucy" play in reconstruction of human evolutionary history?

17. In what ways do the genera *Australopithecus* and *Homo* differ?

18. When did the different species of *Homo* appear, and how did their culture differ?

For Further Thought Many zoologists consider mammalian teeth, rather than endothermy, hair, or lactation, the characteristic most responsible for the success of mammals. Explain why this could be true.

■ Selected References

See also general references on page 432.

Cameron, D. W., and C. P. Groves. 2004. Bones, stones, and molecules. Burlington, Massachusetts, Elsevier Academic Press. *Comprehensive synthesis of human evolution that highlights speciation and extinction in the human lineage.*

Feldhamer, G. A., L. C. Drickamer, S. H. Vessey, J. F. Merritt, and C. Krajewski. 2007. Mammalogy: adaptation, diversity, and ecology, ed. 3. Baltimore, Johns Hopkins University Press. *Modern, well-illustrated textbook.*

Gibbons, A. 2009. A new kind of ancestor: *Ardipithecus* unveiled. Science **326:**36–40. *This, and eleven other articles in this issue describe the anatomy and life of "Ardi."*

Green, R. E., and 55 others. 2010. A draft sequence of the Neandertal genome. Science **328:**701–722. *Neandertal genes are in living Europeans and Asians, evidence of limited interbreeding between neandertals and modern humans.*

Grzimek's encyclopedia of mammals (vol. 1–5). 1990. New York, McGraw-Hill Publishing Company. *Valuable source of information on all mammalian orders.*

Macdonald, D., ed. 1984. The encyclopedia of mammals. New York, Facts on File Publications. *Coverage of all mammalian orders and families, enhanced with fine photographs and color artwork.*

Nowak, R. M. 1999. Walker's mammals of the world, ed. 6. Baltimore, The Johns Hopkins University Press. *The definitive illustrated reference work on mammals, with descriptions of all extant and recently extinct species.*

Preston-Mafham, R., and K. Preston-Mafham. 1992. Primates of the world. New York, Facts on File Publications. *A small "primer" with high-quality photographs and serviceable descriptions.*

Rismiller, P. D., and R. S. Seymour. 1991. The echidna. Sci. Am. **294:**96–103 (Feb.). *Recent studies of this fascinating monotreme have revealed many secrets of its natural history and reproduction.*

Suga, N. 1990. Biosonar and neural computation in bats. Sci. Am. **262:**60–68 (June). *How the bat nervous system processes echolocation signals.*

Templeton, A. R. 2002. Out of Africa again and again. Nature **416:**45–51. *A comprehensive analysis of molecular genetic data indicating that humans have evolved as a cohesive lineage for the past 1.7 million years, with multiple migrations of populations out of Africa.*

Wilson, D. E., and D. Reeder, eds. 2005. Mammal species of the world: a taxonomic and geographic reference, 3. ed. Baltimore, Johns Hopkins University Press. *An authoritative compilation of the world's mammal species.*

■ Custom Website

The *Animal Diversity* Online Learning Center is a great place to check your understanding of chapter material. Visit www.mhhe.com/hickmanad6e for access to key terms, quizzes, and more! Further enhance your knowledge with Web links to chapter-related material.

Explore live links for these topics:

Classification and Phylogeny of Animals
Class Mammalia
Marine Mammals
Cat Dissections
Rat Dissections

Fetal Pig Dissections
Vertebrate Laboratory Exercises
Conservation Issues Concerning Mammals

General References

These references pertain to groups covered in more than one chapter. They include a number of very valuable field manuals that aid in identification, as well as general texts.

Barrington, E. J. W. 1979. Invertebrate structure and function, ed. 2. New York, John Wiley & Sons, Inc. *Excellent account of function in major invertebrate groups.*

Brusca, R. C., and G. J. Brusca. 2003. Invertebrates, ed. 2. Sunderland, Massachusetts, Sinauer Associates, Inc. *Invertebrate text organized around the bauplan ("body plan") concept—structural range, architectural limits, and functional aspects of a design—for each phylum. Includes cladistic analysis of phylogeny for most groups.*

Cracraft, J., and M. J. Donoghue (eds.). 2004. Assembling the tree of life. New York, Oxford University Press. *This volume summarizes much of the current work on the Tree of Life project, an attempt to uncover the evolutionary relationships for all living taxa. The early chapters of the book discuss the relevance of the Tree of Life project for society. The overview and introductory sections are very readable, and these are paired with results of recent phylogenetic analyses for all domains of life.*

Fotheringham, N. 1980. Beachcomber's guide to Gulf Coast marine life. Houston, Texas, Gulf Publishing Company. *Coverage arranged by habitats. No keys, but common forms that occur near shore can be identified.*

Gosner, K. L. 1979. A field guide to the Atlantic seashore: invertebrates and seaweeds of the Atlantic coast from the Bay of Fundy to Cape Hatteras. The Peterson Field Guide Series. Boston, Houghton Mifflin Company. *A helpful aid for students of invertebrates found along the northeastern coast of the United States.*

Halstead, B. W. 1992. Dangerous aquatic animals of the world: a color atlas: with prevention, first aid, and emergency treatment procedures. Princeton, New Jersey, Darwin Press.

Excellent coverage of threats to humans by dangerous aquatic animals and how to avoid fateful encounters.

Humann, P., and N. DeLoach. 2002. Reef creature identification. Florida, Caribbean, Bahamas. Jacksonville, Florida, New World Publications. *Excellent field guide to aid identification of Atlantic reef invertebrates except corals.*

Hyman, L. H. 1940–1967. The invertebrates, 6 vols. New York, McGraw-Hill Book Company. *Informative discussions on the phylogenies of most invertebrates are treated in this outstanding series of monographs. Volume I contains a discussion of the colonial hypothesis of the origin of metazoa, and volume 2 contains a discussion of the origin of bilateral animals, body cavities, and metamerism.*

Kaplan, E. H. 1988. A field guide to southeastern and Caribbean seashores: Cape Hatteras to the Gulf Coast, Florida, and the Caribbean. A. Peterson Field Guide Series. Boston, Houghton Mifflin Company. *More than just a field guide, this comprehensive book is filled with information on the biology of seashore animals: complements Gosner's field guide, which covers animals north of Cape Hatteras.*

Kardong, K. V. 2012. Vertebrates: comparative anatomy, function, evolution, ed. 6. New York, McGraw-Hill. *Evolution of vertebrate structure.*

Kozloff, E. N. 1996. Marine invertebrates of the Pacific Northwest. Seattle, University of Washington Press. *Contains keys for many marine groups.*

Lane, R. P., and H. W. Crosskey. 1993. Medically important insects and arachnids. London, Chapman and Hall. *The most up-to-date medical entomology text available.*

Mebs, D. 2002. Venomous and poisonous animals. Stuttgart, Germany, Medpharm Scientific Publishers. *Fascinating coverage of animals dangerous to humans. Highlights include color photos of every species and medical treatment descriptions.*

Meglitsch, P. A., and F. R. Schram. 1991. Invertebrate zoology, ed. 3. New York, Oxford University Press. *Schram's thorough revision of the older Meglitsch text. Includes cladistic treatments.*

Morris, R. H., D. P. Abbott, and E. C. Haderlie. 1980. Intertidal invertebrates of California. Stanford, Stanford University Press. *An essential reference on the most important invertebrates of the intertidal zone in California. Contains 900 color photographs.*

Nielsen, C. 2001. Animal evolution: interrelationships of the living phyla. New York, Oxford University Press. *Cladistic analysis of morphology is used to develop sister-group relationships of the living Metazoa. An advanced but essential reference.*

Pechenik, J. A. 2009. Biology of the invertebrates, ed. 6. Dubuque, McGraw-Hill Companies, Inc. *Up-to-date text, including recognition of superphyla Ecdysozoa and Lophotrochozoa.*

Pough, F. H., J. B. Heiser, and W. N. McFarland. 2009. Vertebrate life, ed. 8. Upper Saddle River, New Jersey, Prentice Hall. *Vertebrate morphology, physiology, ecology, and behavior cast in a cladistic framework.*

Ricketts, E. F., J. Calvin, and J. W. Hedgpeth (revised by D. W. Phillips). 1985. Between Pacific tides, ed. 5. Stanford, Stanford University Press. *A revision of a classic work in marine biology. It stresses the habits and habitats of the Pacific coast invertebrates, and the illustrations are revealing. It includes an excellent, annotated systematic index and bibliography.*

Roberts, L. S., and J. Janovy, Jr. 2005. Gerald D. Schmidt and Larry S. Roberts' Foundations of parasitology, ed. 7. Dubuque, McGraw-Hill Publishers. *Highly readable and up-to-date account of parasitic protozoa, worms, and arthropods.*

Ruppert, E. E., R. S. Fox, and R. D. Barnes. 2004. Invertebrate zoology, ed. 7. Florence, Brooks/Cole Publishing. *Authoritative, detailed coverage of invertebrate phyla.*

432

Smith, D. G. 2001. Pennak's freshwater invertebrates of the United States, ed. 4. New York, John Wiley & Sons, Inc. *Contains keys for identification of freshwater invertebrates with brief accounts of each group. Indispensable for freshwater biologists.*

Smith, D. L., and K. B. Johnson. 1996. A guide to marine coastal plankton and marine invertebrate larvae. Dubuque, Iowa, Kendall/Hunt Publishing Company. *Valuable manual for identification of marine plankton, which is usually not covered in most field guides.*

Tudge, C. 2000. The variety of life; a survey and celebration of all the creatures that have ever lived. New York, Oxford University Press. *This well-written book might be called a celebration of systematics, the discipline, as the author says, that introduces us to the creatures themselves. A fine introduction that explains cladistics is followed by chapters that describe … almost everything!*

Valentine, J. V. 2004. On the origin of phyla. Chicago, Illinois, University of Chicago Press. *A thorough coverage of paleontological, morphological, and molecular evidence on relationships among the animal phyla and the evolutionary origins of animal body plans.*

Willmer, P., G. Stone, and I. A. Johnston. 2005. Environmental physiology of animals. Malden, Massachusetts, Blackwell Publishers. *A comparative treatment of the physiological ecology of animals.*

Glossary

This glossary lists definitions, pronunciations, and derivations of the most important recurrent technical terms, units, and names (excluding taxa) used in this text.

A

abiotic (ā´bī-ät´ik) (Gr. *a*, without, + *biōtos*, life, livable) Characterized by the absence of life.

abomasum (ab´ō-mā´səm) (L. *ab*, from, + *omasum*, paunch) Fourth and last chamber of the stomach of ruminant mammals.

aboral (ab-o´rəl) (L. *ab*, from, + *os*, mouth) Region of an animal opposite the mouth.

acanthodians (a´kan-thō´dē-əns) (Gr. *akantha*, prickly, thorny) Group of jawed fishes, characterized by large spines in their fins, from Lower Silurian to Lower Permian.

acclimatization (ə-klī´mə-tə-zā-shən) (L. *ad*, to, + Gr. *klima*, climate) Gradual physiological adjustment of an organism in response to relatively long-lasting environmental changes.

acoelomate (ā-sēl´ə-māt´) (Gr. *a*, not, + *koilōma*, cavity) Without a coelom, as in flatworms and proboscis worms.

acontium (ə-kän´chē-əm), pl. **acontia** (Gr. *akontion*, dart) Threadlike structure bearing nematocysts located on mesentery of sea anemone.

adaptation (L. *adaptatus*, fitted) Anatomical structure, physiological process, or behavioral trait that evolved by natural selection and improves an organism's ability to survive and leave descendants.

adaptive radiation Evolutionary diversification that produces numerous ecologically disparate lineages from a single ancestral one, especially when this diversification occurs within a short interval of geological time.

adaptive zone A characteristic reaction and mutual relationship between environment and organism ("way of life") demonstrated by a group of evolutionarily related organisms.

adductor (ə-duk´tər) (L. *ad*, to, + *ducere*, to lead) A muscle that draws a part toward a median axis, or a muscle that draws the two valves of a mollusc shell together.

adipose (ad´ə-pōs) (L. *adeps*, fat) Fatty tissue; fatty.

aerobic (ā-rō´bik) (Gr. *aēr*, air, + *bios*, life) Oxygen-dependent form of respiration.

afferent (af´ə-rənt) (L. *ad*, to, + *ferre*, to bear) Adjective meaning leading or bearing toward some organ—for example, nerves conducting impulses toward the brain or blood vessels carrying blood toward an organ; opposed to efferent.

age structure An accounting of the ages of individuals in a population at a particular time and place.

agnathan (ag-nā´-thən) (Gr. *a*, without, + *gnathos*, jaw) A jawless fish of the paraphyletic superclass Agnatha of the phylum Chordata.

alate (ā´lāt) (L. *alatus*, wing) Winged.

albumin (al-bū´mən) (L. *albuman*, white of egg) Any of a large class of simple proteins that are important constituents of vertebrate blood plasma and tissue fluids and also present in milk, whites of eggs, and other animal substances.

allantois (ə-lan´tois) (Gr. *allas*, sausage, + *eidos*, form) One of the extraembryonic membranes of the amniotes that functions in respiration and excretion in birds and nonavian reptiles and plays an important role in the development of the placenta in most mammals.

allele (ə-lēl´) (Gr. *allēlōn*, of one another) Alternative forms of genes coding for the same trait, and situated at the same locus in homologous chromosomes.

allelic frequency An estimation of the proportion of gametes produced in a population (gene pool) that contains a particular allelic form of a particular gene.

allopatric (Gr. *allos*, other, + *patra*, native land) In separate and mutually exclusive geographical regions.

allopatric speciation The hypothesis that new species are formed by dividing an ancestral species into geographically isolated subpopulations that evolve **reproductive barriers** between them through independent evolutionary divergence from their common ancestor.

altricial (al-tri´shəl) (L. *altrices*, nourishers) Referring to young animals (especially birds) having the young hatched in an immature, dependent condition.

alula (al´yə-lə) (L. dim of *ala*, wing) The first digit or thumb of a bird's wing, much reduced in size.

alveoli (al-vē´ə-lī) Pockets or spaces bounded by membrane or epithelium.

ambulacra (am´byə-lak´rə) (L. *ambulare*, to walk) In echinoderms, radiating grooves where podia of the water-vascular system characteristically project outside the organism.

ameboid movement (ə-mē´boid) (Gr. *amoibē*, change, + *oid*, like) Cellular locomotion by protrusion of cytoplasm to form pseudopodia.

amensalism (ā-men´səl-iz´əm) An asymmetric competitive interaction between two species in an ecological community in which only one of the species is affected.

amictic (ā-mik´tic) (Gr. *a*, without, + *miktos*, mixed or blended) Pertaining to female rotifers, which produce only diploid eggs that cannot be fertilized, or to the eggs produced by such females; contrasts with **mictic.**

ammocoetes (am-ə-sēt´ēz) (Gr. *ammos,* sand, + *koitē,* bed) The filter-feeding larval stage of lampreys.

amnion (am´nē-än) (Gr. *amnion,* membrane around the fetus) The innermost of the extraembryonic membranes forming a fluid-filled sac around the embryo in amniotes.

amniote (am´nē-ōt) Having an amnion; as a noun, an animal that develops an amnion in embryonic life; refers collectively to nonavian reptiles, birds, and mammals. Adj., **amniotic.**

amphid (am´fəd) (Gr. *amphidea,* anything that is bound around) One of a pair of anterior sense organs in certain nematodes.

amplexus (am-plek´səs) (L., embrace) The embrace of frogs or toads in which males fertililze eggs as they leave the female's body.

ampulla (am-pūl´ə) (L., flask) Membranous vesicle; dilation at one end of each semicircular canal containing sensory epithelium; muscular vesicle above the tube foot in water-vascular system of echinoderms.

anadromous (an-ad´rə-məs) (Gr. *anadromos,* running upward) Refers to fishes that migrate up streams from the sea to spawn.

anaerobic (an´ə-rō´bik) (Gr. *an,* not, + *aēr,* air, + *bios,* life) Not dependent on oxygen for respiration.

analogy (L. *analogous,* ratio) Similarity of function but not of origin.

anamniote (an´am-nē-ōt) A vertebrate that lacks an amniotic membrane around the embryo. Includes fishes and amphibians.

anapsids (ə-nap´səds) (Gr. *an,* without, + *apsis,* arch) Extinct amniotes in which the skull lacks temporal openings.

ancestral character state The condition of a taxonomic character inferred to have been present in the most recent common ancestor of a taxonomic group being studied cladistically.

androgenic gland (an´drō-jen´ək) (Gr. *anēr,* male, + *gennaein,* to produce) Gland in Crustacea that causes development of male characteristics.

annulus (an´yəl-əs) (L., ring) Any ringlike structure, such as superficial rings on leeches.

antenna (L., sail yard) A sensory appendage on the head of arthropods, or the second

pair of two such pairs of structures in crustaceans.

antennal gland Excretory gland of Crustacea located in the antennal metamere.

anterior (L. comparative of *ante,* before) The head of an organism or (as the adjective) toward that end.

anthropoid (an´thrə-poyd) (Gr. *anthrōpos,* man, + *eidos,* form) Resembling humans, especially the great apes.

aperture (ap´ər-chər) (L. *apertura* from *aperire,* to uncover) An opening; the opening into the first whorl of a gastropod shell.

apical (ā´pə-kl) (L. *apex,* tip) Pertaining to the tip or apex.

apical complex A certain combination of organelles found in the protozoan phylum Apicomplexa.

apocrine (ap´ə-krən) (Gr. *apo,* away, + *krinein,* to separate) Applies to a type of mammalian sweat gland that produces a viscous secretion by breaking off a part of the cytoplasm of secreting cells.

apodeme (ap´ə-dēm) A protrusion on the inside of the cuticle of certain arthropods (crustaceans and insects) to which the muscles attach.

apopyle (ap´ə-pīl) (Gr. *apo,* away from, + *pylē,* gate) In sponges, opening of the radial canal into the spongocoel.

aposematic (ap-ə-si-mat´ik) A conspicuous condition that serves as a warning; for example, the bright colors of a Monarch butterfly's wings warn potential predators that the butterfly is distasteful.

arboreal (är-bōr´ē-əl) (L. *arbor,* tree) Living in trees.

archaeocyte (ar´kē-ə-sīt) (Gr. *archaios,* beginning, + *kytos,* hollow vessel) Ameboid cells of varied function in sponges.

archenteron (ärk-en´tə-rän) (Gr. *archē,* beginning, + *eneron,* gut) The main cavity of an embryo in the gastrula stage; it is lined with endoderm and represents the future digestive cavity.

archosaur (är´kə-sor) (Gr. *archōn,* ruling, + *sauros,* lizard) A clade of diapsid vertebrates that includes the living crocodiles and birds and the extinct pterosaurs and dinosaurs.

Aristotle's lantern Masticating apparatus of some sea urchins.

artiodactyl (är´tē-ō-dak´təl) (Gr. *artios,* even, + *daktylos,* toe) One of an order of mammals with two or four digits on each foot.

asconoid (Gr. *askos,* bladder) Simplest form of sponges, with canals leading directly from the outside to the interior.

asexual Without distinct sexual organs; not involving formation of gametes.

asymmetric competition See **amensalism.**

atoke (ā´tōk) (Gr. *a,* without, + *tokos,* offspring) Anterior, nonreproductive part of a marine polychaete, as distinct from the posterior, reproductive part (epitoke) during the breeding season.

atrium (ā´trē-əm) (L. *atrium,* vestibule) One of the chambers of the heart. Also, the tympanic cavity of the ear. Also, the large cavity containing the pharynx in tunicates and cephalochordates.

auricle (aw´ri-kl) (L. *auricula,* dim. of *auris,* ear) One of the less muscular chambers of the heart; atrium; the external ear, or pinna; any earlike lobe or process.

auricularia (ə-rik´u-lar´ē-ə) (L. *auricula,* a small ear) Type of larva found in Holothuroidea.

autogamy (aw-täg´ə-mē) (Gr. *autos,* self, + *gamos,* marriage) Condition in which the gametic nuclei produced by meiosis fuse within the same organism that produced them to restore the diploid number.

autotomy (aw-tät´ə-mē) (Gr. *autos,* self, + *tomos,* a cutting) Detachment of a part of the body by the organism itself.

autotroph (aw´tō-trōf) (Gr. *autos,* self, + *trophos,* feeder) Organism that makes its organic nutrients from inorganic raw materials.

autotrophic nutrition (Gr. *autos,* self, + *trophia,* denoting nutrition) Nutrition characterized by the ability to use simple inorganic substances to synthesize more complex organic compounds, as in green plants and some bacteria.

average effect A quantitative-genetic parameter estimating the incremental contribution of each copy of a particular allele to the mean value of a particular organismal phenotype (such as height or weight) in the population being studied. Average effect is calculated from measurements of the population frequencies of all genotypes containing the allele, and the mean deviations of each genotypic class from the mean value of the phenotype in the population as a whole.

axial (L. *axis,* axle) Relating to the axis, or stem; on or along the axis.

axocoel (aks´ə-sēl) (Gr. *axon,* axle, + *koilos,* hollow) Anterior coelomic compartment in echinoderms; corresponds to protocoel.

axolotl (ak´sə-lot´l) (Nahuatl *atl*, water, + *xolotl*, doll, servant, spirit) Salamanders of the species *Ambystoma mexicanum*, which do not metamorphose and which retain aquatic larval characteristics throughout adulthood.

axon (ak´sän) (Gr. *axōn*) Elongate extension of a neuron that conducts impulses away from the cell body and toward the synaptic terminals.

axoneme (aks´ə-nēm) (L. *axis*, axle, + Gr. *nēma*, thread) The microtubules in a cilium or flagellum, usually arranged as a circlet of nine pairs enclosing one central pair; also, the microtubules of an axopodium.

axopodium (ak´sə-pō´dē-əm) (Gr. *axon*, an axis, + *podion*, small foot) Long, slender, more or less permanent pseudopodium in certain amebas. (Also called an **axopod.**)

axostyle (aks´ō-stīl) Tubelike organelle in some flagellate protozoa, extending from the area of the kinetosomes to the posterior end, where it often protrudes.

B

basal body Also called kinetosome or blepharoplast, a cylinder of nine triplets of microtubules found at the base of a flagellum or cilium; same structure as a centriole.

basal disc Aboral attachment site on a cnidarian polyp.

basis, basipodite (bā´səs, bā-si´pō-dīt) (Gr. *basis*, base, + *pous, podos*, foot) The distal or second joint of the protopod of a crustacean appendage.

benthos (ben´thōs) (Gr., depth of the sea) Organisms that live on the floor of a sea or lake. Adj., **benthic.** Also, the submerged substrate itself.

Bilateria (bī´lə-tir´ē-ə) (L. *bi-*, two, + *latus*, side) Bilaterally symmetrical animals.

binary fission Mode of asexual reproduction in which the animal splits into two approximately equal offspring.

binomial nomenclature The Linnean system of naming species in which the first word is the name of the genus (first letter capitalized) and the second word is the specific epithet (uncapitalized), usually an adjective modifying the name of the genus. Both of these words are written in italics.

biogenetic law A statement postulating a characteristic relationship between

ontogeny and phylogeny. Examples include Haeckel's law of **recapitulation** and Von Baer's law that general characteristics (those shared by many species) appear earlier in ontogeny than more restricted ones; neither of these statements is universally true.

biogeochemical cycle A description of the flow of elementary matter, such as carbon or phosphorus, through the component parts of an ecosystem and its **abiotic** environment, including the amount of an element present at the various stages of a **food web.**

biological species concept A reproductive community of populations (reproductively isolated from others) that occupies a specific niche in nature.

bioluminescence Method of light production by living organisms; usually, certain proteins (luciferins), in the presence of oxygen and an enzyme (luciferase), are converted to oxyluciferins with liberation of light.

biomass (Gr. *bios*, life, + *maza*, lump or mass) The weight of total living organisms or of a species population per unit of area.

biome (bī´ōm) (Gr. *bios*, life, + *ōma*, abstract group suffix) Complex of plant and animal communities characterized by climatic and soil conditions; the largest ecological unit.

biosphere (Gr. *bios*, life, + *sphaira*, globe) That part of earth containing living organisms.

biotic (bī-ät´ik) (Gr. *biōtos*, life, livable) Of or relating to life.

bipinnaria (L. *bi*, double, + *pinna*, wing, + *aria*, like or connected with) Free-swimming, ciliated, bilateral larva of the asteroid echinoderms; develops into the brachiolaria larva.

biradial symmetry Type of radial symmetry in which only two planes passing through the oral-aboral axis yield mirror images because some structure is paired.

biramous (bī-rām´əs) (L. *bi*, double, + *ramus*, a branch) Adjective describing appendages with two distinct branches; contrasts with uniramous, unbranched.

blastocoel (blas´tə-sēl) (Gr. *blastos*, germ, + *koilos*, hollow) Cavity of the blastula.

blastomere (Gr. *blastos*, germ, + *meros*, part) An early cleavage cell.

blastopore (Gr. *blastos*, germ, + *poros*, passage, pore) External opening of the archenteron in the gastrula.

blastula (Gr. *blastos*, germ, + *ula*, dim) Early embryological state of many animals; consists of a hollow mass of cells.

blepharoplast (blə-fä´rə-plast) (Gr. *blepharon*, eyelid, + *plastos*, formed) See **basal body.**

book gill Respiratory structure of aquatic chelicerates (Arthropoda) in which many thin, blood-filled gills are layered like the pages of a book. Gas exchange occurs as seawater passes between each pair of gills.

book lung Respiratory structure of terrestrial chelicerates (Arthropoda) in which many thin-walled air pockets extend into a blood-filled chamber in the abdomen.

BP Before the present.

brachial (brāk´ē-əl) (L. *brachium*, forearm) Referring to the arm.

branchial (brank´ē-əl) (Gr. *branchia*, gills) Referring to gills.

brown bodies Remnants of the lophophore and digestive tract of a degenerating adult ectoproct left behind in the chamber as a new lophophore and digestive tract are formed.

buccal (buk´əl) (L. *bucca*, cheek) Referring to the mouth cavity.

budding Reproduction in which the offspring arises as a small outgrowth from the parent. Failure of the offspring to separate from the parent leads to colony formation.

bursa, pl. **bursae** (M.L. *bursa*, pouch, purse made of skin) A saclike cavity. In ophiuroid echinoderms, pouches opening at bases of arms and functioning in respiration and reproduction (genitorespiratory bursae).

C

caecilian (si-sil´-yən) Denotes any member of the amphibian order Gymnophiona (also called Apoda).

calyx (kā´-liks) (L., bud cup of a flower) Any of various cup-shaped zoological structures.

capitulum (ka-pi´chə-ləm) (L., small head) Term applied to small, headlike structures of various organisms, including projection from body of ticks and mites carrying mouthparts.

carapace (kar´ə-pās) (F. from Sp. *carapacho*, shell) Shieldlike plate covering the cephalothorax of certain crustaceans; dorsal part of the shell of a turtle.

carinate (kar´ə-nāt) (L. *carina*, keel) Having a keel; in particular, the flying birds with a keeled sternum for the insertion of flight muscles; contrasts with **ratite.**

carnivore (kar´nə-vōr´) (L. *carnivorus*, flesh-eating) A mammal of the order

Carnivora. Also, any organism that eats animals. Adj., **carnivorous.**

carrying capacity The maximum number of individuals of a particular species that can persist under specified environmental conditions.

cartilage (L. *cartilago;* akin to L. *cratis,* wickerwork) Translucent elastic tissue that forms most of the skeleton of embryos, very young vertebrates, and adult cartilaginous fishes, such as sharks and rays; in adult amniotes, much of it is converted into bone.

caste (kast) (L. *castus,* pure, separated) One of the polymorphic forms within an insect society; each caste has its specific duties, such as queen, worker, soldier, and others.

catadromous (kə-tad´rə-məs) (Gr. *kata,* down, + *dromos,* a running) Refers to fishes that migrate from fresh water to the ocean to spawn.

catastrophic species selection Differential survival among species during a time of mass extinction based on character variation that permits some species but not others to withstand severe environmental disturbances, such as those caused by an asteroid impact.

caudal (käd´l) (L. *cauda,* tail) Constituting, belonging to, or relating to a tail.

cecum (sē´kəm) (L. *caecus,* blind) A blind pouch at the beginning of the large intestine; any similar pouch.

cellulase (sel´ū-lās) (L. *cella,* small room) Enzyme that cleaves cellulose; synthesized only by bacteria and some protists.

cellulose (sel´ū-lōs) (L. *cella,* small room) Chief polysaccharide constituent of the cell wall of green plants and some fungi; an insoluble carbohydrate $(C_6H_{10}O_5)_n$ that is converted to glucose by hydrolysis.

centriole (sen´trē-ōl) (Gr. *kentron,* center of a circle, + L. *ola,* small) Minute cytoplasmic organelle usually found in the centrosome and considered the active division center of the animal cell; organizes spindle fibers during mitosis and meiosis. Same structure as basal body or kinetosome.

cephalization (sef´ə-li-zā´shən) (Gr. *kephalē,* head) The evolutionary process by which specialization, particularly of the sensory organs and appendages, became localized in the head end of animals.

cephalothorax (sef´ə-lä-thō´raks) (Gr. *kephalē,* head, + thorax) A body division found in many Arachnida and higher

Crustacea in which the head is fused with some or all of the thoracic segments.

cerata (sə-ra´tə) (Gr. *keras,* a horn, bow) Dorsal processes on some nudibranchs for gas exchange.

cercaria (ser-kar´ē-ə) (Gr. *kerkos,* tail, + L. *aria,* like or connected with) Tadpolelike juveniles of trematodes (flukes).

cerebellum (ser-ə-bel´əm) (L. *cereb,* brain). A region of the hindbrain that processes spatial orientation information and coordinates muscle movement.

cervical (sər´və-kəl) (L. *cervix,* neck) Relating to a neck; usually refers to vertebrae in the neck region, or to the neck (cervix) of a mammalian uterus.

character (kar´ik-tər) A component of phenotype (including specific molecular, morphological, behavioral, or other features) used by systematists to diagnose species or higher taxa, or to evaluate phylogenetic relationships among different species or higher taxa, or relationships among populations within a species.

character displacement Differences in morphology or behavior within a species caused by competition with another species; characteristics typical of one species differ according to whether the other species is present in or absent from a local community.

chela (kēl´ə) (Gr. *chēlē,* claw) Pincerlike claw.

chelicera (kə-lis´ə-rə), pl. **chelicerae** (Gr. *chēlē,* claw, + *keras,* horn) One of a pair of the most anterior head appendages on members of subphylum Chelicerata.

chelipeds (kēl´ə-peds) (Gr. *chēlē, claw,* + L. *pes,* foot) Pincerlike first pair of legs in most decapod crustaceans; specialized for seizing and crushing.

chemoautotroph (kē-mō-aw´tō-trōf) (Gr. *chemeia,* transmutation, + *autos,* self, + *trophos,* feeder) An organism utilizing inorganic compounds as a source of energy.

chitin (kī´tən) (Fr. *chitine,* from Gr. *chitōn,* tunic) A hard substance that forms part of the cuticle of arthropods and occurs sparingly in certain other invertebrates; a nitrogenous polysaccharide insoluble in water, alcohol, dilute acids, and the digestive juices of most animals.

chloragogen cells (klōr´ə-gog-ən) (Gr. *chlōros,* light green, + *agōgos,* a leading, a guide) Modified peritoneal cells, greenish or brownish, clustered around the digestive tract of certain annelids;

apparently they aid in elimination of nitrogenous wastes and in food transport.

chlorophyll (klō´rə-fil) (Gr. *chlōros,* light green, + *phyllōn,* leaf) Green photosynthetic pigment found in cyanobacteria, plants, and some flagellates.

chloroplast (Gr. *chlōros,* light green, + *plastos,* molded) An organelle in plant cells that contains chlorophyll and is the physical location of photosynthesis; occurs also in photosynthetic unicellular eukaryotes and *Volvox.*

choanoblast (kō-an´ə-blast) (Gr. *choanē,* funnel, + *blastos,* germ) One of several cellular elements within the syncytial tissue of a hexactinellid sponge.

choanocyte (kō-an´ə-sīt) (Gr. *choanē,* funnel, + *kytos,* hollow vessel) One of the flagellate collar cells that line cavities and canals of sponges.

choanoflagellate (kō-an´ə-fla-jel´āt) Any member of a protozoan clade having a single flagellum surrounded by a column of microvilli; some form colonies, and all are included within the larger clade of opisthokonts.

chorioallantoic placenta (kō´rē-ō-al´an-tō´-ic) (Gr. *chorion,* skin, + *allas,* sausage) Type of placenta that occurs in placental mammals and some marsupials; a structure modified from the embryonic chorionic and allantoic membranes through which materials are exchanged between the embryo and the mother.

chorion (kō´rē-on) (Gr. *chorion,* skin) The outer layer of the double membrane that surrounds the embryo of reptiles, birds, and mammals; in mammals, it contributes to the placenta.

choriovitelline placenta (kor´ē-ō-vi´təl-ən) (Gr. *chorion,* skin, + *vittel,* yolk of an egg) An often transitory placenta that forms during the early developmental stages of marsupials and placental mammals. Also called the "yolk sac placenta," it is formed from the yolk sac and chorionic membrane of the embryo.

chromatophore (krō-mat´ə-fōr) (Gr. *chrōma,* color, + *herein,* to bear) Pigment cell, usually in the dermis, in which usually the pigment can be dispersed or concentrated.

chromosomal theory of inheritance The well-established principle, initially proposed by Sutton and Boveri in 1903–1904, that nuclear chromosomes are the physical bearers of genetic material in eukaryotic organisms. It is the foundation for modern evolutionary genetics.

bat/āpe/ärmadillo/herring/fēmale/finch/līce/crocodile/crōw/cȯin/cōre/duck/ūnicorn/tǘna/ə indicates unaccented vowel sound "uh" as in mammal, fishes, cardinal, heron, vulture/stress as in bi-ol´o-gy, bi´o-log´i-cal

chrysalis (kris´ə-lis) (L., from Gr. *chrysos*, gold) The pupal stage of a butterfly.

cilium (sil´ē-əm), pl. **cilia** (L., eyelid) A hairlike, vibratile organellar process found on many animal cells. Cilia may be used for moving particles along the cell surface or, in ciliated unicellular forms, for locomotion.

cirrus (sir´əs) (L., curl) Hairlike tuft on an insect appendage; locomotor organelle of fused cilia; male copulatory organ of some invertebrates. Cirri can refer to tufts of fused cilia in members of the Ciliophora.

clade (klād) (Gr. *klados*, branch) A taxon or other group consisting of a particular ancestral lineage and all of its descendants, forming a distinct branch on a cladogram or phylogenetic tree.

cladistics (klad-is´təks) (Gr. *klados*, branch, sprout) A system of arranging taxa by analysis of primitive and derived characteristics so that the arrangement reflects phylogenetic relationships.

cladogram (klād´ə-gram) (Gr. *klados*, branch, + *gramma*, letter) A branching diagram showing the pattern of sharing of evolutionarily derived characters among species or higher taxa.

clasper Digitiform projection on the medial side of the pelvic fins of male chondrichthians and some placoderms; used as an intromittent organ to transfer sperm to the female reproductive tract.

clitellum (kli-tel´əm) (L. *clitellae*, pack-saddle) Thickened, saddlelike portion of certain midbody segments of many oligochaetes and leeches.

cloaca (klō-ā´kə) (L., sewer) Posterior chamber of the digestive tract in many vertebrates, receiving feces and urogenital products. In certain invertebrates, a terminal portion of the digestive tract that serves also as a respiratory, excretory, or reproductive duct.

cloning (klō´ning) Production of genetically identical organisms by asexual reproduction.

cnida (nī´də) (Gr. *knidē*, nettle). Stinging or adhesive organelles formed within cnidocytes in phylum Cnidaria; nematocysts are a common type.

cnidoblast (nī´də-blast) (Gr. *knidē*, nettle, + Gr. *blastos*, germ) A cnidocyte is called a cnidoblast during the time when a cnida is forming within it.

cnidocil (nī´dō-sil) (Gr. *knidē*, nettle, + L. *cilium*, hair) Triggerlike spine on a nematocyst.

cnidocyte (nī´dō-sīt) (Gr. *knidē*, nettle, + *kytos*, hollow vessel) Modified interstitial cell that holds the cnida.

coccidian (kok-sid´ē-ən) (Gr. *kokkis*, kernel, grain) Intracellular protozoan parasite belonging to a class within phylum Apicomplexa; the organism causing malaria is an example.

cochlea (kōk´lē-ə) (L., snail, from Gr. *kochlos*, a shellfish) A tubular cavity of the inner ear containing the essential organs of hearing; occurs in crocodiles, birds, and mammals; spirally coiled in mammals.

cocoon (kə-kün´) (Fr. *cocon*, shell) Protective covering of a resting or developmental stage, sometimes including its contents; for example, the cocoon of a moth or the protective covering for the developing embryos in some annelids.

coelenteron (sē-len´tər-on) (Gr. *koilos*, hollow, + *enteron*, intestine) Internal cavity of a cnidarian; gastrovascular cavity; archenteron.

coelom (sē´lōm) (Gr. *koilōma*, cavity) The body cavity in triploblastic animals, lined with mesodermal peritoneum.

coelomate (sē´lōm-āt) Animals that possess a **coelom;** also called eucoelomate.

coelomoduct (sē-lō´mə-dukt) (Gr. *koilos*, hollow, + *ductus*, a leading) Duct that carries gametes or excretory products (or both) from the coelom to the exterior.

cohesion species concept The most inclusive population of individuals having the potential for phenotypic cohesion through intrinsic cohesion mechanisms; a refinement of the **evolutionary species concept** emphasizing population-genetic processes.

cohort (kō´hort) All organisms of a population born within a specified time interval.

collagen (käl´ə-jən) (Gr. *kolla*, glue, + *genos*, descent) Tough, fibrous protein occurring in vertebrates as the chief constituent of collagenous connective tissue; also occurs in invertebrates—for example, the cuticle of nematodes.

collar bodies Extensions of choanoblasts bearing flagellated collars in hexactinellid sponges.

collar cells Cells having a single flagellum surrounded by a ring of microvilli. Sponge choanocytes are collar cells, as are choanoflagellates, but collar cells also occur outside these taxa.

collencyte (käl´ən-sīt) (Gr. *kolla*, glue, + *kytos*, hollow vessel) Type of cell in sponges that secretes fibrillar collagen.

colloblast (käl´ə-blast) (Gr. *kolla*, glue, + *blastos*, germ) A glue-secreting cell on the tentacles of ctenophores.

comb plate One of the plates of fused cilia arranged in rows for ctenophore locomotion.

commensalism (kə-men´səl-iz´əm) (L. *com*, together with, + *mensa*, table) A relationship in which one individual lives close to or on another and benefits, and the host is unaffected; often symbiotic.

common descent Darwin's theory that all forms of life are derived from a shared ancestral population through a branching of evolutionary lineages.

community (L. *communitas*, community, fellowship) See **ecological community** and **reproductive community**

comparative biochemistry Studies of the structures of biological macromolecules, especially proteins and nucleic acids, and their variation within and among species to reveal homologies of macromolecular structure.

comparative cytology or **karyology** Studies of the structures of chromosomes within and among species to reveal homologies of chromosomal structure.

comparative method Use of patterns of similarity and dissimilarity among species or populations to test hypotheses of character **homology** and to infer phylogenetic relationships; use of phylogeny to examine evolutionary processes and history.

comparative morphology Studies of organismal form and its variation within and among species to reveal homologies of organismal characters.

competition Some degree of overlap in ecological niches of two populations in the same community, such that both depend on the same food source, shelter, or other resources, and negatively affect each other's survival.

competitive exclusion An ecological principle stating that two species whose niches are very similar cannot coexist indefinitely in the same community; one species is driven to extinction by competition between them.

conjugation (kon´jū-gā´shən) (L. *conjugare*, to yoke together) Temporary union of two ciliate protozoa while they are exchanging chromatin material and undergoing nuclear phenomena leading to binary fission. Also, formation of cytoplasmic bridges between bacteria for transfer of plasmids.

conodont (kōn´ə-dänt) (Gr. *kōnos*, cone, + *odontos*, tooth) Toothlike element from a Paleozoic animal now considered an early marine chordate.

conspecific (L. *com*, together, + species) Of the same species.

consumer (kən-sū´-mer) Organism whose energy and matter are acquired by eating other organisms, which may be **primary producers, herbivores,** or **carnivores.**

contractile vacuole Clear, fluid-filled cell vacuole in protozoa and a few animals; collects water and releases it to the outside in a cyclical manner, for osmoregulation and some excretion.

control That part of a scientific experiment to which the experimental variable is not applied, but similar to the experimental group in all other respects.

convexity The property of a taxonomic group that a path can be drawn between any two members on a cladogram or phylogenetic tree without leaving the group. **Monophyletic** and **paraphyletic** groups are convex, whereas polyphyletic groups are not.

coprophagy (ko-prō´fā´-jē) (Gr. *kopros,* dung, + *phagein,* to eat) Feeding on dung or excrement as a normal behavior among animals; reingestion of feces.

copulation (Fr., from L. *copulare,* to couple) Sexual union to facilitate fertilization of eggs by sperm.

coral bleaching Corals become white and brittle after expelling their mutualist zooxanthellae in response to increased ocean temperatures.

corneum (kor´nē-əm) (L. *corneus,* horny) Outermost epithelial layer of dead, keratinized cells. Also called stratum corneum.

corona (kə-rō´nə) (L., crown) Head or upper portion of a structure; ciliated disc on anterior end of rotifers.

corpora allata (kor´pə-rə əl-la´tə) (L. *corpus,* body, + *allatum,* aided) Endocrine glands in insects that produce juvenile hormone.

cortex (kor´teks) (L., bark) The outer layer of a structure.

coxa, coxopodite (kox´ə, kox´-ə pō´-dīt) (L. *coxa,* hip, + Gr. *pous, podos,* foot) The proximal joint of an insect or arachnid leg; in crustaceans, the proximal joint of the protopod.

Cretaceous extinction A **mass extinction** that occurred 65 million years ago in which 76% of existing species, including all dinosaurs, became extinct, marking the end of the Mesozoic era.

crop Region of the esophagus specialized for storing food.

crystalline style Rod containing digestive enzymes present in the stomach of a bivalve.

ctene (tēn) (Gr. *kteis, ktenos,* comb) Fused cilia that form a comb plate in members of phylum Ctenophora.

ctenidia (te-ni´dē-ə) (Gr. *kteis,* comb). Comblike structures, especially gills of molluscs; also applied to comb plates of Ctenophora.

ctenoid scales (tēn´oid) (Gr. *kteis, ktenos,* comb) Thin, overlapping dermal scales of teleost fishes; exposed posterior margins have fine, toothlike spines.

cuticle (kū´ti-kəl) (L. *cutis,* skin) Protective, noncellular, organic layer secreted by the external epithelium (hypodermis) of many invertebrates. In vertebrates, the term refers to the epidermis or outer skin.

Cuvierian tubules (*Cuvier,* 19th-century French comparative vertebrate anatomist) Sticky, often toxic elongate internal organs of holothurians expelled to entangle potential predators; these can be regenerated.

cyanobacteria (sī-an´ō-bak-ter´ē-ə) (Gr. *kyanos,* a dark-blue substance, + *bakterion,* dim. of *baktron,* a staff) Photosynthetic prokaryotes, also called blue-green algae, cyanophytes.

cycloid scales (sī´kloid) (Gr. *kyklos,* circle) Thin, overlapping dermal scales of teleost fishes; posterior margins are smooth.

cynodonts (sin´ə-dänts) (Gr. *kynodōn,* canine tooth) Group of mammal-like carnivorous synapsids of the Upper Permian and Triassic periods.

cyst (sist) (Gr. *kystis,* a bladder, pouch) Resistant, quiescent stage of an organism, usually with a secreted wall.

cysticercus (sis´tə-ser´kəs) (Gr. *kystis,* bladder, + *kerkos,* tail) Type of juvenile tapeworm having an invaginated and introverted scolex contained in a fluid-filled bladder.

cystid (sis´tid) (Gr. *kystis,* bladder) In an ectoproct, the dead secreted outer parts plus the adherent underlying living layers.

cytopharynx (Gr. *kytos,* hollow vessel, + *pharynx,* throat) Short, tubular gullet in ciliate protozoa.

cytoplasm (sī´tə-plasm) (Gr. *kytos,* hollow vessel, + *plasma,* mold) The living matter of the cell, excluding the nucleus.

cytoproct (sī´tə-prokt) (Gr. *kytos,* hollow vessel, + *prōktos,* anus) Site on a protozoan where indigestible matter is expelled.

cytopyge (sī´tə-pīj) (Gr. *kytos,* hollow vessel, + *pyge,* rump or buttocks) In some protozoa, localized site for expulsion of wastes.

cytostome (sī´tə-stōm) (Gr. *kytos,* hollow vessel, + *stoma,* mouth) The cell mouth in many protozoa.

D

dactylozooid (dak-til´ə-zō-id) (Gr. *dakos,* bite, sting, + *tylos,* knob, + *zōon,* animal) Polyp of a colonial hydroid specialized for defense or killing food.

Darwinism Theory of evolution emphasizing common descent of all living organisms, gradual change, multiplication of species, and natural selection.

data, sing. datum (Gr. *dateomai,* to divide, cut in pieces) The results of a scientific experiment, or descriptive observations, upon which a conclusion is based.

deciduous (də-sij´ü-wəs) (L. *deciere,* to fall off) Shed at the end of a growing period.

decomposer (dē´-kəm-pō´zər) **Consumer** that breaks organic matter into soluble components available to plants at the base of the food web; most are bacteria or fungi.

deduction (L. *deductus,* led apart, split, separated) Reasoning from the general to the particular, from given premises to their necessary conclusion.

definitive host The host in which sexual reproduction of a symbiont occurs; if no sexual reproduction, then the host in which the symbiont becomes mature and reproduces; contrasts with **intermediate host.**

deme (dēm) (Gr., populace) A local population of closely related animals.

demography (də-mäg´rə-fē) (Gr. *demos,* people, + *graphy*) The properties of the rate of growth and the age structure of populations.

density-dependent Biotic environmental factors, such as predators and parasites, whose effects on a population vary according to the number of organisms in the population.

density-independent Abiotic environmental factors, such as fires, floods, and temperature changes, whose effects on a population are unaffected by the number of organisms in the population.

deposit feeders Aquatic organisms that consume detritus and small organisms in soil and other sediments.

derived character state Condition of a taxonomic character inferred by cladistic analysis to have arisen within a taxon being examined cladistically rather than having been inherited from the most recent common ancestor of all members of the taxon.

dermal (Gr. *derma*, skin) Pertaining to the skin; cutaneous.

dermal ostia (Gr. *derma*, skin, + L. *ostium*, door) Incurrent pores in a sponge.

dermis The inner, sensitive mesodermal layer of skin; corium.

determinate cleavage Type of cleavage, usually spiral, in which the fate of the blastomeres is determined very early in development; see **mosaic development.**

detorsion (L. *de*, down, from, + *torquere*, to twist) Developmental process in post-veliger stage of certain gastropod molluscs whereby orientation of internal organs resulting from torsion is altered to resemble a state prior to torsion; anus and mantle cavity are posterior after detorsion.

detritus (də-trī´tus) (L., that which is rubbed or worn away) Any fine particulate debris of organic or inorganic origin.

Deuterostomia (dū´tə-rō-stō´mē-ə) (Gr. *deuteros*, second, secondary, + *stoma*, mouth) A group of higher phyla in which cleavage is indeterminate and ancestrally radial. The endomesoderm is enterocoelous, and the mouth is derived away from the blastopore. Includes Echinodermata, Chordata, and Hemichordata; contrasts with **Protostomia.**

diapause (dī´ə-pawz) (Gr. *diapausis*, pause) Period of arrested development in the life cycle of insects and certain other animals in which physiological activity is very low and the animal is highly resistant to unfavorable external conditions.

diaphragm (dī´ə-fram) (Gr. *dia*, separate, + *phragm*, partition) A sheetlike muscle that separates the thoracic and abdominal cavities of mammals. Contraction of this muscle draws air into the lungs.

diapsids (dī-ap´səds) (Gr. *di*, two, + *apsis*, arch) Amniotes in which the skull bears two pairs of temporal openings; includes nonavian reptiles and birds. The temporal openings are evolutionarily lost in turtles.

dictyosome (dik´tē-ə-sōm) (Gr. *diction*, to throw, + *sōma*, body) A part of the secretory system of endoplasmic reticulum in protozoa; also called Golgi bodies.

diffusion (L. *diffusion*, dispersion) Movement of particles or molecules from an area of high concentration of the particles or molecules to an area of lower concentration.

digitigrade (dij´ə-tə-grād) (L. *digitus*, finger, toe, + *gradus*, step, degree) Walking on the digits, with the heel of the foot raised; contrasts with **plantigrade** in describing mammalian locomotion.

dimorphism (dī-mor´fizm) (Gr. *di*, two, + *morphē*, form) Existence within a species of two distinct forms according to color, sex, size, organ structure, or behavior. Occurrence of two kinds of zooids in a colonial organism.

dioecious (dī-esh´əs) (Gr. *di*, two, + *oikos*, house) Having male and female organs in separate individuals.

diphycercal (dif´i-sər´kəl) (Gr. *diphyēs*, twofold, + *kerkos*, tail) A tail that tapers to a point, as in lungfishes; vertebral column extends to tip without upturning.

diphyodont (dī-fi´ə-dänt) (Gr. *diphyēs*, twofold, + *odous*, tooth) Having successive sets of deciduous and permanent teeth.

diploblastic (di´plə-blas´tək) (Gr. *diploos*, double, + *blastos*, bud) Organism with two germ layers, endoderm and ectoderm.

diploid (dip´lóid) (Gr. *diploos*, double, + *eidos*, form) Having the somatic (double, or 2n) number of chromosomes or twice the number characteristic of a gamete of a given species.

direct development A postnatal ontogeny primarily featuring growth in size rather than a major change in body shape or organs present; a life history that lacks **metamorphosis.**

directional selection Natural selection that favors one extreme value of a continuously varying trait and disfavors other values.

disruptive selection Natural selection that simultaneously favors two different extreme values of a continuously varying trait but disfavors intermediate values.

distal (dis´tal) Farther from the center of the body than a reference point.

DNA barcoding Technique for identifying organisms to species using sequence information of a standard gene present in all animals. The mitochondrial gene encoding cytochrome *c* oxidase I (*COI*) is often used.

domain (dō-mān´) An informal taxonomic rank above the Linnean kingdom; Archaea, Bacteria, and Eucarya are ranked as domains.

dorsal (dor´səl) (L. *dorsum*, back) Toward the back, or upper surface, of an animal.

double circulation A blood-transport system having a distinct pulmonary circuit of blood vessels separate from the circuit of blood vessels serving the remainder of the body.

dual-gland adhesive organ Organs in the epidermis of most turbellarians, with three cell types: viscid and releasing gland cells and anchor cells.

E

eccrine (ek´rən) (Gr. *ek*, out of, + *krinein*, to separate) Applies to a type of mammalian sweat gland that produces a watery secretion.

ecdysiotropin (ek-dī´-sē-ə-trō´pən) (Gr. *ekdysis*, to strip off, escape + *tropos*, a turn, change) Hormone secreted in brain of insects that stimulates prothoracic gland to secrete molting hormone. Prothoracicotropic hormone; brain hormone.

ecdysis (ek-dī´-sis) (Gr. *ekdysis*, to strip off, escape) Shedding of outer cuticular layer; molting, as in insects or crustaceans.

ecdysone (ek-dī´sōn) (Gr. *ekdysis*, to strip off) Molting hormone of arthropods; stimulates growth and ecdysis; produced by prothoracic glands in insects and Y-organs in crustaceans.

ecdysozoan protostome (ek-dī´-sō-zō´ən prō´tə-stōm) (Gr. *ekdysis*, to strip off, escape, + *zōon*, animal. Gr. *protos*, first, + *stoma*, mouth) Any member of a clade within Protostomia whose members shed the cuticle as they grow; includes arthropods, nematodes, and several smaller phyla.

ECM (extracellular matrix) A layer composed of proteoglycans, linking proteins, and fibrous proteins, such as collagen; the layer is secreted by cells and has structural and functional roles (e.g., cell orientation, adhesion, communication, filtering of materials); the matrix is sometimes called a basal lamina or a basal membrane.

ecological community An assemblage of species that are associated in a common area and interact with each other in a self-regulating relationship.

ecological pyramid A quantitative measurement of a **food web** in terms of amount of **biomass,** numbers of **organisms,** or energy at each of the different **trophic** levels present (**producers, herbivores,** first-level **carnivores,** higher-level carnivores).

ecology (Gr. *oikos*, house, + *logos*, discourse) Part of biology that concerns the relationship between organisms and their environment.

ecosystem (ek´ō-sis-təm) (eco[logy] from Gr. *oikos*, house, + system) An ecological unit comprising both the biotic communities and the nonliving (abiotic) environment, which interact to produce a stable system.

ectoderm (ek´tō-derm) (Gr. *ektos*, outside, + *derma*, skin) Outer layer of cells of an early embryo (gastrula stage); one of the germ layers, also sometimes used to include tissues derived from ectoderm.

ectognathous (ek´tō-nā´thəs) (Gr. *ektos*, outside, without, + *gnathos*, jaw) Derived character shared by most insects in which mandibles and maxillae are not in pouches.

ectolecithal (ek´tō-les´ə-thəl) (Gr. *ektos*, ouside, + *ekithos*, yolk) Yolk for nutrition of the embryo contributed by cells that are separate from the egg cell and are combined with the zygote by envelopment within the eggshell.

ectoparasite (ek´tō-par´ə-sīt) **Parasite** that resides on the outside surface of its host organism; contrasts with **endoparasite.**

ectoplasm (ek´tō-pla-zm) (Gr. *ektos*, outside, + *plasma*, form) The cortex of a cell or that part of cytoplasm just under the cell surface; contrasts with **endoplasm.**

ectothermic (ek´tō-therm´ic) (Gr. *ektos*, outside, + *thermē*, heat) Having a variable body temperature derived from heat acquired from the environment; contrasts with **endothermic.**

efferent (ef´ə-rənt) (L. *ex*, out, + *ferre*, to bear) Leading or conveying away from some organ—for example, nerve impulses conducted away from the brain, or blood conveyed away from an organ; contrasts with **afferent.**

egestion (ē-jes´chən) (L. *egestus*, to discharge) Act of dispelling indigestible or waste matter from the body by any normal route.

elephantiasis (el-ə-fən-tī´ə-səs) Disfiguring condition caused by chronic infection with filarial worms *Wuchereria bancrofti* and *Brugia malayi.*

Eltonian pyramid An **ecological pyramid** showing numbers of organisms at each of the **trophic** levels.

embryogenesis (em´brē-ō-jen´ə-səs) (Gr. *embryon*, embryo, + *genesis*, origin) The origin and development of the embryo; embryogeny.

encystment Process of **cyst** formation.

endemic (en-dem´ik) (Gr. *en*, in, + *demos*, populace) Geographically restricted to a certain region or country; native to a restricted area; not introduced.

endoderm (en´də-dərm) (Gr. *endon*, within, + *derma*, skin) Innermost germ layer of an embryo, forming the primitive gut; also may refer to tissues derived from endoderm.

endognathous (en´dō-nā-thəs) (Gr. *endon*, within, + *gnathous*, jaw) Ancestral character in insects (orders Diplura, Collembola, and Protura) in which the mandibles and maxillae are located in pouches.

endolecithal (en´dō-les´ə-thəl) (Gr. *endon*, within, + *lekithos*, yolk) Yolk for nutrition of the embryo incorporated into the egg cell itself.

endoparasite (en´dō-par´ə-sīt) Parasite that resides inside the body of its host organism; contrasts with **ectoparasite.**

endoplasm (en´dō-pla-zm) (Gr. *endon*, within, + *plasma*, mold or form) The portion of cytoplasm that immediately surrounds the nucleus.

endopod, endopodite (en´dō-päd, en-dop´ə-dīt) (Gr. *endon*, within, + *pous, podos*, foot) Medial branch of a biramous crustacean appendage.

endoskeleton (Gr. *endon*, within, + *skeletos*, hard) Skeleton or supporting framework within the living tissues of an organism; contrasts with **exoskeleton.**

endostyle (en´dō-stīl) (Gr. *endon*, within, + *stylos*, a pillar) Ciliated groove(s) in the floor of the pharynx of tunicates, cephalochordates, and larval lampreys; used for accumulating and moving food particles to the stomach.

endosymbiosis A symbiosis in which the symbiont lives inside its host; origin of eukaryotes, in which one prokaryote (symbiont) came to live inside another prokaryote (host), and symbionts eventually became organelles, such as mitochondria, of the host.

endothermic (en´dō-therm´ik) (Gr. *endon*, within, + *thermē*, heat) Having a body temperature determined by heat derived from an animal's own oxidative metabolism; contrasts with **ectothermic.**

energy budget Economic analysis of the energy used by an organism, partitioned into **gross productivity, net productivity,** and **respiration.**

enterocoel (en´tər-ō-sēl´) (Gr. *enteron*, gut, + *koilos*, hollow) Type of coelom formed by the outpouching of a mesodermal sac from the endoderm of the primitive gut.

enteron (en´tə-rän) (Gr., intestine) The digestive cavity.

entognathy (en´tə-nā-thē) Bases of mouthparts enclosed within the head capsule in insect orders Collembola, Diplura and Protura.

entomology (en´tə-mol´ə-jē) (Gr. *entoma*, an insect, + *logos*, discourse) Study of insects.

ephyra (ef´ə-rə) (Gr. *Ephyra*, Greek city) Juvenile medusa budded from a strobilating polyp in class Scyphozoa, phylum Cnidaria.

epidermis (ep´ə-dər´məs) (Gr. *epi*, on, upon, + *derma*, skin) The outer, nonvascular layer of skin of ectodermal origin; in invertebrates, a single layer of ectodermal epithelium.

epipod, epipodite (ep´ə-päd, e-pi-pō´-dīt) (Gr. *epi*, on, upon, + *pous, podos*, foot) Lateral process on the protopod of a crustacean appendage, often modified as a gill.

epistome (ep´i-stōm) (Gr. *epi*, on, upon, + *stoma*, mouth) Flap over the mouth in some lophophorates bearing the protocoel.

epithelium (ep´i-thē´lē-um) (Gr. *epi*, on, upon, + *thēlē*, nipple) Cellular tissue covering a free surface or lining a tube or cavity.

epitoke (ep´i-tōk) (Gr. *epitokos*, fruitful) Posterior part of a marine polychaete when swollen with developing gonads during the breeding season; contrasts with **atoke.**

erythrocyte (ə-rith´rō-sīt) (Gr. *erythros*, red, + *kytos*, hollow vessel) Red blood cell; has hemoglobin to carry oxygen from lungs or gills to tissues. During formation in mammals, erythrocytes lose their nuclei, whereas those of other vertebrates retain the nuclei.

estivation (es´tə-vā´shen) (L. *aestivates*, to spend the summer) A state of dormancy during the summer when temperatures are high, food is scarce and/ or dehydration threatens. Metabolism and breathing rate decrease.

estrous cycle Periodic episodes of estrus, or "heat," when females of most mammalian species become sexually receptive.

estrus (es´trəs) (L. *oestrus*, gadfly, frenzy) The period of heat, or rut, especially of a female mammal during ovulation. Associated with maximum sexual receptivity.

eukaryotic, eucaryotic (ū´ka-rē-ot´ik) (Gr. *eu*, good, true, + *karyon*, nut, kernel) Organisms whose cells characteristically contain a membrane-bound nucleus or nuclei; contrasts with **prokaryotic.**

eumetazoan (ü´-met-ə-zō´ən) (Gr. *eu*, good, true, + *meta*, after, + *zōon*, animal) Any multicellular animal with distinct germ layers that form true tissues; animals beyond the cellular grade of organization.

bat/āpe/ärmadillo/herring/fēmale/finch/līce/crocodile/crōw/cȯin/cōre/duck/ūnicorn/tüna/ə indicates unaccented vowel sound "uh" as in mammal, fishes, cardinal, heron, vulture/stress as in bi-ol´o-gy, bi´o-log´i-cal

euryhaline (ū-rə-hā´lı̄n) (Gr. *eurys,* broad, + *hals,* salt) Able to tolerate wide ranges of saltwater concentrations.

euryphagous (yə-rif´ə-gəs) (Gr. *eurys,* broad, + *phagein,* to eat) Eating a large variety of foods.

eurytopic (ū-rē-tŏp´ik) (Gr. *eurys,* broad, + *topos,* place) Describes an organism with a wide environmental range.

eutely (ū´tē-lē) (Gr. *euteia,* thrift) Condition of a body composed of a constant number of cells or nuclei in all adult members of a species, as in rotifers, acanthocephalans, and nematodes.

evagination (ē-vaj-ə-nā´shən) (L. *e,* out, + *vagina,* sheath) An outpocketing from a hollow structure.

evolution (L. *evolvere,* to unfold) All changes in the characteristics and diversity of life on earth throughout its history.

evolutionary sciences Empirical investigation of ultimate causes in biology using the comparative method.

evolutionary species concept A single lineage of ancestral-descendant populations that maintains its identity from other such lineages and has its own evolutionary tendencies and historical fate; differs from the **biological species concept** by explicitly including a time dimension and including asexual lineages.

evolutionary taxonomy System of classification, formalized by George Gaylord Simpson, that groups species into Linnean higher taxa representing a hierarchy of distinct adaptive zones; such taxa may be **monophyletic** or **paraphyletic** but not **polyphyletic.**

exaptation (ek-sap´-tā´shən) Evolutionary cooption of an organismal or molecular character for a biological role unrelated to the character's evolutionary origin. Bird feathers are considered an exaptation for flight because they originated prior to avian flight but provided utility for flight following its origin; contrasts with **adaptation;** bird feathers are considered an adaptation for the biological role of thermoregulation.

exopod, exopodite (ex´ə-päd, ex-äp´ō-dı̄t) (Gr. *exō,* outside, + *pous, podos,* foot) Lateral branch of a biramous crustacean appendage.

exoskeleton (ek´sō-skel´ə-tən) (Gr. *exō,* outside, + *skeletos,* hard) Supporting structure secreted by ectoderm or epidermis; external, not enveloped by living tissue, as opposed to **endoskeleton;** in vertebrates, a supporting structure formed within the integument.

experiment (L. *experiri,* to try) A trial made to support or to disprove a hypothesis.

experimental method General procedure for testing hypotheses by predicting how a biological system will respond to a disturbance, making the disturbance under controlled conditions, and then comparing the observed results with the predicted ones.

experimental sciences Empirical investigation of proximate causes in biology using the **experimental method.**

extrinsic factor An environmental variable that influences the biological properties of a population, such as observed number of individuals or rate of growth.

extrusome (eks´trə-sōm) (L. *extrusus,* driven out, + *soma,* body) Any membrane-bound organelle used to extrude something from a cell.

eyelid (ı̄´-lid) A thin surface of skin and muscle that can be closed to protect an eye from light, abrasion, and/or desiccation. Appears in many but not all terrestrial vertebrates.

F

filipodium (fi´li-pō´dē-əm) (L. *filum,* thread, + Gr. *pous, podos,* foot) Type of pseudopodium that is very slender and may branch but does not rejoin to form a mesh.

filter feeding Any feeding process by which particulate food is filtered from water in which it is suspended.

fission (L. *fissio,* a splitting) Asexual reproduction by division of the body into two or more parts.

fitness Degree of adjustment and suitability for a particular environment. Genetic fitness is the relative contribution of a genotype to the next generation; organisms with high genetic fitness are those favored by natural selection.

flagellum (flə-jel´əm), pl. **flagella** (L., a whip) Whiplike organelle of locomotion.

flame cell Specialized hollow excretory or osmoregulatory structure composed of one or several small cells containing a tuft of flagella (the "flame") and situated at the end of a minute tubule; connected tubules ultimately open to the outside. See **protonephridium.**

fluke (O.E. *flōc,* flatfish) Member of platyhelminth classes Trematoda or Monogenera. Also, certain of the flatfishes (order Pleuronectiformes).

food chain Movement of energy from plant compounds to organisms that eat plants,

then to other organisms that eat the plant feeders, and possibly further through a linear series of organisms that feed and are then eaten by others. Food chains connect and branch to form **food webs.**

food vacuole A digestive organelle in the cell.

food web An analysis relating species in an ecological community according to how they acquire nutrition, such as by fixing atmospheric carbon (**producers**), consuming producers (**herbivores**), consuming herbivores (first-level **carnivores**), or consuming carnivores (higher-level carnivores).

fossil (fos´əl) Any remains or impression of an organism from a past geological age that has been preserved by natural processes, usually by mineralization in the earth's crust.

fossorial (fä-sōr´ē-əl) (L. *fossor,* digger) Characterized by digging or burrowing.

fouling Contamination of feeding or respiratory areas of an organism by excrement, sediment, or other matter. Also, accumulation of sessile marine organisms on the hull of a boat or ship so as to impede its movement through water.

fovea (fō´vē-ə) (L., a small pit) Small pit or depression, especially the fovea centralis, a small, rodless pit in the retina of some vertebrates; a point of acute vision.

frog (fräg´) Denotes any member of the amphibian order Anura (also called Salientia).

frontal plane Plane parallel to the main axis of the body and at right angles to the sagittal plane.

fundamental niche A variety of roles potentially performed by an organism or population in an ecological community; limits on such roles are set by the intrinsic biological attributes of an organism or population. See also **niche** and **realized niche.**

furcula (fur´cū-la) (L. *furc,* fork) Fused clavicles of birds and some dinosaurs.

funnel The tube from which a jet of water exits the mantle cavity of a cephalopod mollusc.

G

gamete (ga´mēt, gə-mēt´) (Gr. *gamos,* marriage) A mature haploid sex cell; usually male and female gametes can be distinguished. An egg or a sperm.

gametic meiosis Meiosis that occurs during formation of the gametes, as in humans and other metazoa.

gametocyte (ga-mēt´ə-sı̄t) (Gr. *gametēs,* spouse, + *kytos,* hollow vessel) The mother cell of a gamete; an immature gamete.

ganglion (gang´lē-ən), pl. **ganglia** (Gr., little tumor) Aggregation of nerve tissue containing nerve cells.

ganoid scales (ga´noyd) (Gr. *ganos*, brightness) Thick, bony, rhombic scales of some bony fishes; not overlapping.

gastrocoel (gas´trō-sēl) (Gr. *gastēr*, stomach, + *koilos*, hollow) Embryonic cavity forming in gastrulation that becomes the adult gut; also called an archenteron.

gastrodermis (gas´trō-dər´mis) (Gr. *gastēr*, stomach, + *derma*, skin) Lining of the digestive cavity of cnidarians.

gastrovascular cavity (Gr. *gastēr*, stomach, + L. *vasculum*, small vessel) Body cavity in certain lower invertebrates that functions in both digestion and circulation and has a single opening serving as both mouth and anus.

gastrozooid (gas´trō-zō-id) (Gr. *gastēr*, stomach, + *zoon*, animal). The feeding polyp of a hydroid, a hydranth.

gastrula (gas´trə-lə) (Gr. *gastēr*, stomach, + L. *ula*, dim.) Embryonic stage, usually cap- or sac-shaped, with walls of two layers of cells surrounding a cavity (archenteron) with one opening (blastopore).

gemmule (je´mūl) (L. *gemma*, bud, + *ula*, dim.) Asexual, cystlike reproductive unit in freshwater sponges; formed in summer or autumn and capable of overwintering.

gene (Gr. *genos*, descent) The part of a chromosome that is the hereditary determiner and is transmitted from one generation to another. Specifically, a gene is a nucleic acid sequence (usually DNA) that encodes a functional polypeptide or RNA sequence.

gene pool Collection of all the alleles of all the genes in a population.

genetic drift Change in allelic frequencies by chance processes in the evolution of animals. In small populations, one allele may drift to fixation, becoming the only representative of that gene locus.

genotype (jēn´ə-tīp) (Gr. *genos*, offspring, + *typos*, form) The genetic constitution, expressed and latent, of an organism; the total set of genes present in the cells of an organism; contrasts with **phenotype.**

genus (jē´nus), pl. **genera** (L., race) Group of related species with the taxonomic rank between family and species.

germinative zone The site immediately following the scolex on the body of a mature tapeworm where new proglottids are produced.

germ layer In an animal embryo, one of three basic layers (ectoderm, endoderm, mesoderm) from which the various organs and tissues arise in the multicellular animal.

germovitellarium (jer´mə-vit-əl-ar´ē-əm) (L. *germen*, a bud, offshoot, + *vitellus*, yolk) Closely associated ovary (germarium) and yolk-producing structure (vitellarium) in rotifers.

germ plasm The germ cells of an organism, as distinct from the somatoplasm; the hereditary material (genes) of the germ cells.

gestation (je-stā´shən) (L. *gestare*, to bear) Period during which offspring are carried in the uterus.

gizzard (giz´erd) A modified stomach with thick muscular walls that grinds food, often with the aid of stones.

glochidium (glō-kid´ē-əm) (Gr. *glochis*, point, + *idion*, dimin. suffix) Bivalved larval stage of freshwater mussels.

glycogen (glī´kə-jən) (Gr. *glykys*, sweet, + *genes*, produced) A polysaccharide constituting the principal form in which carbohydrate is stored in animals; animal starch.

gnathobase (nath´ə-bās´) (Gr. *gnathos*, jaw, base) Median basic process on certain appendages in some arthropods, usually used for biting or crushing food.

gnathostomes (nath´ə-stōmz) (Gr. *gnathos*, jaw, + *stoma*, mouth) Vertebrates with jaws.

gonad (gō´nad) (N.L. *gonas*, a primary sex organ) Organ that produces gametes (ovary in the female and testis in the male).

gonangium (gō-nan´jē-əm) (N.L. *gonas*, primary sex organ, + *angeion*, dimin. of vessel) Reproductive zooid of hydroid colony (Cnidaria).

gonoduct (gän´ə-dukt) (Gr. *gonos*, seed, progeny, + duct) Duct leading from a gonad to the exterior.

gonophore (gon´ə-for) (Gr. *gonos*, seed, progeny, + *phoros*, bearer). Sexual reproductive structure developing from reduced medusae in some hydrozoans; it may be retained on the colony or released.

gonopore (gän´ə-pōr) (Gr. *gonos*, seed, progeny, + *poros*, an opening) Genital pore in many invertebrates.

grade (L. *gradus*, step) Level of organismal complexity or adaptive zone characteristic of a taxonomic group.

gradualism (graj´ə-wal-iz´əm) A component of Darwin's evolutionary theory postulating that evolution occurs by the temporal accumulation of small, incremental changes by populations, usually across very long periods of geological time; it opposes claims that evolution can occur by mutations having large and discontinuous phenotypic effects.

granular glands Integumentary structures of modern amphibians associated with secretion of defensive compounds.

green gland Excretory gland of certain Crustacea; the antennal gland.

gregarine (gre-ga-rin´) (L. *gregarious*, belonging to a herd or flock) Protozoan parasites belonging to class Gregarinea within phylum Apicomplexa; these organisms infect the guts or body cavities of invertebrates.

gross productivity Measurement of the total energy assimilated by an organism.

ground substance The matrix in which connective tissue fibers are embedded.

growth rate The proportion by which a population changes in numbers of individuals at a given time by reproduction and possibly by immigration.

guild (gild) (M.E. *gilde*, payment, tribute) Species of a local community that partition resources through character displacement to avoid niche overlap and competition, such as Galápagos finch communities whose component species differ in beak size for specializing on different-sized seeds.

gynecophoric canal (gī´nə-kə-fōr´ik) (Gr. *gynē*, woman, + *pherein*, to carry) Groove in male schistosomes (certain trematodes) that carries the female.

H

habitat (L. *habitare*, to dwell) The place where an organism normally lives or where individuals of a population live.

halter (hal´tər), pl. **halteres** (hal-ti´rēz) (Gr., leap) In Diptera, small, club-shaped structure on each side of the metathorax representing the hind wings; thought to be sense organ for balancing; also called a balancer.

haplodiploidy (Gr. *haploos,* single, + *diploos,* double, + *eidos,* form) Reproduction in which haploid males are produced parthenogenetically and diploid females develop from fertilized eggs.

haploid (Gr. *haploos,* single) The reduced, or n, number of chromosomes, typical of gametes, as opposed to the diploid, or 2n, number found in somatic cells. In certain groups, some mature organisms have a haploid number of chromosomes.

Hardy-Weinberg equilibrium Mathematical demonstration that the Mendelian hereditary process does not change the populational frequencies of alleles or genotypes across generations, and that change in allelic or genotypic frequencies requires factors such as natural selection, genetic drift in finite populations, recurring mutation, migration of individuals among populations, and nonrandom mating.

hemal system (hē´məl) (Gr. *haima,* blood) System of small vessels in echinoderms; function unknown.

hemimetabolous (he´mē-mə-ta´bə-ləs) (Gr. *hēmi,* half, + *metabolē,* change) Refers to gradual metamorphosis during development of insects without a pupal stage; contrasts with **holometabolous.**

hemocoel (hē´mə-sēl) (Gr. *haima,* blood, + *koilos,* hollow) Main body cavity of arthropods; may be subdivided into sinuses, through which blood flows.

hemoglobin (Gr. *haima,* blood, + L. *globulus,* globule) Iron-containing respiratory pigment occurring in vertebrate red blood cells and in blood plasma of many invertebrates; a compound of an iron porphyrin heme and globin proteins.

hemolymph (hē´mə-limf) (Gr. *haima,* blood, + L. *lympha,* water) Fluid composed of blood plasma and lymph residing in the coelom or hemocoel of invertebrates having open circulatory systems.

herbivore ([h]erb´ə-vōr´) (L. *herba,* green crop, + *vorare,* to devour) Any organism subsisting on plants. Adj., **herbivorous.**

hermaphrodite (hər-maf´rə-dīt) (Gr. *hermaphroditos,* containing both sexes; from Greek mythology, Hermaphroditos, son of Hermes and Aphrodite) An organism with both male and female functional reproductive organs. **Hermaphroditism** often denotes an aberration in species that normally have separate male and female individuals; **monoecism** implies that this is the normal condition for the species.

heterocercal (het´ər-o-sər´kəl) (Gr. *heteros,* different, + *kerkos,* tail) In some fishes, a tail with the upper lobe larger than the lower, and the end of the vertebral column somewhat upturned in the upper lobe, as in sharks; contrasts with **homocercal.**

heterochrony (het´ə-rō-krōn-ē) (Gr. *heteros,* different, + *chronos,* time) Evolutionary change in the relative time of appearance or rate of development of characteristics from ancestor to descendant.

heterodont (het´ə-rō-dänt) (Gr. *heteros,* different, + *odous,* tooth) Having teeth differentiated into incisors, canines, and molars for different purposes in mammals; contrasts with **homodont.**

heterolobosea (het´ə-rō-lo-bo´sē-ə) (Gr. *heteros,* other, different, + *lobos,* lobe). Protozoan clade in which most members can assume both ameboid and flagellate forms.

heterostracans (het´ə-rō-strā´-kanz) (Gr. *heteros,* different, + *ostrakon,* shell) Group of extinct fishes with dermal armor and no jaws or paired fins; known from the Ordovician to Devonian periods.

heterotroph (het´ə-rō-trōf) (Gr. *heteros,* different, + *trophos,* feeder) Organism that obtains both organic and inorganic raw materials from the environment in order to live; includes most animals and microorganisms and those plants that do not have photosynthesis.

heterozygous Describes an organism in which homologous chromosomes contain different allelic forms (often dominant and recessive) of a gene; derived from a zygote formed by union of gametes of dissimilar allelic constitution.

hexamerous (hek-sam´ər-əs) (Gr. *hex,* six, + *meros,* part) Six parts; specifically, symmetry based on six or multiples thereof.

hibernation (L. *hibernus,* wintry) Condition, especially of mammals, of passing the winter in a torpid state in which the body temperature drops nearly to freezing and the metabolism drops close to zero.

hierarchical system Scheme that arranges organisms into a series of taxa of increasing inclusiveness, as illustrated by Linnean classification.

histology (hi-stäl´-ə-jē) (Gr. *histos,* web, tissue, + *logos,* discourse) Study of the microscopic anatomy of tissues.

holometabolous (hō´lō-mə-ta´bə-ləs) (Gr. *holo,* complete, + *metabolē,* change) Complete metamorphosis during insect development from larval to pupal to adult stages; contrasts with **hemimetabolous.**

holophytic nutrition (hō-lō-fit´ik) (Gr. *holo,* whole, + *phyt,* plant) Occurs in green plants and certain protozoa and involves synthesis of carbohydrates from carbon dioxide and water in the presence of light, chlorophyll, and certain enzymes.

holozoic nutrition (hō-lō-zō´ik) (Gr. *holo,* whole, + *zoikos,* of animals) Type of nutrition involving ingestion of liquid or solid organic food particles.

homeobox (hō´mē-ō-box) (Gr. *homolos,* like, resembling, + L. *buxus,* boxtree [used in the sense of enclosed, contained]) A highly conserved 180-base-pair sequence found in homeotic genes, regulatory sequences of protein-coding genes that regulate development.

homeothermic (hō´mē-ō-thər´mik) (Gr. *homeo,* alike, + *thermē,* heat) Having a nearly uniform body temperature, regulated independent of the environmental temperature; "warm-blooded," **endothermic.**

homeotic genes (hō-mē-ät´ik) (Gr. *homolos,* like, resembling) Genes, identified through mutations, that give developmental identity to specific body segments.

home range The area over which an animal ranges in its activities. Unlike territories, home ranges are not defended.

hominid (häm´ə-nid) (L. *homo, hominis,* man) A member of the family Hominidae, which includes chimpanzees, gorillas, humans, orangutans, and extinct forms descended from their most recent common ancestor.

hominoid (häm´ə-noyd) Relating to the Hominoidea, a superfamily of primates to which the great apes and humans belong.

homocercal (hō´mə-ser´kəl) (Gr. *homos,* same, common, + *kerkos,* tail) A tail with the upper and lower lobes symmetrical and the vertebral column ending near the middle of the base, as in most teleost fishes; contrasts with **heterocercal.**

homodont (hō´mō-dänt) (Gr. *homos,* same, + *odous,* tooth) Having all teeth similar in form; contrasts in mammals with **heterodont.**

homology (hō-mäl´ə-jē) (Gr. *homologos,* agreeing) Similarity of parts or organs of different organisms caused by evolutionary derivation from a corresponding part or organ in a common ancestor, and usually having a similar embryonic origin. May also refer to a matching pair of chromosomes. Serial homology is the correspondence in the same individual of repeated structures having the same origin and development, such as the appendages of arthropods. Adj., **homologous.**

homoplasy (hō´mō-plā´sē) Phenotypic similarity among characteristics of different species or populations (including molecular, morphological, behavioral, or other features) that does not accurately represent patterns of common evolutionary descent (= nonhomologous similarity); it is produced by evolutionary parallelism, convergence, and/or reversal, and is revealed by incongruence among different characters on a cladogram or phylogenetic tree.

hyaline (hī´ə-lən) (Gr. *hyalos,* glass) Adj., glassy, translucent. Noun, a clear, glassy, structureless material occurring in cartilage, vitreous bodies, mucin, and glycogen, for example.

hydatid cyst (hī-da´təd) (Gr. *hydatis,* watery vesicle) Type of cyst formed by juveniles of certain tapeworms (*Echinococcus*) in their vertebrate hosts.

hydranth (hī´dranth) (Gr. *hydōr,* water, + *anthos,* flower) Nutritive zooid of hydroid colony.

hydrocoel (hī´drə-sēl) (Gr. *hydōr,* water, + *koilos,* hollow) Second or middle coelomic compartment in echinoderms; left hydrocoel gives rise to water-vascular system.

hydrocoral (Gr. *hydōr,* water, + *korallion,* coral) Certain members of the cnidarian class Hydrozoa that secrete calcium carbonate, resembling true corals.

hydrogenosomes (hī-drə-jen´ə-sōmz) Small organelles in certain anaerobic protozoa that produce molecular hydrogen as an end product of energy metabolism.

hydroid The polyp form of a cnidarian as distinguished from the medusa form. Formerly, hydrozoans of the discontinued, paraphyletic order Hydroida.

hydrostatic skeleton A mass of fluid or plastic parenchyma enclosed within a muscular wall to provide the support necessary for antagonistic muscle action; for example, parenchyma in acoelomates and perivisceral fluids in pseudocoelomates serve as hydrostatic skeletons.

hydrothermal vent A submarine hot spring; seawater seeping through the sea floor is heated by magma and expelled back into the sea through a hydrothermal vent.

hyperosmotic (hī´pər-oz-mo´-tik) (Gr. *hyper,* over, + *ōsmos,* impulse). Describes a solution that contains a greater concentration of dissolved particles than another solution to which it is compared; gains water through a selectively permeable membrane from a solution containing fewer particles; contrasts with **hypoosmotic.**

hyperparasitism (hī´pər-par´ə-si-ti´zm) A parasite itself parasitized by another parasite.

hypodermis (hī´pə-dər´mis) (Gr. *hypo,* under, + L. *dermis,* skin) The cellular layer lying beneath and secreting the cuticle of annelids, arthropods, and certain other invertebrates.

hypoosmotic (hī´pō-oz-mo´-tik) (Gr. *hypo,* under, + *ōsmos,* impulse). Describes a solution that contains a lesser concentration of dissolved particles than another solution to which it is compared; loses water through a selectively permeable membrane from a solution containing more particles; contrasts with **hyperosmotic.**

hypostome (hī´pə-stōm) (Gr. *hypo,* under, + *stoma,* mouth) Name applied to a structure in various invertebrates (such as mites and ticks), located at the posterior or ventral area of the mouth; elevation supporting the mouth of a hydrozoan.

hypothesis (hī´po-thə´-sis) (Gr. *hypothesis,* foundation, supposition) A statement or proposition that can be tested by observation or experiment.

hypothetico-deductive method The central procedure of scientific inquiry in which a postulate is advanced to explain a natural phenomenon and then subjected to observational or experimental testing that potentially could reject the postulate.

I

ichthyosaur (ik´thē-ō-sor) (Gr. *ichthyo,* fish, + *saur,* lizard) Aquatic, Mesozoic reptiles characterized by a porpoiselike body, but having a vertical tail and large eyes.

immediate cause See **proximate cause.**

inbreeding The tendency among members of a population to mate preferentially with close relatives.

indeterminate cleavage Type of embryonic development in which the fate of the blastomeres is not determined very early as to tissues or organs, as

occurs, for example, in echinoderms and vertebrates.

indigenous (in-dij´ə-nəs) (L. *indigna,* native) Pertains to organisms that are native to a particular region; not introduced.

induction (L. *inducere, inductum,* to lead) Reasoning from the particular to the general; deriving a general statement (hypothesis) based on individual observations. In embryology, the alteration of cell fates caused by interaction with neighboring cells.

infraciliature (in´frə-sil´ē-ə-chər) (L. *infra,* below, + *cilia,* eyelashes) The organelles just below the cilia in ciliate protozoa.

inheritance of acquired characteristics The discredited Lamarckian notion that organisms, by striving to meet the demands of their environments, obtain new adaptations and pass them by heredity to their offspring.

instar (inz´tär) (L., form) Stage in the life of an insect or other arthropod between molts.

integument (in-teg´ū-mənt) (L. *integumentum,* covering) An external covering or enveloping layer.

intermediary meiosis Meiosis that occurs neither during gamete formation nor immediately after zygote formation, resulting in both haploid and diploid generations, such as in foraminiferan protozoa.

intermediate host A host in which some development of a symbiont occurs, but in which maturation and sexual reproduction do not occur; contrasts with **definitive host.**

internal nostrils Palatal structures connecting the nasal cavity and throat in lungfishes and tetrapod vertebrates; used for olfaction and/or breathing when the mouth is closed.

interstitial (in´tər-sti´shəl) (L. *inter,* among, + *sistere,* to stand) Situated in the interstices or spaces between structures such as cells, organs, or grains of sand.

intracellular (in-trə-sel´yə-lər) (L. *intra,* inside, + *cellula,* chamber) Occurring within a body cell or within body cells.

intrinsic growth rate Exponential growth rate of a population; the difference between the density-independent components of the birth and death rates of a natural population with stable age distribution.

intrinsic rate of increase See **intrinsic growth rate.**

introvert (L. *intro*, inward, + *vertere*, to turn) The anterior narrow portion that can be withdrawn (introverted) into the trunk of a sipunculid worm.

iteroparity (i´tər-o-pā´ri-tē´) A life history in which individual organisms of a population normally reproduce more than one time before dying; contrasts with **semelparity.**

J

Jacobson's organ (*Jacobson*, 19th-century Danish surgeon and anatomist) Chemosensory organ in the roof of the mouth of many terrestrial vertebrates; the tongue transfers pheromones and food-derived chemicals to this organ, also called the vomeronasal organ.

juvenile hormone Hormone produced by the corpora allata of insects; among its effects are maintenance of larval or nymphal characteristics during development.

K

karyology (ker´-ē-ol´ə-jē) See **comparative cytology**

keratin (ker´ə-tən) (Gr. *kera*, horn, + *in*, suffix of proteins) A scleroprotein found in epidermal tissues and modified into hard structures such as horns, hair, and nails.

keystone species Species (typically a predator) whose removal leads to reduced species diversity within the community.

kinetoplast (kī-nēt´ə-plast) (Gr. *kinētos*, moving, + *plastos*, molded, formed) Cellular organelle that functions in association with a kinetosome at the base of a flagellum; presumed to be derived from a mitochondrion.

kinetosome (kin-et´ə-sōm) (Gr. *kinētos*, moving, + *sōma*, body) The self-duplicating granule at the base of a flagellum or cilium; similar to centriole; also called a basal body or blepharoplast.

L

labium (lā´bē-əm) (L., a lip) The lower lip of an insect formed by fusion of the second pair of maxillae. Also part of human female external genitalia.

labrum (lā´brəm) (L., a lip) The upper lip of insects and crustaceans situated above or

in front of the mandibles; also denotes the outer lip of a gastropod shell.

labyrinthodont (lab´ə-rin´thə-dänt) (Gr. *labyrinthos*, labyrinth, + *odous, odontos*, tooth) Group of Paleozoic amphibians containing the temnospondyls and the anthracosaurs.

lachrymal glands (lak´rə-məl) (L. *lacrimia*, tear) Structures in terrestrial vertebrates that secrete tears to lubricate the eyes.

lacunar system Netlike set of circulatory canals filled with fluid in an acanthocephalan.

Lamarckism Hypothesis, as expounded by Jean-Baptiste de Lamarck, of evolution by acquisition during an organism's lifetime of characteristics that are transmitted to offspring.

lamella (lə-mel´ə) (L. dim. of *lamina*, plate) One of the two plates forming a gill in a bivalve mollusc. One of the thin layers of bone laid concentrically around an osteon (Haversian canal). Any thin, platelike structure.

larva (lar´və), pl. **larvae** (L., a ghost) An immature stage that is quite different from the adult.

lateral (L. *latus*, the side, flank) Of or pertaining to the side of an animal; a *bilateral* animal has two sides.

lateral-line system Sensory organ that detects water vibrations; consists of neuromast organs in canals and grooves on the head and sides of the body of fishes and some amphibians.

lek (lek) (Sw., play, game) Area where animals assemble for communal courtship display and mating.

lemniscus (lem-nis´kəs) (L., ribbon) One of a pair of internal projections of the epidermis from the neck region of Acanthocephala, functioning in fluid control in the protrusion and invagination of the proboscis.

lepidosaurs (lep´ə-dō-sors) (L. *lepidos*, scale, + *sauros*, lizard) A group of diapsid reptiles that appeared in the Permian period and that includes the modern snakes, lizards, and tuataras.

leptocephalus (lep´tə-sef´ə-ləs), pl. **leptocephali** (Gr. *leptos*, thin, + *kephalē*, head) Transparent, ribbonlike migratory larva of the European or American eel.

limiting resource A particular source of nutrition, energy, or living space whose scarcity is causally associated with a population having fewer individuals than otherwise expected in a particular environment.

lineage (lin´əj) An unbranched succession of ancestral-descendant populations in

evolutionary time. Lineages related to others through branching of ancestral lineages form a phylogenetic tree. In molecular evolution, a succession of ancestral-descendant DNA molecules traced through an organismal genealogy or phylogeny.

lobopodium (lō´bə-pō´dē-əm) (Gr. *lobos*, lobe, + *pous, podos*, foot) Blunt, lobelike pseudopodium.

lobosea (lə-bō´sē-ə) (Gr. *lobos*, lobe) A protozoan clade comprising amebas with lobopodia.

lophophore (lōf´ə-fōr) (Gr. *lophos*, crest, + *phoros*, bearing) Tentacle-bearing ridge or arm within which is an extension of the coelomic cavity in lophophorate animals (ectoprocts, brachiopods, and phoronids).

lophotrochozoan protostome (lō´fō-trō´kō-zō´ən) (Gr. *lophos*, crest, + *trochos*, wheel, + *zōon*, animal) Any member of a clade within Protostomia whose members generally possess either a trochophore larva or a lophophore; examples are annelids, molluscs, and ectoprocts.

lorica (lor´ə-kə) (L. *lorica*, corselet) A secreted, protective covering, as in phylum Loricifera.

lymph (limf) (L. *lympha*, water) The interstitial (intercellular) fluid in the body; also, the fluid in the lymphatic space.

M

macroevolution (L. *makros*, long, large, + *evolvere*, to unfold) Evolutionary change on a grand scale, encompassing the origin of novel designs, evolutionary trends, adaptive radiation, and mass extinction.

macrogamete (mak´rə-gam´ēt) (Gr. *makros*, long, large, + *gamos*, marriage) The larger of the two gamete types in a heterogametic species, considered the female gamete.

macronucleus (ma´krō-nū´klē-əs) (Gr. *makros*, long, large, + *nucleus*, kernel) The larger of the two kinds of nuclei in ciliate protozoa; controls all cell function except reproduction.

madreporite (ma´drə-pōr´īt) (Fr. *madrépore*, reef-building coral, + *ite*, suffix for some body parts) Sievelike structure that is the intake of the water-vascular system of echinoderms.

malaria (mə-lar´-ē-ə) (It. *malaria*, bad air) Disease marked by periodic chills, fever, anemia, and other symptoms; caused by *Plasmodium* spp.

Malpighian tubules (mal-pig´ē-ən) (Marcello Malpighi, Italian anatomist, 1628–94) Blind tubules opening into the hindgut of nearly all insects and some myriapods and arachnids and functioning primarily as excretory organs.

mandible (L. *mandibula*, jaw) One of the lower jaw bones in vertebrates; one of the head appendages in arthropods.

mantle Soft extension of the body wall in certain invertebrates—for example, brachiopods and molluscs—which usually secretes a shell; thin body wall of tunicates.

manubrium (mə-nü´brē-əm) (L., handle) The portion projecting from the oral side of a jellyfish medusa, bearing the mouth; oral cone; presternum or anterior part of sternum; handlelike part of malleus of ear.

marsupial (mär-sü´pē-əl) (Gr. *marsypion*, little pouch) One of the pouched mammals of the subclass Metatheria.

mass extinction Relatively short interval of geological time in which a large portion (75–95%) of existing species or higher taxa are eliminated nearly simultaneously.

mastax (mas´təx) (Gr., jaws) Pharyngeal mill of rotifers.

matrix (mā´triks) (L. *mater,* mother) The extracellular substance of a tissue, or that part of a tissue into which an organ or process is set.

maxilla (mak-sil´ə) (L. dim. of *mala*, jaw) One of the upper jawbones in vertebrates; one of the head appendages in arthropods.

maxilliped (mak-sil´ə-ped) (L. *maxilla*, jaw, + *pes,* foot) One of the pairs of head appendages located just posterior to the maxilla in crustaceans; a thoracic appendage that has become incorporated into the feeding mouthparts.

medial (mē´dē-əl) Situated, or occurring, in the middle.

medulla (mə-dül´ə) (L., marrow) The inner portion of an organ in contrast to the cortex, or outer portion. Also, hindbrain.

medusa (mə-dü´-sə) (Gr. mythology, female monster with snake-entwined hair) A jellyfish, or the free-swimming stage that reproduces sexually in the life cycle of cnidarians.

Mehlis glands (me´ləs) Glands of uncertain function surrounding the junction of yolk duct, oviduct, and uterus in trematodes and cestodes.

meiosis (mī-ō´səs) (Gr. from *meioun,* to make small) The nuclear changes by means of which the chromosomes are reduced from the diploid to the haploid number; in animals, usually occurs in the last two divisions in the formation of the mature egg or sperm.

melanin (mel´ə-nin) (Gr. *melas*, black) Black or dark-brown pigment found in plant or animal structures.

membranelle Tiny, membrane-like structure; may be formed by fused cilia.

merozoite (me´rə-zō´īt) (Gr. *meros*, part, + *zōon*, animal) A very small trophozoite at the stage just after cytokinesis has been completed in multiple fission of a protozoan.

mesenchyme (me´zn-kīm) (Gr. *mesos*, middle, + *enchyma*, infusion) Embryonic connective tissue; irregular or amebocytic cells often embedded in gelatinous matrix.

mesocoel (mēz´ō-sēl) (Gr. *mesos*, middle, + *koilos*, hollow) Middle body coelomic compartment in some deuterostomes, anterior in lophophorates; corresponds to hydrocoel in echinoderms.

mesoderm (me´zə-dərm) (Gr. *mesos*, middle, + *derma*, skin) The third germ layer, formed in the gastrula between the ectoderm and endoderm; gives rise to connective tissues, muscle, urogenital and vascular systems, and the peritoneum.

mesoglea (mez´ō-glē´ə) (Gr. *mesos*, middle, + *glia*, glue) The layer of jellylike or cement material between the epidermis and gastrodermis in cnidarians and ctenophores.

mesohyl (me´zō-hil) (Gr. *mesos*, middle, + *hyle*, a wood) Gelatinous matrix surrounding sponge cells; also called mesoglea, mesenchyme.

mesonephros (me-zō-nef´rōs) (Gr. *mesos*, middle, + *nephros*, kidney) The middle of three pairs of embryonic renal organs in vertebrates. Functional kidney of embryonic amniotes; its collecting duct is a Wolffian duct. Adj., **mesonephric.**

mesosome (mez´ə-sōm) (Gr. *mesos*, middle, + *sōma*, body) The portion of the body in lophophorates and some deuterostomes that contains the mesocoel.

metacercaria (me´tə-sər-ka´rē-ə) (Gr. *meta*, after, + *kerkos*, tail, + L. *aria*, connected with) Fluke juvenile (cercaria) that has lost its tail and become encysted.

metacoel (met´ə-sēl) (Gr. *meta*, after, + *koilos*, hollow) Posterior coelomic compartment in some deuterostomes and lophophorates; corresponds to somatocoel in echinoderms.

metamere (met´ə-mēr) (Gr. *meta*, after, + *meros*, part) A repeated body unit along the longitudinal axis of an animal; a somite or segment.

metamerism (mə-ta´-mə-ri´zəm) (Gr. *meta*, between, after, + *meros*, part) Being composed of serially repeated parts (metameres); serial segmentation.

metamorphosis (Gr. *meta*, after, + *morphē*, form, + *osis*, state of) Sharp change in form during postembryonic development— for example, tadpole to frog, or larval insect to adult.

metanephridium (me´tə-nə-fri´di-əm) (Gr. *meta*, after, + *nephros*, kidney) Type of tubular nephridium with the inner open end draining the coelom and the outer open end discharging to the exterior.

metanephros (me´tə-ne´frōs) (Gr. *meta*, between, among, after, + *nephros*, kidney) Vertebrate kidney formed from the most posterior of three embryonic regions capable of forming renal organs; functional kidney of adult amniotes; drained by ureter. Adj., **metanephric.**

metapopulation dynamics The structure of a large population that comprises numerous semi-autonomous subpopulations, termed demes, with some limited movement of individuals among demes. Demes of a metapopulation are often geographically distinct.

metasome (met´ə-sōm) (Gr. *meta*, after, behind, + *sōma*, body) The portion of the body in lophophorates and some deuterostomes that contains the metacoel.

metazoan (met-ə-zō´ə) (Gr. *meta*, after, + *zōon*, animal) Animal; contrasts with **protozoan.**

microevolution (mī´krō-ev-ə-lü-shən) (L. *mikros,* small, + *evolvere*, to unfold) A change in the gene pool of a population across generations.

microfilariae (mīk´rə-fil-ar´ē-ē) (Gr. *mikros*, small, + L. *filum*, a thread) Born alive, partially developed juvenile filarial worms (phylum Nematoda).

microgamete (mīk´rə-ga´-mēt) (Gr. *mikros*, small, + *gamos*, marriage) The smaller of the two gamete types in a heterogametic species, considered the male gamete.

micron (μ) (mī´-krän) (Gr. neuter of *mikros*, small) One-thousandth of a millimeter; about 1/25,000 of an inch. Now largely replaced by micrometer (μ).

microneme (mī´krə-nēm) (Gr. *mikros*, small, + *nēma*, thread) One type of structure forming the apical complex in phylum Apicomplexa; slender and elongate, leading to the anterior; thought to function in host-cell penetration.

micronucleus The smaller of two nuclei in ciliated protozoa; controls the reproductive functions of these organisms.

microsporidian (mī´krō-spo-rid´ē-ən) (Gr. *micros*, small, + *spora*, seed, + *idion*, dim. suffix) Any member of a protozoan clade comprising intracellular parasites with a distinctive spore morphology.

microthrix See **microvillus.**

microtubule (Gr. *mikros*, small, + L. *tubule*, pipe) Long, tubular cytoskeletal element with an outside diameter of 20 to 27 nm. Microtubules influence cell shape and play important roles during cell division.

microvillus (Gr. *mikros*, small, + L. *villus*, shaggy hair) Narrow, cylindrical cytoplasmic projection from epithelial cells; microvilli form the brush border of several types of epithelial cells. Also, microvilli with unusual structure cover the surface of cestode tegument (also called **microthrix** [pl. **microtriches**]).

mictic (mik´tik) (Gr. *miktos*, mixed or blended) Pertaining to haploid eggs of rotifers or the females that lay such eggs.

mimic (mim´ik) (Gr. *mimicus*, imitator) Species whose morphological or behavioral characteristics copy those of another species because those characteristics deter shared predators.

mimicry Evolution by natural selection of similar forms by different species, such as the sharing of warning signals that discourage common predators. In Batesian mimicry, a species tasteful to a predator deceptively copies warning signals of a distasteful species. In Müllerian mimicry, two or more distasteful species evolve common warning signals to avoid a common predator.

miracidium (mīr´ə-sid´ē-əm) (Gr. *meirakidion*, youthful person) Minute, ciliated larval stage in the life of flukes.

mitochondrion (mīt´ō-kän´drē-ən) (Gr. *mitos*, a thread, + *chondrion*, dim. of *chondros*, corn, grain) Cellular organelle in which aerobic metabolism occurs.

mitosis (mī-tō´səs) (Gr. *mitos*, thread, + *osis*, state of) Nuclear division in which chromosomal material is equally divided both qualitatively and quantitatively between the two resulting nuclei; ordinary cell division.

model (mod´l) (Fr. *modèle*, pattern) A species whose morphological or behavioral characteristics are copied by another species because those characteristics deter shared predators.

modular (moj´ə-lər) Describes the structure of a colony of genetically identical organisms that are physically associated and produced asexually by cloning.

molting Shedding of the outer cuticular layer; see **ecdysis.**

monoecious (mə-nē´shəs) (Gr. *monos*, single, + *oikos*, house) Having both male and female gonads in the same organism, usually denoting that this is the typical condition of a species; hermaphroditic.

monogamy (mə-näg´ə-mē) (Gr. *monos*, single, + *gamos*, marriage) The condition of having a single mate at any one time. Adj., **monogamous.**

monophyly (män´ō-fī´lē) (Gr. *monos*, single, + *phyle*, tribe) Condition occurring when a taxon or other group of organisms contains the most recent common ancestor of the group and all of its descendants. Adj., **monophyletic.**

monotreme (mä´nō-trēm) (Gr. *monos*, single, + *trēma*, hole) Egg-laying mammal of the order Monotremata.

morphogenesis (mor´fə-je´nə-səs) (Gr. *morphē*, form, + *genesis*, origin) Development of the architectural features of organisms; formation and differentiation of tissues and organs.

morphology (Gr. *morphē*, form, + *logos*, discourse) The science of structure. Includes cytology, the study of cell structure; histology, the study of tissue structure; and anatomy, the study of gross structure.

mosaic development Embryonic development characterized by independent differentiation of each part of the embryo; determinate cleavage.

mucus (mū´kəs) (L. *mucus*, nasal mucus) Viscid, slippery secretion rich in mucins produced by secretory cells such as those in mucous membranes. Adj., **mucous.**

multiple fission Mode of asexual reproduction in some protistans in which the nuclei divide more than once before cytokinesis occurs.

multiplication of species The Darwinian theory that the evolutionary process generates new species through a branching of evolutionary lineages derived from an ancestral species.

mutation (mū-tā´shən) (L. *mutare*, to change) Stable and abrupt change of a gene; the heritable modification of a character.

mutualism (mū´chə-wə-li´zəm) (L. *mutuus*, lent, borrowed, reciprocal) Type of interaction in which two different species derive benefit from their association, sometimes necessary for both species to survive; often symbiotic.

myocyte (mī´ə-sīt) (Gr. *mys*, muscle, + *kytos*, hollow vessel) Contractile cell (pinacocyte) in sponges.

myofibril (Gr. *mys*, muscle, + L. dim. of *fibra*, fiber) Contractile filament within muscle or muscle fiber.

myomere (mī´ə-mēr) (Gr. *mys*, muscle, + *meros*, part) A muscle segment of successive segmental trunk musculature.

myotome (mī´ə-tōm) (Gr. *mys*, muscle, + *tomos*, cutting) A voluntary muscle segment in cephalochordates and vertebrates; that part of a somite destined to form muscles; the muscle group innervated by a single spinal nerve.

N

nacre (nā´kər) (F., mother-of-pearl) Innermost, lustrous layer of a mollusc shell, secreted by mantle epithelium. Adj., **nacreous.**

nares (na´rēz), sing. **naris** (L., nostrils) Openings into the nasal cavity, both internally and externally, in the head of a vertebrate.

natural selection The interactions between organismal character variation and the environment that cause differences in rates of survival and reproduction among varying organisms in a population; leads to evolutionary change if variation is heritable.

nauplius (naw´plē-əs) (L., a kind of shellfish) Free-swimming microscopic larval stage of certain crustaceans, with three pairs of appendages (antennules, antennae, and mandibles) and a median eye. Characteristic of ostracods, copepods, barnacles, and some others.

nekton (nek´tən) (Gr. neuter of *nēktos*, swimming) Term for actively swimming organisms, essentially independent of wave and current action; contrasts with **plankton.**

nematocyst (ne-mat´ə-sist´) (Gr. *nēma*, thread, + *kystis*, bladder) Stinging organelle of cnidarians.

neo-Darwinism (nē´ō´där´wə-niz´əm) A modified version of Darwin's evolutionary theory that eliminates elements of the Lamarckian inheritance of acquired characteristics and pangenesis that were present in Darwin's formulation; this theory originated with August

Weissmann in the late nineteenth century and, after incorporating Mendelian genetic principles, has become the currently favored version of Darwinian evolutionary theory.

neopterygian (nē-äp´tə-rij´ē-ən) (Gr. *neos,* new, + *pteryx,* fin) Any of a large group of bony fishes that includes most modern species.

nephridium (nə-frid´ē-əm) (Gr. *nephridios,* of the kidney) One of the segmentally arranged, paired excretory tubules of many invertebrates, notably the annelids. In a broad sense, any tubule specialized for excretion and/or osmoregulation; with an external opening and with or without an internal opening.

nephron (ne´frän) (Gr. *nephros,* kidney) Functional unit of a vertebrate kidney comprising Bowman's capsule, an enclosed glomerulus, and the attached uriniferous tubule.

nephrostome (nef´rə-stōm) (Gr. *nephros,* kidney, + *stoma,* mouth) Ciliated, funnel-shaped opening of a nephridium.

nested hierarchy Ordering of species into a series of increasingly more inclusive clades according to the taxonomic distribution of synapomorphies.

net productivity The energy stored by an organism, equal to the energy assimilated **(gross productivity)** minus the energy used for metabolic maintenance **(respiration).**

neural crest Populations of ectodermally derived embryonic cells that differentiate into many skeletal, neural, and sensory structures; unique to vertebrates.

neural spine A dorsal projection of a vertebra that functions as a muscle attachment site.

neuroglia (nur´ə-glē´ə) (Gr. *neuron,* nerve, + *glia,* glue) Tissue supporting and filling the spaces between the nerve cells of the central nervous system.

neuromast (Gr. *neuron,* sinew, nerve, + *mastos,* knoll) Cluster of sense cells on or near the surface of a fish or amphibian that is sensitive to vibratory stimuli and water current.

neuron (Gr., nerve) A nerve cell.

neuropodium (nur´ə-pō´dē-əm) (Gr. *neuron,* nerve, + *pous, podos,* foot) Lobe of parapodium nearer the ventral side in polychaete annelids.

neurosecretory cell (nur´ə-sə-krē´to-rē) Any cell (neuron) of the nervous system that produces a hormone.

niche (nich´) The role of an organism, population, or species in an ecological community comprising its usage of resources, its unique way of life, and its relationship to other biotic and abiotic factors.

niche overlap A comparison of two species that quantifies the proportion of each species' resources also utilized by the other species.

notochord (nō´tə-kord´) (Gr. *nōtos,* back, + *chorda,* cord) Elongated cartilaginous cellular cord, enclosed in a sheath, that forms the primitive axial skeleton of chordate embryos, adult cephalochordates, and jawless vertebrates.

notopodium (nō´tə-pō´dē-əm) (Gr. *nōtos,* back, + *pous, podos,* foot) Lobe of parapodium nearer the dorsal side in polychaete annelids.

nucleolus (nü-klē´ə-ləs) (dim. of L. *nucleus,* kernel) Deeply staining body within the nucleus of a cell and containing ribosomal RNA; nucleoli are specialized portions of certain chromosomes that carry multiple copies of the genes encoding ribosomal RNA and where ribosomal RNA is actively synthesized.

nucleoplasm (nü´klē-ə-pla´zəm) (L. *nucleus,* kernel, + Gr. *plasma,* mold) Protoplasm of nucleus, as distinguished from cytoplasm.

nucleus (nü´klē-əs) (L. *nucleus,* a little nut, the kernel) The organelle in eukaryotes that contains the chromatin and is bounded by a double membrane (nuclear envelope).

nurse cells Single cells or layers of cells surrounding or adjacent to other cells or structures for which the nurse cells provide nutrients or other molecules (for example, for insect oocytes or *Trichinella* spp. juveniles).

nymph (L. *nympha,* nymph, bride) Immature stage (following hatching) of a hemimetabolous insect that lacks a pupal stage.

O

ocellus (ō-sel´əs) (L. dim. of *oculus,* eye) A simple eye or eyespot in many types of invertebrates.

octomerous (ok-tom´ər-əs) (Gr. *oct,* eight, + *meros,* part) Eight parts; specifically, symmetry based on eight.

odontophore (ō-don´tə-for´) (Gr. *odous,* tooth, + *pherein,* to carry) Tooth-bearing

organ in molluscs, including the radula, radular sac, muscles, and cartilages.

olfactory epithelium A specialized chemosensory surface tissue inside the nasal cavities of aquatic and terrestrial vertebrates.

omasum (ō-ma´səm) (L., paunch) The third compartment of the stomach of a ruminant mammal.

ommatidium (ä´mə-tid´ē-əm) (Gr. *omma,* eye, + *idium,* small) One of the optical units of the compound eye of arthropods.

omnivore (äm´nə-vōr´) (L. *omnis,* all, + *vorare,* to devour) Animal whose diet includes a variety of animal and plant material.

oncosphere (än´kō-sfiər) (Gr. *onkinos,* a hook, + *sphaira,* ball) Rounded larva common to all cestodes; bears hooks.

ontogeny (on-to´jə-nē) (Gr. *ontos,* being, + *geneia,* act of being born, from *genēs,* born) The course of development of an individual from egg to senescence.

oocyst (ō´ə-sist) (Gr. *ōion,* egg, + *kystis,* bladder) Cyst formed around zygote of malarial parasite and related organisms.

oocyte (ō´ə-sīt) (Gr. *ōion,* egg, + *kytos,* hollow) Stage in formation of ovum, just preceding first meiotic division (primary oocyte) or just following first meiotic division (secondary oocyte).

ookinete (ō-ə-kī´nēt) (Gr. *ōion,* egg, + *kinein,* to move) Motile zygote of malarial parasites.

operculum (ō-per´kū-ləm) (L., cover) The gill cover in bony fishes; keratinized plate in some snails.

opisthaptor (ō´pəs-thap´tər) (Gr. *opisthen,* behind, + *haptein,* to fasten) Posterior attachment organ of a monogenetic trematode.

opisthokont (ō-pis´thō-kont) (G. *opisthen,* behind, + *kontos,* a pole) Any member of the eukaryotic clade comprising fungi, microsporidians, choanoflagellates, and animals; if present, flagellated cells possess a single posterior flagellum.

opisthonephros (ō-pis´-thō-nef´-rōs) (Gr. *opisth,* back + *nephros,* kidney) A kidney that develops from the middle and posterior portions of the nephrogenic region of vertebrates and is drained by the Wolffian duct or accessory ducts. Functional kidney of most adult anamniotes (fishes and amphibians). Adj., **opisthonephric.**

oral disc The end of a cnidarian polyp bearing the mouth.

oral lobe Flaplike extension of the mouth of a scyphozoan medusa that aids in feeding.

organelle (Gr. *organon*, tool, organ, + L. *ella*, dimin. suffix) Specialized part of a cell; a subcellular structure that performs functions analogous to organs of multicellular animals.

organism (or´-gə-niz´-əm) A biological individual composed of one or more cells, tissues, and/or organs whose parts are interdependent in producing a collective physiological system. Organisms of the same species may form **populations.**

osculum (os´kū-ləm) (L. *osculum*, a little mouth) Excurrent opening in a sponge.

osmoregulation Maintenance of proper internal salt and water concentrations in a cell or in the body of a living organism; active regulation of internal osmotic pressure.

osmosis (oz-mō´sis) (Gr. *ōsmos*, act of pushing, impulse) A flow of solvent (usually water) through a semipermeable membrane.

osmotroph (oz´mə-trōf) (Gr. *ōsmos*, a thrusting, impulse, + *trophē*, to eat) Heterotrophic organism that absorbs dissolved nutrients.

osphradium (os-frā´dē-əm) (Gr. *osphradion*, strong smell) Sense organ in aquatic snails and bivalves that tests incoming water.

ossicles (L. *ossiculum*, small bone) Small separate pieces of echinoderm endoskeleton. Also, tiny bones of the middle ear of vertebrates.

osteostracans (os-tē-os´trə-kəns) (Gr. *osteon*, bone, + *ostrakon*, shell) Group of jawless, extinct fishes with dermal armor and pectoral fins from the Silurian and Devonian periods.

ostium (L., door) Opening.

ostracoderm (os-trak´ō-derm) (Gr. *ostrakon*, shell, + *derma*, skin) A paraphyletic group of extinct, jawless fishes with dermal armor known from the late Cambrian to Devonian periods.

otolith (ōt´ə-lith´) (Gr. *ous, otos*, ear, + *lithos*, stone) Calcareous concretions in the membranous labyrinth of the inner ear of lower vertebrates or in the auditory organ of certain invertebrates.

outgroup In phylogenetic systematic studies, a species or group of species closely related to but not included within a taxon whose phylogeny is being studied, and used to polarize variation of characters and to root the phylogenetic tree.

outgroup comparison Method for determining the polarity of a character in cladistic analysis of a taxonomic group. Character states found within the group being studied are judged ancestral if they occur also in related taxa outside the study group (= outgroups); character states that occur only within the taxon being studied but not in outgroups are judged to have been derived evolutionarily within the group being studied.

oviger (ō´vi-jər) (L. *ovum*, egg, + *gerere*, to bear) Leg that carries eggs in pycnogonids.

oviparity (ō´və-pa´rət-ē) (L. *ovum*, egg, + *parere*, to bring forth) Reproduction in which eggs are released by the female; development of offspring occurs outside the maternal body. Adj., **oviparous** (ō-vip´ə-rəs).

ovipositor (ō´ve-poz´ət-ər) (L. *ovum*, egg, + *positor*, builder, placer, + *or*, suffix denoting agent or doer) In many female insects, a structure at the posterior end of the abdomen for laying eggs.

ovoviviparity (ō´vo-vī-və-par´ə-tē) (L. *ovum*, egg, + *vivere*, to live, + *parere*, to bring forth) Reproduction in which eggs develop within the maternal body without additional nourishment from the parent and hatch within the parent or immediately after laying. Adj., **ovoviviparous** (ō-vo-vī-vip´ə-rəs).

ovum (L. *ovum*, egg) Mature female germ cell (egg).

P

paedomorphosis (pē-dō-mor´fə-səs) (Gr. *pais*, child, + *morphē*, form) Displacement of ancestral juvenile features to later stages of the ontogeny of descendants.

pangenesis (pan-jen´ə-sis) (Gr. *pan*, all, + *genesis*, descent) Darwin's discredited hypothesis that hereditary factors are produced by individual body parts according to the organism's use of those parts, and collected in the germ cells for transmission to offspring.

papilla (pə-pil´ə), pl. **papillae** (L., nipple) Small, nipplelike projection. A vascular process that nourishes the root of a hair, feather, or developing tooth.

papula (pa´pü-lə), pl. **papulae** (L., pimple) Respiratory processes on skin of sea stars; also, pustules on skin.

parabasal bodies Cellular organelles similar to Golgi bodies, presumed to function as part of the secretory system in endoplasmic reticulum.

parabronchi (par-ə-bron´kī) (Gr. *para*, beside, + *bronchos*, windpipe) Fine air-conduction pathways of a bird lung.

paraphyly (per´ə-fī´lē) (Gr. *para*, before, + *phyle*, tribe) The condition that a taxon or other group of organisms contains the most recent common ancestor of all members of the group but excludes some descendants of that ancestor; shows **convexity.** Adj., **paraphyletic.**

parapodium (pe´rə-pō´dē-əm) (Gr. *para*, beside, + *pous, podos*, foot) One of the paired lateral processes on each side of most segments in polychaete annelids; variously modified for locomotion, respiration, or feeding.

parasite (par´ə-sīt) Organism that lives physically on or in, and at the expense of, another organism.

parasitism (per´ə-sit´iz-əm) (Gr. *parasitos*, from *para*, beside, + *sitos*, food) The condition of an organism living in or on another organism (host) at whose expense the parasite is maintained; destructive symbiosis.

parasitoid (per´ə-si´-tȯid) Organism that is a typical parasite early in its development but that finally kills the host during or at the completion of development; refers to many insect parasites of other insects.

parenchyma (pə-ren´kə-mə) (Gr., anything poured in beside) In simpler animals, a spongy mass of vacuolated mesenchyme cells filling spaces between viscera, muscles, or epithelia; in some, the cells are cell bodies of muscle cells. Also, the specialized tissue of an organ as distinguished from the supporting connective tissue.

parenchymula (pə´rən-kīm´yə-lə) (Gr. *para*, beside, + *enchyma*, infusion) Flagellated, solid-bodied larva of some sponges.

parietal (pä-rī´-ə-təl) (L. *paries*, wall) Something next to, or forming part of, a wall of a structure.

parsimony (pär´sə-mō´nē´) (L. *parsus*, to spare). A general methodological principle that the simplest hypothesis capable of explaining observations is the best working hypothesis and should be tested first before investigating more complex hypotheses. In phylogenetic systematics, this principle involves using the phylogenetic tree that requires the smallest amount of evolutionary change as the best working hypothesis of phylogenetic relationships.

parthenogenesis (pär´thə-nō-gen´ə-sis) (Gr. *parthenos*, virgin, + L. from Gr.

genesis, origin) Unisexual reproduction involving the production of young by females not fertilized by males; common in rotifers, cladocerans, aphids, bees, ants, and wasps. A parthenogenic egg may be diploid or haploid.

pecten (pek´-tən) (L., comb) Any of several types of comblike structures on various organisms; for example, a pigmented, vascular, and comblike process that projects into the vitreous humor from the retina at a point of entrance of the optic nerve in the eyes of all birds and many reptiles.

pectoral (pek´tə-rəl) (L. *pectoralis,* from *pectus,* the breast) Of or pertaining to the breast or chest, to the pectoral girdle, or to a pair of keratinized shields of the plastron of certain turtles.

pedalium (pə-dal´ē-əm) (Gr. *pedalion,* a prop, rudder) The flattened, bladelike base of a tentacle or group of tentacles in the cnidarian class Cubozoa.

pedal laceration Asexual reproduction in sea anemones; a form of fission.

pedicel (ped´ə-sel) (L. *pediculus,* little foot) A small or short stalk or stem. In insects, the second segment of an antenna or the waist of an ant.

pedicellaria (ped´ə-sə-ler´ē-ə) (L. *pediculus,* little foot, + *aria,* like or connected with) One of many minute, pincerlike organs on the surface of certain echinoderms.

pedipalps (ped´ə-palps´) (L. *pes, pedis,* foot, + *palpus,* stroking, caress) Second pair of appendages of arachnids.

peduncle (pē-dun´kəl) (L. *pedunculus,* dim. of *pes,* foot) A stalk. Also, a band of white matter joining different parts of the brain.

pelage (pel´ij) (Fr., fur) Hairy covering of mammals.

pelagic (pə-laj´ik) (Gr. *pelagos,* the open sea) Occupying or moving through water rather than the underlying substrate; contrasts with **benthic.**

pellicle (pel´ə-kəl) (L. *pellicula,* dim. of *pelis,* skin) Thin, translucent, secreted envelope covering many protozoa.

pelycosaur (pel´ə-kō-sor) (Gr. *pelyx,* basin, + *sauros,* lizard) Any of a group of Permian synapsids characterized by homodont dentition and sprawling limbs.

pen A flattened, flexible internal support in a squid; a remnant of the ancestral shell.

pentadactyl (pen-tə-dak´təl) (Gr. *pente,* five, + *daktylos,* finger) Having five digits, or five fingerlike parts, to the hand or foot.

perennibranchiate (pə-rən´ə-brank´ē-ət) (L. *perennis,* throughout the year, + Gr. *branchia,* gills) Having permanent gills, relating especially to certain paedomorphic salamanders.

periostracum (pe-rē-äs´trə-kəm) (Gr. *peri,* around, + *ostrakon,* shell) Outer keratinized layer of a mollusc shell.

peripheral (pə-ri´fər-əl) (Gr. *peripherein,* to move around) Structure or location distant from center, near outer boundaries.

periproct (per´ə-präkt) (Gr. *peri,* around, + *prōktos,* anus) Region of aboral plates around the anus of echinoids.

perisarc (per´ə-särk) (Gr. *peri,* around, + *sarx,* flesh) Sheath covering the stalk and branches of a hydroid.

perissodactyl (pə-ris´ə-dak´təl) (Gr. *perissos,* odd, + *daktylos,* finger, toe) Pertaining to an order of ungulate mammals with an odd number of digits.

peristomium (per´ə-stō´mē-əm) (Gr. *peri,* around, + *stoma,* mouth) One of two parts forming the annelid head; it bears the mouth.

peritoneum (per´ə-tə-nē´əm) (Gr. *peritonaios,* stretched around) The membrane that lines the coelom and covers the coelomic viscera.

Permian extinction A mass extinction that occurred 245 million years ago in which 96% of existing species became extinct, marking the end of the Paleozoic era.

perpetual change The most basic theory of evolution, stating that the living world is neither constant nor cycling, but is always undergoing irreversible modification through time.

phagocyte (fag´ə-sīt) (Gr. *phagein,* to eat, + *kytos,* hollow vessel) Any cell that engulfs and devours microorganisms or other particles.

phagocytosis (fag´ə-sī-tō´səs) (Gr. *phagein,* to eat, + *kytos,* hollow vessel) The engulfment of a particle by a phagocyte or a protozoan.

phagosome (fa´gə-sōm) (Gr. *phagein,* to eat, + *sōma,* body) Membrane-bound vessel in cytoplasm containing food material engulfed by phagocytosis.

phagotroph (fag´ə-trōf) (Gr. *phagein,* to eat, + *trophē,* food) Heterotrophic organism that ingests solid particles for food.

pharynx (far´inks), pl. **pharynges** (Gr. *pharynx,* gullet) The part of the digestive tract between the mouth cavity and the esophagus that, in vertebrates, is common to both the digestive and the respiratory tracts. In cephalochordates, the gill slits open from it.

phenetic taxonomy (fə-ne´tik) (Gr. *phaneros,* visible, evident) Uses overall similarity to classify organisms into taxa; contrasts with classification based explicitly on reconstruction of phylogeny.

phenotype (fē´nə-tīp) (Gr. *phainein,* to show) The visible or expressed characteristics of an organism; influenced by the genotype in interaction with environmental conditions.

pheromone (fer´ə-mōn) (Gr. *pherein,* to carry, + *hormōn,* exciting, stirring up) Chemical substance released by one organism that influences the behavior or physiological processes of another organism.

photoautotroph (fō-tō-aw´tō-trōf) (Gr. *phōtos,* light, + *autos,* self, + *trophos,* feeder) Organism requiring light as a source of energy for making organic nutrients from inorganic raw materials.

photosynthesis (fō-tō-sin´thə-sis) (Gr. *phōs,* light, + *synthesis,* action or putting together) The synthesis of carbohydrates from carbon dioxide and water in chlorophyll-containing cells exposed to light.

phototaxis (fō-tō-tak´sis) (Gr. *phōtos,* light, + *taxis,* arranging, order) A taxis in which light is the orienting stimulus. Involuntary tendency for an organism to turn toward (positive) or away from (negative) light.

phototrophs (fō-tō-trōfs) (Gr. *phōtos,* light, + *trophē,* nourishment) Organisms capable of using CO_2 in the presence of light as a source of metabolic energy.

phyletic gradualism A model of evolution in which morphological evolutionary change is continuous and incremental and occurs mainly within unbranched species or lineages over long periods of geological time; contrasts with **punctuated equilibrium.**

phylogenetic species concept An irreducible (basal) cluster of organisms, diagnosably distinct from other such clusters, and within which there is a parental pattern of ancestry and descent.

phylogenetic systematics See **cladistics.**

phylogenetic tree A diagram whose branches represent evolutionary lineages; depicts the common descent of species or higher taxa.

phylogeny (fī´läj´ə-nē) (Gr. *phylon,* tribe, race, + *geneia,* origin) The origin and diversification of any taxon, or the evolutionary history of its origin and diversification, usually presented as a dendrogram.

phylum (fī´ləm), pl. **phyla** (N.L. from Gr. *phylon,* race, tribe) A chief taxonomic category, between kingdom and class, into which are grouped organisms of common descent that share a fundamental pattern of organization.

physiology (L. *physiologia,* natural science) Branch of biology covering the organic processes and phenomena of an organism or any of its parts or a particular body process.

phytophagous (fi-täf´ə-gəs) (Gr. *phyton,* plant, + *phagein,* to eat). Feeding on plants.

pinacocyte (pin´a-kō-sīt´) (Gr. *pinax,* tablet, + *kytos,* hollow vessel) Flattened cells comprising dermal epithelium in sponges.

pinna (pin´ə) (L., feather, sharp point) The external ear. Also, a feather, wing, fin, or similar part.

pinocytosis (pin´o-sī-tō´sis, pīn´o-sī-tō´sis) (Gr. *pinein,* to drink, + *kytos,* hollow vessel, + *osis,* condition) Acquisition of fluid by a cell in which the plasma membrane invaginates and pinches off to form small vesicles.

placenta (plə-sen´tə) (L., flat cake) The vascular structure, both embryonic and maternal, through which a mammalian embryo and fetus are nourished while in the uterus; also used to denote analogous but nonhomologous structures in some elasmobranchs, bony fishes, and squamates.

placoderms (plak´ə-dərmz) (Gr. *plax,* plate, + *derma,* skin) A group of heavily armored jawed fishes of the Lower Devonian to Lower Carboniferous periods.

placoid scale (pla´kȯid) (Gr. *plax, plakos,* tablet, plate) Type of scale found in cartilaginous fishes, composed of a basal plate of dentine embedded in the skin and a backward-pointing spine tipped with enamel.

plankton (plank´tən) (Gr. neuter of *planktos,* wandering) The passively floating animal and plant life of a body of water; contrasts with **nekton.**

plantigrade (plan´tə-grād´) (L. *planta,* sole, + *gradus,* step, degree) Pertaining to mammals that walk on the whole surface of the foot (for example, humans and bears); contrasts with **digitigrade.**

planula (plan´yə-lə) (N.L. dim. from L. *planus,* flat) Free-swimming, ciliated larval type of cnidarians; usually flattened and ovoid, with an outer layer of ectodermal cells and an inner mass of endodermal cells.

plasma membrane (plaz´mə) (Gr. *plasma,* a form, mold) Living, external, limiting, protoplasmic structure that regulates the exchange of nutrients across the cell surface.

plasmid (plaz´məd) (Gr. *plasma,* a form, mold) Small circle of DNA that a bacterium may carry in addition to its genomic DNA.

plastron (plas´trən) (Fr. *plastron,* breast plate) Ventral body shield of turtles; structure in corresponding position in certain arthropods; thin film of gas retained by epicuticle hairs of aquatic insects.

pleura (plü´rə) (Gr., side, rib) The membrane that lines each half of the thorax and covers the lungs.

pneumostome (nü´-mo-stōm) An external opening to the internal modified mantle cavity that functions as a lung in terrestrial gastropods.

podium (pō´dē-əm) (Gr. *pous, podos,* foot) A footlike structure—for example, the tube foot of echinoderms.

poikilothermic (pȯi-ki´lə-thər´mik) (Gr. *poikilos,* variable, + thermal) Pertaining to animals whose body temperature is variable and fluctuates with that of the environment; cold-blooded; contrasts with **homeothermic.**

polarity (Gr. *polos,* axis) In systematics, the ordering of alternative states of a taxonomic character from ancestral to successively derived conditions in an evolutionary transformation series. In developmental biology, the tendency for the axis of an ovum to orient corresponding to the axis of the mother. Also, condition of having opposite poles; differential distribution of gradation along an axis.

Polian vesicles (pō´lē-ən) (From G. S. Poli, 1746–1825, Italian naturalist) Vesicles opening into a ring canal in most asteroids and holothuroids.

polyandry (pol´ē-an´drē) (Gr. *polys,* many, + *anēr,* man) Having more than one male mate at one time.

polygamy (pə-lig´ə-mē) (Gr. *polys,* many, + *gamos,* marriage) Having more than one mate at one time.

polygyny (pə-lij´ə-nē) (Gr. *polys,* many, + *gynē,* woman) Having more than one female mate at one time.

polymorphism (pä´lē-mor´fi-zəm) (Gr. *polys,* many, + *morphē,* form) The presence in a species of more than one structural type of individual; genetic variation in a population.

polyp (päl´əp) (Fr. *polype,* octopus, from L. *polypus,* many-footed) The sessile stage in the life cycle of cnidarians.

polyphyletic (pä´lē-fī-let´-ik) (Gr. *polys,* many, + *phylon,* tribe) Derived from more than one ancestral source; contrasts with monophyletic and paraphyletic.

polyphyly (pä´lē-fī´lē) (Gr. *polys,* full, + *phylon,* tribe) The condition that a taxon or other group of organisms does not contain the most recent common ancestor of all members of the group, implying that it has multiple evolutionary origins; such groups are not valid as formal taxa and are recognized as such only through error.

polyphyodont (pä-lē-fī´ə-dänt) (Gr. *polyphyes,* manifold, + *odous,* tooth) Having several sets of teeth in succession.

polypide (pä´lē-pīd) (L. *polypus,* polyp) An individual or zooid in a colony, specifically in ectroprocts, that has a lophophore, digestive tract, muscles, and nerve centers.

polytypic (pol-ē-tip´-ik). Denotes a species that contains two or more taxonomically designated subspecies; taxonomic recognition of subspecies is controversial and rejected by many taxonomists.

population (L. *populus,* people) A group of organisms of the same species inhabiting a specific geographical locality.

porocyte (pō´rə-sīt) (Gr. *porus,* passage, pore, + *kytos,* hollow vessel) Type of cell in asconoid sponges through which water enters the spongocoel.

portal system (L. *porta,* gate) System of large veins beginning and ending with a bed of capillaries; for example, the hepatic portal and renal portal systems in vertebrates.

positive assortative mating Tendency of an individual to mate preferentially with others whose phenotypes are similar to its own.

posterior (L., latter) Situated at or toward the rear of the body; in bilateral forms, the end of the main body axis opposite the head.

preadaptation The possession of a trait that coincidentally predisposes an organism for survival in an environment different from those encountered in its evolutionary history.

precocial (prē-kō´shəl) (L. *praecoquere,* to ripen beforehand) Referring (especially) to birds whose young are covered with down and are able to walk when newly hatched.

predaceous, predacious (prē-dā´shəs) (L. *praedator,* a plunderer; *praeda,* prey) Living by killing and consuming other animals; predatory.

predation (prə-dā´shən) Interaction between species in an ecological community in which members of one species (prey) serve as food for another species (**predator**).

predator (pred´ə-tər) (L. *praedator,* a plunderer; *praeda,* prey) Organism that preys on other organisms for its food.

prehensile (prē-hen´səl) (L. *prehendere,* to seize) Adapted for grasping.

primary producer Species whose members begin **productivity** by acquiring energy and matter from **abiotic** sources, as for example when plants synthesize sugars from water and carbon dioxide using solar energy (see **photosynthesis**).

primate (prī-māt) (L. *primus,* first) Any mammal of the order Primates, which includes the tarsiers, lemurs, marmosets, monkeys, apes, and humans.

primitive (L. *primus,* first) Primordial; ancient; little evolved; characteristics closely approximating those possessed by early ancestral types.

proboscis (prō-bäs´əs) (Gr. *pro,* before, + *boskein,* feed) A snout or trunk. Also, a tubular sucking or feeding organ with the mouth at the end, as in planarians, leeches, and insects. Also, the sensory and defensive organ at the anterior end of certain invertebrates.

producers (L. *producere,* to bring forth) Organisms, such as plants, able to produce their own food from inorganic substances.

production In ecology, the energy accumulated by an organism that becomes incorporated into new biomass.

productivity (prō´duk-tiv´-ət-ē) Property of a biological system measured by the amount of energy and/or materials that it incorporates.

proglottid (prō-glät´əd) (Gr. *proglōttis,* tongue tip, from *pro,* before, + *glōtta,* tongue, + *id,* suffix) Portion of a tapeworm containing a set of reproductive organs; usually corresponds to a segment.

prokaryotic, procaryotic (pro-kar´ē-ät´ik) (Gr. *pro,* before, + *karyon,* kernel, nut) Not having a membrane-bound nucleus or nuclei. Prokaryotic cells characterize Bacteria and Archaea.

pronephros (prō-nef´rōs) (Gr. *pro,* before, + *nephros,* kidney) Most anterior of three pairs of embryonic renal organs of vertebrates; functional only in adult hagfishes and larval fishes and amphibians; vestigial in mammalian embryos. Adj., **pronephric.**

prosimian (prō-sim´ē-ən) (Gr. *pro,* before, + L. *simia,* ape) Any member of a group of arboreal primates including lemurs, tarsiers, and lorises but excluding monkeys, apes, and humans.

prosopyle (prōs´-ə-pīl) (Gr. *prosō,* forward, + *pylē,* gate) Connections between the incurrent and radial canals in some sponges.

prostomium (prō-stō´mē-əm) (Gr. *pro,* before, + *stoma,* mouth) In most annelids and some molluscs, that part of the head located in front of the mouth.

protein (prō´tēn, prō´tē-ən) (Gr. *protein,* from *proteios,* primary) A macromolecule consisting of carbon, hydrogen, oxygen, and nitrogen and usually containing sulfur; composed of chains of amino acids joined by peptide bonds; present in all cells.

protein polymorphism Occurrences of allelic variants in the amino acid sequences of proteins within a population or species. Separation of allelic variants by protein electrophoresis provided an early method for quantifying the amount of genetic variation present in natural populations.

prothoracic glands Glands in the prothorax of insects that secrete the hormone ecdysone.

prothoracicotropic hormone See **ecdysiotropin.**

protist (prō´-tist) (Gr. *protos,* first) A member of the discontinued, paraphyletic kingdom Protista, generally considered to include the protozoa and eukaryotic algae.

protocoel (prō-tō-sēl) (Gr. *prōtos,* first, + *koilos,* hollow) The anterior coelomic compartment in some deuterostomes; corresponds to the axocoel in echinoderms.

protocooperation Mutually beneficial interaction between organisms in which the interaction is not physiologically necessary to the survival of either.

protonephridium (prō-tō-nə-frid´ē-əm) (Gr. *prōtos,* first, + *nephros,* kidney) Primitive osmoregulatory or excretory organ consisting of a tubule terminating internally with a flame bulb or solenocyte; the unit of a flame bulb system.

proton pump Active transport of hydrogen ions (protons) across an inner mitochondrial membrane during cellular respiration.

protopod, protopodite (prō´-tō-päd, prō´-tō-pō-dīt) (Gr. *prōtos,* first, + *pous, podos,* foot) Basal portion of crustacean appendage, containing coxa and basis.

protostome (prō´tō-stōm) (Gr. *protos,* first, + *stoma,* mouth) A member of the group Protostomia. Protostome taxa have recently been divided into ecdysozoan protostomes and lophotrochozoan protostomes.

Protostomia (prō-tō-stō´mē-ə) (Gr. *prōtos,* first, + *stoma,* mouth) A clade in which cleavage is determinate, the mouth is derived at or near the blastopore, and a coelom (when present) is formed by proliferation of mesodermal bands that later split (schizocoely). Phyla within the clade are divided between two subgroups: Lophotrochozoa and Ecdysozoa. Lophotrochozoans, exemplified by annelids and molluscs, share spiral cleavage and derive the mesoderm from a particular blastomere (called 4d). Ecdysozoans (arthropods and related taxa) have a unique cleavage pattern and do not form mesoderm from the 4d cell. Contrasts with **Deuterostomia.**

protozoan A common name often used to denote unicellular eukaryotes; contrasts with "metazoan" (= animal), and also excludes fungi and plants.

proventriculus (pro´ven-trik´ū-ləs) (L. *pro,* before, + *ventriculum,* ventricle) In birds, the glandular stomach between the crop and gizzard. In insects, a muscular dilation of the foregut armed internally with chitinous teeth.

proximal (L. *proximus,* nearest) Situated toward or near the point of attachment; opposite of distal, distant.

proximate cause (L. *proximus,* nearest, + *causa*) The factors that underlie the functioning of a biological system at a particular place and time, including those responsible for metabolic, physiological, and behavioral functions at the molecular, cellular, organismal, and population levels. Immediate cause.

pseudocoelom (sü´də-sē-lōm) (Gr. *pseudēs,* false, + *koilos,* hollow) Body cavity not lined with peritoneum and not a part of the blood or digestive systems; embryonically derived from the blastocoel.

pseudopodium (sü´də-pō´dē-əm) (Gr. *pseudēs*, false, + *podion*, small foot, + *eidos*, form) A temporary cytoplasmic protrusion extended from a protozoan or ameboid cell and serving for locomotion or for engulfing food.

punctuated equilibrium Model of evolution in which morphological evolutionary change is discontinuous, being associated primarily with discrete, geologically instantaneous events of speciation that lead to phylogenetic branching; morphological evolutionary stasis characterizes species between episodes of speciation; contrasts with **phyletic gradualism.**

pupa (pū´pə) (L., girl, doll, puppet) Inactive quiescent state of the holometabolous insects. It follows the larval stages and precedes the adult stage.

pygidium (pə-jid´ē-əm) (Gr. *pyge*, rump, buttocks, + *idion*, dim. ending) Posterior region of a segmented animal bearing the anus.

pygostyle (pī´gō stīl) (Gr. *pygo*, rump + *styl*, piller) A bone at the end of a bird's vertebral column, formed from fused caudal vertebrae.

Q

queen In entomology, a reproductive female in a colony of social insects such as bees, ants, and termites; distinguished from workers, nonreproductive females, and soldiers.

R

radial canals Canals along the ambulacra radiating from the ring canal of echinoderms. Also, choanocyte-lined canals in syconoid sponges.

radial cleavage Embryonic development in which early cleavage planes are symmetrical to the polar axis, each blastomere of one tier lying directly above the corresponding blastomere of the next layer; see **indeterminate cleavage.**

radial symmetry A morphological condition in which the parts of an animal are arranged concentrically around an oral-aboral axis, and more than one imaginary plane through this axis yields halves that are mirror images of each other.

Radiata (rā´dē-ä´tə) (L. *radius*, ray) Phyla showing radial symmetry, specifically Cnidaria and Ctenophora.

radiole (rā´dē-ōl) (L. *radiolus*, dim. of *radius*, ray, spoke of a wheel) Featherlike structure extending from head of some sedentary polychaetes; used in feeding on suspended particles.

radula (ra´jə-lə) (L., scraper) Rasping tongue found in most molluscs.

ratite (ra´tīt) (L. *ratis*, raft) Referring to birds having an unkeeled sternum; contrasts with **carinate.**

realized niche The role actually performed by an organism or population in its ecological community at a particular time and place as constrained by both its intrinsic biological attributes and its particular environmental conditions. See also **niche** and **fundamental niche.**

recapitulation Summarizing or repeating; hypothesis that an individual repeats its phylogenetic history in its development.

redia (rē´dē-ə), pl. **rediae** (rē´dē-ē) (from Francesco Redi, 1626–97, Italian biologist) Larval stage in the life cycle of flukes; it is produced by a sporocyst larva, and in turn gives rise to many cercariae.

regulative development Progressive determination and restriction of initially totipotent embryonic material.

reproductive barrier (L. *re*, + *producere*, to lead forward; M.F. *barriere*, bar) The factors that prevent one sexually propagating population from interbreeding and exchanging genes with another population.

reproductive community A general criterion for the species category shared to some degree by all formal species concepts is that species constitute a reproductively bounded population or lineage of populations that does not freely merge with others in nature.

resource (rē´so(ə)rs) Available source of nutrition, energy, or space in which to live.

respiration (L. *respiratio*, breathing) Gas interchange between an organism and its surrounding medium. In the cell, the release of energy by the oxidation of food molecules.

rete mirabile (rē´tē mə-rab´ə-lē) (L., wonderful net) Network of small blood vessels so arranged that the incoming blood runs countercurrent to the outgoing blood and thus makes possible efficient exchange between the two bloodstreams. Such a mechanism serves to maintain the high concentration of gases in the fish swim bladder.

reticulopodia (rə-tik´ū-lə-pə-dē´ə) (L. *reticulum*, dim. of *rete*, net, + *podos*, pous*, foot) Pseudopodia that branch and rejoin extensively.

reticulum (rə-tik´yə-ləm) (L. *rete*, dim. *reticulum*, a net) Second stomach of ruminants; a netlike structure.

retortamonad (rə-tort´ə-mō´nad) (L. *retro*, bend backward, + *monas*, single) Any member of a protozoan clade composed of certain heterotrophic flagellates.

rhabdite (rab´dīt) (Gr. *rhabdos*, rod) Rodlike structures in the cells of the epidermis or underlying parenchyma in certain turbellarians. They are discharged in mucous secretions.

rheoreceptor (rē´ō-rē-cep´tor) (Gr. *rheos*, a flowing, + L. *receptus*, accept) Sensory organ of aquatic animals that responds to water current.

rhinophore (rī´nə-fōr) (Gr. *rhis*, nose, + *pherein*, to carry) Chemoreceptive tentacles in some molluscs (opisthobranch gastropods).

rhipidistian (rip-ə-dis´tē-ən) (Gr. *rhipis*, fan, + *histion*, sail, web) Member of a group of Paleozoic lobe-finned fishes.

rhizopodia (rī´zə-pō´dē-ə) (Gr. *rhiza*, root, + *podos*, foot) Branched filamentous pseudopodia made by some amebas.

rhopalium (rō-pā´lē-əm) (N.L. from Gr. *rhopalon*, a club) One of the marginal, club-shaped sense organs of certain jellyfishes; tentaculocyst.

rhoptries (rōp´trēz) (Gr. *thopalon*, club, + *tryō*, to rub, wear out) Club-shaped bodies in Apicomplexa forming one of the structures of the apical complex; open at anterior and apparently functioning in penetration of host cell.

rhynchocoel (rink´ō-sēl) (Gr. *rhynchos*, snout, + *koilos*, hollow) In nemertines, the dorsal tubular cavity that contains the inverted proboscis. It has no opening to the outside.

rostrum (ros´trəm) (L., ship's beak) A snoutlike projection on the head.

rumen (rü´mən) (L., cud) The large first compartment of the stomach of ruminant mammals. Fermentation of cellulose by microorganisms occurs in it.

ruminant (rüm´ə-nənt) (L. *ruminare*, to chew the cud) Cud-chewing artiodactyl mammals with a complex, four-chambered stomach.

S

sagittal plane (saj´ə-təl) (L. *sagitta*, arrow) Pertaining to the median anteroposterior plane that divides a bilaterally symmetrical organism into right and left halves.

salamander (sal´-ə-man´-der). Denotes any member of the amphibian order Urodela (also called Caudata, sometimes with a different content of fossil forms).

saprophagous (sə-prof´ə-gəs) (Gr. *sapros*, rotten, + *phagos*, from *phagein*, to eat) Feeding on decaying matter; saprobic; saprozoic.

saprozoic nutrition (sap´rə-zō´ik) (Gr. *sapros*, rotten, + *zōon*, animal) Animal nutrition by absorption of dissolved salts and simple organic nutrients from surrounding medium; also refers to feeding on decaying matter.

sauropterygians (so-rop´tə-rij´ē-əns) (Gr. *sauros*, lizard, + *pteryginos*, winged) Mesozoic marine reptiles.

scalids (skā-lədz) (Gr. *skalis*, hoe, mattock) Recurved spines on the head of kinorhynchs.

schistosomiasis (shis´tō-sō-mī´-ə-sis) (Gr. *schistos*, divided, + *soma*, body, + *iasis*, a diseased condition) Infection with blood flukes of the genus *Schistosoma*.

schizocoel (skiz´ə-sēl) (Gr. *schizo*, from *schizein*, to split, + *koilos*, hollow) A coelom formed by the splitting of embryonic mesoderm. Noun, **schizocoelomate,** an animal with a schizocoel, such as an arthropod or mollusc. Adj., **schizocoelous.**

schizocoelous mesoderm formation (skiz´ō-sē-ləs) Embryonic formation of the mesoderm as cords of cells between ectoderm and endoderm; splitting of these cords produces the coelomic space.

schizogony (skə-zo´gə-nē) (Gr. *schizein*, to split, + *gonos*, seed) Multiple asexual fission.

sclerite (skle´rīt) (Gr. *sklēros*, hard) A hard, chitinous or calcareous plate or spicule; one of the plates forming the exoskeleton of arthropods, especially insects.

scleroblast (skler´ə-blast) (Gr. *sklēros*, hard, + *blastos*, germ) Amebocyte specialized to secrete a spicule; occurs in sponges.

sclerocyte (skler´ə-sīt) (Gr. *sklēros*, hard, + *kytos*, hollow vessel) Amebocyte in sponges that secretes spicules.

sclerotin (skler´-ə-tən) (Gr. *sklērotēs*, hardness) Insoluble, tanned protein permeating the cuticle of arthropods.

sclerotization (skle´rə-tə-zā´shən) Hardening of the cuticle of arthropods by the formation of stabilizing cross linkages between peptide chains of adjacent protein molecules.

scolex (skō´leks) (Gr. *skōlex*, worm, grub) The holdfast, or so-called head, of a tapeworm; bears suckers and, in some, hooks; posterior to it, new proglottids are differentiated.

scyphistoma (sī-fis´tə-mə) (Gr. *skyphos*, cup, + *stoma*, mouth) The polyp form of a scyphozoan.

sebaceous (sə-bāsh´əs) (L. *sebaceus*, made of tallow) Type of mammalian epidermal gland that produces a fatty substance.

sebum (sē´bəm) (L., grease, tallow) Oily secretion of the sebaceous glands of the skin.

sedentary (sed´ən-ter-ē) Stationary, sitting, inactive; staying in one place.

segmentation Division of the body into discrete segments or metameres; also called **metamerism.**

semelparity (se´məl-pā´ri-tē´) A life history in which individual organisms of a population normally reproduce only one time before dying, although numerous offspring may be produced at the time of reproduction; contrasts with **iteroparity.**

sensillum, pl. **sensilla** (sen-si´ləm) (L., *sensus*, sense) A small sense organ, especially in arthropods.

septum, pl. **septa** (L., fence) A wall between two cavities.

serial homology See **homology.**

serosa (sə-rō´sə) (N.L. from L. *serum*, serum) The outer embryonic membrane of birds and reptiles; chorion. Also, the peritoneal lining of the body cavity.

serous (sir´əs) (L. *serum*, serum) Watery, resembling serum; applied to glands, tissue, cells, or fluid.

serum (sir´əm) (L., whey, serum) The liquid that separates from the blood after coagulation; blood plasma from which fibrinogen has been removed. Also, the clear portion of a biological fluid separated from its particulate elements.

sessile (ses´əl) (L. *sessilis*, low, dwarf) Attached at the base; fixed to one spot.

seta (sēt´ə), pl. **setae** (sē´tē) (L. *bristle*) Needlelike chitinous structure of the integument of annelids, arthropods, and others.

sex ratio The proportion of males versus females in a population at a particular time and place.

sexual selection Charles Darwin's theory that a struggle for mates exists and that characteristics favorable for mating may prevail through reproductive success even if they are not advantageous in the struggle for survival.

siliceous (si-li´shəs) (L. *silex*, flint) Containing silica.

simian (sim´ē-ən) (L. *simia*, ape) Pertaining to monkeys or apes.

sink deme A subpopulation (deme) whose members are drawn disproportionately from other subpopulations of the same species (see **metapopulation dynamics**); for example, a deme occupying an environmentally unstable area whose members are periodically destroyed by climatic changes and then replenished by colonists from other demes when favorable conditions are restored.

sinus (sī´nəs) (L., curve) Cavity or space in tissues or in bone.

siphon (sī´-fən) Tube for directing water flow.

siphonoglyph (sī´fan´ə-glif´) (Gr. *siphōn*, reed, tube, siphon, + *glyphē*, carving) Ciliated furrow in the gullet of sea anemones.

siphuncle (sī´fun-kəl) (L. *siphunculus*, small tube) Cord of tissue running through the shell of a nautiloid, connecting all chambers with the animal's body.

sister taxon (=“sister group”) The relationship between a pair of species or higher taxa that are each other's closest phylogenetic relatives.

solenia (sə-len´ē-ə) (Gr. *solen*, pipe) Channels through the material connecting the polyps in an octocorallian colony (phylum Cnidaria).

soma (sō´mə) (Gr., body) The whole of an organism except the germ cells (germ plasm).

somatic (sō-mat´ik) (Gr. *sōma*, body) Refers to the body—for example, somatic cells in contrast to germ cells.

somatocoel (sə-mat´ə-sēl) (Gr. *sōma*, the body, + *koilos*, hollow) Posterior coelomic compartment of echinoderms; left somatocoel gives rise to oral coelom, and right somatocoel becomes aboral coelom.

somite (sō´mīt) (Gr. *sōma*, body) One of the blocklike masses of mesoderm arranged segmentally (metamerically) in a longitudinal series beside the digestive tube of the embryo; metamere.

sorting Differential survival and reproduction among varying individuals; often confused with natural selection, which is one possible cause of sorting.

source deme A stable subpopulation (deme) that serves differentially as a source of colonists for establishing, joining, or replacing other such subpopulations of the same species (see **metapopulation dynamics**); for example, a deme inhabiting an environmentally stable area whose members routinely establish transitory populations in environmentally unstable nearby areas.

speciation (spē´sē-ā´shən) (L. *species,* kind) The evolutionary process or event by which new species arise.

species (spē´shez, spē´sēz) sing. and pl. (L., particular kind) A group of interbreeding individuals of common ancestry that are reproductively isolated from all other such groups; a taxonomic unit ranking below a genus and designated by a binomial name consisting of its genus and the species name.

species epithet The second, uncapitalized word in the binomial name of a species. It is usually an adjective modifying the first word, the genus into which the species is placed.

species richness The number of different **species** that coexist at a given time and place to form an **ecological community.**

species selection Differential rates of speciation and/or extinction among varying evolutionary lineages caused by interactions among species-level characteristics and the environment.

spermatheca (spər´mə-thē´kə) (Gr. *sperma,* seed, + *thēkē,* case) A sac in the female reproductive organs for the reception and storage of sperm.

spermatophore (spər-ma´-tə-fōr´) (Gr. *sperma, spermatos,* seed, + *pherein,* to bear) Capsule or packet enclosing sperm; produced by males of several invertebrate groups and a few vertebrates.

spicule (spi´kyul) (L. dim. of *spica,* point) One of the minute calcareous or siliceous skeletal bodies found in sponges, radiolarians, soft corals, and sea cucumbers.

spiracle (spi´rə-kəl) (L. *spiraculum,* from *spirare,* to breathe) External opening of a trachea in arthropods. One of a pair of openings on the head of elasmobranchs for passage of water. Exhalant aperture of tadpole gill chamber.

spiral cleavage Type of early embryonic cleavage in which cleavage planes are diagonal to the polar axis, and unequal cells are produced by the alternate clockwise and counterclockwise

cleavage around the axis of polarity; see **determinate cleavage.**

spongin (spun´jin) (L. *spongia,* sponge) Fibrous, collagenous material forming the skeletal network of some sponges.

spongocoel (spun´jō-sēl) (Gr. *spongos,* sponge, + *koilos,* hollow) Central cavity in sponges.

spongocyte (spun´jō-sīt) (Gr. *spongos,* sponge, + *kytos,* hollow vessel) A cell in sponges that secretes spongin.

sporocyst (spō´rə-sist) (Gr. *sporos,* seed, + *kystis,* pouch) Larval stage in the life cycle of flukes; it originates from a miracidium.

sporogony (spor-og´ə-nē) (Gr. *sporos,* seed, + *gonos,* birth) Multiple fission to produce sporozoites after zygote formation.

sporozoite (spo´rə-zō´īt) (Gr. *sporos,* seed, + *zōon,* animal, + *ite,* suffix for body part) Stage in the life history of many sporozoan protozoa; released from oocysts.

stabilizing selection Natural selection that favors average values of a continuously varying trait and disfavors extreme values.

stapes (stā´pēz) (L., stirrup) Stirrup-shaped, innermost bone of the middle ear of mammals; the only ear bone of other tetrapods.

statoblast (sta´-tō-blast) (Gr. *statos,* standing, fixed, + *blastos,* germ) Asexually produced overwintering stage of many freshwater ectoprocts.

statocyst (Gr. *statos,* standing, + *kystis,* bladder) Sense organs of equilibrium; a fluid-filled cellular cyst containing one or more granules (statoliths) used to sense direction of gravity.

stenohaline (sten-ə-ha´līn, -lən) (Gr. *stenos,* narrow, + *hals,* salt) Pertaining to aquatic organisms that have restricted tolerance to changes in environmental saltwater concentration.

stenophagous (stə-näf´ə-gəs) (Gr. *stenos,* narrow, + *phagein,* to eat) Eating few kinds of foods.

stereom (ster´ē-ōm) (Gr. *stereos,* solid, hard, firm) Meshwork structure of endoskeletal ossicles of echinoderms.

sternum (ster´nəm) (L., breastbone) Ventral plate of an arthropod body segment; breastbone of vertebrates.

stigma (Gr. *stigma,* mark, tattoo mark) Eyespot in certain protozoa. Spiracle of certain terrestrial arthropods.

stolon (stō´lən) (L. *stolō, stolonis,* a shoot, or sucker of a plant) Rootlike extension of the body wall giving rise to buds that may develop into new zooids, thus forming a

compound animal in which the zooids remain united by the stolon. Occurs in some colonial anthozoans, hydrozoans, ectoprocts, and ascidians.

stoma (stō´mə) (Gr., mouth) A mouthlike opening.

strobila (strō´bə-lə) (Gr. *strobilē,* lint plug like a pine cone [*strobilos*]) Scyphozoan jellyfish polyp with a stack of ephyrae atop it. Also, the chain of proglottids of a tapeworm.

strobilation (strō´bi-lā´shən) (Gr. *strobilos,* a pine cone) Repeated linear budding of individuals, as in scyphozoan ephyrae (phylum Cnidaria), or sets of reproductive organs in tapeworms (phylum Platyhelminthes).

stroma (strō´mə) (Gr. *stroma,* bedding) Supporting connective tissue framework of an animal organ; filmy framework of red blood corpuscles and certain cells.

subnivean (səb-ni´vē-ən) (L. *sub,* under, below, + *nivis,* snow) Applied to environments beneath snow, which acts as an insulator against colder atmospheric temperature.

survivorship (sər-vī´vər-ship) The proportion of individuals of a cohort or population that persist from one point in their life history, such as birth, to another one, such as reproductive maturity or a specified age.

suspension feeders Aquatic organisms that collect suspended food particles from the surrounding water; particles may be filtered or taken by other methods.

swim bladder Gas-filled sac of many bony fishes used in buoyancy and, in some cases, respiratory gas exchange.

sycon (sī´kon) (Gr. *sykon,* fig) Type of canal system in certain sponges. Sometimes called syconoid.

symbiosis (sim-bī-ōs´əs, sim´bē-ōs´əs) (Gr. *syn,* with, + *bios,* life) The living together of two different species in an intimate relationship. The symbiont always benefits; the host may benefit (mutualism), be unaffected (commensalism), or be harmed (parasitism).

synapomorphy (sin-ap´ə-mōr´fē) (Gr. *syn,* together with, + *apo,* of, + *morphē,* form) Shared, evolutionarily derived character states that are used to recover patterns of common descent among two or more species.

synapsids (si-nap´sədz) (Gr. *synapsis,* contact, union) An amniote lineage comprising the mammals and their closest fossil relatives, having a skull with a single pair of temporal openings.

syncytium (sin-sish´ē-əm) (Gr. *syn*, with, + *kytos*, hollow vessel) A multinucleated cell. Adj., **syncytial**.

syngamy (sin´gə-mē) (Gr. *syn*, with, + *gamos*, marriage) Fertilization of one gamete with another individual gamete to form a zygote; found in most animals that have sexual reproduction.

synsacrum (sin-sa´krəm) (Gr. *syn*, together + sacrum) The pelvic girdle of birds, consisting of fused lumbar, sacral, and caudal vertebrae.

syrinx (sir´inks) (Gr., shepherd's pipe) The vocal organ of birds located at the base of the trachea.

systematics (sis-tə-mat´iks) Science of taxonomy and reconstruction of phylogeny.

systematization (sis-tə-mə-tī-zā´shən) Construction of a hierarchical taxonomy that conveys the structure of common evolutionary descent among species. Each recognized taxon comprises an ancestral population lineage and all of its descendants.

T

tactile (tak´til) (L. *tactilis*, able to be touched, from *tangere*, to touch) Pertaining to touch.

taenidia (tə´nid´-ē-ə) (Gr. *tainia*, ribbon) Spiral thickenings of the cuticle that support tracheae (phylum Arthropoda).

tagma, pl. **tagmata** (Gr. *tagma*, arrangement, order, row) Compound body section of an arthropod resulting from embryonic fusion of two or more segments; for example, head, thorax, abdomen.

tagmatization, tagmosis Organization of the arthropod body into tagmata.

taxon (tak´son), pl. **taxa** (Gr. *taxis*, order, arrangement) Any taxonomic group or entity.

taxonomy (tak-sän´ə-mē) (Gr. *taxis*, order, arrangement, + *nomas*, law) Study of the principles of systematic ordering and naming of organisms.

tegument (teg´ū-ment) (L. *tegumentum*, from tegere, to cover) An integument; specifically, the external covering in cestodes and trematodes.

teleology (tel´ē-äl´ə-jē) (Gr. *telos*, end, + L. *logia*, study of, from Gr. *logos*, word) The philosophical view that natural events are goal-directed and preordained, as opposed to the scientific view of mechanical determinism.

teleost (tē´lē-ost) A clade of advanced ray-finned fishes characterized by a homocercal caudal fin.

telson (tel´sən) (Gr., extremity) Posterior projection of the last body segment in many crustaceans.

temnospondyls (tem-nō-spän´dəls) (Gr. *temnō*, to cut, + *spondylos*, vertebra) A large group of early tetrapods that lived from the Carboniferous to the Triassic.

tergum (ter´gəm) (L., back) Dorsal part of an arthropod body segment.

territory (L. *territorium*, from *terra*, earth) Restricted area preempted by an animal or pair of animals, usually for breeding purposes, and guarded from other individuals of the same species.

test (L. *testa*, shell) A shell or hardened outer covering.

tetrapods (te´trə-pods) (Gr. *tetras*, four, + *pous, podos*, foot) Four-limbed vertebrates; the group includes amphibians, reptiles, birds, and mammals, including taxa that have lost limbs, such as caecilians and snakes.

theory A scientific hypothesis or set of related hypotheses that offers very powerful explanations for a wide variety of related phenomena and serves to organize the scientific investigation of those phenomena.

therapsid (thə-rap´sid) (Gr. *theraps*, an attendant) Extinct Mesozoic synapsid amniotes from which mammals evolved.

thoracic (thō-ra´sək) (L. *thōrax*, chest) Pertaining to the thorax or chest.

Tiedemann's bodies (tēd´ə-mənz) (from F. Tiedemann, German anatomist) Four or five pairs of pouchlike bodies attached to the ring canal of sea stars, apparently functioning in production of coelomocytes.

tissue (ti´shü) (M.E. *tissu*, tissue) An aggregation of cells and cell products organized to perform a common function.

tornaria (tor-na´re-ə) (Gr. *tornos*, compass, circle, wheel) A hemichordate (acorn worm) larva that closely resembles bipinnaria larvae of sea stars.

torsion (L. *torquere*, to twist) A twisting phenomenon in gastropod development that alters the position of the visceral and pallial organs by 180 degrees.

toxicyst (tox´i-sist) (Gr. *toxikon*, poison, + *kystis*, bladder) Structures of predatory ciliate protozoa that, on stimulation, expel a poison to subdue prey.

trabecular reticulum (trə-bek´yə-lər rə-tik´yə-ləm) An extensive syncytial tissue in hexactinellid sponges, bearing choanoblasts and collar bodies and forming flagellated chambers.

trachea (trā´kē-ə) (M.L., windpipe) The windpipe. Also, any of the air tubes of insects.

tracheal system (trāk´ē-əl) (M.L. *trachia*, windpipe) A network of thin-walled tubes that branch throughout the entire body of terrestrial insects; used for respiration.

tracheole (trāk´ē-ōl) (L. *trachia*, windpipe) Fine branches of the tracheal system, filled with fluid, but not shed at ecdysis.

transverse plane (L. *transversus*, across) A plane or section that lies or passes across a body or structure dividing it into cephalic and caudal pieces.

trend A directional change in the characteristic features or patterns of diversity in a group of organisms when viewed over long periods of evolutionary time in the fossil record.

trichinosis (trik-ən-o´səs) Disease caused by infection with the nematode *Trichinella* spp.

trichocyst (trik´ə-sist) (Gr. *thrix*, hair, + *kystis*, bladder) Saclike protrusible organelle in the ectoplasm of ciliates, which discharges as a threadlike weapon of defense.

triploblastic (trip´lō-blas´tik) (Gr. *triploos*, triple, + *blastos*, germ) Pertaining to animals in which the embryo has three primary germ layers—ectoderm, mesoderm, and endoderm.

trochophore (trōk´ə-fōr) (Gr. *trochos*, wheel, + *pherein*, to bear) A free-swimming ciliated marine larva characteristic of most molluscs and certain ectoprocts, brachiopods, and marine worms; an ovoid or pyriform body with a preoral circlet of cilia and sometimes a secondary circlet behind the mouth.

trophallaxis (tröf´ə-lak´səs) (Gr. *trophē*, food, + *allaxis*, barter, exchange) Exchange of food between young and adults, especially certain social insects.

trophic (trō´fək) (Gr. *trophē*, food) Pertaining to nutrition.

trophic level Position of a species in a **food web,** such as producer, herbivore, first-level carnivore, or higher-level carnivore.

trophosome (trŏf´ə-sōm) (Gr. *trophē,* food, + *soma,* body) Organ in pogonophorans bearing mutualistic bacteria; derived from midgut.

trophozoite (trŏf´ə-zō´īt) (Gr. *trophē,* food, + *zōon,* animal) Adult stage in the life cycle of a protozoan in which it is actively absorbing nourishment.

tube feet (podia) Numerous small, muscular, fluid-filled tubes projecting from an echinoderm; part of water-vascular system; used in locomotion, clinging, food handling, and respiration.

tubulin (tü´bū-lən) (L. *tubulus,* small tube, + *in,* belonging to) Globular protein forming the hollow cylinder of microtubules.

tunic (L. *tunica,* tunic, coat) In tunicates, a cuticular, cellulose-containing covering of the body secreted by the underlying body wall.

turbinates (tər´bin-āts) (L. *turbin,* whirling) Highly convoluted bones covered in mucous membrane in the nasal cavity of endotherms that serve to reduce heat and water lost during respiration.

tympanic membrance (tim-pan´ik) (Gr. *tympanon,* drum) The surface that separates the outer and middle ear; also called the eardrum.

typhlosole (tif´lə-sōl´) (Gr. *typhlos,* blind, + *sōlēn,* channel, pipe) A longitudinal fold projecting into the intestine in certain invertebrates such as the earthworm.

typological species concept The discredited, pre-Darwinian notion that species are classes defined by the presence of fixed, unchanging characters (= "essence") shared by all members.

U

ultimate cause (L. *ultimatus,* last, + *causa*) The evolutionary factors responsible for the origin, state of being, or role of a biological system.

umbo (um´bō), pl. **umbones** (um-bō´nēz) (L., boss of a shield) One of the prominences on either side of the hinge region in a bivalve mollusc shell. Also, the "beak" of a brachiopod shell.

undulating membrane Membranous structure on a protozoan associated with a flagellum; on other protozoa, may be formed from fused cilia.

ungulate (un´gū-lət) (L. *ungula,* hoof) Hooved. Noun, any hooved mammal.

uniformitarianism (ū´nə-for´mə-ter´ē-ə-niz´əm) Methodological assumptions that the laws of chemistry and physics have remained constant throughout the history of the earth, and that past geological events occurred by processes that can be observed today.

uniramous (ü´nə-rām´əs) (L. *unus,* one, + *ramus,* a branch) Adjective describing unbranched appendages (phylum Arthropoda).

unitary (ü´nə-ter´ē) Describes the structure of a population in which reproduction is strictly sexual and each organism is genetically distinct from others.

uropod (ū´rə-pod) (Gr. *oura,* tail, + *pous, podos,* foot) Posteriormost appendage of many crustaceans.

V

vacuole (vak´yə-wōl) (L. *vacuus,* empty, + Fr. *ole;* dimin. suffix) A membrane-bound, fluid-filled space in a cell.

valve (L. *valva,* leaf of a double door) One of the two shells of a typical bivalve mollusc or brachiopod.

variation (L. *varius,* various) Differences among individuals of a group or species that cannot be ascribed to age, sex, or position in the life cycle.

velarium (və-la´rē-əm) (L. *velum,* veil, covering) Shelflike extension of the subumbrellar edge in cubozoans (phylum Cnidaria).

veliger (vēl´ə-jər) (L. *velum,* veil, covering) Larval form of certain molluscs; develops from the trochophore and has the beginning of a foot, mantle, and shell.

velum (vē´ləm) (L., veil, covering) A membrane on the subumbrellar surface of jellyfishes of class Hydrozoa. Also, a ciliated swimming organ of the veliger larva.

ventral (ven´trəl) (L. *venter,* belly) Situated on the lower or abdominal surface.

vestige (ves´tij) (L. *vestigium,* footprint) A rudimentary organ that may have been well developed in some ancestor or in the embryo.

vibrissa (vī-bris´ə), pl. **vibrissae** (L., nostril-hair) Stiff hairs that grow from the nostrils or other parts of the face of many mammals and that serve as tactile organs; "whiskers."

villus (vil´əs), pl. **villi** (L., tuft of hair) Small fingerlike, vascular process on the wall of the small intestine. Also, one of the

branching, vascular processes on the embryonic portion of the placenta.

viscera (vis´ər-ə) (L. pl. of *viscus,* internal organ) Internal organs in the body cavity.

visceral (vis´ər-əl) Pertaining to viscera.

viviparity (vī´və-par´ə-tē) (L. *vivus,* alive, + *parere,* to bring forth) Reproduction in which eggs develop within the female body, which supplies nutritional aid; occurs in therian mammals, many non-avian reptiles, and some fishes; offspring are born as juveniles. Adj., **viviparous** (vī-vip´ə-rəs).

W

water-vascular system System of fluid-filled closed tubes and ducts peculiar to echinoderms; used to move tentacles and tube feet that serve variously for clinging, food handling, locomotion, and respiration.

weir (wēr) (Old English *wer,* a fence placed in a stream to catch fish) Interlocking extensions of a flame cell and a collecting tubule cell in some protonephridia.

X

X-organ Neurosecretory organ in eyestalk of crustaceans that secretes molt-inhibiting hormone.

Y

Y-organ Gland in the antennal or maxillary segment of some crustaceans that secretes molting hormone.

Z

zoecium (zō-ē´shē-əm) (Gr. *zōon,* animal, + *oikos,* house) Cuticular sheath or shell of Ectoprocta; also spelled **zooecium.**

zooid (zō-oid) (Gr. *zōon,* animal) An individual member of a colony of animals, such as colonial cnidarians and ectoprocts.

zooxanthella (zō´ə-zan-thəl´ə) (Gr. *zōon,* animal, + *xanthos,* yellow) A minute dinoflagellate alga living in the tissues of many types of marine invertebrates.

zygote (Gr. *zygōtos,* yoked) A fertilized egg.

zygotic meiosis Meiosis that occurs within the first few divisions after zygote formation; thus, all stages in the life cycle other than the zygote are haploid.

Photo Credits

Chapter 1

Figure 1.1a: © William Ober; 1.1b: Cleveland P. Hickman, Jr.; 1.1c: Courtesy Duke University Marine Laboratory; 1.2a: Image # 32662, American Museum of Natural History; 1.2b, 1.3, 1.4: © 2001 The Natural History Museum, London; 1.6a: BAL Portrait of Charles Darwin, 1840 by George Richmond (1809-96) Downe House, Downe, Kent, UK/Bridgeman Art Library, London/New York; 1.6b: © Stock Montage; 1.7-1.8: Cleveland P. Hickman, Jr.; p. 12: © The Natural History Museum, London; 1.10a: © A. J. Copley/Visuals Unlimited; 1.10b: © G. O. Poinar/Oregon State University, Corvallis; 1.10c: © Ken Lucas; 1.10d: © Roberta Hess Poinar; 1.11a: Boehm Photography; 1.12: Cleveland P. Hickman, Jr.; 1.17: Library of Congress; 1.20 (top): © Steve Simonsen; 1.20 (bot): © David Wrobel/Visuals Unlimited; 1.23b: Cleveland P. Hickman, Jr.; 1.24: Courtesy of Storrs Agricultural Experiment Station, University of Connecticut at Storrs; 1.27a-b: © Michael Tweedie/Photo Researchers, Inc.; 1.30: Jack Hollingsworth/Getty Images RF; 1.31: © S. Krasemann/Photo Researchers, Inc.; 1.35: NASA.

Chapter 2

Opener: Cleveland P. Hickman, Jr.; 2.8-2.9 (top): © Noble Proctor/Photo Researchers, Inc.; 2.9 (bot): Cleveland P. Hickman, Jr.; 2.13a (top): © Patti Murray/Animals Animals/Earth Scenes; 2.13a (bot): © Bill Beatty/Animals Animals/Earth Scenes; 2.13b (top and bot), 2.13c: © James L. Castner/Visuals Unlimited; p. 51: © D. Foster/WHOI/Visuals Unlimited.

Chapter 3

Opener: Larry S. Roberts; 3.12a-c, 3.13 (top and bot), 3.14 a-b: © Ed Reschke; 3.14c: Cleveland P. Hickman, Jr.; 3.14d, 3.15 a-c: © Ed Reschke.

Chapter 4

Opener, 4.1: Courtesy of the Library of Congress; 4.2: Image # 334101, American Museum of Natural History; 4.6a: © OSF/M. Cole/Animals Animals/Earth Scenes; 4.6b: © OSF/D. Allen/Animals-Animals/Earth Scenes; 4.8: Courtesy of George W. Byers/University of Kansas.

Chapter 5

Opener: © Michael Abbey/Visuals Unlimited; 5.1a: Punchstock RF; 5.1b: CDC/Dr. Stan Erlandsen and Dr. Dennis Freely; 5.1c: Melba Photo Agency/Punchstock RF; 5.3c: Courtesy of Jerry Niederkorn, Ph. D., UT Southwestern Medical Center, Dallas, Texas; 5.4b: Courtesy Dr. Ian R. Gibbons; 5.5a-c: © Michael Abbey/Visuals Unlimited; 5.19: © Bill Bachman/Photo Researchers, Inc.; 5.22a: Courtesy Gustaf M. Hallegraeff; 5.22b: © A. M. Siegelman/Visuals Unlimited; 5.23: Courtesy of J. and M. Cachon from Lee J. J. Hunter, S.H., and Boves, E.C. (editors) 1985, "An Illustrated Guide to the Protozoa," Society of Protozoologists; Permission by Edna Kaneshiro.

Chapter 6

Opener, 6.3, 6.5, 6.13a-c: Larry S. Roberts.

Chapter 7

Opener: Larry S. Roberts; 7.2a: © William Ober; 7.6b: Thien T. Mai; 7.7: © Rick Harbo; 7.8: © CABISCO/Phototake; 7.9: © CABISCO/Visuals Unlimited; 7.11: © Diane W. Gotshall; 7.13: © OSF/Peter Parks/Animals Animals/Earth Scenes; 7.14a-b: Larry S. Roberts; 7.15: © Rick Harbo; 7.17: Larry S. Roberts; 7.18: © Rick Harbo; 7.20: © Daniel W. Gotshall; 7.21a: © Michael Patrick O'Neill/Photo Researchers, Inc.; 7.21b, 7.23, 7.24a-c, 7.28: Larry S. Roberts; 7.29: Courtesy of Dr. Ernesto Weil, Department of Marine Sciences, University of Puerto Rico; 7.30a: © Andrew J. Martinez/Photo Researchers, Inc.; 7.30b: © Kjell Sandved/Butterfly Alphabet.

Chapter 8

Opener: Larry S. Roberts; 8.4: © James L. Castner/Visuals Unlimited; 8.11a, 8.12: Larry S. Roberts; 8.16: © CABISCO/Visuals Unlimited; 8.18: Larry S. Roberts; 8.19a: © Larry Stepanowicz/Visuals Unlimited; 8.20a, 8.21c: © Martin V. Sorensen; 8.22a: © John Walsh/Photo Researchers, Inc.; 8.23a: Courtesy Dr. Wayne Lord, Department of Biology, University of Central Oklahoma; 8.23b-c: © Inga Sidor.

Chapter 9

Opener: From the collection of Herman Zaiman, "A Presentation of Pictorial Parasites"; 9.2: Courtesy of Dr. Peter Funch, University of Aarhus; 9.3c: Courtesy of Dr. Kerstin Wasson; 9.4b: Larry S. Roberts; 9.5a: © William Ober; 9.6a-b: © Robert Brons/Biological Photo Service; 9.7c: David J. Eisenhour; 9.10: © Stan Elems/Visuals Unlimited.

Chapter 10

Opener: Larry S. Roberts; 10.1a: © Kjell Sandved/Visuals Unlimited; 10.1b-c: © Rick Harbo; 10.1d: © Daniel W. Gotshall; 10.1e: Larry S. Roberts; 10.4b and inset: © Ardea/Spears, David & Sasitorn, DAE/Animals Animals; 10.6b: © Dr. Thurston C. Lacalli, Biology Department, University of Victoria; 10.7: © Kjell Sandved/Visuals Unlimited; 10.8a: © 2006, L. Kuhnz/MBARI; 10.8b: © 2003, L. Kuhnz/MBARI; 10.10: © Daniel Gotshall/Visuals Unlimited; 10.16a-b: © Daniel W. Gotshall; 10.17a-b, 10.18 a-b: Cleveland P. Hickman, Jr.; 10.20a: © Rick Harbo; 10.20b: © William Ober; 10.21, 10.22a: Larry S. Roberts; 10.22b: Cleveland P. Hickman, Jr.; 10.23, 10.27a-b: Larry S. Roberts; 10.28: © Rick Harbo; 10.29: Larry S. Roberts; 10.30b: Courtesy Richard J. Neves; 10.31b: © Dave Fleetham/Tom Stack & Associates; 10.32a: Courtesy M. Butschler, Vancouver Public Aquarium; 10.33: Larry S. Roberts.

Chapter 11

Opener: © Photo Gear RF; 11.4a: © William Ober; 11.4b: Larry S. Roberts; 11.6: Cleveland P. Hickman, Jr.; 11.7: Larry S. Roberts; 11.13b: Cleveland P. Hickman, Jr.; 11.18g: © G.L. Twiest/Visuals Unlimited; 11.20: © Timothy Branning; 11.22, 11.23b: Cleveland P. Hickman, Jr.

Chapter 12

Opener: From the collection of Herman Zaiman, "A Presentation of Pictorial Parasites"; 12.4a: Courtesy of Frances M. Hickman; 12.4b: Larry S. Roberts; 12.5: Photo by E. Pike; slide is made available courtesy of the American Society of Tropical Medicine and Hygiene/ From the collection of Herman Zaiman "A Presentation of Pictorial Parasites"; 12.6: From the collection of Herman Zaiman, "A Presentation of Pictorial Parasites"; 12.7a: © R. Calentine/Visuals Unlimited; 12.7b: Larry S. Roberts; 12.8: From the collection of Herman Zaiman, "A Presentation of Pictorial Parasites"; 12.9: © Robert Calentine, Visuals Unlimited; 12.10: Dr. Andreas Schmidt-Rhaesa; 12.11b: © David Schart/Photo Researchers, Inc.; 12.12b: © Erling Svensen; 12.13b: Courtesy Reinhard Kristensen, Copenhagen; 12.14b: Dr. Morley Read/Photo Researchers, Inc.; 12.15: Courtesy Diane R. Nelson.

Chapter 13

Opener: Larry S. Roberts; 13.1a-b: © A. J. Copley/Visuals Unlimited; 13.8: © Michael Lustbader/Photo Researchers, Inc.; 13.9a: © Wild and Natural/Animals Animals; 13.9b, 13.10: © J. H. Gerard/Nature Press; 13.11: © Matt Meadows/Photo Researchers, Inc.; 13.12a: Centers for Disease Control; 13.12b: Gregory G. Dimijian, M.D./Photo Researchers, Inc.; 13.13a: © Karl H. Switak/Photo Researchers, Inc.; 13.13b: © Ken Lucas/Ardea/Animals Animals; 13.13c: Cleveland P. Hickman, Jr.; 13.14a: Centers for Disease Control; 13.14b: From G. W. Wharton, "Mites and commercial extracts of house dust" in Science 167: 1382-1383. Copyright © 1970 by AAAS. 13.14c: © D. S. Snyder/Visuals Unlimited; 13.15a: © Nature's Images/Photo Researchers, Inc.; 13.16a: Digital Vision/Getty Images RF; 13.25a: © CABISCO/Phototake; 13.26b, 13.26e: Courtesy L. G. Abele, W. Kim, and B. E. Felgenhauer; 13.27a-b: © Rick Harbo; 13.28a: Cleveland P. Hickman, Jr.; 13.29: Larry S. Roberts; 13.30a: © Rick Harbo; 13.30b-c: © Kjell Sandved/Butterfly Alphabet; 13.32a: Cleveland P. Hickman, Jr.; 13.32b-c: © Rick Harbo; 13.32d-e, 13.35a: Larry S. Roberts; 13.36a-b: © Ron West/Nature Photography; 13.37: Joel Santore/Getty Images RF; 13.40 © John D. Cunningham/Visuals Unlimited; 13.41: © Dennis Kunkel Microscopy, Inc./Visuals Unlimited; 13.42: Centers for Disease Control/James Gathaney; 13.43a: Cleveland P. Hickman, Jr.; 13.43b: © J. H. Gerard/Nature Press; 13.45c: © James Castner/Visuals Unlimited;

13.46, 13.49a: Cleveland P. Hickman, Jr.; 13.49b: © J. H. Gerard/Nature Press; 13.49c: © CABISCO/Phototake; 13.50a: Design Pics Inc./Alamy RF; 13.50b: © Ingram Publishing/Fotosearch RF; 13.51: © J. H. Gerard/Nature Press; 13.52: Courtesy James E. Lloyd; 13.53: © Creatas Images/Picture Quest RF; 13.54: © K. Lorensen/Andromedia Productions; 13.55a: © George D. Lepp/Photo Researchers, Inc.; 13.55b: © Kjell B. Sandved/Photo Researchers, Inc.; 13.56: Larry S. Roberts; 13.57a: © USDA; 13.57b: Photo by Scott Bauer/USDA; 13.57c: © Fran Keller; 13.58a: © L. L. Rue, III; 13.58b: © J. S. Sira/Gap Photo/Visuals Unlimited; 13.58c: © J. H. Gerard/Nature Press; 13.59: Digital Vision/Getty Images RF; 13.60: © Mark Smith/Photo Researchers, Inc.; 13.61b: M. Frances Keller; 13.62: © Leroy Simon/Visuals Unlimited; 13.63: © Alex Wild/Visuals Unlimited.

Chapter 14

Opener: © Peter Gier/Visuals Unlimited; 14.2b: Thuesen, E.V. and Bieri, 1987; 14.4a: Larry S. Roberts; 14.4b: © Marty Snyderman/Visuals Unlimited, Inc.; 14.4c: © Brand X Pictures/PunchStock RF; 14.4d: Larry S. Roberts; 14.7d: © Rick Harbo; 14.8: Larry S. Roberts; 14.12a: © William Ober; 14.12b: © The McGraw-Hill Companies Inc./Diane R. Nelson, photographer; 14.15: © Rick Harbo; 14.16a-b: Larry S. Roberts; 14.19a: © Rick Harbo; 14.19b, 14.23: Larry S. Roberts; 14.24b: © William Ober.

Chapter 15

Opener: © Heather Angel/Natural Visions; 15.6: Larry S. Roberts.

Chapter 16

Chapter: Larry S. Roberts; 16.4: © Berthoule-Scott/Jacana/Photo Researchers, Inc.; 16.8: © IFA-Bilderteam/Getty Images; 16.10a: © William Ober; 16.10b: © Comstock Images/PictureQuest RF; 16.16a-b: © John G. Shedd Aquarium 1997/Patrice Ceisel; 16.17a: © James D. Watt/Animals Animals/Earth Scenes; 16.17b: © Francois Odendaal/Biological Photo Service; 16.17c: © Brand X Pictures/PunchStock RF; 16.17d: © Fred McConnaughey/Photo Researchers, Inc.; 16.28: © William Troyer/Visuals Unlimited; 16.30: © Daniel W. Gotshall; 16.31: © Frederick R. McConnaughey.

Chapter 17

Opener: © Gary Mezaros/Visuals Unlimited; 17.4b: © Mark Smith/Photo Researchers, Inc.; 17.5a-b: Courtesy L. Houck; 17.9: Cleveland P. Hickman, Jr.; 17.10: Allan Larson; 17.11a: © Ken Lucas; 17.11b-17.12: Cleveland P. Hickman, Jr.; 17.13: Image # 125617, American Museum of Natural History; 17.14-17.15: Cleveland P. Hickman, Jr.

Chapter 18

Opener: © Ron Magill; 18.6, 18.8: Cleveland P. Hickman, Jr.; 18.10: © John Mitchell/Photo Researchers, Inc.; 8.11: Cleveland P. Hickman, Jr.; 18.12: © John D. Cunningham/Visuals Unlimited; 18.13: © Leonard Lee Rue, III; 18.14: © Craig K. Lorenz/Photo Researchers, Inc.; 18.15: © Paul Freed/Animals Animals/Earth Scenes; 18.16: © Austin J. Stevens/Animals Animals/Earth Scenes; 18.17: Cleveland P. Hickman, Jr.; 18.18: © Joe McDonald/Visuals Unlimited; 18.20: Cleveland P. Hickman, Jr.; 18.22: © Mark Carwardine/Visuals Unlimited; 18.23: Carmela Leszczynski/Animals Animals/Earth Scenes; 18.24a-b: Cleveland P. Hickman, Jr.

Chapter 19

Opener: PhotoDisc RF; 19.1a: Image # 125065, American Museum of Natural History; 19.19a-b: © Leonard Lee Rue, III; 19.20: Cleveland P. Hickman, Jr.; 19.22: © John Gerlach/Visuals Unlimited; 19.23: © Richard R. Hansen/Photo Researchers, Inc.; 19.27-19.30: Cleveland P. Hickman, Jr.

Chapter 20

Opener: © Jeff Lepore/Photo Researchers, Inc.; 20.5: © Leonard Lee Rue, III; 20.6a: © Photo Disc RF; 20.6b: © Corbis RF; 20.7, 20.11: Courtesy Robert E. Treat; 20.12a: Cleveland P. Hickman, Jr.; 20.14: © S. Maslowski/Visuals Unlimited; 20.16: © Kjell Sandved/Visuals Unlimited; 20.17: © Leonard Lee Rue, III; 20.19: © G. Herben/Visuals Unlimited; 20.21: © Zoological Society of San Diego; 20.22a: © Medioimages/Superstock RF; 20.22b: Cleveland P. Hickman, Jr.; 20.23: © Milton H. Tierney, Jr./Visuals Unlimited; 20.25: © John Reader/Photo Researchers, Inc.; 20.27: © Photodisc/Getty Images RF; 20.28: Cleveland P. Hickman, Jr.; 20.29: © John Gerlach/Visuals Unlimited; 20.30: Cleveland P. Hickman, Jr.; 20.32: © William Ober.

Index

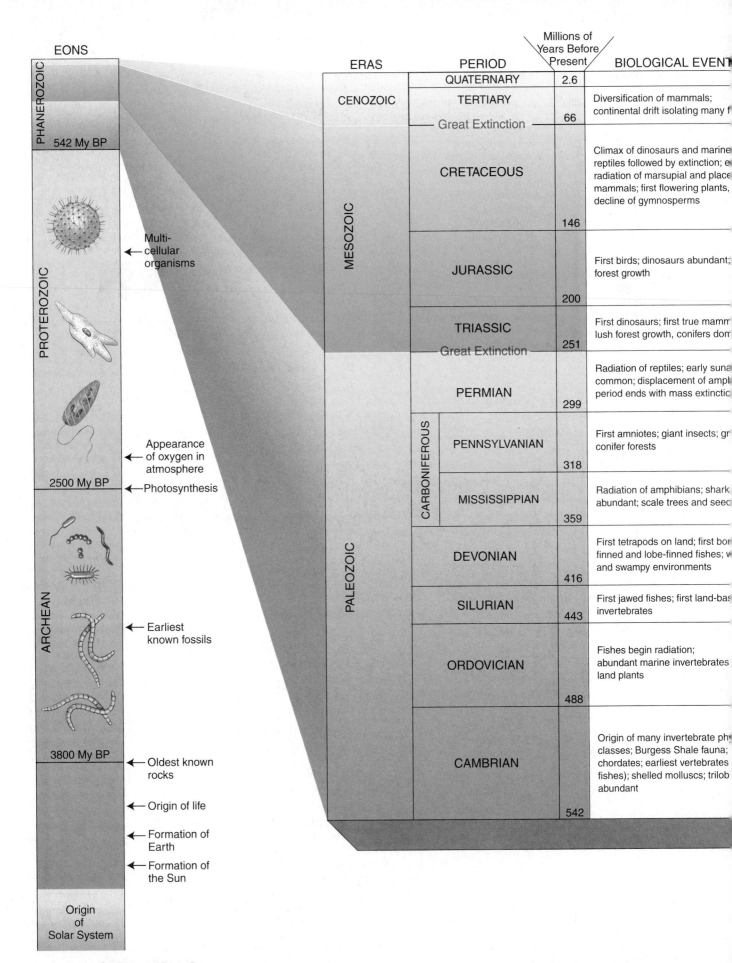

EONS			ERAS	PERIOD	Millions of Years Before Present	BIOLOGICAL EVENT
PHANEROZOIC			CENOZOIC	QUATERNARY	2.6	
				TERTIARY		Diversification of mammals; continental drift isolating many f
				Great Extinction	66	
542 My BP			MESOZOIC	CRETACEOUS		Climax of dinosaurs and marine reptiles followed by extinction; e radiation of marsupial and place mammals; first flowering plants, decline of gymnosperms
					146	
PROTEROZOIC		Multi-cellular organisms		JURASSIC		First birds; dinosaurs abundant; forest growth
					200	
				TRIASSIC		First dinosaurs; first true mamm lush forest growth, conifers dom
					251	
				Great Extinction		
			PALEOZOIC	PERMIAN		Radiation of reptiles; early suna common; displacement of ampl period ends with mass extinctic
					299	
		Appearance of oxygen in atmosphere	CARBONIFEROUS	PENNSYLVANIAN		First amniotes; giant insects; gr conifer forests
2500 My BP		Photosynthesis			318	
				MISSISSIPPIAN		Radiation of amphibians; shark abundant; scale trees and seed
					359	
				DEVONIAN		First tetrapods on land; first bor finned and lobe-finned fishes; w and swampy environments
					416	
ARCHEAN		Earliest known fossils		SILURIAN		First jawed fishes; first land-bas invertebrates
					443	
				ORDOVICIAN		Fishes begin radiation; abundant marine invertebrates land plants
					488	
3800 My BP		Oldest known rocks		CAMBRIAN		Origin of many invertebrate phy classes; Burgess Shale fauna; chordates; earliest vertebrates fishes); shelled molluscs; trilob abundant
		Origin of life				
		Formation of Earth			542	
		Formation of the Sun				
Origin of Solar System						